Hans-Christoph Pape
Roy Sanders • Joseph Borrelli, Jr.
Editors

The Poly-Traumatized Patient with Fractures

A Multi-Disciplinary Approach

Second Edition

Editors
Hans-Christoph Pape
Department of Trauma, Hand,
Plastic and Reconstructive Surgery
Aachen Medical Center
Aachen
Germany

Joseph Borrelli, Jr.
Orthopedic Medicine Specialists
Arlington, TX
USA

Roy Sanders
Florida Orthopaedic Institute
Temple Terrace, FL
USA

ISBN 978-3-662-47211-8   ISBN 978-3-662-47212-5   (eBook)
DOI 10.1007/978-3-662-47212-5

Library of Congress Control Number: 2015949820

Springer Heidelberg New York Dordrecht London
© Springer-Verlag Berlin Heidelberg 2016
This work is subject to copyright. All rights are reserved by the Publisher, whether the whole or part of the material is concerned, specifically the rights of translation, reprinting, reuse of illustrations, recitation, broadcasting, reproduction on microfilms or in any other physical way, and transmission or information storage and retrieval, electronic adaptation, computer software, or by similar or dissimilar methodology now known or hereafter developed.
The use of general descriptive names, registered names, trademarks, service marks, etc. in this publication does not imply, even in the absence of a specific statement, that such names are exempt from the relevant protective laws and regulations and therefore free for general use.
The publisher, the authors and the editors are safe to assume that the advice and information in this book are believed to be true and accurate at the date of publication. Neither the publisher nor the authors or the editors give a warranty, express or implied, with respect to the material contained herein or for any errors or omissions that may have been made.

Printed on acid-free paper

Springer-Verlag GmbH Berlin Heidelberg is part of Springer Science+Business Media (www.springer.com)

Preface

This book focuses on the multidisciplinary management of the patient with multiple injuries. The second edition of this book has been modified according to certain changes in trauma care. Trauma system changes can influence rescue conditions, the patient flow inside the hospital, and outcome. The selection of authors has been maintained as done for the first edition. All of them are experts in their particular fields. In addition, we have sought to include contributions from all over the world, thus respecting the fact that trauma is a global challenge.

Also, the second edition of the book has been expanded in terms of outcome assessment for certain injury types known to be responsible for long-term issues. Among these are bone infections, bone defects, and certain fracture types.

We hope that these changes will help improve trauma care and the challenges yet to come.

Aachen, Germany	Hans-Christoph Pape, MD, FACS
Temple Terrace, FL, USA	Roy Sanders, MD
Arlington, TX, USA	Joseph Borrelli, Jr., MD

Contents

1. **The Impact of Trauma on Society** 1
   Tahira Devji, Farrah Naz Hussain, and Mohit Bhandari

2. **Economic Aspects of Trauma Care** 9
   Tim Mathes, Christoph Mosch, and Michaela Eikermann

3. **Evidence-Based Orthopaedic Trauma Care** 15
   Martin Mangupli and Richard Buckley

4. **Inflammatory Changes and Coagulopathy in Multiply Injured Patients** 23
   Markus Huber-Lang and Florian Gebhard

5. **Pathophysiology of Polytrauma** 41
   Theodoros Tosounidis and Peter V. Giannoudis

6. **Head Injuries: Neurosurgical and Orthopedic Strategies** 55
   Philip F. Stahel and Michael A. Flierl

7. **Soft Tissue Injuries** 65
   Norbert Pallua and Stefan Bohr

8. **Chest Trauma** 87
   Mirjam B. de Jong, Marike C. Kokke, Falco Hietbrink, and Luke P.H. Leenen

9. **Abdominal Injuries: Indications for Surgery** 111
   Clay Cothren Burlew and Ernest E. Moore

10. **Management of Pelvic Ring Injuries** 127
    David J. Hak and Cyril Mauffrey

11. **Urological Injuries in Polytraumatized Patients** 143
    David Pfister and Axel Heidenreich

12. **Fracture Management** 157
    Roman Pfeifer and Hans-Christoph Pape

13. **Mangled Extremity: Management in Isolated Extremity Injuries and in Polytrauma** 169
    Mark L. Prasarn, Peter Kloen, and David L. Helfet

14	**Management of Spinal Fractures** Keith L. Jackson, Michael Van Hal, Joon Y. Lee, and James D. Kang	187
15	**The Management of the Multiply Injured Elderly Patient** Charles M. Court-Brown and N. Clement	201
16	**General Management in the Elderly: Preoperative and ICU** Alain Corcos and Andrew B. Peitzman	219
17	**Polytrauma in Young Children** Achim Braunbeck and Ingo Marzi	231
18	**Fracture Management in the Pregnant Patient** Erich Sorantin, Nima Heidari, Karin Pichler, and Annelie-Martina Weinberg	245
19	**Open Fractures: Initial Management** Michael Frink and Steffen Ruchholtz	261
20	**Vascular Injuries: Indications for Stents, Timing for Vascular and Orthopedic Injuries** Luke P.H. Leenen	277
21	**Management of Articular Fractures** Tak-Wing Lau and Frankie Leung	289
22	**Outcome and Management of Primary Amputations, Subtotal Amputation Injuries, and Severe Open Fractures with Nerve Injuries** William W. Cross III and Marc F. Swiontkowski	307
23	**High-Energy Injuries Caused by Penetrating Trauma** Yoram A. Weil and Rami Mosheiff	329
24	**Management of Traumatic Bone Defects** Richard P. Meinig and Hans-Christoph Pape	343
25	**Acute Soft Tissue and Bone Infections** Lena M. Napolitano	351
26	**Posttraumatic Acute and Chronic Osteomyelitis** John K. Sontich	371
27	**Management of Malunions and Nonunions in Patients with Multiple Injuries** Nicholas Greco, Peter Siska, and Ivan S. Tarkin	387
28	**Psychological Sequelae After Severe Trauma** Bianca Voss, Frank Schneider, and Ute Habel	399
29	**Outcome After Extremity Injuries** Boris A. Zelle	407
30	**Clinical Outcome of Pelvic and Spinal Fractures** Roman Pfeifer, Bilal M. Barkatali, Thomas Dienstknecht, and Hans-Christoph Pape	419

Index 427

The Impact of Trauma on Society

Tahira Devji, Farrah Naz Hussain, and Mohit Bhandari

Contents

1.1 Introduction 1
1.2 The Psychological Implications of Trauma 2
1.3 Chronic Pain and Disability Due to Trauma 3
1.4 Return to Work After Trauma 3
1.5 Global Trends and Perspectives on Trauma 4
Conclusion 5
References 6

T. Devji, BSc, MSc(c), Ms
Clinical Epidemiology and Biostatistics,
McMaster University, 293 Wellington St. N,
Suite 110, Hamilton, ON, Canada
e-mail: devjits@mcmaster.ca

F.N. Hussain, MD
Department of Obstetrics and Gynecology,
Mount Sinai St Luke's,
Roosevelt, NY, USA
e-mail: Fhussain@chpnet.org

M. Bhandari, MD, PhD, FRCSC (✉)
Division of Orthopaedic Surgery,
Department of Surgery, McMaster University,
293 Wellington St N, Suite 110,
Hamilton L8L 8E7, ON, Canada
e-mail: bhandam@mcmaster.ca

1.1 Introduction

Injury has become a major cause of fatality and disability in countries of all economic levels [1]. Nearly 16,000 people die from injuries each day and for each of these fatalities, several thousand individuals survive with permanently disabling injuries [2]. In the United States, trauma-related costs, such as lost wages, medical expenses, insurance administration costs, property damage, and employer costs, exceed $400 billion annually [3]. Despite this massive financial burden, the real cost can only be ascertained when one considers that trauma affects the youngest and most productive members of society [3]. Studies have shown that the functional outcome of trauma patients at 1 year or more following injury is below that of the normal population [4]. Many continue to suffer from residual problems such as long-term physical impairments, disabilities and handicaps that may even impact their ability to fully return to their previous work or way of life [4]. A substantial number of individuals who have suffered orthopaedic trauma may also possess less obvious forms of residual sequelae, such as emotional or psychosocial disabilities [5].

Reintegrating trauma patients who have sustained musculoskeletal injuries into society requires a multidisciplinary approach [6]. Therefore, knowledge about the impact of trauma on society is essential in adopting such an approach to orthopaedic trauma care.

1.2 The Psychological Implications of Trauma

Trauma is sudden and unexpected in nature and can be especially frightening for victims who may have lost their ability to comprehend and adapt to the unfamiliar situation around them [7]. The management of trauma, therefore, requires treatment not only of the immediate physical injuries, but also the behavioural and psychological aspects associated with the event, which can severely impact patient recovery [7]. Patient psychological status after orthopaedic trauma is a common source of complaints from patients and is a clinically relevant outcome [8]. As outcomes research has shifted its focus from physician-derived measures towards patient important outcomes, evidence of psychological distress as a consequence of orthopaedic trauma has come to light [9]. Many studies have reported high rates of psychological distress following trauma and have shown a strong association between psychological status and functional outcomes [9]. Starr et al. [10] surveyed 580 patients who had sustained orthopaedic trauma using the Revised Civilian Mississippi Scale for Posttraumatic Stress Disorder questionnaire. The authors reported 51 % of respondents met the criteria for diagnosis of posttraumatic stress disorder (PTSD) [10]. Moreover, patients with PTSD had significantly higher Injury Severity Scores (ISS) and Extremity Abbreviated Injury Scores (EAIS) [10]. Crichlow et al. [11] interviewed 161 orthopaedic trauma patients and found that the presence of clinically relevant depression was 45 %, as determined by the Beck Depression Inventory (BDIA). The authors also demonstrated a close correlation between the presence of depression and poorer scores on functional outcome measures, such as the Short Musculoskeletal Function Assessment (SMFA) [11].

Not only do psychological problems such as PTSD and depression pose an impact on functional outcomes, they also affect quality of life (QOL) [12]. Measures of QOL provide insight into how a disability may affect an individual's overall well-being, such as an individual's perception of his or her position in life with respect to goals, standards, concerns and expectations [12]. In an observational study investigating the extent of psychological symptoms of 215 patients following orthopaedic trauma, Bhandari et al. [8] reported that one in five met the threshold for psychological distress in all primary dimensions of the SCL-90-R. In particular, phobic anxiety and somatization (i.e. the expression of physical symptoms as a result of emotional or psychological distress) ranked high in comparison to age- and sex-matched population control subjects [8]. In terms of the relationship between psychological problems and patients' health-related quality of life, the authors found that the global severity of psychological symptoms were significantly associated with the Physical Component and Mental Component summary scores of the Medical Outcomes Study 36-item Short Form (SF-36) [8].

Although few other studies in the orthopaedic trauma literature have considered the impact of psychological distress on QOL following the acute phase of injury, studies involving patients with other injuries have come to similar conclusions concerning this relationship. O'Donnell et al. [12] examined the 12-month outcomes of 363 consecutive admissions to a Level I trauma service and found that an individual's acute psychological response (e.g. anxiety and depression) directly predicted QOL, as measured by the WHOQoL-Bref, as well as level of disability. More specifically, anxiety and depression were associated with PTSD, which in turn was associated with lower levels of QOL and functioning [12].

Psychological morbidity is common following orthopaedic trauma and interventions are required to prevent further sequelae. Management of acute pain is of primary importance as injured patients are less likely to respond to psychotherapeutic interventions [7]. In a systematic review and meta-analysis of randomized controlled trials (RCTs) of psychological interventions aimed at preventing or treating PTSD within 3 months of a traumatic event, Roberts et al. [13] found that cognitive behavioural therapy was significantly more effective than waiting list or usual care in reducing traumatic stress symptoms in already symptomatic individuals, and particularly for those patients who met the diagnostic criteria for

PTSD; however, the magnitude of effect varied considerably. In the past decade, collaborative care (CC), a multifaceted disease management strategy (e.g. combined case management, pharmacotherapy, and psychotherapy) has gained support due to the model's comprehensive nature in treatment delivery for patients with medical and psychiatric disorders [14]. In a recent RCT to test a multifaceted CC intervention targeting PTSD, Zatzick et al. [14] failed to demonstrate a reduction in symptoms of PTSD. Also, there was no association of early evaluation and supportive intervention with attenuated PTSD symptoms for patients in the CC condition [14]. Further large-scaled CC trials are required to assess the effectiveness of this intervention strategy with respect to PTSD, functional outcome improvements and cost-effectiveness [14]. From this growing body of research, it is evident that the trauma patient's psychological state is as important as injury severity and physical health for injury recovery and long-term outcomes [12]. For a complete discussion of PTSD and psychological sequelae after severe trauma, please see Chap. 28.

1.3 Chronic Pain and Disability Due to Trauma

Chronic or ongoing pain includes several symptoms and conditions, including acute post-traumatic pain, depression, hostility, anxiety, sleep and rest disturbances [15, 16]. Many trauma patients suffer from long-term impairments, disabilities and handicaps, and at least half of all major trauma patients are left with one or more residual problems [4]. Therefore, understanding the determinants of long-term functional consequences following trauma is important in order to improve the chances of a patient's recovery [4]. Trauma has been proposed as a causal factor or trigger of chronic or persistent pain [15]. Chronic pain affects as many as 50 million Americans and is one of the leading causes of disability among those under the age of 45 [15]. The overall productivity lost due to chronic pain is estimated to be four times more than productivity lost due to lost workdays alone [15, 17]. In a prospective analysis of the prevalence and early predictors of chronic pain in a cohort of severe lower extremity trauma patients, Castillo et al. [15] found that more than a quarter of the study group reported that their pain highly interfered with daily activities. Pain also has other consequences for its victims, including psychological regression [16]. Those who suffer from chronic pain also use five times more health services than the general population [15, 18].

As surgeons, we know that pain is an inevitable result of traumatic injury and the accompanying healing process. However, why do patients continue to endure pain long after they have been treated? The biomedical model of health focuses on pain as the result of a physical injury [19]. This makes it difficult to clinically explain the presence of disability after the pathology related to the injury has healed [19]. Studies focusing on trauma populations suggest that factors during the course of recovery other than the injury are critical to the development of persistent pain and associated functional impairment [19]. Such factors include high initial pain intensity, PTSD, worker's compensation status, education, low recovery expectations and depression [19]. In the aforementioned study by Castillo et al. [15], several early predictors of chronic pain were identified at baseline, including having less than a high school education, having less than a college education, low self-efficacy for return to daily activities, and high levels of alcohol consumption. In addition, high reported acute pain intensity, sleep and rest dysfunction, depression and anxiety at 3 months post-discharge were found to be predictors of chronic pain at 7 years [15].

1.4 Return to Work After Trauma

Return to work is defined as a complete or almost complete return to pre-injury full-time paid employment [20]. While traumatic injury often results in psychological distress and chronic pain for its victims, another burden it poses to society is the long-term impairment of its most productive members and a subsequent loss of working days [21]. Survivors of severe injury are able to achieve a QOL comparable to the normal population once they have returned to their pre-injury occupation [22, 23]. Because it increases an individual's sense of self-worth and personal fulfilment, return to work is indicative of successful social reintegration after major trauma [20]. Return to work is therefore one of the most important methods by which to evaluate treatment outcomes [16, 21]. In the United States, more days are lost to work as a result of chronic pain than any other medical reason [15]. Road traffic injuries in particular are a major cause of trauma and have resulted in greater than 1 million deaths and 50 million injuries worldwide [24]. In 2001, 2.1 million people aged 18–65 were victims of car crashes in the United States [25]. Cumulatively, victims of these crashes lost an estimated 60.8 million days of work [25].

To realize the impact of trauma on society, it is important to consider the factors contributing to lost productivity among survivors of injury [16, 21]. Factors contributing to a delayed return to work include injury severity, pre-injury characteristics of the patient (i.e. socioeconomic status, self-efficacy, health habits, social support with respect to the home and workplace), characteristics of the pre-injury occupation (i.e. white- versus blue-collar work, physical demands, tenure, job satisfaction and flexibility), motivation to work, receipt of disability compensation and baseline measures of physical functioning, pain, anxiety and depression [16, 19, 21]. Patients are especially delayed from returning to work if they have significant physical disabilities, psychosocial impairments, cognitive impairments or changes to their personality [20]. Recovery times can be lengthy, even taking longer than a year in certain cases [20]. Patients may also be unable to return to their pre-injury job due to the replacement of their previous roles [20]. These factors can render a return to pre-injury work status challenging for many victims of trauma and can therefore pose a significant financial and social burden to victims as well as their families [19].

The focus of return to work following victims of major trauma has been on returning to paid work, and thus evidence concerning achievement of satisfactory levels of unpaid activities while returning to full-time employment is scarce in the literature [26]. van Erp et al. [26] looked at four specific domains of unpaid work in patients who sustained a major trauma and were full-time employed at the time of the incident. The strongest predictors of limitations in these unpaid work items (i.e. household work, shopping, caring for children, and odd jobs around the house) were the percentage of permanent impairment, followed by the level of participation (return to work) [26]. Co-morbidity, lower limb fractures and female gender were also determinants of limitations in unpaid activities. Resuming paid work following major trauma showed not to be associated with reductions in unpaid work, that is, the hypothesis that limited total energy levels may cause resumed paid work to mitigate participation in unpaid activities, could not be confirmed. Individuals who return to full-time employment seem to experience few or no reductions in all four types of unpaid activities, whereas those who return to part-time work or not at all also experience several limitations in unpaid work [26]. The results of this study underline the public health implications of major trauma in the setting of lost productivity in paid work, but also, the direct impact on personal environments and satisfaction in the private lives of trauma victims [26]. Future research concerning return to work should emphasize full-time workers and also part-time workers, as this cohort contributes substantially to the economy. The effectiveness of interventions that may increase return to work and patient satisfaction in trauma victims should also be a future directive of research in this area [26].

1.5 Global Trends and Perspectives on Trauma

Unintentional injuries, including road traffic injuries, drowning, burns, poisoning and falls, are an increasingly significant public health issue, constituting for 6.6 % of global mortality [27]. The growing burden of trauma is disproportionately concentrated in low- and middle-income countries (LMICs) [27, 28]. According to the World Health Organization, in 2004 over 91 % of unintentional injury-related deaths and 94 % of disability-adjusted life years were lost in LMICs [27]. The worldwide rate of unintentional injuries is 61 per 100,000 population per year, with the highest rate in the Southeast Asian region (80 per 100,000) and lowest in the American region (39 per 100,000) [27]. Road traffic injuries are a leading cause of global morbidity and account for the largest proportion of unintentional injury deaths (33 %), followed by falls (11 %) and drowning (10 %) [27]. There are apparent regional differences in the mechanism of certain injuries. For instance, the distribution of road traffic deaths by road user group varies across countries of different income levels. A review conducted by Naci et al. [29] reported that road traffic fatalities among pedestrians are estimated to be 45 % in low-income countries, compared with 29 % in middle-income countries and 18 % in high-income countries. Sixty-three percent of road traffic fatalities in high-income countries are among motorized four wheelers, 40 % in middle-income countries and 34 % in low-income countries. The global costs for road traffic injuries was estimated to be over US $518 billion and over US $65 billion in LMICs [30]. Most traumatic injury cost estimates relate to road traffic injuries; however, there is a critical gap in comprehensive estimates regarding the global economic burden of unintentional injuries [27].

A consistent theme identified in trauma and injury research is the paucity of researchers and research institutions in LMICs. Despite the disproportionately large burden of injury on LMICs, these countries are least likely to have effective surveillance systems implemented for monitoring injury trends [27]. For instance, several developing countries rely on hospital-based death-reporting systems, which undercount deaths occurring outside of hospitals, and in turn underestimate injuries [27]. Furthermore, only countries with highly developed health infrastructures have national surveillance in place for nonfatal injuries, making data collection efforts on surviving trauma victims a sizeable challenge [27]. Thus, surveillance systems and population-based data are crucial for research and prevention efforts, as well as attracting the attention of policy makers and community leaders [27].

Ultimately, the greatest burden of injury falls on those countries with the weakest evidence to guide and implement intervention strategies, the scarcest resources, and the least developed infrastructure to effect change [27]. Thus, research investments and acquisition for injury prevention and control, especially in LMICs, should concentrate on core areas such as, epidemiology, interventions, economic analysis, social sciences and policy [27]. Further insight is required regarding the causes, extent and nature of trauma and injury risks relevant to LMICs [28]. Efforts should be centred on improvements in diagnosis, treatment and innovative intervention strategies for management of acute and long-term sequelae of traumatic injuries [28]. Studies focused on understanding of the context in which trauma and injury occur in LMICs, development and evaluation of public education strategies and effectiveness of primary prevention will be beneficial for prevention of unintentional injuries on a global scale [28]. Defining the economical impact of injury-related costs on society and underlining the cost-effectiveness of interventions to local and national policy makers is necessary to influence policies pertaining to injury prevention and control [27]. Understanding the attitudes and perceptions of people surrounding the burden of trauma and injury causation is imperative for public engagement and the success of interventions [27].

Conclusion

Although much of this chapter focused on trauma victims themselves as members of society, a final thought to consider is the impact of trauma on the families of victims. Having someone close become seriously injured can be an immense source of psychological stress for family members [7]. Many may exhibit the behaviour of 'hovering', which is defined as an initial sense of confusion, distress and uncertainty prior to seeing the patient and understanding the diagnosis and prognosis [7, 31]. It can also be difficult for relatives to cope with their sudden change in role and status in the life of a loved one experiencing trauma [7]. Feelings of isolation from other family members, financial constraints and transportation concerns may also surface. Such problems are only amplified by a lack of medical knowledge [7]. Hence, comprehensive trauma services should consider providing support to family members alongside severely damaged patients. Provision of comprehensive care of trauma patients is essential. While experiencing trauma, patients become lost in an unfamiliar and threatening situation. Many become dependent, losing control over their environment and personal well-being. During the injury, treatment and recovery procedures, and for years afterward, patients can experience immense psychological and emotional distress, chronic pain and resultant productivity loss. Although trauma can happen to anyone, its tendency to affect individuals during their youngest and most productive years poses a significant impact on society [3]. Therefore, knowledge about this impact is imperative in adopting an interdisciplinary approach to orthopaedic trauma care. It should be recognized that trauma is a global health burden, which is disproportionately concentrated in low- and middle-income countries. Countries with poorly established health infrastructures have not placed a high priority on injury prevention as a public health issue [27]. Thus, there is a large demand for investment in relevant research for injury prevention and control, especially in developing countries [27].

References

1. Mock C, Quansah R, Krishnan R, Arreola-Risa C, Rivara F. Strengthening the prevention and care of injuries worldwide. Lancet. 2004;363(9427):2172–9.
2. Krug EG, Sharma GK, Lozano R. The global burden of injuries. Am J Public Health. 2000;90(4):523–6.
3. Surgeons ACO. Advanced trauma life support. 7th ed. Chicago: American College of Surgeons; 2004.
4. Holtslag HR, van Beeck EF, Lindeman E, Leenen LP. Determinants of long-term functional consequences after major trauma. J Trauma. 2007;62(4):919–27.
5. Ponsford J, Hill B, Karamitsios M, Bahar-Fuchs A. Factors influencing outcome after orthopedic trauma. J Trauma. 2008;64(4):1001–9.
6. Spiegel DA, Gosselin RA, Coughlin RR, Kushner AL, Bickler SB. Topics in global public health. Clin Orthop Relat Res. 2008;466(10):2377–84.
7. Mohta M, Sethi AK, Tyagi A, Mohta A. Psychological care in trauma patients. Injury. 2003;34(1):17–25.
8. Bhandari M, Busse JW, Hanson BP, Leece P, Ayeni OR, Schemitsch EH. Psychological distress and quality of life after orthopedic trauma: an observational study. Can J Surg. 2008;51(1):15–22.
9. Starr AJ. Fracture repair: successful advances, persistent problems, and the psychological burden of trauma. J Bone Joint Surg Am. 2008;90 Suppl 1:132–7.
10. Starr AJ, Smith WR, Frawley WH, Borer DS, Morgan SJ, Reinert CM, et al. Symptoms of posttraumatic stress disorder after orthopaedic trauma. J Bone Joint Surg Am. 2004;86-a(6):1115–21.
11. Crichlow RJ, Andres PL, Morrison SM, Haley SM, Vrahas MS. Depression in orthopaedic trauma patients. Prevalence and severity. J Bone Joint Surg Am. 2006;88(9):1927–33.
12. O'Donnell ML, Creamer M, Elliott P, Atkin C, Kossmann T. Determinants of quality of life and role-related disability after injury: impact of acute psychological responses. J Trauma. 2005;59(6):1328–34; discussion 34–5.
13. Roberts NP, Kitchiner NJ, Kenardy J, Bisson JI. Systematic review and meta-analysis of multiple-session early interventions following traumatic events. Am J Psychiatry. 2009;166(3):293–301.
14. Zatzick D, Roy-Byrne P, Russo J, Rivara F, Droesch R, Wagner A, et al. A randomized effectiveness trial of stepped collaborative care for acutely injured trauma survivors. Arch Gen Psychiatry. 2004;61(5):498–506.
15. Castillo RC, MacKenzie EJ, Wegener ST, Bosse MJ. Prevalence of chronic pain seven years following limb threatening lower extremity trauma. Pain. 2006;124(3):321–9.
16. MacKenzie EJ, Morris Jr JA, Jurkovich GJ, Yasui Y, Cushing BM, Burgess AR, et al. Return to work following injury: the role of economic, social, and job-related factors. Am J Public Health. 1998;88(11):1630–7.
17. Blyth FM, March LM, Nicholas MK, Cousins MJ. Chronic pain, work performance and litigation. Pain. 2003;103(1–2):41–7.
18. Von Korff M, Dworkin SF, Le Resche L. Graded chronic pain status: an epidemiologic evaluation. Pain. 1990;40(3):279–91.
19. Clay FJ, Newstead SV, Watson WL, Ozanne-Smith J, Guy J, McClure RJ. Bio-psychosocial determinants of persistent pain 6 months after non-life-threatening acute orthopaedic trauma. J Pain. 2010;11(5):420–30.
20. Holtslag HR, Post MW, van der Werken C, Lindeman E. Return to work after major trauma. Clin Rehabil. 2007;21(4):373–83.

21. MacKenzie EJ, Bosse MJ, Kellam JF, Pollak AN, Webb LX, Swiontkowski MF, et al. Early predictors of long-term work disability after major limb trauma. J Trauma. 2006;61(3):688–94.
22. Hou WH, Tsauo JY, Lin CH, Liang HW, Du CL. Worker's compensation and return-to-work following orthopaedic injury to extremities. J Rehabil Med. 2008;40(6):440–5.
23. Post RB, van der Sluis CK, Ten Duis HJ. Return to work and quality of life in severely injured patients. Disabil Rehabil. 2006;28(22):1399–404.
24. Sharma BR. Road traffic injuries: a major global public health crisis. Public Health. 2008;122(12):1399–406.
25. Ebel BE, Mack C, Diehr P, Rivara FP. Lost working days, productivity, and restraint use among occupants of motor vehicles that crashed in the United States. Inj Prev. 2004;10(5):314–9.
26. van Erp S, Holtslag HR, van Beeck EF. Determinants of limitations in unpaid work after major trauma: a prospective cohort study with 15 months follow-up. Injury. 2014;45:629–34.
27. Chandran A, Hyder AA, Peek-Asa C. The global burden of unintentional injuries and an agenda for progress. Epidemiol Rev. 2010;32(1):110–20.
28. Hofman K, Primack A, Keusch G, Hrynkow S. Addressing the growing burden of trauma and injury in low- and middle-income countries. Am J Public Health. 2005;95(1):13–7.
29. Naci H, Chisholm D, Baker TD. Distribution of road traffic deaths by road user group: a global comparison. Inj Prev. 2009;15(1):55–9.
30. Norton R, Hyder AA, Bishai D, Peden M. Unintentional injuries. In: Jamison DT, Breman JG, Measham AR, Alleyne G, Claeson M, Evans DB, et al., editors. Disease control priorities in developing countries. Washington, DC: World Bank/The International Bank for Reconstruction and Development/The World Bank Group; 2006.
31. Jamerson PA, Scheibmeir M, Bott MJ, Crighton F, Hinton RH, Cobb AK. The experiences of families with a relative in the intensive care unit. Heart Lung. 1996;25(6):467–74.

Economic Aspects of Trauma Care

Tim Mathes, Christoph Mosch, and Michaela Eikermann

Contents

2.1 Introduction 9
2.2 Cost of Illness 10
2.3 Prevention 11
2.4 Economic Evaluation of Health Technologies in Trauma Care 11
2.5 Economic Aspects of Multiple/Severe Injuries in Hospital Care ... 12
References 13

2.1 Introduction

Worldwide each year about 5.8 million people die from trauma and injuries, which accounts for 10 % of all deaths. Hence, injuries are one of the most frequent causes of deaths (Fig. 2.1) and the trend is rinsing. In many countries, injuries are the main cause for death of young and middle-aged people [1, 2]. Furthermore, injuries are a main cause for lifelong disabilities. It is estimated that 11.2 of all disability adjusted life years (DALYS) are caused by injuries [3, 4].[1] The pandemic character [5] of injuries by itself and the fact that mostly working aged adults are affected shows clearly the burden for national economies that are caused by injuries.

The deaths caused by injuries and the concomitant loss of productivity are just the tip of the iceberg. In addition to injuries that result in death, there are even more injuries that are associated with health care costs of the survivors because of long hospitalizations, rehabilitation and/or lifelong need for care because of disability [6, 7].

T. Mathes (✉) • C. Mosch • M. Eikermann
Institute for Research in the Operative Medicine (IFOM), Private University Witten/Herdecke gGmbH, Ostmerheimer Str. 200, 51109 Cologne, Germany
e-mail: tim.mathes@uni-wh.de; christoph.mosch@uni-wh.de; michaela.eikermann@uni-wh.de

[1] DALY: Sum of the Years of Life Lost (YLL: number of deaths x standard life expectancy at age of death in years) due to premature mortality in the population and the Years Lost due to Disability (YLD: number of incident cases x disability weight x average duration of the case until remission or death (years)) (http://www.who.int/healthinfo/global_burden_disease/metrics_daly/en/).

© Springer-Verlag Berlin Heidelberg 2016
H.-C. Pape et al. (eds.), *The Poly-Traumatized Patient with Fractures: A Multi-Disciplinary Approach*, DOI 10.1007/978-3-662-47212-5_2

2.2 Cost of Illness

Like any other illness, each trauma that is treated in primary care facilities, emergency departments or hospitals involves the use of medical resources. Most of the trauma care is very resource intensive because many different services and medical disciplines are involved (Fig. 2.2) partly over a very long period. The direct medical costs include cross-sectorally the treatment costs by the physician, nursing costs, pharmaceuticals, medical products and often intensive care measures like mechanical ventilation.

The largest portion of costs for trauma treatment arise in the first year after occurrence (e.g. USA $76,210 per patient) [9]. In trauma care, most direct cost arise in the hospitalization period. Especially, the inpatient treatment of multiple trauma is associated with high consumption of medical resources (e.g. UK £20.742 per patient, Germany € 21.866 per patient) [10, 11].

Even after 1 year, the treatment of the long-term consequences of major trauma like pain treatment or rehabilitation can cause high direct medical costs [12, 13]. Thus, for example in the USA, even in the year post discharge, multiple trauma is associated with cost per patient of $78,577 [11].

The direct medical costs often make up only a small part of the total costs [14]. Because trauma often results in absenteeism, disability and premature death of young and middle-aged people, the costs due to loss of productivity play a mature role for the national economy in the total cost that arise by trauma and are more important than in other conditions were mainly older people are affected [2, 15]. In major trauma only about 60 % of the people return to full-time employment [16]. In the USA alone, the indirect cost because of productivity loss due to injuries are about $326 billion per annum, which is about four times more as much as the direct medical costs in the treatment of these patients [17]. These estimates do not include the intangible cost that can be associated with trauma. Also if there is no disability that prevents

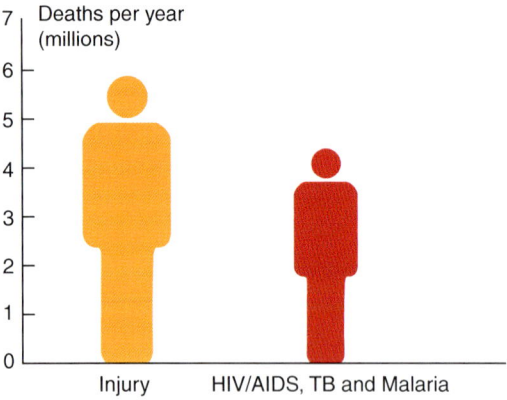

Fig. 2.1 The scale of the problem [1]

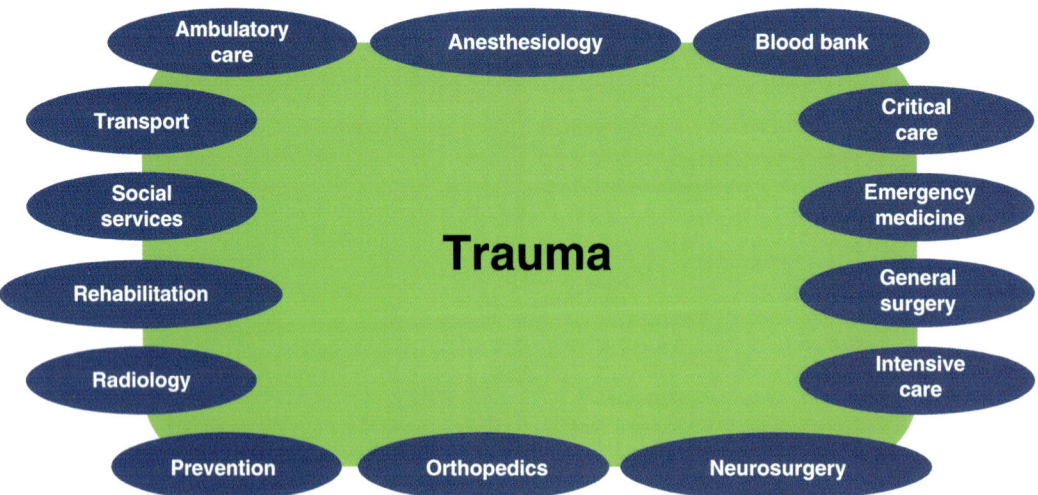

Fig. 2.2 Services and disciplines involved in trauma care (Modified from [8])

the patients from work, there are intangible costs for patients due to reduced quality of life [14].

All together, injuries make up a large proportion of the gross national product of national economies. The World Health Organization (WHO) estimates that the cost for traffic injuries by itself amount between 1 and 2 % of the gross national product in low- and high-income countries, respectively [18]. Apart from the costs of injuries for the society, if the main wage earner is affected by the incapacity to work, the resulting loss of income may mean a poor financial situation for all persons living in the household.

2.3 Prevention

Nearly 25 % of the injuries are caused by road traffic injuries. Road traffic injuries and also other unintentional injuries (e.g. falls) are preventable in most cases [1]. Therefore, injuries should be considered like diseases (such as cancer) as public health problems that are often preventable and respond to targeted interventions [18]. Although the improving in trauma care in western countries the prevention has a particular importance considering that many injuries are preventable plus the high cost for trauma care and loss of productivity. For middle- and low-income countries, it is estimated that only if the prevention effort and trauma care is increased so that the mortality caused by injuries is on the same level as in high-income countries, the economic benefit is about $250 billion.

2.4 Economic Evaluation of Health Technologies in Trauma Care

Economic evaluations compare the costs and benefits (e.g. quality of life, mortality) of a certain health technology[2] [19], which means that

[2] The application of organized knowledge and skills in the form of devices, medicines, vaccines, procedures and systems developed to solve a health problem and improve quality of life. It is used interchangeably with health-care technology.

the inputs and outputs are assessed at the same time. An incremental cost-effectiveness ratio (ICER) is calculated based on the differences of the effects/benefits and the difference of costs between the compared health technologies (Eq. 2.1) [20].

$$\text{ICER} = \frac{\text{Cost}_a - \text{Cost}_b}{\text{Effects}_a - \text{Effects}_b} \quad (2.1)$$

The results are associated with different decisions depending on differences between costs and the differences between effects of the compared health technologies (Fig. 2.3). Under the assumption that the existing standard of care is mostly replaced only by health technologies with higher benefits, in particular, the decision in the case of higher cost and higher benefits is difficult because it implies that a decision about the value of an additional unit of benefit has to be made. This value for an additional unit of benefit should reflect the individual value of the respective society [22].

Economic evaluation of health technologies for trauma care is difficult because the health technologies are often more complex, compared to drug therapy [23] (see Fig. 2.1). The challenge in evaluating complex interventions arises by the variation of the active elements (e.g. staff characteristics) of the intervention. Therefore, it is difficult to specify the health technology, to identify the effective components and to replicate the health technology in different (health care) settings [24].

There is also the fact that trauma has different clinical pictures and disease patterns. Thus, the cost or cost-effectiveness can vary widely depending on many different factors like age, injury severity score or involved body parts [10, 25, 26].

The consequences of trauma often affect a long time period or even the whole lifetime of patients. To avoid misleading results, the economic evaluation has to cover a sufficient time horizon to ensure that all benefits (increased quality of life) and costs associated with the health technology are captured [27]. For that reason, long follow-ups can be necessary for appropriate economic evaluations. However, in clinical practice, long-term follow-ups are challenging or

Fig. 2.3 Cost-effectiveness plan [21]

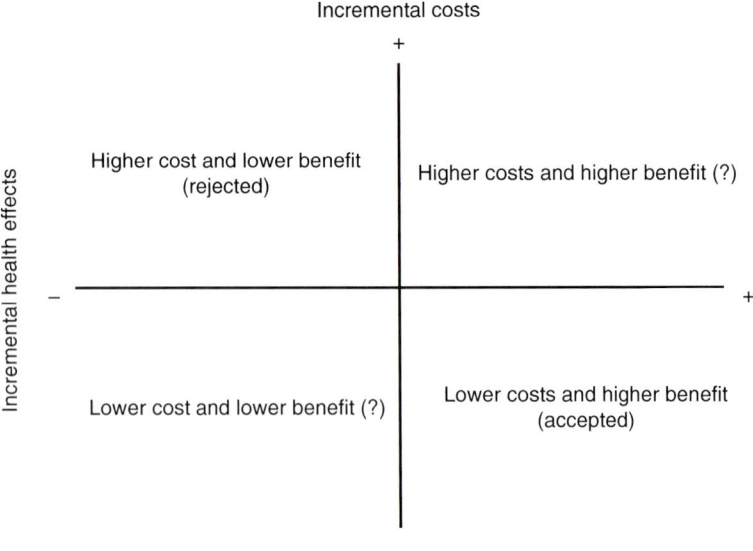

even not feasible. In such cases, the trial-based results should be extrapolated.

The recovery of working ability is of particular importance in trauma care in order to avoid productivity loss. Thus, valid estimates of the cost for the society should incorporate the possible savings due to reduced productive loss that are attributable to the health technology because this can lead to cost-effective results, even if the health technology causes substantial direct medical costs. Especially economic evaluations of health technologies that aim among other on the recovery of working ability like rehabilitation should consider the productivity loss. However, often the cost for trauma services is high [13, 28]. Consequently, also if the indirect costs are incorporated, the benefit of the health technology has to be large to reach cost-effectiveness [28].

A further consequence of the complexity and need for long observation periods is that the health technology has to be evaluated with an observational design because experimental designs are not always practicable [23]. The mentioned difficulties in the evaluation are probably one reason that many of the applied health technologies are not evaluated sufficiently [13].

2.5 Economic Aspects of Multiple/Severe Injuries in Hospital Care

As a consequence of the differences in clinical pictures and disease patterns, the complex medical challenges and the variety of involved disciplines in the inpatient treatment of multiple or severe injured/poly-traumatized patients, it is necessary to find suitable payment modes for the hospitals and their single units. In health care systems with reimbursement to the hospitals by daily charges (and if any capital costs), it is almost uncomplicated to reward every clinical department in an appropriate scope. In contrast to that settings with an allocation of public financed or insurance-related budgets and a compensation of hospital services based on diagnosis/case-specific lump-sum rewards (e.g. German OR Australian diagnosis related groups (DRG)) require agreements and budget transfers within the hospitals itself ("income splitting"). The fact that the remuneration bases (apart from patients and services characteristics like the duration of mechanical ventilation) primarily on the individual discharge diagnosis has as a consequence that mostly the discharging unit receives the major part of the funds. In order to distribute this amount adequately to the involved departments and disciplines, it demands a labour-intensive internal cost allocation. Due to missing market prices in hospitals, there are particular demands to quantify the rendered deliverables and services monetarily [29, 30]. To calculate the actual direct medical and non-medical costs, it is necessary to use the *bottom-up approach*. Different from the *top-down approach* (which implies the total cost of care and calculates the "average" arithmetic mean by dividing the total cost by the number of the affected patients), the bottom-up approach needs a detailed and complete itemization of services and costs with a sophisticated acquisition, weighting and transparent depiction of all services based on the respective personnel and tangible use of resources [31, 32]. Moreover, this practice allows a structured overview of all processes which are factually imputable to the treated patients.

An elementary determinant affecting the cost coverage of the case-based lump-sum compensation is the correct classification into the respective diagnosis-related case group. The heterogeneity of these patients and the concomitant difficulty to diversify the respective (diagnosis-related) case group impede a demand-actuated remuneration [33]. Additional to the actual treatment, there are high extra costs for specialized hospitals with trauma centres which are equipped to treat multiple/severely injured patients who usually appear as an emergency on 24 h a day without any warning and a need of a broad range of clinical skills and resources. Already the provision of separately necessary structures regarding staff, facilities and (technical) equipment increases the intrahospital costs per patient immensely [34, 35]. Besides high expenditures especially in the shock and operating room as well as in the intensive care unit, for high-priced medications and an outstanding overall length of stay, this fact can be seen as one important reason why the assigned lump-sum compensations may not cover the expenses of the hospitals in health care systems with a reimbursement through DRGs. Different German studies show a significant and enduring funding gap of barely 4300–13,000 € per treated severely injured patient [10, 35–38] so that in consequence hospitals with particular trauma centres get into financial hardship or are incentived to select severe injured patients and to cut the treatment of such patients back.

References

1. World Health Organization (WHO). Injuries and violence – the facts. Geneva: WHO; 2010.
2. Centers for Disease Control and Prevention National Center for Injury Prevention and Control (NCIPC). Injury: the leading cause of death among persons 1–44, 14 Apr 2014. Available from: http://www.cdc.gov/injury/overview/leading_cod.html.
3. Murray CJL, Vos T, Lozano R, Naghavi M, Flaxman AD, Michaud C, et al. Disability-adjusted life years (DALYs) for 291 diseases and injuries in 21 regions, 1990? 2010: a systematic analysis for the Global Burden of Disease Study 2010. Lancet. 2012;380(9859):2197–223.
4. World Health Organization (WHO). Metrics: disability-adjusted life year (DALY), 14 Apr 2014. Available from: http://www.who.int/healthinfo/global_burden_disease/metrics_daly/en/.
5. Murray CJL, Lopez AD. Alternative projections of mortality and disability by cause 1990–2020: Global Burden of Disease Study. Lancet. 1997;349(9064):1498–504.
6. Bederman SS, Murnaghan O, Malempati H, Lansang E, Wilkinson M, Johnston E, et al. In-hospital mortality and surgical utilization in severely polytraumatized patients with and without spinal injury. J Trauma. 2011;71(4):E71–8. Epub 2011/03/15.
7. Rivara FP, Dicker BG, Bergman AB, Dacey R, Herman C. The public cost of motorcycle trauma. JAMA. 1988;260(2):221–3.
8. San Francisco General Hospital and Trauma Center. Trauma care, University of San Francisco, 14 Apr 2014. Available from: http://sfgh.ucsf.edu/trauma-care.
9. Weir S, Salkever DS, Rivara FP, Jurkovich GJ, Nathens AB, Mackenzie EJ. One-year treatment costs of trauma care in the USA. Expert Rev Pharmacoecon Outcomes Res. 2010;10(2):187–97. Epub 2010/04/14.
10. Rösch M, Klose T, Leidl R, Gebhard F, Kinzl L, Ebinger T. Kostenanalyse der Behandlung polytraumatisierter Patienten. Unfallchirurg. 2000;103(8):632–9.
11. Sikand M, Williams K, White C, Moran CG. The financial cost of treating polytrauma: implications for tertiary referral centres in the United Kingdom. Injury. 2005;36(6):733–7.
12. Parsons B, Schaefer C, Mann R, Sadosky A, Daniel S, Nalamachu S, et al. Economic and humanistic burden of post-trauma and post-surgical neuropathic pain among adults in the United States. J Pain Res. 2013;6:459–69. Epub 2013/07/05.
13. Khan F, Amatya B, Hoffman K. Systematic review of multidisciplinary rehabilitation in patients with multiple trauma. Br J Surg. 2012;99(S1):88–96.

14. Haeusler J-MC, Tobler B, Arnet B, Huesler J, Zimmermann H. Pilot study on the comprehensive economic costs of major trauma: consequential costs are well in excess of medical costs. J Trauma Acute Care Surg. 2006;61(3):723–31. doi:10.1097/01.ta.0000210453.70742.7f.
15. Ringburg AN, Polinder S, van Ierland MC, Steyerberg EW, van Lieshout EM, Patka P, et al. Prevalence and prognostic factors of disability after major trauma. J Trauma. 2011;70(4):916–22. Epub 2010/11/04.
16. Holtslag HR, Post MW, van der Werken C, Lindeman E. Return to work after major trauma. Clin Rehabil. 2007;21(4):373–83.
17. Corso P, Finkelstein E, Miller T, Fiebelkorn I, Zaloshnja E. Incidence and lifetime costs of injuries in the United States. Inj Prev. 2006;12(4):212–8. Epub 2006/08/05.
18. World Health Organization (WHO). World report on road traffic injury prevention. Geneva: WHO; 2004.
19. Drummond MF, Stoddart GL, Torrance GW. Methods for the economic evaluation of health care programmes. 3rd ed. Oxford: Oxford University Press; 2005. 396 p.
20. World Health Organization (WHO). Health technology assessment, Cited 14 Apr 2014. Available from: http://www.who.int/medical_devices/assessment/en/.
21. Mathes T, Walgenbach M, Antoine S-L, Pieper D, Eikermann M. Methods for systematic reviews of health economic evaluations: a systematic review, comparison, and synthesis of method literature. Med Decis Making. 2014;34:826–40.
22. Coast J. Is economic evaluation in touch with society's health values? BMJ. 2004;329(7476):1233–6.
23. Craig P, Dieppe P, Macintyre S, Michie S, Nazareth I, Petticrew M. Developing and evaluating complex interventions: the new Medical Research Council guidance. BMJ. 2008;337:a1655.
24. Shiell A, Hawe P, Gold L. Complex interventions or complex systems? Implications for health economic evaluation. BMJ. 2008;336(7656):1281–3.
25. Morris S, Ridley S, Lecky FE, Munro V, Christensen MC. Determinants of hospital costs associated with traumatic brain injury in England and Wales. Anaesthesia. 2008;63(5):499–508. Epub 2008/04/17.
26. MacKenzie EJ, Weir S, Rivara FP, Jurkovich GJ, Nathens AB, Wang W, et al. The value of trauma center care. J Trauma. 2010;69(1):1–10. Epub 2010/07/14.
27. Weinstein MC, O'Brien B, Hornberger J, Jackson J, Johannesson M, McCabe C, et al. Principles of good practice for decision analytic modeling in health-care evaluation: report of the ISPOR Task Force on Good Research Practices–modeling studies. Value Health. 2003;6(1):9–17. Epub 2003/01/22.
28. Delgado MK, Staudenmayer KL, Wang NE, Spain DA, Weir S, Owens DK, et al. Cost-effectiveness of helicopter versus ground emergency medical services for trauma scene transport in the United States. Ann Emerg Med. 2013;62(4):351–64.e19. Epub 2013/04/16.
29. Conrad HJ. Controlling im Krankenhaus: Controlling als Instrument zur Sicherung des wirtschaftlichen Erfolges von Krankenhäusern. Kulmbach: Baumann-Fachverl; 2008.
30. Hesse S, Boyke J, Zapp W. Innerbetriebliche Leistungsverrechnung im Krankenhaus: Verrechnungskonstrukte und Wirkungen für Management und Controlling. Wiesbaden: Springer Fachmedien; 2014.
31. Busch H-P. Interne Leistungsverrechnung im Krankenhaus am Beispiel eines "Profitcenters". Das Krankenhaus. 2006;12:1109–17.
32. Rowell D, Connelly L, Webber J, Tippett V, Thiele D, Schuetz M. What are the true costs of major trauma? J Trauma. 2011;70(5):1086–95. Epub 2011/03/12.
33. Flohe S, Buschmann C, Nabring J, Merguet P, Luetkes P, Lefering R, et al. Definition of polytrauma in the German DRG system 2006. Up to 30% "incorrect classifications". Unfallchirurg. 2007;110(7):651–8. Epub 2007/07/10. Polytraumadefinition im G-DRG-System 2006. Bis zu 30% "Fehlgruppierungen".
34. Taheri PA, Butz DA, Lottenberg L, Clawson A, Flint LM. The cost of trauma center readiness. Am J Surg. 2004;187(1):7–13. Epub 2004/01/07.
35. Schmelz A, Ziegler D, Beck A, Kinzl L, Gebhard F. Costs for acute, stationary treatment of polytrauma patients. Unfallchirurg. 2002;105(11):1043–8. Epub 2002/10/29. Akutstationare Behandlungskosten polytraumatisierter Patienten.
36. Garving C, Santosa D, Bley C, Pape HC. Cost analysis of emergency room patients in the German diagnosis-related groups system: a practice relevant depiction subject to clinical parameters. Unfallchirurg. 2014;117:716–22. Epub 2013/08/10. Kostenanalyse von Schockraumpatienten im DRG-System : Eine praxisnahe Abbildung in Abhangigkeit klinischer Parameter.
37. Juhra C, Franz D, Roeder N, Vordemvenne T, Raschke MJ. Classification of severely injured patients in the G-DRG System 2008. Unfallchirurg. 2009;112(5):525–32. Epub 2009/03/17. Abbildung des schwer verletzten Patienten im G-DRG-System 2008.
38. Grotz M, Schwermann T, Lefering R, Ruchholtz S, Graf v d Schulenburg JM, Krettek C, et al. DRG reimbursement for multiple trauma patients – a comparison with the comprehensive hospital costs using the German trauma registry. Unfallchirurg. 2004;107(1):68–75. Epub 2004/01/30. DRG-Entlohnung beim Polytrauma. Ein Vergleich mit den tatsachlichen Krankenhauskosten anhand des DGU-Traumaregisters.

Evidence-Based Orthopaedic Trauma Care

3

Martin Mangupli and Richard Buckley

Contents

3.1	**Origins**	15
3.2	**Present State**	16
3.3	**Evidence-Based Approach**	16
3.3.1	Ask	17
3.3.2	Acquire	17
3.3.3	Appraise	18
3.3.4	Apply	18
3.3.5	Act	18
3.4	**Challenges**	19
3.5	**Future Directions**	19
3.6	**Models for Success in Orthopaedic Trauma Care**	19
3.7	**Origins**	19
3.8	**Formalization and Funding**	20
3.9	**Commitment**	20
3.10	**Research Coordinators**	20
3.11	**Biannual Meetings**	21
3.12	**Operationalization of the EBM Approach**	21
	Conclusion	22
	References	22

M. Mangupli, MD
Division of Orthopaedic Trauma Surgery,
Sanatorio Allende, Hipolito Yrigoyen 384,
Córdoba 5000, Argentina
e-mail: manguplim@hotmail.com

R. Buckley, MD, FRCS (✉)
Division of Orthopaedic Trauma Surgery,
Department of Surgery, Foothills Medical Centre,
University of Calgary, AC 144A, 0490 McCaig
Tower, Calgary, AB T2N 5A1, Canada
e-mail: buckclin@ucalgary.ca

3.1 Origins

Knowledge and new information is widely available in more than 100 orthopaedic journals published around the world. They provide evidence-based orthopaedic surgery, with the emphasis increasing every year. While it would be disingenuous to contend that the inception of evidence-based medicine occurred in its entirety at any discrete time, two key points are widely recognized as holding significant importance. In 1967, Professor David L. Sackett founded Canada's first department of clinical epidemiology at McMaster University and developed the "Hierarchy of Evidence" (Fig. 3.1). In essence, this step placed greater value and emphasis on research that limits bias and confounding variables through elements of design and methodology. This model formed the foundation of evidence-based philosophy and remains one of its pillars today [1, 2].

Secondly, and years later in 1990, Professor Gordon Guyatt coined the term "evidence-based medicine" (EBM) in a document for applicants to the internal medicine residency program at

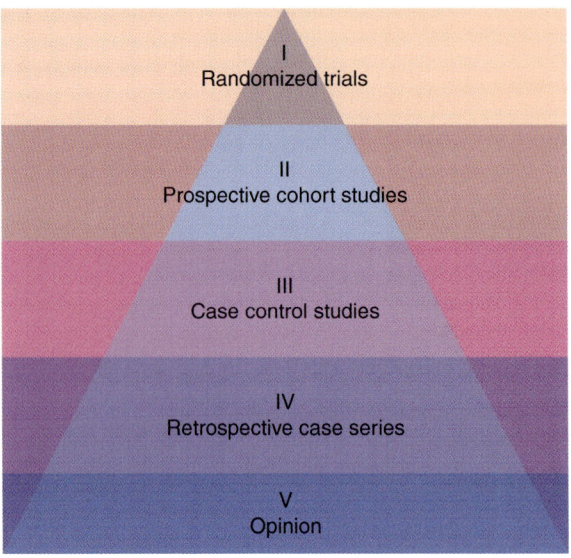

Fig. 3.1 Hierarchy of evidence

McMaster University. A year later, he introduced the term to the academic literature and it was defined as "an attitude of enlightened skepticism towards the application of diagnostic, therapeutic and prognostic technologies" [3]. Moving forward, the definition of EBM evolved to "the conscientious, explicit, and judicious use of the current best evidence in making decisions about the care of individual patients". In a practical sense, it is the use of the best evidence in the literature combined with the preferences of patients and the expertise of the treating physician to facilitate medical decision making for an individual patient integrating clinical judgement and the patient's values [4–6].

3.2 Present State

Evidence-based practice is ongoing continuous change with the appearance of new research, new technology, new ideas or even a mix of the old and the new put together in novel ways. A July 2013 Pubmed search of the term "evidence-based medicine" produced five citations before 1993 and 87,391 to date. The days when opinions of experts can effectively counter an evidence-based document are vanishing; decision making relies more on randomized controlled trials (RCTs) because they have demonstrated the potential to produce the most valid conclusions minimizing the effects of bias [7]. In the orthopaedic literature, historically, much of the evidence came from uncontrolled case series. Bhandari reported that as few as 3 % of the studies in the Journal of Bone and Joint Surgery (American Volume) (JBJS) were RCTs and over 70 % were level IV or V studies in 2000. Most recently, Hanzlik et al. reviewed 1,058 articles that had been published in JBJS over a 30-year period (1975–2005). They assessed levels of evidence for each study and found that the proportion of level I studies increased to 21 % in 2005 concluding that there was a favourable improvement in the number of high-quality studies in orthopaedic surgery [8, 9]. Also the number of systematic reviews and meta-analyses that have been published has substantially increased in the last two decades showing the high regard for the most valid information in the evidence hierarchy [10, 11]. This will provide the most advanced orthopaedic care for the patient as well as the best and most efficient way in which to do so.

3.3 Evidence-Based Approach

Although the methodological quality has improved over time, in 2008, 68 % of published studies still had methodological flaws [12].

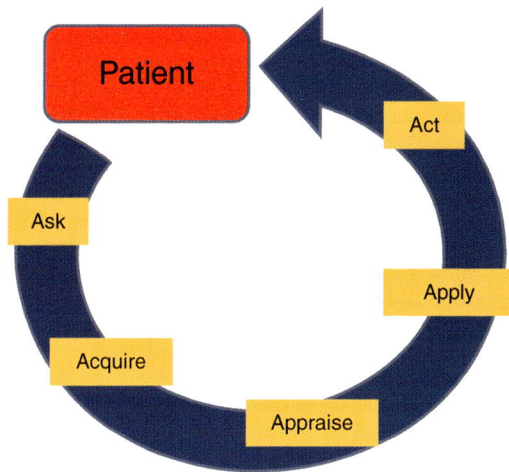

Fig. 3.2 Evidence cycle

Therefore, the information may be more plentiful, but it still of questionable quality. As evidence-based skills have been refined over time, an approach has been produced that allows a clear and concise framework from which to work and better use the available evidence [1, 13, 14]. This approach has come to be known as the "Evidence Cycle" and consists of the five As (Fig. 3.2):

*A*sk (formulate a relevant question)
*A*cquire (conduct an efficient literature search)
*A*ppraise (critically appraise the available evidence)
*A*pply (determine applicability of best evidence to the clinical situation)
*A*ct (use clinical expertise to integrate the best available evidence with the clinical circumstance and the patients' values)

3.3.1 Ask

In order to obtain a relevant answer, it is necessary to begin with an appropriate question. Conceptually, as health-care practitioners, we ask two types of questions [1]. Those related to physiology, pathophysiology, epidemiology, and disease/condition progression are "background" questions and are associated with complementary knowledge. Once practitioners have a thorough understanding of the background information, they should then begin to ask "foreground" questions including vital issues such as screening, diagnosis, prognosis, treatment alternatives, and possible outcomes. The answers will direct them in the management of specific aspects of patient care and have a major impact in clinical decision-making [1, 15]. It is important to address a question so it is answerable. For that, it has to be objective and clearly stated including a clearly identified group or condition, and intervention or specific issue, a comparison point, and an outcome or result. To help recall these features, the "PICO" approach is useful:

*P*opulation
*I*ntervention
*C*omparison
*O*utcome

So, the question "How should closed tibial shaft fractures be treated?" could be developed through the PICO approach. The question would become, "In adolescents with non displaced closed midshaft tibial fracture with intact fibula (*population*), does operative treatment with intramedullary nailing (*Intervention*) versus non-operative treatment with casting (*Comparison*) provide a better outcome reducing the risk of angular deformity (*Outcome*)?".

3.3.2 Acquire

The ability to thoroughly and efficiently search for literature pertaining to the question is necessary to make well-informed clinical decisions. There are presently numerous electronic databases with powerful search engines necessary to deal with the ever-expanding volume of studies and trials. Pubmed (www.pubmed.com) and Google Scholar (www.scholar.google.com) are two quality and free search engines. As well, most academic institutions and many professional organizations have made available medical librarian services that can greatly increase the ease and efficiency of searching the literature. However, no one involved in daily clinical work is capable of reading all of these new publications. Therefore, several approaches have

been developed by readers and providers of knowledge; therefore, health practitioners can gain knowledge on a given topic in a reasonable timeframe [16]. Conceptually, evidence sources can be considered to fall into one of the following groups:
- *Preappraised*: abstracts or guidelines
- *Summarized*: systematic reviews or meta-analyses
- *Primary studies*: individual studies [17]

Preappraised sources may be useful to busy practising clinicians because this type of information has undergone a filtering process to include only those studies of higher quality, regularly updated, so that the evidence we access through these resources is current. These recommendations are the result of consensus meetings and although not all guidelines are evidence based, most are based on systematic reviews or randomized trials. Summarized sources in the form of systematic reviews or meta-analyses are also valuable in addressing specific questions, but when this type or resource is not available, primary studies may be searched.

3.3.3 Appraise

An evidence-based approach to a clinician's practice relies on an awareness of the evidence upon which the practice is based as well as the strength of inference and the degree of certainty permitted by the evidence [14]. A critical appraisal of the available evidence determines its significance and applicability to the clinical situation in question. Assigning level of evidence can be a rapid approach to evaluating study quality, by determining the following:
- Primary question of the study
- Study type (therapeutic, prognostic, diagnostic, economic, or decision analysis)
- Level of evidence I–V [18]

When assigning levels of evidence, greater agreement exists between reviewers trained in epidemiology; however, those without training still demonstrate high levels of agreement. Therefore, the GRADE system is an example of a

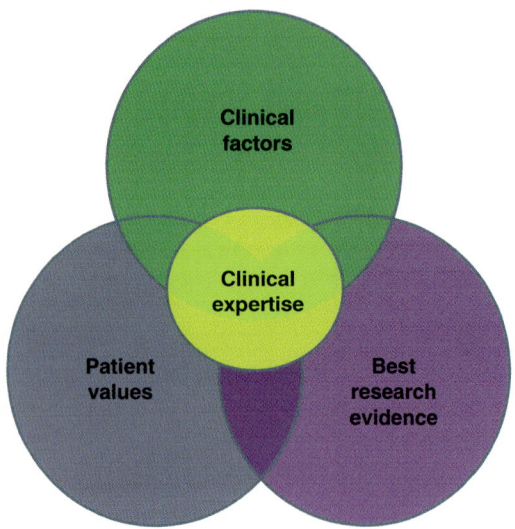

Fig. 3.3 Model of evidence-based practice

thorough and validated model being widely adopted as a standard [19, 20].

3.3.4 Apply

For conceptual purposes, evidence-based practice has been refined as the *conscientious* use of the *current best evidence* in making *health-care decisions* given the *clinical circumstances*. Implicit in this are the following components (Fig. 3.3):
- *Conscientious* – requires clinical expertise
- *Current best evidence* – hierarchy of evidence
- *Health-care decisions* – patient values
- *Clinical circumstances* – factors pertinent to the situation [17]

In essence, clinical judgement must be exercised in deciding how to apply the evidence in a balanced fashion to individual patients given their circumstances and preferences.

3.3.5 Act

The evidence cycle begins and ends with the patient. A patient issue induces a question thereby initiating the cycle that concludes with acting on that patient issue.

3.4 Challenges

Evidence-based practice is not the blind transference of study results into clinical applications but the integration of the results from the best evidence with the clinical circumstances and patient values as guided by clinical expertise. To do so, evidence-based medicine requires education. Although physicians today may have received some training on this topic during medical school, there are still many people who have received no formal training. The most elementary form of evidence-based education is on levels of evidence because knowing about the quality of a study and how to evaluate it might also motivate better research [21]. Another challenge resides in how medical literature searches have become more complex because of the vast amount of published material. It is hard to keep up with the new publications (more than 7,300 citations added weekly in more than 3,800 biomedical journals) turning it into a daunting task. Also, in orthopaedic surgery, as an experience-based science, the opinion of leaders, experienced surgeons and some organizations still have an impact in decision making; therefore, a commitment to evidence-based medicine needs greater support of the results of well-designed studies encouraging high-quality research and further publications, even if those results are negative or unpopular. It means valuing evidence over opinion and supporting data collection for quality improvement and physician accountability [21].

3.5 Future Directions

Although there is an overwhelming amount of orthopaedic knowledge available, not all orthopaedic surgeons know how to use it efficiently and acquire information with the most valid evidence. It is also recognized that many issues related to orthopaedic trauma care will never be subjected to randomized trials because of the rarity of the event or the unique ethical or logistical limitation of the clinical circumstances. Therefore, most of these situations will most likely continue to be addressed with level 2 and 3 studies with acceptance as the highest level of available evidence. The attitude to an awareness of evidence-based orthopaedics remains at a high level; young surgeons should be encouraged to incorporate this philosophy into patient care from the early stages of training, teaching more about EBM, and instituting critical appraisal of the evidence. As the world becomes functionally smaller, multicentre and international trials are becoming increasingly feasible and will strengthen our foundation of literature and body of knowledge.

3.6 Models for Success in Orthopaedic Trauma Care

Beyond understanding, valuing, and utilizing evidence-based principles, studying and practising orthopaedic trauma surgery is the advanced act of contributing to the available body of knowledge in the field [22]. Evidence-based principles must reside as a core principle, so that the field is on solid ground as it is advanced by its researchers and thought leaders. This may be accomplished in many ways, but important lessons may be learned by exploring one model of success where the whole has been recognized as greater than the sum of its part.

3.7 Origins

The Canadian Orthopaedic Trauma Society (COTS) [22] has been successful in producing a number of multicentre randomized trials in the field of orthopaedic trauma surgery. The humble beginnings of this pioneering group can be tracked back to a social meeting in 1990 between three collegial academic orthopaedic trauma surgeons from different centres discussing a clinical problem. Although this initial meeting did not produce a study of merit, it more importantly produced an appreciation for the potential held by the collaboration and communication between centres. COTS has now grown to consist of over 85 members from different academic centres meeting at least biannually and contributing to

randomized trials. The functional basis of the group began with meetings for a study on the management of intra-articular fractures of the calcaneus [23]. This group helped not only in contributing patients, but also in the development of the study design and protocol. This early venture lead to the acquisition of funding and also in more interested colleagues who joined to consider the prospect of creating a bigger group to conduct randomized controlled trials. At least 29 studies have been written in different journals or are current research projects, discussing difficult topics such as the following:
- The best implants for intertrochanteric and geriatric hip fractures
- Operative or non-operative treatment for displaced fractures of the distal clavicle
- Operative or non-operative treatment of displaced isolated ulnar shaft fractures
- Replacement or ORIF of distal humerus fractures
- Reamed or unreamed IM nails for femoral shaft fractures
- Treatment of distal radius fractures
- Management of calcaneus fractures and prediction of subtalar fusions
- Operative or non-operative treatment of acute acromioclavicular joint dislocation
- Operative or non-operative treatment of unstable lateral malleolar fractures
- Management of bicondylar tibial plateau fractures

COTS has continued to deal with the toughest fracture questions that have not yet been answered and for that the multicentre, randomized collaborative teamwork approach has been essential.

3.8 Formalization and Funding

Decisions regarding how to formalize and legitimize the group were required. It was necessary to choose either an independent existence or one under the umbrella of a pre-existing association. It was decided for both legal and funding reasons to stay within the Canadian Orthopaedic Association (COA) and use the Canadian Orthopaedic Fund as the research fund depot. In essence, pre-existing infrastructure was utilized in keeping with its mandate and to the mutual benefit of both COTS and COA. This was not only efficient and cost-effective, but its legitimacy allowed a more aggressive approach to pursuing grants and research funding from various sources (i.e. peer reviewed, association, community, industry).

3.9 Commitment

Similar goals and interests of the involved members are not enough to ensure the functional success of such a group. Despite the requirement for every project to be led by a single surgeon, each protocol has to be discussed, sometimes for years, and reviewed by the entire COTS group, so much so that some of the group felt that the acronym (COTS) should stand for "Compromising Orthopaedic Trauma Surgeons" [23]. This effectively leads to the motivation for as many centres as possible to be involved in each and every study while embracing an "all for one, one for all" philosophy. The universal acceptance of a negotiated protocol followed by dedication to the protocol for the length of the study are the keys to success of the group in producing practice changing trials. In addition, a proactive approach has been taken toward the future membership of the group. To facilitate the recruitment and mentorship of subsequent generation of active members, COTS annually makes available a Young Investigator Grant for principal investigators less than 40 years of age in an effort to enthuse and motivate the next generation of COTS investigators.

3.10 Research Coordinators

The investigators are ultimately responsible for all aspects of a study. However, they delegate some duties to trained and experienced professionals called "Research Coordinators". These are the people who manage the day-to-day aspects of a study. They are involved in one or more aspects of research, including, but not limited to, data collection, analysis, or monitoring;

recruitment and enrolment of study subjects; protection of subjects and their rights in conjunction with institutional review boards (ethics review boards); development of informed consents; reporting of adverse events; development of case report forms; grant and budget development; report preparation; education of other health-care professionals, patients or families about research studies and protocol requirements; and dissemination of study results. The importance of including research coordinators as team members within COTS cannot be understated; therefore, they are recognised as associated members of COTS. In 2006, their increased autonomy and ownership in their roles began to reflect as they began to meet in conjunction with the COTS and OTA annual meetings addressing issues such as how to improve enrolment, the design of data forms, data acquisition and control, website information, and updates. The meetings also provide a forum for study. The coordinator's ability to solve day-to-day issues quickly and among themselves engenders a seamless environment in which to run these studies [23].

3.11 Biannual Meetings

Attendance is promoted by scheduling the meetings at national and international orthopaedic meetings. The regularity of the meetings is required to maintain team rapport and the enthusiasm required to complete medium and long-term protocols. As mentioned before, the development of a research protocol is a collaborative effort hinging on compromise. Proposed protocol questions are appraised on its merits of suitability and feasibility in the following manner:
- Is that question worth answering?
- Is there controversy or debate?
- Is there sufficient interest among surgeons?
- Does a large enough study population exist in combined centres to allow completion?
- Study design – is a RCT the best choice?

If the protocol question passes the group screen, then the investigator completes the requisite literature search, study design, inclusion and exclusion criteria, outcomes to be measured, power analysis, and estimation of time of completion. Following this, it is submitted again to the group for review of the appropriateness. At this point, application for funding begins with the advantage of going forth validated by the previous successes of the group associated with the production of high-level studies.

3.12 Operationalization of the EBM Approach

COTS has met with great success using the strategies outlines above leading to multiple studies being presented at national and international meetings as well many papers being published in top peer-reviewed journals. The academic centres in Canada have all been able to obtain excellent coordinators to enable them to design and complete the clinical studies. COTS has been fortunate to have extremely cooperative colleagues at all the centres who have been able to direct the patients toward individual studies. The patient population across the country is relatively stable allowing for long-term follow-up in most studies. Trauma centres are widely separated, which ensures that patients have little alternative but to return to their own centre for follow-up care, and they are not easily lost, with long-term follow-up (2 years) often reaching 85 %. The support of the Canadian Orthopaedic Association has allowed the COTS group to use the annual meeting as a place to meet and present their academic work. This has proved to be a positive situation for both the organization and the COTS group. The guidance and support from the Canadian Orthopaedic Association in the early years has allowed COTS to take complete control of academic trauma education in the country. These face-to-face biannual meetings held in conjunction with the Canadian Orthopaedic Association in June and OTA in October remain the stabilizing structure of the organization. Partners in industry have also supported COTS with unrestricted educational grants. This was particularly helpful in the initial years. However, peer-reviewed funding at local, national and international levels has sustained the group. The initiative of all members of COTS has allowed keeping the momentum going as they move from older to younger orthopaedic trauma surgeons [23]. All members treat fractures on a day-to-day basis, and everyone recruits and follows up patients, showing the power of collaboration and compromise among the members.

Conclusion

Evidence-based medicine has evolved from the need of solving clinical problems. In contrast to the traditional paradigm of clinical practice, evidence-based medicine acknowledges that intuition, clinical experience, and pathophysiologic rationale are not sufficient for making the best clinical decisions. Although evidence-based medicine recognizes the importance of clinical experience, it includes the evaluation of evidence from clinical research and the integration of patients' values, preferences, and actions for best clinical decision-making. We, as surgeons, need to recognize the need for evidence and use it. For that, we sometimes have to admit that we do not know the answer and look for the best one available, or sometimes admit there is a better answer for a clinical issue than ours. Individual and collective efforts in the field of orthopaedic trauma care will be more efficient, valued, and successful by utilizing and embracing evidence-based skills and principles.

References

1. Guyatt GH, Rennie D, The Evidence-Based Medicine Working Group. User's guides to the medical literature: a manual for evidence-based clinical practice. 2nd ed. Chicago: AMA Press; 2008.
2. Sackett DL, Haynes RB, Guyatt GH, et al. Clinical epidemiology: a basic science for clinical medicine. Boston: Little Brown; 1991.
3. Guyatt GH. Evidence-based medicine. ACP J Club. 1991;114:A16.
4. Bhandari M, Tornetta III P, Guyatt GH. Glossary of evidence-based orthopaedic terminology. Clin Orthop Relat Res. 2003;413:158–63.
5. Sackett DL, Rosenberg WM, Gray JA, et al. Evidence-based medicine: what it is and what it isn't. BMJ. 1996;312:71–2.
6. Bederman S, Wright JG. Randomized trials in surgery: how far have we come? J Bone Joint Surg Am. 2012;94 Suppl 1(E):2–6.
7. Shore BJ, Nasreddine AY, Kocher MS. Overcoming the funding challenge: the cost of randomized controlled trials in the next decade. J Bone Joint Surg Am. 2012;94 Suppl 1(E):101–6.
8. Bhandari M, Richards RR, Sprague S, Schemitsch EH. The quality of reporting randomized controlled trials in the Journal of Bone and Joint Surgery from 1988 through 2000. J Bone Joint Surg Am. 2002;84:388–96.
9. Hanzlik S, Mahabir RC, Baynosa RC, Khiabani KT. Levels of evidence in research published in the Journal of Bone and Joint Surgery (American volume) over the last thirty years. J Bone Joint Surg Am. 2009;91:425–8.
10. Bhandari M, Morrow F, Kulkarni AV, Tornetta 3rd P. Metaanalyses in orthopaedic surgery. A systematic review of their methodologies. J Bone Joint Surg Am. 2001;83(1):15–24.
11. Gagnier JJ, Kellam PJ. Reporting and methodological quality of systematic reviews in the orthopaedic literature. J Bone Joint Surg Am. 2013;95:e771–7.
12. Dijkman BG, Tornetta 3rd P, Bhandari M, et al. Twenty years of meta analyses in orthopaedic surgery: has quality kept up with the quantity? J Bone Joint Surg Am. 2010;92(1):48–57.
13. Poolman RW, Kerkhoffs GM, Struijs PA, et al. Don't be misled by the orthopaedic literature: tips for critical appraisal. Acta Orthop Scand. 2007;78:162–71.
14. Sackett DL, Richardson WS, Rosenberg W, et al. Evidence-based medicine. How to practice and teach EBM. 2nd ed. London: Churchill Livingstone; 2000.
15. Petrisor BA, Bhandari M. Principles of teaching evidence-based medicine. Injury. 2006;37:335–9.
16. Goldhahn S, Audige L, Helfet D, Hanson B. Pathways to evidence-based knowledge in orthopaedic surgery: an international survey of AO course participants. Int Orthop. 2005;29:59–64.
17. Bhandari M, Joensson A. Clinical research for surgeons. New York: Thieme; 2009.
18. Wright JG. A practical guide to assigning levels of evidence. J Bone Joint Surg Am. 2007;89:1128–30.
19. Bhandari M, Swiontkowski MF, Einhorn TA, et al. Interobserver agreement in the application of levels of evidence to scientific papers in the American volume of the Journal of Bone and Joint Surgery. J Bone Joint Surg Am. 2004;86-A:1717–20.
20. Atkins D, Best D, Briss PA, et al. Grading quality of evidence and strength of recommendations. BMJ. 2004;328:1490.
21. Henly B, Turkelson C, Jacobs J, Haralson RH. Evidence-based medicine, the quality initiative and P4P: performance or paperwork? J Bone Joint Surg Am. 2008;90:2781–90.
22. Leighton R, Trask K. The Canadian orthopaedic trauma society: a model for success in orthopaedic research. Injury. 2009;40:1131–6.
23. Buckley R, Leighton R, Trask K. The Canadian Orthopaedic Trauma Society. J Bone Joint Surg Br. 2011;93(6):722–5.

Inflammatory Changes and Coagulopathy in Multiply Injured Patients

Markus Huber-Lang and Florian Gebhard

Contents

4.1	Introduction	23
4.2	Damage-Associated Molecular Patterns	24
4.3	Acute-Phase Reaction	24
4.4	Immune Response After Multiple Injury	26
4.5	Activation and Dysfunction of the Serine Protease Systems	26
4.5.1	The Coagulation System: Coagulopathy	26
4.5.2	The Complement System: Complementopathy	27
4.5.3	The Kallikrein-Kinin System	28
4.6	Cytokines	29
4.6.1	Pro-inflammatory Cytokines	29
4.6.2	Anti-inflammatory Cytokines	30
4.7	Reactive Oxygen Species (ROS)	31
4.8	Cells Implicated in Multiple Trauma	32
4.8.1	Neutrophils	32
4.8.2	Monocytes/Macrophages	33
4.8.3	Natural Killer Cells	33
4.9	Mechanisms of the Development of Organ Dysfunction	33
4.9.1	Severity of Initial Injury (First Hit)	33
4.9.2	Two-Hit Theory	34
4.9.3	Ischemia/Reperfusion Injury	34
4.9.4	Barrier Breakdown	35
Conclusion		35
References		36

M. Huber-Lang, MD, PhD (✉) • F. Gebhard, MD, PhD
Department of Orthopedic Trauma, Hand, Plastic and Reconstruction Surgery, Center for Surgery, Ulm University, Albert-Einstein-Allee 23, 89081 Ulm, Germany
e-mail: markus.huber-lang@uniklinik-ulm.de; florian.gebhard@uniklinik-ulm.de

4.1 Introduction

Multiple trauma results in a significant blood loss and accumulation of necrotic and/or devitalized tissue in an ischemic-hypoxic environment, both of which will become the origin of coagulatory and inflammatory changes. The inflammatory response after polytrauma is a major part of the host's molecular danger response. The acute posttraumatic phase of inflammation consists of two rather synchronically mounted columns: the pro-inflammatory response (systemic inflammatory response syndrome, SIRS) and the anti-inflammatory response (compensatory anti-inflammatory response syndrome, CARS) [1]. SIRS includes changes in the heart rate, respiratory rate, temperature regulation, and immune cell activation (Table 4.1) [2]. In the natural course of the inflammatory response after trauma, the balance of the pro- and anti-inflammatory response is in equilibrium, which maintains the biological homeostasis and induces controlled regeneration processes, enabling the

Table 4.1 Diagnostic criteria for systemic inflammatory response syndrome (SIRS)

Parameter	Values
Temperature	<36 °C (96.8 °F) or >38 °C (100.4 °F)
Heart rate	>90 min
Respiratory rate	>20 breaths/min or $PaCO_2$ < 32 mmHg (4.3 kPa)
White blood cell count	>12,000 mm³ or <4000 mm³ or the presence of >10 % immature neutrophils (band forms)

SIRS can be diagnosed when two or more of these criteria are present

patient to recover normally without significant complications. However, the excessive inflammatory response after trauma seems to simultaneously and rapidly involve the induction of innate (both pro- and anti-inflammatory mediators) and suppression of adaptive immunity [1, 3, 4] all of which decisively contribute to the development of the early multi-organ dysfunction syndrome (MODS). Furthermore, a prolonged and dysregulated immune-inflammatory state is associated with delayed recovery and complications, especially the development of late MODS. Based on improved intensive care and organ support, there is often a progress to the clinically evident persistent inflammation, immune suppression, and catabolism syndrome (PICS) which might have replaced the late MODS, but still is associated with a poor outcome, appearing as "silent death" [5].

The steps of an inflammatory reaction to trauma involve fluid phase mediators (cytokines, chemokines, coagulation- and complement activation products, oxygen radicals, eicosanoids, and nitric oxide (NO)) and cellular effectors (neutrophils, monocytes/macrophages, and endothelial cells) that translate the trauma-induced signals into cellular responses. These factors are closely interrelated and interconnected by up-regulatory and down-regulatory mechanisms. The combination of these factors may cause severe SIRS, acute respiratory distress syndrome (ARDS) and sepsis, acute kidney injury (AKI), progressing to MODS, depending on the type of injured tissue, the surgical and anesthesiological management after injury, age, gender, genetics, and most importantly, underlying comorbidities and physical conditions (exogenous and endogenous factors) (Fig. 4.1).

4.2 Damage-Associated Molecular Patterns

Patient survival after severe trauma requires an adequate molecular and cellular danger response. The injured tissues release cytosolic molecules (e.g., ATP), organelles (e.g., mitochondria), histones, nucleosomes, DNA, RNA, matrix, and membrane fragments, all functioning as damage-associated molecular patterns (DAMPs). Furthermore, damage of external and internal barriers (e.g., skin, gut-blood barrier, air-blood barrier, brain-blood barrier) facilitates invasion of microorganisms, resulting in additional exposure to microorganisms-derived pathogen-associated molecular patterns (PAMPs). After multiple injury, the immune system of the injured patient is exposed to both DAMPs (also termed alarmins) and PAMPs, which are summarized as danger-associated molecular patterns [6]. The "3-R-challenge" for the innate and adaptive immune system is to recognize, respond to, and resolve the "molecular danger". For recognition of the damage, there are effective fluid-phase "master alarm systems", such as the coagulation and complement cascade, and effective cellular "danger sensors", such as the pattern recognition receptors (PRR). These systems transfer the damage/danger signals to the cells which in turn mount an acute phase reaction and inflammatory response to resolve the damaged tissue load [7].

4.3 Acute-Phase Reaction

Within an hour after trauma, inflammation resulting from tissue injury induces an increase in plasma concentration of a number of liver-derived proteins (the acute phase proteins, APP). Pro-inflammatory cytokines (IL-1β, TNF, IL-6) released locally by Kupffer cells can systemically influence other cell types such as hepatocytes to synthesize more APPs. Proactive APPs,

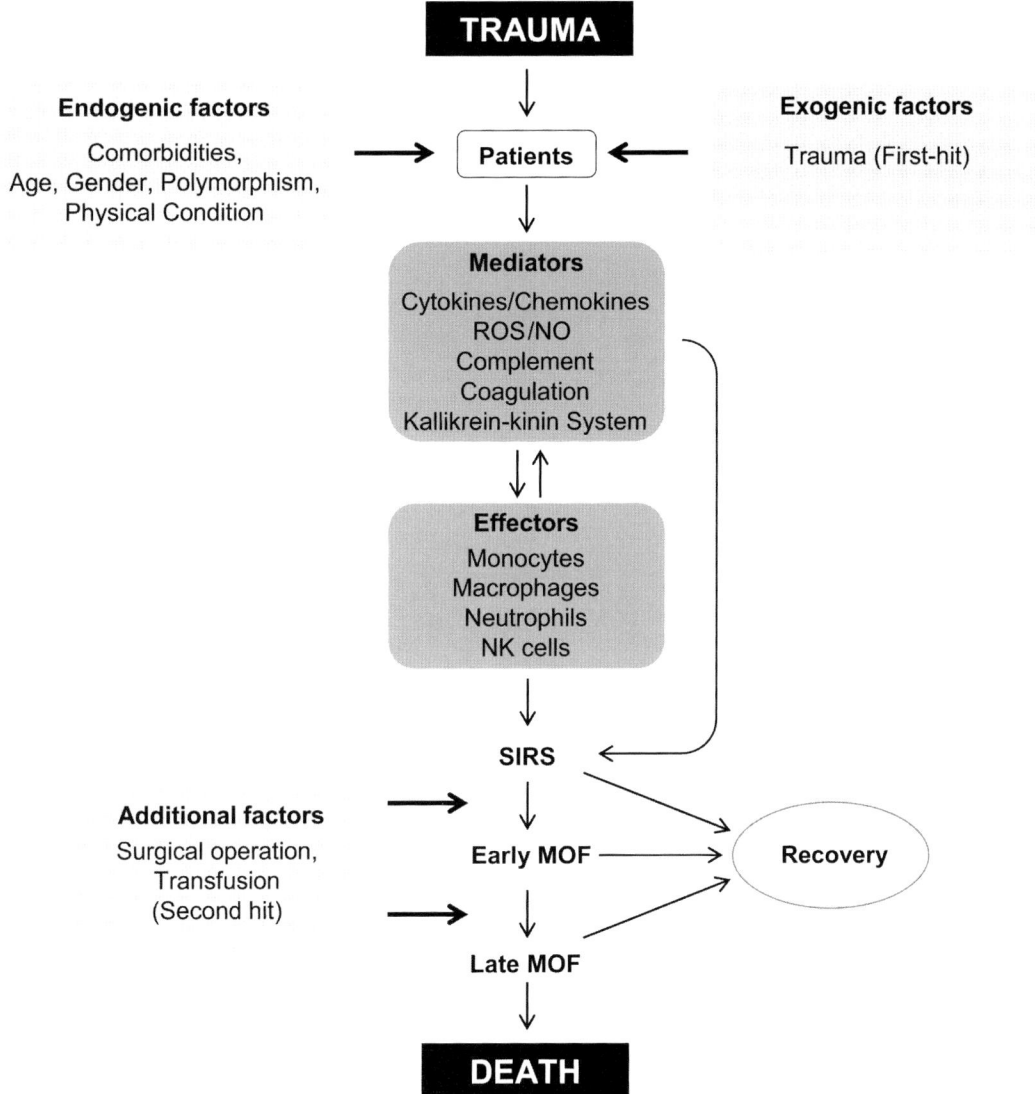

Fig. 4.1 Trauma-induced systemic inflammatory response syndrome (SIRS) and complications; *NO* nitric oxide, *ROS* reactive oxygen species, *NK cells* natural killer cells, *MOF* multiple organ failure

such as C-reactive protein (CRP), procalcitonin (PCT), serum amyloid A (SAA), complement activation products (C3a, C5a), activated coagulation proteins (FVIIa, FXa, FIIa), proteinase inhibitors, and metal-binding proteins, are increased during this phase [8], whereas the production of inhibitory APPs, such as albumin, high-density lipoprotein (HDL), protein C, protein S, and ATIII are decreased [9, 10].

Plasma concentrations of CRP are normally below 10 mg/l [11]. Hepatic synthesis of CRP is regulated mainly by IL-6. Serum levels of CRP can be detected about 12 h after systemic detection of IL-6. Clinically, the plasma levels of CRP are relatively non-specific and may not correlate with injury severity and are not predictive of post-traumatic complications such as infections [12]. In the context of trauma, it is also still unclear whether the native pentameric or the denatured monomeric form of CRP is responsible for the CRP-induced cellular effects [13].

PCT is physiologically produced in the thyroid gland as the precursor molecule of calcitonin [10]. During sepsis, stimulation by endotoxins or pro-inflammatory cytokines such as IL-1β or TNF dramatically increases the serum levels of PCT up to 1000-fold [14]. In trauma patients, PCT has been proposed as a practical biomarker for predicting posttraumatic complications such as severe SIRS, sepsis, and MODS [14–17].

4.4 Immune Response After Multiple Injury

The biological immune response after trauma was considered in the past to be divided into an early innate phase and a late adaptive response. However, since multiple intensive interactions between both systems are known (e.g., via the complement cascade), a spatial- or time-dependent discrimination of both systems in regard to pathomechanistic changes after multiple injury is irrational. Both immune mechanisms contribute to effective recognition, activation, discrimination, regulation, and eradication of invading damage- and pathogen-associated signals [18]. Nevertheless, the innate immune response represents the "first line of defense", consisting of a barrier against exogenous non-self antigens and microorganisms. This includes the integrity of epithelial and mucosal cells: skin, respiratory tract, alimentary tract, urogenital tract, brain, and conjunctiva. Exogenous pathogens that escape the first barrier are rapidly recognized and removed by the multiple components of innate immune cells such as neutrophils, monocytes/macrophages, natural killer cells, and dendritic cells [19]. The innate immune response is closely accompanied by the specifically acquired immune response after the trauma impact. The adaptive immune response is conducted by the interaction of antigen-presenting cells (APCs), dendritic cells, monocytes/macrophages, T-lymphocytes, and B-lymphocytes. The APCs capture invading pathogens and create peptide-MHC (major histocompatibility complex) protein complexes. T-lymphocytes recognize the peptide-MHC protein complex via T-cells expressing antigen-binding receptors (TCRs) and are thereby activated. In turn, activated T-lymphocytes release cytokines to activate and amplify further cells of the immune system. T-helper lymphocytes (CD4+ T cells) differentiate into two phenotypes according to the cytokine release, the Th1 and Th2 lymphocytes. Th1 cells promote the pro-inflammatory response through the release of IL-2, TNF, and interferon-γ (IFN-γ), while Th2 cells produce anti-inflammatory cytokines (IL-4, IL-5, and IL-10), which suppress macrophage activity [10]. Attention has been focused on the Th1/Th2-ratio. IL-12 secreted from monocytes/macrophages promotes the differentiation of Th1 cells by increasing the production of IFN-γ [20, 21]. Several studies have shown that a suppressed IL-12, IL-2, and IFN-γ, and elevated IL-4 are observed after major trauma, which correlated with a shift of the Th1/Th2 ratio towards the Th2-type pattern [22, 23]. This imbalance in Th1/Th2-type cytokine response (from pro- to anti-inflammation) is not only a compensatory response but also increases the risk of infection by immune suppression [20]. However, other reports do not support this view and question the clinical relevance of the Th1/Th2-shift after major tissue injury [24, 25].

4.5 Activation and Dysfunction of the Serine Protease Systems

4.5.1 The Coagulation System: Coagulopathy

Bleeding is a leading cause of death following polytrauma, and acute trauma-induced coagulopathy (ATIC) increases both the risk and severity of bleeding. Clinically, there are several routine laboratory parameters which are indicative of coagulopathy development (Table 4.2). Around one third of severe polytrauma patients are already coagulopathic upon arrival in the emergency room [27] and coagulopathy belongs

Table 4.2 Clinical parameters for acute trauma-induced coagulopathy

Trauma-induced coagulopathy	
TT	>15 s [26]
Prothrombin Time Test (Quick)/INR	<70 % [27]/>1.2 [28]
PT	<18 s [26]
aPTT	>60 s [26]
Platelets	<100,000 µl [27]

Modified from Maegele et al. [27], Brohi et al. [26], Greuters et al. [28]

together with acidosis and hypothermia to the "lethal triad" of polytrauma. Thus, an important diagnostic and therapeutic strategy has been developed proposed as the "STOP bleeding campaign" [29] that addresses three major aspects of coagulopathy: fast detection and stopping of relevant bleeding sources; estimation and resuscitation of the lost blood volume; and rapid monitoring for coagulopathic conditions.

The major mechanism of activation of the coagulation cascade following trauma is via the extrinsic coagulation system [30]. The extrinsic cascade mediates inflammation by tissue factor (TF). Exposure of the FVII to TF (e.g., from injured cells) results in the conversion of FVII to FVIIa. The FVIIa-TF-complexes activate FX to FXa, and FXa converts prothrombin to thrombin (FIIa). Thrombin activates FV, FVIII, and FXI, which results in enhanced thrombin formation. Thrombin also cleaves fibrinogen, and the fibrin clot is formed following polymerization and stabilization. In normal conditions, small amounts of TF are exposed to the circulating blood. However, under pathophysiological conditions, TF is upregulated on the surface of neutrophils, macrophages, and endothelial cells. Endotoxin, activated complement (C5a), and cytokines (IL-1β, TNF) induce TF expression [31]. TF is highly thrombogenic, and its upregulation often results in hypercoagulability, leading to an increased tendency of thrombosis [32, 33]. Another phylogenetically ancient activation pathway is the rather unknown FSAP (FVII activating protease) pathway that is activated by an autocatalytic mechanism promoted by factors released by necrotic or post-apoptotic cells such as nucleic acids, nucleosomes, and polyamines. FSAP can regulate coagulation and fibrinolysis by activating Factor VII and pro-urokinase, respectively. In polytrauma patients, an early and robust activation of FSAP is seen which in turn contributes to the activation of both, the coagulation and complement system [34].

In addition, coagulation mediators (FVIIa, FXa, and FIIa) elicit inflammation with expression of TNF, cytokines, adhesion molecules (MCP-1, ICAM-1, VCAM-1, selectins, etc.), and growth factors (e.g., VEGF) [33]. Inhibitors to prevent a hypercoagulable state include antithrombin III (ATIII), protein C, protein S and TF pathway inhibitor (TFPI). ATIII inhibits FIXa, FXa, and thrombin. TFPI suppresses the activity of TF/FVIIa/FXa complexes [35]. Protein C is activated by the thrombin-thrombomodulin complex on endothelial cells, and activated protein C, in combination with free protein S, cleaves and inactivates FV and FVIII [36]. Therapeutically intervening with the production and/or activity of inhibitors could help to improve outcome by mitigating complications such as ARDS.

For example, the CRASH2 trial has recently revealed that early application of tranexamic acid (a synthetic derivative of the amino acid lysine) that inhibits fibrinolysis by blocking the lysine binding sites on plasminogen significantly reduces the risk of death in bleeding trauma patients [37].

4.5.2 The Complement System: Complementopathy

Almost synchronically to the coagulation response, there is an activation of the complement cascade immediately after multiple trauma [38, 39]. The complement system consists of more than 30 proteins. In the resting state, complement proteins circulate as inactive forms in plasma. The activation of the complement system can occur through four pathways (alternative, classical, lectin, and coagulation paths). The classical pathway of complement is activated by antigen-antibody complexes (immune-globulin M or G) or CRP. The alternative pathway is activated by bacterial products such as lipopolysaccharides (LPS). The lectin pathway is initiated by lectin binding to mannose, glucose, or other sugars of microorganisms. Upon activation of the complement system, there is a generation of biologically active peptides. The cleavage of the central complement components C3 and C5 to the anaphylatoxins C3a and C5a, respectively, also induces the formation of opsonins and the membrane attack complexes (MAC, C5b-9) [40, 41]. Early after polytrauma, serum levels of the complement activation products C3a and C5a are significantly elevated and correlate with the severity of the injury (e.g., traumatic brain injury), septic complications, and mortality [27, 38]. The circulating soluble MAC is also enhanced within the first hours after polytrauma but almost not detectable between 4 and 48 h after polytrauma [38, 39]. Regulation of complement activation and protection against complement-mediated tissue destruction is provided by a selection of soluble- and membrane-bound complement regulatory proteins (CRegs). The expression profile of CRegs on leukocytes is specifically altered post polytrauma: CD46 (membrane co-factor protein) is significantly reduced in neutrophils, monocytes, and lymphocytes. In contrast, CD55 (decay accelerating factor) seems to be increased on neutrophils early after trauma. A delayed up-regulation of CD55 has been observed in monocytes from trauma patients. An initial enhancement of CD59 (MAC inhibitor) expression was measured in neutrophils and monocytes at the time of admission. Remarkably, C5a receptor (C5aR), CD59 and CD46 expression on neutrophils reversely correlated with injury severity [42]. The anaphylatoxins C3a and C5a mainly play pro-inflammatory roles, which include the recruitment and activation of phagocytic cells (polymorphonuclear cells, PMNs), monocytes/macrophages, the enhancement of the hepatic acute-phase reaction, stimulation of the release of vasoactive mediators (such as histamine), and promoting the adhesion of leukocytes to endothelial cells and their permeation through injured tissues. C5b forms a complex by the consecutive binding of proteins C6–C9, culminating in the formation of the MAC (C5b-9), which leads to the formation of pores in the cellular membrane causing lysis and death of the target cells [43]. Furthermore, the inflammatory response of complement activation leads to the production of free oxygen radicals and arachidonic acid metabolites and cytokines.

The complement cascade bridges innate and adaptive immunity for defense against microbial pathogens. However, excessive consumption of complement proteins may also cause tissue damage of the host after trauma. Within the first 24 h after multiple injuries, there is a massive reduction in complement hemolytic activity (CH50), which recovers only around 5 days after trauma, and can be used to discriminate between lethal and non-lethal outcome. This trauma-induced reduction of global complement function is referred to as trauma-induced "complementopathy" in analogy with "coagulopathy", both of which significantly participate to the impairment of the innate immune response after polytrauma (Fig. 4.2).

4.5.3 The Kallikrein-Kinin System

The kallikrein-kinin system involves a cascade of plasma proteases and is related to the complement and clotting cascade (intrinsic activation) [44]. This contact system consists of plasma proteins factor XII (Hageman factor; FXII), prekallikrein, high molecular weight kininogen (HMWK), and FXI. Contact with negatively charged surfaces such as foreign bodies or the membrane fragments of stimulated platelets activates FXII [44]. The active protein FXIIa converts prekallikrein into the proteolytic enzyme kallikrein, which in turn cleaves the plasma glycoprotein precursor HMWK to form bradykinin [45]. Bradykinin increases vascular permeability and causes dilation of blood vessels by its action on smooth muscle cells. In turn, as a positive feedback loop, kallikrein itself accelerates the conversion of FXII to FXIIa. Kallikrein can also activate fibrinolysis to counterbalance the clotting cascade activated by FXIIa. Furthermore, kallikrein also exhibits chemotactic activity, converting C3 and C5 into the chemoattractant products C3a and C5a, respectively [46].

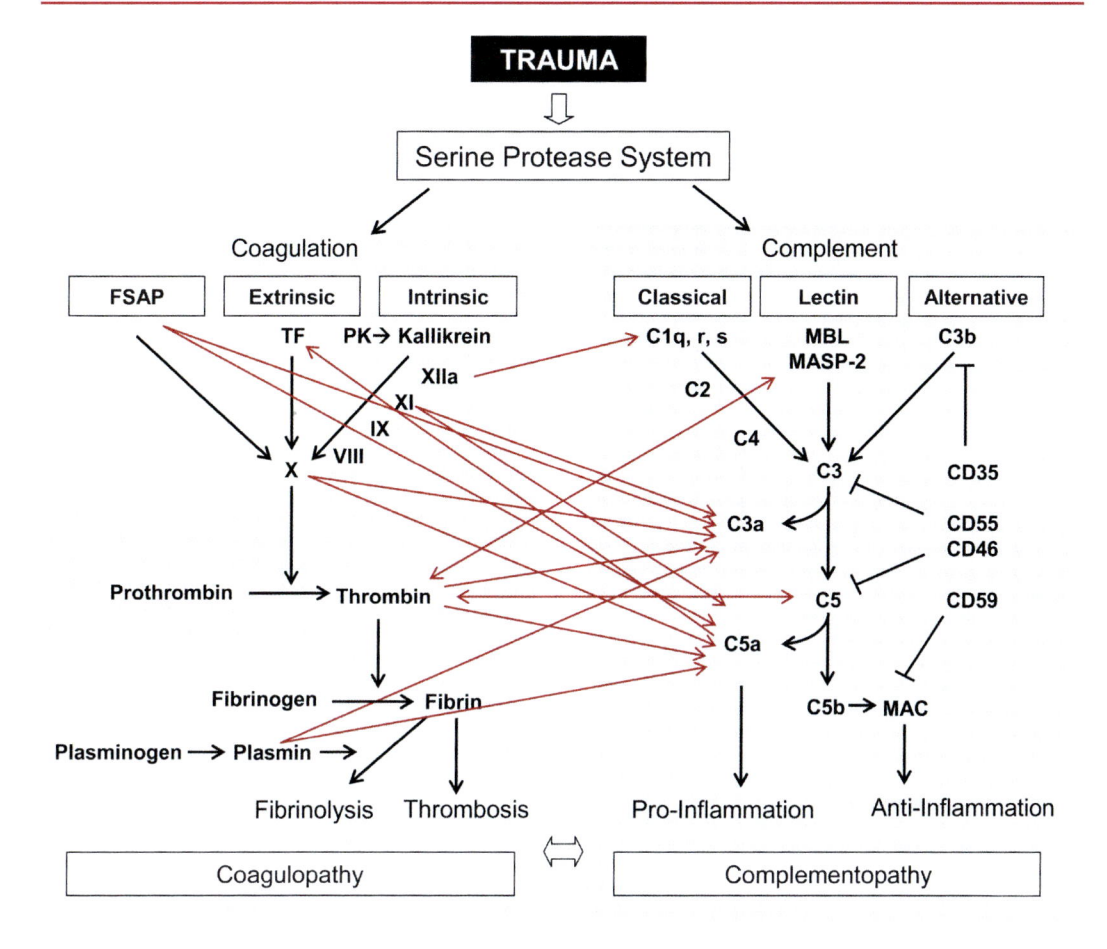

Fig. 4.2 Posttraumatic activation of the serine protease system; *PK* prekallikrein, *TF* tissue factor, *FSAP* Factor VII activating protease, *MBL* mannose-binding lectin, *MASP-2* mannose-associated serine protease-2, *MAC* membrane attack complex

4.6 Cytokines

4.6.1 Pro-inflammatory Cytokines

Pro-inflammatory cytokines play key local and systemic roles as intercellular messengers to initiate, amplify, and perpetuate the inflammatory response after trauma (Table 4.3). Cytokines are produced by many cell types in all organs. They have multiple targets and act in a pleiotropic manner. Early after trauma, production and release of pro-inflammatory cytokines such as IL-1β, TNF, IL-6, and IL-8 is initiated by monocytes and macrophages. IL-1β and TNF as well as IL-6 and IL-8 are released early after polytrauma [3, 47] and predominantly function as pro-inflammatory mediators to repair damaged tissue. The release of IL-1β and TNF is mainly stimulated by bacterial endotoxins or other microbial products, immune complexes, and a variety of inflammatory stimuli. Upon release, IL-1β and TNF usually return to baseline levels within 4 h. TNF increases the activity of neutrophils and monocytes by activating the underlying endothelium. TNF promotes the expression and release of adhesion molecules such as ICAM1 or E-selectin, and increases the permeability of endothelial cells, which facilitates neutrophil migration into the damaged tissue [48]. Some studies have proposed TNF as a valid serum marker for complications after

Table 4.3 Features of the major pro-inflammatory cytokines

Cytokine/chemokine	Source	Function
TNF	Monocytes/macrophages, mast cells, T lymphocytes, epithelial cells	Upregulation of adhesion molecules and secretion of cytokines, chemokines, and NO by endothelial cells
		Acute-phase response
		Fever
IL-1β	Monocytes/macrophages, mast cells, T lymphocytes, endothelial cells, some epithelial cells	Similar to TNF
IL-6	Monocytes/macrophages, T lymphocytes, endothelial cells	Acute-phase response
		T and B lymphocyte proliferation
		Prognostic marker of complications (SIRS, sepsis, MOF) after trauma
IL-8	Macrophages, neutrophils, endothelial cells, T lymphocytes, mast cells	Chemotaxis
		Leukocyte activation
		Diagnostic marker for AIDS

trauma. However, the results are inconsistent and to date, no data is available indicating whether TNF correlates to the severity of trauma or trauma outcome [49–54]. Many different cell types produce IL-6: In addition to immune cells such as monocytes, macrophages, neutrophils, T cells, and B cells, it is also produced by endothelial cells, smooth muscle cells, and fibroblasts. IL-6 upregulates the hepatic acute-phase response, stimulating generation of C-reactive protein (CRP), procalcitonin, serum amyloid A, fibrinogen, α1-antitrypsin, and complement activation products (e.g., C5a), which then promote neutrophil activation. There is strong evidence that serum IL-6 level correlates with the severity of trauma, trauma pattern (especially in combination with chest trauma), and the risk of subsequent ARDS, MOF, and lethal outcome [47, 55]. Therefore, IL-6 may be considered as a clinically relevant and feasible parameter to estimate the severity of injury and prognosis after trauma [56, 57]. In addition, for patients requiring second or subsequent surgeries following trauma, IL-6 may prove to be an important biological marker in deciding the correct timing of surgery. In trauma patients with high initial levels of IL-6 (>500 pg/dL), it is recommended to delay secondary procedures for more than 4 days [58]. The chemokine IL-8 is secreted by monocytes/macrophages, neutrophils, and endothelial cells.

After trauma, serum levels of IL-8 are elevated within 24 h. Its production following trauma stimulates leukocyte recruitment to the injured and inflammation site. Plasma levels of IL-8 correlate with the subsequent development of ARDS and MOF [57, 59–61].

4.6.2 Anti-inflammatory Cytokines

IL-10 is mainly synthesized by T lymphocytes and monocytes/macrophages. It is the pivotal role of IL-10 to inhibit the production of monocyte/macrophage-derived TNF, IL-6, IL-8, and free oxygen radicals [62]. IL-10 plasma levels are proportional to the severity of trauma and to posttraumatic complications [63–67] (Table 4.4). In addition to its pro-inflammatory role, IL-6 also has anti-inflammatory properties. As an immunoregulatory cytokine, IL-6 stimulates macrophages to release anti-inflammatory cytokines such as IL-1 receptor antagonists and soluble TNF receptors [8]. Moreover, IL-6 induces macrophages to release prostaglandin E2 (PGE2), the most powerful endogenous immune suppressant. PGE2 regulates the synthesis of TNF and IL-1β by macrophages and induces the release of IL-10 [68–70].

Overall, it has to be emphasized that almost all cytokines may not act strictly in either a pro- or

4 Inflammatory Changes and Coagulopathy in Multiply Injured Patients

Table 4.4 Features of the major anti-inflammatory cytokines

Cytokine/chemokine	Source	Function
IL-4	Th2 lymphocytes	B cell class switch
IL-6	See Table 4.3	Reduction of TNF and IL-1 synthesis
		Release of IL-1 Ra and sTNF-Rs
IL-10	Monocytes/macrophages, T lymphocytes	Inhibited secretion of pro-inflammatory cytokines and ROS production
		Reduced adhesion molecule expression
		Enhanced B lymphocyte survival, proliferation, and antibody production
		Levels correlated with injury severity and outcome

Fig. 4.3 Posttraumatic pro- and anti-inflammatory immune response; *PICS* persistent inflammation, immune suppression, and catabolism syndrome

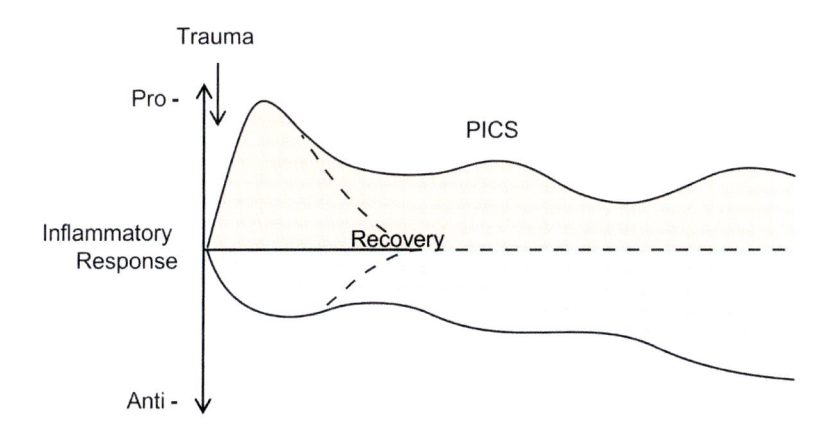

anti-inflammatory manner, but rather may exhibit a "janus-faced behavior" depending on the underlying tissue, local environment, and trauma conditions. Furthermore, the categorized pro- and anti-inflammatory cytokines follow not a specific temporal pattern but are rather synchronically and rapidly generated and released [1, 3], mounting the overall inflammatory response. When the simultaneous cytokine response is excessive, prolonged, and dysregulated, this may lead to severe complications, such as organ dysfunctions [1] or persistent inflammation, immunosuppression, and catabolism syndrome (PICS) [5] (Fig. 4.3).

4.7 Reactive Oxygen Species (ROS)

Reactive oxygen species are released by leukocytes after exposure to pro- and anti-inflammatory cytokines, chemokines, complement factors, and bacterial products. There are several mechanisms of ROS production: mitochondrial oxidation, metabolism of arachidonic acid, activation of nicotin-adenine-dinucleotide-phosphate (NADPH) oxidase, and activation of xanthine oxidase. With ischemia and subsequent reperfusion, reintroduced molecular oxygen reacts with hypoxanthine and xanthine oxidase generated as the result of ATP consumption during the ischemia phase to generate superoxide anions ($•O_2^-$). Superoxide anions are further reduced to hydrogen peroxide (H_2O_2) by superoxide dismutase (SOD). The initial ROS (superoxide anion and H_2O_2) are relatively low-energy oxygen radicals and are not considered to cause high levels of cytotoxicity [71]. The most detrimental agents of the ROS are hydroxyl radicals ($•OH$) which are generated from superoxide anions and H_2O_2 by the Haber-Weiss reaction: $•O_2^- + H_2O_2 \rightarrow •OH + OH^- + O_2$ or from H_2O_2 by the Fenton reaction in the presence of iron ($LFe^{II}(H_2O_2) \rightarrow LFe^{III} + OH + OH^-$) (Fig. 4.4). ROS cause lipid peroxidation, cell membrane disintegration,

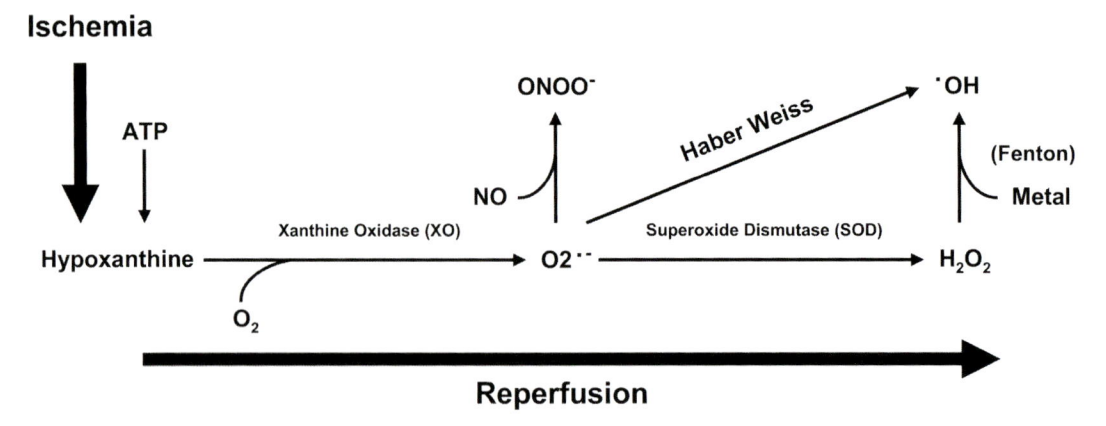

Fig. 4.4 Production of reactive oxygen species (ROS)

and DNA damage to endothelial and parenchymal cells [72, 73]. Furthermore, ROS secreted by polymorphonuclear leukocytes (PMN) induce cytokines, chemokines [74], heat shock protein (HSP) [75], and adhesion molecules (p-selectin, ICAM-1) [76] leading to cell and tissue damage.

4.8 Cells Implicated in Multiple Trauma

4.8.1 Neutrophils

Early after severe tissue trauma, neutrophils migrate along the chemoattractant gradient of complement activation products, interleukins, and ROS to the site of tissue damage and to remote organ tissue. Neutrophil mobilization is important for wound healing and protection against invading microorganisms, but their immigration to remote organ tissue contributes to SIRS [77]. Neutrophil migration is composed of four steps: The first step, generation of leukocyte selectins (e.g., L-selectins) and E- and P-selectins on the endothelium is induced by anaphylatoxins (e.g., C5a), cytokines (e.g., IL-6), and toxins [78]. These adhesion molecules are responsible for the rolling of neutrophils. The second step involves expression of integrins on neutrophils such as CD11 and CD18, and intercellular adhesion molecules (ICAM-1) and vascular cell adhesion molecules (VCAM-1) on the surface of endothelial cells, all of which are strongly induced by C5a [79–81]. The interaction of these upregulated molecules activate neutrophils to reinforce the contact between neutrophils and endothelial cells (sticking). In the next step, migration and accumulation into tissues occur, mediated by chemokines and complement anaphylatoxins. To migrate through cellular barriers, neutrophils undergo significant deformational changes to permeate through small cellular gaps with the help of locally released matrix metalloproteinases. In the final step, activation of neutrophils occurs to protect against dangerous molecules, microorganisms, and cells. Neutrophils utilize a large arsenal for forming the "first line of defense" after trauma: chemotaxis, phagocytosis, oxidative burst reaction with release of ROS and myeloperoxidase (MPO), generation of NO, leukotriens, platelet-activating factor (PAF), tissue factor (TF), proteases, and multiple pro-inflammatory cytokines. However, the active substances released from neutrophils may not only harm the invading microorganisms or injured cells but also healthy host cells, especially since neutrophils become "long-lived" after trauma by significant inhibition of neutrophil apoptosis. Thus, neutrophils after trauma function as "friend and foe".

4.8.2 Monocytes/Macrophages

Monocytes/macrophages and neutrophils play a central role for the innate host defense, tissue repair, and remodelling, and for the intermediaries to the antigen-specific adaptive immune response. Monocytes are circulating precursors of macrophages. Monocytes migrate into the different tissues (liver, spleen, lungs, etc.) even in absence of local inflammation and become tissue macrophages. When monocytes/macrophages are activated by various phagocytotic events in response to trauma, they regulate the activation of T and B lymphocytes, which induce antigen presentation by the major histocompatibility complex II (MHC II). Monocytes/macrophages also release chemokines, cytokines (IL-6, TNF, IL-10, IL-12, TGF-β), and various growth factors (fibroblast growth factor [FGF], epidermal growth factor [EGF], and platelet-derived growth factors [PDGF]) that initiate the formation of new extracellular matrix and promote angiogenesis and generation of new tissue at the site of injury. The functional phenotype shifts from a pronounced pro-inflammatory M1 type to a more anti-inflammatory and regenerative M2-type macrophage. The monocyte/macrophage cellular response after minor trauma embodies several beneficial effects for the host. However, major trauma induces massive monocyte/macrophage activation. In this state, the effects of the monocyte/macrophage response become systemic and may also induce detrimental effects. Systemically, the macrophage-modulated immune response influences microcirculation, metabolism, and triggering and progression of remote organ injury. Deactivation of monocytes and decreased expression of MHC II on their surface are observed after major trauma correlating with the severity of injury [82].

4.8.3 Natural Killer Cells

Natural killer (NK) cells are antigen-non-specific lymphocytes that recognize pathogen-associated molecular patterns (PAMPs) of invading microorganisms [83] as well as damaged, transformed, or virus-infected host cells [84]. Since they are not dependent on pre-sensitization [85] to mediate their cytotoxic effects and to release excessive amounts of pro- and anti-inflammatory cytokines within minutes of stimulation, NK cells are regarded as part of the "first line of defense" [86]. Their ability to release immune-modulatory cytokines may provide important regulatory functions during immune response, especially following severe injury. However, studies addressing the role of human NK cells after severe tissue trauma are rare and contradictory. Some studies revealed an increase of NK cells in the early stage after severe trauma [77], whereas NK cell function is greatly depressed by traumatic injury. However, there was no correlation between the NK cell count or activity and injury severity [85, 87]. Concerning the effect of plasma samples from trauma patients on the cytotoxic activity of healthy NK cells in vitro, it has been shown that incubation times of more than 40 h lead to suppressed NK cell function, suggesting that posttraumatic immune suppression is associated with suppression of NK cell activity [85]. Vice versa, murine experiments have collectively shown that NK cells as a key source of interferon γ exert harmful pro-inflammatory effects in the posttraumatic immune response and during the pathogenesis of sepsis [88, 89]. In support, early depletion of NK cells results in reduction of liver IL-6 expression and a 50 % improved survival rate in a murine polytrauma model. Lymphocyte apoptosis in spleen as well as neutrophil infiltration into lungs and liver is also attenuated [88]. Furthermore, in various mouse models of sepsis, depletion of NK cells leads to improved survival [89, 90] suggesting that early posttraumatic activation of NK cells promotes amplification of the inflammatory response, and the subsequent loss of cellular functions might contribute to immune suppression manifested in later stages after trauma [87].

4.9 Mechanisms of the Development of Organ Dysfunction

4.9.1 Severity of Initial Injury (First Hit)

The initial trauma insult activates an inflammatory cascade that stimulates the host immune system. Massive initial trauma impact (first hit) causes severe

SIRS. In this situation, the overwhelming production and release of pro- and anti-inflammatory mediators result in rapid MODS and early death.

An initial trauma insult of lower severity induces a moderate state of SIRS/CARS. In this instance, inflammatory and immune cells undergo some "priming". However, some patients develop posttraumatic complications, such as sepsis, AKI, ARDS, and MODS. The development of these complications is regulated by various exogenous and endogenous factors. Among these factors, it is important to understand the relationship between the biological changes and the anatomical region of initial injury. The central nervous system is a rich source of inflammatory mediators. Traumatic brain injuries (TBI) with the disruption of the blood-brain barrier (BBB) allow immune cells to migrate into the subarachnoid space, leading to an accumulation of leukocytes from the periphery [10, 91–93]. Trauma to the chest area, particularly lung contusions, leads to an early increase in plasma mediators, which is associated with systemic inflammatory and anti-inflammatory reactions, such as pneumonia, ARDS, and MODS [94–96]. Patients with severe soft tissue injuries to the extremities with resulting hemorrhagic shock or severe muscle crush syndrome are at risk of developing more serious remote organ injury (e.g., AKI). Ischemia/reperfusion injury (I/R) leads to the production of large quantities of ROS. Femoral fractures with soft tissue injuries usually result in alteration of hemodynamic parameters such as increased cardiac output, tachycardia, decreased systemic vascular resistance, and decreased hepatic blood flow [97]. Long bone fractures and unstable pelvic fractures are characterized by high blood loss and are associated with severe soft tissue injury, which initiate both a local and systemic inflammatory response [65, 98–102]. These bodies of evidence suggest that the initial trauma itself predisposes trauma patients to posttraumatic complications.

4.9.2 Two-Hit Theory

Traumatized patients who survive the initial injury ("first hit") may still be at risk of death from sepsis and multiple organ failure. Secondary insults following the initial injury amplify the systemic inflammatory response and upset the balance of pro- and anti-inflammatory mediators, pro- and anti-coagulatory factors, pro- and anti-apoptotic events, and pro- and anti-regenerative processes. Secondary insults ("second hits") are compounded by endogenous and exogenous factors. Endogenous secondary insults include respiratory distress, cardiovascular instability, ischemia and reperfusion injury, and infection. Exogenous secondary insults include surgical and anesthesiological interventions [103–105], blood transfusions, and – not to forget – missed injuries.

Clinical studies have revealed that orthopedic surgical intervention can also cause major changes in the inflammatory response, and these changes are in proportion to the magnitude of surgery. For instance, femoral nailing induces an increase in systemic plasma levels of IL-6 and IL-10. In these patients, human leukocyte antigen-DR expression on monocytes is reduced as well [106, 107]. Furthermore, reamed femoral nailing appears to be associated with greater impairment of immune reactivity than un-reamed nailing [107].

Blood transfusions are a paramount therapy in the management of trauma/hemorrhagic shock patients. However, various studies have demonstrated that blood transfusions are associated with infection, SIRS, ARDS, and MODS after trauma [108–113], also representing a "second hit" for the multiply injured patient.

4.9.3 Ischemia/Reperfusion Injury

Ischemia/reperfusion (I/R) injury is a common and important event in clinical situations such as trauma, hemorrhagic shock, cardiac arrest (hypoxemia, hypotension of systemic tissue), contusions, lacerations, vascular injuries, and compartment syndrome (increased pressure in a preformed anatomical compartment with resulting hypoperfusion and hypoxemia of local tissue). Inadequate microvascular flow results in the activation of leukocytes and converts local endothelial cells into a pro-inflammatory and pro-thrombotic phenotype. I/R injury consists of two specific stages. During the first stage of ischemia and hypoxemia, oxygen and nutrients are deprived from tissues temporarily by the disruption of blood supply. During the ischemic phase, the lack of oxygen leads to decreased production as well as consumption of adenosine triphosphate (ATP). As consumption of ATP continues, it is degraded into adenosine diphosphate (ADP) and adenosine monophosphate (AMP), which is further degraded to inosine and hypoxanthine [114]. ATP depletion leads to an alteration in intercellular calcium and sodium concentration. It also results in the activation of cytotoxic enzymes such as proteases or phospholipases, all cumulating to reversible or irreversible cellular damage. The second stage of reperfusion is the revascularization or reestablished supply of oxygen to the ischemic tissue. The hallmark of the reperfusion phase is the generation of by-products of neutrophil activation, which induces secondary tissue damage and organ dysfunction. On reperfusion with the reintroduction of molecular oxygen into the ischemic tissue, oxygen reacts with leukocytes and endothelial cells promoting the generation of reactive oxygen species and platelet-activation factor. The interactions of neutrophils and endothelial cells have been shown to contribute to massive interstitial edema caused by microvascular capillary leakage after reperfusion injury.

4.9.4 Barrier Breakdown

The ischemia and reperfusion injury with ATP depletion is a major cause for breakdown of physiological organ-blood barriers, such as blood-brain, blood-gut, and blood-alveolus barrier. Broken barriers characterized by diffuse microvascular leakage and tissue edema are thought to be main drivers of bacterial translocation (BT) and sepsis [115]. Bacterial translocation is defined as the phenomenon of both viable and nonviable bacteria as well as their products (bacterial cell wall components, LPS, and peptidoglycan) crossing the intestinal barrier to external sites such as the mesenteric lymph nodes, liver, and spleen. BT occurs as a result of a loss of integrity of the gut barrier function after trauma, hemorrhagic shock, and burns [116], and may be associated with post-traumatic complications [117, 118]. Although most data on BT and its complications have shown consistent results in animal models of hemorrhagic shock, trauma, and severe burns, its importance in humans is questionable, with variable results in clinical studies. In addition, it is still debatable whether BT is an important pathophysiologic event or simply an epiphenomenon of severe disease [119].

Conclusion

Following trauma, acute inflammatory reactions may be triggered by infections (bacterial, viral, fungal, parasitic) and microbial toxins, or by any of several molecules released from necrotic tissue (HMGB1, hyaluronic acid, etc.). Pattern recognition receptors (PRRs), including toll-like receptors, can detect these stimuli and trigger a signaling pathway that leads to the production of various mediators. In the acute phase of trauma, vasodilatation is induced by vasodilatatory mediators (NO, prostaglandins), quickly followed by increased permeability of the microvasculature. Vasodilatation and extravasation of plasma result in hemoconcentration, facilitating the peripheral migration of neutrophils. Neutrophil migration from the blood stream into interstitial tissue is divided into several steps, which are mediated by endothelial cell adhesion molecules, cytokines produced by monocytes/macrophages and various other cells, chemokines, the complement system, and arachidonic acid. Migrated neutrophils produce several mediators such as neutral protease, reactive oxygen species (ROS), lipids (leukotriene, PAF), and tissue factor (TF). These mediators act as secondary tissue damage mediators and pro-coagulatory factors depending on the degree of initial injury as well as additional insults. During inflammation, the plasmaic cascade, consisting of the complement cascade, the kallikrein-kinin system, and the coagulation cascade, is activated by toxins and inflammatory mediators. Activation of the complement system induces generation and depletion of complement activation products, causing an increase in vascular permeability, chemotaxis, opsonization, activation of the coagulation cascade, and trauma-induced complementopathy. Excessive activation of the coagulation system results in a hypercoagulable state, leading to an acute trauma-induced coagulopathy (ATIC). Activation of the kallikrein-kinin system results in kinins with vasoactive properties. In addition to its role in stimulating inflammation, the immune system (innate and adaptive) is a main driver for the barrier breakdown, clinically evident as diffuse microvascular leakage syndrome and organ failure. The exact knowledge of the pathophysiological changes after polytrauma is a prerequisite for effective, targeted, and patient-tailored future therapies to support the immune and organ functions after severe tissue trauma.

References

1. Xiao W, Mindrinos MN, Seok J, et al. A genomic storm in critically injured humans. J Exp Med. 2011;208:2581–90.
2. Bone RC, Balk RA, Cerra FB, et al. Definitions for sepsis and organ failure and guidelines for the use of innovative therapies in sepsis. The ACCP/SCCM Consensus Conference Committee. American College of Chest Physicians/Society of Critical Care Medicine. Chest. 1992;101:1644–55.
3. Gebhard F, Bruckner UB, Strecker W, et al. Untersuchungen zur systemischen posttraumatischen Inflammation in der Frühphase nach Trauma, Hefte zu der Unfallchirurg, vol. 276. Berlin/Heidelberg/New York: Springer; 2000. p. 276.
4. Adib-Conquy M, Cavaillon JM. Compensatory anti-inflammatory response syndrome. Thromb Haemost. 2009;101:36–47.
5. Gentile LF, Cuenca AG, Efron PA, et al. Persistent inflammation and immunosuppression: a common syndrome and new horizon for surgical intensive care. J Trauma Acute Care Surg. 2012;72:1491–501.
6. Matzinger P. The danger model: a renewed sense of self. Science. 2002;296:301–5.
7. Gebhard F, Huber-Lang M. Polytrauma–pathophysiology and management principles. Langenbecks Arch Surg. 2008;393:825–31.
8. Lin E, Calvano SE, Lowry SF. Inflammatory cytokines and cell response in surgery. Surgery. 2000;127:117–26.
9. Gruys E, Toussaint MJ, Niewold TA, et al. Acute phase reaction and acute phase proteins. J Zhejiang Univ Sci B. 2005;6:1045–56.
10. Keel M, Trentz O. Pathophysiology of polytrauma. Injury. 2005;36:691–709.
11. el Hassan BS, Peak JD, Whicher JT, et al. Acute phase protein levels as an index of severity of physical injury. Int J Oral Maxillofac Surg. 1990;19:346–9.
12. Gosling P, Dickson GR. Serum c-reactive protein in patients with serious trauma. Injury. 1992;23:483–6.
13. Taylor KE, van den Berg CW. Structural and functional comparison of native pentameric, denatured monomeric and biotinylated C-reactive protein. Immunology. 2007;120:404–11.
14. Castelli GP, Pognani C, Cita M, et al. Procalcitonin as a prognostic and diagnostic tool for septic complications after major trauma. Crit Care Med. 2009;37:1845–9.
15. Mimoz O, Benoist JF, Edouard AR, et al. Procalcitonin and C-reactive protein during the early posttraumatic systemic inflammatory response syndrome. Intensive Care Med. 1998;24:185–8.
16. Uzzan B, Cohen R, Nicolas P, et al. Procalcitonin as a diagnostic test for sepsis in critically ill adults and after surgery or trauma: a systematic review and meta-analysis. Crit Care Med. 2006;34:1996–2003.
17. Wanner GA, Keel M, Steckholzer U, et al. Relationship between procalcitonin plasma levels and severity of injury, sepsis, organ failure, and mortality in injured patients. Crit Care Med. 2000;28:950–7.
18. Lenz A, Franklin GA, Cheadle WG. Systemic inflammation after trauma. Injury. 2007;38:1336–45.
19. Pillay J, Hietbrink F, Koenderman L, et al. The systemic inflammatory response induced by trauma is reflected by multiple phenotypes of blood neutrophils. Injury. 2007;38:1365–72.
20. Trinchieri G. Interleukin-12 and the regulation of innate resistance and adaptive immunity. Nat Rev Immunol. 2003;3:133–46.
21. Watford WT, Moriguchi M, Morinobu A, et al. The biology of IL-12: coordinating innate and adaptive immune responses. Cytokine Growth Factor Rev. 2003;14:361–8.
22. Decker D, Schondorf M, Bidlingmaier F, et al. Surgical stress induces a shift in the type-1/type-2 T-helper cell balance, suggesting down-regulation of cell-mediated and up-regulation of antibody-mediated immunity commensurate to the trauma. Surgery. 1996;119:316–25.
23. Spolarics Z, Siddiqi M, Siegel JH, et al. Depressed interleukin-12-producing activity by monocytes correlates with adverse clinical course and a shift toward Th2-type lymphocyte pattern in severely injured male trauma patients. Crit Care Med. 2003;31:1722–9.
24. Heizmann O, Koeller M, Muhr G, et al. Th1- and Th2-type cytokines in plasma after major trauma. J Trauma. 2008;65:1374–8.
25. Wick M, Kollig E, Muhr G, et al. The potential pattern of circulating lymphocytes TH1/TH2 is not altered after multiple injuries. Arch Surg. 2000;135:1309–14.
26. Brohi K, Singh J, Heron M, et al. Acute traumatic coagulopathy. J Trauma. 2003;54:1127–30.

27. Maegele M, Lefering R, Yucel N, et al. Early coagulopathy in multiple injury: an analysis from the German Trauma Registry on 8724 patients. Injury. 2007;38:298–304.
28. Greuters S, van den Berg A, Franschman G, et al. Acute and delayed mild coagulopathy are related to outcome in patients with isolated traumatic brain injury. Crit Care. 2011;15:R2.
29. Rossaint R, Bouillon B, Cerny V, et al. The STOP the bleeding campaign. Crit Care. 2013;17:136.
30. Chu AJ. Blood coagulation as an intrinsic pathway for proinflammation: a mini review. Inflamm Allergy Drug Targets. 2010;9:32–44.
31. Ritis K, Doumas M, Mastellos D, et al. A novel C5a receptor-tissue factor cross-talk in neutrophils links innate immunity to coagulation pathways. J Immunol. 2006;177:4794–802.
32. Abraham E. Coagulation abnormalities in acute lung injury and sepsis. Am J Respir Cell Mol Biol. 2000;22:401–4.
33. Chu AJ. Tissue factor mediates inflammation. Arch Biochem Biophys. 2005;440:123–32.
34. Kanse SM, Gallenmueller A, Zeerleder S, et al. Factor VII-activating protease is activated in multiple trauma patients and generates anaphylatoxin C5a. J Immunol. 2012;188:2858–65.
35. Riddel Jr JP, Aouizerat BE, Miaskowski C, et al. Theories of blood coagulation. J Pediatr Oncol Nurs. 2007;24:123–31.
36. Rigby AC, Grant MA. Protein S: a conduit between anticoagulation and inflammation. Crit Care Med. 2004;32:S336–41.
37. Shakur H, Roberts I, Bautista R, et al. Effects of tranexamic acid on death, vascular occlusive events, and blood transfusion in trauma patients with significant haemorrhage (CRASH-2): a randomised, placebo-controlled trial. Lancet. 2010;376:23–32.
38. Burk AM, Martin M, Flierl MA, et al. Early complementopathy after multiple injuries in humans. Shock. 2012;37:348–54.
39. Hecke F, Schmidt U, Kola A, et al. Circulating complement proteins in multiple trauma patients–correlation with injury severity, development of sepsis, and outcome. Crit Care Med. 1997;25:2015–24.
40. Fosse E, Pillgram-Larsen J, Svennevig JL, et al. Complement activation in injured patients occurs immediately and is dependent on the severity of the trauma. Injury. 1998;29:509–14.
41. Mollnes TE, Fosse E. The complement system in trauma-related and ischemic tissue damage: a brief review. Shock. 1994;2:301–10.
42. Amara U, Kalbitz M, Perl M, et al. Early expression changes of complement regulatory proteins and C5A receptor (CD88) on leukocytes after multiple injury in humans. Shock. 2010;33:568–75.
43. Mastellos D, Lambris JD. Complement: more than a 'guard' against invading pathogens? Trends Immunol. 2002;23:485–91.
44. Sugimoto K, Hirata M, Majima M, et al. Evidence for a role of kallikrein-P6nin system in patients with shock after blunt trauma. Am J Physiol. 1998;274:R1556–60.
45. Joseph K, Kaplan AP. Formation of bradykinin: a major contributor to the innate inflammatory response. Adv Immunol. 2005;86:159–208.
46. Amara U, Flierl MA, Rittirsch D, et al. Molecular intercommunication between the complement and coagulation systems. J Immunol. 2010;185:5628–36.
47. Gebhard F, Pfetsch H, Steinbach G, et al. Is interleukin 6 an early marker of injury severity following major trauma in humans? Arch Surg. 2000;135:291–5.
48. Dinarello CA. Proinflammatory cytokines. Chest. 2000;118:503–8.
49. Ayala A, Perrin MM, Meldrum DR, et al. Hemorrhage induces an increase in serum TNF which is not associated with elevated levels of endotoxin. Cytokine. 1990;2:170–4.
50. Rabinovici R, John R, Esser KM, et al. Serum tumor necrosis factor-alpha profile in trauma patients. J Trauma. 1993;35:698–702.
51. Rhee P, Waxman K, Clark L, et al. Tumor necrosis factor and monocytes are released during hemorrhagic shock. Resuscitation. 1993;25:249–55.
52. Roumen RM, Hendriks T, van der Ven-Jongekrijg J, et al. Cytokine patterns in patients after major vascular surgery, hemorrhagic shock, and severe blunt trauma. Relation with subsequent adult respiratory distress syndrome and multiple organ failure. Ann Surg. 1993;218:769–76.
53. Stylianos S, Wakabayashi G, Gelfand JA, et al. Experimental hemorrhage and blunt trauma do not increase circulating tumor necrosis factor. J Trauma. 1991;31:1063–7.
54. Zingarelli B, Squadrito F, Altavilla D, et al. Role of tumor necrosis factor-alpha in acute hypovolemic hemorrhagic shock in rats. Am J Physiol. 1994;266:H1512–5.
55. Biffl WL, Moore EE, Moore FA, et al. Interleukin-6 in the injured patient. Marker of injury or mediator of inflammation? Ann Surg. 1996;224:647–64.
56. Pape HC, Tsukamoto T, Kobbe P, et al. Assessment of the clinical course with inflammatory parameters. Injury. 2007;38:1358–64.
57. Partrick DA, Moore FA, Moore EE, et al. Jack A. Barney Resident Research Award winner. The inflammatory profile of interleukin-6, interleukin-8, and soluble intercellular adhesion molecule-1 in postinjury multiple organ failure. Am J Surg. 1996;172:425–9.
58. Pape HC, van Griensven M, Rice J, et al. Major secondary surgery in blunt trauma patients and perioperative cytokine liberation: determination of the clinical relevance of biochemical markers. J Trauma. 2001;50:989–1000.
59. DeLong Jr WG, Born CT. Cytokines in patients with polytrauma. Clin Orthop Relat Res. 2004;(422):57–65.

60. Donnelly SC, Strieter RM, Kunkel SL, et al. Interleukin-8 and development of adult respiratory distress syndrome in at-risk patient groups. Lancet. 1993;341:643–7.
61. Pallister I, Dent C, Topley N. Increased neutrophil migratory activity after major trauma: a factor in the etiology of acute respiratory distress syndrome? Crit Care Med. 2002;30:1717–21.
62. Oswald IP, Wynn TA, Sher A, et al. Interleukin 10 inhibits macrophage microbicidal activity by blocking the endogenous production of tumor necrosis factor alpha required as a costimulatory factor for interferon gamma-induced activation. Proc Natl Acad Sci U S A. 1992;89:8676–80.
63. Armstrong L, Millar AB. Relative production of tumour necrosis factor alpha and interleukin 10 in adult respiratory distress syndrome. Thorax. 1997;52:442–6.
64. Donnelly SC, Strieter RM, Reid PT, et al. The association between mortality rates and decreased concentrations of interleukin-10 and interleukin-1 receptor antagonist in the lung fluids of patients with the adult respiratory distress syndrome. Ann Intern Med. 1996;125:191–6.
65. Giannoudis PV, Smith RM, Perry SL, et al. Immediate IL-10 expression following major orthopaedic trauma: relationship to anti-inflammatory response and subsequent development of sepsis. Intensive Care Med. 2000;26:1076–81.
66. Neidhardt R, Keel M, Steckholzer U, et al. Relationship of interleukin-10 plasma levels to severity of injury and clinical outcome in injured patients. J Trauma. 1997;42:863–70.
67. Pajkrt D, Camoglio L, Tiel-van Buul MC, et al. Attenuation of proinflammatory response by recombinant human IL-10 in human endotoxemia: effect of timing of recombinant human IL-10 administration. J Immunol. 1997;158:3971–7.
68. Opal SM, DePalo VA. Anti-inflammatory cytokines. Chest. 2000;117:1162–72.
69. Phipps RP, Stein SH, Roper RL. A new view of prostaglandin E regulation of the immune response. Immunol Today. 1991;12:349–52.
70. Tilg H, Trehu E, Atkins MB, et al. Interleukin-6 (IL-6) as an anti-inflammatory cytokine: induction of circulating IL-1 receptor antagonist and soluble tumor necrosis factor receptor p55. Blood. 1994;83:113–8.
71. Sasaki M, Joh T. Oxidative stress and ischemia-reperfusion injury in gastrointestinal tract and antioxidant, protective agents. J Clin Biochem Nutr. 2007;40:1–12.
72. Cristofori L, Tavazzi B, Gambin R, et al. Early onset of lipid peroxidation after human traumatic brain injury: a fatal limitation for the free radical scavenger pharmacological therapy? J Investig Med. 2001;49:450–8.
73. Kong SE, Blennerhassett LR, Heel KA, et al. Ischaemia-reperfusion injury to the intestine. Aust N Z J Surg. 1998;68:554–61.
74. Remick DG, Villarete L. Regulation of cytokine gene expression by reactive oxygen and reactive nitrogen intermediates. J Leukoc Biol. 1996;59:471–5.
75. Schreck R, Rieber P, Baeuerle PA. Reactive oxygen intermediates as apparently widely used messengers in the activation of the NF-kappa B transcription factor and HIV-1. EMBO J. 1991;10:2247–58.
76. Gasic AC, McGuire G, Krater S, et al. Hydrogen peroxide pretreatment of perfused canine vessels induces ICAM-1 and CD18-dependent neutrophil adherence. Circulation. 1991;84:2154–66.
77. Hua R, Chen FX, Zhang YM, et al. Association of traumatic severity with change in lymphocyte subsets in the early stage after trauma. Zhonghua Wei Zhong Bing Ji Jiu Yi Xue. 2013;25:489–92.
78. Zallen G, Moore EE, Johnson JL, et al. Circulating postinjury neutrophils are primed for the release of proinflammatory cytokines. J Trauma. 1999;46:42–8.
79. Law MM, Cryer HG, Abraham E. Elevated levels of soluble ICAM-1 correlate with the development of multiple organ failure in severely injured trauma patients. J Trauma. 1994;37:100–9.
80. Seekamp A, Jochum M, Ziegler M, et al. Cytokines and adhesion molecules in elective and accidental trauma-related ischemia/reperfusion. J Trauma. 1998;44:874–82.
81. Simon SI, Green CE. Molecular mechanics and dynamics of leukocyte recruitment during inflammation. Annu Rev Biomed Eng. 2005;7:151–85.
82. Ayala A, Ertel W, Chaudry IH. Trauma-induced suppression of antigen presentation and expression of major histocompatibility class II antigen complex in leukocytes. Shock. 1996;5:79–90.
83. Chalifour A, Jeannin P, Gauchat JF, et al. Direct bacterial protein PAMP recognition by human NK cells involves TLRs and triggers alpha-defensin production. Blood. 2004;104:1778–83.
84. Lanier LL. NK cell recognition. Annu Rev Immunol. 2005;23:225–74.
85. Morrison G, Cunningham-Rundles S, Clowes Jr GH, et al. Augmentation of NK cell activity by a circulating peptide isolated from the plasma of trauma patients. Ann Surg. 1986;203:21–4.
86. Lodoen MB, Lanier LL. Natural killer cells as an initial defense against pathogens. Curr Opin Immunol. 2006;18:391–8.
87. Joshi P, Hauser CJ, Jones Q, et al. Mechanism of suppression of natural killer cell activity in trauma patients. Res Commun Mol Pathol Pharmacol. 1998;101:241–8.
88. Barkhausen T, Frerker C, Putz C, et al. Depletion of NK cells in a murine polytrauma model is associated with improved outcome and a modulation of the inflammatory response. Shock. 2008;30:401–10.
89. Chiche L, Forel JM, Thomas G, et al. The role of natural killer cells in sepsis. J Biomed Biotechnol. 2011;2011:986491.
90. Badgwell B, Parihar R, Magro C, et al. Natural killer cells contribute to the lethality of a murine model of

Escherichia coli infection. Surgery. 2002;132:205–12.
91. Ghirnikar RS, Lee YL, Eng LF. Inflammation in traumatic brain injury: role of cytokines and chemokines. Neurochem Res. 1998;23:329–40.
92. Morganti-Kossmann MC, Satgunaseelan L, Bye N, et al. Modulation of immune response by head injury. Injury. 2007;38:1392–400.
93. Schmidt OI, Heyde CE, Ertel W, et al. Closed head injury–an inflammatory disease? Brain Res Brain Res Rev. 2005;48:388–99.
94. Knoferl MW, Liener UC, Perl M, et al. Blunt chest trauma induces delayed splenic immunosuppression. Shock. 2004;22:51–6.
95. Perl M, Gebhard F, Bruckner UB, et al. Pulmonary contusion causes impairment of macrophage and lymphocyte immune functions and increases mortality associated with a subsequent septic challenge. Crit Care Med. 2005;33:1351–8.
96. Strecker W, Gebhard F, Perl M, et al. Biochemical characterization of individual injury pattern and injury severity. Injury. 2003;34:879–87.
97. Schirmer WJ, Schirmer JM, Townsend MC, et al. Femur fracture with associated soft-tissue injury produces hepatic ischemia. Possible cause of hepatic dysfunction. Arch Surg. 1988;123:412–5.
98. Giannoudis PV, Pape HC, Cohen AP, et al. Review: systemic effects of femoral nailing: from Kuntscher to the immune reactivity era. Clin Orthop Relat Res. 2002;(404):378–86.
99. Hauser CJ, Joshi P, Zhou X, et al. Production of interleukin-10 in human fracture soft-tissue hematomas. Shock. 1996;6:3–6.
100. Hauser CJ, Zhou X, Joshi P, et al. The immune microenvironment of human fracture/soft-tissue hematomas and its relationship to systemic immunity. J Trauma. 1997;42:895–903.
101. Pape HC, Schmidt RE, Rice J, et al. Biochemical changes after trauma and skeletal surgery of the lower extremity: quantification of the operative burden. Crit Care Med. 2000;28:3441–8.
102. Perl M, Gebhard F, Knoferl MW, et al. The pattern of preformed cytokines in tissues frequently affected by blunt trauma. Shock. 2003;19:299–304.
103. Angele MK, Chaudry IH. Surgical trauma and immunosuppression: pathophysiology and potential immunomodulatory approaches. Langenbecks Arch Surg. 2005;390:333–41.
104. Flohe S, Flohe SB, Schade FU, et al. Immune response of severely injured patients–influence of surgical intervention and therapeutic impact. Langenbecks Arch Surg. 2007;392:639–48.
105. Ni CN, Redmond HP. Cell response to surgery. Arch Surg. 2006;141:1132–40.
106. Giannoudis PV, Smith RM, Bellamy MC, et al. Stimulation of the inflammatory system by reamed and unreamed nailing of femoral fractures. An analysis of the second hit. J Bone Joint Surg Br. 1999;81:356–61.
107. Smith RM, Giannoudis PV, Bellamy MC, et al. Interleukin-10 release and monocyte human leukocyte antigen-DR expression during femoral nailing. Clin Orthop Relat Res. 2000;(373):233–40.
108. Malone DL, Dunne J, Tracy JK, et al. Blood transfusion, independent of shock severity, is associated with worse outcome in trauma. J Trauma. 2003;54:898–905.
109. Moore EE, Johnson JL, Cheng AM, et al. Insights from studies of blood substitutes in trauma. Shock. 2005;24:197–205.
110. Moore FA, Moore EE, Sauaia A. Blood transfusion. An independent risk factor for postinjury multiple organ failure. Arch Surg. 1997;132:620–4.
111. Sauaia A, Moore FA, Moore EE, et al. Early predictors of postinjury multiple organ failure. Arch Surg. 1994;129:39–45.
112. Shander A. Emerging risks and outcomes of blood transfusion in surgery. Semin Hematol. 2004;41:117–24.
113. Silliman CC, Moore EE, Johnson JL, et al. Transfusion of the injured patient: proceed with caution. Shock. 2004;21:291–9.
114. Nakao A, Kaczorowski DJ, Sugimoto R, et al. Application of heme oxygenase-1, carbon monoxide and biliverdin for the prevention of intestinal ischemia/reperfusion injury. J Clin Biochem Nutr. 2008;42:78–88.
115. Goldenberg NM, Steinberg BE, Slutsky AS, et al. Broken barriers: a new take on sepsis pathogenesis. Sci Transl Med. 2011;3:88ps25.
116. Macintire DK, Bellhorn TL. Bacterial translocation: clinical implications and prevention. Vet Clin North Am Small Anim Pract. 2002;32:1165–78.
117. Fukushima R, Alexander JW, Gianotti L, et al. Bacterial translocation-related mortality may be associated with neutrophil-mediated organ damage. Shock. 1995;3:323–8.
118. Nieuwenhuijzen GA, Goris RJ. The gut: the 'motor' of multiple organ dysfunction syndrome? Curr Opin Clin Nutr Metab Care. 1999;2:399–404.
119. Lichtman SM. Bacterial [correction of baterial] translocation in humans. J Pediatr Gastroenterol Nutr. 2001;33:1–10.

Pathophysiology of Polytrauma

Theodoros Tosounidis and Peter V. Giannoudis

Contents

5.1	Introduction	41
5.2	Initial Response	42
5.3	Inflammatory Response	43
5.4	Clinical Course and Appropriate Actions	46
5.5	Clinical Course and Immunomarkers	48
References		51

T. Tosounidis, MD (✉) • P.V. Giannoudis, MD, FRCS
Trauma and Orthopaedic Surgery,
Leeds General Infirmary, A Floor Clarendon Wing,
Leeds General Infirmary, Great George Street,
Leeds LS1 3EX, UK
e-mail: ttosounidis@yahoo.com;
pgiannoudi@aol.com

5.1 Introduction

Recent advances in seminal fields of medicine have refined our contemporary understanding of the pathophysiology of polytraumatized patient and allowed the application of novel treatment strategies in the management of patients with multiple injuries. Although the definition of the polytraumatized patient is still an issue of discussion [1], it is nowadays common knowledge that the cascade of physiologic adaptations to obtain and maintain homeostasis after severe trauma is a multifaceted phenomenon that mainly involves the cardiorespiratory and immune systems. Recent technical advances in the field of genotyping have contributed to detection of polymorphisms and haplotypes of genes related to inflammatory molecules. A promising and growing body of evidence is emerging with regard to genetic predisposition of complications after major trauma. Identification of patients prone to developing complications based in their genome profile will probably allow tailoring the diagnosis and therapeutic interventions on an individual basis. Nevertheless the results of this research have not yet been translated to everyday clinical practice.

The cardiovascular adaptation to trauma leads to early clinically observed changes better described as the three phases of hypodynamic flow, the hyperdynamic flow, and the recovery [2, 3]. Stress reaction to trauma involves the

© Springer-Verlag Berlin Heidelberg 2016
H.-C. Pape et al. (eds.), *The Poly-Traumatized Patient with Fractures:
A Multi-Disciplinary Approach*, DOI 10.1007/978-3-662-47212-5_5

activation of the immune system and the development of both the systemic inflammatory response syndrome (SIRS) and the counter anti-inflammatory response syndrome (CARS). Under ideal circumstances a fine balance between those two immune reactions is maintained and the recovery is uneventful. On the contrary an exaggerated SIRS might lead to adult respiratory distress syndrome (ARDS), multiple organ dysfunction (MOD), or even death while a decompensated CARS could contribute to immunosuppression and early sepsis [4, 5]. The rationale of contemporary management of the polytrauma patient has been based on the understanding of the implications of actions and interventions early in the course of trauma that might potentially alter the equilibrium of these two immune reactions. In selected group of patients being at high risk of complications, Damage control surgery has replaced the early total care as the conceptual framework of polytrauma management [2, 6]. The classification of polytrauma patients according to their physiology and response after prehospital and emergency resuscitation is currently used to guide the most appropriate course of action minimizing the impact of the second hit to the physiology of the patient thus providing the grounds for decreased mortality and posttraumatic complications [6]. This approach represents a paradigm shift in the management of polytraumatized patients that is based on the evidence provided by understanding the pathophysiologic mechanisms involved in the setting of polytrauma.

5.2 Initial Response

The local tissue damage (fractures, soft tissue injury), primary organ injury (lung, head), acidosis, hypoxia, and pain perception trigger the activation of local and systematic reactions in order to control hemorrhage and the function of vital organs [7]. All together represent the traumatic load posed to the organism after the initial injury.

The initial major threats during the first hours after injury include hypovolemia due to hemorrhage, hypoxia, and hypercapnia due to direct lung injury and indirect lung insult, hypothermia, and the direct and indirect traumatic brain injuries. The resuscitation efforts during the initial approach of the polytrauma are limited only to life saving procedures as these are described by the ATLS protocol. A reliable method to characterize the patients' physiology at the scene of injury is to classify the degree of shock [8]. Specific attention should be paid in certain categories of patients such as the very young and the athletes who can sufficiently compensate shock before rapid and possibly fatal deterioration.

The current pre-hospital at scene management is based on the "scoop and run" perspective [9] which relies on a rapid assessment of the injuries, securing the airway, controlling the major bleeding, and supporting the circulation until the patient arrives to hospital. In addition to ATLS, standardized life-saving procedures and hospital protocols based on local resources have been established over the years assisting the clinicians with the decision-making process. The importance of the trauma network and the effective collaboration between the central major trauma center, peripheral hospitals, and ambulance services is of paramount importance [10].

The so-called end points of resuscitation include stable hemodynamics with no requirement of inotropic support, stable oxygen saturation, lactate level less than 2 mmol/L, no coagulation disturbances, normal temperature, urinary output greater than 1 mL/kg/h [6]. These parameters provide an indirect estimation or intravascular volume restoration and oxygen delivery and consumption by the tissues. The end points of resuscitation are currently used to physiologically classify the patient and dictate the most appropriate course of action [6]. Of note is the fact that the decreased base deficit and the elevated lactate levels at the time of admission still remain an invaluable index in reflecting mortality in blunt trauma patients as this was demonstrated in a recent prognostic study that included 2269 patients with Injury Severity Score >12 [11]. On the other hand newer indices such as the shock index (Heart Rate/Systolic Blood Pressure) and new markers such as Shock Index × age, Systolic Blood Pressure/age, Maximum Heart

Rate (220 − age) − Heart Rate, and Heart Rate/maximum HR seem to better correlate with mortality at 48 h and might represent better triage tools in some trauma patients, that is, patients without head and or spine injury and no prehospital intubation or cardiac arrest [12].

It is contemporary knowledge that trauma-induced coagulopathy (TIC) is a major key factor in the pathophysiological derangements that occurs after severe trauma and significantly contributes to severe hemorrhage. Almost one third of the patients with severe trauma suffer from coagulopathy [13] and its management is of paramount importance in the success of the resuscitation efforts. The above term pertains to acute intrinsic coagulopathy and the mechanisms involved in its pathophysiology are thought to be more complex than the simplistic explanation of "dilutional coagulopathy" [14–16]. Several anticoagulation mechanisms such as the thrombin-thrombomodulin-protein C anticoagulant system [17, 18] and decreased activated protein C [18, 19] are considered to be involved in the pathogenesis of TIC. Additionally fibrinolysis [20, 21] and platelet dysfunction even in the degree of a mild decrease in their aggregation adversely affect coagulation in trauma [22, 23]. The endothelial damage due to systemic inflammation and trauma-induced complementopathy seem also to play a significant role in the development of coagulation disturbances [24, 25]. Moreover, hemodilution resulting from resuscitation with high volume of crystalloids [26] is correlated to defective coagulation, which in case of severe hypothermia and acidosis can be fatal [27]. The triad of acidosis, coagulopathy, and hypothermia is known to be well associated to very increased mortality and its presence denotes that the patient is very close to death ("in extremis" patient). Consequently, prevention of heat loss and bleeding control along with reversal of coagulopathy constitute the clinical management priorities [28]. In this context, the concept of permissive or "damage control resuscitation" with permissive hypotension has evolved to minimize the detrimental effects of hemorrhage in trauma patients [29].

The immediate central nervous system response after major trauma is mainly driven by the activation of the neuroendocrine axis. Pain and fear, the by-products of metabolism, that cross the blood-brain barrier and brain injury itself are the basic stimuli for the activation of this axis. The hypothalamus and subsequently the sympathetic-adrenal system are activated. In addition stimuli from aortic and carotid receptors trigger the renin-angiotensin system in an effort to control blood pressure through vasoconstriction and increased heart rate [7, 30]. At the same time the organism enters a reduced metabolic state in order to minimize the energy expenditure [7, 30, 31].

5.3 Inflammatory Response

Nowadays, it is well recognized that major trauma induces an intense immuno-inflammatory response. The magnitude of this response depends on the initial trauma load, the pain stimuli, the systemic and local release of pro-inflammatory cytokines, the age, the sex as well as the genetic make-up of the patient. The activation of various cells such as polymorphonuclear leukocytes (PMNL), monocytes, lymphocytes, natural killer (NK), and parenchymal cells leads to dysfunction of the endothelial membrane of almost every vital organ and the development of SIRS [7, 32, 33]. The microenvironment theory [34] describes the interactions between PMNL and endothelial cells facilitated by the expression of adhesion molecules. When firm adhesion is established then the PMNL can extravasate and induce remote organ injury [34]. This injury affects not only the tissues at the site of injury but the endothelium of vital organs and especially the lung. Activated neutrophils migrate to the site of injury. Vascular endothelial damage and increased endothelial permeability may occur leading to generalized hypoxemia causing further sequestration and priming of neutrophils and macrophages facilitating activation of the coagulation, complement and the prostaglandin system [35].

Apart from the aforementioned systemic early innate response, paracrine action of locally produced inflammatory mediators plays a significant role. Prostaglandins and thromboxanes from damaged endothelial membranes, as well as histamine, bradykinin, and kallidin from interstitial mast cells, are locally produced and can magnify capillary permeability and local tissue edema. At the same time the cascade of these events can be amplified from the dissemination of these mediators to the peripheral bloodstream [35].

In the early phase, major trauma also triggers the release of signaling molecules called alarmins that mainly play a role in the activation of innate immune response without the presence of a bacterial focus. The alarmins are released factors, antigens, and cell debris from the traumatized/dead tissue. Alarmins are also chemoattractants and activators of antigen presenting cells (APCs) [36]. They belong to the so-called damage associated molecular patterns (DAMPs) [37] that include the alarmins and the pathogen associated molecular patterns (PAMPs). PAMPs represent inflammatory molecules of microbial origin recognized by the immune system as foreign due to their peculiar molecular patterns. Alarmins act as "danger molecules" and are actively secreted from the dead cells in the site of injury and passively released from cells that are in the process of imminent cellular death or apoptosis (Fig. 5.1) [38, 39]. The "danger molecules," that is, endogenous alarmins and exogenous PAMPs collectively interact with pattern recognition receptors (PRRs) and upon their detection from immune system cells trigger inflammatory, chemotactic, antimicrobial, and adaptive immune cell reactions [40]. In a recent study [41], the presence of mitochondrial DAMPs in circulation after trauma was evident and the common pathway of immune system response that results in a sepsis-like state was demonstrated. Antibacterial peptides, S100, heat-shock proteins and high-mobility group box 1 (HMGB1), with the latter being the most important are some of the molecules included in the family of alarmins [42]. Our knowledge about these molecules has substantiated over the last years and their role in the development of the "aseptic SIRS" and the pathogenesis of multiple organ dysfunction syndrome (MODS) is under ongoing investigation. In fact blockade of HMGB1 in animal models of trauma has been shown to decrease the inflammatory response and to improve outcomes [43, 44]. Nevertheless, DAMPs targeting strategies that could potentially be used in clinical practice have not been yet evolved for the patients with SIRS [45]. From the alarmin receptors, the toll-like receptor 4 (TLR4) seems to represent the most important one [40]. It has been demonstrated that the alarmin HMGB1 is released early after severe trauma and shock, activates the TLR4, and is associated with traumatic coagulopathy and other systemic inflammatory markers. Moreover, elevated levels of HMGB1 are correlated to increase morbidity and mortality [45, 46]. Furthermore, it has been shown that increased peripheral blood levels of histone-complex DNA fragments, that is, extracellular nucleic constituents released from damaged tissue after trauma play a role in endothelial damage, hypercoagulable state, and inflammation in the early course of trauma [47].

Cytokines are polypeptides that are produced from a variety of cells such as monocytes/macrophages and T-helper lymphocytes. Interleukin-1 (IL-1), IL-6, IL-8, IL-10 and tumor necrosis factor (TNF) are cytokines that transmit signals between cells thus enhancing their communication and playing an important role in the development of SIRS and MODS. In particular, TNF activates cells such as NK-cell and macrophages and induces apoptosis [48]. It leads to thromboxane A2, prostaglandin, selectin, platelet activation factor, and intracellular adhesion molecules production. It exerts its effects via remote and local action. Up-to-date effective inhibition of TNF has not been successful although blocking it might work in septic patients [49].

Interleukin 1 (IL-1) is another cytokine involved in signaling during major trauma. Its secretion pathway has not been fully understood so far. It induces T-cell and macrophage application and activates a cascade that leads to transcription of many different pro-inflammatory cytokines [35]. Interleukin 6 (IL-6) is the most extensively studied cytokine that is promptly detectable after major trauma (within hours).

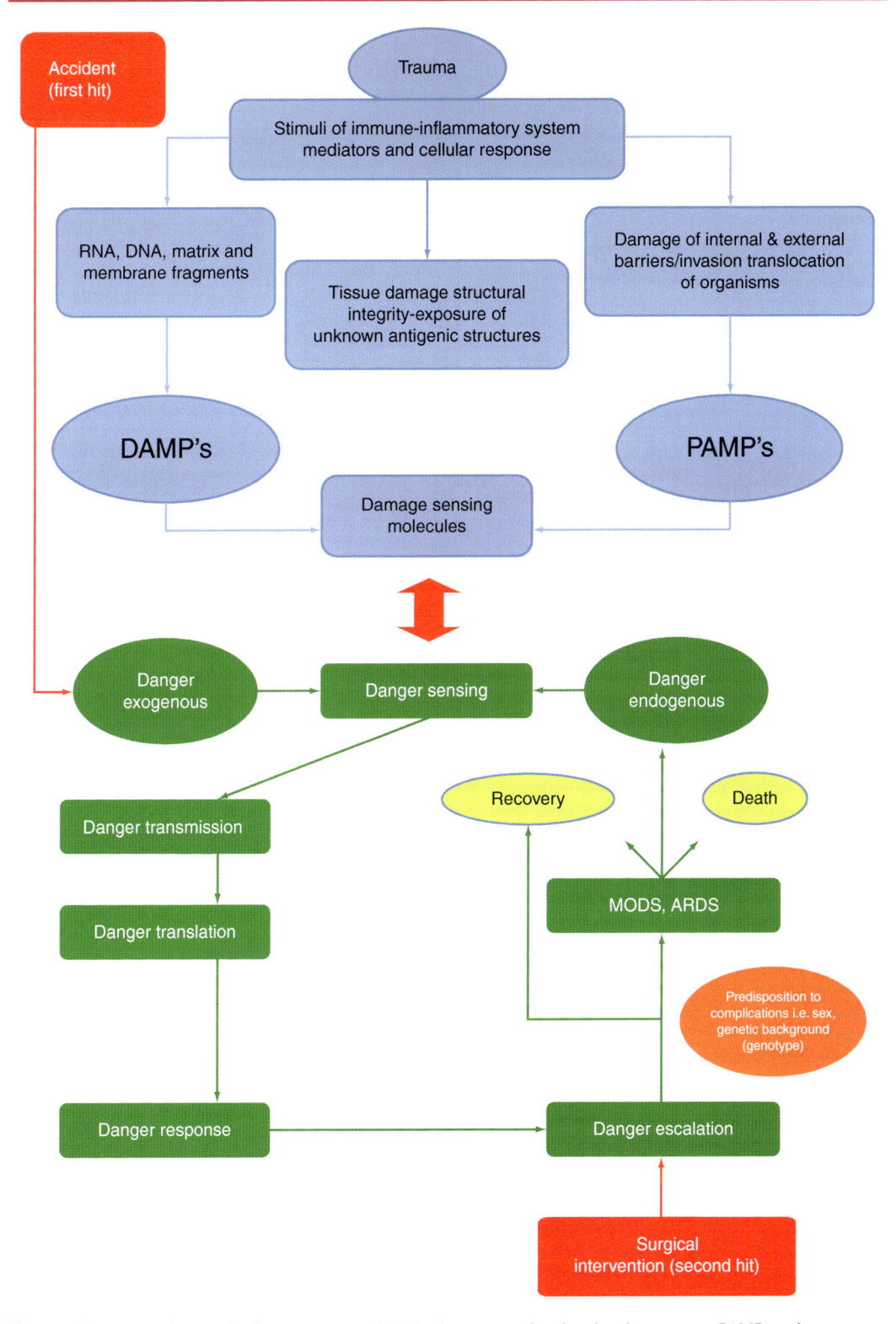

Fig. 5.1 Danger sensing mechanisms at trauma. *DAMPs* danger associated molecular patterns, *PAMPs* pathogen associated molecular patterns

Its plasma half-life and the consistent pattern of expression have established it as the most widely studied pro-inflammatory molecule [50]. It regulates growth and differentiation of lymphocytes, and activates NK-cells and neutrophils. At the same time it inhibits the apoptosis of neutrophils having therefore a role both as a pro-inflammatory and anti-inflammatory protein [51, 52]. In animal models it has been shown that blockade of IL-6 increases survival [53]. It has also been proved that a certain cut-off of 200 pg/dl could effectively be used as a diagnostic and predictive means of SIRS and later complications in the clinical setting [50]. IL-8 belongs to chemotactic cytokines which are called chemokines and act as chemoattractants. Depending on its concentration gradient IL-8 can act as an angiogenic factor and a very effective chemmoattractant. It activates the neutrophills as well as lymphocytes, monocytes, endothelial cells, and fibroblasts [54].

The physiologic response to trauma is a multifaceted phenomenon that can be influenced and modified by several different variables. It has been shown to be gender dependent and the role of sex hormones in the course of postinjury immune response is now accepted. In animal models, males and ovariectomized females exhibit a more intense alteration in immune function following hemorrhage after trauma [55]. Females have demonstrated a relative better response to traumatic shock and hemorrhage compared to males [56, 57]. In a retrospective analysis of 43,394 patients Haider et al. [58] demonstrated an increased survival rate among severely injured females (ISS > 16 and systolic blood pressure <90 mmHg) between 13 and 64 years of age who sustained shock after trauma. Asimilar difference has not in mortality was observed in younger and elderly patients, that is, patients not affected by sex hormones a finding that suggests the possible contribution of hormones to sex-based mortality after shock following trauma. Furthermore, a recent large retrospective analysis of 244,371 adult patients with blunt trauma and ISS > 16 revealed that Asian females had a 40 % lower risk of mortality relative to Asian males suggesting that in Asian race there is a sex-based outcome difference [59]. These findings underpin the racial and gender-based differences in outcome after significant trauma.

In recent years there is a growing body of evidence that posttraumatic complications could be influenced by the genetic background (genotype) of each patient and that genotyping might be helpful in detecting complication prone polytrauma patients [33, 60–63]. The application of genomic studies to practice is a promising evolving field that will probably affect our perspectives and management of polytrauma in the near future and will allow a more personalized approach to our patients [63, 64]. Bronkhorst et al. [65] performed genotyping of single-nucleotide polymorphisms in Toll-like receptor and cluster of differentiation 14 (CD14) genes in 219 polytraumatized patients and detected that the presence of a specific type of TLR2 genotype increased the risk of developing gram-positive infection and SIRS. Similarly, specific single nucleotide polymorphisms in the lectin pathway have been found to correlate with increased risk of developing sepsis, SIRS, and septic shock in polytrauma patients [66]. Additionally, three single nucleotide polymorphisms and haplotypes were found to be associated with increased production of high-mobility group box protein 1 in the peripheral blood and the increased risk of MODS and sepsis in trauma patients was suggested [67]. An overview of the most recent studies investigating the association of specific gene polymorphism with the development of complications in polytrauma patients is presented in Table 5.1.

5.4 Clinical Course and Appropriate Actions

The extent of the inflammatory response is mainly dependent on the magnitude of the traumatic load during injury. This response (SIRS) can be very intense due to the initial injury (first hit) or can be exaggerated from actions and intervention during treatment (second hit) [27, 43]. Any additional interventional (e.g., massive transfusions) or surgical (e.g., prolonged operations, operations with severe tissue damage) load, represents an exogenous hit. Furthermore antigenic

5 Pathophysiology of Polytrauma

Table 5.1 Recent studies investigating the association of specific gene polymorphism with the development of complications in polytrauma patients

Authors	Year	Gene	Number of patients	Conclusion
Bronkhorst et al. [66]	2013	MBL2, MASP2, and FCN2 in the lectin pathway	219	SNPs in the lectin pathway predispose polytrauma patients for SIRS, infectious complication, septic shock, and positive culture findings
Bronkhorst et al. [65]	2013	TLR/CD14	219	SNPs in TRL2 T-16934A and TRL9 T-1486C increased, decreased the risk of developing posttraumatic infectious complications, respectively. Early genotyping might be useful to detect the patients prone to develop complications
Zeng et al. [68]	2012	RAGE	728	The rs1800625 polymorphism is potentially useful to estimate the risk of developing multiple organ failure and sepsis in polytrauma patients
Zeng et al. [69]	2012	MD-2	726	The rs11465996 polymorphism is potentially useful to estimate the risk of developing multiple organ failure and sepsis in polytrauma patients
Zeng et al. [70]	2012	LBP	787	The rs2232618 polymorphism is potentially useful to estimate the risk of developing multiple organ failure and sepsis in polytrauma patients
Zeng et al. [67]	2012	HMGB1	556	The rs2249825 and the haplotype TCG could be used to estimate the risk of developing multiple organ failure and sepsis in polytrauma patients
Zhang et al. [71]	2011	NLRP3	718	The rs2027432 and rs12048215 polymorphisms are potentially useful to estimate the risk of developing multiple organ failure and sepsis in polytrauma patients
Chen et al. [72]	2011	TLR9	557	The rs187084 and rs352162 could be used to estimate the risk of developing multiple organ failure and sepsis in polytrauma patients
Chen et al. [73]	2011	TRL2	410	The rs3804099 and the haplotype ATT could potentially be used to estimate the risk of sepsis and multiple organ failure in polytrauma patients
Gu et al. [74]	2011	IL-4 promoter	308	IL-4 -589 T/C polymorphism potentially affects t-helper balance and susceptibility to infection
Chen et al. [75]	2010	TLR4	303	TLR4/2242 polymorphism might be used to estimate the relevant risk and organ dysfunction in polytrauma patients
Hildebrand et al. [76]	2009	CALCA	137	Polymorphisms of CALCA had no significant impact in complication development
Christie et al. [77]	2008	MYLK	237	Specific polymorphisms found to be associated with the development of acute lung injury

load from infections, ischemia/reperfusion injuries, acidosis, respiratory or cardiovascular distress, add an endogenous hit. An uncontrolled inflammatory response may lead to remote organ damage primarily in the lung leading to the development of adult respiratory distress syndrome (ARDS), MODS, and potentially death. At the same time CARS is evolving. If this hypoinflammation state is overwhelming, it may lead to immune-suppression that is responsible for the subsequent septic complications [78, 79]. An uneventful clinical course indicates that a fine balance between these extreme reactions of the immune system has prevailed.

Staging of the physiological status of the patient after the initial assessment and life-saving procedures dictates the sequence and priorities of any further actions. The patient may be classified in one of four categories: stable, borderline, unstable, and extremis [23, 48]. Stable patients have no immediate life-threatening injuries and do not need inotropic support to become hemodynamically stable. Borderline patients have been stabilized during the initial period but the type of their injuries makes them vulnerable to further rapid deterioration. Unstable patients have not achieved the end points of resuscitation and are hemodynamically unstable. Extremis patients usually suffer from the lethal triad and require inotropic support. These patients are very "sick" and usually they succumb as a result of their injuries. Extreme vigilance is required in specific patient groups such as the children, young adults, and athletes since shock can be initially compensated until rapid deterioration occurs. The clinical condition of the patient in any given time reflects a stage in ongoing evolving immune inflammatory reactions. If the magnitude of the initial trauma is well tolerated and physiological markers of stress are not abnormal, early implementation of definite care can be performed with subsequent uneventful recovery. If the initial injury is of great magnitude then hemorrhage control takes priority and temporary stabilization of musculoskeletal injuries utilizing external fixators is performed, in order to minimize the second hit insult and to protect the organism from an exaggerated SIRS, which might lead to ARDS, MODS, or even death. Secondary definitive treatment and reconstruction procedures can be performed when the clinical condition of the patient allows. The rationale behind any intervention is to eliminate the extent of the "second hit" whenever possible [27, 48, 57]. This staged approach minimizes the degree of surgical insult to the patient who is in an unstable equilibrium after major trauma. The management of these patients can be divided into four stages. During the acute phase only the resuscitation and life-saving procedures are performed. After the initial resuscitation and during the primary stabilization period major extremity injuries, arterial injuries and compartment syndromes are managed with DCO. In the secondary period the patient is reassessed constantly and appropriate actions are taken. Major procedures (second hit) are not justified due to the additional burden that may exert to the already compromised patient's immunological status. Subsequently, between days 5–10 the so-called period of "window of opportunity" definite fracture treatment can be performed [80]. Thereafter, any complex reconstruction procedures can be planned accordingly [6] (Fig. 5.2). Although concerns about longer hospital stay and cost implications have been raised, this approach has definitely modified the perceptions and daily practice of the orthopedic trauma surgeons [81].

5.5 Clinical Course and Immunomarkers

From the above described theory of "two" or "multiple hits" is becoming evident that monitoring the patient's status and clinical course via a scoring system of inflammation would be useful in both guiding our clinical decisions with regard to therapeutic intervention and predicting the possible outcome and complications in the setting of polytrauma. Various attempts have been made and are ongoing to describe the degree of the inflammatory response [82–89].

Immunomonitoring is a term used to describe the value of monitoring the inflammatory markers that are released and can be clinically measured

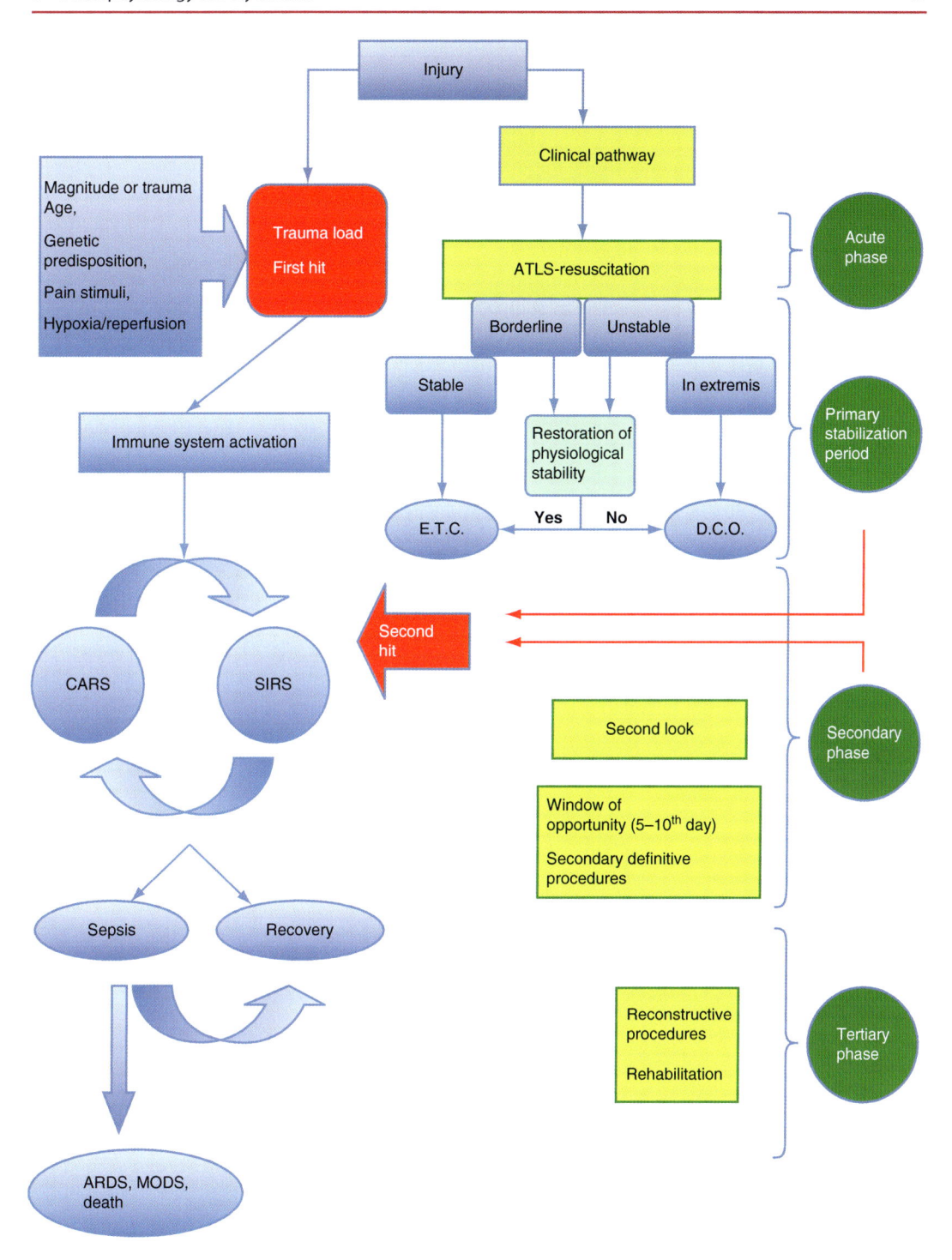

Fig. 5.2 The immune response after trauma, its correlation to clinical status of the patient and the appropriate management in any given phase. *ETC* early total care, *DCO* damage control orthopedics, *ARDS* adult respiratory distress syndrome, *MODS* multiple organ dysfunction syndrome, *ATLS* advanced trauma life support, *SIRS* systemic inflammatory response syndrome, *CARS* counter anti-inflammatory response syndrome

in the setting of polytrauma. The necessity of "immunovigilance" and its possible clinical implications became clearer during the last few years. Until recently we could only draw indirect information regarding the inflammatory status of the patient mainly from clinical markers such as fluid balance [33, 82], lactate and base deficit [90]. However, as our understanding of the complex mechanisms involved in the immune response after trauma has expanded and as our technical ability to measure various molecular mediators has improved a new era in documenting the evolving physiological status of the traumatized patient at the molecular level has been established.

The markers of immune reactivity that may have clinical utility are the acute phase reactants (liposaccharide-binding protein, C-reactive protein, precalcitonin), the markers of mediator activity (TNF-a, IL-1, IL-10, IL-6, IL-8) and the markers of cellular activity (human leukocyte antigen) [91]. While the first category has been proven to be nonspecific for trauma, there is evidence that molecules from the other two categories may have some predictive value.

More specifically TNF-a was one from the first markers that was studied. It has been correlated with poorer outcome in multiple traumatized patients in the intensive care unit but nowadays is not considered a reliable predictive index for the clinical course of inflammation in trauma unless sepsis is present [92]. The clinical utility of IL-1 and IL-10 has not been effectively supported so far [33]. The expression of major histocompatibility complex antigens (MHC class II) at the mononuclear cells of the peripheral blood has been shown to be associated to morbidity due to sepsis after trauma [93]. Many other circulating molecules have been described as potential predictors of the clinical course including the serum amyloid A, procalcitonin, C3 complement, and haptoglobin [94–96]. It appears that a continuously high level or a second rise in their values is correlated with complications and MODS, respectively [82].

Continuous monitoring is more reliable in the case of the pro-inflammatory cytokines and especially in the case of IL-6. The relatively persistent pattern of expression and the long plasma half-life have established IL-6 as the most clinically useful molecule [50]. High values have been correlated to adverse outcome after early surgery [97, 98]. IL-6 is considered to be of prognostic value for systemic inflammatory response, sepsis, and multiple organ failure [82]. IL-6 and SIRS have been correlated to new injury severity score (NISS) and to each other. A numerical value of 200 pg/dL has been proven to be of diagnostic documentation of a SIRS state [50, 99]. In a recent review of the published literature [100] about the clinical implications of Interleukin-6, the positive relationship of the extent of its elevation to the severity/extent of trauma as well as the correlation of its elevation to posttraumatic complications was evident. The authors concluded that further research is needed in order to elucidate the genetic polymorphism related to IL-6 as well as its pathophysiologic role. Large sample population, a sufficient size control group and serial measurements of IL-6 are needed and emphasis in the early posttraumatic period has been recommended. The relatively recent discovery of the alarmins (danger signaling molecules subcategorized as DAMPs and PAMPs) seems to be promising for their use as a predictive marker but up to date there are no powerful studies to support that. On the other hand the characterization and quantification of endothelial injury after trauma has been attempted to be correlated with the inflammatory and clinical status of the traumatized patient. The molecules that are released from the injured endothelium and are measurable in plasma are mainly the selectins (L-, P-, E-selectin), the vascular adhesion molecules, the thrombomodulin and the vW-factor. L-selectin has been shown to be positively related to the prognosis of potential complication after major trauma but definite conclusion cannot be drawn as yet [101].

Finally, the completion of the human genome project has open novel avenues in the clinical setting for the investigation of the genetic make-up of the patient and how this could influence the physiological responses and outcome [102]. Currently there is evidence to support the involvement of various polymorphic variants of genes in determining the posttraumatic course [103]. Single nucleotide polymorphisms result in different immune responses to trauma and might in the future guide an individualized approach to diagnosis and interventions in specific patient groups [63]. Although such an approach appears to be promising, results from different studies have not been reproducible because of the ethnic admixture, variable linkage disequilibrium, and genotype misclassification [103–105]. Further genome-wide and sufficiently powered studies are needed to provide more robust evidence about the contribution of genes in determining the clinical outcome of patients [63]. The need of translation research is also of paramount importance until novel perspectives in the polytrauma pathophysiology find their implementation in clinical practice.

References

1. Butcher N, Balogh ZJ. The definition of polytrauma: the need for international consensus. Injury. 2009;40 Suppl 4:S12–22.
2. Giannoudis PV, Dinopoulos H, Chalidis B, Hall GM. Surgical stress response. Injury. 2006;37 Suppl 5:S3–9.
3. Smith RM, Giannoudis PV. Trauma and the immune response. J R Soc Med. 1998;91:417–20.
4. Aosasa S, Ono S, Mochizuki H, Tsujimoto H, Osada S, Takayama E, et al. Activation of monocytes and endothelial cells depends on the severity of surgical stress. World J Surg. 2000;24:10–6.
5. Ono S, Aosasa S, Tsujimoto H, Ueno C, Mochizuki H. Increased monocyte activation in elderly patients after surgical stress. Eur Surg Res. 2001;33:33–8.
6. Giannoudis PV. Surgical priorities in damage control in polytrauma. J Bone Joint Surg. 2003;85:478–83.
7. Keel M, Trentz O. Pathophysiology of polytrauma. Injury. 2005;36:691–709.
8. Pape HC, Tornetta 3rd P, Tarkin I, Tzioupis C, Sabeson V, Olson SA. Timing of fracture fixation in multitrauma patients: the role of early total care and damage control surgery. J Am Acad Orthop Surg. 2009;17:541–9.
9. Smith RM, Conn AK. Prehospital care – scoop and run or stay and play? Injury. 2009;40 Suppl 4:S23–6.
10. Kanakaris NK, Giannoudis PV. Trauma networks: present and future challenges. BMC Med. 2011;9:121.
11. Ouellet JF, Roberts DJ, Tiruta C, Kirkpatrick AW, Mercado M, Trottier V, et al. Admission base deficit and lactate levels in Canadian patients with blunt trauma: are they useful markers of mortality? J Trauma Acute Care Surg. 2012;72:1532–5.
12. Bruijns SR, Guly HR, Bouamra O, Lecky F, Lee WA. The value of traditional vital signs, shock index, and age-based markers in predicting trauma mortality. J Trauma Acute Care Surg. 2013;74:1432–7.
13. Moore EE, Moore FA, Fabian TC, Bernard AC, Fulda GJ, Hoyt DB, et al. Human polymerized hemoglobin for the treatment of hemorrhagic shock when blood is unavailable: the USA multicenter trial. J Am Coll Surg. 2009;208:1–13.
14. Frith D, Brohi K. The pathophysiology of trauma-induced coagulopathy. Curr Opin Crit Care. 2012;18:631–6.
15. White NJ. Mechanisms of trauma-induced coagulopathy. Hematology. 2013;2013:660–3.
16. Maegele M, Spinella PC, Schochl H. The acute coagulopathy of trauma: mechanisms and tools for risk stratification. Shock. 2012;38:450–8.
17. Brohi K, Cohen MJ, Ganter MT, Matthay MA, Mackersie RC, Pittet JF. Acute traumatic coagulopathy: initiated by hypoperfusion: modulated through the protein C pathway? Ann Surg. 2007;245:812–8.
18. Brohi K, Cohen MJ, Ganter MT, Schultz MJ, Levi M, Mackersie RC, et al. Acute coagulopathy of trauma: hypoperfusion induces systemic anticoagulation and hyperfibrinolysis. J Trauma. 2008;64:1211–7; discussion 7.
19. Cohen MJ, Call M, Nelson M, Calfee CS, Esmon CT, Brohi K, et al. Critical role of activated protein C in early coagulopathy and later organ failure, infection and death in trauma patients. Ann Surg. 2012;255:379–85.
20. Tauber H, Innerhofer P, Breitkopf R, Westermann I, Beer R, El Attal R, et al. Prevalence and impact of abnormal ROTEM(R) assays in severe blunt trauma: results of the 'Diagnosis and Treatment of Trauma-Induced Coagulopathy (DIA-TRE-TIC) study'. Br J Anaesth. 2011;107:378–87.
21. Ives C, Inaba K, Branco BC, Okoye O, Schochl H, Talving P, et al. Hyperfibrinolysis elicited via thromboelastography predicts mortality in trauma. J Am Coll Surg. 2012;215:496–502.
22. Kutcher ME, Redick BJ, McCreery RC, Crane IM, Greenberg MD, Cachola LM, et al. Characterization of platelet dysfunction after trauma. J Trauma Acute Care Surg. 2012;73:13–9.
23. Solomon C, Traintinger S, Ziegler B, Hanke A, Rahe-Meyer N, Voelckel W, et al. Platelet function following trauma. A multiple electrode aggregometry study. Thromb Haemost. 2011;106:322–30.
24. Gebhard F, Huber-Lang M. Polytrauma–pathophysiology and management principles. Langenbecks Arch Surg. 2008;393:825–31.
25. Ostrowski SR, Johansson PI. Endothelial glycocalyx degradation induces endogenous heparinization in patients with severe injury and early traumatic coagulopathy. J Trauma Acute Care Surg. 2012;73:60–6.
26. Hubetamann B, Lefering R, Taeger G, Waydhas C, Ruchholtz S. Influence of prehospital fluid resuscitation on patients with multiple injuries in hemorrhagic shock in patients from the DGU trauma registry. J Emerg Trauma Shock. 2011;4:465–71.

27. Hess JR, Brohi K, Dutton RP, Hauser CJ, Holcomb JB, Kluger Y, et al. The coagulopathy of trauma: a review of mechanisms. J Trauma. 2008;65:748–54.
28. Cirocchi R, Abraha I, Montedori A, Farinella E, Bonacini I, Tagliabue L, et al. Damage control surgery for abdominal trauma. Cochrane Database Syst Rev. 2013;3:CD007438.
29. Schochl H, Grassetto A, Schlimp CJ. Management of hemorrhage in trauma. J Cardiothorac Vasc Anesth. 2013;27:S35–43.
30. Hill AG, Hill GL. Metabolic response to severe injury. Br J Surg. 1998;85:884–90.
31. Plank LD, Hill GL. Sequential metabolic changes following induction of systemic inflammatory response in patients with severe sepsis or major blunt trauma. World J Surg. 2000;24:630–8.
32. Cipolle MD, Pasquale MD, Cerra FB. Secondary organ dysfunction. From clinical perspectives to molecular mediators. Crit Care Clin. 1993;9:261–98.
33. Giannoudis PV. Current concepts of the inflammatory response after major trauma: an update. Injury. 2003;34:397–404.
34. Hietbrink F, Koenderman L, Rijkers G, Leenen L. Trauma: the role of the innate immune system. World J Emerg Surg. 2006;1:15.
35. Lenz A, Franklin GA, Cheadle WG. Systemic inflammation after trauma. Injury. 2007;38:1336–45.
36. Oppenheim JJ, Yang D. Alarmins: chemotactic activators of immune responses. Curr Opin Immunol. 2005;17:359–65.
37. Bianchi ME. DAMPs, PAMPs and alarmins: all we need to know about danger. J Leukoc Biol. 2007;81:1–5.
38. Raucci A, Palumbo R, Bianchi ME. HMGB1: a signal of necrosis. Autoimmunity. 2007;40:285–9.
39. El Mezayen R, El Gazzar M, Seeds MC, McCall CE, Dreskin SC, Nicolls MR. Endogenous signals released from necrotic cells augment inflammatory responses to bacterial endotoxin. Immunol Lett. 2007;111:36–44.
40. Stoecklein VM, Osuka A, Lederer JA. Trauma equals danger–damage control by the immune system. J Leukoc Biol. 2012;92:539–51.
41. Zhang Q, Raoof M, Chen Y, Sumi Y, Sursal T, Junger W, et al. Circulating mitochondrial DAMPs cause inflammatory responses to injury. Nature. 2010;464:104–7.
42. Pugin J. Dear SIRS, the concept of "alarmins" makes a lot of sense! Intensive Care Med. 2008;34:218–21.
43. Levy RM, Mollen KP, Prince JM, Kaczorowski DJ, Vallabhaneni R, Liu S, et al. Systemic inflammation and remote organ injury following trauma require HMGB1. Am J Physiol Regul Integr Comp Physiol. 2007;293:R1538–44.
44. Sawa H, Ueda T, Takeyama Y, Yasuda T, Shinzeki M, Nakajima T, et al. Blockade of high mobility group box-1 protein attenuates experimental severe acute pancreatitis. World J Gastroenterol. 2006;12:7666–70.
45. Hirsiger S, Simmen HP, Werner CM, Wanner GA, Rittirsch D. Danger signals activating the immune response after trauma. Mediators Inflamm. 2012;2012:315941.
46. Cohen MJ, Brohi K, Calfee CS, Rahn P, Chesebro BB, Christiaans SC, et al. Early release of high mobility group box nuclear protein 1 after severe trauma in humans: role of injury severity and tissue hypoperfusion. Crit Care. 2009;13:R174.
47. Johansson PI, Windelov NA, Rasmussen LS, Sorensen AM, Ostrowski SR. Blood levels of histone-complexed DNA fragments are associated with coagulopathy, inflammation and endothelial damage early after trauma. J Emerg Trauma Shock. 2013;6:171–5.
48. DeLong WG Jr, Born CT. Cytokines in patients with polytrauma. Clin Orthop Relat Res. 2004;21:57–65.
49. Abraham E. Why immunomodulatory therapies have not worked in sepsis. Intensive Care Med. 1999;25:556–66.
50. Giannoudis PV, Harwood PJ, Loughenbury P, Van Griensven M, Krettek C, Pape HC. Correlation between IL-6 levels and the systemic inflammatory response score: can an IL-6 cutoff predict a SIRS state? J Trauma. 2008;65:646–52.
51. Lin E, Calvano SE, Lowry SF. Inflammatory cytokines and cell response in surgery. Surgery. 2000;127:117–26.
52. Xing Z, Gauldie J, Cox G, Baumann H, Jordana M, Lei XF, et al. IL-6 is an antiinflammatory cytokine required for controlling local or systemic acute inflammatory responses. J Clin Invest. 1998;101:311–20.
53. Riedemann NC, Neff TA, Guo RF, Bernacki KD, Laudes IJ, Sarma JV, et al. Protective effects of IL-6 blockade in sepsis are linked to reduced C5a receptor expression. J Immunol. 2003;170:503–7.
54. Keane MP, Strieter RM. Chemokine signaling in inflammation. Crit Care Med. 2000;28:N13–26.
55. Choudhry MA, Bland KI, Chaudry IH. Trauma and immune response–effect of gender differences. Injury. 2007;38:1382–91.
56. Choudhry MA, Bland KI, Chaudry IH. Gender and susceptibility to sepsis following trauma. Endocr Metab Immune Disord Drug Targets. 2006;6:127–35.
57. Sperry JL, Minei JP. Gender dimorphism following injury: making the connection from bench to bedside. J Leukoc Biol. 2008;83:499–506.
58. Haider AH, Crompton JG, Chang DC, Efron DT, Haut ER, Handly N, et al. Evidence of hormonal basis for improved survival among females with trauma-associated shock: an analysis of the National Trauma Data Bank. J Trauma. 2010;69:537–40.
59. Sperry JL, Vodovotz Y, Ferrell RE, Namas R, Chai YM, Feng QM, et al. Racial disparities and sex-based outcomes differences after severe injury. J Am Coll Surg. 2012;214:973–80.
60. Bogner V, Kirchhoff C, Baker HV, Stegmaier JC, Moldawer LL, Mutschler W, et al. Gene expression

profiles are influenced by ISS, MOF, and clinical outcome in multiple injured patients: a genome-wide comparative analysis. Langenbecks Arch Surg. 2007;392:255–65.
61. Gundersen Y, Vaagenes P, Thrane I, Bogen IL, Haug KH, Reistad T, et al. Response of circulating immune cells to major gunshot injury, haemorrhage, and acute surgery. Injury. 2005;36:949–55.
62. Liese AM, Siddiqi MQ, Siegel JH, Deitch EA, Spolarics Z. Attenuated monocyte IL-10 production in glucose-6-phosphate dehydrogenase-deficient trauma patients. Shock. 2002;18:18–23.
63. Hildebrand F, Mommsen P, Frink M, van Griensven M, Krettek C. Genetic predisposition for development of complications in multiple trauma patients. Shock. 2011;35:440–8.
64. Matzko ME, Bowen TR, Smith WR. Orthogenomics: an update. J Am Acad Orthop Surg. 2012;20:536–46.
65. Bronkhorst MW, Boye ND, Lomax MA, Vossen RH, Bakker J, Patka P, et al. Single-nucleotide polymorphisms in the Toll-like receptor pathway increase susceptibility to infections in severely injured trauma patients. J Trauma Acute Care Surg. 2013;74:862–70.
66. Bronkhorst MW, Lomax MA, Vossen RH, Bakker J, Patka P, van Lieshout EM. Risk of infection and sepsis in severely injured patients related to single nucleotide polymorphisms in the lectin pathway. Br J Surg. 2013;100:1818–26.
67. Zeng L, Zhang AQ, Gu W, Chen KH, Jiang DP, Zhang LY, et al. Clinical relevance of single nucleotide polymorphisms of the high mobility group box 1 protein gene in patients with major trauma in southwest China. Surgery. 2012;151:427–36.
68. Zeng L, Zhang AQ, Gu W, Zhou J, Zhang LY, Du DY, et al. Identification of haplotype tag single nucleotide polymorphisms within the receptor for advanced glycation end products gene and their clinical relevance in patients with major trauma. Crit Care. 2012;16:R131.
69. Zeng L, Zhang AQ, Gu W, Zhou J, Zhang LY, Du DY, et al. Identification of haplotype tag SNPs within the whole myeloid differentiation 2 gene and their clinical relevance in patients with major trauma. Shock. 2012;37:366–72.
70. Zeng L, Gu W, Zhang AQ, Zhang M, Zhang LY, Du DY, et al. A functional variant of lipopolysaccharide binding protein predisposes to sepsis and organ dysfunction in patients with major trauma. Ann Surg. 2012;255:147–57.
71. Zhang AQ, Zeng L, Gu W, Zhang LY, Zhou J, Jiang DP, et al. Clinical relevance of single nucleotide polymorphisms within the entire NLRP3 gene in patients with major blunt trauma. Crit Care. 2011;15:R280.
72. Chen KH, Zeng L, Gu W, Zhou J, Du DY, Jiang JX. Polymorphisms in the toll-like receptor 9 gene associated with sepsis and multiple organ dysfunction after major blunt trauma. Br J Surg. 2011; 98:1252–9.
73. Chen KH, Gu W, Zeng L, Jiang DP, Zhang LY, Zhou J, et al. Identification of haplotype tag SNPs within the entire TLR2 gene and their clinical relevance in patients with major trauma. Shock. 2011;35:35–41.
74. Gu W, Zeng L, Zhang LY, Jiang DP, Du DY, Hu P, et al. Association of interleukin 4–589T/C polymorphism with T(H)1 and T(H)2 bias and sepsis in Chinese major trauma patients. J Trauma. 2011;71: 1583–7.
75. Chen K, Wang YT, Gu W, Zeng L, Jiang DP, Du DY, et al. Functional significance of the Toll-like receptor 4 promoter gene polymorphisms in the Chinese Han population. Crit Care Med. 2010;38:1292–9.
76. Hildebrand F, Kalmbach M, Kaapke A, Krettek C, Stuhrmann M. No association between CALCA polymorphisms and clinical outcome or serum procalcitonin levels in German polytrauma patients. Cytokine. 2009;47:30–6.
77. Christie JD, Ma SF, Aplenc R, Li M, Lanken PN, Shah CV, et al. Variation in the myosin light chain kinase gene is associated with development of acute lung injury after major trauma. Crit Care Med. 2008;36:2794–800.
78. Bone RC. Sir Isaac Newton, sepsis, SIRS, and CARS. Crit Care Med. 1996;24:1125–8.
79. Schroder O, Laun RA, Held B, Ekkernkamp A, Schulte KM. Association of interleukin-10 promoter polymorphism with the incidence of multiple organ dysfunction following major trauma: results of a prospective pilot study. Shock. 2004;21:306–10.
80. Stahel PF, Heyde CE, Wyrwich W, Ertel W. Current concepts of polytrauma management: from ATLS to "damage control". Orthopade. 2005;34:823–36.
81. Giannoudis PV, Giannoudi M, Stavlas P. Damage control orthopaedics: lessons learned. Injury. 2009;40 Suppl 4:S47–52.
82. Pape HC, Tsukamoto T, Kobbe P, Tarkin I, Katsoulis S, Peitzman A. Assessment of the clinical course with inflammatory parameters. Injury. 2007;38:1358–64.
83. Bochicchio GV, Napolitano LM, Joshi M, McCarter Jr RJ, Scalea TM. Systemic inflammatory response syndrome score at admission independently predicts infection in blunt trauma patients. J Trauma. 2001; 50:817–20.
84. Donnelly SC, MacGregor I, Zamani A, Gordon MW, Robertson CE, Steedman DJ, et al. Plasma elastase levels and the development of the adult respiratory distress syndrome. Am J Respir Crit Care Med. 1995;151:1428–33.
85. Giannoudis PV, Smith RM, Windsor AC, Bellamy MC, Guillou PJ. Monocyte human leukocyte antigen-DR expression correlates with intrapulmonary shunting after major trauma. Am J Surg. 1999;177:454–9.
86. Harwood PJ, Giannoudis PV, van Griensven M, Krettek C, Pape HC. Alterations in the systemic inflammatory response after early total care and damage control procedures for femoral shaft fracture in severely injured patients. J Trauma. 2005;58: 446–52; discussion 52–4.

87. Nast-Kolb D, Waydhas C, Gippner-Steppert C, Schneider I, Trupka A, Ruchholtz S, et al. Indicators of the posttraumatic inflammatory response correlate with organ failure in patients with multiple injuries. J Trauma. 1997;42:446–54; discussion 54–5.
88. Partrick DA, Moore EE, Moore FA, Biffl WL, Barnett Jr CC. Release of anti-inflammatory mediators after major torso trauma correlates with the development of postinjury multiple organ failure. Am J Surg. 1999;178:564–9.
89. Wanner GA, Keel M, Steckholzer U, Beier W, Stocker R, Ertel W. Relationship between procalcitonin plasma levels and severity of injury, sepsis, organ failure, and mortality in injured patients. Crit Care Med. 2000;28:950–7.
90. Rossaint R, Cerny V, Coats TJ, Duranteau J, Fernandez-Mondejar E, Gordini G, et al. Key issues in advanced bleeding care in trauma. Shock. 2006;26:322–31.
91. Sears BW, Stover MD, Callaci J. Pathoanatomy and clinical correlates of the immunoinflammatory response following orthopaedic trauma. J Am Acad Orthop Surg. 2009;17:255–65.
92. Riche F, Panis Y, Laisne MJ, Briard C, Cholley B, Bernard-Poenaru O, et al. High tumor necrosis factor serum level is associated with increased survival in patients with abdominal septic shock: a prospective study in 59 patients. Surgery. 1996;120:801–7.
93. Ayala A, Ertel W, Chaudry IH. Trauma-induced suppression of antigen presentation and expression of major histocompatibility class II antigen complex in leukocytes. Shock. 1996;5:79–90.
94. Casl MT, Coen D, Simic D. Serum amyloid A protein in the prediction of postburn complications and fatal outcome in patients with severe burns. Eur J Clin Chem Clin Biochem. 1996;34:31–5.
95. Dehne MG, Sablotzki A, Hoffmann A, Muhling J, Dietrich FE, Hempelmann G. Alterations of acute phase reaction and cytokine production in patients following severe burn injury. Burns. 2002;28:535–42.
96. Spies M, Wolf SE, Barrow RE, Jeschke MG, Herndon DN. Modulation of types I and II acute phase reactants with insulin-like growth factor-1/binding protein-3 complex in severely burned children. Crit Care Med. 2002;30:83–8.
97. Guillou PJ. Biological variation in the development of sepsis after surgery or trauma. Lancet. 1993;342:217–20.
98. Pape HC, van Griensven M, Rice J, Gansslen A, Hildebrand F, Zech S, et al. Major secondary surgery in blunt trauma patients and perioperative cytokine liberation: determination of the clinical relevance of biochemical markers. J Trauma. 2001;50:989–1000.
99. Barber RC, Chang LY, Purdue GF, Hunt JL, Arnoldo BD, Aragaki CC, et al. Detecting genetic predisposition for complicated clinical outcomes after burn injury. Burns. 2006;32:821–7.
100. Jawa RS, Anillo S, Huntoon K, Baumann H, Kulaylat M. Interleukin-6 in surgery, trauma, and critical care part II: clinical implications. J Intensive Care Med. 2011;26:73–87.
101. Giannoudis PV, Tosounidis TI, Kanakaris NK, Kontakis G. Quantification and characterisation of endothelial injury after trauma. Injury. 2007;38:1373–81.
102. Hildebrand F, Pape HC, van Griensven M, Meier S, Hasenkamp S, Krettek C, et al. Genetic predisposition for a compromised immune system after multiple trauma. Shock. 2005;24:518–22.
103. Giannoudis PV, van Griensven M, Tsiridis E, Pape HC. The genetic predisposition to adverse outcome after trauma. J Bone Joint Surg Br. 2007;89:1273–9.
104. Abraham E. Host defense abnormalities after hemorrhage, trauma, and burns. Crit Care Med. 1989;17:934–9.
105. Nadel S. Helping to understand studies examining genetic susceptibility to sepsis. Clin Exp Immunol. 2002;127:191–2.

Head Injuries: Neurosurgical and Orthopedic Strategies

6

Philip F. Stahel and Michael A. Flierl

Contents

6.1	Introduction..	55
6.2	Pathophysiology of Head Injury......................	56
6.3	The "Deadly Duo": Hypoxia and Hypotension..	56
6.4	Clinical Assessment and Management...........	58
6.5	Strategies of Fracture Fixation in Head-Injured Patients..................................	59
Conclusion...		60
References...		61

P.F. Stahel, MD, FACS (✉)
Department of Orthopaedics, Denver Health Medical Center, University of Colorado, School of Medicine, 777 Bannock Street, Denver, CO 80204, USA

Department of Neurosurgery, Denver Health Medical Center, University of Colorado, School of Medicine, Denver, CO, USA
e-mail: philip.stahel@dhha.org

M.A. Flierl
Department of Orthopaedic Surgery, Denver Health Medical Center, University of Colorado, School of Medicine, Denver, CO 80204, USA
e-mail: michael.a.flierl@gmail.com

6.1 Introduction

Traumatic brain injury (TBI) represents the leading cause of death in the trauma patient and is associated with dramatic long-term neurological sequelae among survivors [1–4]. One of the central aspects of our current understanding of the pathophysiology of TBI is that the extent of neurological injury is not solely determined by the traumatic impact itself, but rather evolves over time [5]. The evolution of secondary brain injury is characterized by a complex cascade of molecular and biochemical reactions to the initial trauma which occur as a consequence of complicating processes initiated by the primary traumatic impact [6–8]. These events trigger an acute inflammatory response within the injured brain, leading to development of cerebral edema, breakdown of the blood-brain barrier (BBB), and leakage of neurotoxic molecules from the peripheral blood stream into the subarachnoid space of the injured brain [9–13]. Ultimately, the extent of secondary brain injury, characterized by neuroinflammation, ischemia/reperfusion injuries, cerebral edema, intracranial hemorrhage, and intracranial hypertension, represents the main determinant for the poor outcome of head-injured patients [14, 15]. In addition, iatrogenic factors, such as permissive hypotension, prophylactic hyperventilation, overzealous volume resuscitation, and inappropriate timing and technique of associated fracture fixation may contribute to a deterioration of secondary brain injury [13, 16–19]. Despite recent advances in basic and clinical research and improved neurointensive care, no specific pharmacological therapy is currently available which may attenuate or prevent the development of secondary brain injuries [20]. Due to the complex underlying pathophysiology and the high vulnerability of the injured brain to "2nd hit" insults, it is imperative to closely coordinate the timing and surgical priorities for the management of associated injuries in head-injured patients.

6.2 Pathophysiology of Head Injury

The primary brain injury is a result of mechanical forces applied to skull at the time of impact, whereas secondary brain injury evolves over time and cannot be detected on initial CT imaging studies [21]. Evidence of secondary brain injury has been found on autopsy in 70–90 % of all fatally head-injured patients [22, 23]. Secondary brain injury is initiated by a trauma-induced, host-mediated inflammatory response within the intracranial compartment, and is aggravated by hypoxia, metabolic acidosis, cerebral fat emboli from the fracture site, injury-triggered activation of the coagulation system, and development of cerebral edema [6, 14, 17]. The immunological and pathophysiological sequelae of TBI are highly complex, and involve numerous brain-derived proinflammatory mediators, such as cytokines, chemokines, complement anaphylatoxins, excitatory molecules, electrolyte disturbances, and blood-derived leukocytes which are migrating across the BBB [11, 24, 25].

The resulting complex neuroinflammatory network leads to a proinflammatory environment with brain edema and brain tissue destruction by leukocyte-released proteases, lipases, and reactive oxygen species [26, 27]. In addition, these events culminate in the break-down of the BBB and allow neurotoxic circulating molecules to enter the brain. As a result, the traumatized brain is highly susceptible to secondary injuries caused by intracerebral inflammation, as well as systemic neurotoxic molecules, which are normally blocked under physiological conditions (Fig. 6.1).

In TBI patients who have sustained concomitant extracerebral trauma to the musculoskeletal system, a profound *systemic* inflammatory response is triggered in parallel, involving cytokines/chemokines, complement activation products, the coagulation system, stress hormones, neuronal signaling, and numerous inflammatory cells [28].

The treating surgeon has to be aware of the neuropathology of TBI as well as the systemic inflammatory invents when deciding on the optimal management approach in this challenging patient population, as inappropriate treatment may result in an iatrogenic secondary insult to the brain.

6.3 The "Deadly Duo": Hypoxia and Hypotension

Episodes of hypoxia and hypotension represent the main independent predictive factors for poor outcome after severe brain injury [8, 29]. In a landmark article published in 1993, Chesnut et al. analyzed the impact of hypotension, as defined as a systolic blood pressure (SBP) <90 mmHg, either during the resuscitation phase ("early") or in the ICU ("late"), on the outcome of head-injured patients prospectively entered into the Traumatic Coma Data Bank (TCDB) [15]. Early hypotension occurred in 248 of 717 patients (34.6 %) and was associated with a doubling of postinjury mortality from 27 to 55 % [15]. Late hypotension occurred in 156 of 493 patients (31.6 %), of which 39 patients (7.9 %) had combined early and late hypotensive episodes. For 117 patients with an exclusive hypotensive episode occurred in the ICU, 66 % either died or survived in a vegetative state, as defined by a Glasgow Outcome Scale (GOS) score of 1 or 2 points [15]. The authors furthermore determined that mortality is drastically increased in combination with hypotension (SBP <90 mmHg) and hypoxia (PaO$_2$ ≤60 mmHg) [7]. A different study by Elf et al. confirmed the notion, that severe secondary insults occur during the neurointensive care period in more than 35 % of all head-injured

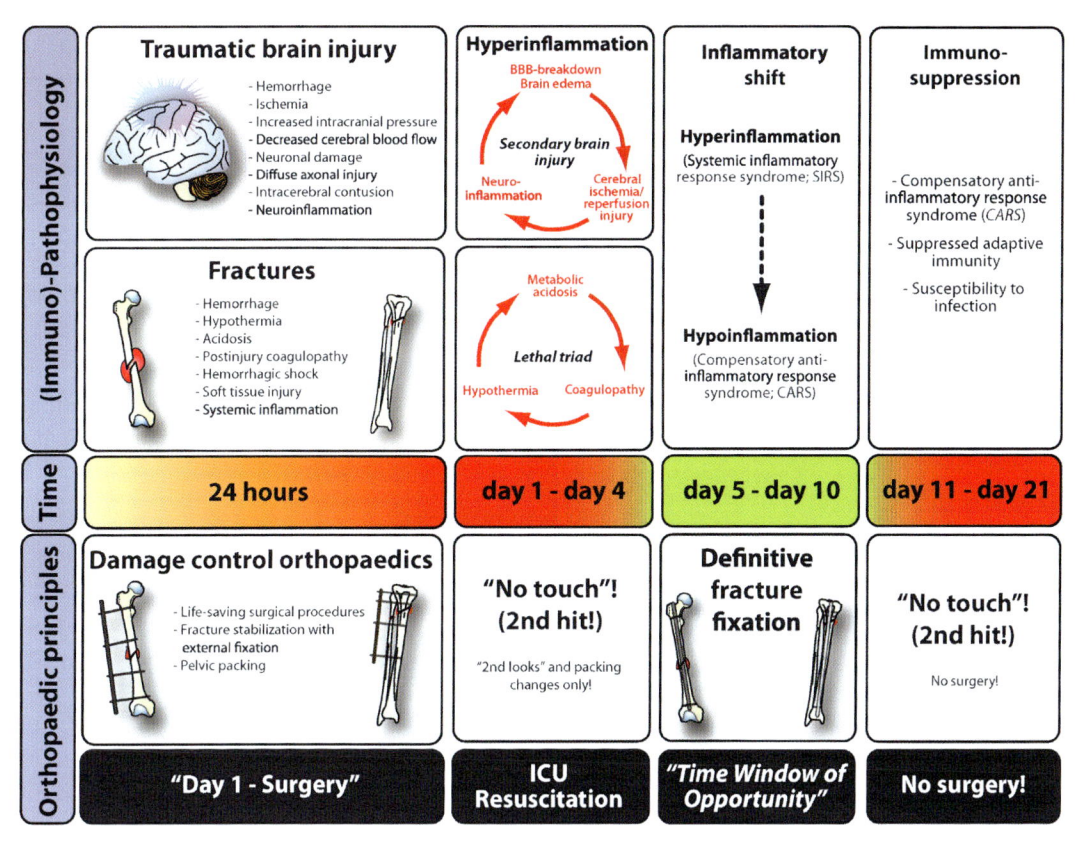

Fig. 6.1 Schematic of priorities in the management of associated orthopedic injuries in patients with severe head injuries, based on the understanding of the underlying immunological pathophysiology

patients, including episodes of hypoxia, hypotension, elevated intracranial pressure (ICP) and decreased cerebral perfusion pressure (CPP) [14].

The prevention of hypoxemia and hypotension represents the "key" parameter for avoiding secondary insults to the injured brain and improving outcomes of TBI patients [29, 30]. National guidelines by the *Brain Trauma Foundation* mandate that blood pressure and oxygenation be monitored in all head-injured patients, and advocate to maintain a systolic blood pressure >90 mmHg and a PaO_2 >60 mmHg, respectively [31]. This notion is of particular importance in view of the ongoing debate on the controversial concept of "permissive hypotension" in patients with traumatic hemorrhage from penetrating or blunt torso injuries [32, 33]. The strategy of "permissive hypotension" is mainly based on a landmark article from the 1990s advocating a modified prehospital resuscitation concept for hypotensive patients with penetrating torso injuries, by delaying fluid resuscitation until arrival in the operating room [34]. This proactive concept is certainly intuitive from the perspective that traditional resuscitation with aggressive fluid administration may lead to increased hydrostatic pressure and displacement of blood clots, a dilution of coagulation factors, and an undesirable hypothermia in critically injured patients [35]. However, in light of the vulnerability of the injured brain to secondary insults mediated by hypoxia and hypotension during the early postinjury period, the concept of hypotensive resuscitation, which has seen an unjustified expansion from penetrating to blunt trauma, in absence of high level evidence [32, 36], appears contraindicated for patients with traumatic brain injuries [33, 37].

6.4 Clinical Assessment and Management

Head-injured patients are initially assessed and resuscitated according to the American College of Surgeons' *Advanced Trauma Life Support* (ATLS®) protocol [35]. The severity of head injury is diagnosed by the combination of (1) mechanism of trauma, the (2) clinical/neurological status, and (3) imaging by computed tomography (CT) scan. The neurologic status is assessed after stabilization of vital functions [38]. The level of consciousness is rapidly evaluated by the Glasgow Coma Scale (GCS), which grades the severity of TBI as mild (GCS 14/15), moderate (GCS 9–13), and severe (GCS 3–8) [21]. The postresuscitation GCS score is of clinical importance due to the significant correlation with patient outcome [21]. A head CT should be obtained under the following circumstances: (1) altered level of consciousness with GCS <14 (moderate or severe brain injury); (2) abnormal neurological status; (3) differences in pupil size or reactivity; (4) suspected skull fracture; (5) intoxicated patients; and should be repeated whenever the patient's neurologic status deteriorates [21].

Elevated intracranial pressure (ICP) above 15-20 mmHg has been associated with poor outcomes after severe TBI [39]. Monitoring of ICP by indwelling catheters is recommended under the following conditions [40–43]:
1. *Severe* TBI (GCS ≤8) and abnormal admission CT scan
2. *Severe* TBI (GCS ≤8) with normal CT scan, but prolonged coma >6 h
3. Surgical evacuation of intracranial hematomas
4. Neurological deterioration (GCS ≤8) in patients with initially mild or moderate extent of TBI
5. Head-injured patients requiring prolonged mechanical ventilation, for example, for management of associated extracranial injuries, unless the initial CT scan is normal

The indications and benefits of emergency craniotomy or decompressive craniectomy are beyond the scope of this chapter, and the reader is deferred to the pertinent peer-reviewed literature [44–46].

Maintenance of an adequate cerebral perfusion pressure (CPP) is recommended above 70–80 mmHg, which is calculated as the mean arterial pressure (MAP) minus ICP [39, 41, 47]. This notion reflects on the imperative not to allow any period of hypotension in head-injured patients, as discussed above [29, 37]. In addition to the outlined dangers of hypoxemia and hypotension, hypercarbia, and hypoglycemia should be strictly avoided or rapidly corrected to minimize the risk of developing secondary brain injuries [14]. Hyperosmolar therapy with mannitol or hypertonic saline is recommended for reduction of cerebral edema and increased ICP, and in patients displaying clinical signs of trans-tentorial herniation, progressive neurological deterioration, or bilaterally dilated and nonreactive pupils [48]. However, the routine use of osmotherapy for management of brain edema represents a topic of heavy debate [49–51]. Similarly, the concept of therapeutic hypothermia for patients with severe head injuries remains controversial [46, 51, 52]. This noninvasive modality of neuroprotection has been investigated for decades in patients with head injuries, cerebrovascular stroke, cardiac arrest, and spinal cord injury [53]. The underlying rationale of moderately lowering the patient's body temperature is aimed at slowing down the acute inflammatory processes in the injured CNS, and to reduce the extent of traumatic and ischemic tissue injury [54]. Interestingly, the historic euphoria in the 1990s for applying therapeutic hypothermia to patients with severe head injuries [55] was revoked later on in additional validation studies, and the debate on the appropriateness of cooling down the injured brain remains unresolved until present [52, 56]. Despite increased understanding of the pathophysiology of secondary brain injury, the pharmacological "golden bullet" for treating TBI patients and preventing or reducing incidence of secondary cerebral insults has not yet been identified [20]. However, there is unequivocal consensus that the use of steroids is considered obsolete and contraindicated for patients with traumatic brain injuries, since the failure of the large-scale "CRASH" trial was published in 2004 [57, 58].

6.5 Strategies of Fracture Fixation in Head-Injured Patients

Head-injured patients with associated orthopedic injuries represent a vulnerable population due to the high risk of "2nd hit" insults, particularly in presence of femur shaft fractures [17]. The benefits of early definitive fracture stabilization in multiply injured patients are well described and include early unrestricted mobility in conjunction with a decreased "antigenic load" related to stress, pain, and systemic inflammation [13, 59, 60]. Clearly, the question regarding the "optimal" timing and modality of long bone fracture fixation in patients with associated head injuries remains a topic of ongoing discussion and debate [18, 61–64]. Even though the benefits of early femur fracture stabilization have been unequivocally demonstrated in Dr. Bone's landmark study more than 20 years ago [65], not all multiply injured patients are able to tolerate early definitive fracture fixation due to hemodynamic instability, refractory hypoxemia, or intracranial hypertension [62]. Impressively, experimental studies in sheep showed that femoral reaming and nailing leads to increased ICP levels above 15 mmHg in models of hemorrhagic shock/resuscitation with or without associated traumatic brain injury [19, 66]. A clinical study in 33 blunt trauma patients with TBI revealed that early definitive fracture fixation within 24 h was associated with adverse neurological outcomes and increased mortality, associated with early episodes of hypoxia and hypotension, compared to TBI patients whose orthopedic injuries were stabilized definitively at a later timepoints (>24 h) [67]. A larger 10-year study on 61 patients with severe TBI revealed that early femur fracture fixation within <24 h is associated with an increased incidence of secondary brain injury, related to significantly increased rates of hypotension and decreased CPP <70 mmHg [68]. These data were corroborated by a different study analyzing changes in ICP and CPP in 17 patients with severe head injuries undergoing reamed intramedullary nailing of associated femur fractures [69]. The authors showed that the CPP dropped below a minimal threshold of 75 mmHg intraoperatively during the fracture fixation in all patients, with an average decrease in CPP of Δ18 mmHg [69]. The decrease in CPP was attributed to intraoperative episodes of systemic hypotension, and patients with early femoral nailing within 24 h had statistically significant lower CPP values than the rest of the cohort [69].

Overall, there is unequivocal evidence – both from experimental animal studies and from clinical trials in patients with severe TBI – that the early (<24 h) definitive fixation of associated femur shaft fractures in head-injured patients leads to significant adverse effects, including intraoperative episodes of hypotension, increases in ICP and critical decreases in CPP, all of which ultimately constitute preventable "2nd hits" and contribute to secondary brain injury and poor long-term outcomes (Fig. 6.1).

Consequently, alternative strategies to provide early fracture stabilization of long bones, while avoiding the risk of "early total care", have been proposed, including skeletal traction and "damage control" external fixation [70]. The concept of "damage control" surgery was extended beyond its initial applications in abdominal and thoracic trauma, to the initial management of major fractures in the severely injured, particularly in presence of associated head injuries [62, 71]. The principal is to provide early fracture stabilization by external fixation as a bridge to definitive fracture care once the patient is physiologically stable, and the injured brain less vulnerable to iatrogenic "2nd hit" insults [17]. The delayed conversion from external fixation to intramedullary nailing of femur shaft fractures is considered safe once the ICP has normalized and/or patients are awake, oriented, and fully resuscitated [35]. In other words, the second procedure related intramedullary reaming and nailing of long bone fractures should be performed outside of "priming" window, once the postinjury hyperinflammatory response has subsided

Treatment options:

(1) Skeletal traction:
- Temporary relative stability
- Inability of patient positioning in ICU
- Motion at fracture site, perpetuating "antigenic load" and elevated ICP

(2) "Damage control" external fixation:
- Temporary relative stability
- Unlimited patient positioning in ICU
- Reduction of "antigenic load"
- Fast procedure
- Safe conversion to IMN for up to 14 days

(3) Reamed intramedullary nailing:
- Definitive stabilization
- Intraoperative hypoxia and hypotension
- Elevated ICP and decreased CPP
- Exacerbation of secondary brain injury

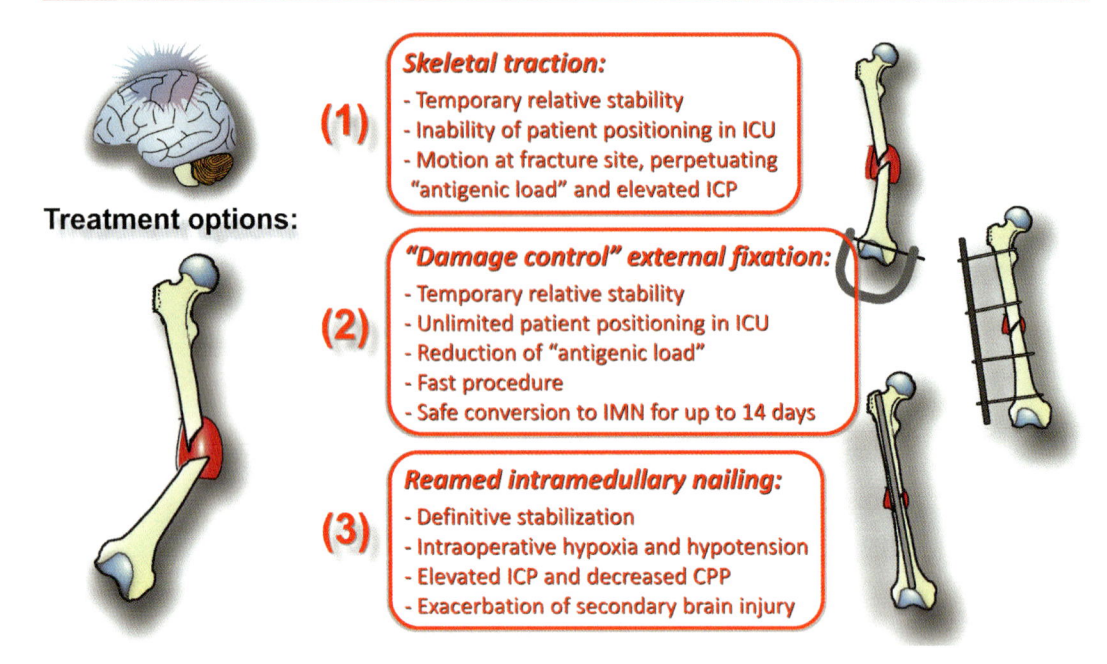

Fig. 6.2 Risks and benefits of distinct management strategies for acute immobilization of femoral shaft fractures in head-injured patients

(Fig. 6.1). When compared to early total care, the "damage control" approach with delayed conversion to definitive care has been shown to decrease the initial operative time and intraoperative blood loss without increasing the risk of procedure related complications such as infection and nonunion [72, 73].

The risks and benefits of distinct modalities for acute management of femur shaft fractures in head-injured patients, namely (1) skeletal traction [70], (2) "damage control" external fixation [71, 72], and (3) "early total care" by reamed intramedullary nail fixation [69] are depicted in Fig. 6.2.

Conclusion

Head-injured patients with associated long bone fractures represent a very vulnerable patient population [17]. These patients have a high risk of sustaining secondary cerebral insults related to hypotension, increased ICP, and decreased CPP, all of which contribute to increased mortality and adverse neurological outcomes [19, 66–69]. The subspecialties involved in the early management of multiply injured patients with head injuries and associated long bone fractures include ED physicians, trauma surgeons, neurosurgeons, and orthopedic surgeons. They all should be on the same page in terms of understanding the underlying pathophysiology of TBI and the time-dependent vulnerability of the injured brain to iatrogenic "2nd hit" insults [17, 21].

When the patient with combined orthopedic and neurosurgical injuries is evaluated in the emergency department, several questions need to be answered. A rapid neurologic exam must be performed to assess the severity of brain injury. A noncontrast craniocerebral CT scan is obtained as the first-line adjunctive diagnostic work-up in stable patients. An ICP monitor (either fiberoptic or ventricular) may be placed in the ED if the patient is too hemodynamically unstable to justify a trip to the CT scanner.

Any patient with a suspected brain injury who needs to be taken to the operating room and will be unable to undergo follow up neurologic examination needs to have ICP monitoring. The exact

ICP threshold of when not to proceed to the operating room is unknown, though sustained pressures beyond 15–20 mmHg should be an indication to proceed to the ICU for resuscitation. Any patient with a progressively worsening neurological exam is also at high risk as is the patient with unexplained changes in ICP. Hypoxia and hypotension significantly increase mortality in the patient with brain injury.

Despite recent advances from basic research and clinical studies [74], the current literature remains conflicting in terms of identifying a clear-cut management strategy for timing and modality of fracture fixation in severely head-injured patients [17, 18, 61, 64, 67, 68]. This notion emphasizes the pressing need for well-designed prospective controlled multicenter trails aimed at comparing the standard treatment strategies for initial management of long bone fractures in patients with severe head injuries (Fig. 6.2).

Until higher level evidence-based recommendations are available, the clinical approach for the management of this vulnerable cohort of patients must be based on the basic principle of "*do not further harm*" by applying simple measures of "damage control" – when in doubt – which respect the underlying pathophysiology of traumatic brain injury and the hyperinflammatory response of the combination of multiple critical injuries [13]. We recommend the following specific management strategy for associated orthopedic injuries in head-injured patients, based on a combination of empiric experience and review of the available pertinent literature in the field:

1. *"Damage control orthopedics"* by spanning external fixation in all patients with *severe* TBI (GCS ≤8, intracranial pathology on CT scan, including cerebral edema, midline shift, sub-/epidural bleeding, or open head injuries).
2. Optional *"damage control orthopedics"* in all patients with *moderate* TBI (GCS 9–13), or patients with GCS of 14/15 with "minor" intracranial pathology on CT scan (e.g., traumatic subarachnoid hemorrhage that warrants observation only). Concomitant neurosurgical procedures may be performed at the same time as DCO, for example, an emergency craniotomy.
3. *No additional operations* (2nd hit) in patients with refractory intracranial hypertension or unexplained deterioration in neurologic exam.
4. *Conversion from external to internal fixation* in TBI patients who recovered from a comatose state and are awake and alert (GCS 13–15), or comatose patients with a stable ICP (<20 mmHg) and CPP in a normal range (>80 mmHg) for more than 48 h.
5. *"Early total care"* for long bone fractures all patients with *mild* TBI (GCS 14/15) and normal initial craniocerebral CT scan.
6. *Temporary skeletal traction* as a valid adjunct for patients *"in extremis"*, that is, in severe protracted traumatic-hemorrhagic shock and coagulopathy, who are unsafe to be taken to the operating room until adequately resuscitated.

References

1. Zgaljardic DJ, Durham WJ, Mossberg KA, Foreman J, Joshipura K, Masel BE, Urban R, Sheffield-Moore M. Neuropsychological and physiological correlates of fatigue following traumatic brain injury. Brain Inj. 2014;28:389–97.
2. Tanev KS, Pentel KZ, Kredlow MA, Charney ME. PTSD and TBI co-morbidity: scope, clinical presentation and treatment options. Brain Inj. 2014;28(3):261–2703.
3. Sullivan-Singh SJ, Sawyer K, Ehde DM, Bell KR, Temkin N, Dikmen S, Williams RM, Hoffman JM. Comorbidity of pain and depression among persons with traumatic brain injury. Arch Phys Med Rehabil. 2014;95:1100–5.
4. Andruszkow H, Urner J, Deniz E, Probst C, Grun O, Lohse R, Frink M, Hildebrand F, Zeckey C. Subjective impact of traumatic brain injury on long-term outcome at a minimum of 10 years after trauma- first results of a survey on 368 patients from a single academic trauma center in Germany. Patient Saf Surg. 2013;7(1):32.
5. Stahel PF, Flierl MA. Closed head injury. In: Smith WR, Stahel PF, editors. Management of musculoskeletal injuries in the trauma patient. New York: Springer; 2014. p. 297–304.
6. Bayir H, Clark RS, Kochanek PM. Promising strategies to minimize secondary brain injury after head trauma. Crit Care Med. 2003;31(1 Suppl):S112–7.
7. Chesnut RM, Marshall LF, Klauber MR, Blunt BA, Baldwin N, Eisenberg HM, Jane JA, Marmarou A, Foulkes MA. The role of secondary brain injury in

determining outcome from severe head injury. J Trauma. 1993;34(2):216–22.
8. Chesnut RM. Secondary brain insults after head injury: clinical perspectives. New Horiz. 1995;3(3):366–75.
9. Weckbach S, Hohmann C, Braumueller S, Denk S, Klohs B, Stahel PF, Gebhard F, Huber-Lang MS, Perl M. Inflammatory and apoptotic alterations in serum and injured tissue after experimental polytrauma in mice: distinct early response compared with single trauma or "double-hit" injury. J Trauma Acute Care Surg. 2013;74(2):489–98.
10. Burk AM, Martin M, Flierl MA, Rittirsch D, Helm M, Lampl L, Bruckner U, Stahl GL, Blom AM, Perl M, et al. Early complementopathy after multiple injuries in humans. Shock. 2012;37:348–54.
11. Bellander BM, Olafsson IH, Ghatan PH, Bro Skejo HP, Hansson LO, Wanecek M, Svensson MA. Secondary insults following traumatic brain injury enhance complement activation in the human brain and release of the tissue damage marker S100B. Acta Neurochir. 2011;153(1):90–100.
12. Stahel PF, Barnum SR. The role of the complement system in CNS inflammatory diseases. Expert Rev Clin Immunol. 2006;2:445–56.
13. Probst C, Mirzayan MJ, Mommsen P, Zeckey C, Tegeder T, Geerken L, Maegele M, Samii A, van Griensven M. Systemic inflammatory effects of traumatic brain injury, femur fracture, and shock: an experimental murine polytrauma model. Mediators Inflamm. 2012;2012:136020.
14. Elf K, Nilsson P, Enblad P. Prevention of secondary insults in neurointensive care of traumatic brain injury. Eur J Trauma. 2003;29:74–80.
15. Chesnut RM, Marshall SB, Piek J, Blunt BA, Klauber MR, Marshall LF. Early and late systemic hypotension as a frequent and fundamental source of cerebral ischemia following severe brain injury in the Traumatic Coma Data Bank. Acta Neurochir Suppl. 1993;59:121–5.
16. Stahel PF, Moore EE, Schreier SL, Flierl MA, Kashuk JL. Transfusion strategies in postinjury coagulopathy. Curr Opin Anaesthesiol. 2009;22:289–98.
17. Flierl MA, Stoneback JW, Beauchamp KM, Hak DJ, Morgan SJ, Smith WR, Stahel PF. Femur shaft fracture fixation in head-injured patients: when is the right time? J Orthop Trauma. 2010;24(2):107–14.
18. Giannoudis PV, Veysi VT, Pape HC, Krettek C, Smith MR. When should we operate on major fractures in patients with severe head injuries? Am J Surg. 2002;183:261–7.
19. Mousavi M, Kolonja A, Schaden E, Gabler C, Ehteshami JR, Vecsei V. Intracranial pressure-alterations during controlled intramedullary reaming of femoral fractures: an animal study. Injury. 2001;32(9):679–82.
20. Beauchamp K, Mutlak H, Smith WR, Shohami E, Stahel PF. Pharmacology of traumatic brain injury: where is the "golden bullet"? Mol Med. 2008;14 (11–12):731–40.
21. Stahel PF, Smith WR. Closed head injury. In: Bland KI, Sarr MG, Büchler MW, Csendes A, Garden OJ, Wong J, editors. Trauma surgery – handbooks in general surgery. London: Springer; 2011. p. 83–101.
22. Badri S, Chen J, Barber J, Temkin NR, Dikmen SS, Chesnut RM, Deem S, Yanez ND, Treggiari MM. Mortality and long-term functional outcome associated with intracranial pressure after traumatic brain injury. Intensive Care Med. 2012;38(11):1800–9.
23. Finfer SR, Cohen J. Severe traumatic brain injury. Resuscitation. 2001;48(1):77–90.
24. Schmidt OI, Heyde CE, Ertel W, Stahel PF. Closed head injury – an inflammatory disease? Brain Res Rev. 2005;48(2):388–99.
25. Helmy A, Guilfoyle MR, Carpenter KL, Pickard JD, Menon DK, Hutchinson PJ. Recombinant human interleukin-1 receptor antagonist in severe traumatic brain injury: a phase II randomized control trial. J Cereb Blood Flow Metab. 2014;34:845–51.
26. Kelso ML, Gendelman HE. Bridge between neuroimmunity and traumatic brain injury. Curr Pharm Des. 2014;20:4284–98.
27. Schwulst SJ, Trahanas DM, Saber R, Perlman H. Traumatic brain injury-induced alterations in peripheral immunity. J Trauma Acute Care Surg. 2013;75(5):780–8.
28. Neher MD, Weckbach S, Flierl MA, Huber-Lang MS, Stahel PF. Molecular mechanisms of inflammation and tissue injury after major trauma–is complement the "bad guy"? J Biomed Sci. 2011;18:90.
29. Stahel PF, Smith WR, Moore EE. Hypoxia and hypotension, the "lethal duo" in traumatic brain injury: implications for prehospital care. Intensive Care Med. 2008;34(3):402–4.
30. Geeraerts T, Friggeri A, Mazoit JX, Benhamou D, Duranteau J, Vigue B. Posttraumatic brain vulnerability to hypoxia-hypotension: the importance of the delay between brain trauma and secondary insult. Intensive Care Med. 2008;34:551–60.
31. The Brain Trauma Foundation. Guidelines for the management of severe traumatic brain injury. J Neurotrauma. 2007;24 Suppl 1:S7–13.
32. Curry N, Davis PW. What's new in resuscitation strategies for the patient with multiple trauma? Injury. 2012;43:1021–8.
33. Rossaint R, Bouillon B, Cerny V, Coats TJ, Duranteau J, Fernández-Mondéjar E, Hunt BJ, Komadina R, Nardi G, Neugebauer E, et al. Management of bleeding following major trauma – an updated European guideline. Crit Care. 2010;14:R52.
34. Bickell WH, Wall MJJ, Pepe PE, Martin RR, Ginger VF, Allen MK, Mattox KL. Immediate versus delayed fluid resuscitation for hypotensive patients with penetrating torso injuries. N Engl J Med. 1994;331:1105–9.
35. Stahel PF, Smith WR, Moore EE. Current trends in resuscitation strategy for the multiply injured patient. Injury. 2009;40 Suppl 4:S27–35.
36. Pieracci FM, Biffl WL, Moore EE. Current concepts in resuscitation. J Intensive Care Med. 2012;27:79–96.

37. Bratton SL, Chestnut RM, Ghajar J, McConnell Hammond FF, Harris OA, Hartl R, Manley GT, Nemecek A, Newell DW, Rosenthal G, et al. Guidelines for the management of severe traumatic brain injury. I. Blood pressure and oxygenation. J Neurotrauma. 2007;24 Suppl 1:S7–13.
38. Stahel PF. Pupil evaluation in addition to Glasgow Coma Scale components in prediction of traumatic brain injury and mortality. Br J Surg. 2012;99 Suppl 1:131.
39. Karamanos E, Teixeira PG, Sivrikoz E, Varga S, Chouliaras K, Okoye O, Hammer P. Intracranial pressure versus cerebral perfusion pressure as a marker of outcomes in severe head injury: a prospective evaluation. Am J Surg. 2014;208:363–71.
40. Stover JF, Steiger P, Stocker R. Need for intracranial pressure monitoring following severe traumatic brain injury. Crit Care Med. 2006;34:1582–3.
41. Bratton SL, Chestnut RM, Ghajar J, McConnell Hammond FF, Harris OA, Hartl R, Manley GT, Nemecek A, Newell DW, Rosenthal G, et al. Guidelines for the management of severe traumatic brain injury. VIII. Intracranial pressure thresholds. J Neurotrauma. 2007;24 Suppl 1:S55–8.
42. Bratton SL, Chestnut RM, Ghajar J, McConnell Hammond FF, Harris OA, Hartl R, Manley GT, Nemecek A, Newell DW, Rosenthal G, et al. Guidelines for the management of severe traumatic brain injury. VII. Intracranial pressure monitoring technology. J Neurotrauma. 2007;24 Suppl 1: S45–54.
43. Bratton SL, Chestnut RM, Ghajar J, McConnell Hammond FF, Harris OA, Hartl R, Manley GT, Nemecek A, Newell DW, Rosenthal G, et al. Guidelines for the management of severe traumatic brain injury. VI. Indications for intracranial pressure monitoring. J Neurotrauma. 2007;24 Suppl 1: S37–44.
44. Chen SH, Chen Y, Fang WK, Huang DW, JHuang KC, Tseng SH. Comparison of craniotomy and decompressive craniectomy in severely head-injured patients with acute subdural hematoma. J Trauma. 2011;71: 1632–6.
45. Lazaridis C, Czosnyka M. Cerebral blood flow, brain tissue oxygenation, and metabolic effects of decompressive craniectomy. Neurocrit Care. 2011; 16:478–84.
46. Honeybul S. An update on the management of traumatic brain injury. J Neurosurg Sci. 2011;55:343–55.
47. Bratton SL, Chestnut RM, Ghajar J, McConnell Hammond FF, Harris OA, Hartl R, Manley GT, Nemecek A, Newell DW, Rosenthal G, et al. Guidelines for the management of severe traumatic brain injury. IX. Cerebral perfusion thresholds. J Neurotrauma. 2007;24 Suppl 1:S59–64.
48. Ropper AH. Hyperosmolar therapy for raised intracranial pressure. N Engl J Med. 2012;367:746–52.
49. Grande PO, Romner B. Osmotherapy in brain edema: a questionable therapy. J Neurosurg Anesthesiol. 2012;24:407–12.
50. Bulger EM, Hoyt DB. Hypertonic resuscitation after severe injury: is it of benefit? Adv Surg. 2012;46: 73–85.
51. Blissitt PA. Controversies in the management of adults with severe traumatic brain injury. AACN Adv Crit Care. 2012;23:188–203.
52. Kramer C, Freeman WD, Larson JS, Hoffman-Snyder C, Wellik KE, Dermaerschalk BM, Wingerchuck DM. Therapeutic hypothermia for severe traumatic brain injury: a critically appraised topic. Neurologist. 2012;18:173–7.
53. Kuffler DP. Maximizing neuroprotection: where do we stand? Ther Clin Risk Manag. 2012;8:185–94.
54. Straus D, Prasad V, Munoz L. Selective therapeutic hypothermia: a review of invasive and noninvasive techniques. Arq Neuropsiquiatr. 2011;69(6):981–7.
55. Marion DW, Penrod LE, Kelsey SF, Obrist WD, Kochanek PM, Palmer AM, Wisniewski SR, DeKosky ST. Treatment of traumatic brain injury with moderate hypothermia. N Engl J Med. 1997;336(8):540–6.
56. Urbano LA, Oddo M. Therapeutic hypothermia for traumatic brain injury. Curr Neurol Neurosci Rep. 2012;12(5):580–91.
57. Sauerland S, Maegele M. A CRASH landing in severe head injury. Lancet. 2004;364:1291–2.
58. Bratton SL, Chestnut RM, Ghajar J, McConnell Hammond FF, Harris OA, Hartl R, Manley GT, Nemecek A, Newell DW, Rosenthal G, et al. Guidelines for the management of severe traumatic brain injury. XV. Steroids. J Neurotrauma. 2007; 24 Suppl 1:S91–5.
59. Keel M, Trentz O. Pathophysiology of polytrauma. Injury. 2005;36:691–709.
60. Stahel PF, Smith WR, Moore EE. Role of biological modifiers regulating the immune response after trauma. Injury. 2007;38(12):1409–22.
61. Velmahos GC, Arroyo H, Ramicone E, Cornwell 3rd EE, Murray JA, Asensio JA, Berne TV, Demetriades D. Timing of fracture fixation in blunt trauma patients with severe head injuries. Am J Surg. 1998;176: 324–9.
62. Scalea TM. Optimal timing of fracture fixation: have we learned anything in the past 20 years? J Trauma. 2008;65(2):253–60.
63. Stahel PF, Ertel W, Heyde CE. Traumatic brain injury: impact on timing and modality of fracture care [German]. Orthopade. 2005;34(9):852–64.
64. Starr AJ, Hunt JL, Chason DP, Reinert CM, Walker J. Treatment of femur fracture with associated head injury. J Orthop Trauma. 1998;12(1):38–45.
65. Bone LB, Johnson KD, Weigelt J, Scheinberg R. Early versus delayed stabilization of femoral fractures: a prospective randomized study. J Bone Joint Surg Am. 1989;71:336–40.
66. Lehmann U, Reif W, Hobbensiefken G, Seekamp A, Regel G, Sturm JA, Dwenger A, Schweitzer G, Mann D, Ellerbeck M, et al. Effect of primary fracture management on craniocerebral trauma in polytrauma. An animal experiment study [German]. Unfallchirurg. 1995;98(8):437–41.

67. Jaicks RR, Cohn SM, Moller BA. Early fracture fixation may be deleterious after head injury. J Trauma. 1997;42(1):1–5; discussion 5–6.
68. Townsend RN, Lheureau T, Protech J, Riemer B, Simon D. Timing fracture repair in patients with severe brain injury (Glasgow Coma Scale score <9). J Trauma. 1998;44(6):977–82; discussion 982–3.
69. Anglen JO, Luber K, Park T. The effect of femoral nailing on cerebral perfusion pressure in head-injured patients. J Trauma. 2003;54(6):1166–70.
70. Scannell BP, Waldrop NE, Sasser HC, Sing RF, Bosse MJ. Skeletal traction versus external fixation in the initial temporization of femoral shaft fractures in severely injured patients. J Trauma. 2010;68(3):633–40.
71. Harwood PJ, Giannoudis PV, van Griensven M, Krettek C, Pape HC. Alterations in the systemic inflammatory response after early total care and damage control procedures for femoral shaft fracture in severely injured patients. J Trauma. 2005;58(3):446–52; discussion 452–4.
72. Scalea TM, Boswell SA, Scott JD, Mitchell KA, Kramer ME, Pollak AN. External fixation as a bridge to intramedullary nailing for patients with multiple injuries and with femur fractures: damage control orthopedics. J Trauma. 2000;48(4):613–21; discussion 621-3.
73. Nowotarski PJ, Turen CH, Brumback RJ, Scarboro JM. Conversion of external fixation to intramedullary nailing for fractures of the shaft of the femur in multiply injured patients. J Bone Joint Surg Am. 2000;82(6):781–8.
74. Helmick K, Baugh L, Lattimore T, Goldman S. Traumatic brain injury: next steps, research needed, and priority focus areas. Mil Med. 2012;177(8 Suppl):86–92.

Soft Tissue Injuries

7

Norbert Pallua and Stefan Bohr

Contents

7.1	Introduction	65
7.2	**Challenges of Soft Tissue Injury Associated with Fractures**	**67**
7.2.1	Principles and Classifications	67
7.2.2	Hypoxia and Perfusion-related Complications	68
7.2.3	Life Threatening Early Wound Infections	69
7.3	**Surgical Management**	**70**
7.3.1	Acute Surgical Management	70
7.3.2	Post-acute Surgical Phase	70
7.3.3	Early-Intent Defect Coverage	71
7.3.4	Peripheral Nerve System	75
References		**81**

N. Pallua, MD (✉) • S. Bohr, MD
Department of Plastic and Hand Surgery – Burn Center, University Clinics RWTH Aachen, Aachen, Germany
e-mail: npallua@ukaachen.de; sbohr@ukaachen.de

7.1 Introduction

Clinically relevant causes of soft tissue injury can be roughly divided into their cause, i.e. thermally, chemically or mechanically induced. All of these demonstrate different characteristics and require specific clinical management. However, a shared feature of any type of relevant tissue injury is the loss of tissue homeostasis caused by (sub-) lethal damage. An initial inflammatory phase (Fig. 7.1) is drastically enhanced by blood borne cells of innate immunity, predominantly polymorphonuclear leucocytes (PMNs), which will home in onto the site of injury [1].

This may lead to a significant secondary tissue necrosis. It also induces tissue ischaemia, e.g. macro- or microvascular thrombosis, hypoperfusion, haemorrhage or relevant tissue swelling. A systemic acute-phase-response is largely defined by various humoral factors secreted by the liver and innate immune cells [2]. Typically within 72 h, a zone of demarcation within injured tissues has been established that essentially divides vital from non-vital tissue. Early on surgical debridement can provide a crucial role in mitigating a systemic inflammatory response (SIR) by reducing both the necrotic and microbial load of wounds. With sufficient perfusion established, wounds will soon enter a proliferative phase of wound healing primarily designed to achieve wound closure through wound contraction, newly formed connective tissue components and

© Springer-Verlag Berlin Heidelberg 2016
H.-C. Pape et al. (eds.), *The Poly-Traumatized Patient with Fractures: A Multi-Disciplinary Approach*, DOI 10.1007/978-3-662-47212-5_7

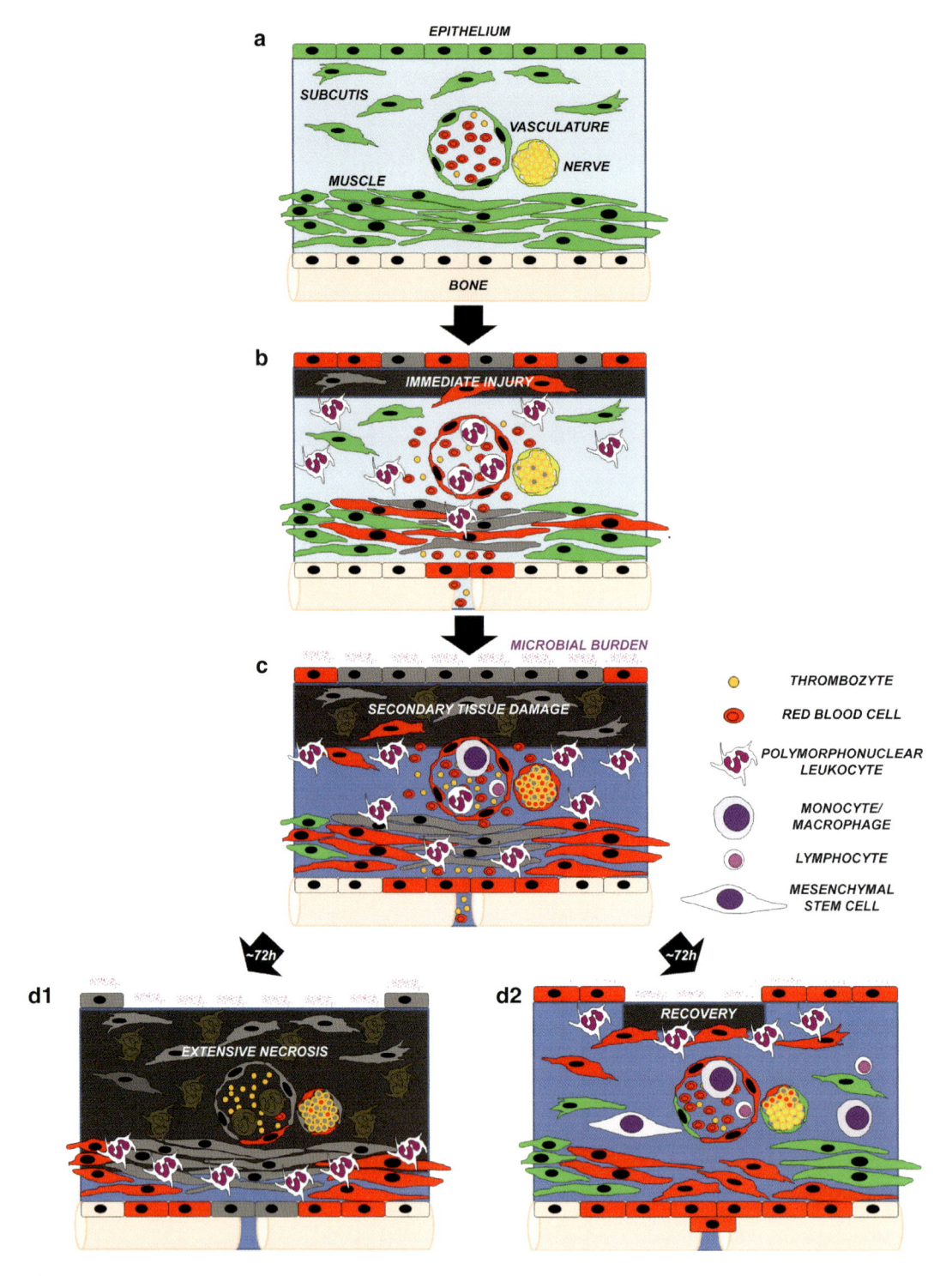

Fig. 7.1 *Schematic*: The dynamics of the initial response of various tissues following injury is often characterized by highly significant secondary cellular damage due to protracted ischaemia and a pro-inflammatory environment. (**a**) Tissue homeostasis as defined by the absence of hypoxia, acidosis or inflammation. (**b**) Relevant trauma will cause immediate cellular death (*grey cells and areas*) and haemorrhage. Also within few hours, pro-inflammatory activation of local cell populations (*red cells*) as well as the arrival of blood borne innate immune cells, predominately

re-epithelialization, often at the cost of tissue function. Thus, scar tissue formation is a key feature of healing wounds in a remodelling phase. Clinical experience has shown that any measure that will shorten the recovery phase following tissue injury impairs functional and aesthetical outcome. Also, there has been a novel understanding of the inflammatory response. It clearly demonstrates that an abatement of inflammation is not just a decrease in pro-inflammatory stimuli. Successful wound healing requires an active, highly regulated process designed to counteract negative effects of prolonged pro-inflammatory signalling. A new type of phospholipid mediators have been described (resolvins and protectins) [3] and appear to affect pro-inflammatory and potentially harmful actions of polymorphonuclear leucocytes. Another important aspect has been the identification of various cellular phenotypes able to modulate pro-inflammatory responses. These enable tissue recovery and regeneration, most notably mesenchymal derived stem cell [4] and subpopulations of macrophages [5]. Cleary, one goal of future treatment strategies following relevant soft tissue injury should be to modulate an innate inflammatory response.

7.2 Challenges of Soft Tissue Injury Associated with Fractures

7.2.1 Principles and Classifications

Background In general, any type of fracture is also associated with varying degrees of soft tissue injury. For the orthopaedic surgeon, accompanying soft tissue trauma is the most relevant factor to determine feasibility and indications for open vs. closed fracture reduction and fixation approaches. Over the last decades the principle of an anatomic fracture reduction has been changed to biological fixation techniques [6]. The term biological refers to leaving a zone of fracture unexposed to avoid further compromise surrounding soft tissue perfusion. Unstable fractures, maintain a state of soft tissue inflammation evidenced by continued swelling, pain and immobility.

Thus, from a plastic surgeon's perspective, surgical incisions required for fracture reduction should be critically assessed, especially regarding the possibly of raising pedicled (e.g. arterial perforator-based) local tissue flaps later on.

Classifications Various grading systems have been suggested with the intent of guiding clinical decision making, both in choice of surgical techniques as well as timing of reconstructive measures with an additional prognostic factor. Gustilo & Anderson classified 'open fractures' into three major types with additional subgroups [7]. Tscherne & Oestern et al. developed a more encompassing classification with a stronger focus on soft tissue injury which also includes 'closed fractures' [8]. In part for outcome comparability, more elaborate classification systems which account for various additional factors such as the extent of skin contusion, muscle injury, vascular and nerve injury et cetera in a check-list format have been continuously revised in recent years including the *Hannover Fracture Scale (HFS)* [9] and the *AO/OTA Fracture and Dislocation Classification* [10]. In addition, an appreciation of the 'transferred energy' leading to the observed pattern of injury can be highly useful in anticipating soft tissue recovery vs. the need for early intent

Fig. 7.1 (continued) of neutrophils, occurs. (**c**) Usually within 72 h post injury a progressive secondary tissue necrosis along with thrombosis and widespread inflammatory infiltrates is observed ('second hit'). Also, loss of epithelial barrier function favours a microbial colonization of wounds increasing the risk of relevant infection. (**d1**) If tissue necrosis is extensive, it will become a burden requiring surgical debridement and eventually reconstructive measures. Strong evidence suggests that certain aspects of a pro-inflammatory innate immune response can prove detrimental to tissue survival by promoting, e.g. thrombosis or pro-apoptotic pathways. (**d2**) In contrast, if tissue necrosis is limited, a pro-inflammatory response will gradually give way to a recovery phase, characterized by the arrival of various, yet not well-defined cellular phenotypes on the scene. *Conclusion*: On top of the initial trauma significant secondary tissue damage commonly occurs. This opens a 'window of opportunity' for various strategies aimed at mitigating this secondary tissue damage, e.g. the modulation of an innate immune response

reconstructive measures. Typically, a simple fall will generate forces around 100 (Ft/Lb, lb$_f$=joule), a skiing accident up to 500 lb$_f$, a gun projectile up to 2,000 lb$_f$, and a motor vehicle accident at 18 miles/h (30 km/h) up to 100,000 lb$_f$.

7.2.2 Hypoxia and Perfusion-related Complications

7.2.2.1 Tissue Hypoxia [11]

All molecular processes necessary for sustained cell survival require a steady generation of high energy transfer compounds, most notably of adenosine triphosphate (ATP). Depending on their metabolic activity, an eukaryotic cell poses limited capabilities of regenerating ATP in a non-oxygen–dependent anaerobic fashion. Prolonged hypoxia leads to metabolic decoupling (mitochondrial PaO$_2$ <0.1–1 mmHg) and accumulation of acidic metabolites which in turn promote inflammation and tissue necrosis. Thus, a continuous monitoring of adequate oxygen supply to tissues in general following trauma is mandatory. Ischaemia (lack of perfusion), arterial Hypoxia (lack of lung dependent respiration), anaemia (lack of blood oxygen transport capacity) and intoxication (e.g. metabolic acidosis, CO-binding of haemoglobin, MetHb formation) have to be identified as different causes of tissue hypoxia. Limitations of compensatory mechanisms are reached in case of an acute drop of Hb values <10 g/dl, an arterial PaO$_2$ <40 mmHg, a venous PaO$_2$ <40 mmHg and oxygen HB-saturation values <85 %. From a surgical view point, control of haemorrhage outweighs all decision making immediately followed by re-establishment of arterial blood supply by means of vascular re-anastomosis and autologous or alloplastic grafts.

7.2.2.2 Compartment Syndrome

Sufficient tissue perfusion depends on an *effective capillary perfusion pressure* (P_{eff}), defined as the difference (simplified) between a *hydrostatic perfusion pressure* (ΔP) to a *colloid osmotic tissue pressure* ($\Delta \pi$) which in turn depends on a pressure gradient between a capillary (P_c, π_c) and an interstitial (P_i, π_i) space:

$$P_{eff} = \Delta P - \Delta \pi = (P_c - P_i) - (\pi_c - \pi)$$
$$= \text{Starling equation}$$

P_{eff} typically ranges from 30 mmHg (4 kPa) at the post-arteriole entry level to −10 mmHg (−1.33 kPa) at the post-capillary venule level. In the poly-traumatized patient virtually all determinants of tissue perfusion, e.g. hydrostatic pressure, plasma protein levels will undergo significant changes resulting in overall interstitial fluid retention. A P_{eff} above 30 mmHg usually results in an insufficient fluid return into the capillary system and thus soft tissue swelling. Soft tissue swelling itself will strongly increase P_i, and thus decrease ΔP and P_{eff}. Although controversial [12], an estimate of P_i can be directly measured using a puncture cannula attached to a pressure device. Pressure values above 35 mmHg are considered pathological by most authors and values above a systemic diastolic pressure inevitably will lead to severe tissue ischaemia. Muscles, especially of the lower leg are typically enclosed in strong fibrous sheets (syn.: fascia, loge, compartment) which will limit soft tissue swelling and thus are most susceptible to a relevant rise in P_i, clinically termed *compartment syndrome* [13].

7.2.2.3 Venous Tissue Congestion and Thrombosis

Insufficient venous return is an often underestimated cause of serious sequels in the poly-traumatized patient. Venous congestion with a disturbed blood circulation between a superficial and a deep, inter-muscular vascular system, especially of the legs is a common feature accompanying post-traumatic tissue swelling. Although mechanisms of thrombosis formation are complex [14], three major contributing factors, originally defined by Virchow (1856), remain valuable in guiding a clinical rational: (1) post-traumatic state of heightened coagulability, (2) decreased venous velocity and (3) traumatic vascular lesions.

For vascular repair following traumatic vascular transection, based on a literature overview, venous re-anastomosis using standard end-to-end suture techniques can be considered the gold standard opposed to end-to-side techniques. Regarding post-surgery anti-coagulative therapy following free-flap tissue transfer, to the authors' knowledge, various recommendations [15] but no widely evaluated and accepted guidelines exist. However, according to a survey by Xipoleas et al. [16], around 85 % of members of the *American Society of Plastic Surgeons* routinely use anti-coagulative therapy post-operatively, either low-fractionated heparin and/or aspirin. To the authors' knowledge, no meta-analysis exists which supports the notion of clinically relevant anti-thrombotic effects of colloidal, e.g. dextrane-based i.v. therapy.

Regarding a generally increased risk of thrombotic events of poly-traumatized patients, the incidence of unnoticed *below-knee deep venous thrombosis* (BKDVT) has been stated between 40 and 80 % according to AWMF-consensus guidelines [17]. These result in an overall rate of about 10 % of above-knee propagation. Anti-thrombotic therapy, e.g. using low-fractionated heparin, hirudin, Danaparoid®, Fondaparinux®, Rivaroxaban® but not aspirin is considered as an effective prophylaxis. With heparin-derivatives, there is a need for a regular screening for *heparin-induced thrombocytopenia* (HIT I/II) and renal function. Activation of the 'muscular pump' by physiotherapy, active or passive early-functional movement of joints, compression stockings and various measures designed to reduce swelling have been shown to be effective in clinical level three studies or lower.

7.2.2.4 Decubitus

Peri- [18] and post-operative immobility exposes the poly-traumatized to a high, avoidable risk of pressure ulcers caused by prolonged periods of B.E.: ischaemic tissue compression around typical anatomical landmarks. These include the occipital region, shoulder blades, elbow region, sacral region, ischial tuberosity, trochanter major, fibula head and the heel. Repositioning measures every 2 h to avoid prolonged, localized tissue compression has been shown to significantly decrease the incidence of pressure sores [19]. Thus, measures of appropriate pressure distribution throughout periods of prolonged immobility is a mandatory part of patients' care with national guidelines for diagnosis (stage I-IV) and management (NPUAP [20] or EPUAP) being revised on a regular basis. If manifested, surgical debridement and secondary defect coverage offers many challenges in terms of peri- and post-operative care due to a relatively high risk of relapse [21]. Additional surgical measures such as colostomy [22] prior to attempting a definite defect closure of sacral decubiti in order to reduce a bacterial load of wounds have been widely promoted.

7.2.3 Life Threatening Early Wound Infections

Even in skin abrasions, a significant rise in bacterial swab colony forming units is to be expected within 24 h following trauma [23]. Thus, in the poly-traumatized patient with fractures, an initial microbial swab evaluation is useful for the anticipation of (1) infections caused by environmental pathogens. These include *Bacillus, Clostridium, Corynebacter, Pseudomonas or Actinobacter* species, (2) fungal infections, e.g. *Aspergillus* species, (3) early onset bacterial infections, e.g. *Streptococcus* and *Staphylococcus* species or (4) pre-existing multi-drug resistant bacteria strain carriers, e.g. *methicillin-resistant staphylococcus* (MRSA). A switch from initially present body surface colonization or environmental wound contamination to nosocomial, hospital acquired bacterial strains (e.g. *Staphylococcus, Escherichia coli, Proteus, Enterococcus, Bacteroides* species) can be typically expected within 72 h post trauma, again advocating early definite surgical closure of wounds. However, to the authors' knowledge, related to a risk assessment for possible wound infection, no evidence-based data exists which dictates delayed vs. primary wound closure [24]. Overall, *staphylococcus species* remain the most common cause of surgical wound infections [25].

Live threatening infections related to initial wound contamination are typically characterized by a sudden onset and rapid progression of soft tissue inflammation and necrosis. Thus, surgical decision making often is solely based on clinical presentation alone. They include: (1) *Gas gangrene* [26] (e.g. *Clostridium perfringens*), more defined by tissue-lytic toxin effects (e.g. *alpha toxin*) rather than bacterial load, can be anticipated in patients with an initial wound contamination with soil as well as with a diabetic, alcoholic and peripheral artery disease (PAD) predisposition. Here, the avoidance of ischemic wound conditions is paramount. A fulminant course of tissue necrosis, intra-tissue gas formation and emergency gram-staining should motivate early and extensive surgical debridement or amputation along with a supportive therapy, e.g. hyperbaric oxygen therapy and antibiotics. With manifestation in the genital region, usually a mixed infection with aerobic and anaerobic bacteria is present, termed Fournier's gangrene (1883). (2) *Necrotizing fasciitis* [27] (e.g. *Streptococcus pyogenes, Staphylococcus aureus, Clostridium perfringens, Bacteroides fragilis, Aeromonas hydrophilia*) shares features with gas gangrene in that toxin effects will cause a sudden, fulminant progression of disease on a subcutaneous, epifascial plane but initial soft tissue reaction is often bland with only minor swelling or redness that does not correlate well with severe pain in the conscious patient. Early surgical intervention is mandatory with diagnosis usually confirmed by tissue histology. (3) *Tetanus* [28] (*Clostridium tetani*), due to extensive passive antitoxin vaccination programmes is a rare condition in developed countries. However, because of its potential lethal cause; checking up-to-date vaccination remains a mandatory part of any initial wound management. (4) *Botulism* [29] (*Clostridium botulinum*), a lethal, toxin-defined disease is avoided by sufficient initial wound decontamination and mitigated by early diagnosis with passive vaccination. (5) *Erysipelas* [30] (*Streptococcus pyogenes*) caused by a superficial, non-pyogenic infection of upper layers of the skin by streptolysin O/S exotoxin expressing strains can be readily treated following early diagnosis using antibiotics combined with antiseptic dressing regimes. A more severe, 'bullous' from that often requires secondary defect coverage can be distinguished based on the appearance of large areas of toxic epidermolysis [31].

7.3 Surgical Management

In contrast to an initial acute phase following trauma with a focus on fostering soft tissue recovery and preventing secondary tissue damage, both on a local and a systemic level, in a post-acute phase clinical reevaluation mandates a timely surgical approach of defects or soft tissue insufficiencies.

7.3.1 Acute Surgical Management

Radical surgical debridement of contaminated or critically damaged, non-viable tissue represents a widely accepted, mandatory part of any surgical first intent strategy. Following initial stabilization of vital body functions, both through intensive care management and surgical control of haemorrhage or acute brain damage, a close multidisciplinary approach is required.

Acute phase surgical measures comprise (1) the avoidance of prolonged ischaemia by securing sufficient arterial perfusion and venous return through re-anastomosis, evacuation of hematoma or decompression of muscular compartments; (2) decontamination of wounds using techniques such as excision of wound edges according to Friedrich (1889) [32], bursectomy, jet-lavage [33], versa-jet [34], open aseptic wound management [34] or the application of occlusive, vacuum-based dressings [35]; (3) additional measures to ensure a viable environment for the primary or secondary reconstruction of nerves, tendons or ligaments.

Soft tissue injury to the upper and lower extremities has been shown to be most susceptible to early infectious complications opposed to soft tissue injuries of the head, neck and genital region [36]. In general, primary wound closure should be attempted within 6 h following trauma which also coincides with an expected peak in tissue swelling.

7.3.2 Post-acute Surgical Phase

Comparable to an initial surgical debridement of traumatized soft tissue, a surgically motivated 'second look' strategy should occur. Given the

7.3.3 Early-Intent Defect Coverage

Background and Principles of Defect Coverage A widely accepted guideline regarding a timely closure of wounds and defect coverage can be summed up by the 'reconstructive ladder' [43] concept (Fig. 7.2) in which the complexity of soft tissue injury directs the choice of surgical options. A more recent 'reconstructive triangle' [44] concept (Fig. 7.2) reflects a certain change in philosophy with more patient-centred view on certain defined goals of treatment and means to achieve them. Accordingly, decision making on how soft tissue injury and defect closure should be managed is based on an evaluation of overall *safety, function* and *aesthetic form*. This also includes two-timed approaches where early defect closure is achieved by technically simple measures such as split-skin grafting followed by more complex procedures such as functional myoplasty later on when the poly-traumatized patient has entered a rehabilitatory phase of treatment.

In general, if primary wound closure is not feasible, defect coverage can be achieved by two major principles.

First, by allowing the formation of granulation tissue which will lead to stable and well perfused

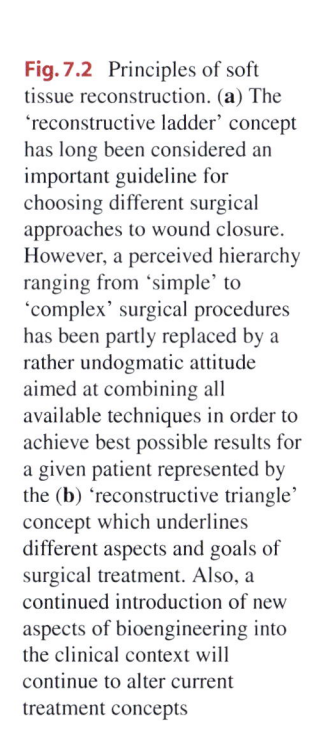

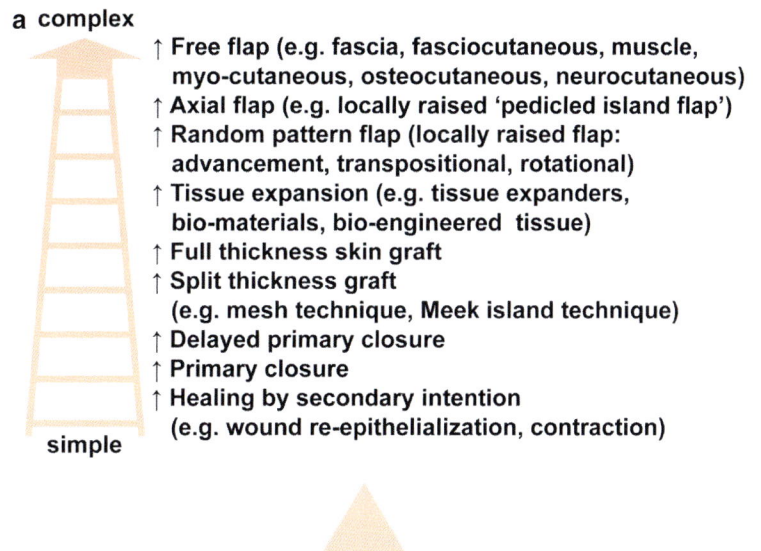

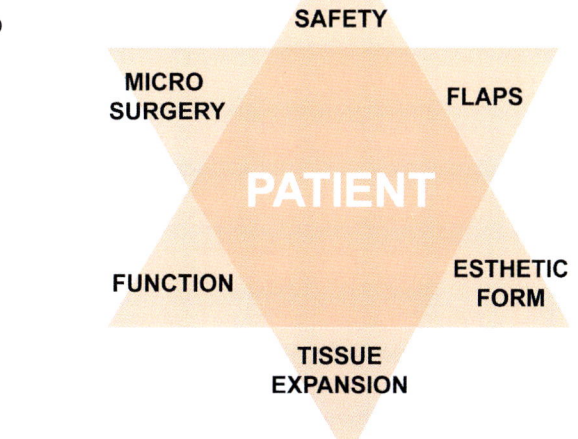

Fig. 7.2 Principles of soft tissue reconstruction. (**a**) The 'reconstructive ladder' concept has long been considered an important guideline for choosing different surgical approaches to wound closure. However, a perceived hierarchy ranging from 'simple' to 'complex' surgical procedures has been partly replaced by a rather undogmatic attitude aimed at combining all available techniques in order to achieve best possible results for a given patient represented by the (**b**) 'reconstructive triangle' concept which underlines different aspects and goals of surgical treatment. Also, a continued introduction of new aspects of bioengineering into the clinical context will continue to alter current treatment concepts

a complex

↑ Free flap (e.g. fascia, fasciocutaneous, muscle, myo-cutaneous, osteocutaneous, neurocutaneous)
↑ Axial flap (e.g. locally raised 'pedicled island flap')
↑ Random pattern flap (locally raised flap: advancement, transpositional, rotational)
↑ Tissue expansion (e.g. tissue expanders, bio-materials, bio-engineered tissue)
↑ Full thickness skin graft
↑ Split thickness graft
   (e.g. mesh technique, Meek island technique)
↑ Delayed primary closure
↑ Primary closure
↑ Healing by secondary intention
   (e.g. wound re-epithelialization, contraction)

simple

b

SAFETY
MICRO SURGERY
FLAPS
PATIENT
FUNCTION
ESTHETIC FORM
TISSUE EXPANSION

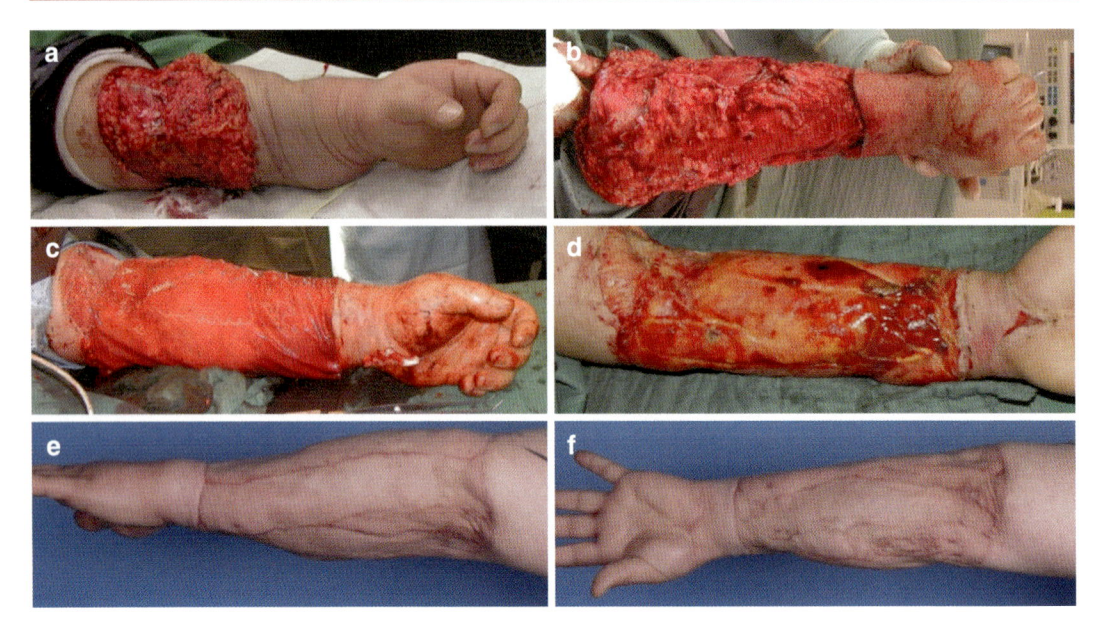

Fig. 7.3 *Clinical Case*: This 24-year-old patient was overrun by a motor vehicle and sustained (**a**) a degloving injury of the left lower arm (**b**) requiring debridement of the resulting soft tissue necrosis. (**c**) Defect coverage was performed using a collagen-glycosaminoglycan-based biodegradable matrix wound dressing (Integra™) which (**d**) when integrated acted as dermal substitute suitable for autologous split-skin grafting. This allowed the formation of (**e**, **f**) a levelled, stable new soft tissue sheath with unhindered sliding of underlying muscle and tendon

wound(beds) which in turn can then be closed using various techniques of autologous skin grafting. This approach is usually limited to soft tissue defects that will not leave bone or tendons exposed. However, in recent years, development and application techniques regarding the use of biodegradable skin substitutes [45] such as MatriDerm® [46] or Integra® [47] have matured sufficiently to allow for coverage of exposed bone, muscle or tendons with stable and functionally satisfying results (Fig. 7.3).

Secondly, any type of surgical measure that will 'move' a defined block of tissue from one area to another defines defect coverage using flaps. A block of tissue that does not incorporate a defined vascular pattern is called a 'random pattern' flap, as opposed to the 'axial pattern' or 'island' flaps with an identifiable vascular pedicle. Flaps with a vascular pedicle are most versatile since they allow for a safe mobilization of large tissue blocks both locally and distally as 'free flaps'. A list of flaps commonly used for defect coverage is given in Table 7.1. The authors' recommendation for defect coverage using flaps in defined anatomical areas is given in Fig. 7.4.

Random pattern flaps are essentially synonymous with locally mobilized skin flaps. Standard skin flaps used for defect closure are further sub-categorized based on the underlying principle of mobilization as either *advancement*, *transpositional* or *rotational*. A key consideration of local tissue mobilization for defect closure is that unidirectional loss of tissue mobility (through defect) is best compensated for by tissue mobilization along a main vector that is perpendicular to a corresponding vector of reduced tissue mobility. Also, as a rule of thumb, the ratio between length and width of a randomly patterned flap should not exceed 3:1 for sufficient flap perfusion. This ratio can be modulated using, e.g. a 'bridge' flap technique, effectively creating two pedicles.

A detailed study of human vascular anatomy underlying skin perfusion such as performed by C. Manchot (1889) and M. Salmon (1936) proved to be essential for the design of larger skin or fascio-cutaneous flaps. Human vascular anatomy (opposed to, e.g. rodent models) rarely allow larger cutaneous flaps to be raised based on a vascular pedicle that runs along the subcutaneous

Table 7.1 Shown: List of commonly used flaps for soft tissue coverage associated with trauma

Flap-name (abbreviation)	Flap-type+	Flap-size++	Comments+++	Literature
Upper extremity				
Anti-cubital perforator (AC)	FC	s	Local	[63]
Brachio-radialis muscle (BR)	M, MC	s	Local	[64]
Deltoideus perforator (DAP)	FC	s	Local	
Distal ulnar perforator (dUP)	FC	m	Local, "Becker flap"	[65]
Interossea posterior artery (IP)	FC	s	Local	[66]
Radial artery forearm (RAF)	FC	m	"Chinese flap"	[67–69]
Ulnar artery forearm (UAF)	FC	m	Local and free	[70, 71]
Upper arm perforator (Lateral, medial, anterior, posterior UA)	FC	m	Local and free	[72–76]
Lower extremity				
Antero-lateral thigh perforator (ALTP)	FC	l	Local and free	[77–81]
Biceps femoris muscle (BF)	M, MC	m	Local	[82]
Distal lateral thigh (DLT)	FC	m	Local and free	
Distally based sural (DBS)	FC	s	Local	[83, 84]
Dorsalis pedis artery (DPA)	FC	s	Local	[85–87]
Extensor digitorum brevis muscle (EDB)	M	s	Local	[88, 89]
Fibular osteocutaneous (FOC)	OC	m	Free	[90–92]
Flexor digitorum communis muscle (FDC)	M	s	Local	[93, 94]
Gastrocnemius muscle (GCM)	M	s	Local, medial or lateral head	[95–97]
Gluteus maximus muscle (GM)	M, MC	l	Local and free	[98–100]
Gracilis muscle (G)	M, MC	m	Local and free	[101, 102]
Inferior gluteal perforator (iGAP)	FC	m	Local	[103–105]
Lateral supramalleolar (LSM)	FC	m	Local, "fibular/peroneal artery perforator"	[106]
(Reverse) medial plantar artery (RMP/MP)	FC	s	Local	[107–111]
Peroneus brevis muscle (PB)	M	s	Local	[112–114]
(Distal-medial thigh) saphenus (DMTS)	FC	s	Local, 'saphena neuro-cutaneous'	[115, 116]
Soleus muscle (SM)	M	s	Local	[117–119]
Superior gluteal perforator (sGAP)	FC	m	Local	[100]
Tensor fasciae latae muscle (TFL)	MC, FC	m	Local and free	[120]
Tibialis anterior muscle (TAM)	M	s	Local	[121, 122]
Vastus lateralis muscle (VL)	M, MC	m	Local	[123, 124]
Thorax, abdomen, pelvis				
(Para)scapular (PS/S)	FC, OC	l	Local and free	[125–128]
(Deep) inferior epigastric artery (perforator) (DIEP/IE)	FC	m	Local and free	[129–132]
Iliacus muscle (IM)	M	l	Local	[133]
Inguinal flap (IF)	FC, OC	l	"Groin flap"	[134–137]
Latissimus dorsi muscle (LD)	M, MC	l	Local and free	[138–142]
Omentum major (OM)		l	Local and free	[143–146]
Pectoralis major muscle (PM)	M, MC		Local and free	[147–150]
(Trans-) rectus abdominis muscle (TRAM/RAM)	M, MC	m	Local and free	[151–154]
Serratus anterior muscle (fascia) (SA/SAF)	M, F, OC	l	Free	[155–158]
Superior epigastric artery (SE)	M, MC	s	Local	[159, 160]
Supraclavicular island flap/anterior supraclavicular artery perforator (SIF/aSAP)	FC	l	Local	[161, 162]
Free-style perforator flaps	FC		Local	[163–165]

Additional specifications are given (+) according to the type of tissue that is transferred (e.g. muscle), (++) typical size or (+++) the applicability for free vs. local tissue transfer. In addition, quotes for original work related to specific flaps are listed
M: muscle, *MC*: musculo-cutaneous, *F*: fascia, *FC*: fascio-cutaneous, *OC*: osteo-cutaneous, *small(s)* ≤20 cm^2, *medium(m)* ≤50 cm^2, *large(l)* >50 cm^2

Defect Area	Zone
upper arm, elbow, lower arm, wrist	1
shoulder, neck, upper thorax	2
lower thorax, abdomen, pelvis	3
gluteal area, hip, thigh	4
knee, lower leg	5
ankle, foot	6

Local Flap
AC, BR, dUP, IP, RAF, UAF, UA
aSAP/SIF, PM, LD, DAP, SE
(T)RAM), (D)IEP, LD, IF, IM, iGAP, sGAP, G, GM, OM
G, GM, IF, RAM, sGAP, iGAP, BF, TFL, VL
DLT, DBS, DMTS, LSM, PB, SM, TAM, GCM, FDC, EHL, PB
DBS, DPA, EDB, LSM, RMP/MP

Free Flap
UA, RAF, ALTP, G, PS/S, SAF, IF (distant flap)
ALTP, PS/S, LD, SA, (T)RAM, (D)IEP, G
LD, ALTP, PS/S, SA/SAF, G
LD, ALTP, PS/S, G
ALTP, PS/S, LD, G, TFL, RAF, UAF, SA/SAF
ALTP, PS/S, G, LD, UA, RAF, SA/SAF, TFL, IF, RAM, OM

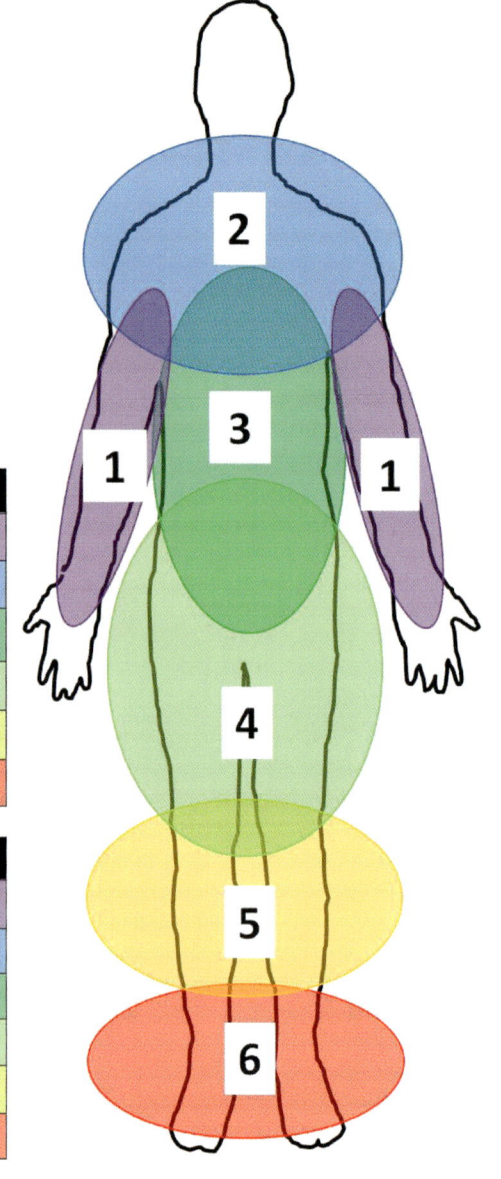

Fig. 7.4 *Shown*: A comprehensive listing of local and free tissue transfer for defined anatomical regions recommended by the authors (for abbreviations see Table 7.1)

plane. Instead, skin perfusion is predominantly ensured by 'perforators' which originate from a deeper muscle plane. Thus, the vascular anatomy of skin is closely linked to that of muscle. Regarding free transfer of muscle tissue, a highly valuable classification system of muscular artery perfusion was introduced by Mathes and Nahai [48]. Nowadays, so-called perforator flaps play a dominant role whenever coverage of soft tissue defects of significant size and depth is required. As already mentioned, perforator flaps are essentially fascio-cutaneous flaps where a substantial area of skin and underlying fat tissue are dependent on a vascular pedicle that usually perforates perpendicular to the surface plane of the flap. Design and surgical preparation of a perforator flap requires a concise knowledge of a perforator's relation to the vascular system of an underlying muscle. As elaborated by Cormack and Lamberty [49], a perforator-based flap pedicle might require transmuscular, septal, subfascial preparation techniques. Knowledge of perforator anatomy is also invaluable in raising osteo-myo-cutaneous flaps. In order to minimize donor site co-morbidity due to flap raising, muscle flaps are usually reserved for large and deep defects requiring a certain volume and an increased 'mechanical resistance' of the flap. Also, commonly used muscle flaps do cause only minor functional deficits. A 'monitor skin island' as part of a muscular flap is still considered the most practical way of monitoring adequate flap perfusion despite various alternative approaches or monitor devices available on the market (e.g. O2C™ [50]).

Fracture Management and Soft Tissue Coverage Early definite fracture reduction and stabilization regarding weight bearing bones, highly instable fractures or open fractures is widely accepted [51, 52]. In addition, immediate definite soft tissue coverage using flaps within 24 h post trauma has been recommended. The rationale behind this approach can be justified by three major clinical determinates.

First, accompanying soft tissue will result in considerable soft tissue swelling and inflammation which will in turn temporarily compromise tissue perfusion.

Second, relevant microbial colonization of wounds usually occurs within an interval of ~24–72 h post trauma.

Third, systemic effects of injury as well of trauma to other organ systems such as the brain, lung or liver exert a well documented heightened vulnerability of patients related to surgical management in an interval ranging from day 1 to several days post trauma. However, Byrd et al. [53] reported excellent results regarding flap coverage of fracture-associated soft tissue defects within a 1–6 days interval post trauma. 'Best' results here refer to (1) number of surgical procedures, (2) expedited fracture consolidation vs. non-union, (3) lower rates of osteomyelitic complications or (4) overall hospitalization time. Most recently, Harrison et al. [54] concluded that no evidence exist that even delayed (>21 days post trauma) defect coverage in the upper extremity will result in a higher incidence of infection, bony non-union or flap loss.

One remaining problem here however is 'pushing the limits of microsurgery'. With diameters of arterial perforators typically in the range of 0.5–2 mm, mean arterial pressures and the avoidance of vasospasms become of paramount importance for the survival of flaps. Thus, the overall condition of patients, e.g. requirements of catecholamine pressure agents has to be included into flap-surgery planning. In addition, on the venous return side, flaps also require adequate anti-coagulative therapy, thus increasing the risk of continued haemorrhage in the initial phase of trauma. The argument of microbial contamination and subsequent contamination is a valid one. However, the proper use of negative pressure dressings is highly effective in bridging the interval between radical initial debridement, definite fracture stabilization and soft tissue defect coverage [55]. In conclusion, in the authors' opinion a combination of various surgical techniques and a well timed and if necessary repeated surgery in order to achieve an optimal functional and aesthetical outcome is the most viable approach when it comes to complex tissue trauma and defects. Further aspects of defect coverage are discussed with the presented cases (Figs. 7.5, 7.6 and 7.7).

7.3.4 Peripheral Nerve System

Various degrees of nerve injury ranging from *neuropraxia* (self limited contusion), to *axonotmesis* (severe contusion), to *neurotmesis* (complete discontinuity) as elaborated by Seddon et al. [56] and Sunderland [57] can be expected in the poly-traumatized patient with fractures. Careful and repeated clinical re-evaluation, e.g. of Tinel's sign in order to identify and monitor

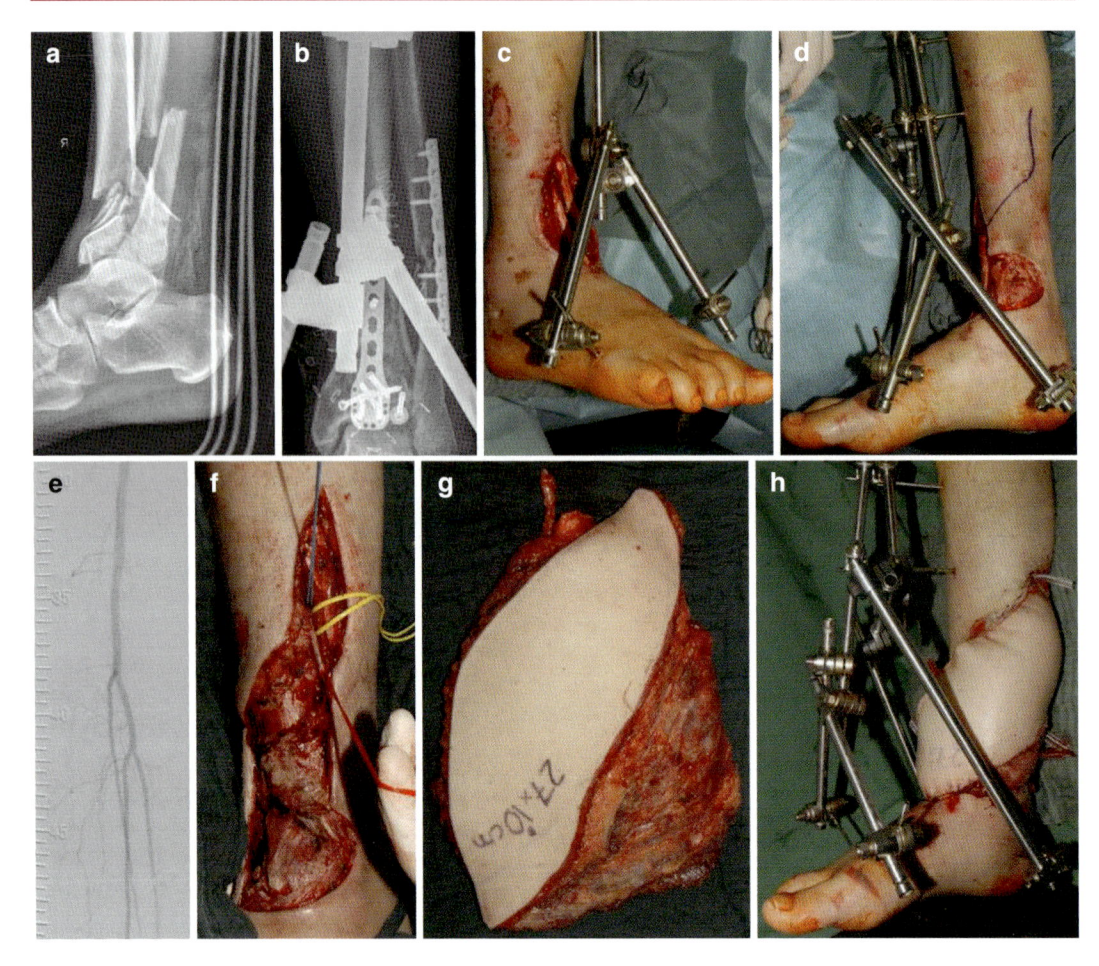

Fig. 7.5 *Clinical Case*: Fall from height of a 56-year-old patient, heavy smoker, which resulted in bilateral distal tibial pilon fractures. (**a**) Initial X-ray of the right lower ankle region. (**b**) Immediate fracture reduction followed by combined internal and external fixation techniques was performed. Due to the extent of soft tissue injury, primary wound closure was not feasible and temporary wound coverage was performed using vacuum-based dressing techniques. *Problem*: Following secondary debridement, post-traumatic soft tissue recovery over several days still did not allow for a secondary wound closure resulting in a defect (**c**) anteriorly and (**d**) medially to the ankle joint. *Solution*: Pre-operatively, the vascular status of the right leg using (**e**) digital subtraction angiography allowed choosing (**f**) a favourable sight of microvascular anastomosis to the posterior tibial artery above fracture level. Considering the size and location of the defect with the exposure of tendon and joint structures along with metal implants; defect coverage was then performed by (**g**, **h**) free latissimus dorsi muscle flap transfer. *Considerations*: Complex fracture situations near joints, despite surgical strategies of immediate anatomical reconstruction are commonly associated with prolonged soft tissue swelling, partial necrosis and fibrosis resulting in relevant defects with a high risk of joint infection, pseudoarthrosis or osteomyelitis. The rational of free tissue transfer in this case using a free myocutaneous flap opposed to other flap options was based on criteria such as defect size, required length of pedicle and safety regarding flap resistance to ischaemia, joint movement, infection or foreseeable repeated joint surgery

recovery of possibly accompanying nerve lesions in the conscious patient is usually sufficient to direct further diagnostic and therapeutic measures. Guidelines for decision making regarding delayed surgical nerve repair has been suggested by Brenner et al. [58]. However, in the unconscious patient high-energy trauma, the localization of factures (e.g. humerus shaft fractures, proximal

7 Soft Tissue Injuries 77

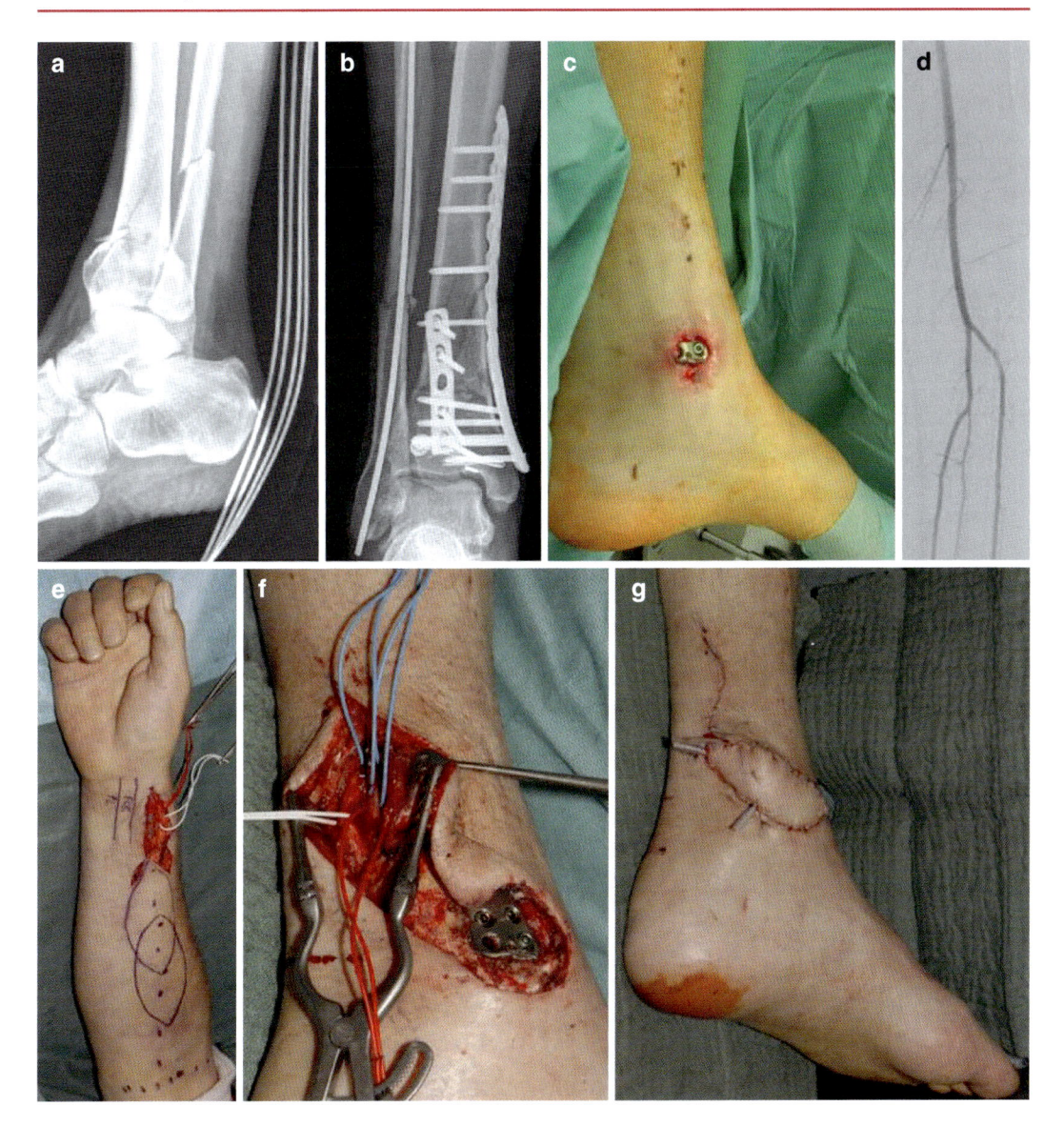

Fig. 7.6 *Clinical Case*: The same patient as in Fig. 7.5 demonstrated a less displaced tibial pilon fracture on (**a**) an initial X-ray of the left lower ankle region. Here, associated soft tissue swelling was allowed to recover over several days and (**b**) definite fracture fixation was achieved using a surgical strategy based on implants and techniques that allow for minimized surgical incision and thus bony exposure. *Problem*: Secondary soft tissue necrosis of (**c**) the medial malleolar region occurred with exposure of implant. *Solution*: Again following (**d**) digital subtraction angiography of the left lower leg; defect coverage was performed by raising a (**e**, **f**) fascio-cutaneous radial artery forearm flap ('chinese flap'; short or long pedicle). *Considerations*: This case demonstrates remaining limits of state-of-the-art implants and current strategies designed to minimize additional soft tissue injury associated with fracture reduction and fixation. The choice of defect coverage using free vs. local tissue transfer in this case was based on flap safety since a local flap solution, e.g. dorsalis pedis artery *or* distally based sural flap in the authors experience would have been associated with a heightened risk of partial necrosis considering local post-injury vascular fibrosis and smoking-related ischaemia

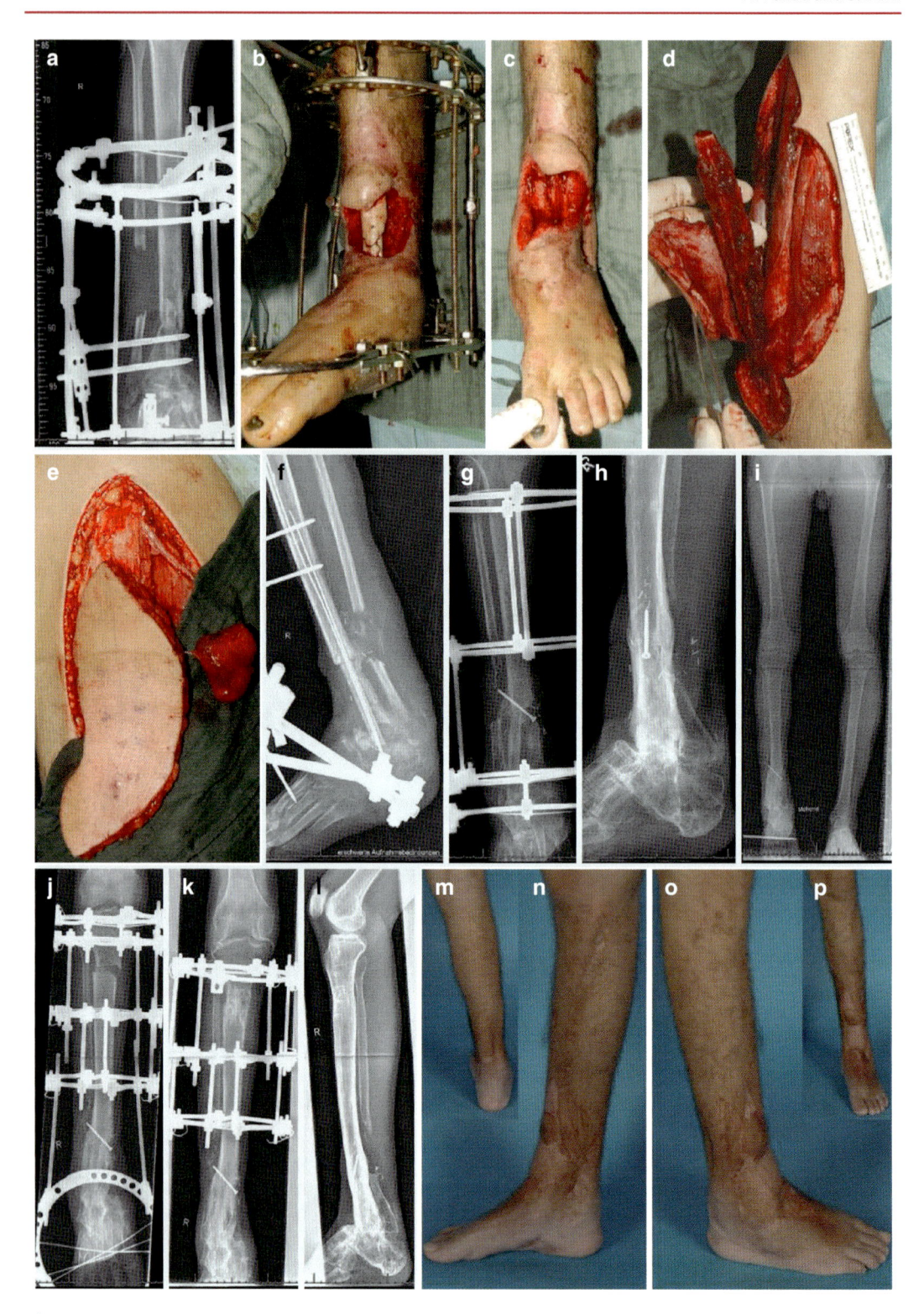

Fig. 7.7 *Clinical Case*: This 19-year-old patient sustained a crush injury of the right lower leg during a traffic accident. *Problem*: The clinical picture upon admission included failed attempts of free tissue transfer as well as cross-leg flap surgery performed in the patient's home country resulting in (+) significant bone defects, (++)

tibia and fibula fractures), joint dislocation, highly instable or displaced or open fractures and pelvic or spinal fractures mandate advanced diagnostic measures such as MRI-imaging, NCV, EMG early on. Whenever possible during an initial exploration of wounds or during open fracture reduction, primary suture of relevant transected nerves should be attempted using microsurgical techniques, e.g. surgical loupes, microscopes, < 8–0/non-resorbable/monofilament sutures. Anatomical reconstruction can be achieved with perineural or fascicular + epineural suture techniques whereas epineural sutures are usually sufficient for non-mixed type nerves. As a rule, an environment free of tension and significant fibrosis within or around a nerve structure is crucial for its recovery and long-term function.

7.3.4.1 Plexus Injury

Motorcycle accidents together with falls from heights are among the most common causes of injuries to the cervical plexus. In the polytraumatized patient, plexus lesions are often incomplete and caused by indirect or blunt trauma with imaging techniques biased by concomitant hematoma and soft tissue reaction. Thus, a close follow-up reevaluation is usually required. The evacuation of hematoma is performed as an emergency procedure if a progressive plexus lesion is suspected. Also, ultrasound examination has become an invaluable tool here for guiding surgical decision making. Integrity of the neural structures should be confirmed using microsurgical preparation techniques. In the case of discontinuity, it might be advisable to perform nerve transplant procedures at a delayed stage with subsided SIR and secured soft tissue coverage. In the meantime, physical therapy and other measures of rehabilitation are crucial in order to extend a 'surgical window of opportunity' and to improve the overall clinical outcome. Among others, electromyography is an important diagnostic follow-up tool for indication and timing of plexus surgery. Despite a lack of studies with higher levels of evidence, the outcome of plexus lesions that do not show signs of recovery within 3 months following injury can be considered poor. Also, reconstructive plexus surgery should be performed within an interval of 12–18 month post injury in order to antedate irreversible muscle atrophy [59]. In addition, various techniques of functional muscle transfer, especially in the upper extremity have been described [60].

7.3.4.2 Peripheral Nerve Injury

Surgical measures to prevent or mitigate nerve damage include the initial open or closed fracture reduction along with nerve decompression, nerve transposition or end-to-end of severed nerves. Any nerve reconstructive measure should be performed in a manner allowing for tension-free-suturing based on principles established by Millesi et al. [61]. In general, peripheral nerve lesions characterized by a mechanical nerve discontinuity without defect should be reconstructed in a primary intent approach if (1) a

---

Fig. 7.7 (continued) chronically exposed and thus nonvital and infected bone, (+++) a status following resection of the talus bone. (**a**) X-ray upon admission of the right lower ankle region. (**b**) Extensive debridement with removal of the Ilizarov fixator resulted in (**c**) an extensive combined bone and soft tissue defect. *Solution*: A coordinated, interdisciplinary approach was first aimed at defect coverage with arthrodesis of the angle region. To achieve this, a combined free (**d**) osteocutanous fibular transplant of the left leg and an (**e**) anterior lateral thigh perforator ALTP-flap with an additional 'muscle plombage' (*chimeric* flap) was raised with combined (**f**) internal and external fixation. Four months later, an additional (**g**) non-vascularized iliac crest bone graft proximal to the original fibular transplant was performed along with a switch to an external ring fixator. At 16 month post the initial surgery, complete (**h**) fracture consolidation with full weight bearing and without signs of infections was observed. However, a resulting shortening of the (**i**) right compared to the left leg indicated a (**j-l**) corrective distraction-osteotomy performed at the proximal tibial level of the right leg using an Ilizarov external fixator system. (m-p). Post-operative aspect of the right leg at 23 months. *Considerations*: This case demonstrates (+) that the initial assessment and management of soft tissue trauma is a key factor to successful fracture treatment and that (++) free soft tissue transfer can be highly successful in avoiding amputation vs. a functional reconstruction of extremities

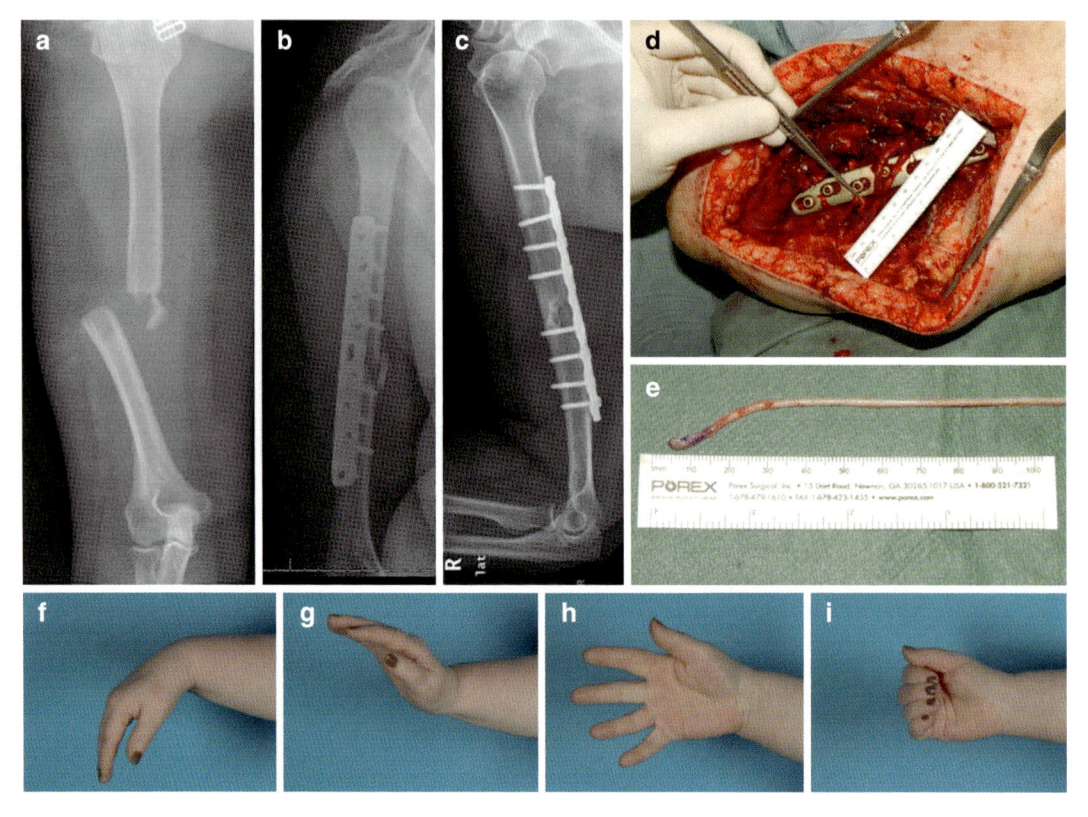

Fig. 7.8 *Clinical Case*: This 58-year-old female patient sustained a highly displaced fracture of the right humerus at mid-shaft level. *Problem*: The initial clinical examination revealed a radial nerve lesion, thus requiring an open surgical approach. Surgical exploration demonstrated a ruptured radial nerve. Shown: X-ray at (**a**) admission, (**b**) post-operatively, (**c**) consolidated. *Solution*: Following LDCDP-Plate fracture stabilization the nerve defect zone was bridged using a (**d**) double-barrelled (**e**) suralis nerve transplant. (**f–i**) At 16 months post-operatively progressive reinnervation of functional lower arm muscle groups was observed. *Considerations*: This case demonstrates that the awareness of typical nerve lesions associated with fractures should direct surgical decision making, e.g. closed vs. open fracture reduction techniques, especially in the unconscious patient. Early reconstruction, if feasible, of nerve continuity will commonly result in superior clinical outcomes

viable and sufficient soft tissue bed is guaranteed and (2) definite internal fracture fixation can be achieved. Autologous nerve grafts that bridge defects should be oriented in reverse in order to ensure that regenerating fibres will only be diverted into fibres that bridge the whole defect. Commonly used donor nerves are non-motory nerves such as antebrachial cutaneous nerves, the saphenus nerve or the sural nerve. Alternatively, end-to-side nerve repair [62], in which the distal stump of a transected nerve is attached to the side of an uninjured donor nerve, has been suggested as a technique for repair of peripheral nerve injuries in situations where reconnection to or bridging of the proximal nerve stump is not feasible. This can also be a temporary measure, where the uninjured nerve acts as a 'chaperone' until wound conditions allow for nerve grafts. However, with end-to-side repair of motor nerves, significant injury (by incision) of the donor nerve is necessary in order to be effective. Further aspects of peripheral nerve injury are discussed with the presented cases (Figs. 7.8 and 7.9).

7 Soft Tissue Injuries

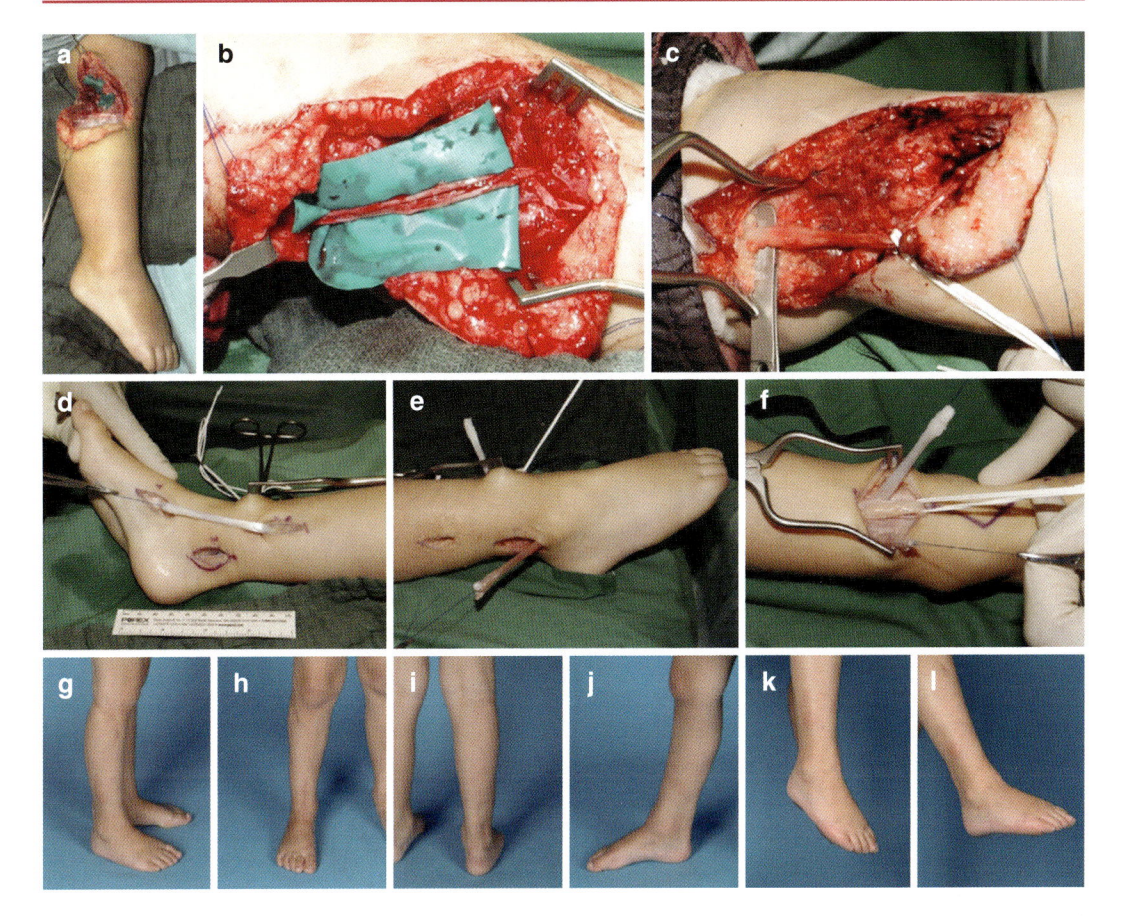

Fig. 7.9 *Clinical Case*: This 3-year-old girl sustained a deep soft tissue injury of the right lower leg caused by a stirring gear. *Problem*: Upon referral, clinical infected wound conditions and a complete peroneal nerve lesion was apparent. (**a**) Initial surgical debridement, co-adaption of muscle, marking of nerve stumps and vacuum wound dressings followed by (**b**) delayed bridging suralis nerve transplant. Due to a protracted delay of signs of reinnervation together with a positive Tinel's sign, (**c**) excision of proximal stump neurinoma with nerve co-adaptation was indicated 6 months later. However, functional recovery of peroneal nerve dependent muscle groups was still insufficient 8 months later and (**d–f**) transposition of the posterior tibial tendon onto the anterior tibial and long peroneal tendon was performed with (**g–l**) adequate foot elevation at 6 months post-operatively. *Considerations*: Soft tissue infection and fibrosis are highly significant impairments to reconstructive surgical measures following nerve injury and often require salvage operations such as transposition tenoplasty or (free) functional muscle transfer

References

1. Martin P, Leibovich SJ. Inflammatory cells during wound repair: the good, the bad and the ugly. Trends Cell Biol. 2005;15:599–607.
2. Keel M, Trentz O. Pathophysiology of polytrauma. Injury. 2005;36:691–709.
3. Serhan CN, Petasis NA. Resolvins and protectins in inflammation resolution. Chem Rev. 2011;111:5922–43.
4. Bernardo ME, Fibbe WE. Mesenchymal stromal cells: sensors and switchers of inflammation. Cell Stem Cell. 2013;13:392–402.
5. Murray PJ, Wynn TA. Protective and pathogenic functions of macrophage subsets. Nat Rev Immunol. 2011;11:723–37.
6. Perren SM. Evolution of the internal fixation of long bone fractures. The scientific basis of biological internal fixation: choosing a new balance between stability and biology. J Bone Joint Surg Br. 2002;84:1093–110.
7. Kim PH, Leopold SS. In brief: Gustilo-Anderson classification. [corrected]. Clin Orthop Relat Res. 2012;470:3270–4.
8. Tscherne H, Oestern HJ. A new classification of soft-tissue damage in open and closed fractures (author's transl). Unfallheilkunde. 1982;85:111–5.

9. Krettek C, Seekamp A, Kontopp H, Tscherne H. Hannover Fracture Scale '98–re-evaluation and new perspectives of an established extremity salvage score. Injury. 2001;32:317–28.
10. Meling T, Harboe K, Arthursson AJ, Soreide K. Steppingstones to the implementation of an inhospital fracture and dislocation registry using the AO/OTA classification: compliance, completeness and commitment. Scand J Trauma Resusc Emerg Med. 2010;18:54.
11. Eltzschig HK, Eckle T. Ischemia and reperfusion–from mechanism to translation. Nat Med. 2011;17:1391–401.
12. Nelson JA. Compartment pressure measurements have poor specificity for compartment syndrome in the traumatized limb. J Emerg Med. 2013;44:1039–44.
13. Mabvuure NT, Malahias M, Hindocha S, Khan W, Juma A. Acute compartment syndrome of the limbs: current concepts and management. Open Orthop J. 2012;6:535–43.
14. Noel P, Cashen S, Patel B. Trauma-induced coagulopathy: from biology to therapy. Semin Hematol. 2013;50:259–69.
15. Talbot SG, Pribaz JJ. First aid for failing flaps. J Reconstr Microsurg. 2010;26:513–5.
16. Xipoleas G, Levine E, Silver L, Koch RM, Taub PJ. A survey of microvascular protocols for lower-extremity free tissue transfer I: perioperative anticoagulation. Ann Plast Surg. 2007;59:311–5.
17. Struijk-Mulder MC, Ettema HB, Verheyen CC, Buller HR. Comparing consensus guidelines on thromboprophylaxis in orthopedic surgery. J Thromb Haemost. 2010;8:678–83.
18. O'Brien DD, Shanks AM, Talsma A, Brenner PS, Ramachandran SK. Intraoperative Risk Factors Associated With Postoperative Pressure Ulcers in Critically Ill Patients: a Retrospective Observational Study. Crit Care Med. 2014;42(1):40–7.
19. Still MD, Cross LC, Dunlap M, Rencher R, Larkins ER, et al. The turn team: a novel strategy for reducing pressure ulcers in the surgical intensive care unit. J Am Coll Surg. 2013;216:373–9.
20. The NPUAP Dual Mission Conference: reaching consensus on staging and deep tissue injury. Ostomy Wound Manage. 2005;51:34. PMID: 16089057.
21. Levine SM, Sinno S, Levine JP, Saadeh PB. An evidence-based approach to the surgical management of pressure ulcers. Ann Plast Surg. 2012;69:482–4.
22. de la Fuente SG, Levin LS, Reynolds JD, Olivares C, Pappas TN, et al. Elective stoma construction improves outcomes in medically intractable pressure ulcers. Dis Colon Rectum. 2003;46:1525–30.
23. DeBoard RH, Rondeau DF, Kang CS, Sabbaj A, McManus JG. Principles of basic wound evaluation and management in the emergency department. Emerg Med Clin North Am. 2007;25:23–39.
24. Eliya-Masamba MC, Banda GW. Primary closure versus delayed closure for non bite traumatic wounds within 24 hours post injury. Cochrane Database Syst Rev. 2013;(10): CD008574.
25. Owens CD, Stoessel K. Surgical site infections: epidemiology, microbiology and prevention. J Hosp Infect. 2008;70 Suppl 2:3–10.
26. Jeavons RP, Dowen D, Rushton PR, Chambers S, O'Brien S. Management of Significant and Widespread, Acute Subcutaneous Emphysema: Should We Manage Surgically or Conservatively? J Emerg Med. 2014;46(1):21–7.
27. Lancerotto L, Tocco I, Salmaso R, Vindigni V, Bassetto F. Necrotizing fasciitis: classification, diagnosis, and management. J Trauma Acute Care Surg. 2012;72:560–6.
28. Miyagi K, Shah AK. Tetanus prophylaxis in the management of patients with acute wounds. J Plast Reconstr Aesthet Surg. 2011;64:e267–9.
29. Taylor SM, Wolfe CR, Dixon TC, Ruch DS, Cox GM. Wound botulism complicating internal fixation of a complex radial fracture. J Clin Microbiol. 2010;48:650–3.
30. Karppelin M, Siljander T, Vuopio-Varkila J, Kere J, Huhtala H, et al. Factors predisposing to acute and recurrent bacterial non-necrotizing cellulitis in hospitalized patients: a prospective case–control study. Clin Microbiol Infect. 2010;16:729–34.
31. Zimmerman JL, Shen MC. Rhabdomyolysis. Chest. 2013;144:1058–65. PMID: 24008958.
32. Schmitt W. P. L. Friedrich and the problem of wound infection (author's transl). Zentralbl Chir. 1978;103:65–9.
33. Ketterl R, Leitner A, Wittwer W. Programmed debridement, combined with jet lavage in extensive hand infections. Langenbecks Arch Chir Suppl Kongressbd. 1997;114:562–5.
34. Fraccalvieri M, Serra R, Ruka E, Zingarelli E, Antoniotti U, et al. Surgical debridement with VERSAJET: an analysis of bacteria load of the wound bed pre- and post-treatment and skin graft taken. A preliminary pilot study. Int Wound J. 2011;8:155–61.
35. Siegel HJ, Herrera DF, Gay J. Silver negative pressure dressing with vacuum-assisted closure of massive pelvic and extremity wounds. Clin Orthop Relat Res. 2014;472(3):830–5. PMID: 23813240.
36. Di Benedetto C, Bruno A, Bernasconi E. Surgical site infection: risk factors, prevention, diagnosis and treatment. Rev Med Suisse. 2013;9:1832–4, 36–39.
37. Schlag G, Redl H, Bahrami S. SIRS (Systemic Inflammatory Response Syndrome) following trauma and during sepsis. Anasthesiol Intensivmed Notfallmed Schmerzther. 1994;29:37–41. PMID: 8142569.
38. Steigbigel NH. Effect of treatment with low doses of hydrocortisone and fludrocortisone on mortality in patients with septic shock. Curr Infect Dis Rep. 2003;5:363–4. PMID: 13678564.
39. Baranov D, Neligan P. Trauma and aggressive homeostasis management. Anesthesiol Clin. 2007;25:49–63, viii. PMID: 17400155.

40. Landucci F, Mancinelli P, De Gaudio AR, Virgili G. Selenium supplementation in critically ill patients: A systematic review and meta-analysis. J Crit Care. 2014;29(1):150–6. PMID: 24135013.
41. Biesalski HK, McGregor GP. Antioxidant therapy in critical care–is the microcirculation the primary target? Crit Care Med. 2007;35:S577–83. PMID : 17713412.
42. Gebhard F, Huber-Lang M. Polytrauma–pathophysiology and management principles. Langenbecks Arch Surg. 2008;393:825–31.
43. Levin LS. The reconstructive ladder. An orthoplastic approach. Orthop Clin North Am. 1993;24:393–409.
44. Mathes SJ. Chest wall reconstruction. Clin Plast Surg. 1995;22:187–98.
45. Iorio ML, Shuck J, Attinger CE. Wound healing in the upper and lower extremities: a systematic review on the use of acellular dermal matrices. Plast Reconstr Surg. 2012;130:232S–41.
46. Haik J, Weissman O, Hundeshagen G, Farber N, Harats M, et al. Reconstruction of full-thickness defects with bovine-derived collagen/elastin matrix: a series of challenging cases and the first reported post-burn facial reconstruction. J Drugs Dermatol. 2012;11:866–8.
47. Shores JT, Hiersche M, Gabriel A, Gupta S. Tendon coverage using an artificial skin substitute. J Plast Reconstr Aesthet Surg. 2012;65:1544–50.
48. Mathes SJ, Nahai F. Classification of the vascular anatomy of muscles: experimental and clinical correlation. Plast Reconstr Surg. 1981;67:177–87.
49. Cormack GC, Lamberty BG. A classification of fascio-cutaneous flaps according to their patterns of vascularisation. Br J Plast Surg. 1984;37:80–7.
50. Forst T, Hohberg C, Tarakci E, Forst S, Kann P, et al. Reliability of lightguide spectrophotometry (O2C) for the investigation of skin tissue microvascular blood flow and tissue oxygen supply in diabetic and nondiabetic subjects. J Diabetes Sci Technol. 2008;2:1151–6.
51. Godina M. Early microsurgical reconstruction of complex trauma of the extremities. Plast Reconstr Surg. 1986;78:285–92.
52. Hertel R, Lambert SM, Muller S, Ballmer FT, Ganz R. On the timing of soft-tissue reconstruction for open fractures of the lower leg. Arch Orthop Trauma Surg. 1999;119:7–12.
53. Byrd HS, Spicer TE, Cierney 3rd G. Management of open tibial fractures. Plast Reconstr Surg. 1985;76:719–30.
54. Harrison BL, Lakhiani C, Lee MR, Saint-Cyr M. Timing of traumatic upper extremity free flap reconstruction: a systematic review and progress report. Plast Reconstr Surg. 2013;132:591–6.
55. Ingargiola MJ, Daniali LN, Lee ES. Does the application of incisional negative pressure therapy to high-risk wounds prevent surgical site complications? A systematic review. Eplasty. 2013;13, e49.
56. Seddon HJ, Medawar PB, Smith H. Rate of regeneration of peripheral nerves in man. J Physiol. 1943;102:191–215.
57. Sunderland S. A classification of peripheral nerve injuries producing loss of function. Brain. 1951;74:491–516.
58. Brenner MJ, Moradzadeh A, Myckatyn TM, Tung TH, Mendez AB, et al. Role of timing in assessment of nerve regeneration. Microsurgery. 2008;28:265–72.
59. Shin AY, Spinner RJ, Steinmann SP, Bishop AT. Adult traumatic brachial plexus injuries. J Am Acad Orthop Surg. 2005;13:382–96.
60. Fischer JP, Elliott RM, Kozin SH, Levin LS. Free function muscle transfers for upper extremity reconstruction: a review of indications, techniques, and outcomes. J Hand Surg Am. 2013;38:2485–90.
61. Millesi H. Interfascicular nerve grafting. Orthop Clin North Am. 1981;12:287–301.
62. Dvali LT, Myckatyn TM. End-to-side nerve repair: review of the literature and clinical indications. Hand Clin. 2008;24:455–60, vii.
63. Becker C, Gilbert A. The cubital flap. Ann Chir Main. 1988;7:136–42.
64. Gilbert A, Restrepo J. The brachioradial muscle: anatomy and use as a muscular rotation flap. Ann Chir Plast. 1980;25:72–5.
65. Becker C, Gilbert A. The ulnar flap. Handchir Mikrochir Plast Chir. 1988;20:180–3.
66. Masquelet AC, Penteado CV. The posterior interosseous flap. Ann Chir Main. 1987;6:131–9.
67. Masquelet AC. Anatomy of the radial forearm flap. Anat Clin. 1984;6:171–6.
68. Masquelet AC. The Chinese flap. Presse Med. 1984;13:43.
69. Yang GF. Free grafting of a lateral brachial skin flap. Zhonghua Wai Ke Za Zhi. 1983;21:272–4.
70. Lovie MJ, Duncan GM, Glasson DW. The ulnar artery forearm free flap. Br J Plast Surg. 1984;37:486–92.
71. Guimberteau JC, Goin JL, Panconi B, Schuhmacher B. The reverse ulnar artery forearm island flap in hand surgery: 54 cases. Plast Reconstr Surg. 1988;81:925–32.
72. Chaput B, Gandolfi S, Ho Quoc C, Chavoin JP, Garrido I, Grolleau JL. Reconstruction of cubital fossa skin necrosis with radial collateral artery perforator-based propeller flap (RCAP). Ann Chir Plast Esthet. 2014;59(1):65–9. PMID: 23891106.
73. Morrison CS, Sullivan SR, Bhatt RA, Chang JT, Taylor HO. The pedicled reverse-flow lateral arm flap for coverage of complex traumatic elbow injuries. Ann Plast Surg. 2013;71:37–9.
74. Masquelet AC, Rinaldi S, Mouchet A, Gilbert A. The posterior arm free flap. Plast Reconstr Surg. 1985;76:908–13.
75. Maruyama Y, Takeuchi S. The radial recurrent fasciocutaneous flap: reverse upper arm flap. Br J Plast Surg. 1986;39:458–61.
76. Kaplan EN, Pearl RM. An arterial medial arm flap – vascular anatomy and clinical applications. Ann Plast Surg. 1980;4:205–15.
77. Begue T, Masquelet AC, Nordin JY. Anatomical basis of the anterolateral thigh flap. Surg Radiol Anat. 1990;12:311–3.

78. Baek SM. Two new cutaneous free flaps: the medial and lateral thigh flaps. Plast Reconstr Surg. 1983;71:354–65.
79. Maruyama Y, Ohnishi K, Takeuchi S. The lateral thigh fascio-cutaneous flap in the repair of ischial and trochanteric defects. Br J Plast Surg. 1984;37:103–7.
80. Javaid M, Cormack GC. Anterolateral thigh free flap for complex soft tissue hand reconstructions. J Hand Surg Br. 2003;28:21–7.
81. Lamberty BG, Cormack GC. The antecubital fascio-cutaneous flap. Br J Plast Surg. 1983;36:428–33.
82. Baker DC, Barton Jr FE, Converse JM. A combined biceps and semitendinosus muscle flap in the repair of ischial sores. Br J Plast Surg. 1978;31:26–8.
83. Dabernig J, McGill D, Shilov B, Dabernig W, Schaff J. The suralis pendulum flap. Plast Reconstr Surg. 2006;117:1071–3.
84. Schepler H, Sauerbier M, Germann G. The distally pedicled suralis flap for the defect coverage of posttraumatic and chronic soft-tissue lesions in the "critical" lower leg. Chirurg. 1997;68:1170–4.
85. Ohmori K, Harii K. Free dorsalis pedis sensory flap to the hand, with microneurovascular anastomoses. Plast Reconstr Surg. 1976;58:546–54.
86. Robinson DW. Microsurgical transfer of the dorsalis pedis neurovascular island flap. Br J Plast Surg. 1976;29:209–13.
87. McCraw JB, Furlow Jr LT. The dorsalis pedis arterialized flap. A clinical study. Plast Reconstr Surg. 1975;55:177–85.
88. Massin P, Romana C, Masquelet AC. Anatomic basis of a pedicled extensor digitorum brevis muscle flap. Surg Radiol Anat. 1988;10:267–72.
89. Crocker AD, Moss AL. The extensor hallucis brevis muscle flap. J Bone Joint Surg Br. 1989;71:532. PMID: 2722953.
90. Taylor GI, Miller GD, Ham FJ. The free vascularized bone graft. A clinical extension of microvascular techniques. Plast Reconstr Surg. 1975;55:533–44.
91. Yajima H, Tamai S, Ono H, Kizaki K, Yamauchi T. Free vascularized fibula grafts in surgery of the upper limb. J Reconstr Microsurg. 1999;15:515–21.
92. Harrison DH. The osteocutaneous free fibular graft. J Bone Joint Surg Br. 1986;68:804–7.
93. Durand S, Sita-Alb L, Ang S, Masquelet AC. The flexor digitorum longus muscle flap for the reconstruction of soft-tissue defects in the distal third of the leg: anatomic considerations and clinical applications. Ann Plast Surg. 2013;71:595–9.
94. Hartrampf Jr CR, Scheflan M, Bostwick 3rd J. The flexor digitorum brevis muscle island pedicle flap: a new dimension in heel reconstruction. Plast Reconstr Surg. 1980;66:264–70.
95. Atchabahian A, Masquelet AC. The distally based medial gastrocnemius flap: case report and anatomic study. Plast Reconstr Surg. 1996;98:1253–7.
96. Feldman JJ, Cohen BE, May Jr JW. The medial gastrocnemius myocutaneous flap. Plast Reconstr Surg. 1978;61:531–9.
97. Arnold PG, Mixter RC. Making the most of the gastrocnemius muscles. Plast Reconstr Surg. 1983;72:38–48.
98. Ger R. The coverage of vascular repairs by muscle transposition. J Trauma. 1976;16:974–8.
99. Ger R. The technique of muscle transposition in the operative treatment of traumatic and ulcerative lesions of the leg. J Trauma. 1971;11:502–10.
100. Scheflan M, Nahai F, Bostwick 3rd J. Gluteus maximus island musculocutaneous flap for closure of sacral and ischial ulcers. Plast Reconstr Surg. 1981;68:533–8.
101. Harii K, Ohmori K, Torii S. Free gracilis muscle transplantation, with microneurovascular anastomoses for the treatment of facial paralysis. A preliminary report. Plast Reconstr Surg. 1976;57:133–43.
102. Holle J, Worseg A, Kuzbari R, Wuringer E, Alt A. The extended gracilis muscle flap for reconstruction of the lower leg. Br J Plast Surg. 1995;48:353–9.
103. Becker H. The distally-based gluteus maximus muscle flap. Plast Reconstr Surg. 1979;63:653–6.
104. Le-Quang C. Two new free flaps developed from aesthetic surgery II. The inferior gluteal flap. Aesthetic Plast Surg. 1980;4:159–68.
105. Erk Y, Spira M, Parsa FD, Stal S. A modified gluteus maximus musculocutaneous free flap based on the inferior gluteal vessels. Ann Plast Surg. 1983;11:344–6.
106. Masquelet AC, Beveridge J, Romana C, Gerber C. The lateral supramalleolar flap. Plast Reconstr Surg. 1988;81:74–81.
107. Masquelet AC, Gilbert A, Restrepo J. The plantar flap in reconstructive surgery of the foot. Presse Med. 1984;13:935–6.
108. Hamm JC, Stevenson TR, Mathes SJ. Knee joint salvage utilising a plantar musculocutaneous island pedicle flap. Br J Plast Surg. 1986;39:249–54.
109. Chai Y, Ma X, Lin C, Wang K, Pan Y, Chen Y. The reverse medialis pedis flap for coverage of forefoot skin defects. Zhonghua Zheng Xing Wai Ke Za Zhi. 2002;18:27–8.
110. Shanahan RE, Gingrass RP. Medial plantar sensory flap for coverage of heel defects. Plast Reconstr Surg. 1979;64:295–8.
111. Reading G. Instep island flaps. Ann Plast Surg. 1984;13:488–94.
112. Eren S, Ghofrani A, Reifenrath M. The distally pedicled peroneus brevis muscle flap: a new flap for the lower leg. Plast Reconstr Surg. 2001;107:1443–8.
113. Lyle WG, Colborn GL. The peroneus brevis muscle flap for lower leg defects. Ann Plast Surg. 2000;44:158–62.
114. Eyssel M, Dresing K. The peroneus brevis muscle flap-plasty. A simple procedure for covering fibular soft tissue defects after osteosynthesis. Unfallchirurg. 1989;92:85–91.
115. Malikov S, Valenti P, Masquelet AC. Neurocutaneous flap of the anteromedial aspect of the thigh with a distal pedicle. Anatomical study and clinical application. Ann Chir Plast Esthet. 1999;44:531–40.

116. Acland RD, Schusterman M, Godina M, Eder E, Taylor GI, Carlisle I. The saphenous neurovascular free flap. Plast Reconstr Surg. 1981;67:763–74.
117. Dumont CE, Masquelet AC. A reverse triangular soleus flap based on small distal communicating arterial branches. Ann Plast Surg. 1998;41:440–3.
118. Levante S, Masquelet AC. Coverage of chronic osteomyelitis of the ankle and the foot using a soleus muscle island flap, vascularized with retrograde flow on the posterior tibial artery. Ann Chir Plast Esthet. 2009;54:523–7.
119. Tobin GR. Hemisoleus and reversed hemisoleus flaps. Plast Reconstr Surg. 1985;76:87–96.
120. Nahai F, Silverton JS, Hill HL, Vasconez LO. The tensor fascia lata musculocutaneous flap. Ann Plast Surg. 1978;1:372–9.
121. Ebraheim NA, Madsen TD, Humpherys B. The tibialis anterior used as a local muscle flap over the tibia after soft tissue loss. J Trauma. 2003;55:959–61.
122. Hirshowitz B, Moscona R, Kaufman T, Har-Shai Y. External longitudinal splitting of the tibialis anterior muscle for coverage of compound fractures of the middle third of the tibia. Plast Reconstr Surg. 1987;79:407–14.
123. Auregan JC, Begue T, Tomeno B, Masquelet AC. Distally-based vastus lateralis muscle flap: a salvage alternative to address complex soft tissue defects around the knee. Orthop Traumatol Surg Res. 2010;96:180–4.
124. Bovet JL, Nassif TM, Guimberteau JC, Baudet J. The vastus lateralis musculocutaneous flap in the repair of trochanteric pressure sores: technique and indications. Plast Reconstr Surg. 1982;69:830–4.
125. Gilbert A, Teot L. The free scapular flap. Plast Reconstr Surg. 1982;69:601–4.
126. Hamilton SG, Morrison WA. The scapular free flap. Br J Plast Surg. 1982;35:2–7.
127. Baker SR, Sullivan MJ. Osteocutaneous free scapular flap for one-stage mandibular reconstruction. Arch Otolaryngol Head Neck Surg. 1988;114:267–77.
128. Nassif TM, Vidal L, Bovet JL, Baudet J. The parascapular flap: a new cutaneous microsurgical free flap. Plast Reconstr Surg. 1982;69:591–600.
129. Grinsell D, Saravolac V, Whitaker IS. Pre-expanded bipedicled deep inferior epigastric artery perforator (DIEP) flap for paediatric lower limb reconstruction. J Plast Reconstr Aesthet Surg. 2012;65:1603–5.
130. Van Landuyt K, Blondeel P, Hamdi M, Tonnard P, Verpaele A, Monstrey S. The versatile DIEP flap: its use in lower extremity reconstruction. Br J Plast Surg. 2005;58:2–13.
131. Blondeel N, Vanderstraeten GG, Monstrey SJ, Van Landuyt K, Tonnard P, Matton G. The donor site morbidity of free DIEP flaps and free TRAM flaps for breast reconstruction. Br J Plast Surg. 1997;50:322–30.
132. Lantieri L, Serra M, Baruch J. Preservation of the muscle in the use of rectus abdominis free flap in breast reconstruction: from TRAM to DIEP (Deep inferior epigastric perforator) flap. Ann Chir Plast Esthet. 1997;42:156–9.
133. Medalie DA, Llull R, Heckler F. The iliacus muscle flap: an anatomical and clinical evaluation. Plast Reconstr Surg. 2011;127:1553–60.
134. Zheng HP, Zhuang YH, Zhang ZM, Zhang FH, Kang QL. Modified deep iliac circumflex osteocutaneous flap for extremity reconstruction: anatomical study and clinical application. J Plast Reconstr Aesthet Surg. 2013;66:1256–62.
135. Lister GD, McGregor IA, Jackson IT. The groin flap in hand injuries. Injury. 1973;4:229–39.
136. McGregor IA, Jackson IT. The groin flap. Br J Plast Surg. 1972;25:3–16.
137. Acland RD. The free iliac flap: a lateral modification of the free groin flap. Plast Reconstr Surg. 1979;64:30–6.
138. Olivari N. The latissimus flap. Br J Plast Surg. 1976;29:126–8.
139. May Jr JW, Gallico 3rd GG, Jupiter J, Savage RC. Free latissimus dorsi muscle flap with skin graft for treatment of traumatic chronic bony wounds. Plast Reconstr Surg. 1984;73:641–51.
140. May Jr JW, Lukash FN, Gallico 3rd GG. Latissimus dorsi free muscle flap in lower-extremity reconstruction. Plast Reconstr Surg. 1981;68:603–7.
141. Tobin GR, Schusterman M, Peterson GH, Nichols G, Bland KI. The intramuscular neurovascular anatomy of the latissimus dorsi muscle: the basis for splitting the flap. Plast Reconstr Surg. 1981;67:637–41.
142. Bostwick 3rd J, Scheflan M, Nahai F, Jurkiewicz MJ. The "reverse" latissimus dorsi muscle and musculocutaneous flap: anatomical and clinical considerations. Plast Reconstr Surg. 1980;65:395–9.
143. Tropet Y, Brientini JM, Vichard P. Free-flap transfer of the omentum to the lower leg. Ann Chir Plast Esthet. 1990;35:338–41.
144. Clement RW, Young VL, Marsh JL. Use of the greater omentum as a vascular supply for free-flap transfer. Plast Reconstr Surg. 1984;74:131–2.
145. Abbes M, Demard F, Richelme H, Valicioni J. The value of the delta-pectoral flap, transplantation of the great omentum and hyperbaric oxygen therapy in the management of severe buccopharyngostoma (4 cases) (author's transl). Ann Chir. 1975;29:761–7.
146. Martin FF, Le Quang C, Texier M, Bonnet F, Dufourmentel C. Treatment of radionecrosis of the wrist using a great omentum flap and skin graft. Ann Chir Plast. 1974;19:247–9.
147. Ariyan S. The pectoralis major myocutaneous flap. A versatile flap for reconstruction in the head and neck. Plast Reconstr Surg. 1979;63:73–81.
148. Tobin GR. Pectoralis major muscle-musculocutaneous flap for chest-wall reconstruction. Surg Clin North Am. 1989;69:991–1006.
149. Morain WD, Geurkink NA. Split pectoralis major myocutaneous flap. Ann Plast Surg. 1980;5:358–61.

150. McGregor IA. A "defensive" approach to the island pectoralis major myocutaneous flap. Br J Plast Surg. 1981;34:435–7.
151. Drever JM. The lower abdominal transverse rectus abdominis myocutaneous flap for breast reconstruction. Ann Plast Surg. 1983;10:179–85.
152. Drever JM. The epigastric island flap. Plast Reconstr Surg. 1977;59:343–6.
153. Pennington DG, Pelly AD. The rectus abdominis myocutaneous free flap. Br J Plast Surg. 1980;33:277–82.
154. Bunkis J, Walton RL, Mathes SJ. The rectus abdominis free flap for lower extremity reconstruction. Ann Plast Surg. 1983;11:373–80.
155. Ulrich D, Fuchs P, Bozkurt A, Pallua N. Free serratus anterior fascia flap for reconstruction of hand and finger defects. Arch Orthop Trauma Surg. 2010;130:217–22.
156. Ozcelik D, Ugurlu K, Turan T. Reconstruction of the replanted hand with latissimus dorsi muscle and serratus anterior fascia combined flap. J Reconstr Microsurg. 2003;19:153–6.
157. Whitney TM, Buncke HJ, Alpert BS, Buncke GM, Lineaweaver WC. The serratus anterior free-muscle flap: experience with 100 consecutive cases. Plast Reconstr Surg. 1990;86:481–90; discussion 491.
158. Moscona RA, Ullmann Y, Hirshowitz B. Free composite serratus anterior muscle – rib flap for reconstruction of severely damaged foot. Ann Plast Surg. 1988;20:167–72.
159. De la Plaza R, Arroyo JM, Vasconez LO. Upper transverse rectus abdominis flap: the flag flap. Ann Plast Surg. 1984;12:410–8.
160. Miller LB, Bostwick 3rd J, Hartrampf Jr CR, Nahai F. The superiorly based rectus abdominis flap: predicting and enhancing its blood supply based on an anatomic and clinical study. Plast Reconstr Surg. 1988;81:713–24.
161. Pallua N, Machens HG, Rennekampff O, Berger A. The fasciocutaneous supraclavicular artery island flap for releasing postburn mentosternal contractures. Plast Reconstr Surg. 1997;99:1878–84; discussion 1885–76.
162. Pallua N, Wolter TP. The anterior supraclavicular artery perforator (a-SAP) flap: a new pedicled or free perforator flap based on the anterior supraclavicular vessels. J Plast Reconstr Aesthet Surg. 2013;66:489–96.
163. Lecours C, Saint-Cyr M, Wong C, Bernier C, Mailhot E, Tardif M, Chollet A. Freestyle pedicle perforator flaps: clinical results and vascular anatomy. Plast Reconstr Surg. 2010;126:1589–603.
164. D'Arpa S, Cordova A, Pignatti M, Moschella F. Freestyle pedicled perforator flaps: safety, prevention of complications, and management based on 85 consecutive cases. Plast Reconstr Surg. 2011;128:892–906.
165. Bhat S, Shah A, Burd A. The role of freestyle perforator-based pedicled flaps in reconstruction of delayed traumatic defects. Ann Plast Surg. 2009;63:45–52.

Chest Trauma

8

Mirjam B. de Jong, Marike C. Kokke,
Falco Hietbrink, and Luke P.H. Leenen

Contents

8.1	**Introduction**	87
8.1.1	Incidence	88
8.1.2	Trauma Mechanism and Pathophysiology	88
8.1.3	Classification	88
8.2	**Diagnostics**	89
8.2.1	Chest Radiography	89
8.2.2	Computed Tomography of the Chest	89
8.2.3	Ultrasonography	90
8.2.4	Bronchoscopy	91
8.3	**Acute Life-Threatening Thoracic Injuries**	91
8.3.1	Tension Pneumothorax	91
8.3.2	Open Pneumothorax	92
8.3.3	Massive Hemothorax	92
8.3.4	Pulmonary Contusion	92
8.3.5	Cardiac Tamponade	93
8.4	**Potentially Life-Threatening Injuries**	93
8.4.1	Simple Pneumothorax	93
8.4.2	Hemothorax	94
8.4.3	Tracheobronchial Tree Rupture	94
8.4.4	Traumatic Aortic Disruption	95
8.4.5	Traumatic Diaphragmatic Injury	95
8.4.6	Esophageal Rupture	96
8.5	**Bone Injuries**	97
8.5.1	Rib Fractures	97
8.5.2	Sternal Fractures	98
8.5.3	Scapular Fractures	98
8.6	**Lung Injuries**	98
8.6.1	Pulmonary Lacerations	98
8.6.2	Pulmonary Contusion	99
8.6.3	Pulmonary Herniation	100
8.7	**Cardiac Injuries**	100
8.7.1	Myocardial Contusion	100
8.7.2	Coronary Artery Injury	100
8.7.3	Cardiac Rupture	101
8.8	**Fracture Treatment in Patients with Concomitant Thoracic Trauma**	101
8.9	**Complications**	102
8.9.1	Acute Respiratory Distress Syndrome	102
8.9.2	Chylothorax	103
8.9.3	Pleural Empyema	104
8.9.4	Persistent Air Leakage	104
8.10	**Operative Techniques**	104
8.11	**References**	105

M.B. de Jong (✉) • M.C. Kokke • F. Hietbrink
L.P.H. Leenen
Department of Traumasurgery, University Medical
Center Utrecht, Heidelberglaan 100, Utrecht 3584 CX,
The Netherlands
e-mail: m.b.dejong-33@umcutrecht.nl;
m.c.kokke@umcutrecht.nl; f.hietbrink@umcutrecht.nl;
l.p.h.leenen@umcutrecht.nl

8.1 Introduction

Thoracic trauma is one of the major burdens in poly-traumatized patients. The mechanisms have changed throughout the years, as vehicles have become available that allow high speed traveling, along with changes in passive car safety.

8.1.1 Incidence

Thoracic trauma is a significant cause of morbidity. In persons younger than 40 years, traumatic injury is even the most common cause of death. Thoracic injuries are responsible for 25 % of deaths in this population [1].

Trauma deaths due to chest injury occur in 76 % in the first day, of which 38 % takes place in the first hour [2]. The majority of patients who dies from pulmonary complications will die more than 10 days after trauma. The so-called golden hour for thoracic trauma in which accurate treatment is required to prevent mortality is still very important. Less than 10 % of the blunt chest injuries and 15–30 % of the penetrating chest injuries require an operation. Most chest injuries can be managed non-operatively. A systematic approach like Advanced Trauma Life Support® provided by the American College of Surgeons is the most well known and most used system [3].

According to the principles of ATLS®, thoracic injuries are separated mainly in two main categories: acute life-threatening and potentially life-threatening injuries (Table 8.1).

8.1.2 Trauma Mechanism and Pathophysiology

Chest trauma is mostly related to automobile or pedestrian accidents and commonly results in chest wall injuries like rib fractures. The pain associated with these injuries can make breathing difficult and this may compromise ventilation. This can be further aggravated by pulmonary contusion, which leads to even more difficulty in respiration. Shunting and dead space ventilation produced by these injuries can also impair oxygenation. Space-occupying conditions include pneumothorax, hemothorax, and hemopneumothorax. These interfere with oxygenation and ventilation by compressing otherwise healthy lung parenchyma. At a cellular level, lung contusion induces an inflammatory response signified by primed polymorph neutrophil granulocytes (PMNs) in blood and tissue [4].

Operative treatment is rarely necessary in blunt thoracic injuries although the advent of several plating systems for rib fixation increases intervention rates (see also rib fractures). Most blunt thoracic injuries can be treated with supportive measures and simple interventional procedures such as chest drainage. Traumatic asphyxia results from a severe blunt injury of the thorax. Patients present with cyanosis of the head and neck, subconjunctival hemorrhage, periorbital ecchymosis, petechiae of the head and neck, and occasionally neurologic symptoms. Factors implicated in the development of these striking physical characteristics include thoraco-abdominal compression after deep inspiration against a closed glottis. This results in venous hypertension in the valveless cervicofacial venous system. Other injuries caused by blunt thoracic trauma are diaphragmatic injuries, pneumothorax, hemothorax, blunt tracheal injuries, bronchial injuries, esophageal injuries, cardiac injuries, and injuries to the major thoracic veins or thoracic duct. These injuries and their management will be described in this chapter.

8.1.3 Classification

In the poly-traumatized patient chest trauma is only one part of all injuries. The evaluation of injury severity and the prediction of outcome is one of the most important functions of scoring systems. Several scoring systems for the classification of blunt thoracic trauma have been developed. The Abbreviated Injury Scale (AIS) is a prognostic scoring system allocating a severity score to every injury of the different body regions

Table 8.1 Thoracic injuries [3]

Acute life threatening	Tension pneumothorax
	Open pneumothorax
	Flail chest and pulmonary contusion
	Massive hemothorax
	Cardiac tamponade
Potentially life threatening thoracic injuries	Simple pneumothorax
	Hemothorax
	Pulmonary contusion
	Tracheobronchial tree rupture
	Blunt cardiac injury
	Traumatic aortic disruption
	Traumatic diaphragmatic injury
	Esophageal rupture

Table 8.2 Thoracic trauma severity score according to Pape et al. [8]

Grade	PO$_2$/FiO$_2$	Rib fractures	Pulmonary contusion	Pleural lesion	Age (years)	Points
0	>400	0	None	None	<30	0
I	300–400	1–3 unilateral	1 lobe unilateral	Pneumothorax	30–40	1
II	200–300	4–6 unilateral	1 lobe bilateral or 2 lobes unilateral	Hemothorax/hemopneumothorax unilateral	41–54	2
III	150–200	>3 bilateral	<2 lobes bilateral	Hemothorax/hemopneumothorax bilateral	55–70	3
IV	<150	Flail chest	≥2 lobes bilateral	Tension pneumothorax	>70	5

(Head, face, neck, thorax, abdomen, spine, upper extremity, lower extremity, external and other trauma). High scores are associated with a lower probability of survival. The AIS is an anatomical scoring system for injury severity assessment of different body regions [5, 6].

Most of the thoracic trauma scores are based on pathological-anatomical changes. The Thoracic Trauma Severity Score (TTS score) seems to be the most suitable for severity assessment and prediction of outcome in poly-traumatized patients with blunt chest injuries [7].

The TTS score is based on five anatomical and physiological parameters: pO$_2$/FiO$_2$, rib fractures, pulmonary contusion, pleural lesions, and age. Each parameter is assigned a value of 0–5. The TTS score ranges from 0 to 25 and with increasing values, a more severe thoracic trauma can be assumed (Table 8.2) [8].

8.2 Diagnostics

Thoracic trauma may result in a variety of different injuries. A prompt assessment of correct diagnosis and severity assessment of thoracic trauma is crucial for the further treatment of thoracic lesions itself and concomitant injuries. There are several diagnostic tools for diagnosis and severity assessment of thoracic trauma.

8.2.1 Chest Radiography

The supine anteroposterior (AP) chest radiography is the initial examination of choice in patients with thoracic trauma. Because of the supine position of the trauma patient in the emergency room there is no lateral view available and therefore limited information can be gained from chest radiography. The chest X-ray is used as a first screening method during the evaluation of the trauma patient at the emergency room. The availability of CT scan, even in the emergency room, lead to a reduction in the need for plain films [9]. Although the CT scan is significantly more effective in detecting thoracic injuries, chest radiography still is recommended by Advanced Trauma Life Support protocol [3]. Especially in unstable patients a chest radiograph is still useful as it is the quickest way to rule out (tension) pneumothorax and hemothorax. CT scan is still more time consuming and requires considerable radiation exposure. Transferring an unstable patient from the emergency room to the radiology suite provides unnecessary risk. Overuse of CT scans can lead to inappropriate delays in patient care [10].

However, in the stable patient with suspicion of blunt thoracic injuries and an indication for chest CT scan, skipping the chest radiograph should be considered [11, 12].

In certain cases physicians even should not wait for a chest radiograph to confirm clinical suspicion. The classic example is hyperresonant note on percussion and the absence of breath sounds over the affected hemithorax combined with signs of hemodynamic compromise, which can be found in patients with tension pneumothorax. This should be immediately decompressed before obtaining a chest radiograph.

8.2.2 Computed Tomography of the Chest

The use of CT for thoracic trauma evaluation has increased dramatically in the past 15 years. Chest CT scan is superior in identifying and visualizing

```
┌─────────────────────────────────────────────────┐
│ Blunt trauma patient > 14 years, need for       │
│ chest imaging at initial assessment             │
└─────────────────────────────────────────────────┘
                        │
                        ▼
┌─────────────────────────────────────────────────┐
│ 1. Age > 60 years                               │
│ 2. Rapid deceleration mechanism: fall > 20 ft   │         ┌──────────────────────────────┐
│    (6 m) or motor vehicle crash > 40 mph        │────────▶│ ≥1 criteria present: cannot  │
│    (> 64 km/h)                                  │         │ exclude intrathoracic injury │
│ 3. Chest pain                                   │         └──────────────────────────────┘
│ 4. Intoxication                                 │
│ 5. Abnormal alertness/mental status             │
│ 6. Distracting painful injury                   │
│ 7. Tenderness to chest wall palpitation         │
└─────────────────────────────────────────────────┘
                        │
                        ▼
┌─────────────────────────────────────────────────┐
│ All criteria absent very low risk for           │
│ intrathoracic injury and chest imaging is       │
│ not indicated.                                  │
└─────────────────────────────────────────────────┘
```

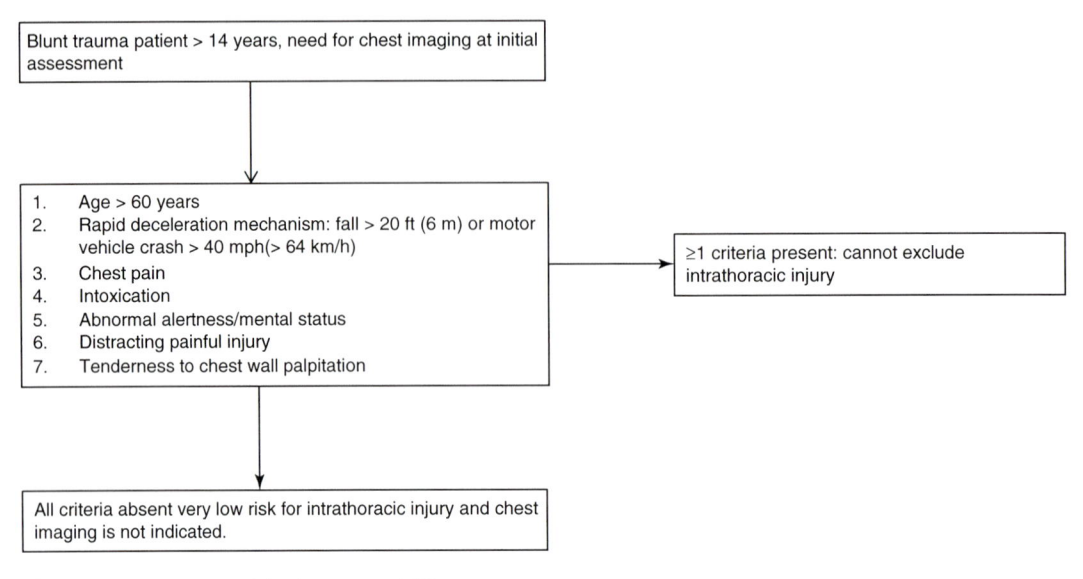

Fig. 8.1 Nexus chest decision instrument [15]

injuries like pulmonary contusions, pneumothorax, hemothorax, vascular injuries and fractures. In about 18–82 % of the patients with a normal chest X-ray additional injuries are found on chest CT scan [11–13].

The marked increase in the number of occult injuries diagnosed on a chest CT scan in patients with blunt thoracic trauma was not, however, accompanied by a similar increase in therapeutic intervention [14]. The disadvantages of a chest CT scan are exposure to radiation, costs and CT scan being more time consuming than a plain chest X-ray. Therefore it is important not to make a chest CT scan routinely but only when significant injuries are suspected.

Using the Nexus (National Emergency X-radiography Utilization study) Chest decision instrument might be helpful in decision-making in patients suffering from blunt thoracic trauma. The sensitivity and negative predictive value for thoracic injury seen on chest imaging was 98.8 and 98.5 % respectively [15]. This Nexus chest decision instrument is meant for all blunt trauma patients over 14 years old who, by initial assessment, may need chest imaging to rule out intrathoracic injury. The criteria used are age >60 years, rapid deceleration mechanism defined as fall >20 ft (>6 m) or motor vehicle crash >40 mph (>64 km/h), chest pain, intoxication, abnormal alertness/mental status, distracting painful injury, and tenderness to chest wall palpation. If all criteria are absent there is a very low risk for intrathoracic injury and chest imaging is not indicated. If one or more criteria are present, intrathoracic injury cannot be excluded and chest imaging should be done (Fig. 8.1) [15].

8.2.3 Ultrasonography

8.2.3.1 Transthoracic Ultrasonography

The Focused Abdominal Sonography for Trauma (FAST) is a rapid ultrasound examination performed in the emergency room. Except for abdominal injuries some thoracic injuries as hemothorax, pneumothorax, and blood in the pericardium can be reliably diagnosed with a sensitivity of 93–96 %. Most trauma patients are usually managed in the supine position with spinal immobilization, which underestimates the prevalence of thoracic lesion on chest X-ray. Especially in unstable high-risk patients thoracic ultrasonography as a bedside diagnostic modality is a better diagnostic test than clinical examination and chest X-ray together [16, 17]. However in the evaluation of pneumothorax the accuracy is not sustained over time, probably as a result of the formation of intrapleural adhesions [18]. As another disadvantage, subcutaneous emphysema precludes an accurate diagnosis by ultrasound.

8.2.3.2 Transesophageal Echocardiography (TEE)

In the workup of possible blunt rupture of thoracic aorta, transesophageal echocardiography has sensitivity and specificity up to 93–96 % in diagnosing a thoracic aorta rupture [19]. TEE also may help define intracardiac anatomy, function, and injuries like cardiac valve injury or traumatic rupture of the interatrial or interventricular septum. The TEE has a better sensitivity and specificity than the transthoracic echocardiography (TTE) for depicting aortic injury, pericardial effusion, myocardial contusion, atrial laceration, and cardiac valve injury [20]. However, the use of the TEE may be limited in patients with severe trauma and hypotension, or head, neck, and spine injuries [21].

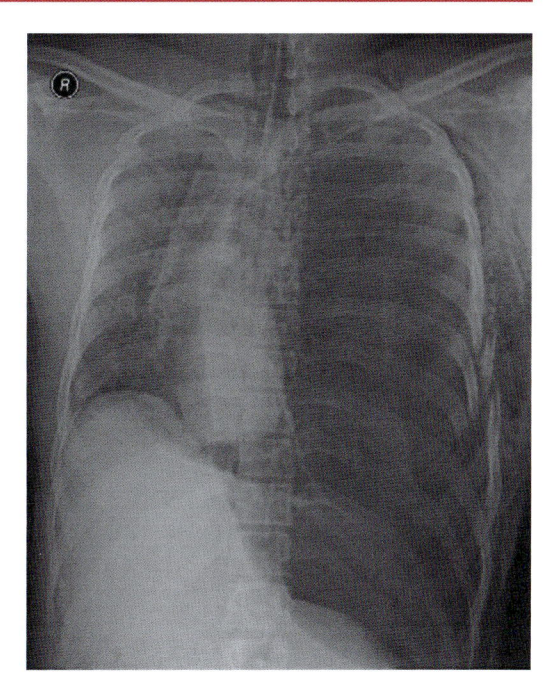

Fig. 8.2 Tension pneumothorax

8.2.4 Bronchoscopy

Fiber optic or rigid bronchoscopy is performed in thoracic trauma patients with suspicion for tracheobronchial injuries. Both techniques have high sensitivity for the diagnosis of these injuries. Bronchoscopy can be used in detecting tracheobronchial lesions, supraglottic injuries, bleeding, and lung contusions. Bronchoscopy can also be of therapeutic use for removing secretions and preventing the formation of atelectasis. Bronchoscopy is rarely used in the primary treatment of patient with thoracic trauma, but a few days after initial trauma it can be useful.

8.3 Acute Life-Threatening Thoracic Injuries

8.3.1 Tension Pneumothorax

A tension pneumothorax occurs when a pneumothorax permits entry but no exit of air from the thoracic cavity (Fig. 8.2). This results in increase of the air in the pleural cavity but leads to collapse of the ipsilateral lung and compression of the intrathoracic structures on the contralateral side. Although needle decompression in the midclavicular line is the recommended method of initial treatment, the patency of this procedure has been a subject of debate. In a porcine model of tension pneumothorax, 58–64 % of the needle placement procedures failed in adequate decompression, compared to a 100 % success rate in thoracostomy tube placement [22].

In the acute clinical setting, a success rate of 59 % has been documented, while in the remaining 41 % the needle did not reach the pleural cavity [23]. Because of these flaws in needle thoracocentesis, blunt dissection and digital decompression should be the first step. When a chest tube cannot directly be placed, the incision is made in the midclavicular line in the second intercostal space, while in a later setting, a formal chest tube can be placed in the 4th intercostal space in the mid-axillary line [24]. In most cases chest tube placement is performed according the ATLS® guidelines [3]. An incision is made in the 4th or 5th intercostal space on the anterior axillary line, after which the pleura is bluntly opened. A large diameter (24–32 French) tube is inserted and placed dorso-cranially [25]. A canister with water seal is connected to the tube and wall suction is initiated. Preferably, prophylactic antibiotics are given; however, this should not delay the placement of a chest tube in an emergency setting [26].

8.3.2 Open Pneumothorax

An open pneumothorax occurs when a pneumothorax is associated with a chest wall defect (Fig. 8.3). During inspiration, air is sucked into the pleural cavity due to the negative intrathoracic pressure. When the diameter of the external wound is over 2/3 of the diameter of the bronchial tree then the air prefers to go through the wound. The wound has to be treated by a venting bandage. This can be applied using commercially available seals [27], or by applying a bandage, which is taped on three sides, allowing air to be vented out, but seals the cavity during inspiration [28].

8.3.3 Massive Hemothorax

A hemothorax is defined as blood in the interpleural space. This occurs in up to 40 % of patients with blunt thoracic trauma. Bleeding is caused by parenchymal injuries, rib fractures, laceration of intercostal or internal mammary artery. Furthermore, hemothorax can be a life-threatening condition when caused by bleeding from the heart or hilar vessels.

All trauma patients with a hemothorax should undergo chest tube placement. In case of gross drain output, a second chest tube is placed promptly. Immediate surgery for massive hemothorax is mandatory when the patient's physiology is unstable (persistent blood transfusion required), regardless of the numbers of initial chest tube output. Furthermore, when >1500 ml in the first 24 h is evacuated, it should prompt surgical intervention [29]. Injuries that are often found when massive hemothorax is present are bleeding from the azygos vein, the mammary artery, laceration of the hilar vessels, severe pulmonary tissue laceration, or dissection of the aorta. When having a massive hemothorax it should not be forgotten to re-infuse the lost blood by using the cell-saver.

8.3.4 Pulmonary Contusion

The most common injury after thoracic trauma is pulmonary contusion. It occurs in 30–75 % of all patients [30]. A severe lung contusion (Figs. 8.4 and 8.5) can be life threatening because of the destruction of alveolar architecture of the lung and intramural bleeding, prohibiting diffusion over the alveolar membrane, leading to severe hypoxia. The lung contusion will further be dealt with in Sect. 8.6.2.

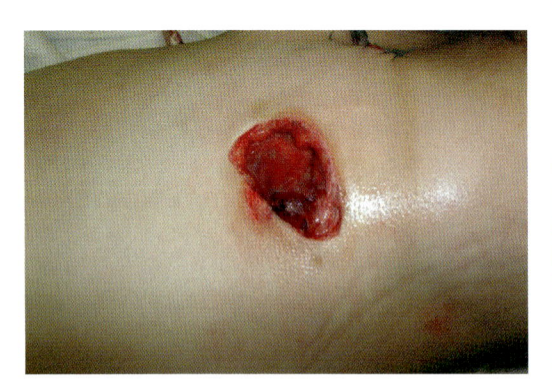

Fig. 8.3 Open pneumothorax

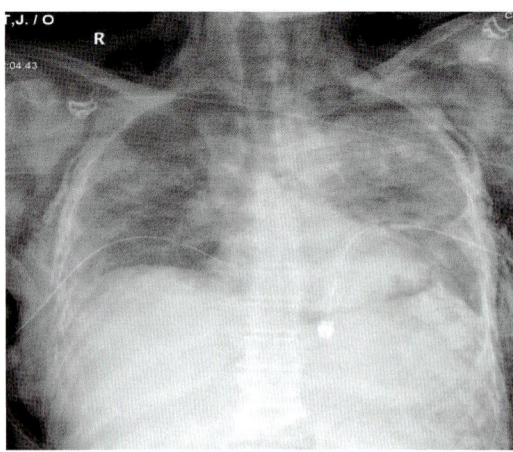

Fig. 8.4 Pulmonary contusion

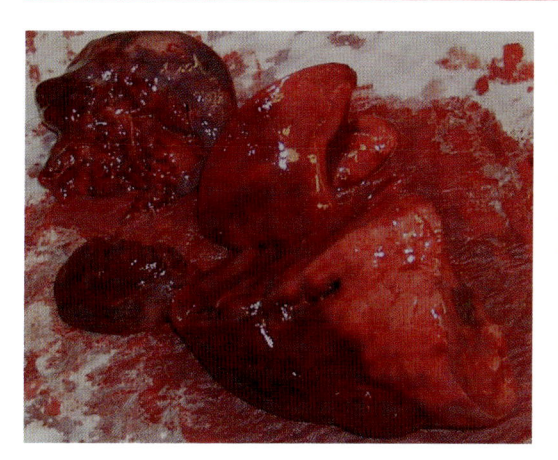

Fig. 8.5 Lung tissue after pulmonary contusion

8.3.5 Cardiac Tamponade

A pericardial tamponade mostly occurs after penetrating trauma, but it also present in about 1 % of blunt chest trauma patients. It develops because of bleeding into the pericardial sac, either from an injury to the heart or from coronary or aorta lesion [31]. Immediate pericardiocentesis is indicated for restoration of normal cardiovascular function. Although successful outcome has been documented in pericardiocentesis as the sole procedure, in a patient with severe hemodynamic instability, the procedure is only to be used as a bridge to surgery or transfer to a definitive care facility [32]. An alternative is the performance of a subxyphoidal window to evacuate the blood from the pericardium. The definitive treatment consists of thoracotomy, repair of the injury causing the bleeding and adequate evacuation of the blood from the pericardium.

8.4 Potentially Life-Threatening Injuries

Besides the life-threatening injuries, which will lead immediately to death if left untreated, there are potentially life-threatening injuries in patients with thoracic trauma (Table 8.1). During the first minutes of trauma resuscitation these injuries can often be missed. So it is of great importance to have a high index of suspicion depending on the trauma mechanism and treat these injuries immediately to prevent further deterioration and eventually death. Additional imaging like a chest X-ray will help you in further diagnosis, however some of these injuries can easily be missed by conventional radiology alone (Sect. 8.2.1) [33].

8.4.1 Simple Pneumothorax

A simple pneumothorax occurs as a result of air entrapment into the pleural cavity between the two pleural layers (visceral and parietal) and will cause a (partial) collapse of the lung and thereby compromising oxygenation and ventilation on the affected side. The air leakage is often caused by a lung laceration after blunt thoracic trauma, but damage to the lung by rib fractures or penetrating injury can also account for this phenomenon.

Diminished breath sounds and hyperresonance to percussion over the affected hemithorax indicates the presence of a pneumothorax. In stable patients an additional chest X-ray or even a CT scan in the case of an occult pneumothorax is necessary to demonstrate the diagnosis. An occult pneumothorax is a pneumothorax that was not suspected clinically, nor was evident on the plain radiograph, but rather identified on CT scan or ultrasound. When adequate follow-up is provided (by means of ultrasound or chest X-rays), occult pneumothorax does not require chest tube drainage [34]. Even on positive pressure ventilation, conservative treatment of an occult pneumothorax can be successful and can reduce the length of hospital stay, given an adequate follow-up by ultrasound or chest radiographs [35–37].

The treatment of a pneumothorax will consist of a tube thoracostomy to release the air and thereby reexpand the collapsed lung. This tube should be placed according to the ATLS® recommendations [3]. Although some authors have stated the drainage of a simple traumatic pneumothorax in patients without other injuries can be done with a pigtail, we strongly suggest that this procedure should be reserved for a very selected patient population [38]. Since the majority of trauma patients with a pneumothorax (especially in blunt trauma) have concomitant injuries, which will often lead to persistent air leak and/or hemothorax, these patients require a formal chest tube. If left untreated a simple pneumothorax can convert into a tension pneumothorax and this certainly needs prompt intervention (see Sect. 8.3.1).

8.4.2 Hemothorax

A hemothorax occurs in up to 40 % of patients with blunt thoracic trauma. As a result of lung laceration, which damages the lung parenchyma, blood can enter the pleural cavity thereby causing a hemothorax. Both bleeding from an intercostal vessel or internal mammary vessel can also contribute to a hemothorax. Depending on the bleeding source and severity, the hemodynamical status of the patient will be influenced and should be treated accordingly. A hemothorax should be treated with a tube thoracostomy to evacuate the accumulated blood. The chest tube production should be monitored closely to recognize a massive hemothorax directly. No further immediate surgical intervention is necessary as a hemothorax is often self-limiting unless there is ongoing bleeding or there consists a massive hemothorax (see Sect. 8.3.3). By draining the intrathoracic hemorrhage the lung can reexpand and the formation of fibrous adhesions is prevented which reduces the risk of a pleural empyema and restrictive pulmonary disease [31, 39].

If a retained or persistent hemothorax is present a VATS is necessary to remove the clotted blood. The VATS procedure should be done within the first 3–7 days after trauma, in order to reduce the chance on conversion to thoracotomy and decrease the risk of infection [29]. Intrapleural thrombolytic therapy has only limited use and should not be considered as standard of care [40].

8.4.3 Tracheobronchial Tree Rupture

Injuries to the tracheobronchial tree (trachea or major bronchus) are rare and most patients die before they reach the hospital. They are frequently caused by blunt trauma which causes compression

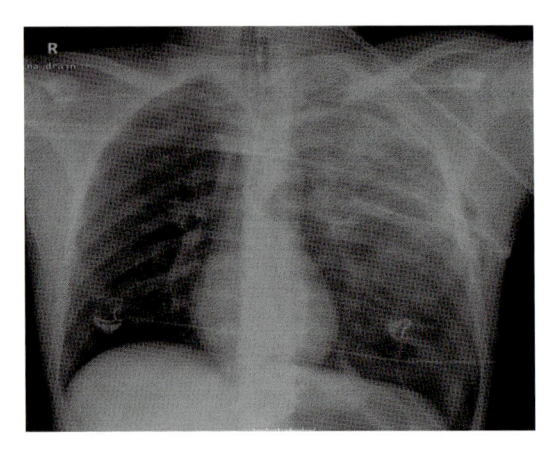

Fig. 8.6 Tube passes right-sided bronchial rupture

of the trachea between the sternum and vertebrae, or by a rapid deceleration trauma. Patients are in severe respiratory distress with coughing, stridor, or an altered voice and present with hemoptysis, massive subcutaneous emphysema, or associated pneumothorax.

A delay in diagnoses leads to a high mortality, even if the patient reaches the hospital alive. So if tracheobronchial injury is suspected immediate treatment is required. The first step is to establish a patent airway by endotracheal intubation. If the endotracheal tube can be managed distal to the tracheal injury, it can prevent a massive air leak (Fig. 8.6). Advantage is that there is no positive pressure in the injured lung; usually healing will appear without surgical intervention. A fiberoptic bronchoscopy can be used as a diagnostic adjunct and if a bronchial injury is diagnosed the tube can be placed in the contralateral bronchus. As the trachea and main bronchus are in the proximity of the great vessels and esophagus, associated injuries must be suspected and treated accordingly. In both tracheal and bronchial disruption further surgical repair is mandatory. When performing a thoracotomy it is possible to intubate directly in the ruptured bronchus (Fig. 8.7).

Tracheal lesions due to blunt trauma usually appear as transverse tears between cartilaginous tracheal rings or longitudinal tears in the posterior tracheal membrane. In tracheal injuries surgical repair is required in order to ensure airway continuity. This can be done by primary suturing with absorbable sutures or by the resection of several tracheal rings and re-anastomosis.

Fig. 8.7 Intubation through left bronchus

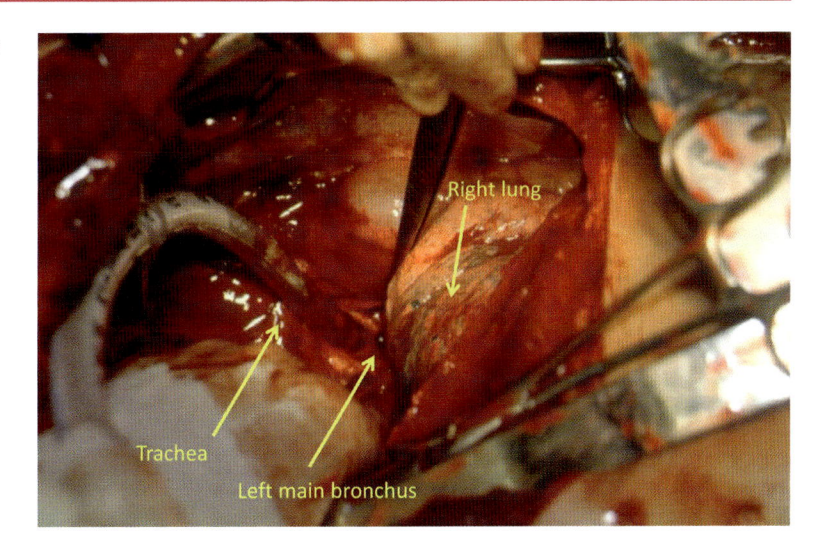

Once this has been done, autogenous tissue is wrapped around the reconstructed trachea. In the neck, all strap muscles can be used for this procedure, while in the chest the intercostal muscles, serratus anterior, latissimus, or pericardial patches can be used.

8.4.4 Traumatic Aortic Disruption

Aortic injury caused by blunt trauma is mostly lethal at the scene; however, those that have only an intimal tear reach the hospital alive and so treatment can be established. In blunt trauma the shearing forces due to rapid deceleration will cause a partial laceration of the aortic wall near the ligamentum arteriosum. This will result in a contained rupture of the aorta (Fig. 8.8). Diagnosis is difficult as specific clinical signs are absent. Together with the trauma mechanism and a high index of suspicion an additional chest X-ray may reveal abnormalities like deviation of the trachea, widened mediastinum, or presence of an apical cap [3]. A CT angiogram of the aorta is more accurate and will confirm the diagnosis and the extent of the injury. Therapy consists of maintaining the mean arterial blood pressure around 60 mmHg, which will reduce the risk of rupture. Thereby it is possible to delay the nowadays often-used endovascular repair of the aortic injury (Figs. 8.9 and 8.10) while treating other severe associated injuries [41, 42].

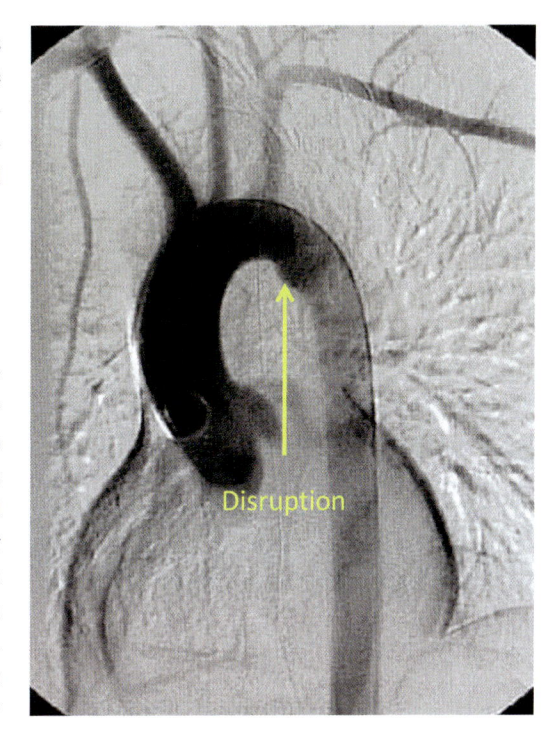

Fig. 8.8 Traumatic aortic disruption

8.4.5 Traumatic Diaphragmatic Injury

Blunt torso injury produces large tears and predominantly occurs at the left hemidiaphragm. These tears can lead to herniation of intra-abdominal organs, which can be diagnosed on

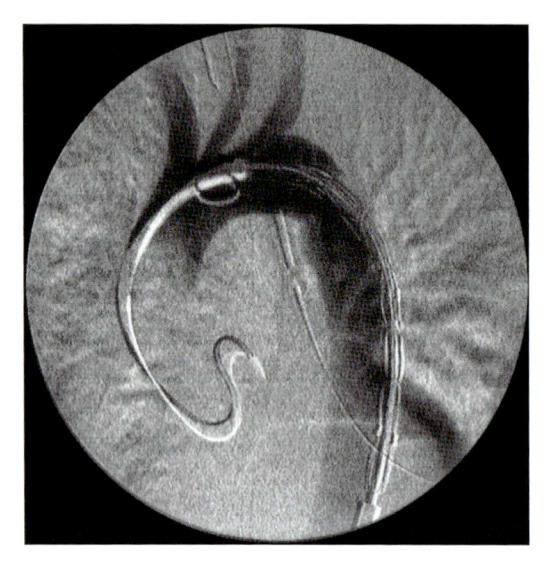

Fig. 8.9 Stent positioning in aorta

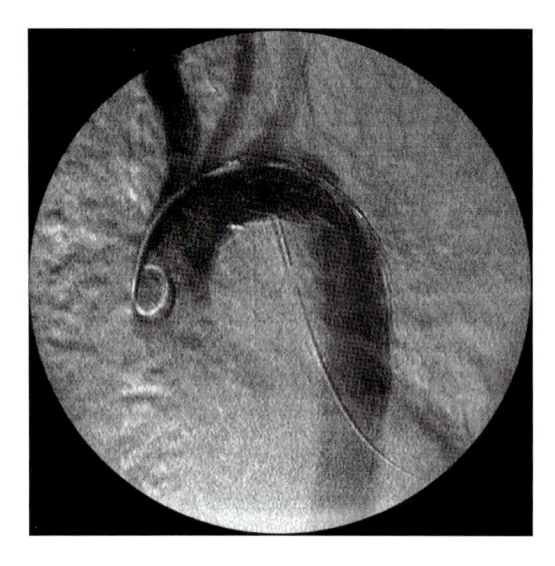

Fig. 8.10 Stent in aorta after deployment

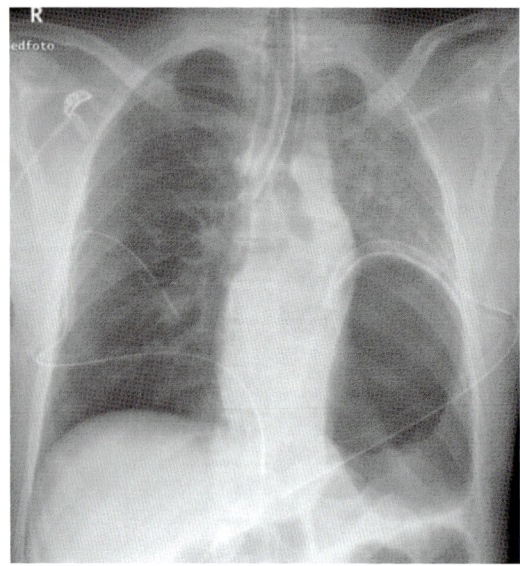

Fig. 8.11 Ruptured diaphragm

chest X-ray (Fig. 8.11). Diagnosis however is often hampered by the fact that the multiply injured patients are treated with positive pressure ventilation, which prevents dislocation of abdominal organs into the thorax. Nowadays early CT scanning can reveal the discrete changes that go with diaphragmatic injury. Many of the diaphragmatic injuries however are diagnosed during an emergency laparotomy or thoracotomy for associated intra-abdominal or intrathoracic injuries. The treatment exists of direct repair. Only in the minority of cases a mesh is necessary.

8.4.6 Esophageal Rupture

Blunt esophageal injuries are very rare. They are caused by a sudden increase in the intra-abdominal pressure, for example by a blow in the upper abdomen. Gastric contents will eject in the esophagus causing a rise in intraluminal pressure. Pressure rise can lead to a tear in the esophagus with leaking of content into the mediastinum.

Patients present with clinical signs like subcutaneous emphysema, pneumomediastinum, pneumothorax, or intra-abdominal free air. Time between trauma and definitive treatment may influence the outcome by developing esophageal injury related complications [43].

If diagnosed early, the majority of the patients can be treated with primary surgical repair and additional drainage of the mediastinum becomes necessary. Nowadays an endoscopic esophageal stent is a possible alternative [44]. If left undiagnosed patients often present with fever and signs of systemic sepsis caused by mediastinitis at a later stage.

8.5 Bone Injuries

8.5.1 Rib Fractures

Rib fractures are the most common thoracic injuries and occur in 10 % of all trauma patients and approximately 30 % of patients with significant chest trauma [45]. Fractures of the first and second rib suggest severe thoracic trauma. These ribs provide a protection of vital structures like brachial plexus and vessels (subclavian artery and vein). Ribs 4–10 are most frequently involved. The mechanism is often due to direct forces on the chest wall. With fractures of ribs 8–12 the presence of intra-abdominal injuries should be considered.

Physical signs of rib fractures include local tenderness and sometimes crepitus over the site of the fracture. Rib fractures may also by an indicator for other significant intrathoracic injuries. Elderly patients and patients with osteoporosis or osteopenia have an increased risk of number and severity of fractures. This is in contrast to children where higher forces are needed to cause fractures, because the chest wall is more pliable and compliant. The most common symptom of rib fractures is pain, which makes it difficult to breathe adequately. Up to 30 % of the patients with rib fractures develop a pneumonia; the older the patient the higher this percentage [46, 47]. The greater the number of fractured ribs the higher the mortality and morbidity [45]. Up to 10 % mortality is reported in patients with more than 4 rib fractures; this increases to 34 % in patients with 8 or more fractures [48]. It is also known that patients with more than 4 rib fractures after the age of 45 have an increased risk of adverse outcomes [48–51].

A flail chest can be defined as fractures of four or more consecutive ribs in two or more places resulting in paradoxical movement of the chest wall during respiration. Paradoxical movement of the chest can increase the work and pain involved with breathing. In most patients the severity and extent of the lung injury determines the clinical course and the requirement of mechanical ventilation. Patients with flail chest have a significant higher need for mechanical ventilation. Although the recovery of mechanical ventilation ensured a better result in the treatment of flail chest, it is also responsible for several ventilation related complications.

Management of rib fractures involves pain control and adequate oxygenation and ventilation possibly using positive pressure ventilation when necessary. In patients with a flail chest, non-operative treatment leads to a mortality of 25–51 and 27–70 % develop pneumonia [52]. External stability by means of operative fixation is an alternative treatment of multiple rib fractures in order to avoid mechanical ventilation (Figs. 8.12 and 8.13). The goal of operative therapy is to improve respiratory mechanics, reduce pain, and prevent pulmonary restriction associated with significant chest wall deformities. Current indication for operative fixation is the presence of a flail chest [53] which is associated with reduction in duration of mechanical ventilation, complications associated with prolonged mechanical ventilation, length of stay in the hospital and mortality [54].

Other indications are patients with rib fractures who, notwithstanding good pain management, are still in pain, have chest wall deformity, or have one or more symptomatic non-union rib fractures. Age over 45 years and more than four rib fractures seem to be important factors in determining outcome of patients with multiple

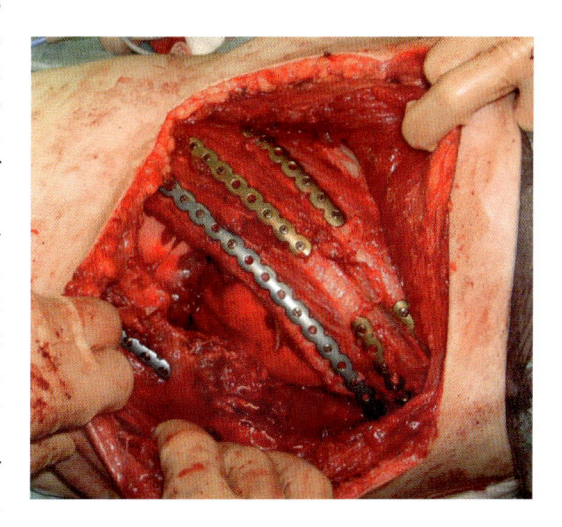

Fig. 8.12 Rib fiation: clinical scenaroi

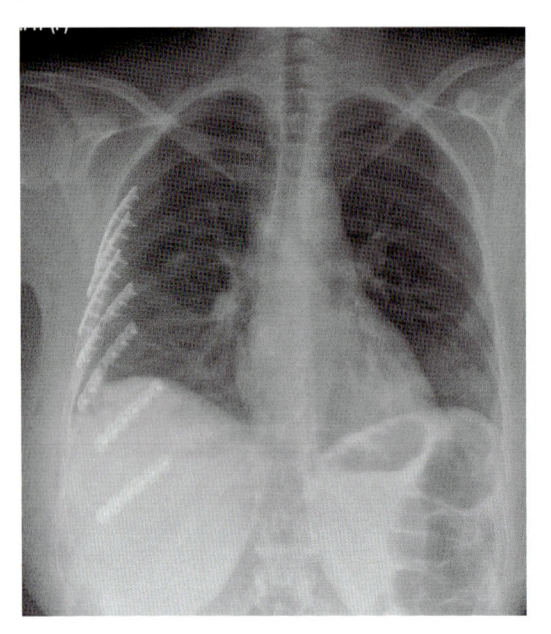

Fig. 8.13 Rib fixation: xray post op

rib fractures. Therefore an operative approach of patients older than 45 years with four or more rib fractures should be considered.

8.5.2 Sternal Fractures

Sternal fractures are present in up to 8 % of the admissions after blunt thoracic trauma and motor vehicle crashes [55]. Before the use of the seatbelt a sternal fracture was a marker of high-energy trauma. With the mandatory use of a seatbelt, the survival after motor vehicle crashes increased together with a rise in the incidence of the sternal fracture, also called the typical "seat belt injury" [56].

The typical sternal fracture is a transverse fracture located in the upper and midportions of the sternal body. The symptoms consist of localized tenderness, swelling, and deformity. A sternal fracture can be diagnosed by a lateral view because a sternal fracture is rarely apparent on the anteroposterior chest film. The highest sensitivity is reached by the chest CT scan is.

As in all thoracic fractures, sternal fractures are often associated with more serious occult injuries. Underlying myocardial injury is not uncommon. Treatment of sternal fractures is similar to that for rib fractures. It consists primarily of pain control and appropriate pulmonary hygiene. Patients with isolated, stable sternal fractures that have normal radiographic findings and normal electrocardiograms can be treated as outpatients [57].

When the sternal fracture is severely displaced open reduction and internal fixation by a midline incision should be done. Various techniques are described, including wire suturing and the placement of plates and screws. Although there are several pre-contoured plates available for this aim, the less massive plates employed for rib fixation can also be used. In the presence of a flail chest a different approach can be followed to fixate both ribs and sternum.

8.5.3 Scapular Fractures

Fractures of the scapula are uncommon; they occur due to a high-energy dissipation. These patients usually have associated injuries (61 %) with higher treatment priority. The associated injuries reported most frequently are rib fractures including pneumothorax, hemothorax, pulmonary and spinal injuries.

A patient with a scapular fracture typically presents with the arm adducted along the body. In physical examination, swelling, crepitus, ecchymosis, and local tenderness may be present. Active range of motion is restricted in all directions. With the presence of a scapular fracture, arterial injury and/or brachial plexopathy should also be considered. Most fractures occur in body (30 %) and neck (25 %) and can be treated non-operatively. In contrast, displaced intraarticular fractures of glenoid mostly need operative fixation.

8.6 Lung Injuries

8.6.1 Pulmonary Lacerations

Pulmonary laceration can be the result of penetrating chest trauma. But also blunt injury, which causes penetration due to rib fractures or torn

Table 8.3 Classification of pulmonary lacerations according to Wagner et al. [59]

Type	Mechanism of injury	Appearance on CT
1	Compression rupture	Air filled or air-fluid level in intraparenchymal cavity. Linear tear (when rupture through visceral pleura)
2	Compression shear	Paravertebral laceration
3	Rib penetration	Small peripheral cavity. Small peripheral linear radiolucency
4	Adhesion tear	Only seen at surgery of autopsy

lung tissue as a result of shearing forces can lead to lacerations of the lung parenchyma. Lung lacerations are characterized by the disruption of the pulmonary architecture, which will cause air or blood leakage. If this ruptures through the visceral pleura it will lead to a pneumothorax, hemothorax, or both. When blood or air becomes entrapped in the lung parenchyma a traumatic cyst develops which will be called pulmonary hematoma or pneumatocele, respectively [58].

The only classification system for pulmonary lacerations published until now is from Wagner et al. (Table 8.3). They describe 4 types of lacerations based on CT findings or mechanism of injury: compression rupture, compression shear, rib penetration, and adhesion tears [59]. Treatment is often non-operative; however, depending on the grade and location, sometimes surgical treatment is necessary. A thoracotomy with preservation of the lung is the primary goal in combination with wedge resection and segmentectomy if required. In case of penetrating injury caused by rib fractures with a through and through tract, a pulmonary tractotomy can be performed (see Sect. 8.10, Figs. 8.14 and 8.15).

8.6.2 Pulmonary Contusion

The most common injury after thoracic trauma is pulmonary contusion. It occurs in 30–75 % of all patients [30]. It arises after severe blunt impact with chest wall injury, due to shearing forces in deceleration trauma or after penetrating injury,

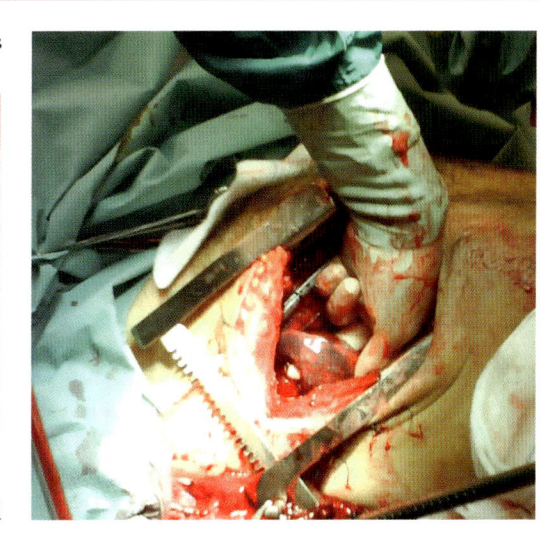

Fig. 8.14 Tractotomy

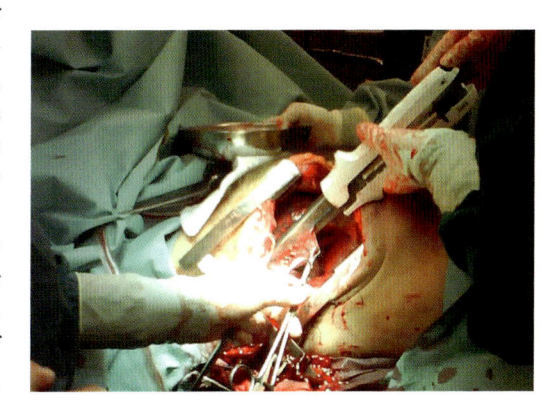

Fig. 8.15 Tractotomy: stapling with stapling device

especially gunshot wound with high energy missiles. In adults, pulmonary contusion is often associated with other injuries whereas in children, as their chest is pliable, it can be found in isolation. The lung is affected due to the direct trauma and damage to the parenchyma causing extravasation of blood and edema in the alveolar space (Figs. 8.4 and 8.5). It occurs mostly on the peripheral lung parenchyma and in contrast to aspiration pneumonia it does not stick to the anatomic pulmonary segments. The patient often present with respiratory distress or failure. On the initial chest X-ray, significant pulmonary contusion will be apparent; however, only approximately half of the abnormalities are detected at the time of admission [60]. Additional CT scans will show the extent of the contusion. Depending on the severity of the pulmonary contusion, resulting in hypoxia and hypercarbia, optimization of oxygenation by ventilatory support may be necessary. After 5–7 days the pathophysiological changes of pulmonary contusion resolve but the recovery of the patient will depend on associated injuries or the appearance of complications.

8.6.3 Pulmonary Herniation

Herniation of the lung through a traumatic event is rare [61]. Patients who are involved in motor vehicle accidents can suffer multiple rib fractures by the compression forces of their seat belt. If those rib fractures originate at the costochondral sternal junction and are dislocated they can cause a ventral chest wall defect. The protrusion of lung parenchyma and pleural membranes through the thoracic cage defect results in pulmonary herniation. As the anterior wall has minimal soft tissue support, this location is more prone to herniation although it may occur even in other areas. It can also be caused by chondrocostal or clavicle sternal dislocation. Patients can be asymptomatic or will have clinical signs like respiratory distress, thoracic ecchymosis caused by the safety belt or subcutaneous emphysema [62]. Furthermore an obvious soft bulging mass, which changes in size with the respiratory cycle, may be present. A CT scan of the chest will reveal the diagnosis but even a standard chest X-ray can show some signs of herniation. The treatment is surgical and may vary from open reduction and internal fixation to using a mesh to close the defect.

8.7 Cardiac Injuries

The exact incidence of cardiac injuries in varying degree of severity is unknown but is reported to be between 16 and 76 % [63]. Blunt cardiac injury (BCI) occurs when the heart is crushed between the sternum and thoracic vertebrae mostly in motor vehicle accidents; this injury can also occur after a fall from height, crush injuries, or even in sports trauma with a direct blow to the chest. These type of injuries can result in myocardial contusion, cardiac rupture, coronary artery injury, or valvular disruption [3].

To diagnose BCI after thoracic trauma is difficult, but to rule out BCI is important especially in patients without further associated injuries who do not require monitoring or even admission to the hospital. The Eastern Association recommends ECG and troponin evaluation and states that a normal ECG with the addition of a normal troponine I has a negative predictive value for BCI of 100 % [64].

8.7.1 Myocardial Contusion

Myocardial contusion is the most common of BCI with an incidence ranging from 3 to 56 % [65]. The patients complain of chest pain; however, as this is a common symptom after chest trauma, the differentiation between a musculoskeletal origin and myocardial contusion or even infarction is challenging. Intramyocardial hemorrhage, edema, and necrosis of muscle cells after myocardial contusion can cause a similar increase in serum troponine, due to loss of membrane integrity, as seen in acute myocardial infarction. ECG changes may show non-specific abnormalities, conduction disorders, or arrhythmias but these changes may be the result of non-cardiac factors like hypoxia or anemia. Over the last decade the CT technology has developed and improved overall sensitivity and specificity. A Multidetector CT with ECG gated capabilities might be able to differ between traumatic or ischemic injury in selected patients [64]. Echocardiographic evaluation, preferably esophageal, can reveal motility and contractility disorders in these cases.

8.7.2 Coronary Artery Injury

As a result of blunt chest trauma, injury to the coronary vessels might appear. This can consist of dissection, intimal tear, thrombus, vessel spasm, vessel rupture, or embolism. Due to its anterior position and the proximity of the chest wall, the coronary artery LAD, which is in the most vulnerable anatomic position, is affected mostly after blunt cardiac trauma. Secondary to this injury myocardial infarction can develop. Therefore in patients complaining of acute chest pain without pre-existent angina pectoris, clinical suspicion must arise. An ECG must be performed to rule out coronary artery injury in an early phase.

Late presentation will consist of single vessel coronary disease in young patients without atherosclerosis disease as cause for angina pectoris or myocardial infarction. Treatment should consist of acute PTCA and stenting as often intimal tears or dissections are present. Thrombolytic therapy is contraindicated as this will worsen the case [66].

8.7.3 Cardiac Rupture

The incidence of cardiac rupture after blunt thoracic trauma is rare and is reported to be 0.16–2 % [67]. Most patients die at the scene in consequence of acute cardiac tamponade. There are several etiologic mechanisms described. Due to compression forces the atria or ventricle, at times of maximal filling status, may tear. Furthermore a rapid deceleration of the heart may cause a rupture at the junction between the atria and the vena cava or pulmonary veins [68]. Patients present with signs of cardiac tamponade and if other causes of hypotension are ruled out (tension pneumothorax, abdominal bleeding) a high index of suspicion must exist. In case of an additional pericardium laceration, massive hemothorax and exsanguination – due to loss of tamponading effect – will result. In certain cases emergency thoracotomy can be lifesaving; however, even in emergency thoracotomy, survival rates are limited.

8.8 Fracture Treatment in Patients with Concomitant Thoracic Trauma

In severely injured patients, damage control surgery is the current standard. This treatment algorithm is primarily developed for patients with massive abdominal hemorrhage. The surgical procedure focuses on bleeding control and limitation of contamination. Only the most necessary procedures are performed and the patient is transported to the intensive care unit as soon as possible. Definitive surgical procedures are postponed until the lethal triad of acidosis, hypothermia, and coagulopathy is corrected [69]. In instable patients, interventions should be rapid and minimally traumatic to the patient. The primary focus is hemorrhage control and other life saving measures. Complex reconstructive work is delayed until the patient is better able to withstand the additional surgical trauma. This approach was readily adopted in patients with pelvic injuries [70].

Damage control surgery is developed to counter the homeostatic complications arising from hypovolemic shock. In addition, severely injured patients suffer immunological disturbances as well [71]. Surgery functions as a second trauma and increases the alterations in the immunological response [71]. This second hit is deemed the underlying mechanism for the development of organ failure, frequently affecting the pulmonary tissue. By minimizing the burden of surgery, an attempt is made to attenuate the inflammatory response and reduce the incidence of organ failure. Damage control orthopedics is used as the current strategy to limit the surgical hit in severely injured patients, in contrast to the early total care principles, in which the patient is treated to the full extent in the first session [72–75].

Currently the two concepts are both used in the clinical setting for the fixation of fractures. "Early total care" (ETC) is used in patients who are deemed stable, while "damage control orthopedics" (DCO) is the treatment of choice in patients who are unstable. Early total care consists of immediate repair by complex operative procedures. In contrast, based on the concept of damage control orthopedics long bone fractures are stabilized by external fixation, which is later converted to intramedullary nailing or plate fixation. Early fracture fixation has been described to be essential to avoid pulmonary complications in multi-trauma patients, such as infection and pulmonary dysfunction [76, 77]. However, ETC gives an increased incidence of pulmonary

Table 8.4 Borderline patients according to Pape et al. [81]

ISS >40
Hypothermia <35 °C
Multiple trauma with ISS >20 and AIS$_{chest}$ >2
Multiple trauma with abdominal/pelvic injury (AIS >2) and shock (RR$_{systol}$ <90 mmHg)
Bilateral lung contusion in chest radiography or CT
Pulmonary artery pressure (PAP) >24 mmHg
Increase of PAP >6 mmHg during femoral nailing

Table 8.5 Features of ARDS [86]

Bilateral pulmonary infiltrates on chest X-ray
PaO$_2$/FiO$_2$ ratio of 200 mmHg or less
Absence of clinical evidence of left atrial hypertension

failure in severely injured patients [78, 79]. In these cases, DCO might be a more suitable approach, keeping in mind the increased percentage of complications at the fracture site, such as delayed union and infectious complications [80]. In conclusion, stable patients can undergo ETC, while unstable patients or patients in extremis undergo DCO. For so-called "borderline" patients however, a clear cut answer is still pending (Table 8.4) [81].

Patients with concomitant thoracic injuries are at increased risk for the development of pulmonary complications when long bone fractures are treated with intramedullary nailing. In postmortem studies large amounts of neutrophils are found in the lungs of patients who die of organ failure. These patients did not have infectious problems in the lungs and thus it is thought that the damage to the pulmonary tissue is caused by the neutrophils [82, 83]. This is supported by the increase in circulating levels of cytokines and activation of circulating neutrophils in patients undergoing intramedullary nailing of a femur fracture [81, 84, 85]. On the contrary, this alteration in inflammatory response is not seen during intramedullary reaming of a tibia fracture.

Patients who are in extremis or are unstable undergo damage control, whether it is thoracic/abdominal surgery or orthopedic surgery. When damage control is applied in the correct "borderline" patients, the incidence of acute lung injury (ALI), acute respiratory distress syndrome (ARDS), sepsis, and multiple organ failure (MOF) is decreased. However, when damage control is applied in patients who would have been stable enough to undergo ETC, more adverse events are seen in the staged approach group [78]. Which patients are at increased risk and who should undergo ETC/DCO is still the subject of research. Several immunological parameters have been discovered to aid the treating physician, and new drugs to modify the inflammatory response are being tested. However, future prospective randomized studies are needed to increase the sensitivity and specificity of parameters to identify those patients who might benefit from DCO concept of fracture care.

8.9 Complications

8.9.1 Acute Respiratory Distress Syndrome

Acute respiratory distress syndrome (ARDS) is a syndrome of inflammation and increased permeability that is associated with clinical, radiological, and physiological abnormalities, which usually develops over 4–48 h and persists for days or weeks (Table 8.5) [86].

The most important risk factors for the development of ARDS are Injury Severity Score (ISS) and the presence of pulmonary contusion [87, 88]. Other risk factors described are transfusion requirement and hypotension on admission. ARDS is associated with complex changes in the lung, manifested by an early exudative phase and followed by proliferative and fibrotic phases. The pathogenesis of ARDS is described in Table 8.6. The treatment of ARDS is supportive care, including optimized mechanical ventilation, nutritional support, manipulation of fluid balance, and prevention of intervening medical complications. All patients with acute respiratory insufficiency require ventilatory support in order to minimize the risks of endobronchial mucus plugging, pneumonia, and atelectasis. The main aim of mechanical ventilation is to maintain adequate oxygenation and ventilation while preventing ventilator-induced lung injury and maintaining adequate tissue perfusion.

Table 8.6 Pathogenesis of acute respiratory distress syndrome [4]

Cellular mechanisms:
Macrophage activation
Neutrophil recruitment and activation
Endothelial injury
Platelet aggregation and degranulation
Plasma protein activation
Alveolar epithelial injury
Tissue responses
Increased pulmonary microvascular permeability
Microvascular thrombosis
Intraalveolar and interstitial edema
Intraalveolar fibrin deposition
Altered pulmonary vasomotor tone
Pathophysiology
Hypoxemia
Decreased pulmonary compliance
Increased shunt fraction
Decreased functional residual capacity
Increased work of breathing

With the use of mechanical ventilation, we know that next to the pulmonary contusion mechanical ventilation can also induce an inflammatory response [4]. Positive end expiratory pressure (PEEP) ventilation maintains PaO_2 above 60 mmHg and is considered to be effective in patients with ARDS. The use of PEEP can partially correct the ventilation-perfusion mismatching in the lung and improve oxygenation. Low tidal volume ventilation reduces mortality compared with high tidal volume ventilation, but can lead to respiratory acidosis. There is no evidence for the use of nitric oxide, corticosteroids, or nursing in a prone position in ARDS [89]. Alternative techniques like permissive hypercapnea, inverse ratio ventilation, and high frequency ventilation are used to protect the lung and prevent more ventilator-induced lung injury. However, early use of mechanical ventilation cannot prevent from developing ARDS.

8.9.2 Chylothorax

Traumatic chylothorax is a rare complication following thoracic trauma and is usually due to penetrating trauma or iatrogenic, secondary to operative procedures. There is a disruption of the thoracic duct. After blunt thoracic trauma a chylothorax is a seldom seen complication. The most common form of blunt injury to the thoracic duct is produced by hyperextension of the spine with rupture of the duct just above the diaphragm in the right thorax. The thoracic duct enters the thorax through the aortic hiatus and travels up just at the right side of the spine. Approximately at the level of the fifth or sixth thoracic vertebra the thoracic duct crosses posterior to the aorta and the aortic arch into the left posterior mediastinum. Therefore a left-sided chylothorax is found in case of ruptures of the upper part of the thoracic duct, whereas right-sided chylothorax is seen in injury of lower levels. The color of the pleural fluid seen is chylous and has a white and milky aspect. However the color of the pleural fluid is not always indicative of a chylothorax. Pleural fluid may not appear chylous if the patient is fasting or the pleural fluid is mixed with blood. An easy test to perform is giving the patient fatty liquid, such as whipped cream, which induces the production of chylous fluids. The diagnosis of chylothorax can be confirmed by the presence of chylomicrons in the pleural fluid. Other characteristics of the fluid are pH 7.4–7.8, lymphocyte predominance in cell count of a specific gravity of 1.012 or higher [90].

A chylothorax may also be suspected if the pleural fluid to serum triglyceride ratio is more than 1 and a pleural fluid to serum cholesterol ratio is less than 1 [91]. The treatment for chylothorax usually starts with total peripheral nutrition or medium chain triglycerides instead of a normal diet combined with a chest tube. Conservative treatment had a success rate up to 88 % [92]. Surgical intervention gives better results than conservative management when the daily production exceeds 1 l for a period more than 5 days [93].

This can be done by an open thoracotomy or video-assisted thoracoscopic surgery (VATS) to ligate or clip the thoracic duct. When the thoracic duct cannot be found, pleurectomy can be done. Newer techniques described are percutaneous CT guided drainage, percutaneous embolization, and robotic surgery [94, 95].

8.9.3 Pleural Empyema

An empyema has been defined as a loculated collection of pus within the pleural cavity. Common etiologies are post-pneumonic, post-resection, and post-traumatic. Untreated post-traumatic empyema results in a restrictive ventilator deficit and atelectasis. Several factors may contribute to a potentially higher risk for empyema in the trauma population. Potential causes for post-traumatic empyema include iatrogenic infection of the pleural space during chest tube placement, direct infection resulting from penetrating injuries of the thoracic cavity, secondary infection of the pleural cavity from associated intra-abdominal organ injuries with diaphragmatic disruption, secondary infection of undrained or inadequately drained hemathorax, hemotogenous or lymphatic spread of subdiaphragmatic empyema resulting from post-traumatic pneumonia, pulmonary contusion or acute respiratory distress syndrome. Almost 27 % of the patients with a retained hemathorax will develop an empyema [96].

Management of thoracic empyema includes decortication by a thoracotomy. Thoracoscopy seems to be an effective method also in selected patients if performed early. Several studies are performed to prove the evidence of the use of prophylactic antibiotics in chest tube placement. Patients who require a chest tube after penetrating injury might benefit from prophylactic antibiotics. Further research is needed to prove the benefit of the use of antibiotics in blunt thoracic injury [26].

8.9.4 Persistent Air Leakage

If the air leakage is large or persistent without re-expansion of the lung a tracheobronchial injury or deep parenchymal injury should be suspected.

A persistent pneumothorax is arbitrarily defined as failure to seal an air leak and achieve full lung expansion within 72 h of chest tube placement. We suggest performing a VATS (video assisted thoracic surgery) when a persistent pneumothorax is present after 72 h, as the cause of this persistent air leak is often deep parenchymal injury [97]. Another temporary solution in the acute setting is to ventilate both lungs separately or making the tube pass the defect by pushing it little deeper.

8.10 Operative Techniques

For the hemodynamically unstable patient a left anterolateral thoracotomy in the 4th intercostal space is the first approach to be applied in supine position (Fig. 8.16). With this approach you still can extend to the other side crossing the sternum, and the abdomen is also reachable without repositioning the patient. To make is easier you can shove a rolled sheet behind the scapula in order to lift and medially rotate the patient a little. In just three strokes using knife, skin and subcutaneous tissue, pectoralis and serratus muscle, intercostal muscle, and thoracic cavity can be reached. The posterior mediastinum cannot be reached with this approach. A double-lumen endotracheal tube placed by an experienced anesthesiologist gives an advantage; however, it can be time consuming with an inexperienced anesthesiologist. In that case a blocker can give the solution after a rapid single lumen intubation.

The anterolateral approach can easily be extended towards the other side by going through the sternum by using a Gigli saw or just big scissors.

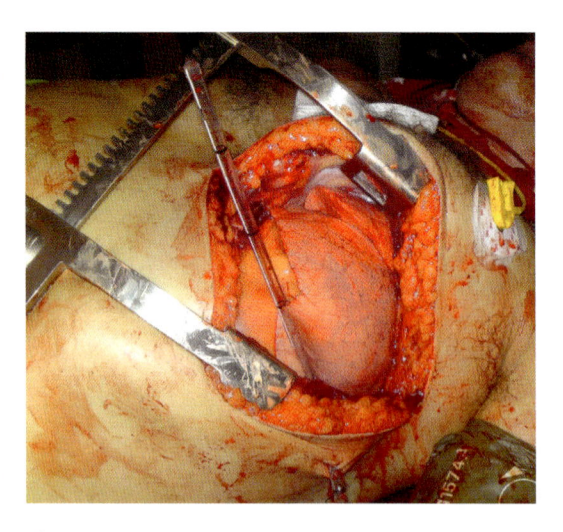

Fig. 8.16 Anterolateral thoracotomy

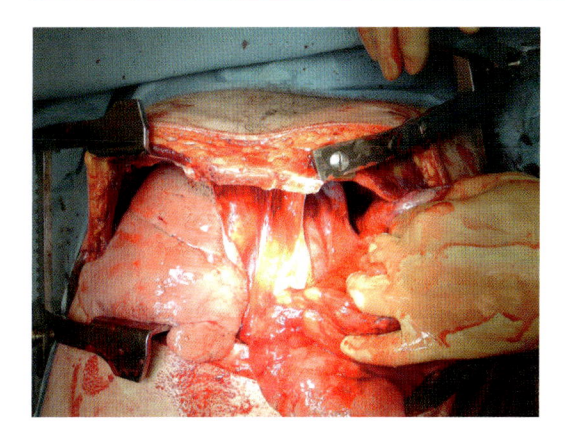

Fig. 8.17 Clamshell thoracotomy

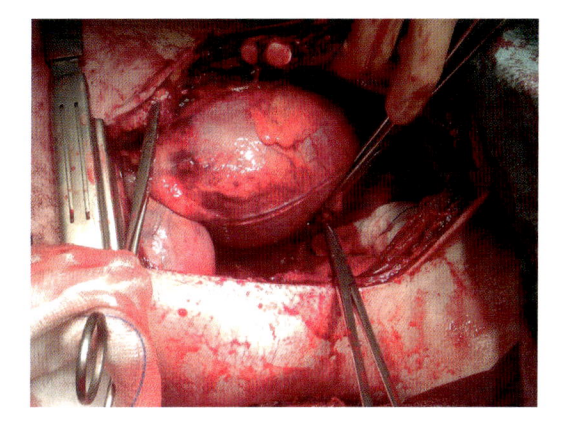

Fig. 8.18 Hilar cross clamping

The bilateral anterolateral approach combined with a transverse sternotomy results in the "clamshell" incision, the largest incision commonly used in thoracic surgery (Fig. 8.17). By crossing the sternum, always look for and take care of the internal mammary artery, which can cause severe bleeding once the patients is not instable anymore.

Once you are inside the chest, first cut the inferior pulmonary ligament so the lung can be manipulated more easily. In case of a massive bleeding from a central lung injury try to stop the bleeding by manual pressure first. If this is not effective enough than clamp the pulmonary hilum (Fig. 8.18). Realize that it might be tricky because you often cannot see what you are doing within the restricted workspace provided by an anterolateral thoracotomy.

It might be clear that there are more operative options in peripheral pulmonary lesions than in central injuries close to the hilum. The tractotomy is a very useful technique for fixing through-and-through lung injuries. You can either remove injured lung when it is peripherally located or you can perform tractotomy which is lung sparing. With a staple device the injured tract inside the lung can be opened and connected with the lung surface and bleeding vessels ligated (Figs. 8.14 and 8.15).

An anatomic resection is seldom necessary. A median sternotomy is a good approach for pericardial penetrating wounds or wound close to the sternum. The internal mammary artery, the heart, and even both pulmonary hilar structures can be reached. The peripheral pulmonary structures and the posterior mediastinum cannot be accessed. It should be realized that this approach is of limited use in patients with a massive hemothorax where the bleeding structure is still not known. Choosing the wrong incision can give a lot of trouble.

References

1. LoCicero 3rd J, Mattox KL. Epidemiology of chest trauma. Surg Clin North Am. 1989;69(1):15–9.
2. Lansink KWW, Gunning AC, Leenen LPH. Cause of death and time of death distribution of trauma patients in a level I trauma centre in the Netherlands. Eur J Trauma Emerg Surg. 2013;39:375–83.
3. American College of Surgeons Committee on trauma. Advanced trauma life support for doctors, student course manual. 8th ed. Chicago: American College of Surgeons Committee on Trauma; 2008.
4. van Wessem KJ, Hennus MP, van Wagenberg L, Koenderman L, Leenen LP. Mechanical ventilation increases the inflammatory response induced by lung contusion. J Surg Res. 2013;183(1):377–84.
5. Eren S, Balci AE, Ulku R, Cakir O, Eren MN. Thoracic firearm injuries in children: management and analysis of prognostic factors. Eur J Cardiothorac Surg. 2003;23(6):888–93.
6. Oestern HJ, Kabus K. The classification of the severely and multiply injured – what has been established? Chirurg. 1997;68(11):1059–65.
7. Mommsen P, Zeckey C, Andruszkow H, Weidemann J, Frömke C, Puljic P, van Griensven M, Frink M, Krettek C, Hildebrand F. Comparison of different thoracic trauma scoring systems in regards to prediction

of post-traumatic complications and outcome in blunt chest trauma. J Surg Res. 2012;176(1):239–47.
8. Pape HC, Remmers D, Rice J, Ebisch M, Krettek C, Tscherne H. Appraisal of early evaluation of blunt chest trauma: development of a standardized scoring system for initial clinical decision making. J Trauma. 2000;49(3):496–504.
9. Berg EE, Chebuhar C, Bell RM. Pelvic trauma imaging: a blinded comparison of computed tomography and roentgenograms. J Trauma. 1996;41:994–8.
10. Costello P, Dupuy D, Ecker C, TEllo R. Spiral CT of the thorax with small volumes of contrast material: a comparative study. Radiology. 1992;183:663–6.
11. Traub M, Stevenson M, McEvoy S, Briggs G, Lo SK, Leibman S, Joseph T. The use of chest computed tomography versus chest X-ray in patients with major blunt trauma. Injury. 2007;38(1):43–7.
12. Cook AD, Klein JS, Rogers FB, Osler TM, Shackford SR. Chest radiographs of limited utility in the diagnosis of blunt traumatic aortic laceration. J Trauma. 2001;50(5):843–7.
13. Trupka A, Waydhas C, Hallfeldt KK, Nast-Kolb D, Pfeifer KJ, Schweiberer L. Value of thoracic computed tomography in the first assessment of severely injured patients with blunt chest trauma: results of a prospective study. J Trauma. 1997;43(3):405–11; discussion 411–2.
14. Kea B, Gamarallage R, Vairamuthu H, Fortman J, Lunney K, Hendey GW, Rodriguez RM. What is the clinical significance of chest CT when the chest x-ray result is normal in patients with blunt trauma? Am J Emerg Med. 2013;31(8):1268–73.
15. Rodriguez RM, Anglin D, Langdorf MI, Baumann BM, Hendey GW, Bradley RN, Medak AJ, Raja AS, Juhn P, Fortman J, Mulkerin W, Mower WR. NEXUS chest: validation of a decision instrument for selective chest imaging in blunt trauma. JAMA Surg. 2013;148(10):940–6.
16. Hyacinthe AC, Broux C, Francony G, Genty C, Bouzat P, Jacquot C, Albaladejo P, Ferretti GR, Bosson JL, Payen JF. Diagnostic accuracy of ultrasonography in the acute assessment of common thoracic lesions after trauma. Chest. 2012;141(5):1177–83.
17. Wilkerson RG, Stone MB. Sensitivity of bedside ultrasound and supine anteroposterior chest radiographs for the identification of pneumothorax after blunt trauma. Acad Emerg Med. 2010;17(1):11–7.
18. Dente CJ, Ustin J, Feliciano DV, Rozycki GS, Wyrzykowski AD, Nicholas JM, Salomone JP, Ingram WL. The accuracy of thoracic ultrasound for detection of pneumothorax is not sustained over time: a preliminary study. J Trauma. 2007;62(6):1384–9.
19. Restrepo CS, Gutierrez FR, Marmol-Velez JA, Ocazionez D, Martinez-Jimenez S. Imaging patients with cardiac trauma. Radiographics. 2012;32(3): 633–49.
20. Karalis DG, Victor MF, Davis GA, McAllister MP, Covalesky VA, Ross Jr JJ, Foley RV, Kerstein MD, Chandrasekaran K. The role of echocardiography in blunt chest trauma: a transthoracic and transesophageal echocardiographic study. J Trauma. 1994;36(1):53–8.
21. Chirillo F, Totis O, Cavarzerani A, Bruni A, Farnia A, Sarpellon M, Ius P, Valfrè C, Stritoni P. Usefulness of transthoracic and transoesophageal echocardiography in recognition and management of cardiovascular injuries after blunt chest trauma. Heart. 1996;75(3): 301–6.
22. Martin M, Satterly S, Inaba K, Blair K. Does needle thoracostomy provide adequate and effective decompression of tension pneumothorax? J Trauma Acute Care Surg. 2012;73(6):1412–7.
23. Harcke HT, Mabry RL, Mazuchowski EL. Needle thoracentesis decompression: observations from postmortem computed tomography and autopsy. J Spec Oper Med. 2013;13(4):53–8.
24. Fitzgerald M, Mackenzie CF, Marasco S, Hoyle R, Kossmann T. Pleural decompression and drainage during trauma reception and resuscitation. Injury. 2008;39(1):9–20.
25. Inaba K, Lustenberger T, Recinos G, Georgiou C, Velmahos GC, Brown C, Salim A, Demetriades D, Rhee P. Does size matter? A prospective analysis of 28–32 versus 36–40 French chest tube size in trauma. J Trauma Acute Care Surg. 2012;72(2):422–7.
26. Bosman A, de Jong MB, Debeij J, van den Broek PJ, Schipper IB. Systematic review and meta-analysis of antibiotic prophylaxis to prevent infections from chest drains in blunt and penetrating thoracic injuries. Br J Surg. 2012;99(4):506–13.
27. Butler FK, Dubose JJ, Otten EJ, Bennett DR, Gerhardt RT, Kheirabadi BS, Gross KR, Cap AP, Littlejohn LF, Edgar EP, Shackelford SA, Blackbourne LH, Kotwal RS, Holcomb JB, Bailey JA. Management of open pneumothorax in tactical combat casualty care: TCCC guidelines change 13–02. J Spec Oper Med. 2013;13(3):81–6.
28. Mattox KL, Allen MK. Systematic approach to pneumothorax, haemothorax, pneumomediastinum and subcutaneous emphysema. Injury. 1986;17(5): 309–12.
29. Mowery NT, Gunter OL, Collier BR, Diaz Jr JJ, Haut E, Hildreth A, Holevar M, Mayberry J, Streib E. Practice management guidelines for management of hemothorax and occult pneumothorax. J Trauma. 2011;70(2):510–8.
30. Simon B, Ebert J, Bokhari F, Capella J, Emhoff T, Hayward 3rd T, Rodriguez A, Smith L, Eastern Association for the Surgery of Trauma. Management of pulmonary contusion and flail chest: an Eastern Association for the Surgery of Trauma practice management guideline. J Trauma Acute Care Surg. 2012;73(5 Suppl 4):S351–61.
31. Waydhas C. Thoracic trauma. Unfallchirurg. 2000;103(10):871–89; quiz 890, 910.
32. Lee TH, Ouellet JF, Cook M, Schreiber MA, Kortbeek JB. Pericardiocentesis in trauma: a systematic review. J Trauma Acute Care Surg. 2013;75(4):543–9.
33. Aukema TS, Beenen LF, Hietbrink F, Leenen LP. Initial assessment of chest X-ray in thoracic trauma

patients: awareness of specific injuries. World J Radiol. 2012;4(2):48–52.
34. Oveland NP, Lossius HM, Wemmelund K, Stokkeland PJ, Knudsen L, Sloth E. Using thoracic ultrasonography to accurately assess pneumothorax progression during positive pressure ventilation: a comparison with CT scanning. Chest. 2013;143(2):415–22.
35. Wilson H, Ellsmere J, Tallon J, Kirkpatrick A. Occult pneumothorax in the blunt trauma patient: tube thoracostomy or observation? Injury. 2009;40(9):928–31.
36. Moore FO, Goslar PW, Coimbra R, Velmahos G, Brown CV, Coopwood Jr TB, Lottenberg L, Phelan HA, Bruns BR, Sherck JP, Norwood SH, Barnes SL, Matthews MR, Hoff WS, de Moya MA, Bansal V, Hu CK, Karmy-Jones RC, Vinces F, Pembaur K, Notrica DM, Haan JM. Blunt traumatic occult pneumothorax: is observation safe? – results of a prospective, AAST multicenter study. J Trauma. 2011;70(5):1019–23; discussion 1023–5.
37. Kirkpatrick AW, Rizoli S, Ouellet JF, Roberts DJ, Sirois M, Ball CG, Xiao ZJ, Tiruta C, Meade M, Trottier V, Zhu G, Chagnon F, Tien H, Canadian Trauma Trials Collaborative and the Research Committee of the Trauma Association of Canada. Occult pneumothoraces in critical care: a prospective multicenter randomized controlled trial of pleural drainage for mechanically ventilated trauma patients with occult pneumothoraces. J Trauma Acute Care Surg. 2013;74(3):747–54; discussion 754–5.
38. Kulvatunyou N, Erickson L, Vijayasekaran A, Gries L, Joseph B, Friese RF, O'Keeffe T, Tang AL, Wynne JL, Rhee P. Randomized clinical trial of pigtail catheter versus chest tube in injured patients with uncomplicated traumatic pneumothorax. Br J Surg. 2014;101(2):17–22.
39. Trupka A, Nast-Kolb D, Schweiberer L. Thoracic trauma. Unfallchirurg. 1998;101(4):244–58.
40. Oğuzkaya F, Akçali Y, Bilgin M. Videothoracoscopy versus intrapleural streptokinase for management of post traumatic retained haemothorax: a retrospective study of 65 cases. Injury. 2005;36(4):526–90.
41. Demetriades D, Velmahos GC, Scalea TM, Jurkovich GJ, Karmy-Jones R, Teixeira PG, Hemmila MR, O'Connor JV, McKenney MO, Moore FO, London J, Singh MJ, Lineen E, Spaniolas K, Keel M, Sugrue M, Wahl WL, Hill J, Wall MJ, Moore EE, Margulies D, Malka V, Chan LS, American Association for the Surgery of Trauma Thoracic Aortic Injury Study Group. Operative repair or endovascular stent graft in blunt traumatic thoracic aortic injuries: results of an American Association for the Surgery of Trauma Multicenter Study. J Trauma. 2008;64(3):561–70.
42. Hershberger RC, Aulivola B, Murphy M, Luchette FA. Endovascular grafts for treatment of traumatic injury to the aortic arch and great vessels. J Trauma. 2009;67(3):660–71.
43. Smakman N, Nicol AJ, Walther G, Brooks A, Navsaria PH, Zellweger R. Factors affecting outcome in penetrating oesophageal trauma. Br J Surg. 2004;91(11): 1513–9.
44. Sepesi B, Raymond DP, Peters JH. Esophageal perforation: surgical, endoscopic and medical management strategies. Curr Opin Gastroenterol. 2010;26(4): 379–83.
45. Simali M, Türüt H, Topçu S, Gülhan E, Yazici Ü, Kaya S, Taştepe I. A comprehensive analysis of traumatic rib fractures: morbidity, mortality and management. Eur J Cardiothorac Surg. 2003;24:133–8.
46. Bergeron E, Lavoie A, Clas D, Moore L, Ratte S, Tetreault S, Lemaire J, Martin M. Elderly trauma patients with rib fractures are at greater risk of death and pneumonia. J Trauma. 2003;54(3):478–85.
47. Bulger EM, Arneson MA, Mock CN, Jurkovich GJ. Rib fractures in the elderly. J Trauma. 2000;48(6):1040–6; discussion 1046–7.
48. Flagel BT, Luchette FA, Reed RL, et al. Half-a-dozen ribs: the breakpoint for mortality. Surgery. 2005;138: 717–25.
49. Holcomb JB, McMullin NR, Kozar RA, Lygas MH, Moore FA. Morbidity from rib fractures increases after age 45. J Am Coll Surg. 2003;196(4):549–55.
50. Todd SR, McNally MM, Holcomb JB, et al. A multidisciplinary clinical pathway decreases rib fracture-associated infectious morbidity and mortality in high-risk trauma patients. Am J Surg. 2006;192:806–11.
51. Kent R, Woods W, Bostrom O. Fatality risk and the presence of rib fractures. Ann Adv Automot Med. 2008;52:73–82.
52. Tanaka H, Yukioka T, Yamaguti Y, Shimizu S, Goto H, Matsuda H, Shimazaki S. Surgical stabilization of internal pneumatic stabilization? A prospective randomized study of management of severe flail chest patients. J Trauma. 2002;52(4):727–32.
53. Nirula R, Mayberry JC. Rib fracture fixation: controversies and technical challenges. Am Surg. 2010; 76(8):793–802.
54. Leinicke JA, Elmore L, Freeman BD, Colditz GA. Operative management of rib fractures in the setting of flail chest. A systematic review and meta-analysis. Ann Surg. 2013;258(6):914–21.
55. Brookes JG, Dunn RJ, Rogers IR. Sternal fractures: a retrospective analysis of 272 cases. J Trauma. 1993;35:46.
56. Budd JS. Effect of seat belt legislation on the incidence of sternal fractures seen in the accident department. Br Med J. 1985;291:785.
57. Sadaba JR, Oswal D, Munsch CM. Management of isolated sternal fractures: determining the risk of blunt cardiac injury. Ann R Coll Surg Engl. 2000;82(3):162–6.
58. Sangster GP, González-Beicos A, Carbo AI, Heldmann MG, Ibrahim H, Carras-cosa P, Nazar M, D'Agostino HB. Blunt traumatic injuries of the lung parenchyma, pleura, thoracic wall, and intrathoracic airways: multidetector computer tomography imaging findings. Emerg Radiol. 2007;14(5):297–310.
59. Wagner RB, Crawford Jr WO, Schimpf PP. Classification of parenchymal injuries of the lung. Radiology. 1988;167(1):77–82.
60. Cohn SM, Dubose JJ. Pulmonary contusion: an update on recent advances in clinical management. World J Surg. 2010;34(8):1959–70.

61. François B, Desachy A, Cornu E, Ostyn E, Niquet L, Vignon P. Traumatic pulmonary hernia: surgical versus conservative management. J Trauma. 1998;44(1):217–9.
62. Arslanian A, Oliaro A, Donati G, Filosso PL. Posttraumatic pulmonary hernia. J Thorac Cardiovasc Surg. 2001;122(3):619–21.
63. Prêtre R, Chilcott M. Blunt trauma to the heart and great vessels. N Engl J Med. 1997;336(9):626–32.
64. Clancy K, Velopulos C, Bilaniuk JW, Collier B, Crowley W, Kurek S, Lui F, Nayduch D, Sangosanya A, Tucker B, Haut ER, Eastern Association for the Surgery of Trauma. Screening for blunt cardiac injury: an Eastern Association for the Surgery of Trauma practice management guideline. J Trauma Acute Care Surg. 2012;73(5 Suppl 4):S301–6.
65. Sybrandy KC, Cramer MJ, Burgersdijk C. Diagnosing cardiac contusion: old wisdom and new insights. Heart. 2003;89(5):485–9.
66. Christensen MD, Nielsen PE, Sleight P. Prior blunt chest trauma may be a cause of single vessel coronary disease; hypothesis and review. Int J Cardiol. 2006;108(1):1–5.
67. Nan YY, Lu MS, Liu KS, Huang YK, Tsai FC, Chu JJ, Lin PJ. Blunt traumatic cardiac rupture: therapeutic options and outcomes. Injury. 2009;40(9):938–45.
68. Brathwaite CE, Rodriguez A, Turney SZ, Dunham CM, Cowley R. Blunt traumatic cardiac rupture. A 5-year experience. Ann Surg. 1990;212(6):701–4.
69. Rotondo MF, Schwab CW, McGonigal MD, Phillips 3rd GR, Fruchterman TM, Kauder DR, Latenser BA, Angood PA. 'Damage control': an approach for improved survival in exsanguinating penetrating abdominal injury. J Trauma. 1993;35(3):375–82.
70. Giannoudis PV, Pape HC. Damage control orthopaedics in unstable pelvic ring injuries. Injury. 2004;35(7):671–7.
71. Pape HC, van Griensven M, Rice J, Gänsslen A, Hildebrand F, Zech S, Winny M, Lichtinghagen R, Krettek C. Major secondary surgery in blunt trauma patients and perioperative cytokine liberation: determination of the clinical relevance of biochemical markers. J Trauma. 2001;50(6):989–1000.
72. Goris RJ, Gimbrère JS, van Niekerk JL, Schoots FJ, Booy LH. Improved survival of multiply injured patients by early internal fixation and prophylactic mechanical ventilation. Injury. 1982;14(1):39–43.
73. Bone LB, Johnson KD, Weigelt J, Scheinberg R. Early versus delayed stabilization of femoral fractures. A prospective randomized study. J Bone Joint Surg Am. 1989;71(3):336–40.
74. Pape HC, Auf'm'Kolk M, Paffrath T, Regel G, Sturm JA, Tscherne H. Primary intramedullary femur fixation in multiple trauma patients with associated lung contusion – a cause of posttraumatic ARDS? J Trauma. 1993;34(4):540–7.
75. Hildebrand F, Giannoudis P, Kretteck C, Pape HC. Damage control: extremities. Injury. 2004;35(7):678–89.
76. Gray AC, White TO, Clutton E, Christie J, Hawes BD, Robinson CM. The stress response to bilateral femoral fractures: a comparison of primary intramedullary nailing and external fixation. J Orthop Trauma. 2009;23(2):90–7.
77. Robinson CM, Ludlam CA, Ray DC, Swann DG, Christie J. The coagulative and cardiorespiratory responses to reamed intramedullary nailing of isolated fractures. J Bone Joint Surg Br. 2001;83(7):963–73.
78. Pape HC, Rixen D, Morley J, Husebye EE, Mueller M, Dumont C, Gruner A, Oestern HJ, Bayeff-Filoff M, Garving C, Pardini D, van Griensven M, Krettek C, Giannoudis P, EPOFF Study Group. Impact of the method of initial stabilization for femoral shaft fractures in patients with multiple injuries at risk for complications (borderline patients). Ann Surg. 2007;246(3):491–9.
79. Taeger G, Ruchholtz S, Waydhas C, Lewan U, Schmidt B, Nast-Kolb D. Damage control orthopedics in patients with multiple injuries is effective, time saving, and safe. J Trauma. 2005;59(2):409–16.
80. Nowotarski PJ, Turen CH, Brumback RJ, Scarboro JM. Conversion of external fixation to intramedullary nailing for fractures of the shaft of the femur in multiply injured patients. J Bone Joint Surg Am. 2000;82(6):781–8.
81. Pape HC, Grimme K, Van Griensven M, Sott AH, Giannoudis P, Morley J, Roise O, Ellingsen E, Hildebrand F, Wiese B, Krettek C, EPOFF Study Group. Impact of intramedullary instrumentation versus damage control for femoral fractures on immunoinflammatory parameters: prospective randomized analysis by the EPOFF Study Group. J Trauma. 2003;55(1):7–13.
82. Hietbrink F, Koenderman L, Rijkers G, Leenen L. Trauma: the role of the innate immune system. World J Emerg Surg. 2006;1:15.
83. Keel M, Trentz O. Pathophysiology of polytrauma. Injury. 2005;36(6):691–709.
84. Pape HC, Bund M, Meier R, Piepenbrock S, von Glinski S, Tscherne H. Pulmonary dysfunction following primary bilateral femoral nailing – a case report. Intensive Care Med. 1999;25(5):547.
85. Hietbrink F, Koenderman L, Leenen LP. Intramedullary nailing of the femur and the systemic activation of monocytes and neutrophils. World J Emerg Surg. 2011;6:34. doi:10.1186/1749-7922-6-34.
86. Bernard GR, Artigas A, Brigham KL. The American-European Consensus Conference on ARDS: definition, mechanisms, relevant outcomes and clinical trial coordination. Am J Respir Crit Care Med. 1994;149:818–24.
87. Miller PR, Croce MA, Kilgo PD, Scott J, Fabian TC. Acute respiratory distress syndrome in blunt trauma: identification of independent risk factors. Am Surg. 2002;68(10):845–50.
88. Becher RD, Colonna AL, Enniss TM, Weaver AA, Crane DK, Martin RS, Mowery NT, Miller PR, Stitzel JD, Hoth JJ. An innovative approach to predict the

development of adult respiratory distress syndrome in patients with blunt trauma. J Trauma Acute Care Surg. 2012;73(5):1229–35.
89. Sharma S. Acute respiratory distress syndrome. Clin Evid (Online). 2010;2010. pii:1511.
90. Pai GP, Bhatti NA, Ellison RG, Rubin JW, Moore HV. Thoracic duct injury from blunt trauma. South Med J. 1984;77(5):667–8.
91. Romero S. Nontraumatic chylothorax. Curr Opin Pulm Med. 2000;6(4):287–91.
92. McCormick 3rd J, Henderson SO. Blunt trauma-induced bilateral chylothorax. Am J Emerg Med. 1999;17(3):302–4.
93. Dugue L, Sauvanet A, Farges O, Goharin A, Le Mee J, Belghiti J. Output of chyle as an indicator of treatment for chylothorax complicating oesophagectomy. Br J Surg. 1998;85(8):1147–9.
94. Seitelman E, Arellano JJ, Takabe K, Barrett L, Faust G, Angus LD. Chylothorax after blunt trauma. J Thorac Dis. 2012;4(3):327–30.
95. Itkin M, Kucharczuk JC, Kwak A, Trerotola SO, Kaiser LR. Nonoperative thoracic duct embolization for traumatic thoracic duct leak: experience in 109 patients. J Thorac Cardiovasc Surg. 2010;139(3): 584–9.
96. DuBose J, Inaba K, Demetriades D, Scalea TM, O'Connor J, Menaker J, Morales C, Konstantinidis A, Shiflett A, Copwood B, AAST Retained Hemothorax Study Group. Management of post-traumatic retained hemothorax: a prospective, observational, multicenter AAST study. J Trauma Acute Care Surg. 2012;72(1):11–22; discussion 22–4; quiz 316.
97. Ahmed N, Jones D. Video-assisted thoracic surgery: state of the art in trauma care. Injury. 2004;35(5):479–89. Review. Billeter AT, Druen D, Franklin GA, Smith JW, Wrightson W, Richardson JD. Video-assisted thoracoscopy as an important tool for trauma surgeons: a systematic review. Langenbecks Arch Surg. 2013; 398(4):515–23.

Abdominal Injuries: Indications for Surgery

Clay Cothren Burlew and Ernest E. Moore

Contents

9.1 Initial Evaluation of the Injured Patient 111
9.1.1 Primary Survey 111
9.1.2 Secondary Survey 112

9.2 **Imaging for Abdominal Injuries** 113

9.3 **Penetrating Injuries** 115

9.4 **Blunt Abdominal Trauma** 118
9.4.1 Liver and Spleen 118
9.4.2 Pancreatic Injuries 120
9.4.3 Bowel Injuries 121
9.4.4 Genitourinary 121

9.5 **Postinjury Complications Requiring Abdominal Exploration** 122

9.6 **Collaboration in the Multiply Injured Patient** 123

References 124

C.C. Burlew, MD, FACS (✉)
E.E. Moore, MD, FACS, MCCM
Department of Surgery, Denver Health
Medical Center and the University
of Colorado Denver School of Medicine,
777 Bannock Street, MC 0206, Denver,
CO 80204, USA
e-mail: clay.cothren@dhha.org

9.1 Initial Evaluation of the Injured Patient

9.1.1 Primary Survey

The initial management of seriously injured patients consists of the primary survey, concurrent resuscitation, the secondary survey, diagnostic evaluation, and definitive care as promulgated by the Advanced Trauma Life Support (ATLS) course of the American College of Surgeons Committee on Trauma [1]. The first step in patient management in the emergency department (ED) is the "ABCs" (*A*irway with cervical spine protection, *B*reathing, and *C*irculation) of the primary survey, and evaluating the patient's response to resuscitation. At this point in the patient's evaluation, any episode of hypotension (defined as a SBP less than 90 mmHg) is assumed to be caused by hemorrhage until proven otherwise. Blood pressure and pulse should be measured manually at least every 5 min in patients with significant blood loss until normal vital signs are restored.

Patients with hemodynamic instability and either a penetrating abdominal wound or intraabdominal hemorrhage noted on FAST exam should undergo urgent laparotomy. In patients with gunshot wounds to the chest or abdomen, a chest and abdominal film, with radiopaque markers at the wound sites, should be obtained to determine trajectory of the bullet or location of a retained fragment. For example, a patient with a gunshot wound to the upper abdomen should have a chest radiograph to ensure the bullet did not traverse the diaphragm causing intrathoracic injury. If a patient has a penetrating weapon remaining in place, this should *not* be removed in the ED as it could be tamponading a lacerated blood vessel. The surgeon should extract the offending instrument in the controlled environment of the operating room, ideally once an incision has been made with adequate exposure.

In a patient without either of these clear operative indications who has persistent hypotension, one should systematically evaluate the five potential sources of blood loss: scalp, chest, abdomen, pelvis, and extremities. Thoracoabdominal trauma should be evaluated with a combination of chest radiograph, FAST, and pelvic radiograph. If the FAST is negative and no other source of hypotension is obvious, DPA should be entertained [2]. This is a diagnostic measure that can easily be performed in the trauma bay. Patients with high energy mechanisms should have their pelvis wrapped with a sheet until radiography can be done. Transport of a hypotensive patient out of the ED for computed tomographic (CT) scan evaluation may be hazardous; monitoring is compromised and the environment is suboptimal to deal with acute problems. If the DPA is positive, with greater than 10 cc of frank blood aspirated, the patient should be emergently transported to the operating room for laparotomy.

9.1.2 Secondary Survey

For hemodynamically stable trauma patients, further evaluation for abdominal injuries is systematically performed. Abdominal examination includes inspection for abdominal wall abrasions or ecchymosis, and assessing for distension, rigidity, tenderness, or rebound. Drugs, alcohol, and head and spinal cord injuries, however, can render the physical examination unreliable. Patients with evidence of peritonitis on examination should undergo laparotomy. Digital rectal examination is performed to evaluate for sphincter tone, presence of blood, rectal perforation, or a high-riding prostate; this is particularly critical in patients with a suspected pelvic fracture or a transpelvic gunshot wound. Vaginal examination with a speculum should also be done in women with pelvic fractures to exclude a laceration, which would classify the fracture as an open fracture.

Adjuncts to the physical exam include vital sign and ECG monitoring, repeat FAST, laboratory measurements, and radiographs. A nasogastric tube (NGT) should be inserted in all intubated patients to decrease the risk of gastric aspiration, but may not be indicated in the awake patient. NGT evaluation of stomach contents for blood may suggest occult gastroduodenal injury or the course of the NGT on chest film may suggest a diaphragm injury. A Foley catheter should be inserted in patients unable to void to decompress the bladder, obtain a urine specimen, and monitor urine output. Gross hematuria demands evaluation of the genitourinary system for injury. Foley catheter placement should be deferred until urologic evaluation in patients with signs of urethral injury: blood at the meatus, perineal or scrotal hematomas, or a high-riding prostate.

Blunt abdominal trauma is initially evaluated by FAST exam in most major trauma centers, and this has largely supplanted DPL (Fig. 9.1) [3]. The advantage of FAST is that it is noninvasive, portable, rapid, and easily repeatable over the patient's ED course. Any of the three standard views of the abdomen – right and left upper quadrants, and pelvis – can reveal intraabdominal fluid, which is presumed to be hemorrhage unless the patient has liver disease with known ascites (Fig. 9.2). Although ultrasound is sensitive for detecting intraperitoneal fluid greater than 400 mL, it does not reliably determine the source of hemorrhage or extent of solid organ injuries, and does have limitations in some settings such as subcutaneous emphysema, morbid obesity, and retroperitoneal hemorrhage from pelvic fractures [4–6]. As noted above, FAST is not 100 % sensitive; therefore, DPA is warranted in hemodynamically unstable patients without a defined source of blood loss to rule out massive hemoperitoneum. The FAST exam should be repeated to verify the patient has not developed hemoperitoneum. Occasionally, the first ultrasound views

9 Abdominal Injuries: Indications for Surgery

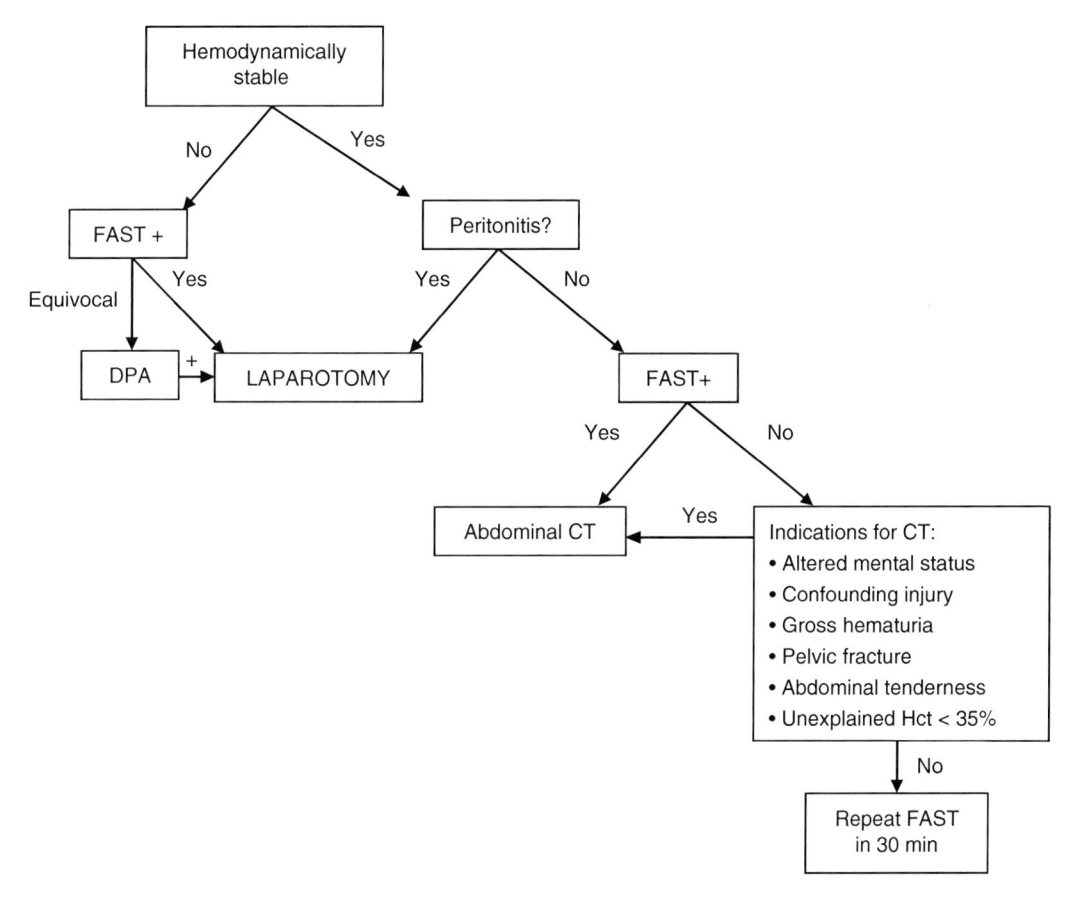

Fig. 9.1 Algorithm for the initial evaluation of the patient with suspected blunt abdominal trauma (*FAST* focused abdominal sonography for trauma, *DPA* diagnostic peritoneal aspirate, *CT* computed tomography, *Hct* hematocrit)

of the abdomen may be normal; repeating the ultrasound as a second snapshot in time is critical for patients at high risk for injury. Patients with fluid on FAST exam, considered a "positive FAST," who do not have immediate indications for laparotomy and are hemodynamically stable, undergo CT scan to quantify their injuries.

9.2 Imaging for Abdominal Injuries

Based upon mechanism, location of injuries identified on physical examination, screening radiographs, and the patient's overall condition, additional diagnostic studies may be indicated. Selective radiographs are done early in the patient's ED evaluation. For patients with severe blunt trauma, chest and pelvic radiographs should be obtained. Historically, a lateral cervical spine radiograph was also obtained, hence the reference to *the big three* films, but currently patients preferentially undergo CT scanning of the spine rather than plain film radiography. Since its initial use in the early 1980s, CT scanning has become a routine part of the trauma evaluation for abdominal injury. With multi-slice helical scanning, the entire torso can be scanned in under 5 min. Patients with a positive FAST, who do not have immediate indications for laparotomy and are hemodynamically stable, undergo CT scan to quantify their injuries. Additionally, patients with persistent abdominal tenderness, significant abdominal wall trauma, distracting injuries, or altered mental status should undergo CT imaging. Although the majority of abdominal penetrating injuries that violate the

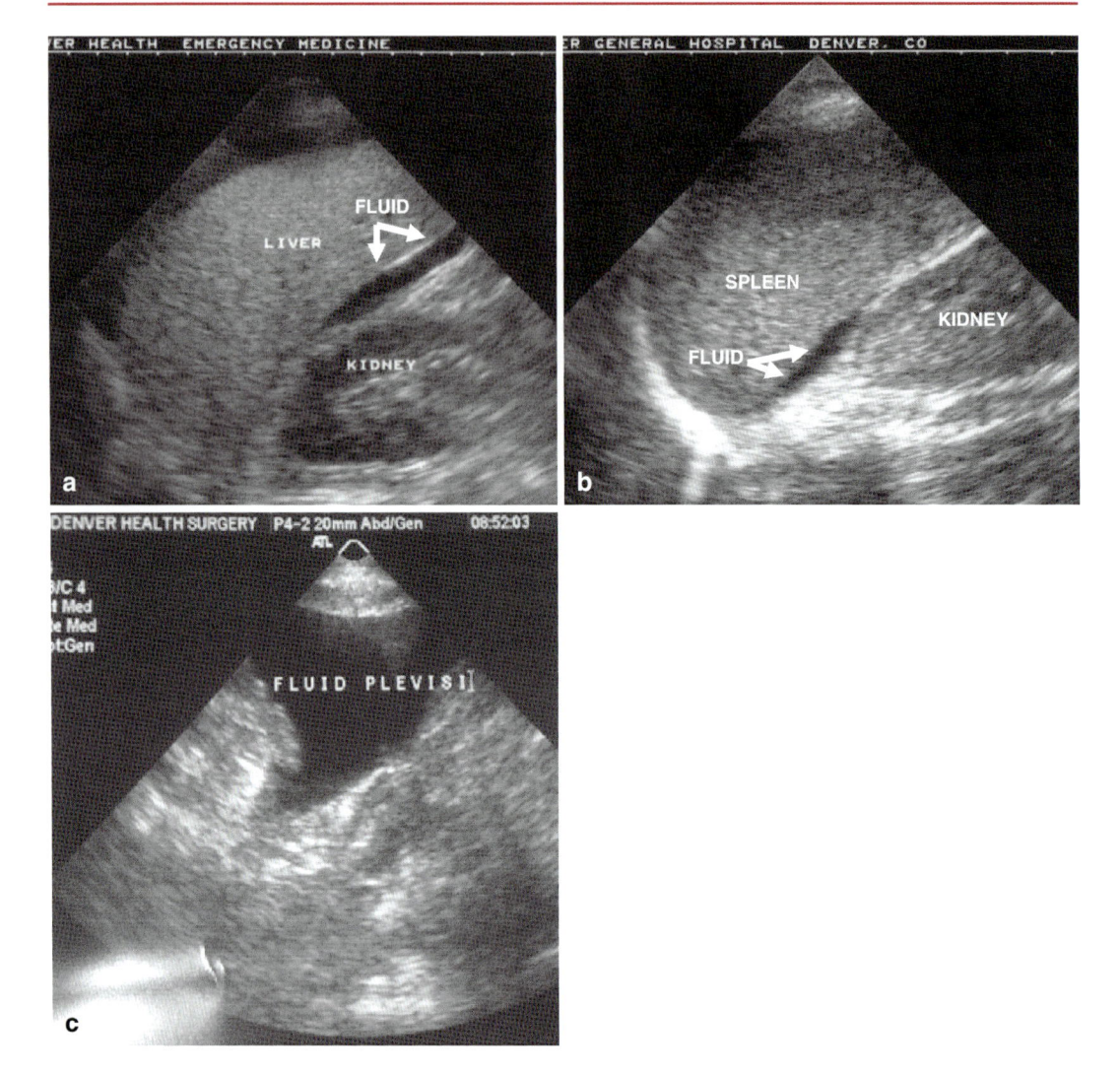

Fig. 9.2 FAST imaging detects intraabdominal hemorrhage. Hemorrhage is presumed when there is a fluid stripe visible between the right kidney and liver (**a**), left kidney and spleen (**b**), or in the pelvis (**c**)

peritoneum require laparotomy, the exception is penetrating trauma isolated to the right upper quadrant. In hemodynamically stable patients with the trajectory of penetrating trauma confined to the liver by CT scan, nonoperative observation is an option [7, 8].

CT scanning is excellent for identifying injuries of the solid organs (liver, spleen, kidney). If a diaphragmatic injury is not clearly identified on ED radiograph, CT scan can also be used to delineate these injuries, particularly with sagittal or coronal reconstructions (Fig. 9.3). Despite the increasing diagnostic accuracy of multi-slice CT scanners, CT still has limited sensitivity for identification of intestinal injuries. Bowel injury is suggested by findings of thickened bowel wall, "streaking" in the mesentery, free fluid without associated solid organ injury, or free intraperitoneal air [9].

The American Association for the Surgery Trauma (AAST) developed a grading scale to provide a uniform definition of solid organ injuries based upon the magnitude of anatomic disruption (Table 9.1) [10]. Solid organ injury

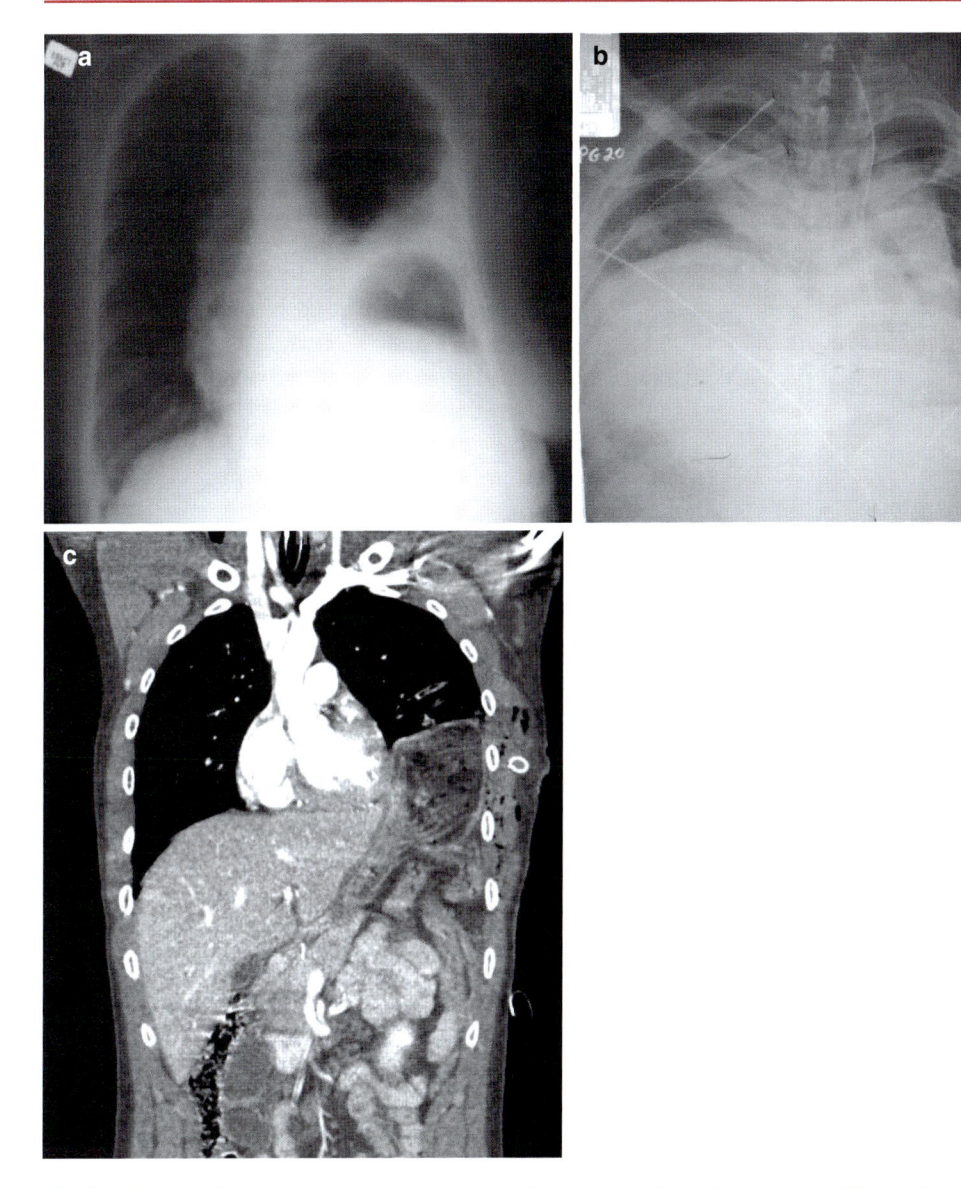

Fig. 9.3 Left diaphragm ruptures are evident with the gastric bubble located in the left hemithorax (**a**), while right-sided ruptures present with the appearance of an elevated hemidiaphragm (**b**). CT scanning may be used in questionable cases to better identify the injury (**c**)

grading permits effective transfer of information between treating physicians, and predicts failure rates and complication rates of nonoperative management (NOM). In addition to grading the injury, specific findings that should be noted on CT scan include contrast extravasation (i.e., a "blush"), the amount of intraabdominal hemorrhage, and the presence of pseudoaneurysms (Fig. 9.4).

9.3 Penetrating Injuries

The diagnostic approach differs between penetrating and blunt abdominal trauma. As a rule, minimal evaluation is required prior to laparotomy for gunshot or shotgun wounds that violate the peritoneal cavity, because over 90 % of patients have significant internal injuries. Anterior truncal GSWs between the fourth intercostal space and the pubic symphysis, whose trajectory by x-ray or entrance/exit wound indicates peritoneal penetration, should undergo operative exploration (Fig. 9.5). GSWs to the back or flank are more difficult to evaluate because of the retroperitoneal location of the injured abdominal organs. Triple-contrast CT scan can delineate the trajectory of the bullet and identify peritoneal violation or retroperitoneal entry, but may miss specific injuries [11]. Similarly, in obese patients if the GSW is thought to be tangential through the subcutaneous tissues, CT scan can delineate the tract and exclude peritoneal violation. Laparoscopy is another option to assess peritoneal penetration, and is followed by laparotomy to repair injuries if found. If in doubt, it is always safer to explore the abdomen than to equivocate, but a period of close observation in the patient with a reliable examination and hemodynamic stability may be considered.

In contrast to GSWs, SWs that penetrate the peritoneal cavity are less likely to injure intraabdominal organs. Anterior abdominal SWs (from costal margin to inguinal ligament and bilateral

Table 9.1 AAST solid organ injury grading scales

	Subcapsular hematoma	Laceration
Liver injury grade		
I	<10 % surface area	<1 cm in depth
II	10–50 % surface area	1–3 cm
III	>50 % or >10 cm	>3 cm
IV	25–75 % of a hepatic lobe	
V	>75 % of a hepatic lobe	
VI	Hepatic avulsion	
Spleen injury grade		
I	<10 % surface area	<1 cm in depth
II	10–50 % surface area	1–3 cm
III	>50 % surface area	>3 cm
IV	>25 % devascularization	Hilar injury
V	Shattered spleen	

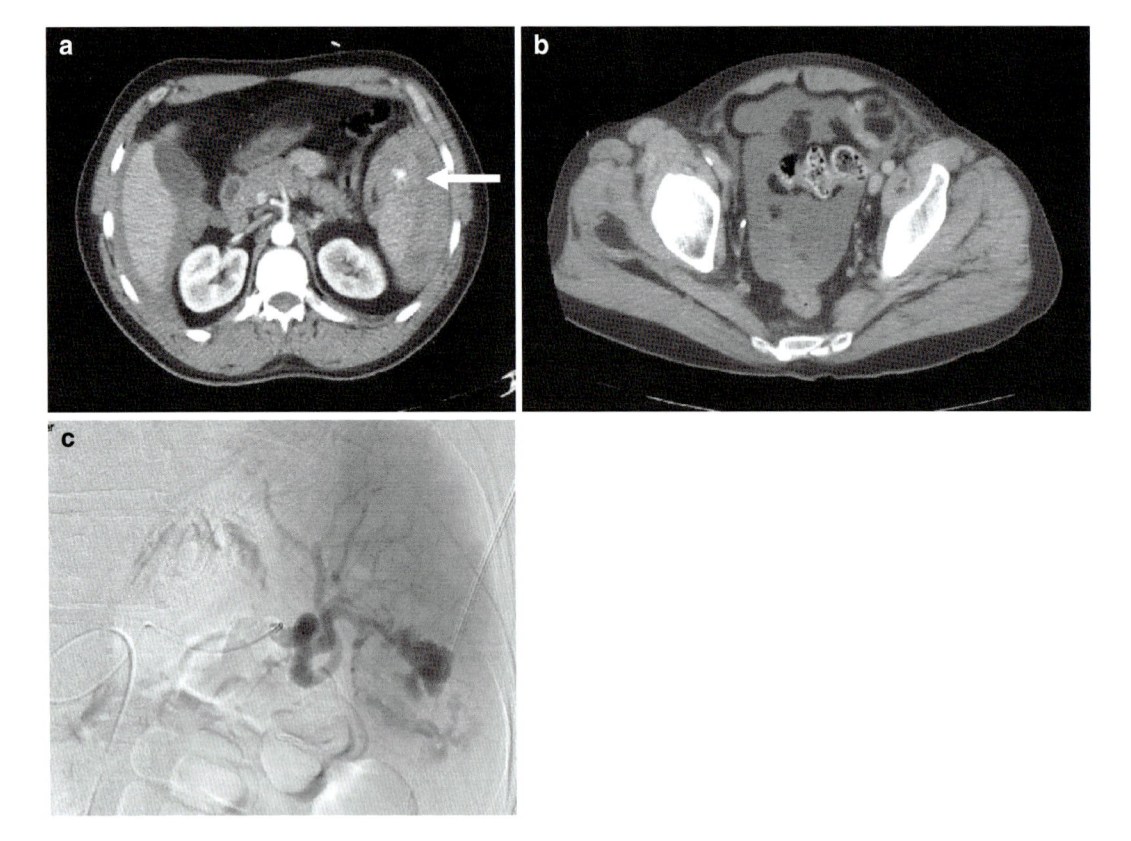

Fig. 9.4 Findings on imaging that are associated with failure of NOM for splenic injuries: contrast extravasation or "blush" (*arrow*) (**a**), intraabdominal hemorrhage extending into the pelvis (**b**), and pseudoaneurysms (**c**)

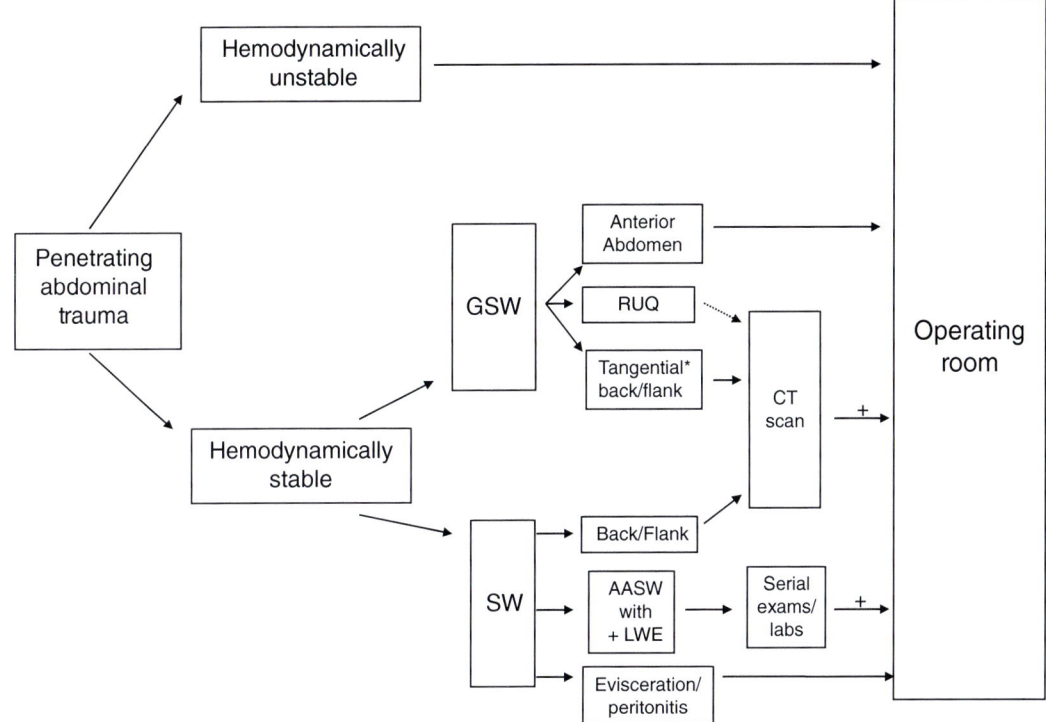

9 Abdominal Injuries: Indications for Surgery

Fig. 9.5 Algorithm for the evaluation of penetrating abdominal injuries (*GSW* gunshot wound, *SW* stab wound, *RUQ* right upper quadrant, *AASW* anterior abdominal stab wound, *LWE* local wound exploration)

*Tangential GSWs may also be evaluated with diagnostic laparoscopy.
** A positive local wound exploration is defined as violation of the posterior fascia.

mid-axillary lines) should be explored under local anesthesia in the ED to determine if the fascia has been violated. Injuries that do not penetrate the peritoneal cavity do not require further evaluation, and the patient is discharged from the ED. Patients with fascial penetration must be further evaluated for intraabdominal injury, as there is up to a 50 % chance of requiring laparotomy. The optimal diagnostic approach remains debated between serial examination, diagnostic peritoneal lavage (DPL), and CT scanning; the most recent evidence supports serial examination and laboratory evaluation [12].

Abdominal SWs of three body regions require a unique diagnostic approach: thoracoabdominal SWs, right upper quadrant SWs, and back/flank SWs. Occult injury to the diaphragm must be ruled out in patients with SWs to the lower chest. Diagnostic laparoscopy or DPL can be used to exclude diaphragmatic injury. Patients undergoing DPL evaluation have different laboratory value cut-offs than those traditional values formerly used for abdominal stab wounds (Table 9.2). A RBC count of more than 10,000/μL is considered positive, and an indication for laparotomy while patients with a DPL RBC count between 1,000/μL and 10,000/μL should

Table 9.2 A positive diagnostic peritoneal lavage following trauma is defined by specific laboratory values

Laboratory study	Positive value	
	AASW	TSW
White blood cell (WBC)	>500 cells/μL	>500 cells/μL
Red blood cell (RBC)	>100,000 cells/μL	>10,000 cells/μL
Amylase	>19 IU/L	>19 IU/L
Alkaline phosphatase	>2 IU/L	>2 IU/L
Bilirubin	>0.1 mg/dL	0.1 mg/dL

AASW anterior abdominal stab wounds, *TSW* thoracoabdominal stab wounds

undergo laparoscopy or thoracoscopy. A RBC count of less than 1,000/μL is considered negative; that is, the red cells are due to the procedure itself. Alternatively, laparoscopy may be preferred in those who cannot tolerate a DPL or those with evidence of a hemothorax or pneumothorax on chest radiograph. SWs to the flank and back should undergo triple-contrast CT to assess for retroperitoneal injuries of the colon, duodenum, and urinary tract [11] (Fig. 9.6).

Although not universally embraced, selected patients with penetrating injuries to the right upper quadrant may be candidates for NOM [13–16]. Patients must have a CT scan that documents confinement of the injury to the liver (Fig. 9.7). Additionally, the patient must be hemodynamically stable, have a reliable physical examination without evidence of peritonitis (i.e., cannot have depressed mental status), and not require blood products. Patients should be admitted for serial examination and hemoglobin monitoring; any alteration should prompt laparotomy. Violation of the diaphragm has the risk of a biliopleural fistula. An alternative approach is to perform laparoscopy to confirm trajectory of the missile or knife, and to repair the diaphragm. In addition to avoiding the morbidity of a laparotomy, the success of NOM for penetrating trauma has resulted in decreased hospital stays, lower transfusion requirements, and diminished abdominal infection rates.

9.4 Blunt Abdominal Trauma

9.4.1 Liver and Spleen

With the advent of CT scanning, NOM of solid organ injuries has replaced routine operative exploration. NOM of blunt solid organ injuries is appropriate in hemodynamically stable patients that do not have overt peritonitis or other indications for laparotomy. High-grade injuries, a large amount of hemoperitoneum, contrast extravasation, and pseudoaneurysms are not absolute contraindications for nonoperative management; however, these patients are at high risk for failure and are more likely to need angioembolization [17–21]. Likewise, there is not a patient age cut-off for the NOM of solid organ injuries. A multidisciplinary approach including angiography with selective angioembolization has improved NOM success rates as well as survival [19, 20, 22]. Over 80 % of patients with liver injuries may be managed nonoperatively.

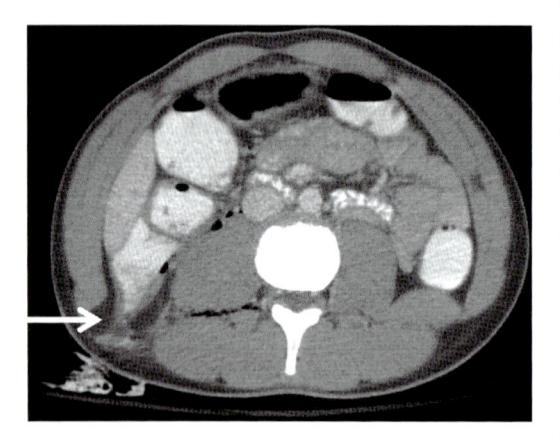

Fig. 9.6 Triple-contrast CT scan of the abdomen can delineate retroperitoneal injuries such as this colon injury (with contrast extravasation) (*arrow*) following a stab wound to the back

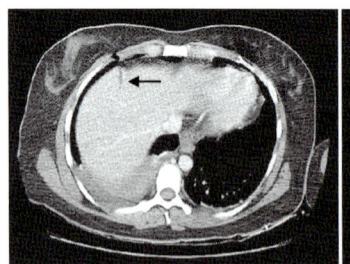

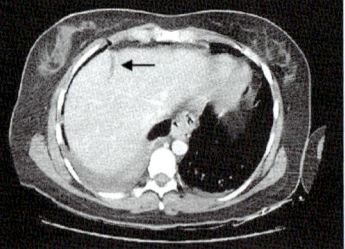

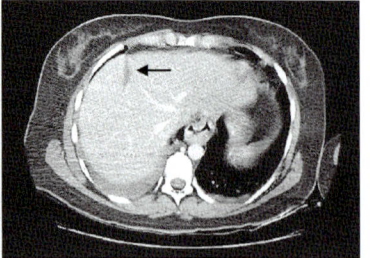

Fig. 9.7 Nonoperative management of penetrating abdominal trauma may be considered if the wound is isolated to the liver as documented by CT scan (*arrow* points to stab wound tract in liver)

Patients who require laparotomy for their liver injuries typically fail NOM in the first 24–48 h [17, 22]. Patients with persistent hemodynamic instability despite red cell transfusions of 4 units in 6 h or 6 units in 24 h should undergo laparotomy. Patients that develop peritonitis following admission should also undergo laparotomy with concern of a missed bowel injury. Of the minority (8 %) of patients that fail NOM, half require operation due to associated injuries (i.e., enteric or pancreatic injuries) while half undergo laparotomy for hepatic-related hemorrhage [17]. Predicting which patients will ultimately require laparotomy has yet to be accomplished. Perhaps not surprisingly, those patients who fail NOM have increasing rates of failure associated with increasing grades of hepatic injury. Series published to date report failure rates of 14 % in grade IV injuries and 23 % in grade V injuries [18]. The amount of hemoperitoneum appears to inversely correlate with successful management; patients with a large amount of hemoperitoneum (i.e., blood extending into the pelvis) are more likely to fail NOM.

An indication for angioembolization to address ongoing hepatic bleeding is transfusion of 4 units of RBCs in 6 h or 6 units of RBCs in 24 h in the hemodynamically stable patient. Hemodynamic instability, however, often requires laparotomy with perihepatic packing for hemostasis. Patients with contrast extravasation identified on CT scanning, indicating arterial hemorrhage, should also be considered as a candidate for hepatic angiography. Originally, evidence of extravasation was an indication for laparotomy; however, the advent of endovascular techniques has resulted in effective hemostasis in selected cases. Angioembolization is particularly helpful in hemodynamically stable patients with contrast pooling within the hepatic parenchyma [19]. Patients with contrast extravasation into the peritoneal cavity are more likely to require laparotomy [20], but cases of successful embolization have been reported [21].

Until the 1970s, splenectomy was considered mandatory for all splenic injuries. Recognition of the immune function of the spleen refocused efforts on splenic salvage in the 1980s [23, 24].

Following success in pediatric patients, NOM of splenic injuries was adopted in the adult population, and has become the prevailing strategy for blunt splenic trauma [25]. NOM of solid organ injuries is pursued in hemodynamically stable patients that do not have overt peritonitis or other indications for laparotomy [26–30]. Similar to liver injuries, there is not an age cut-off for patients for the NOM of splenic injuries [31, 32]. High-grade injuries, a large amount of hemoperitoneum, contrast extravasation, and pseudoaneurysms are not absolute contraindications for NOM; however, these patients are at high risk for failure [33–36]. The identification of contrast extravasation as a risk factor for failure of NOM led to liberal use of angioembolization in an attempt to avoid laparotomy. The true value of angioembolization in splenic salvage has not been rigorously evaluated. Patients with intraparenchymal splenic blushes who are otherwise asymptomatic may be considered for a period of observation rather than empiric angioembolization [37]; it is thought that the contained hemorrhage within the splenic capsule may result in tamponade of the bleeding (Fig. 9.8).

It is clear, however, that 20–30 % of patients with splenic trauma deserves early splenectomy, and that failure of NOM often represents poor patient selection [38, 39]. In adults, indications for prompt laparotomy include initiation of blood transfusion within the first 12 h considered to be secondary to the splenic injury, or hemodynamic instability. In the pediatric population, blood transfusions up to half the patient's blood volume are utilized prior to operative intervention. Following the first 12 postinjury hours, indications for laparotomy are not as black and white. Determination of the patient's age, comorbidities, current physiology, degree of anemia, and associated injuries will determine the use of transfusion alone versus intervention with either embolization or operation. Unlike hepatic injuries, which rebleed in 24–48 h, delayed hemorrhage or rupture of the spleen can occur up to weeks following injury. Overall, nonoperative treatment obviates laparotomy in more than 90 % of cases.

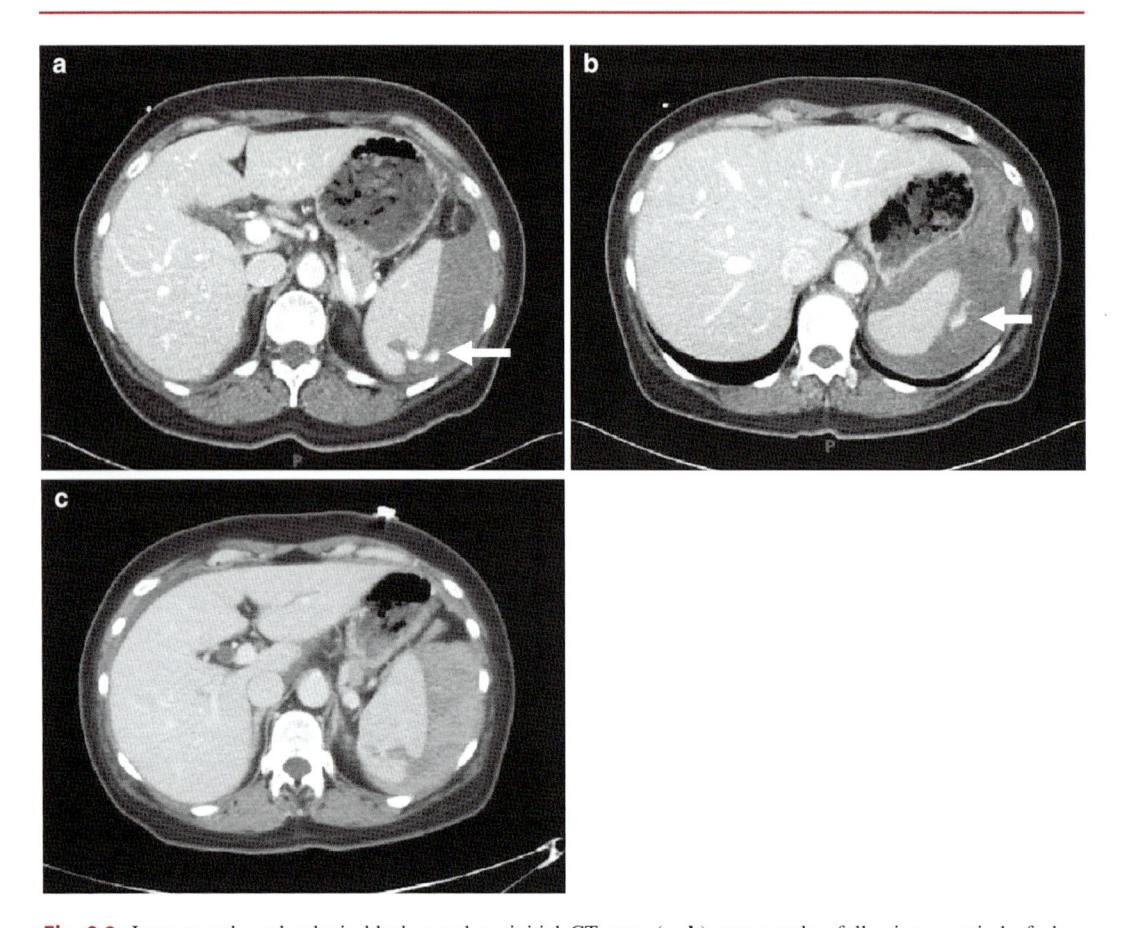

Fig. 9.8 Intraparenchymal splenic blush noted on initial CT scan (**a**, **b**) may resolve following a period of close observation (**c**) (*arrow* points to contrast extravasation)

9.4.2 Pancreatic Injuries

Pancreatic contusions, with or without associated ductal disruption, are difficult to diagnose in patients with blunt abdominal trauma [40]. Patients clearly at risk include those with significant mechanisms including high force, a seatbelt sign on physical examination, or a blow to the epigastrium [41]. The initial CT scan may show nonspecific stranding of pancreas. Associated fluid around the pancreas should prompt further invasive studies such as ERCP or MRCP to rule out a biliary or pancreatic duct injury. With a tentative diagnosis of a pancreatic contusion, one may consider following serial determinations of amylase/lipase; although these lab studies do not have a reliable sensitivity [42], increasing values over time combined with an alteration in clinical exam should prompt a repeat CT scan, duodenal C-loop study, DPL, or an ERCP depending upon the suspected lesion.

Historically, injuries to the pancreas were managed with operative intervention [43]. With the recent evolution of NOM for solid organ injuries, a nonresectional management schema has developed for select pancreatic injuries [44, 45]. Observation of pancreatic contusions, particularly those in the head of the pancreas that may involve ductal disruption, includes serial exams and monitoring of serum amylase. Patients with pancreatic injuries involving the major ducts, originally a strict indication for operative intervention, may be managed with ERCP and stenting in select patients; durability of this approach is currently under investigation [46].

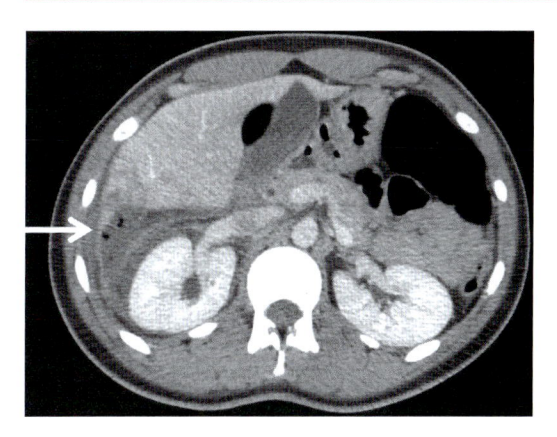

Fig. 9.9 Small bubbles of free air are identified in the right upper quadrant in this patient with an enteric injury (*arrow* point to air bubbles)

the lumen. Clinical exam findings include epigastric pain associated with either emesis or high nasogastric tube (NGT) output; CT scan imaging with oral contrast failing to pass into the proximal jejunum is diagnostic. Patients with suspected associated perforation, suggested by clinical deterioration or imaging with retroperitoneal free air or contrast extravasation, should be explored operatively. NOM includes continuous NGT decompression and nutritional support with total parenteral nutrition (TPN) [56, 57]. A marked drop in NGT output heralds resolution of the hematoma, which typically occurs within 2 weeks; repeat imaging to document these clinical findings is optional. If the patient does not improve clinically or radiographically within 4 weeks, operative evaluation is warranted.

9.4.3 Bowel Injuries

Diagnosing a hollow viscus injury is notoriously difficult [47], and even short delays in diagnoses result in increased morbidity [48, 49]. Findings suggestive of a bowel injury include thickening of the bowel wall, "streaking" in the mesentery, or free intraperitoneal air [9] (Fig. 9.9). If a patient's initial CT scan of the abdomen shows free fluid without evidence of a solid organ injury to explain such fluid, evaluation for a bowel injury should be performed [50–52]. DPL should also be considered in a patient if there is increasing intraabdominal fluid on bedside ultrasound in patients with a solid organ injury but a stable hematocrit, and/or in patients with unexplained clinical deterioration. Particular attention should be paid to elevations in the DPL effluent of bilirubin, alkaline phosphatase, and amylase when pursuing a diagnosis of bowel injury, with specific laboratory values indicating need for laparotomy (Table 9.2) [53, 54]. A rectal injury may be life threatening in patients with pelvic fractures. While some patients have clear findings on physical examination, ranging from hematochezia to overt degloving of the perineum, others may have occult injuries that are missed on initial evaluation in the trauma bay. Flexible or rigid sigmoidoscopy should rule out blood within the canal, clear intestinal perforation, or ischemic mucosa [55].

Following blunt trauma, patients may develop hematomas in the duodenal wall which obstruct

9.4.4 Genitourinary

Over 90 % of all blunt renal injuries are treated nonoperatively. Operative intervention following blunt trauma is limited to renovascular injuries and destructive parenchymal injuries that result in hypotension. The renal arteries and veins are uniquely susceptible to traction injury caused by blunt trauma. As the artery is stretched, the inelastic intima and media may rupture, causing thrombus formation and resultant stenosis or occlusion. The success of renal artery repair approaches 0 % but an attempt is reasonable if the injury is less than 5 h old, or if the patient has a solitary kidney or bilateral injuries [58]. Early CT diagnosis with Interventional Radiology placement of a stent should improve outcomes. Reconstruction of blunt renovascular injuries, however, may be difficult because the injury is typically at the level of the aorta. If repair is not possible within this time frame, leaving the kidney in situ does not necessarily lead to hypertension or abscess formation. The renal vein may be torn or completely avulsed from the vena cava due to blunt trauma. Typically, the large hematoma causes hypotension, leading to operative intervention. The majority of penetrating wounds to kidneys are explored. Renal vascular injuries are common following penetrating trauma, and

Fig. 9.10 A biloma, evident on CT scan (**a**), with an associated right hepatic duct injury evident on ERC (**b**)

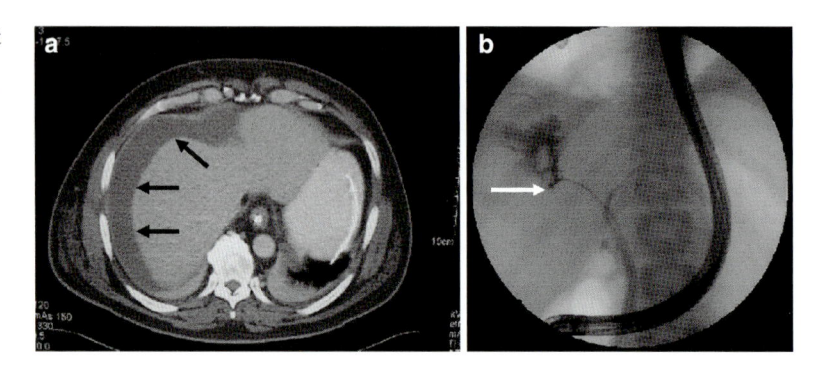

they may be deceptively tamponaded, resulting in delayed hemorrhage. For destructive parenchymal or irreparable renovascular injuries, nephrectomy may be the only option; palpation of a normal contralateral kidney must be performed since unilateral renal agenesis occurs in 0.1 % of patients.

Bladder injuries are subdivided into intraperitoneal versus extraperitoneal location based upon extravasation of contrast on CT cystography. Ruptures or lacerations of the intraperitoneal bladder are operatively closed with a running, single-layer, 3-0 absorbable monofilament suture. Laparoscopic repair is becoming common in patients not requiring laparotomy for other injuries. Extraperitoneal ruptures are treated nonoperatively with bladder decompression for 2 weeks.

9.5 Postinjury Complications Requiring Abdominal Exploration

Following hepatic injuries, the most common complication is a bile leak or biloma, occurring in up to 20 % of patients (Fig. 9.10) [59, 60]. Clinical presentation includes abdominal distension, intolerance of enteral feeds, and elevated liver functions tests. CT scanning effectively diagnoses the underlying problem, and the vast majority is treated with percutaneous drainage and ERCP with sphincterotomy. Occasionally, laparoscopy or laparotomy with drainage of biliary ascites is indicated, particularly if the patient fails to resolve their ileus and fever [61]. Patients undergoing angioembolization for liver trauma must be carefully monitored for hepatic necrosis, and may occasionally require delayed formal hepatic resection.

The most common problem in patients with splenic injuries is delayed bleeding, although as noted previously, the majority fails over an established timeframe. Patients undergoing splenic embolization can fail with rebleeding with 13 % of patients requiring splenectomy [62].

Missed bowel injuries are the most commonly pursued injury, not due to their frequency (less than 5 % of blunt trauma) but rather their associated morbidity. Observation for a missed small or large bowel injury is critical; clinical findings in such patients include a rising white blood cell count, fever, tachycardia, and increasing abdominal pain or frank peritonitis. After repair of bowel injuries, the most common intraabdominal complications are anastomotic failure and abscess. Percutaneous versus operative therapy will be based on the location, timing, and extent of the collection.

An additional postinjury complication that may require laparotomy is the abdominal compartment syndrome (ACS). The ACS is defined as intraabdominal hypertension plus end-organ sequelae (decreased urine output, increased pulmonary inspiratory pressures, decreased cardiac preload, and increased cardiac afterload). The ACS can be due to either intraabdominal injury (primary) or massive resuscitation (secondary). A diagnosis of intraabdominal hypertension cannot reliably be made by physical examination; therefore it is obtained by measuring the intraperitoneal pressure. Organ failure can occur over a wide range of recorded bladder pressures, and there is not a single measurement

9 Abdominal Injuries: Indications for Surgery

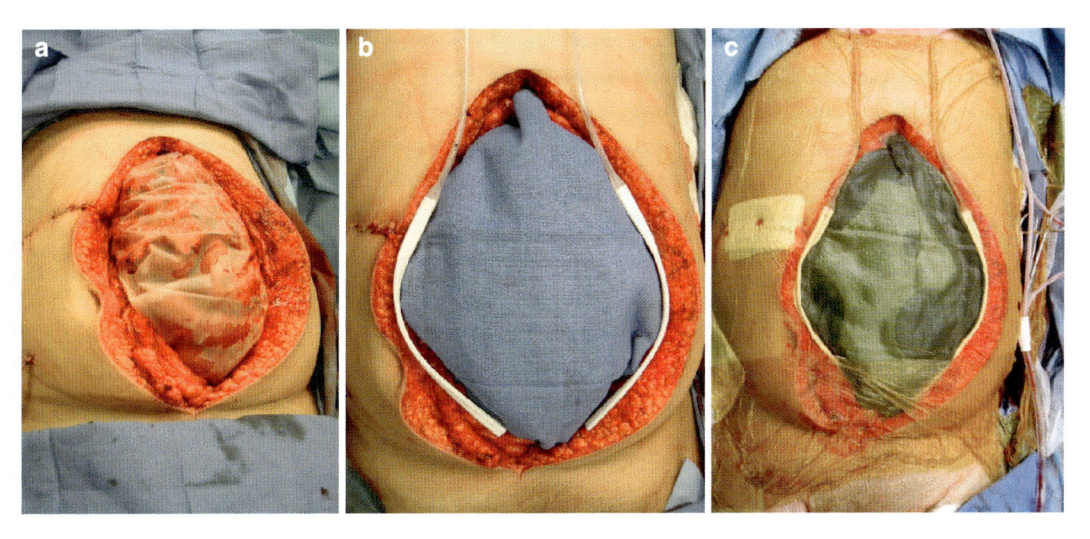

Fig. 9.11 Temporary closure of the abdomen entails covering the bowel with fenestrated 1010 drape (**a**), placement of JP drains and a blue towel (**b**), followed by ioban occlusion (**c**)

of bladder pressure that prompts therapeutic intervention, except >35 mmHg. Rather, emergent decompression is warranted in the patient with intraabdominal hypertension at the level it produces end-organ dysfunction. Laparotomy is performed either in the ICU if the patient is hemodynamically unstable, or in the operating room. ICU bedside laparotomy is easily accomplished, precludes transport in hemodynamically compromised patients, and requires minimal equipment (scalpel, suction, cautery, and abdominal temporary closure dressings). Patients with significant intraabdominal fluid as the primary component of their ACS, rather than bowel or retroperitoneal edema, may be effectively decompressed via a percutaneous drain. This may also be applicable for NOM of major liver injuries. Patients with significant free fluid are identified by bedside ultrasound, and avoid the morbidity of a laparotomy. When laparotomy is required, temporary coverage of the abdominal contents is obtained using a 1010 drape and ioban coverage (Fig. 9.11). Of note, patients can develop recurrent abdominal compartment syndrome despite a widely open abdomen. Therefore, bladder pressures should be monitored every 4 h, with significant increases in pressures alerting the clinician to the possible need for repeat operative decompression.

9.6 Collaboration in the Multiply Injured Patient

Patients with abdominal trauma often have associated fractures of the pelvis and extremities. Early dialogue between the trauma and orthopedic teams is critical to coordinate patient care and optimize outcomes. One illustrative example of this collaboration is the patient with hemodynamic instability and an unstable pelvic fracture. Protocols for care, with the early involvement of both the trauma and orthopedic teams in the trauma bay and in the operating room, has been shown to reduce mortality [63]. In these multiply injured patients, the orthopedic team can stabilize fractures in the ED while the trauma team evaluates the patient for thoracoabdominal trauma and determine the need for operative management. In patients requiring emergent laparotomy who also require intervention for an unstable pelvic fracture, the two teams can operate simultaneously; the trauma team performs the laparotomy while the orthopedic team places an external fixator and performs preperitoneal pelvic packing (PPP) [64, 65]. Alternatively, if the patient's hemodynamic instability is related to the pelvic fracture with concurrent extremity injuries, the trauma team can perform PPP while the orthopedic team places extremity external fixators, washes out open fractures, and performs necessary fasciotomies. Timely communication can ensure appropriate resuscitation and permit simultaneous operations [66, 67].

References

1. American College of Surgeons. Advanced trauma life support. 9th ed. Chicago: American College of Surgeons; 2012.
2. Dolich MO, McKenney MG, Varela JE, et al. 2,576 ultrasounds for blunt abdominal trauma. J Trauma. 2001;50:108.
3. Rozycki GS, Ochsner MG, Schmidt JA, et al. A prospective study of surgeon-performed ultrasound as the primary adjuvant modality for injured patient assessment. J Trauma. 1995;39:492.
4. Branney SW, Wolfe RE, Moore EE, et al. Quantitative sensitivity of ultrasound in detecting free intraperitoneal fluid. J Trauma. 1995;39:375.
5. Ochsner MG, Knudson MM, Pachter HL, et al. Significance of minimal or no intraperitoneal fluid visible on CT scan associated with blunt liver and splenic injuries: a multicenter analysis. J Trauma. 2000;49:505.
6. Rozycki GS, Ballard RB, Feliciano DV, Schmidt JA, Pennington SD. Surgeon-performed ultrasound for the assessment of truncal injuries: lessons learned from 1540 patients. Ann Surg. 1998;228:557.
7. Renz BM, Feliciano DV. Gunshot wounds to the liver. A prospective study of selective nonoperative management. J Med Assoc Ga. 1995;84:275.
8. Demetriades D, Gomez H, Chahwan S, et al. Gunshot injuries to the liver: the role of selective nonoperative management. J Am Coll Surg. 1998;188:343.
9. Malhotra AK, Fabian TC, Katsis SB, et al. Blunt bowel and mesenteric injuries: the role of screening computed tomography. J Trauma. 2000;48:991.
10. Moore EE, Cogbill TH, Jurkovich GJ, et al. Organ injury scaling: spleen and liver. J Trauma. 1995;38:323.
11. Boyle Jr EM, Maier RV, Salazar JD, et al. Diagnosis of injuries after stab wounds to the back and flank. J Trauma. 1997;42:260.
12. Biffl WL, Kaups KL, Pham TN, et al. Validating the Western Trauma Association algorithm for managing patients with anterior abdominal stab wounds: a Western Trauma Association multicenter trial. J Trauma. 2011;71:1494–502.
13. Renz BM, Feliciano DV. Gunshot wounds to the right thoracoabdomen: a prospective study of nonoperative management. J Trauma. 1994;37:737.
14. Demetriades D, Hadjizacharia P, Constantinou C, et al. Selective nonoperative management of penetrating abdominal solid organ injuries. Ann Surg. 2006;244:620.
15. Nance FC, Cohn I. Surgical judgment in the management of stab wounds of the abdomen: a retrospective and prospective analysis based on a study of 600 stabbed patients. Ann Surg. 1969;170:569.
16. Velmahos GC, Constantinou C, Tillou A, et al. Abdominal computed tomographic scan for patients with gunshot wounds to the abdomen selected for nonoperative management. J Trauma. 2005;59:1155.
17. Croce MA, Fabian TC, Menke PG, et al. Nonoperative management of blunt hepatic trauma is the treatment of choice for hemodynamically stable patients. Results of a prospective trial. Ann Surg. 1995;221:744.
18. Malhotra AK, Fabian TC, Croce MA, Gavin TJ, Kudsk KA, Minard G, Pritchard FE. Blunt hepatic injury: a paradigm shift from operative to nonoperative management in the 1990s. Ann Surg. 2000;231:804.
19. Pachter HL, Knudson MM, Esrig B, et al. Status of nonoperative management of blunt hepatic injuries in 1995: a multicenter experience with 404 patients. J Trauma. 1996;40:31.
20. Fang JF, Chen RJ, Wong YC, et al. Classification and treatment of pooling of contrast material on computed tomographic scan of blunt hepatic trauma. J Trauma. 2000;49:1083.
21. Ciraulo DL, Luk S, Palter M, et al. Selective hepatic arterial embolization of grade IV and V blunt hepatic injuries: an extension of resuscitation in the nonoperative management of traumatic hepatic injuries. J Trauma. 1998;45:353.
22. Richardson JD, Franklin GA, Lukan JK, et al. Evolution in the management of hepatic trauma: a 25-year perspective. Ann Surg. 2000;232:324.
23. Feliciano DV, Spjut-Patrinely V, Burch JM, et al. Splenorrhaphy: the alternative. Ann Surg. 1990;211:569.
24. Pickhardt B, Moore EE, Moore FA, et al. Operative splenic salvage in adults: a decade perspective. J Trauma. 1989;29:1386.
25. Richardson JD. Changes in the management of injuries to the liver and spleen. J Am Coll Surg. 2005;200:648.
26. Hurtuk M, Reed RL, Espositio TJ, Davis KA, Luchette FA. Trauma surgeons practice what they preach: the NTDB story on solid organ injury management. J Trauma. 2006;61:243–55.
27. Peitzman AB, Heil B, Rivera L, et al. Blunt splenic injury in adults: multi-institutional study of the Eastern Association for the Surgery of Trauma. J Trauma. 2000;49:177–87.
28. Velmahos GC, Toutouzas KG, Radin R, Chan L, Demetriades D. Nonoperative treatment of blunt injury to solid abdominal organs: a prospective study. Arch Surg. 2003;138:844–51.
29. Pachter HL, Guth AA, Hofstett SR, Spencer FC. Changing patterns in the management of splenic trauma: the impact of nonoperative trauma. Ann Surg. 1998;227:708–17; discussion 717–9.
30. Rojani RR, Claridge JA, Yowler CJ, et al. Improved outcome of adult blunt splenic injury: a cohort analysis. Surgery. 2006;140:625–31; discussion 631–2.
31. Myers JG, Dent DL, Stewart RM, et al. Blunt splenic injuries: dedicated trauma surgeons can achieve a

high rate of nonoperative success in patients of all ages. J Trauma. 2000;48:801–5; discussion 805–6.
32. Harbrecht BG, Peitzman AB, Rivera L, et al. Contribution of age and gender to outcome of blunt splenic injury in adults: multicenter study of the Eastern Association for the Surgery of Trauma. J Trauma. 2001;51:887–95.
33. Goan YG, Huang MS, Lin JM. Nonoperative management for extensive hepatic and splenic injuries with significant hemoperitoneum in adults. J Trauma. 1998;45:360–4; discussion 365.
34. Nwomeh BC, Nadler EP, Meza MP, Bron K, Gaines BA, Ford HR. Contrast extravasation predicts the for operative intervention in children with blunt splenic trauma. J Trauma. 2004;56:537–41.
35. Malhotra AK, Latifi R, Fabian TC, et al. Multiplicity of solid organ injury: influence on management and outcomes after blunt abdominal trauma. J Trauma. 2003;54:925–9.
36. Schurr MJ, Fabian TC, Gavant M, et al. Management of blunt splenic trauma: computed tomographic contrast blush predicts failure of nonoperative management. J Trauma. 1995;39:507–12; discussion 512–3.
37. Burlew CC, Kornblith LZ, Moore EE, Johnson JL, Biffl WL. Blunt trauma induced splenic blushes are not created equal. World J Emerg Surg. 2012;7:8.
38. McIntyre LK, Schiff M, Jurkovich GJ. Failure of nonoperative management of splenic injuries: causes and consequences. Arch Surg. 2005;140:563.
39. Smith HE, Biffl WL, Majercik SD, et al. Splenic artery embolization: have we gone too far? J Trauma. 2006;61:541.
40. Leppaniemi AK, Haapiainen RK. Risk factors of delayed diagnosis of pancreatic trauma. Eur J Surg. 1999;165:1134–7.
41. Arkovitz MS, Johnson N, Garcia VF. Pancreatic trauma in children: mechanisms of injury. J Trauma. 1997;42:49–53.
42. Takishima T, Sugimoto K, Hirata M, Asari Y, Ohwada T, Kakita A. Serum amylase level on admission in the diagnosis of blunt injury to the pancreas: its significance and limitations. Ann Surg. 1997;226:70–6.
43. Patton Jr JH, Lyden SP, Croce MA, et al. Pancreatic trauma: a simplified management guideline. J Trauma. 1997;43:234–9; discussion 239–41.
44. Wales PW, Shuckett B, Kim PC. Long-term outcome after nonoperative management of complete traumatic pancreatic transection in children. J Pediatr Surg. 2001;36:823–7.
45. Jobst MA, Canty TG, Lynch FP. Management of pancreatic injury in pediatric blunt abdominal trauma. J Pediatr Surg. 1999;34:818–23.
46. Lin BC, Liu NJ, Fang JF, Kao YC. Long-term results of endoscopic stent in the management of blunt major pancreatic duct injury. Surg Endosc. 2006;20:1551–5.
47. Fakhry SM, Watts DD, Luchette FA, EAST Multi-Institutional Hollow Viscus Injury Research Group. Current diagnostic approaches lack sensitivity in the diagnosis of perforated blunt small bowel injury: analysis from 275,557 trauma admissions from the EAST multi-institutional HVI trial. J Trauma. 2003;54:295–306.
48. Fakhry SM, Brownstein M, Watts DD, Baker CC, Oller D. Relatively short diagnostic delays (<8 hours) produce morbidity and mortality in blunt small bowel injury: an analysis of time to operative intervention in 198 patients from a multicenter experience. J Trauma. 2000;48:408–14.
49. Niederee MJ, Byrnes MC, Helmer SD, Smith RS. Delay in diagnosis of hollow viscus injuries: effect on outcome. Am Surg. 2003;69:293–8.
50. Ng AK, Simons RK, Torreggiani WC, et al. Intra-abdominal free fluid without solid organ injury in blunt abdominal trauma: an indication for laparotomy. J Trauma. 2002;52:1134–40.
51. Miller PR, Croce MA, Bee TK, Malhortz AK, Fabian TC. Associated injuries in blunt solid organ trauma: implications for missed injury in nonoperative management. J Trauma. 2002;53:238–42.
52. Rodriguez C, Barone JE, Wilbanks TO, Rha CK, Miller K. Isolated free fluid on computed tomographic scan in blunt abdominal trauma: a systematic review of incidence and management. J Trauma. 2002;53:79–85.
53. McAnena OJ, Marx JA, Moore EE. Peritoneal lavage enzyme determinations following blunt and penetrating abdominal trauma. J Trauma. 1991;31:1161–4.
54. Heneman PL, Marx JA, Moore EE, Cantrill SV, Ammons LA. Diagnostic peritoneal lavage: accuracy in predicting necessary laparotomy following blunt and penetrating trauma. J Trauma. 1990;30:1345–55.
55. Velmahos GC, Gomez H, Falabella A, Demetriades D. Operative management of civilian rectal gunshot wounds: simpler is better. World J Surg. 2000;24:114–8.
56. Cogbill TH, Moore EE, Feliciano DV. Conservative management of duodenal trauma: a multicenter perspective. J Trauma. 1990;30:1469–75.
57. Huerta S, Bui T, Porral D, Lush S, Cinat M. Predictors of morbidity and mortality in patients with traumatic duodenal injuries. Am Surg. 2005;71:763–7.
58. Knudson MM, Harrison PB, Hoyt DB, et al. Outcome after major renovascular injuries: a Western trauma association multicenter report. J Trauma. 2000;49:1116.
59. Kozar RA, Moore FA, Cothren CC, et al. Risk factors for hepatic morbidity following nonoperative management: multicenter study. Arch Surg. 2006;141:451–9.
60. Giss SR, Dobrilovic N, Brown RL, Garcia VF. Complications of nonoperative management of pediatric blunt hepatic injury: diagnosis, management, and outcomes. J Trauma. 2006;61:334–9.
61. Goldman R, Zilkowski M, Mullins R, Mayberry J, Deveney C, Trunkey D. Delayed celiotomy for the treatment of bile leak, compartment syndrome, and other hazards of nonoperative management of blunt liver injury. Am J Surg. 2003;185:492–7.
62. Cocanour CS, Moore FA, Ware DN, Marvin RG, Clark JM, Duke JH. Delayed complications of nonoperative

management of blunt adult splenic trauma. Arch Surg. 1998;133:619–24; discussion 624–5.
63. Biffl WL, Smith WR, Moore EE, et al. Evolution of a multidisciplinary clinical pathway for the management of unstable patients with pelvic fractures. Ann Surg. 2001;233(6):843–50.
64. Smith WR, Moore EE, Osborn P, Agudelo JF, Morgan SJ, Parekh AA, Cothren CC. Retroperitoneal packing as a resuscitative technique for hemodynamically unstable patients with pelvic fractures: report of two representative cases and a description of technique.". J Trauma. 2005;59:1510–4.
65. Burlew CC, Moore EE, Stahel PF, et al. Preperitoneal pelvic packing/external fixation with secondary angioembolization: optimal care for life-threatening hemorrhage due to unstable pelvic fractures. J Am Chem Soc. 2011;212(4):628–35.
66. Hak DJ, Smith WR, Suzuki T. Management of hemorrhage in life-threatening pelvic fracture. J Am Acad Orthop Surg. 2009;17(7):447–57.
67. Stahel PF, Smith WR, Moore EE. Current trends in resuscitation strategy for the multiply injured patient. Injury. 2009;40 Suppl 4:S27–35.

Management of Pelvic Ring Injuries

David J. Hak and Cyril Mauffrey

Contents

10.1	**Introduction**	127
10.1.1	Anatomy	127
10.2	**Classification**	129
10.3	**Physical Examination**	133
10.4	**Emergent Treatment/Bony Stabilization**	133
10.4.1	Pelvic Binders	133
10.4.2	Military Antishock Trousers (MAST)	133
10.4.3	Anterior External Fixation	134
10.4.4	C-Clamp	134
10.5	**Hemorrhage Control**	135
10.5.1	Angiography	136
10.5.2	Pelvic Packing	136
10.6	**Novel Resuscitative Strategies**	137
10.7	**Treatment Algorithm**	137
10.8	**Definitive Treatment**	139
10.8.1	Internal Fixation: Anterior Pelvic Ring	139
10.8.2	Internal Fixation: Posterior Pelvic Ring	140
References		141

D.J. Hak, MD, MBA, FACS (✉)
Department of Orthopaedic Surgery,
Denver Health/University of Colorado,
777 Bannock Street, MC 0188,
Denver, CO 80204, USA
e-mail: david.hak@dhha.org

C. Mauffrey, MD, FRCS, FACS
Department of Orthopaedic Surgery,
Denver Health/University of Colorado,
777 Bannock Street, MC 0188,
Denver, CO 80204, USA
e-mail: cyril.mauffrey@dhha.org

10.1 Introduction

10.1.1 Anatomy

The sacrum and two innominate bones (composed of the ilium, ischium, and pubis) are firmly connected by several strong ligaments to form the ring-like structure of the pelvis. Anteriorly, the innominate bones are joined at the pubic symphysis, which consists of a hyaline cartilage articulation with multiple supporting ligaments. Posteriorly, the innominate bones are joined to the sacrum at the sacroiliac joint. The sacroiliac joint consists of an articular portion anteriorly and the fibrous or ligamentous portion posteriorly.

The bones of the pelvis have no intrinsic stability and are stabilized by several strong ligamentous structures (Fig. 10.1). The soft tissue connection at the pubic symphysis consists of fibrocartilage spanning between the two pubic bones and the arcuate ligament inferiorly. The posterior pelvic ring ligaments are critical for pelvic stability. The strongest of these ligaments are the posterior sacroiliac ligaments. These ligaments are made up of short oblique fibers that run from the posterior ridge of the sacrum to the posterosuperior and posteroinferior iliac spines, and longer longitudinal fibers that run from the lateral sacrum to the posterosuperior iliac spine combining with the sacrotuberous ligament. The sacrotuberous ligament is a strong band of tissue that runs from the posterolateral sacrum and dorsal

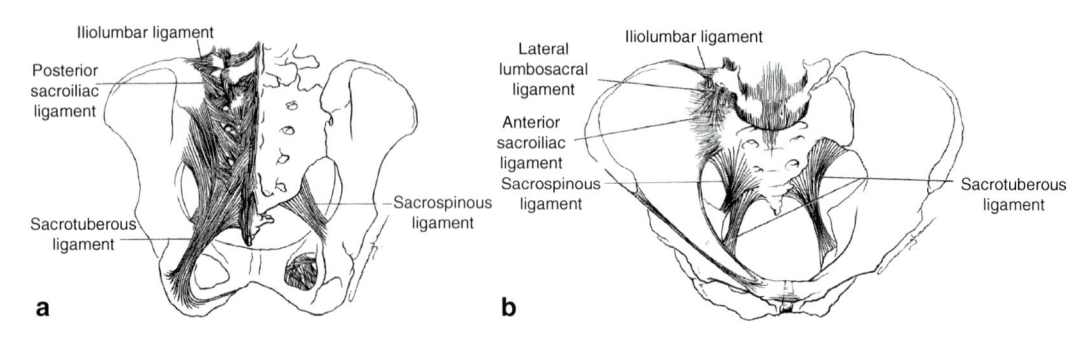

Fig. 10.1 (a) Posterior view of the pelvis showing the strong posterior ligaments which provide critical stability of the pelvic ring. (b) Anterior view of the pelvis showing the important ligamentous structures that stabilize the pelvic ring (From: Tile [1])

aspect of the posterior iliac spine to the ischial tuberosity. The sacrotuberous ligament, along with the posterior sacroiliac ligaments, provides vertical stability to the pelvis. The sacrospinous ligament runs from the lateral edge of the sacrum and coccyx to the sacrotuberous ligament and inserts onto the ischial spine. The iliolumbar ligaments run from the fourth and fifth lumbar transverse processes to the posterior iliac crest; the lumbosacral ligaments run from the fifth lumbar transverse process to the sacral ala. The anterior sacroiliac ligaments are relatively weak compared to the strong posterior sacroiliac ligaments.

Anatomically the pelvis structures can be separated into the true pelvis, located below the iliopectineal line (pelvic brim) and the false pelvis, located above the iliopectineal line. Numerous anatomical structures, including vascular supply for the buttocks and lower extremities, pass between the false and true pelvis. The true pelvis contains the floor of the pelvis along with the urethra, rectum, prostate, and vagina. The false pelvis surrounds the lower intraabdominal contents along with the iliacus muscle.

It is important to understand the location of major blood vessels which lie on the inner wall of the pelvis, since injury to these vessels is commonly associated with severe hemorrhage (Fig. 10.2). An understanding of pelvic anatomy will help the orthopedic surgeon recognize which fracture patterns are more likely to cause direct damage to major vessels and result in significant bleeding. The common iliac artery divides into the external and internal branches. The external iliac artery exits the pelvis anteriorly over the pelvic brim to become the femoral artery. The internal iliac artery lies over the pelvic brim and courses anterior and in close proximity to the sacroiliac joint. The posterior branches of the internal iliac artery include the iliolumbar, superior gluteal, and lateral sacral arteries. The superior gluteal artery, which is the largest branch of the internal iliac artery, courses across the sacroiliac joint in the true pelvis and exits through the greater sciatic notch to supply the gluteus medius, gluteus minimus, and tensor fascia lata muscles. It is the most commonly injured vessel in pelvic fractures with posterior ring disruptions. Anterior branches of the internal iliac artery include the obturator, umbilical, vesical, pudendal, inferior gluteal, rectal, and hemorrhoidal arteries. The inferior gluteal artery exits the pelvis through greater sciatic notch inferior to the piriformis and supplies the gluteus maximus. The pudendal and obturator arteries are adjacent to the pubic rami. In addition to the arteries, there is an associated large venous plexus which drains into the internal iliac vein. Injury to this venous plexus is the major source of hemorrhage in most pelvic fractures.

The neural structures that traverse the pelvis can also be injured in displaced pelvic fractures, leading to long-term morbidity. The sciatic nerve is formed by roots from the lumbosacral plexus (L4, L5, S1, S2, S3) and exits the pelvis deep to the piriformis muscle. The lumbosacral trunk is formed from the anterior rami of L4 and L5, and it crosses anterior to the sacral ala and SI joint. Fractures of sacral ala or dislocations of SI joint

Fig. 10.2 Internal aspect of the pelvis showing the major blood vessels that lie on the inner wall of the pelvis (From Kellam and Browner [2]; Fig. 31.6)

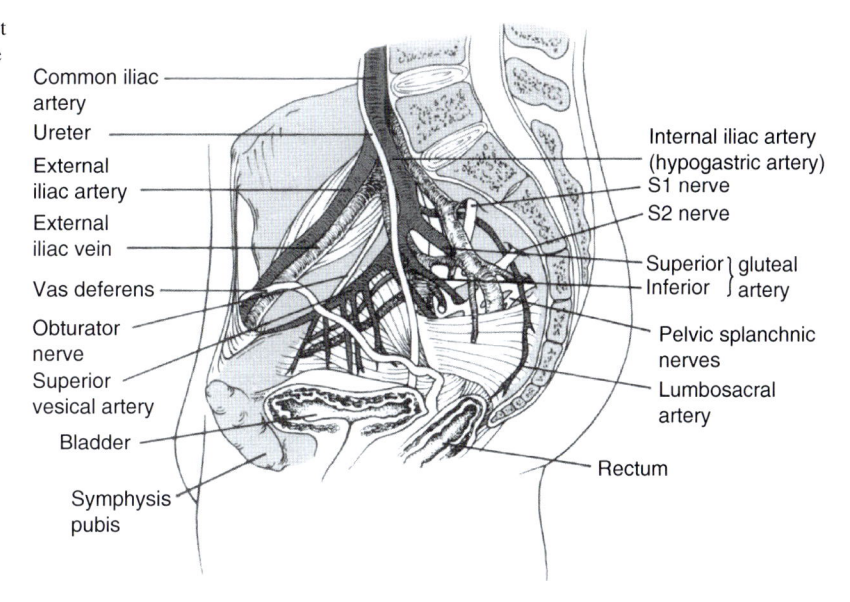

are most likely to injure the lumbosacral trunk. Typical displacement patterns in posterior pelvic fractures include cranial and posterior displacement of the hemipelvis. This may actually decrease the tension on the nerve roots exiting the pelvis posteriorly. More concerning are pelvic injuries with anterior (and caudal) displacement of the hemipelvis, as these displacement patterns potentially put the nerve roots on continued and significant stretch. The L5 nerve root exits below the L5 transverse process and crosses the sacral ala approximately 2 cm medial to the sacroiliac joint. It may be injured in SI joint disruptions and during anterior surgical approaches to the SI joint.

Significant anterior ring disruption can also damage the urethra and/or bladder. The female urethra is short and not rigidly fixed to the pubis or pelvic floor. Since it is more mobile it is less susceptible to injury from shear forces associated with pelvic fractures. The male urethra is less mobile and is making it more susceptible to injury in pelvic fractures. Stricture is the most common long-term complication observed in male patients who have sustained a urethral injury, but impotence may also occur in 25–47 % of patients with urethral rupture and is likely due to associated injury of the parasympathetic nerves (S2–S4). In males the bladder neck is attached to the pubis by puboprostatic ligaments and is contiguous with prostate, whereas in females the bladder lies on pubococcygeal portion of levator ani muscles. The superior and upper posterior portion of the bladder is covered by peritoneum, while the remainder of the bladder is extraperitoneal and covered with loose areolar tissue. Bladder injuries may be caused by a variety of mechanisms including bony spicules from the pubic rami fractures, blunt force injuries causing rupture, or shearing injuries. Intraperitoneal bladder ruptures require operative repair. Extraperitoneal bladder ruptures can usually be managed nonoperatively unless there is a bony spicule invading the bladder. Nonoperative management consists of catheter drainage and broad spectrum antibiotics. Most bladder injuries heal by 3–6 weeks, and a cystogram is obtained prior to catheter removal to confirm bladder healing.

10.2 Classification

Classification systems are useful to aid in decision making and treatment following high-energy pelvic fractures [3, 4]. Several pelvic fracture classification systems have been developed including the Pennal, Letournel, Bucholz, Tile, and Young and Burgess. The two classification

Table 10.1 Tile classification system of pelvic fractures

Type	Description	Mechanism
Stable posterior ring A1	Avulsion fractures (e.g., ASIS, AIIS, or ischial tuberosity)	
A2	Stable iliac wing fracture Minimal fracture displacement of the pelvic ring with intact ligaments (e.g., pubic rami and compression of the sacroiliac joint)	
A3	Isolated anterior ring injuries (e.g., pubic ramus)	
Rotationally unstable & vertically stable B1	"Open book" injuries that cause disruption of the symphysis pubis +/− the sacrospinous and anterior sacroiliac ligaments	
B2	Lateral compression injury causing anterior injury to pubic rami and impaction rather than disruption of the posterior ligament complex B2 has ipsilateral anterior and posterior injuries	
B3	Lateral compression with associated anterior and posterior injuries on the contralateral side	
Vertical shear C1	Unilateral injury with further subdivision according to the nature of posterior injury C1-1 iliac fracture C1-2 sacroiliac dislocation or fracture dislocation C1-3 sacral fracture	
C2	Bilateral injury with one side stable and the other unstable	
C3	Bilateral injury with both sides being unstable	

Images of the pelvic injury are shown in the far *right column*. The *black arrows* show the direction of the vector force causing the injury. The *round jagged red shapes* show the location of ligamentous or bony injury

systems which are commonly used are the Tile classification and the Young and Burgess classification.

The Tile classification primarily describes pelvic instability based on the anterior and posterior injury pattern(s) [5, 6]. Injuries are divided into three broad categories using an ABC classification similar to the AO/OTA classification system. These three main categories are further divided into specific subtypes. Type A injuries are stable pelvic fractures. Type B injuries are rotationally unstable, but vertically stable fractures. Type C injuries are both rotationally and vertically unstable (Table 10.1).

Table 10.2 Young and Burgess classification

Type	Description	Mechanism
Anterior posterior compression (APC) APC I	Slight widening if the symphysis pubis (<2.5 cm) Intact posterior ligamentous complex	
APC II	Widening of the symphysis pubis (>2.5 cm) Anterior widening of the sacroiliac joint (e.g., anterior sacroiliac, sacrotuberous, and sacrospinous ligaments are disrupted) Posterior sacroiliac ligaments are intact	
APC III	Complete disruption of the ipsilateral ligaments (APC II plus posterior sacroiliac ligaments) Hemipelvis instability both rotationally and vertically	
Lateral compression (LC) LC I	Oblique pubic ramus fracture Ipsilateral sacral compression fracture Caused by a direct lateral force	
LC II	Rami fracture Sacral crush injury with either posterior sacroiliac joint disruption or iliac wing fracture Crescent fracture (posterior fracture pattern) Caused by a anteriorly directed lateral force vs. LC I	
LC III	Windswept pelvis (ipsilateral lateral compression and contralateral APC) LC II pattern with continuation of force to the opposite hemipelvis resulting in an external rotation injury of the opposite hemipelvis	
Vertical shear (VS)	Disruption of all the ligamentous structures of the hemipelvis Caused by a vertically directed force Sacroiliac joint disruption or vertical sacral fracture	
Combined mechanism (CM)	High-energy pelvic injuries usually involve fractures caused by more than a single force vector May have combined components of any of the above fracture patterns	

Images of the pelvic injury are shown in the far *right column*. The *black arrows* show the direction of the vector force causing the injury

The Young and Burgess classification is primarily a mechanistic system based on the perceived applied force necessary to produce the injury pattern observed. This classification system should alert the surgeon to common associated injuries, the resuscitation needs of the patient, and may direct clinical care. The pelvic fracture mechanism is categorized into anterior posterior compression (APC), lateral compression (LC), vertical shear (VS), and combined mechanism (CM). Within each category, subtypes indicate the severity of injury (Table 10.2 and Fig. 10.3).

The Young and Burgess pelvic fracture classification has been found to correlate with the pattern of organ injury, resuscitative requirements, and mortality [7, 8]. A rise in mortality has been shown as the APC grade increases, and the APC-III pattern of injuries has been correlated with the greatest 24-h fluid resuscitation requirements.

In a series of 210 consecutive patients with pelvic fractures, Burgess and colleagues reported

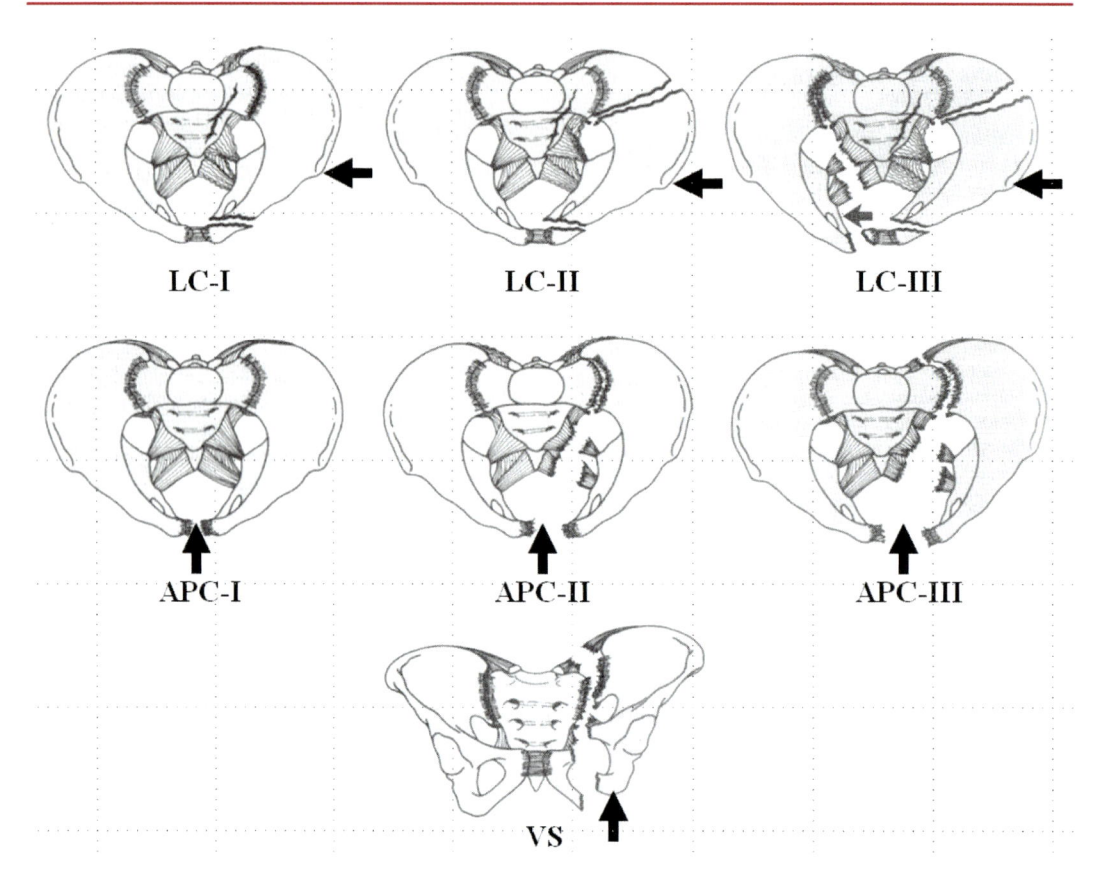

Fig. 10.3 The Young-Burgess classification of pelvic fracture. *LC* lateral compression type pattern, *APC* anteroposterior compression type pattern, *VS* vertical shear type pattern. The *arrow* in each panel indicates the direction of force producing the fracture pattern (From Kellam and Browner [2]; Fig. 31.12)

that transfusion requirements for patients with APC injuries averaged 14.8 units, compared to a mean of 3.6 units for patients with LC injuries, and 5 units for patients with combined mechanism injuries [7]. The overall mortality rate in this series was 8.6 %. A higher mortality rate was seen in the APC (20 %) and CM patterns (18 %), compared to the LC (7 %) and VS (0 %) patterns. Burgess and colleagues noted that exsanguination from pelvic injuries was rare in the lateral compression pattern in which mortality was typically due to other injuries, most commonly a severe closed head injury.

In a study of 343 trauma patients with pelvic fractures, investigators found that as the APC type increased from I to III there was an increasing percentage of injury to the spleen, liver, and bowel [8]. In addition there was an increasing incidence of pelvic vascular injury, retroperitoneal hematoma, shock, sepsis, and acute respiratory distress syndrome. Similarly, as the LC type increased from I to III the authors found an increased incidence of pelvic vascular injury, retroperitoneal hematoma, shock, and 24 h volume needs. Organ injury patterns and mortality in patients with vertical shear injuries were similar to those with high grade APC injuries. Patients with combined mechanisms of injury had an associated injury pattern similar to the lower grades of APC and LC injuries. The pattern of injury in the APC-III was correlated with the greatest 24 h fluid requirements. The investigators also reported major differences in the causes of death between patients with LC patterns compared to APC patterns. Brain injury was the major cause of death in LC injuries, while in APC patterns the most common causes of mortality were shock, sepsis, and ARDS related to massive torso forces.

10.3 Physical Examination

The orthopedic examination of the pelvis should be methodical and complete. An associated limb deformity (shortening or rotation) may be indicative of a pelvic injury with displacement. Hip dislocations should be reduced urgently with a complete neurovascular exam performed before and after reduction. The skin about the entire pelvis should be examined to ensure that there are no associated open wounds. This includes special attention to the perineum and gluteal folds where open fractures frequently occur. A digital rectal examination is required to detect rectal injury and open injuries in this location. In women, a vaginal examination should be performed to rule out an open injury. Manual palpation of the pelvis should be carefully performed and repeated examinations should be avoided.

The potential of an associated urethral injury should be considered in all pelvic fractures with significant anterior ring disruption (APC-II and APC-III type patterns). Signs of potential urethral injury include [9] inability to void despite a full bladder [10], blood at urethral meatus [11], high riding or abnormally mobile prostate [12], elevated bladder on intravenous pyelogram (IVP). However, the absence of meatal blood or a high riding prostate does not rule out urethral injury. If there is a high index of suspicion for genitourinary injury early contrast studies are required. A retrograde urethrogram should be obtained to rule out urethral injury prior to insertion of a urinary catheter, since passing a urinary catheter in the presence of a urethral injury can cause additional iatrogenic injury. Additionally, a cystogram with/without CT can be obtained to rule out bladder injury. However, one should ensure minimal or no contrast dye is remaining in the bladder to prevent problems with intraoperative visualization in patients requiring boney operative intervention.

10.4 Emergent Treatment/Bony Stabilization

It is uncommon for bleeding from a pelvic fracture to be the sole source of blood loss in the multi-injured patient. In fact, massive bleeding from a pelvic fracture alone is uncommon. Nevertheless, pelvic fractures must be considered as a potential source of major bleeding in the hemodynamically unstable patient, particularly when initial attempts to control bleeding from other sources fail to stabilize the patient. Provisional stabilization of the pelvic fracture should occur immediately during the patient's initial evaluation and resuscitation using one of the methods described in the following section.

10.4.1 Pelvic Binders

Circumferential pelvic compression can be easily achieved in the prehospital setting with some form of commercially available pelvic binders, providing early and beneficial pelvic stabilization during transport and resuscitation. In lieu of a commercial binder, a folded sheet wrapped circumferentially around the pelvis can also be used [13] (Fig. 10.4). The use of pelvic binders has been shown to reduce transfusion requirements, length of hospital stay, and mortality in patients with APC injuries [14]. External rotation of the legs is commonly seen in displaced pelvic fractures and forces acting through the hip joint may contribute to pelvic deformity. Correction of lower extremity external rotation can be easily achieved by taping the feet and knees together, which may improve the pelvic reduction provided through use of a pelvic binder.

10.4.2 Military Antishock Trousers (MAST)

In the 1970s and 1980s, military antishock trousers (MAST) were commonly used to provide temporary compression and immobilization of the pelvic ring and lower extremity via pneumatic pressure. Although still useful for stabilization

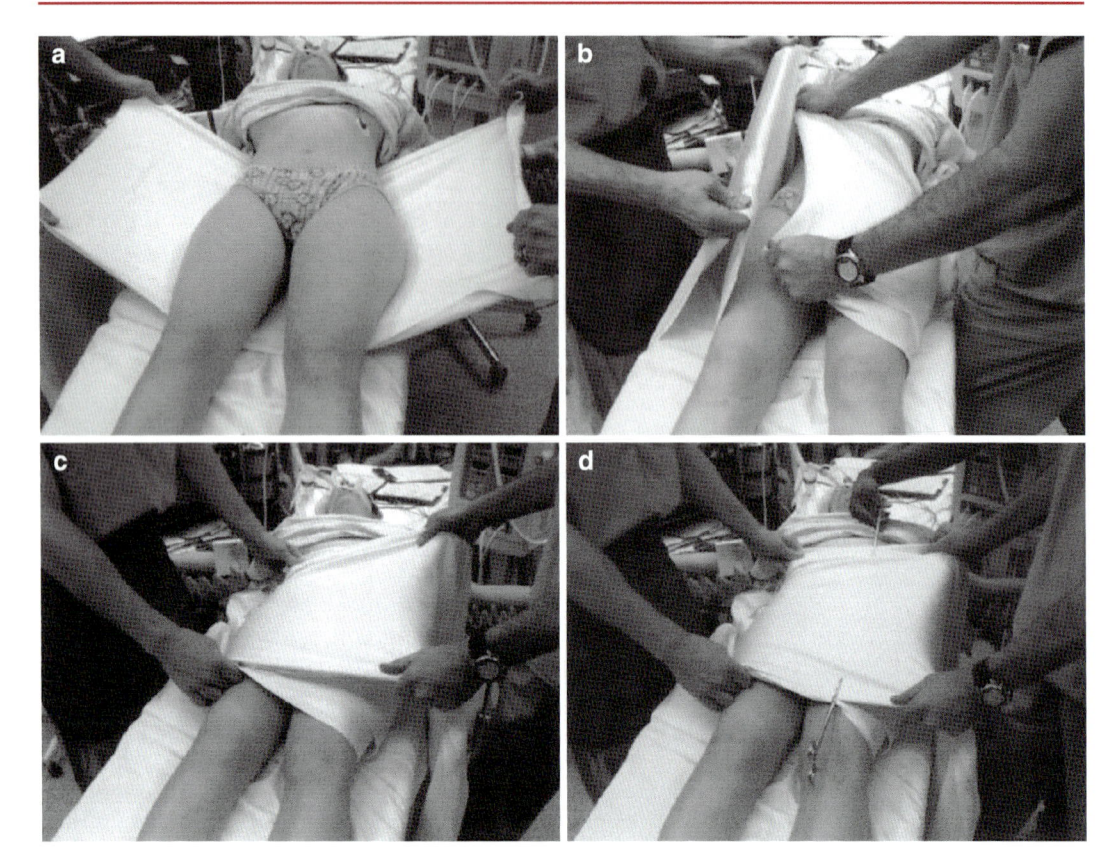

Fig. 10.4 Application of circumferential pelvic antishock sheeting. (**a**) A sheet is folded smoothly to a width of approximately 2 feet and placed beneath the patient's pelvis. (**b**, **c**) The ends of the sheet are crossed in an overlapping manner and pulled taut. (**d**) Clamps are placed proximally and distally to secure the sheet in position (From Routt et al. [13])

of patients with pelvic fractures, MAST has largely been replaced by the use of commercially available pelvic binders. In the past, the use of MAST has been associated with other complications including lower extremity compartment syndrome.

10.4.3 Anterior External Fixation

Several studies have reported a benefit of emergent pelvic external fixation in the resuscitation of the hemodynamically unstable patient with an unstable pelvic fracture [7, 15]. Several factors may contribute to the beneficial effects of external fixation in pelvic fractures. Immobilization helps limit pelvic displacement during patient movements and transfers, decreasing the possibility of clot disruption. In certain patterns (e.g., APC II), reduction of pelvic volume is often achieved by application of the external fixator (Fig. 10.5). Experimental studies have shown that reduction of an APC-II pelvic injury increases the retroperitoneal pressure which may help tamponade venous bleeding [16]. Finally, the apposition of the displaced fracture surfaces can help facilitate the hemostatic pathway to control bony bleeding.

10.4.4 C-Clamp

Standard anterior external pelvic fixation does little to provide posterior pelvic stabilization. This limits the effectiveness of standard anterior external fixators in fracture patterns involving significant posterior disruption or in cases in

10 Management of Pelvic Ring Injuries

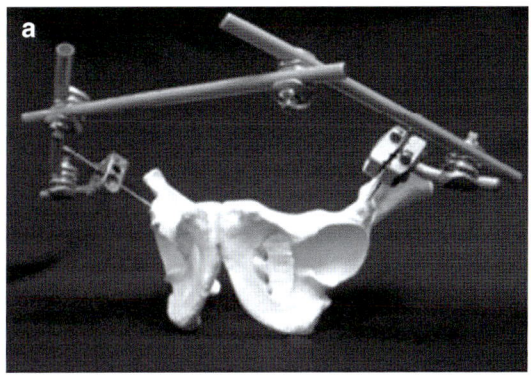

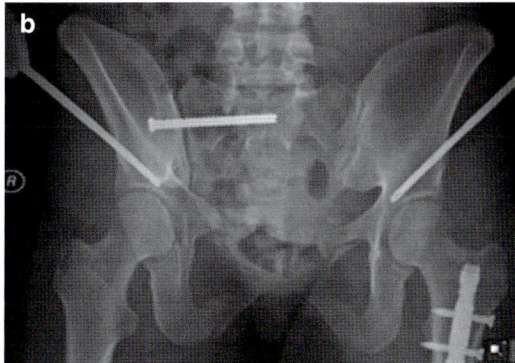

Fig. 10.5 The anterior external fixator can provide good pelvic stability and pins can be placed percutaneously. Additionally, the length of the bars and exact configuration can be adjusted depending on availability, patient body habitus, and surgeon preference. (**a**) Sawbones pelvic model with supra-acetabular pins through the anterior inferior iliac spine. (**b**) Outlet radiograph of a patient with a pelvic ring disruption and supra-acetabular external fixator, with pins traversing the superior aspect of the femoroacetabular joint

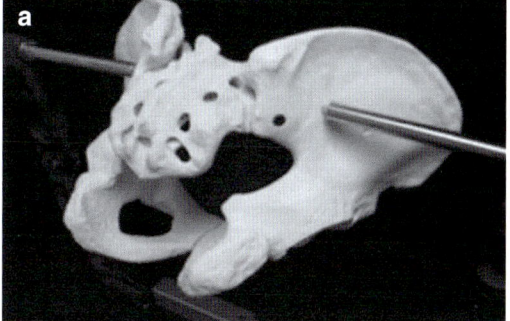

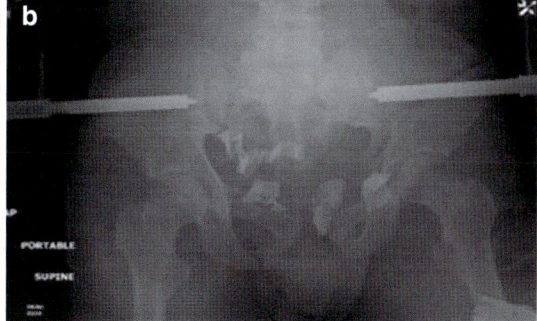

Fig. 10.6 (**a**) C-clamp can be used to provide posterior pelvic stability. The C-clamp pins should be located below the superior aspect of S1 to prevent the SI joint from opening inferiorly. (**a**) Posterior view of a Sawbones pelvic model with placement of C-clamp, note the location of the greater sciatic notch where the superior gluteal neurovascular bundle exits the pelvis. (**b**) Anteroposterior radiograph of a patient with a right SI joint dislocation after application of C-clamp. Of note, patient underwent pelvic packing and also sustained a left iliac wing fracture

which the iliac wing is fractured. The posteriorly applied pelvic C-clamp was developed to address these injury patterns. The C-clamp allows prompt application of a compressive force posteriorly across the sacroiliac joints (Fig. 10.6); however, extreme care must be exercised to avoid iatrogenic injury during its application and generally should be done with fluoroscopic guidance [17]. Additionally, positioning of the pin sites can be a bit more challenging and over reduction of the SI joint or reduction of a comminuted sacral fracture can lead to an iatrogenic sacral nerve root injury. Alternative applications of the C-clamp to the trochanteric region of the femur and to the gluteus medius pillars have also been described as alternative methods of reducing the pelvis in specific circumstances. These methods can be performed more safely without fluoroscopic guidance, but may not be feasible in patients with associated acetabular fractures [10].

10.5 Hemorrhage Control

While stabilization of the bony pelvis is the first stage in hemorrhage control, additional interventions may also be required in selected patients.

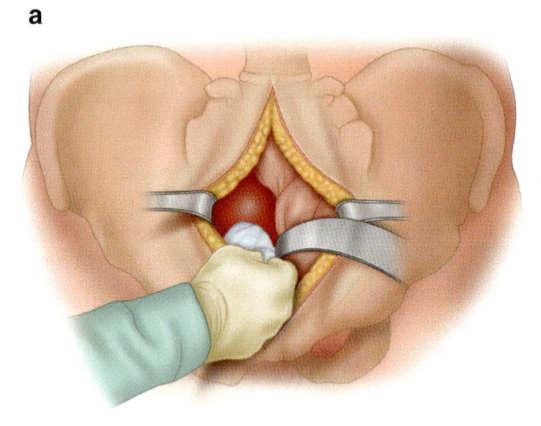

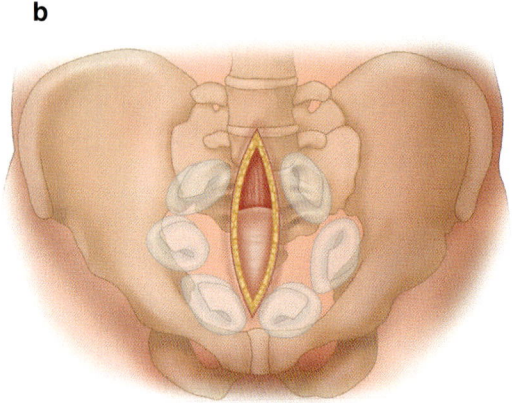

Fig. 10.7 Illustrations demonstrating the retroperitoneal packing technique. (**a**) An 8-cm midline vertical incision is made. The bladder is retracted to one side, and three unfolded lap sponges are packed into the true pelvis (below the pelvic brim) with a forceps. The first is placed posteriorly, adjacent to the sacroiliac joint. The second is placed anterior to the first sponge at a point corresponding to the middle of the pelvic brim. The third sponge is placed in the retropubic space just deep and lateral to the bladder. The bladder is then retracted to the other side, and the process is repeated. (**b**) General location of the six lap sponges following pelvic packing (Adapted from Smith et al. [22])

10.5.1 Angiography

The overall prevalence of patients with pelvic fractures who need embolization is reported to be <10 %. In one review of 162 patients with high-energy pelvic fractures only 8 % underwent angiography. Embolization was more commonly performed in APC and VS patterns (performed in 20 % of cases), but was infrequent in LC patterns (performed in only 1.7 % of cases) [7]. While most pelvic fracture patients do not require angiography, angiographic exploration should be considered in patients with continued hypotension despite pelvic fracture stabilization and aggressive fluid resuscitation. Eastridge et al. reported that 58.7 % of patients with persistent hypotension and a severely unstable pelvic fracture, including APC II, APC III, LC II, LC III, and VS injury patterns, had active arterial bleeding [18]. Miller et al. reported that 67.9 % of patients with pelvic injuries and persistent hemodynamic instability had active arterial bleeding [19].

Early angiography and arterial embolization has been demonstrated to improve patient outcomes [9, 11]. However, it is important to remember that angiography and embolization are not effective in controlling bleeding from venous injuries and bony sites, which represents the predominant source of hemorrhage in high-energy pelvic fractures. Time spent in the angiography suite for hypotensive patients without arterial injury may not contribute to survival. In addition, the aggressive use of angiography is not without consequence and may result in ischemic complications involving the gluteal musculature and subsequent wound healing problems [20].

10.5.2 Pelvic Packing

Pelvic packing was developed as a method to achieve direct hemostasis by controlling venous bleeding resulting from pelvic fractures. Trauma surgeons in Europe have long been advocating exploratory laparotomy followed by pelvic packing [21]. This technique is believed to be especially useful in patients in extremis.

More recently a modified method of pelvic packing, referred to as retroperitoneal packing, has been introduced in North America [22] (Fig. 10.7). In this approach, the intraperitoneal space is not entered, leaving the peritoneum intact to help provide a tamponade effect. Pelvic packing can be performed quickly with minimal blood loss. In one recent series, only 4 of 24 (16.7 %) patients failed to stabilize hemodynamically following pelvic packing and required subsequent embolization, and the authors concluded that packing can quickly control hemorrhage and reduce the need for emergent angiography [23].

10.6 Novel Resuscitative Strategies

The development of posttraumatic coagulopathy following high-energy pelvic fractures is a common and challenging complication to manage. In addition, hypoperfusion and shock lead to the "lethal triad" of acidosis, hypothermia, and coagulopathy further impeding hemostasis [24, 25].

The concept of "permissive hypotension" and earlier blood product administration has recently been implemented in many institutions. By having a systolic blood pressure goal of 80–100 mmHg and limiting crystalloid use you can, in theory, limit the adverse effects of aggressive fluid resuscitation (i.e., promote continued bleeding by increasing the intraluminal pressure at the wound, causing dislodgment of blood clots, dilution of clotting factors, and worsen hypothermia) [3, 4]. However, the optimal blood pressure in trauma is yet to be determined.

Many level I trauma centers have implemented massive transfusion protocols with the goal of improving blood product availability and survival rates. Based mainly on our military experience in Iraq and Afghanistan, several studies have reported on the improved survival rates of major trauma patients receiving a fresh frozen plasma (FFP) and/or platelet (PLT) to red blood cell (RBC) ratio ≥1:2 [4, 26, 27]. Thus, early transfusions with a FFP to RBC ratio of ≥1:2 and 1 unit of plateletpheresis per 5 RBC is now recommended. However, blood product use is not without its risk so goal-directed therapy is encouraged.

One challenge in the management of posttraumatic coagulopathy results from the inadequate real-time monitoring of treatment using traditional laboratory tests (i.e., PT/INR and aPTT). Additionally, they only offer a limited assessment of the coagulation cascade. As a result of these shortcomings, thromboelastography (TEG) use is gaining popularity [12, 28–30]. TEG offers several key advantages over traditional laboratory test and include: bedside (<5 min) results allowing to real-time decisions to be made, it provides clinically relevant results (i.e., clot time, clot strength, and clot lysis rate), and the test is run using whole blood allowing for a more complete coagulation cascade evaluation [28, 30]. For these reasons, TEG-guided transfusions are increasing in popularity.

10.7 Treatment Algorithm

Patients presenting to Denver Health with a high-energy pelvic fracture and hemodynamic instability are initially given 2 L of crystalloid solution (Fig. 10.8). A portable chest radiograph, along with radiographic views of the pelvic and lateral cervical spine, is obtained to rule out a thoracic source of blood loss. A central venous pressure line is placed, and base deficit is measured.

A FAST examination is performed, and, if positive, the patient is taken directly to the operating room for an exploratory laparotomy. A pelvic external fixator is placed, and pelvic packing is performed. If the patient remains hemodynamically unstable they undergo pelvic angiography prior to transfer to the intensive care unit (ICU). If hemodynamic stability is restored, the patient is transferred directly to the ICU. In the ICU the patient receives further fluid resuscitation, is warmed, and attempts are made to normalize the coagulation status. If the patient requires ongoing transfusion while in the ICU, angiographic assessment, if not previously done, should be performed. Recombinant factor VIIa should be considered if the patient is recalcitrant to all other interventions.

If the FAST is negative, transfusion of PRBC is begun in the emergency department. If the patient remains hemodynamically unstable following the second unit of PRBC, they are taken to the operating room for pelvic external fixation and pelvic packing. If the patient remains hemodynamically unstable they undergo pelvic angiography prior to transfer to the intensive care unit (ICU). If hemodynamic stability is restored, the

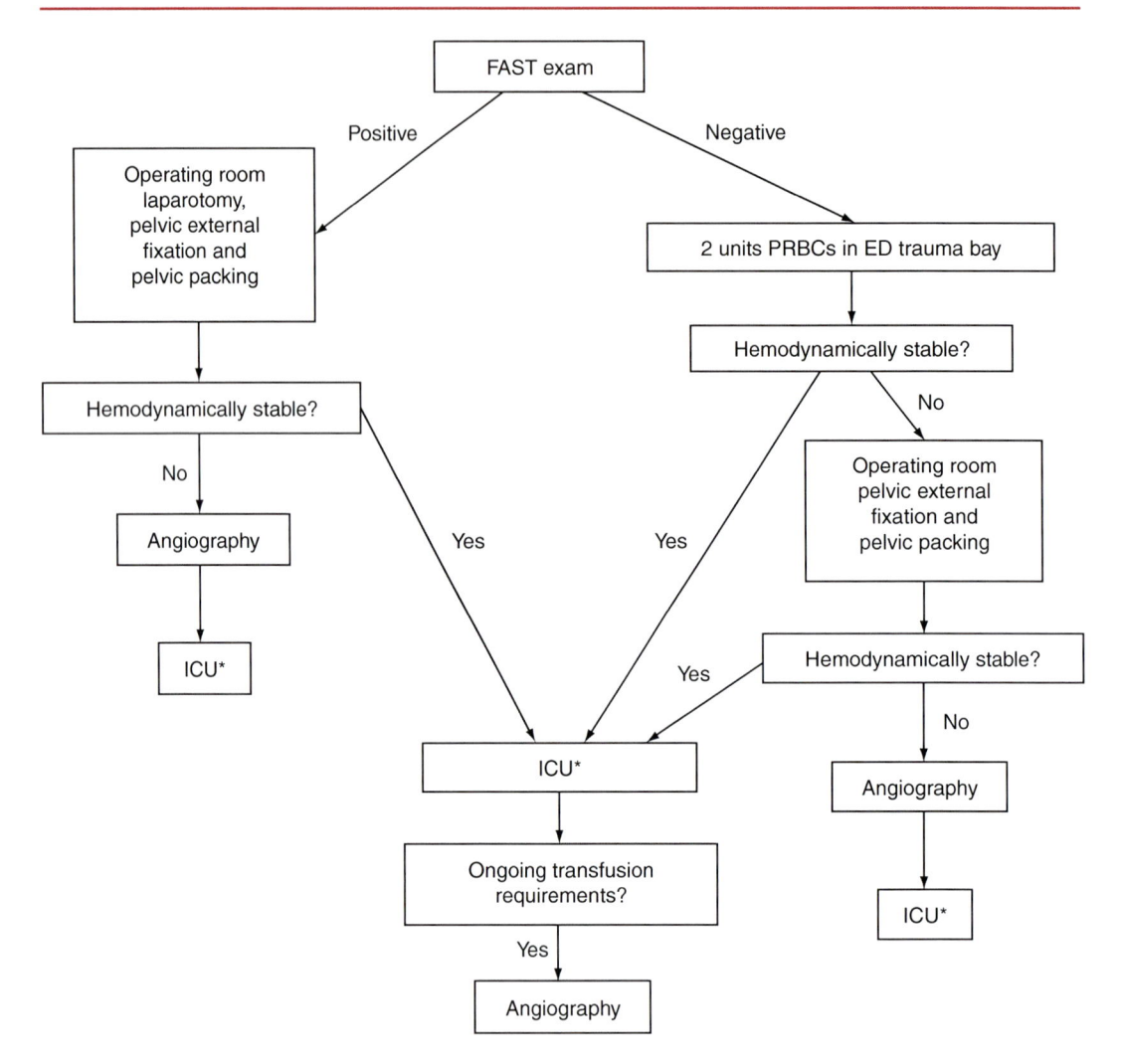

Fig. 10.8 Algorithm for the treatment of patients with pelvic fracture who present with hemodynamic instability. *Patients in whom a laparotomy was not done usually have an abdominal CT scan en route to the intensive care unit (ICU). In the ICU, the patient receives further fluid resuscitation and is warmed; attempts are made to normalize the coagulation status. Recombinant factor VIIa should be considered if the patient is recalcitrant to all other interventions. *FAST* focused abdominal sonography for trauma, *PRBCs* packed red blood cells

patient is transferred directly to the ICU. An abdominal computed tomography scan can be performed at this point in time. If the patient requires ongoing transfusion while in the ICU, angiographic assessment, if not previously done, should be performed.

The experience at Harborview Medical Center has evolved similarly in many respects, especially with regard to resuscitation and ICU management. Additionally, the overall concept of combining pelvic stability with hemorrhage control is adhered to. However, the use of pelvic packing and external fixation is much less commonly performed. Typically, pelvic stability is provided with a circumferentially wrapped sheet while the patient undergoes their initial abdominal and radiographic evaluations. These sheets are left in position for up to 24–48 h if necessary. However, frequent evaluation of the skin is necessary to avoid focal pressure and soft tissue necrosis. For patients with identified intraabdominal pathology requiring exploratory laparotomy, pelvic stability is maintained either by retention of the circumferential pelvic sheet, by application of an external fixator, or by primary percutaneous pelvic stabilization of the posterior and/or anterior pelvic ring as indicated. Individualized care is directed by the fracture pattern. For patients who do not respond to a combination of pelvic stabilization and treatment of any identified intraabdominal sources of bleeding, angiography is typically performed.

Clear and direct communication between the general trauma surgeon, orthopedic trauma specialist, and other care providers, is essential in the management of these severely injured patients. Such communication can help care providers understand each other's concerns and the critical issues which each provider has identified. This communication can lead to improvements in the timing and order of the patient's subsequent diagnostic, interventional, and definitive management.

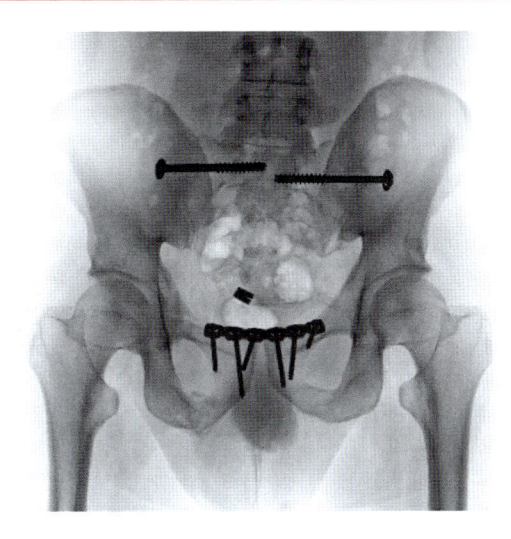

Fig. 10.9 Radiograph following plate fixation of anterior pelvic ring and bilateral percutaneous iliosacral screw fixation of an APC-III pelvic fracture

10.8 Definitive Treatment

10.8.1 Internal Fixation: Anterior Pelvic Ring

Reduction and fixation of a pubic symphysis diastasis may be performed using either a midline incision (extending any prior laparotomy incision), or through a Pfannenstiel incision. A separate Pfannenstiel approach is preferred whenever possible to allow for extension laterally if needed. The midline raphe is identified and dissection occurs between the two bellies of rectus abdominis muscle. The insertion of the rectus is often traumatically avulsed from one of the rami. Surgical release of the rectus from its insertion should be avoided. A Hohman type retractor can be placed beneath the rectus abdominis and over the anterior of the rami to assist with retraction of the rectus and reduction of the hemipelvis. Relaxation of retraction from one side often allows improved retraction and visualization on the opposite side. For "open book" type injuries, a Weber tenaculum is commonly placed anteriorly at the same level of the pubic body to achieve the reduction. Counterforce may need to be applied to correct any flexion or extension deformity of one hemipelvis with respect to the other.

If one hemipelvis is posteriorly displaced an anteriorly directed force may be obtained using Jungbluth pelvic reduction clamp which is applied with screws placed from anterior to posterior in the pubic body.

Several different plate and screw options may be used. Commonly a six-hole 3.5-mm curved reconstruction plate is used (Fig. 10.9). Other options include the use of a two- or four-hole plate with large fragment cortical or cancellous screws. One advantage of a two-hole plate is it permits some mobility, which may be useful in staged fixation when additional posterior reduction is required. However, in a clinical study comparing the use of a two-hole plate to a multi-hole plate fixation construct, investigators found a higher rate of implant failure and a significantly higher rate of pelvic malunion in patients treated with a two-hole symphyseal plate [31]. Locked plate fixation is now available; however, there are no clearly defined benefits over nonlocked plating in the anterior pelvic ring. Double plating has also been described to improve stability if posterior internal fixation cannot be performed and the patient is to be treated definitively with external fixation [32]. This is rarely used, requires a significant anterior soft tissue dissection, and has largely been replaced with a more aggressive approach to fixation of any associated posterior ring injuries.

If there is an associated fracture of the pubic ramus, a longer plate can be used to span across the fracture site. Given the multiple associated soft tissue attachments at the pubic ramus that provide some local stability, the associated rami fractures can sometimes be ignored and the anterior ring is treated with standard symphyseal fixation alone. Alternatively a retrograde ramus screw can be used for internal fixation, but technically is somewhat demanding as the available corridor for screw placement is quite narrow [33].

Based on the success of the anterior external pelvic fixation, a novel anterior pelvic ring fixation technique that uses the same single supra-acetabular pin sites used in traditional external fixation has been developed. This technique uses bilateral single supra-acetabular pedicle screws attached to a subcutaneous connecting rod at the bikini line level [34]. This, in theory, offers the advantages of an anterior external fixator without the risk of pin site infections while improving patient comfort. However as with any new technique complications exist, most common is irritation of the lateral femoral cutaneous nerve but more severe complications have occurred, such as placing the rod too tight on the abdomen or loss of reduction necessitating reoperation [35]. Additionally, this technique can only be used in a semi-elective basis and not in hemodynamically unstable patients.

10.8.2 Internal Fixation: Posterior Pelvic Ring

Injury to the posterior pelvic ring can occur through a dislocation of the sacroiliac joint or through a fracture of the sacrum. These injuries can be addressed through either closed reduction or open reduction and subsequent internal fixation with cannulated or noncannulated screws.

It is important to obtain an anatomical reduction of the SI joint as long-term pain is associated with malreduction. The patient can be positioned either supine or prone, depending on the overall surgical plan and the comfort of the surgeon. A closed reduction can be attempted using a combination of limb traction, a fracture table, or direct manipulation using an external fixator. If an accurate reduction is obtained, percutaneous stabilization of the SI joint with large screws can be performed. When open reduction is required, either a posterior or anterior approach may be used. The posterior approach has been associated with a higher rate of wound healing complications, while the anterior approach has a higher risk of L5 nerve root injury since it runs less than 2 cm medial to the SI joint. However, with careful dissection and strategic posterior incision placement, the soft tissue complications associated with open approaches to the posterior pelvis can be significantly reduced. A combination of direct visualization, palpation of the SI joint, and radiographic evaluation are used to judge the reduction through either approach. Cannulated or noncannulated iliosacral screws can be used following either approach (Fig. 10.9). Another option is following the anterior approach: plate fixation can be used but this is not as strong as iliosacral screws.

Crescent fractures involve a fracture in which a portion of the ilium remains attached to the sacrum. If the intact portion of the ilium is large, the fracture can be reduced through an open posterior approach and fixed with interfragmentary lag screws. Occasionally, if the fracture is quite anterior, an iliac approach may be used. In instances where the fragment is small or the posterior ligaments are injured, then stabilization with iliosacral screws is typically performed.

Posterior transiliac plate fixation may be selected for cases in which there is no available corridor for safe placement of SI screws. Usually a 4.5-mm reconstruction plate is used and tunneled subcutaneously securing fixation to both posterior iliac spines. Postoperative wound complications remain a concern, especially in the presence of a closed internal degloving injury [36].

Alternatively, a trans-sacral bolt/bar can be used to obtain fixation. This technique offers the advantage of having a bar placed posterior to the sacrum transfixing both iliac bones in the most posterior aspect just proximal to the posterior superior iliac spine, resulting in the vertical stress being shared through the entire length of the construct [37, 38] (Fig. 10.10). In addition, compression

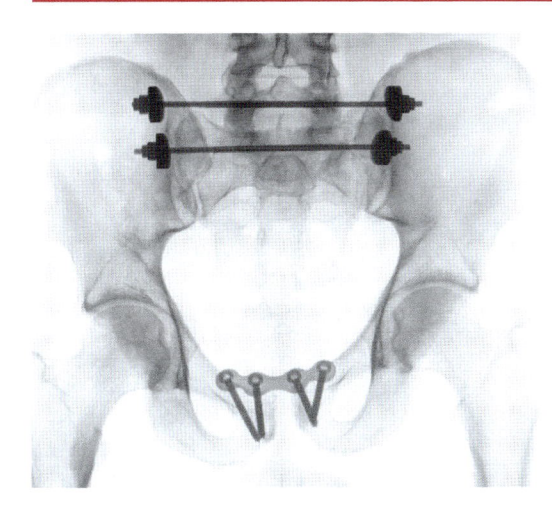

Fig. 10.10 Inlet radiograph of a patient with a pelvic ring injury and unilateral SI joint disruption who underwent anterior pelvic ring plate fixation and dual transiliac sacral bolt fixation

can be obtained through the use of nuts and washer on both ends of the rod, and it can be performed through a minimally invasive approach.

Displaced or unstable fractures of the iliac wing may require fixation through the iliac portion of an ilioinguinal approach. Fixation of the iliac wing can be difficult as the available bone for screw fixations is limited. The iliac wing is very narrow except along crest and as it widens near the acetabulum. Fixation can be accomplished either with plates (on the inner or outer aspect of the ilium), screws (placed between the inner and outer tables of the ilium) or combinations thereof.

References

1. Tile M, editor. Fractures of the pelvic and acetabulum. 2nd ed. Baltimore: Williams & Wilkins; 1995.
2. Kellam JF, Browner BD. Fractures of the pelvic ring: chapter 31. In: Browner BD, Jupiter JB, Levine AM, Trafton PG, editors. Skeletal trauma. WB Saunders: Philadelphia; 1992.
3. Olson SA, Burgess A. Classification and initial management of patients with unstable pelvic ring injuries. Instr Course Lect. 2005;54:383–93.
4. Spahn DR, Bouillon B, Cerny V, et al. Management of bleeding and coagulopathy following major trauma: an updated European guideline. Crit Care. 2013;17(2):R76.
5. Tile M. Pelvic ring fractures: should they be fixed? J Bone Joint Surg Br. 1988;70(1):1–12.
6. Tile M. Acute pelvic fractures: I. Causation and classification. J Am Acad Orthop Surg. 1996;4(3):143–51.
7. Burgess AR, Eastridge BJ, Young JW, et al. Pelvic ring disruptions: effective classification system and treatment protocols. J Trauma. 1990;30(7):848–56.
8. Dalal SA, Burgess AR, Siegel JH, et al. Pelvic fracture in multiple trauma: classification by mechanism is key to pattern of organ injury, resuscitative requirements, and outcome. J Trauma. 1989;29(7):981–1000; discussion 1000–2.
9. Agolini SF, Shah K, Jaffe J, Newcomb J, Rhodes M, Reed 3rd JF. Arterial embolization is a rapid and effective technique for controlling pelvic fracture hemorrhage. J Trauma. 1997;43(3):395–9.
10. Archdeacon MT, Hiratzka J. The trochanteric C-clamp for provisional pelvic stability. J Orthop Trauma. 2006;20(1):47–51.
11. Balogh Z, Caldwell E, Heetveld M, et al. Institutional practice guidelines on management of pelvic fracture-related hemodynamic instability: do they make a difference? J Trauma. 2005;58(4):778–82.
12. Brazzel C. Thromboelastography-guided transfusion Therapy in the trauma patient. AANA J. 2013;81(2):127–32.
13. Routt Jr ML, Falicov A, Woodhouse E, Schildhauer TA. Circumferential pelvic antishock sheeting: a temporary resuscitation aid. J Orthop Trauma. 2002;16(1):45–8.
14. Croce MA, Magnotti LJ, Savage SA, Wood 2nd GW, Fabian TC. Emergent pelvic fixation in patients with exsanguinating pelvic fractures. J Am Coll Surg. 2007;204(5):935–9; discussion 940–2.
15. Riemer BL, Butterfield SL, Diamond DL, et al. Acute mortality associated with injuries to the pelvic ring: the role of early patient mobilization and external fixation. J Trauma. 1993;35(5):671–5; discussion 676–7.
16. Grimm MR, Vrahas MS, Thomas KA. Pressure-volume characteristics of the intact and disrupted pelvic retroperitoneum. J Trauma. 1998;44(3):454–9.
17. Ganz R, Krushell RJ, Jakob RP, Kuffer J. The antishock pelvic clamp. Clin Orthop Relat Res. 1991;267:71–8.
18. Eastridge BJ, Starr A, Minei JP, O'Keefe GE, Scalea TM. The importance of fracture pattern in guiding therapeutic decision-making in patients with hemorrhagic shock and pelvic ring disruptions. J Trauma. 2002;53(3):446–50; discussion 450–1.
19. Miller PR, Moore PS, Mansell E, Meredith JW, Chang MC. External fixation or arteriogram in bleeding pelvic fracture: initial therapy guided by markers of arterial hemorrhage. J Trauma. 2003;54(3):437–43.
20. Yasumura K, Ikegami K, Kamohara T, Nohara Y. High incidence of ischemic necrosis of the gluteal muscle after transcatheter angiographic embolization for severe pelvic fracture. J Trauma. 2005;58(5):985–90.
21. Pohlemann T, Bosch U, Gansslen A, Tscherne H. The Hannover experience in management of pelvic fractures. Clin Orthop Relat Res. 1994;305:69–80.

22. Smith WR, Moore EE, Osborn P, et al. Retroperitoneal packing as a resuscitation technique for hemodynamically unstable patients with pelvic fractures: report of two representative cases and a description of technique. J Trauma. 2005;59(6):1510–4.
23. Cothren CC, Osborn PM, Moore EE, Morgan SJ, Johnson JL, Smith WR. Preperitoneal pelvic packing for hemodynamically unstable pelvic fractures: a paradigm shift. J Trauma. 2007;62(4):834–9; discussion 839–42.
24. Rossaint R, Cerny V, Coats TJ, et al. Key issues in advanced bleeding care in trauma. Shock. 2006;26(4):322–31.
25. Spivey M, Parr MJ. Therapeutic approaches in trauma-induced coagulopathy. Minerva Anestesiol. 2005;71(6):281–9.
26. Holcomb JB, del Junco DJ, Fox EE, et al. The prospective, observational, multicenter, major trauma transfusion (PROMMTT) study: comparative effectiveness of a time-varying treatment with competing risks. JAMA Surg. 2013;148(2):127–36.
27. Holcomb JB, Wade CE, Michalek JE, et al. Increased plasma and platelet to red blood cell ratios improves outcome in 466 massively transfused civilian trauma patients. Ann Surg. 2008;248(3):447–58.
28. Gonzalez E, Pieracci FM, Moore EE, Kashuk JL. Coagulation abnormalities in the trauma patient: the role of point-of-care thromboelastography. Semin Thromb Hemost. 2010;36(7):723–37.
29. Johansson PI, Sorensen AM, Larsen CF, et al. Low hemorrhage-related mortality in trauma patients in a Level I trauma center employing transfusion packages and early thromboelastography-directed hemostatic resuscitation with plasma and platelets. Transfusion. 2013;53(12):3088–99.
30. Johansson PI, Stissing T, Bochsen L, Ostrowski SR. Thrombelastography and tromboelastometry in assessing coagulopathy in trauma. Scand J Trauma Resusc Emerg Med. 2009;17:45.
31. Sagi HC, Papp S. Comparative radiographic and clinical outcome of two-hole and multi-hole symphyseal plating. J Orthop Trauma. 2008;22(6):373–8.
32. Ponson KJ, van Dijke GA H, Joosse P, Snijders CJ, Agnew SG. Improvement of external fixator performance in type C pelvic ring injuries by plating of the pubic symphysis: an experimental study on 12 external fixators. J Trauma. 2002;53(5):907–12; discussion 912–3.
33. Routt Jr ML, Simonian PT, Grujic L. The retrograde medullary superior pubic ramus screw for the treatment of anterior pelvic ring disruptions: a new technique. J Orthop Trauma. 1995;9(1):35–44.
34. Vaidya R, Colen R, Vigdorchik J, Tonnos F, Sethi A. Treatment of unstable pelvic ring injuries with an internal anterior fixator and posterior fixation: initial clinical series. J Orthop Trauma. 2012;26(1):1–8.
35. Vaidya R, Kubiak EN, Bergin PF, et al. Complications of anterior subcutaneous internal fixation for unstable pelvis fractures: a multicenter study. Clin Orthop Relat Res. 2012;470(8):2124–31.
36. Suzuki T, Hak DJ, Ziran BH, et al. Outcome and complications of posterior transiliac plating for vertically unstable sacral fractures. Injury. 2009;40(4):405–9.
37. Mehling I, Hessmann MH, Rommens PM. Stabilization of fatigue fractures of the dorsal pelvis with a trans-sacral bar. Operative technique and outcome. Injury. 2012;43(4):446–51.
38. Vanderschot PM, Broens PM, Vermeire JI, Broos PL. Trans iliac-sacral-iliac bar stabilization to treat bilateral sacro-iliac joint disruptions. Injury. 1999;30(9):637–40.

Urological Injuries in Polytraumatized Patients

11

David Pfister and Axel Heidenreich

Contents

11.1	**Introduction**...	143
11.2	**Renal Trauma**..	143
11.2.1	Clinical Symptoms....................................	144
11.2.2	Imaging Studies	144
11.2.3	Treatment	144
11.3	**Ureteral Trauma**...................................	145
11.3.1	Clinical Symptoms....................................	146
11.3.2	Imaging	146
11.3.3	Management...	146
11.4	**Bladder Trauma**....................................	146
11.4.1	Clinical Symptoms....................................	146
11.4.2	Imaging	148
11.4.3	Treatment	148
11.5	**Urethral Trauma**...................................	150
11.5.1	Clinical Symptoms....................................	152
11.5.2	Radiographic Examination	152
11.5.3	Treatment	153
References...		154

11.1 Introduction

In patients with multiple traumas, urological components are regularly involved approximately in 10 % of the cases [1]. Genitourinary injuries can result in significant morbidity and mortality [1–3]. In general, one has to distinguish between blunt and penetrating injuries to the urogenital organs necessitating an individualized therapeutic approach. According to the exposition of the different organs, the incidence lowers from cranial to caudal with the kidney being the most common injured organ with 1–5 % of trauma cases. In most cases uretral injuries are iatrogenic, whereas about 18 % result in blunt and 7 % in penetrating trauma.

11.2 Renal Trauma

Renal trauma occurs in about 1–5 % of all traumas with blunt trauma accounting for the most common mechanisms of renal injury in about 90 % of the cases [1–7]. While penetrating injuries are less frequent, they tend to be more severe. These result in a higher rate of nephrectomies and are associated with a higher rate of associated organ injuries [8]. Possible indicators for renal trauma are falls, blunt trauma to the flank region, or high speed motor-vehicle accidents [1, 5, 6]. The Committee on the Organ Injury Scaling

D. Pfister • A. Heidenreich (✉)
Department of Urology, RWTH University Aachen,
Pauwelsstr. 30, Aachen 52074, Germany
e-mail: aheidenreich@ukaachen.de

Table 11.1 AAST organ injury severity scale for the kidney

1	Contusion or nonexpanding subcapsular hematoma. No laceration
2	Nonexpanding perirenal hematoma, cortical laceration <1 cm deep w/o extravasation
3	Cortical laceration >1 cm w/o urinary extravasation
4	Laceration: through corticomedullary junction into collecting system or Vascular: segmental renal artery or vein injury with contained hematoma
5	Laceration: shuttered kidney or Vascular: renal pedicle injury or avulsion

of the American Association for Surgery of Trauma (AAST) classified renal injuries as shown in Table 11.1.

11.2.1 Clinical Symptoms

Gross hematuria might be present but it does not correlate with the degree of injury since major injuries such as renal pedicle lacerations or disruption of the ureteropelvic junction may occur without hematuria. Blood transfusion requirements are an indirect indication of the rate of blood loss.

11.2.2 Imaging Studies

Patients with blunt renal, microscopic hematuria, stable vital signs in the absence of deceleration trauma usually do not have to undergo any specific imaging studies [1, 4, 9].

Patients with gross hematuria, penetrating injuries with suspected renal involvement, and instable vital signs must undergo immediate imaging studies (Fig. 11.1) [1, 5]. CT imaging represents the gold standard for radiographic assessment in suspected renal injury because (1) it defines the location and the extent of injuries, (2) detects contusions and devitalized segments, (3) allows for visualization of the entire retroperitoneum, (4) allows for assessment of the renal pedicle, and (5) detects urinary extravasations [1, 5, 9, 10]. Spiral CT scans are advantageous due to shorter scanning times, but do not allow the identification of injuries to the renal collecting system, thereby necessitating the use of delayed CT scans. Angiography is important only for superselective embolization in the management of persisting or delayed hemorrhage.

11.2.3 Treatment

In general, there are several guidelines on treating renal trauma. The management of renal trauma was described in detail by the European Association of Urology [11].

A summary of the various therapeutic approaches is presented in Fig. 11.1. Life-threatening hemodynamic instability or an expanding or pulsatile retroperitoneal hematoma during explorative laparotomy usually represents AAST grade 5 injury and requires immediate surgery [1, 3]. A transperitoneal approach with early occlusion of the pedicle prior to opening of Gerota's fascia is strongly recommended. In patients with avulsion of the renal pedicle close to the aorta or the inferior vena cava, it might be necessary to clamp the major vessels just above and below the renal pedicle to control bleeding and to explore the retroperitoneum. In patients with significant injuries to the vascular pedicle, nephrectomy is the treatment of choice, unless the kidney can be preserved in cases of solitary organ or bilateral injuries. In patients with bleeding from the renal parenchyma due to penetrating injuries, embolization is advised [4, 6, 11, 12].

Persistent bleeding, injuries to the renal collecting system, the renal pelvis, or the ureter with urinary extravasation are relative indications for surgery [1]. Urinary extravasation may be treated by endoluminal stenting and/or placement of a percutaneous nephrostomy. However, surgical reconstruction is advised in the presence of devitalized fragments and associated enteric and pancreatic injuries [13]. Aggressive surgical management for renal lacerations is associated with a 23 % morbidity rate whereas initial nonoperative treatment resulted in an 85 % morbidity rate.

Hemodynamically stable patients with AAST grade 1 and 2 injuries can be managed nonoperatively with supportive care, bed rest, hydration, and prophylactic antibiotics [4–6].

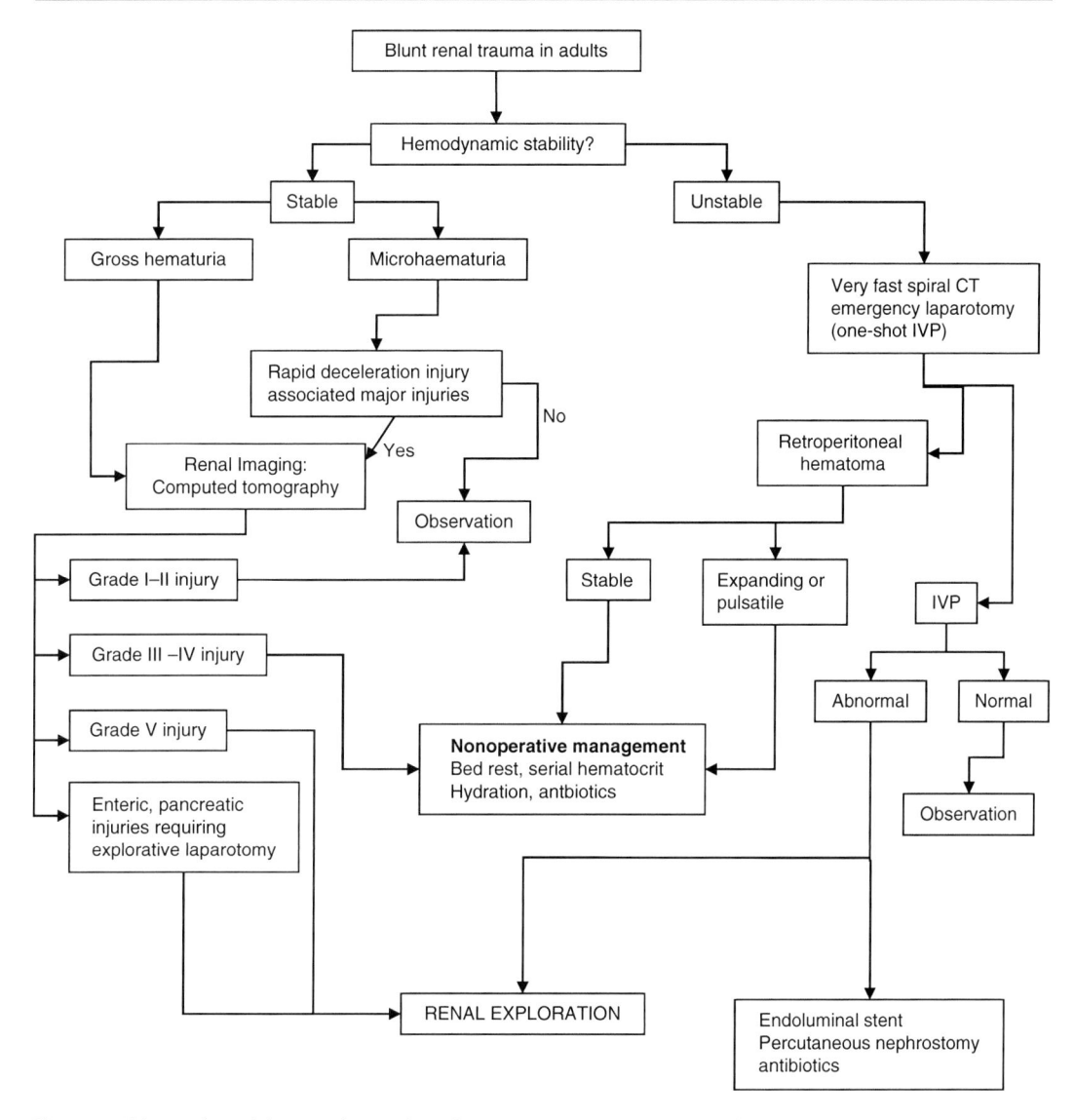

Fig. 11.1 Diagnostic and therapeutic algorithm for suspected blunt renal trauma in adults

Stable patients with renal gunshot injuries or stab wounds must be explored if the renal hilum and the collecting system are involved or if persistent bleeding exists.

In patients with significant renal injuries, postoperative observation is extremely important because a variety of delayed complications may occur within the first 30 days of injury. This includes but is not limited to hemorrhage, urinary fistula, arteriovenous fistula, and pseudoaneurysms [2, 14]. Patients must undergo imaging studies if they develop clinical symptoms such as fever, increasing flank pain, persistent bleeding, and arterial hypertension. As for the primary diagnosis, CT scan of the abdomen is the preferred imaging modality.

11.3 Ureteral Trauma

Trauma to the ureter is rare and it accounts for only about 1 % of all genitourinary injuries. Most commonly, ureteral lesions result from iatrogenic injuries (75 %), and only 7 and 18 % result from blunt and penetrating trauma, respectively. The majority of iatrogenic injuries occur after gynecologic interventions (70–75 %), while about

Table 11.2 Classification of ureteral injury

Grade	Description of injury
I	Hematoma only
II	Laceration <50 % of circumference
III	Laceration >50 % of circumference
IV	Complete tear <2 cm of devascularization
V	Complete tear >2 cm of devascularization

15–20 % result after general surgery and about 10–15 % occur following urologic procedures.

As with all other genitourinary organs, the AAST has classified ureteral injuries according to their severity as indicated in Table 11.2.

11.3.1 Clinical Symptoms

There are no specific clinical symptoms; unspecific symptoms such as meteorism, abdominal distension, and flank pain are caused by retroperitoneal urinoma. Ureteral injury should always be suspected in patients with penetrating abdominal or retroperitoneal injuries, and in patients with blunt deceleration traumas.

11.3.2 Imaging

The most common imaging modality is an intravenous pyelography. It is performed in nearly two-thirds of the patients with suspected ureteral injuries. Typically, IVP demonstrates retroperitoneal extravasation of contrast material. In about 30–50 %, additional retrograde ureteropyelography is performed to verify the location and the extent of the ureteral injury. Small lesions may be managed by the placement of an endoluminal DJ-catheter. In very rare cases, the suspicion of a ureteral injury is based on ultrasound findings of a retroperitoneal fluid collection (urinoma) or a hydronephrosis (Fig. 11.2).

11.3.3 Management

In patients with partial tears of the ureter, the most common, simple, and effective measure is placement of a ureteral stent and/or a percutaneous nephrostomy tube.

If iatrogenic ureteral injuries are detected intraoperatively, an endoluminal DJ stent should be placed with the ureteral laceration being closed by interrupted sutures with a monofil suture. Postoperatively, no drain or suction should be placed in order to prevent the development of a urinary fistula.

Reconstruction of grade 3–5 injuries depends on the anatomic localization of the injury. Usually, grade 3 and 5 injuries can be treated by an end-to-end anastomosis. The anastomosis is reconstructed with absorbable sutures after placement of a ureteral catheter, which can stay in place for about 3–4 weeks. Other surgical options are listed in Table 11.3.

11.4 Bladder Trauma

Bladder injuries are one of the most frequent urological injuries in trauma patients. Among abdominal injuries requiring surgical repair, about 2 % involve the bladder [1, 15, 16]. Blunt trauma accounts for about 65–85 % of bladder ruptures whereas penetrating trauma accounts for only about 14–33 % of all bladder injuries. Bladder ruptures in the setting of blunt traumas are classified as extra- or intraperitoneal, triggering the choice between a conservative approach and a surgical correction. Most commonly, extraperitoneal bladder ruptures occur in about 55 % of the cases, followed by intraperitoneal bladder ruptures in 38 %. Combined injuries are rare (5–8 % of cases). Motor vehicle accidents contribute significantly to bladder rupture by blunt trauma. Seventy to ninety-seven percent of patients with bladder trauma have accompanied pelvic fractures, whereas only 5–30 % of the pelvic fractures are associated with bladder injuries [15–20].

The Committee on the Organ Injury Scaling of the American Association for Surgery of Trauma (AAST) classified bladder injuries as shown in Table 11.4.

11.4.1 Clinical Symptoms

The two most common signs and symptoms for bladder injuries are gross hematuria (80–100 %)

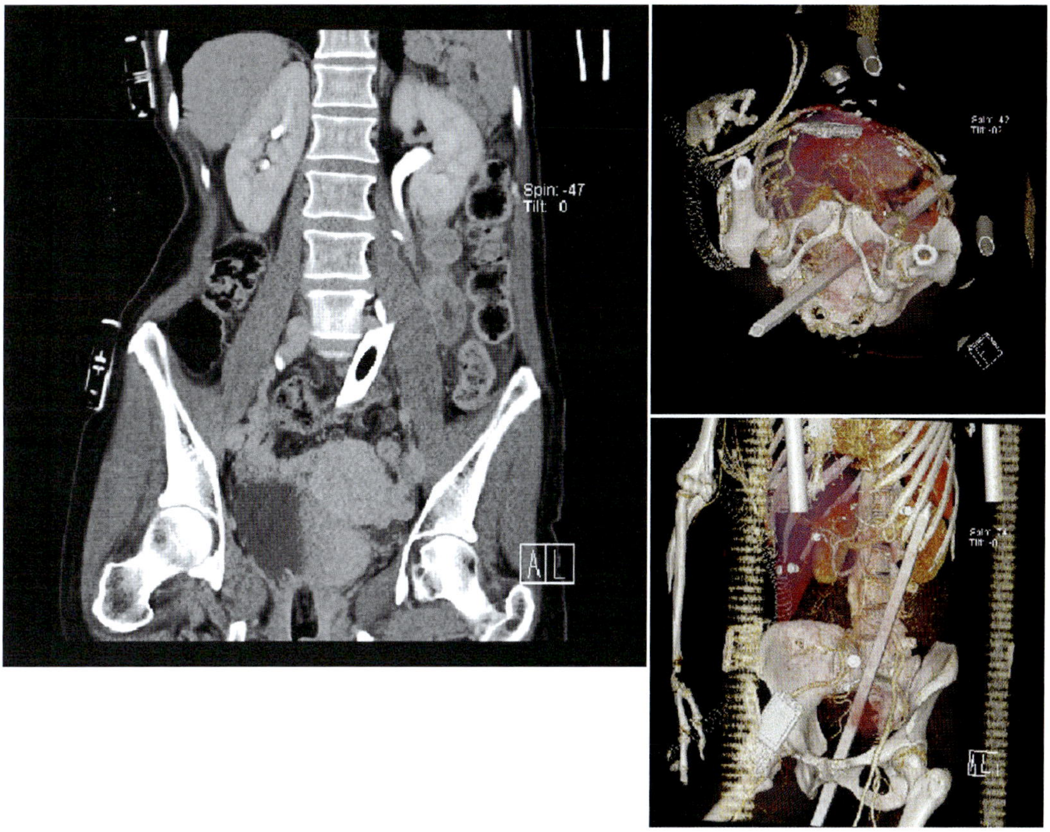

Fig. 11.2 Left ureteral injury with urinoma and hematoma in the small pelvis

Table 11.3 Surgical options to reconstruct ureteral injuries depending on the anatomic level of injury

Level of urethral injury	Options of reconstruction
Upper third	Transuretero-ureterostomy
	Ureterocalycostomy
	Ileal replacement of the ureter
	Percutaneous pyelovesical bypass prosthesis
	Renal autotransplantation
Middle third	Transuretero-ureterostomy
	Boari flap and intravesical reimplantation
	Ileal replacement of the ureter
Lower third	Direct intravesical reimplantation
	Psoas hitch reimplantation
Complete ureteral loss	Ileal replacement (delayed)[a]
	Renal autotransplantation (delayed)[a]
	Percutaneous pyelovesical bypass (delayed)[a]

[a]For urinary drainage, a percutaneous nephrostomy tube should be placed together with occlusion of the ureter by sutures, clips, or occluding catheters

Table 11.4 AAST organ injury severity scale for the bladder

I	Hematoma	Contusion, intramural hematoma
I	Laceration	Partial thickness
II	Laceration	Extraperitoneal bladder wall laceration <2 cm
III	Laceration	Extraperitoneal (>2 cm) or intraperitoneal (<2 cm) bladder wall laceration
IV	Laceration	Intraperitoneal bladder wall laceration >2 cm
V	Laceration	Intraperitoneal or extraperitoneal bladder wall laceration extending into the bladder neck or ureteral orifices

and abdominal tenderness (60–70 %) [15]. Other findings may include the inability to void (rule out: intrapelvic urethral rupture!), bruises over the suprapubic region, and abdominal distension. Depending on the type and extent of associated injuries to the pelvic floor, extravasation of urine

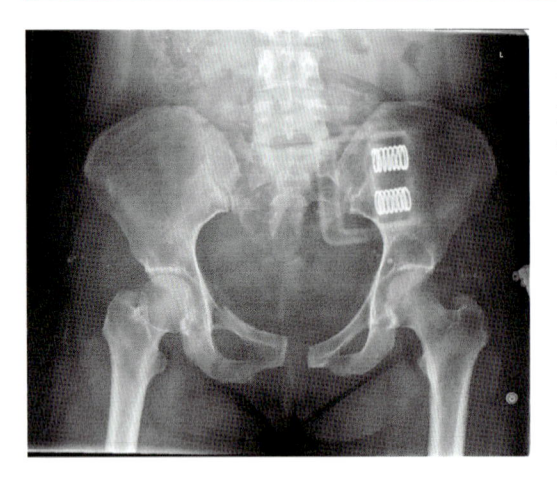

Fig. 11.3 Deceleration trauma after a jump from the third floor

may result in swelling of the perineum, scrotum, thighs, and the anterior abdominal wall.

11.4.2 Imaging

The classic combination of pelvic fracture and gross hematuria requires immediate cystourethrography to rule out urethral and/or bladder ruptures [15, 17, 21, 22]. All patients with pelvic ring fractures and gross hematuria should undergo immediate cystography (Figs. 11.3 and 11.4). Since microscopic hematuria is a relative indicator for significant injury, recommendations for the most appropriate imaging studies are sparse in the literature and in the existing guidelines. Imaging of the bladder may be reserved for those with anterior rami fractures (straddle fractures) or Malgaigne type severe ring disruption (Tile III).

Retrograde cystography in the evaluation of bladder trauma represents the imaging procedure of choice [15, 17, 19–21]. With an adequate filling and postvoid images taken, cystography has an accuracy of 85–100 % in the identification of bladder ruptures. For the highest degree of diagnostic accuracy, the bladder should be filled with at least 350 cc. Bladder rupture may be identified only on the postdrainage film in only about 10 % of patients. Thus, images must always have to include x-rays upon maximal distension and a completely emptied bladder (Fig. 11.5).

Blood at the urethral meatus may be a sign for significant urethral injury. Retrograde urethrography should be performed prior to catheterization of the bladder to exclude associated urethral lesion, which can occur in 10–30 % of the cases [1, 17].

Other imaging studies such as ultrasonography, intravenous pyelography, standard CT scans, or magnetic resonance imaging are inadequate for the evaluation of the bladder and the urethra after trauma [1, 15, 17]. As CT scan is performed in most patients who present with multiple trauma, CT cystography is an excellent substitute for standard cystography. The bladder should be filled with at least 350 cc of dilute (2 %) contrast dye [22].

11.4.3 Treatment

The therapeutic approach to treat any bladder rupture depends on the type of injury, the coexisting injuries, and the condition of the patient (Figs. 11.6 and 11.7)

Most patients with extraperitoneal bladder ruptures may be treated nonoperatively by drainage even in the presence of large extravasations [1, 19, 20, 23]. More than two-thirds of the ruptures resolve within 2 days and almost all within 3 weeks. From the day of catheterization until 3 days after removal of the catheter, antibiotic prophylaxis is recommended.

If a laparotomy is performed for any other reason, extraperitoneal bladder ruptures should be closed by a single layer running suture of 2-0 or 3-0; the bladder is usually drained by a 20 F transurethral catheter before a cystography is performed postoperatively on day 5. Following internal fixation of the pelvic fracture, a direct repair of the extraperitoneal rupture is advised. Concomitant rectal and/or vaginal injuries, open pelvic fractures, the presence of bone fragments in the bladder wall, and entrapment of the bladder wall between bone fragments necessitate immediate surgical repair even in extraperitoneal bladder rupture [1, 15, 16]. Involvement of the bladder neck or the ureteral orifices also requires immediate surgical repair. Bladder neck reconstruction, transurethral placement of an endoluminal catheter, or even ureteral reimplantation (Psoas-Hitch

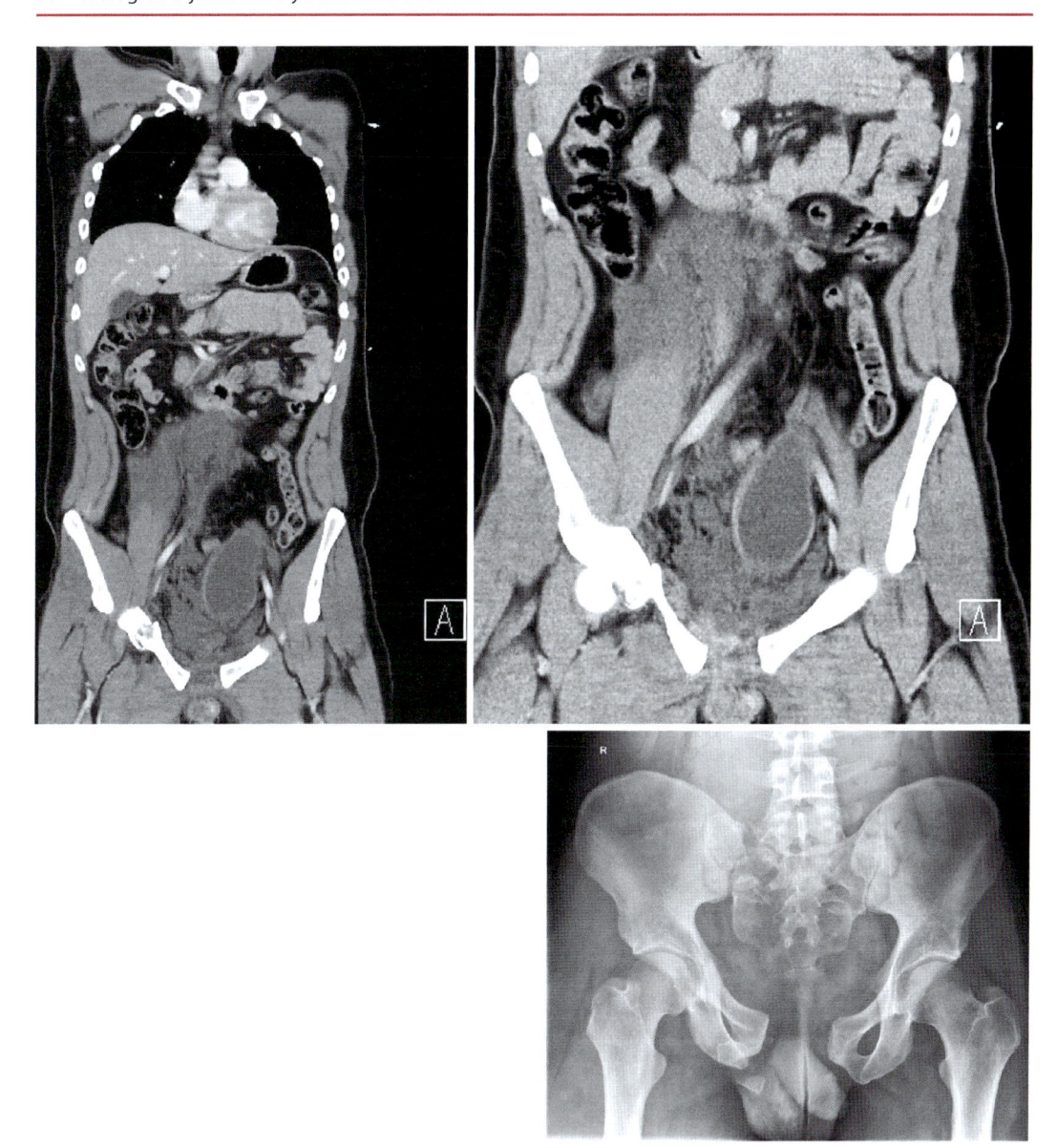

11 Urological Injuries in Polytraumatized Patients

Fig. 11.4 Rupture of the symphysis following a motorcycle accident: hematoma of the small pelvis, cranial dislocation of the bladder due to intrapelvic rupture of the urethra

technique) may be required in cases of severe ureteral orifice damage.

In contrast to an extraperitoneal bladder ruptures, all penetrating and intraperitoneal injuries should undergo immediate surgical repair [1, 15, 16]. In most cases, intraperitoneal bladder perforations are accompanied by other intra-abdominal injuries. Peritonitis might develop because of the urinary leakage. In this scenario, an overlooked bladder perforation may be mimicked by a significant rise in serum creatinine levels due to peritoneal reasorbtion. Antibiotic prophylaxis is administered for 3 days. Standard cystography is feasible on postoperative day 7–10 [17]. A suprapubic catheter is superior to a transurethral catheter for urinary drainage. In case of concomitant rectal or vaginal injuries, the ruptured organs are closed separately in a two-layer technique and a

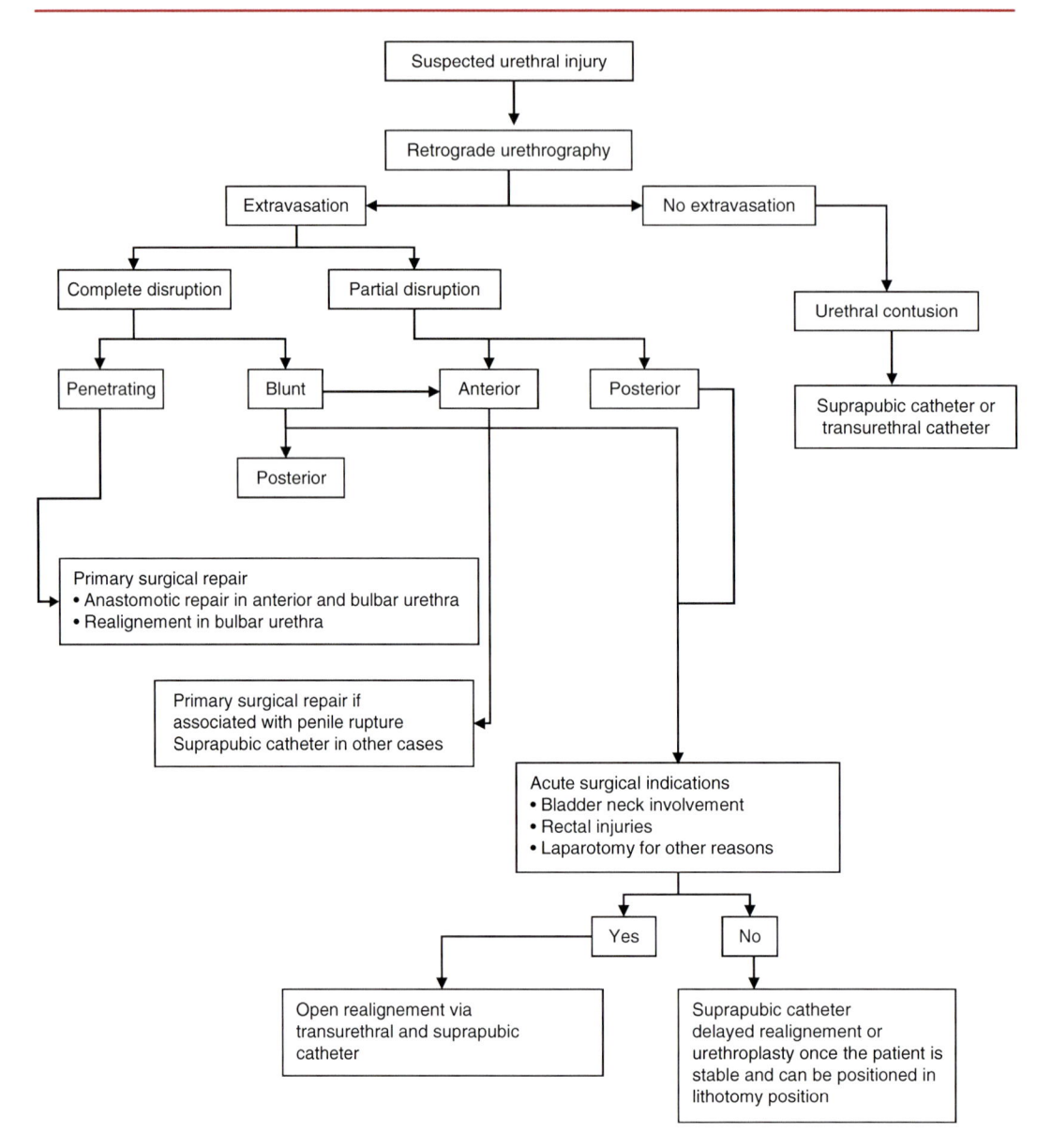

Fig. 11.5 Intraperitoneal bladder perforation with hematoma in the small pelvis after gun shot

11.5 Urethral Trauma

peritoneal flap of a vascularized omentum flap is interposed between bladder, vagina, or rectum.

Urethral injuries occur most commonly in association with pelvic fractures [1, 24, 25]. Unstable diametric pelvic fractures and bilateral ischiopubic rami fractures carry the highest risk of injury to the posterior urethra. In particular, the combination of straddle injuries with diastasis of the sacroiliac joint poses a risk about seven times higher for urethral injuries. The bulbomembranous junction is more vulnerable, as the posterior urethra is fixed at the urogenital diaphragm as well as the puboprostatic ligaments. In children, these are more frequently localized proximally and

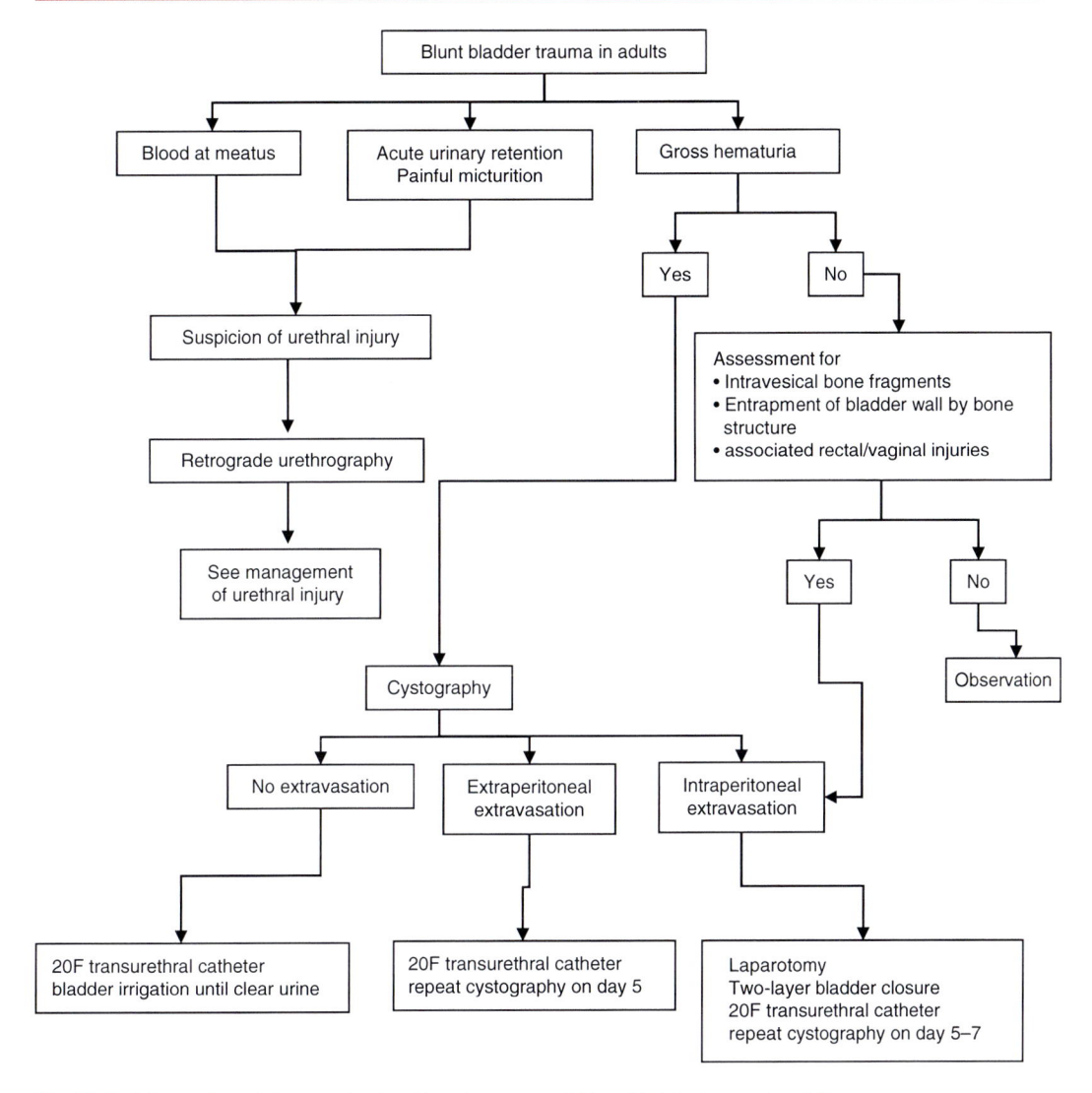

11 Urological Injuries in Polytraumatized Patients

Fig. 11.6 Diagnostic and therapeutic algorithm for suspected blunt bladder trauma in adults

interfere with the bladder neck, as the prostate still is rudimentary. In rare cases, a urethral disruption is diagnosed by the existence of the triad of blood at the external urethral meatus, inability to void, and palpable full bladder. It is usually detected by false catheterization or by the inability to place a transurethral catheter in the emergency department. Additional symptoms may include perineal hematoma and inability to palpate the prostate. In cases of a large pelvic hematoma, the symptom of an impalpable prostate may be misdiagnosed, as the contour of the prostate is smudged. In females with urethral injuries, vulvar edema and blood at the vaginal entrance are among the signs of urethral disorders.

The Committee on Organ Injury Scaling of the American Association for the Surgery of Trauma (AAST) has developed a reliable urethral-injury scaling system (Table 11.5).

No treatment is required for type I and II injuries [1, 24–32]. Usually, types II and III can be managed nonoperatively. A transurethral and a suprapubic catheter are placed. Types IV and V

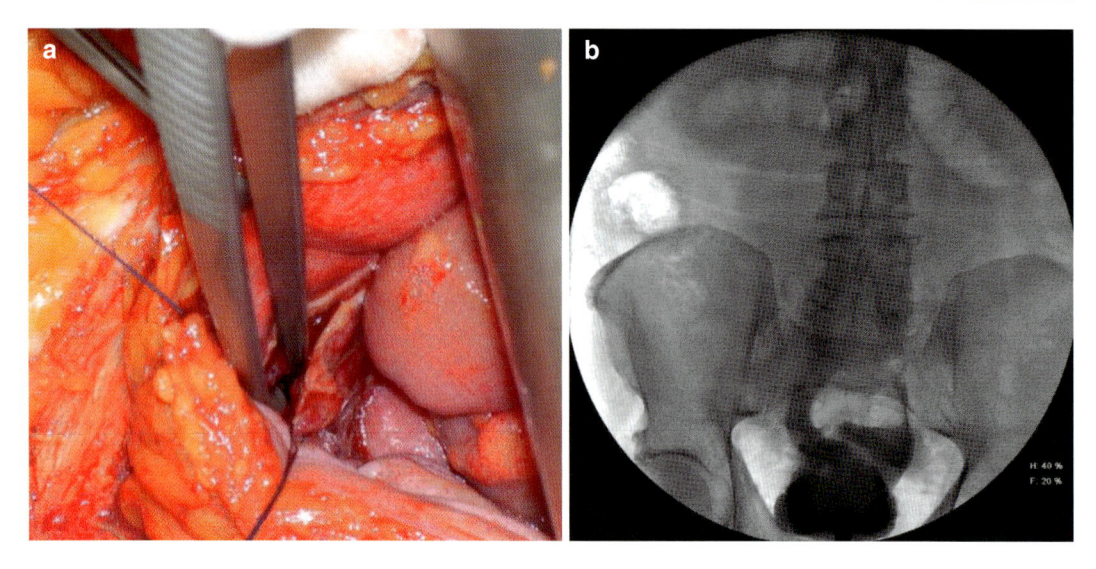

Fig. 11.7 Intraperitoneal bladder rupture after blunt trauma (**b**) with intraoperative situs (**a**)

Table 11.5 AAST organ injury severity scale for the urethra

I	Contusion	Blood at the urethral meatus, normal urethrogram
II	Stretch injury	Elongation of the urethra w/o extravasation on urethrography
III	Partial disruption	Extravasation of contrast at injury site with contrast visualized in the bladder
IV	Complete disruption	Extravasation of contrast at injury site without visualization in the bladder; <2 cm urethral separation
V	Complete disruption	Complete transsection with >2 cm urethral separation, or extension into the prostate or vagina

will require either endoscopic realignment or delayed urethroplasty.

Penetrating injuries to the anterior urethra most commonly derive from gunshots and involve the pendulous and bulbar urethral segments.

11.5.1 Clinical Symptoms

Blood at the meatus is present in about 40–95 % of patients with posterior urethral injuries and in about 75 % of patients with anterior urethral trauma. Its presence should preclude any attempts of urethral manipulation until the entire urethra is adequately imaged. Partial urethral disruption can be very easily transformed into complete urethral disruption due to several attempts of forced transurethral catheterization. In unstable patients, one attempt of transurethral catheterization is justified; if there is any difficulty, a suprapubic tube should be inserted instead. If a urethral injury is suspected, a retrograde urethrogram should be performed.

Gross or microscopic hematuria is a nonspecific clinical sign and the amount of bleeding does not correlate with the extent of injury [1, 25]. Pain during urination or acute urinary retention suggests urethral intrapelvic disruption or temporary spasm of the internal bladder sphincter. Any of the above-mentioned symptoms necessitates immediate radiographic evaluation [31], precludes transurethral manipulation, and prompts placement of a suprapubic catheter for urinary drainage.

Blood at the external urethral meatus is present in more than 80 % of female patients with pelvic fractures and urethral injuries.

11.5.2 Radiographic Examination

When a urethral injury is suspected, immediate retrograde urethrography should be performed

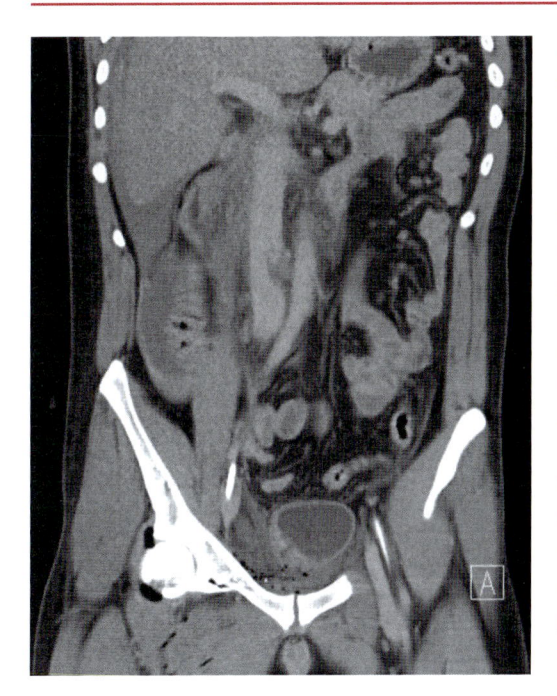

11 Urological Injuries in Polytraumatized Patients

Fig. 11.8 Diagnostic and therapeutic algorithm for suspected blunt urethral injury in male adults

(Fig. 11.8) [1, 25, 31]. In females, direct urethroscopy can be performed. In cases of subsequent urethral strictures, a combined urethrography and cystography can be performed to delineate the pelvic anatomy. Also, magnetic resonance tomography or antegrade cystourethroscopy via the suprapubic tract can be performed to visualize the anatomy of the urethra.

11.5.3 Treatment

Treatment differs with regard to the involvement of the anterior vs. posterior urethra and differs between male and females.

11.5.3.1 Treatment for Urethral Injuries in Males

Type I and II injuries of the anterior urethra can be easily managed by the placement of a transurethral catheter [1]. Type III injuries of the anterior urethra can be managed by the placement of a suprapubic catheter or a transurethral catheter with the advantage of the suprapubic tube avoiding urethral manipulation and diverting urine from the place of injury [24, 25]. In more than 50 % of the cases, spontaneous recanalization occurs; in all other cases, strictures can be managed by internal urethrotomy. Alternatively, delayed urethral reconstructive surgery may be performed with anastomotic urethroplasty or buccal mucosa grafts in strictures <1 cm or longer than 1 cm.

Type IV injuries can be repaired by an end-to-end anastomosis whereas type V injuries should be reconstructed by flap urethroplasty of by buccal mucosa grafts.

In females, most anterior urethral injuries can be sutured primarily from a transvaginal approach [24, 25]. Proximal urethral injuries are best approached transvesically with an optimal view of the bladder neck, the ureteral orifices, and the proximal urethra.

A treatment algorithm for the management of anterior and posterior male urethral injuries is present in Fig. 11.8.

Partial tears or short disruptions of the posterior urethra can be managed in most cases with a suprapubic catheter or a transurethral catheter for about 2 weeks. The majority of injuries heal and the risk of urethral strictures is low.

The management for complete disruption of the posterior urethra is variable [26–32]:

- Immediate open repair in case of any associated injury to the rectum double-layer closure of urethral and rectal lesion and interposition of a flap from the greater omentum.
- Primary endoscopic realignment by antegrade (using the canal of the suprapubic catheter) or retrograde approach.
- Primary open realignment with evacuation of the pelvic hematoma is not recommended; it is associated with frequent postoperative incontinence and impotence.

The most common result of posterior urethral disruption is the development of a short prostatobulbar urethral gap filled with dense fibrotic tissue. Delayed surgical repair of a posterior urethral disrupture should be performed after 3 months. Surgery requires proper positioning of the patient in the lithotomy position. Preoperatively, a retrograde urethrogram and a simultaneous cystogram should be performed to determine the length of the stricture or fibrotic discontinuation of the urethra. If involvement of the bladder neck is suspected, a flexible or rigid urethroscopy is helpful for examining anatomy. In patients who did not undergo primary realignment, the urethral dislocation as well as the length of the defect can be accurately described by MRI. In selected patients with short urethral strictures after realignment of the urethra an endoscopic strategy may follow. In case of complete urethral obstructions, some have favored endoscopic interventions. However, there is a high risk of undermining the urethra and bladder neck and the restructure rate is 80 %. Furthermore, the endoscopic procedure often requires several interventions and long-term repetitive dilatations with recurrent strictures and obliterations.

Usually, long posterior urethral strictures are best managed by an open surgical repair via a perineal approach. The urethra is accessed by a midline or lambda incision. The urethra is then mobilized starting from the beginning of the fibrotic defect to the midscrotum allowing a tension free anastomosis. The scar tissue as well as the fibrotic tissue of the proximal urethra must be excised completely to prevent restrictures. For long strictures, a flap urethroplasty of buccal mucosa grafts is used. Adjunctive maneuvers are infrequently needed. In rare cases, pubectomy can be helpful for cases with extended fibrosis, failed former urethroplasty, or accompanied bladder neck involvement.

Erectile dysfunction is a complication of urethral distraction injuries described in 30–60 % of the patients with pelvic fracture [32]. It is questionable as to whether posttraumatic impotence is a result of the injury itself or because of the surgical management. The frequency of posttreatment erectile dysfunction remains the same, independent of initial therapy (early realignment, open surgery, or no treatment). The overall rate of incontinence, anejaculation, and areflexic bladder is low (2-4 %). Another problem is recurrent urethral strictures, which arise in 15–23 % of patients. Minimally invasive treatment by endoscopic incision of the stricture is often sufficient.

11.5.3.2 Treatment of Urethral Injuries in Females

Vaginal inspection should be performed in every single female patient to assess the extent and localization of the urethral injury and the presence, localization, and extent of potentially associated vaginal injuries. Vaginal injuries are further evaluated with an abdominal CT scan to screen for associated intrapelvic or intraabdominal injuries.

In complete urethral ruptures, immediate surgical repair is recommended to avoid urethrovaginal fistulas and complete urethral obliteration. A complete obliteration with an embedded urethra in scar tissue results in a significantly more complicated surgery with an increased frequency of severe complications. Injuries of the distal urethra can be easily repaired via a transvaginal approach. Injuries of the proximal urethra or the bladder neck are best reconstructed via a retropubic approach. Only in unstable patients should a suprapubic catheter be used and delayed primary reconstruction is justified.

References

1. Lynch TH, Martinez-Pineiro L, Plas E, Serafetinides E, Türkeri L, Santucci RA, Hohenfellner M. EAU guidelines on urological trauma. Eur Urol. 2005;47: 1–15.
2. Starnes M, Demetriades D, Hadjizacharia P, Inaba K, Best C, Chan L. Complications following renal trauma. Arch Surg. 2010;145:377–81.
3. Chow SJ, Thompson KJ, Hartman JF, Wright ML. A 10-year review of blunt renal artery injuries at an urban level I trauma centre. Injury. 2009;40:844–50.
4. Broghammer JA, Fisher MB, Santucci RA. Conservative management of renal trauma: a review. Urology. 2007;70:623–9.
5. Alsikafi NF, Rosenstein DI. Staging, evaluation, and nonoperative management of renal injuries. Urol Clin North Am. 2006;33:13–9.
6. Baverstock R, Simons R, McLoughlin M. Severe blunt renal trauma: a 7-year retrospective review from a provincial trauma centre. Can J Urol. 2001;8:1372.
7. Santucci RA, McAninch JM. Grade IV renal injuries: evaluation, treatment, and outcome. World J Surg. 2001;25:1565–72.
8. Velmahos GC, Demetriades D, Cornwell 3rd EE, Belzberg H, Murray J, Asensio J, Berne TV. Selective management of renal gunshot wounds. Br J Surg. 1998;85:1121–4.

9. Eastham JA, Wilson TG, Ahlering TE. Radiographic evaluation of adult patients with blunt renal trauma. J Urol. 1992;148(2 Pt 1):266–7.
10. Husmann DA, Gilling PJ, Perry MO, Morris JS, Boone TB. Major renal lacerations with a devitalized fragment following blunt abdominal trauma: a comparison between nonoperative (expectant) versus surgical management. J Urol. 1993;150:1774–7.
11. Yeung LL, Brandes SB. Contemporary management of renal trauma: differences between urologists and trauma surgeons. J Trauma. 2011;72:68–77.
12. Umbreit EC, Routh JC, Husmann DA. Nonoperative management of nonvascular grade IV blunt renal trauma in children: meta-analysis and systematic review. Urology. 2009;74:579–82.
13. Mee SL, McAninch JW, Robinson AL, Auerbach PS, Carroll PR. Radiographic assessment of renal trauma: a 10-year prospective study of patient selection. J Urol. 1989;141:1095–8.
14. Malcolm JB, Derweesh IH, Mehrazin R, DiBlasio CJ, Vance DD, Joshi S, Wake RW, Gold R. Nonoperative management of blunt renal trauma: is routine early follow-up imaging necessary? BMC Urol. 2008;8:11.
15. Gomez RG, Ceballos L, Coburn M, Corriere Jr JN, Dixon CM, Lobel B, McAninch J. Consensus statement on bladder injuries. BJU Int. 2004;94:27–32.
16. Dreitlein DA, Suner S, Basler J. Genitourinary trauma. Emerg Med Clin North Am. 2001;19:569–90.
17. Morey AF, Iverson AJ, Swan A, Harmon WJ, Spore SS, Bhayani S, Brandes SB. Bladder rupture after blunt trauma: guidelines for diagnostic imaging. J Trauma. 2001;51:683–6.
18. Paparel P, Badet L, Tayot O, Fessy MH, Bejui J, Martin X. Mechanisms and frequency of urologic complications in 73 cases of unstable pelvic fractures. Prog Urol. 2003;13:54–9.
19. Hsieh CH, Chen RJ, Fang JF, Lin BC, Hsu YP, Kao JL, Kao YC, Yu PC, Kang SC. Diagnosis and management of bladder injury by trauma surgeons. Am J Surg. 2002;184:143–7.
20. Corriere Jr JN, Sandler CM. Mechanisms of injury, patterns of extravasation and management of extraperitoneal bladder rupture due to blunt trauma. J Urol. 1988;139:43–4.
21. Corriere Jr JN, Sandler CM. Management of the ruptured bladder: seven years of experience with 111 cases. J Trauma. 1986;26:830–3.
22. Carlin BI, Resnick MI. Indications and techniques for urologic evaluation of the trauma patient with suspected urologic injury. Semin Urol. 1995;13:9–24.
23. Deck AJ, Shaves S, Talner L, Porter JR. Computerized tomography cystography for the diagnosis of traumatic bladder rupture. J Urol. 2000;164:43–6.
24. Kulkarni SB, Barbagli G, Kulkarni JS, Romano G, Lazzeri M. Posterior urethral stricture after pelvic fracture urethral distraction defects in developing and developed countries, and choice of surgical technique. J Urol. 2010;183:1049–54.
25. Bjurlin MA, Fantus RJ, Mellett MM, Goble SM. Genitourinary injuries in pelvic fracture morbidity and mortality using the National Trauma Data Bank. J Trauma. 2009;67:1033–9.
26. Mundy AR, Andrich DE. Pelvic fracture-related injuries of the bladder neck and prostate: their nature, cause and management. BJU Int. 2010;105(9):1302–8.
27. Koraitim MM. Predictors of surgical approach to repair pelvic fracture urethral distraction defects. J Urol. 2009;182:1435–9.
28. Andrich DE, Day AC, Mundy AR. Proposed mechanisms of lower urinary tract injury in fractures of the pelvic ring. BJU Int. 2007;100:567–73.
29. Ball CG, Jafri SM, Kirkpatrick AW, Rajani RR, Rozycki GS, Feliciano DV, Wyrzykowski AD. Traumatic urethral injuries: does the digital rectal examination really help us? Injury. 2009;40:984–6.
30. Myers JB, McAninch JW. Management of posterior urethral disruption injuries. Nat Clin Pract Urol. 2009;6:154–63.
31. Ingram MD, Watson SG, Skippage PL, Patel U. Urethral injuries after pelvic trauma: evaluation with urethrography. Radiographics. 2008;28:1631–43.
32. Anger JT, Sherman ND, Dielubanza E, Webster GD, Hegarty PK. Erectile function after posterior urethroplasty for pelvic fracture-urethral distraction defect injuries. BJU Int. 2009;104:1126–9.

Fracture Management

12

Roman Pfeifer and Hans-Christoph Pape

Contents

12.1	**Introduction**	157
12.2	**Assessment of the Fracture**	158
12.2.1	Open Fractures	158
12.3	**Assessment of the Severity of Soft Tissue Injury**	159
12.3.1	Soft Tissue Injury in Closed Fractures	159
12.4	**Fracture Treatment**	159
12.4.1	Upper Versus Lower Extremity Injuries	159
12.4.2	Fracture Care in Serial Extremity Fractures	160
12.5	**Stages in Polytrauma**	160
12.5.1	Acute Phase (1–3 h After Admission): Resuscitation/Hemorrhage Control	161
12.5.2	Primary Phase (1–48 h): Stabilization of Fractures	161
12.5.3	Secondary Period (2–10 Days): Regeneration	161
12.5.4	Tertiary Period (Weeks to Months After Trauma): Reconstruction and Rehabilitation	161
12.5.5	Assessment of the Patient	161
12.5.6	Physiology of Staged Treatment	162
12.5.7	Patient Assessment for Initial Definitive Surgery Versus Temporizing Orthopedic Surgery	164
12.6	**Special Situations**	165
12.6.1	Surgical Priorities in the Presence of Additional Head Injuries	165
12.6.2	Surgical Priorities in the Presence of Additional Chest Injuries	165
12.6.3	Surgical Priorities in the Presence of Additional Pelvic Ring Injuries	166
12.6.4	Surgical Priorities Depending on Trauma System	166
References		166

12.1 Introduction

In the multiply injured patient, thorough diagnosis of all fractures is essential to allow for strategic planning of stabilization. Likewise, the assessment of soft tissue injuries is crucial to avoid infectious complications. Most of the fractures in multiply injured patients are amenable to early definitive fixation. Over the years, many authors have argued if an early total care or damage control approach is advantageous. We have developed a concept of safe definitive surgery (SDS) that allows to combine both approaches, depending on the physiologic condition of the patient.

R. Pfeifer
Department of Orthopaedic and Trauma Surgery,
Aachen University Medical Center, Aachen, Germany
e-mail: rpfeifer@ukaachen.de

H.-C. Pape, MD, FACS (✉)
Department of Orthopaedic and Trauma Surgery,
University of Aachen Medicine Center,
Pauwelsstr. 30, Aachen 52074, Germany
e-mail: papehc@aol.com

12.2 Assessment of the Fracture

12.2.1 Open Fractures

The standard classification system for open fractures was described by Gustilo [1]. It is a descriptive classification describing a spectrum of soft tissue injury. The classification is associated with the risk of infection, nonunion, and can help guide fracture management. Type I injuries are lower-energy injuries with a skin defect of 1 cm or less in length. Type II injuries have a large skin defect ranging from 1 to 10 cm. Type III injuries are involved with higher energy and more injury to the soft tissues. In type IIIa injuries, primary soft tissue coverage can be obtained without a flap in contrast to type IIIb injuries associated with severe soft tissue trauma requiring rotational flap or free tissue transfers to obtain soft tissue coverage. Type IIIc open fractures are associated with vascular injuries (see the Chap. 19).

Initial care of open fractures consists of thorough irrigation, debridement, and assessment of the soft tissue damage, followed by fracture stabilization. Exposed bone requires soft tissue coverage which should be performed as soon as possible.

The extent of vascular and nerve damage and the general patient condition are important. In severe soft tissue trauma, the extent of the injury and devitalized tissue may not be completely evident at the time of initial debridement and planned reevaluation is often required.

Amputation versus limb salvage reconstruction of upper and lower extremity fractures associated with severe open injuries remains a question [2]. Time-consuming reconstructive surgery in severely injured patients may increase morbidity and mortality. In cases of pending amputation, the MESS score (Mangled Extremity Severity Score) can be of some help, which provides an objective evaluation [3].

Open fractures with limited soft tissue injury should be stabilized definitively at the time of initial debridement. After the initial debridement, the fracture is stabilized with the most suitable implant and method of fixation.

Open fractures caused by high-energy trauma are usually associated with severe soft tissue damage and commonly combined with extensive bone loss or destruction. This injury requires a graded concept of care. Usually, a temporal fixation strategy is used, if soft tissue coverage of the hardware cannot be achieved. Placement of the external fixator should be considered the definitive stabilization until closure of the wound. The personality of each fracture requires individual treatment. In multiply injured patients, the overall injury severity has to be considered as well as the extent of shock and any initial blood loss.

During initial debridement, all soft tissues should be assessed. If necrotic tissue is left in place, further contamination, bacterial growth, and infection is likely to occur. Sufficient surgical exposure of the injury is essential for adequate assessment.

Special situations include the following:

1. Local soft tissue injury versus degloving

    A degloving injury has to be ruled out or diagnosed properly. The assessment includes the degree of soft tissue laceration and periosteal stripping. Thereby, assessment of osseous vascularity is helpful to decide whether fragments should be maintained or removed.

2. Treatment of Morel-Lavallée lesions (subcutaneous degloving)

    Morel-Lavallé lesions are defined as large subcutaneous tissue degloving injuries induced by shearing forces. This mechanism causes a large underlying hematoma. In contrast to other soft tissue injuries, Morel-Lavallé lesions should *not* be debrided aggressively. Small incisions allow complete evacuation of the hematoma. The cutaneous skin flap is decompressed and has a better chance to survive.

3. Consultation of the plastic surgeon

    Exposed bone and tendons in an area with limited soft tissue coverage often require early treatment with soft tissue flaps. If severe muscle injury or nerve damage is present, muscle or tendon transfer procedures can be performed in a timely fashion to avoid severe disabilities secondary to loss of motion.

In multiply injured patients, there is a higher risk of increasing soft tissue necrosis due to impaired soft tissue perfusion (in posttraumatic edema and increased capillary permeability caused by massive volume resuscitation). Therefore, multiple planned operative revisions have to be scheduled. These "second look" surgeries allow for recurrent assessment of the soft tissues and any additional muscle or skin necrosis. This strategy enables the surgeon to do a timely repeat debridement if required (e.g., with high-pressure irrigation). These operative revisions of soft tissue injuries should be scheduled every 48 h as long as there is an impairment of local perfusion. The traumatic wound should be left open and covered with a synthetic saline-soaked dressing or by vacuum therapy. Local vacuum therapy may save the patient some of the planned "second look" surgeries. It has been shown to be successful in treatment of a variety of wounds including extensive degloving injuries [4, 5]. Subatmospheric pressure on the wound site enhances wound healing, reduces the amount of fluid, and increases local blood flow [6, 7]. These effects have been shown to minimize the risk for wound infection [8].

When definitive internal fixation is possible from the soft tissue point of view, the insertion of stable devices is preferred. In case of shaft fractures of the femur or tibia, the use of intramedullary nails is recommended whenever possible.

For intra-articular open fractures, most surgeons prefer a two-step strategy. Some authors recommend limited internal fixation and gross reduction of severely displaced fragments for soft tissue decompression. The minimally invasive fixation comprises the reconstruction of the joint itself and temporary stabilization with K-wires followed by stabilization with lag screws and adjusting/set screws. Definitive fixation is carried out secondarily following consolidation of the soft tissues.

12.3 Assessment of the Severity of Soft Tissue Injury

12.3.1 Soft Tissue Injury in Closed Fractures

Proper diagnosis and assessment of the true degree of soft tissue damage in closed fractures is crucial. Contusions may raise more therapeutic questions than simple inside-out puncture wounds. Weakening of the skin barrier may be followed by necrosis and infection. Assessment of the severity of a closed fracture helps guide the timing and type of osteosynthesis, (Table 12.1). Early detection and evaluation of neural, vascular, and muscular injuries also affect the overall outcome.

Specific attention has to be dedicated to the presence of compartment syndromes. These should be anticipated when the capillary perfusion pressure is less than intracompartmental pressure. Pain out of proportion in responsive patients is the hallmark indicator. In sedated patients, measurement of intracompartmental pressures is mandatory. Elevated compartment pressures should be treated with emergent fasciotomy.

Table 12.1 Classification of soft tissue injuries in closed fractures [36]

Closed fracture G0: No injury or very minor soft tissue injury. The G0 classification covers simple fractures, i.e., fractures caused by indirect injury mechanisms

Closed fracture G1: Inside-out contusions caused by fracture fragments

Closed fracture G2: Deep, contaminated abrasions or local dermal and muscular contusions. Impending compartment syndrome is usually associated with a G2 lesion. These injuries usually are caused by direct forces that shear off soft tissue and are often associated with moderate to severe fracture types

Closed fracture G3: Extensive skin contusions, muscular disruption, decollement, and obvious compartment syndrome combined with any closed fracture are graded as G3. In this subgroup, severe fracture types and comminuted fractures are usually seen

12.4 Fracture Treatment

12.4.1 Upper Versus Lower Extremity Injuries

In severe open fractures of the upper extremity, certain principles are different from those of the lower extremities. It is widely accepted that surgical management of lower extremities precedes the treatment of upper limb injuries. Moreover, the maintenance of correct length is less important in the treatment of upper extremity fractures.

Severe upper extremity injuries, such as open fractures, compartment syndrome, and concomitant vascular injuries require immediate surgical management. In general, splinting or definitive fixation are more frequently performed in the upper extremity because soft tissue coverage is usually easier.

12.4.2 Fracture Care in Serial Extremity Fractures

The sequence of fracture care in patients with serial extremity injuries is important. Simultaneous treatment of extremity injuries can be achieved if the logistic conditions allow the surgeon to do so. The recommendations for the timing of fixation are summarized as follows:

In serial injuries of the upper extremity, immobilization of humeral shaft fractures is an adequate option unless the injuries are open or if neurovascular injuries require surgical intervention. In forearm fractures, early fixation is advised due to limited soft tissue coverage.

In periarticular fractures, early fixation should be performed if the patient's condition is adequate. If no definitive fixation can be performed and if the patient goes to the OR for other causes, transarticular external fixation (TEF) is preferred over casting. External fixation allows for better stability and assessment of soft tissues. This is of utmost importance due to the risk of compartment syndrome in these injuries.

In serial injuries of the lower extremities, definitive fixation should be achieved whenever possible. In floating knee injuries, retrograde femoral nails and antegrade tibial nails can be placed using the same incision. In unstable patients, closed reduction and transarticular external fixation is performed for temporary fracture stabilization.

In metadiaphyseal and periarticular fractures, the priorities of care are dictated by the degree of soft tissue damage. The orthopedic emergencies that require operative care are as follows:

- Compartment syndrome
- Vascular injuries

Table 12.2 Classification system of complex extremity injuries [37]

Fracture-associated injury	Points
Severe soft tissue damage	2
+ hemorrhagic shock	3
ISS 16–25	1
ISS > 25	2
Neurovascular injury	1
Articular involvement	1

Type of complex extremity fracture	Points	Fracture care
Low risk	1–2	Definitive internal
Moderate risk	3–4	External
High risk	>4	Consider amputation

- Irreducible hip dislocation
- Open fractures

Among the higher priorities are femoral head fractures (Pipkin I-III) and fractures of the talus. Any other periarticular fracture is of lower priority, if no further complication is evident (compartment syndrome, pulse less extremity, or open fracture).

In the care of upper extremity fractures, similar principles are applied. In bilateral fractures, simultaneous treatment should be considered. Both extremities can be draped at the same time. Some parts of the procedure may require operative treatment of only one extremity at the time because of fluoroscopy, or handling issues. If the vital signs of the patient deteriorate during the operation, the second extremity may just be temporarily stabilized using external fixation. The classification system of complex extremity fractures is shown in Table 12.2.

12.5 Stages in Polytrauma

Initial fracture care of severely injured patients requires anticipation of potential problems and decision making about the timing of interventions using a systematic approach [9]. In general, four different phases of the posttraumatic course are separated:

1. Acute phase (1–3 h): resuscitation
2. Primary phase (1–48 h): stabilization

3. Secondary period (2–10 days): regeneration
4. Tertiary period (weeks to months after trauma): reconstruction and rehabilitation

12.5.1 Acute Phase (1–3 h After Admission): Resuscitation/Hemorrhage Control

Initially, the focus of treatment lies in the control of acute life-threatening conditions. Complete patient assessment is required to identify all life-threatening conditions. This involves airway control, thoracocentesis, rapid control of external bleeding, and fluid and/or blood replacement therapy. Decompression in head injuries with impending herniation may outweigh all other measures. Prioritization of the orthopedic injuries is crucial as well. The orthopedic fractures that require immediate surgery are listed above. Spinal and pelvic fractures are covered in different chapters.

12.5.2 Primary Phase (1–48 h): Stabilization of Fractures

The primary phase is the usual time where major extremity injuries are managed. These include acute stabilization of major extremity fractures associated with arterial injuries and compartment syndrome. Fractures can be temporally stabilized by external fixation and the compartments released where appropriate. Systemic complications, such as development of systemic inflammatory response syndrome (SIRS) and acute lung injury (ALI), have to be considered if major musculoskeletal injuries are present.

12.5.3 Secondary Period (2–10 Days): Regeneration

During the secondary phase, the general condition of the patient is stabilized and monitored. In most cases, this implies 2–4 days after trauma. Surgical interventions should be limited to those absolutely required ("second look," debridement) and lengthy procedures should be avoided. Physiological and intensive care scoring systems help monitor the clinical progress.

12.5.4 Tertiary Period (Weeks to Months After Trauma): Reconstruction and Rehabilitation

During the tertiary phase, the patient is able to undergo definitive fracture stabilization. Intensive rehabilitation can help maintain range of motion and improve functional outcomes, social reintegration, and return to work.

12.5.5 Assessment of the Patient

Blunt injuries to extremities and trunk (thorax and abdomen) have been shown to be of immense importance to clinical course of severely injured patients [10]. Patients with multiple blunt traumas have to be assessed for "four pathophysiologic cascades" (hemorrhagic shock, coagulopathy, hypothermia, and soft tissue injuries) in order to avoid life-threatening systemic complications (Fig. 12.1). These cascades have common end point that results in endothelial damage [10].

- Hemorrhagic shock: The systolic blood pressure, dependence on vasopressors, and low urine output are reliable clinical markers of hypovolemia. It must be kept in mind that younger patients are able to compensate for severe shock states but can rapidly decompensate after a period without adequate resuscitation.
- Hypothermia: Several factors are known to affect the development of hypothermia in trauma patients, especially hypovolemia with consecutive centralization of blood circulation and prolonged rescue time. Core temperature below 33 °C has been described to be a critical value [11]. Patients presenting with hypothermia are prone to develop cardiac arrhythmia, cardiac arrest, and coagulopathy.

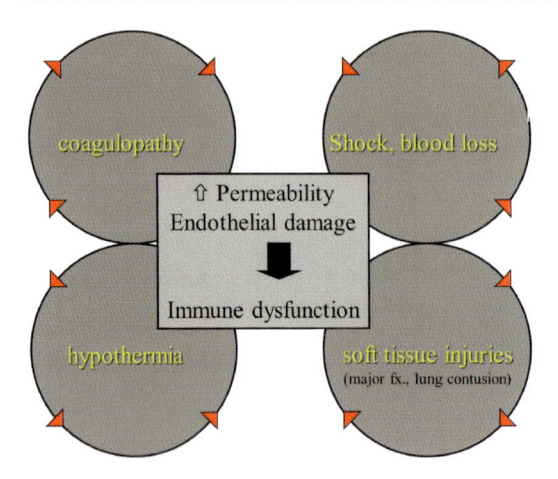

Fig. 12.1 Four vicious cycles demonstrate the pathophysiological cascades. These are known to be associated with the development of posttraumatic immune dysfunction and endothelial damage. The exhaustion of the compensatory mechanisms results in development of systemic complications

- Coagulopathy: Low platelet count is a reliable screening marker for posttraumatic coagulopathy. It can indicate impending disseminated intravascular coagulation. Studies have shown that decreased systemic platelet count (below 90,000) on the first day is associated with multiple organ failure and death [12, 13].
- Soft tissue injury: Major extremity injuries, crush injuries, severe pelvic fractures, and thoracic and abdominal trauma (AIS >2) are included into this category. Severe soft tissue trauma may have additional systemic immunologic effects with consecutive stimulation of immune system and development of the systemic inflammatory response syndrome (SIRS).

12.5.6 Physiology of Staged Treatment

The term "first hit" stands for the initial insult of trauma. The "second hit" refers to the physiologic effect which is impacted by surgical procedures or clinical course. The second hit can enlarge the degree of damage from the primary trauma leading to increased morbidity and mortality [14]. In order to reduce the secondary trauma load, the timing of fracture fixation in polytrauma patients with multiple blunt injuries is based on physiological parameters. All patients should be placed into one of four categories (stable, borderline, unstable, and in extremis) in order to direct the treatment approach. Figure 12.2 demonstrates the *safe definitive surgery* (SDS) concept in treatment of severely injured patient.

12.5.6.1 Stable Condition
Stable patients are hemodynamically stable and respond to initial fluid therapy. Moreover, there is no evidence of respiratory disorders, coagulopathy, hypothermia, and abnormalities of acid base status. Stable patients without comorbidities usually tolerate early definitive fracture fixation [15, 16].

12.5.6.2 Borderline Condition
Borderline conditions are defined as indicated in Table 12.3. In this group of patients, a cautious operative strategy should be used. Additional invasive monitoring should be instituted preoperatively. A low threshold should be used for conversion to a "damage control" approach to the patient management, as detailed below, at the first sign of deterioration.

12.5.6.3 Unstable Condition
Hemodynamically unstable patients are at risk of rapid deterioration, subsequent multiple organ failure, and death. In these patients, a "damage control" approach is required. This entails rapid life-saving surgery only when absolutely necessary and timely transfer to the intensive care unit for further stabilization and monitoring. Temporary stabilization of fractures using external fixation, hemorrhage control, and exteriorization of gastrointestinal injuries is advocated. Complex reconstructive extremity procedures should be delayed until the patient's condition is stabilized and the acute immunoinflammatory response to injury has subsided.

12.5.6.4 In Extremis Condition
These patients have ongoing uncontrolled blood loss. Despite the extensive resuscitation, they remain severely unstable and suffer the effects of four vicious cycles: coagulopathy, shock, hypothermia, and tissue injury. The patients should be transferred directly to the intensive

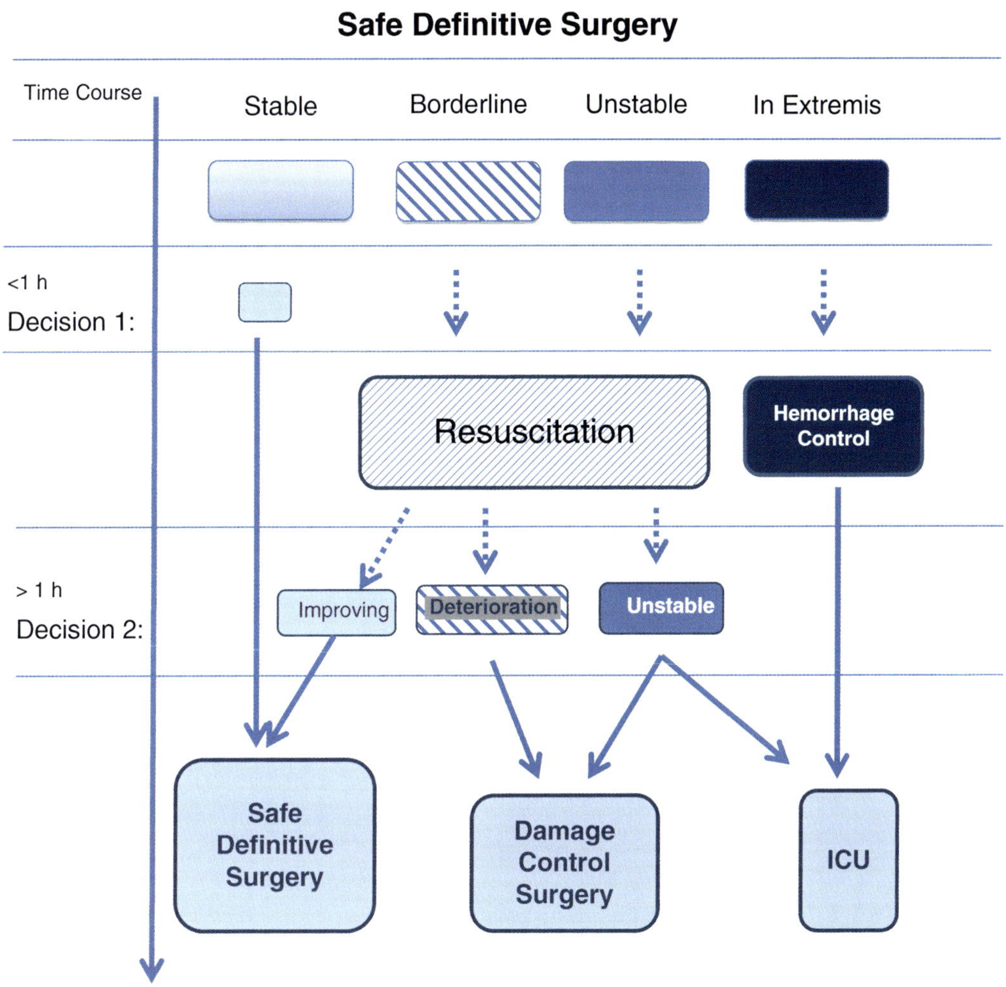

Safe Definitive Surgery

Fig. 12.2 Polytraumatized patients are assessed according to the advanced trauma life support (ATLS) approach. Next, classification (stable, borderline, unstable, in extremis) of the patients is performed using clinical parameters. In "stable" patients, a safe definitive surgery (SDS) strategy can be applied. The patients "in extremis" should be transferred directly to the intensive care unit for invasive monitoring and advanced hematologic, pulmonary, and cardiovascular support. "Borderline" and "unstable" patients are brought to the ICU department for resuscitation. Thereafter, reevaluation of the clinical status is performed. "Unstable" patients and "borderline" patients with secondary deterioration should be treated according to the damage control orthopedics (DCO) concept. Patients with improving conditions can be subjected to safe definitive surgery

Table 12.3 Clinical parameters used to identify patients in uncertain condition, named "borderline"

Factors to identify the borderline patient
Injury severity score > 40
Multiple injuries (ISS > 20) in association with thoracic trauma (AIS > 2)
Multiple injuries in association with severe abdominal or pelvic injury and hemorrhagic shock at presentation (systolic BP < 90 mmHg)
Patients with bilateral femoral fractures
Radiographic evidence of pulmonary contusion
Hypothermia below 35 °C

Usually, at least three of these have to be present to allow for classification as borderline [10]

Table 12.4 Incorporation of existing classification systems for clinical patient assessment

	Parameter	Stable (Grade I)	Borderline (Grade II)	Unstable (Grade III)	In extremis (Grade IV)
Shock	Blood pressure (mmHg)	100 or more	80–100	60–90	<50–60
	Blood units (2 h)	0–2	2–8	5–15	>15
	Lactate levels	Normal range	Around 2.5	>2.5	Severe acidosis
	Base deficit mmol/l	Normal range	No data	No data	>6–8
	ATLS classification	I	II–III	III–IV	IV
Coagulation	Platelet count (µg/ml)	>110,000	90,000–110,000	<70,000–90,000	<70,000
	Factor II and V (%)	90–100	70–80	50–70	<50
	Fibrinogen (g/dl)	>1	around 1	<1	DIC
	D-Dimer	Normal range	Abnormal	Abnormal	DIC
Temperature		<33 °C	33 °C–35 °C	30 °C–32 °C	30 °C or less
Soft tissue injuries	Lung function; PaO_2/FiO_2	350–400	300–350	200–300	<200
	Chest trauma scores; AIS	AIS I or II (e.g., abrasion)	AIS 2 or more (e.g., 2–3 rib fractures)	AIS 3 or more (e.g., serial rib fx. >3)	AIS 3 or more (e.g., unstable chest)
	Chest trauma score; TTS	0	I–II	II–III	IV
	Abdominal trauma (Moore)	<or = II	<or = III	III	III or >III
	Pelvic trauma (AO class)	A type (AO)	B or C	C	C (crush, rollover abd.)
	External (AIS)	AIS I–II (e.g., abrasion)	AIS II–III (e.g., mult. >20 cm tears)	AIS III–IV (e.g., <30 % burn)	(Crush injury, >30 % burn)

Among the parameters, at least three should be met to qualify for a certain category. It is of note that patients that respond to resuscitation qualify for early definitive fracture care, as long as prolonged surgeries are avoided [10]

care unit for invasive monitoring and advanced hematologic, pulmonary, and cardiovascular support. Orthopedic injuries can be stabilized rapidly in the emergency department or intensive care unit using external fixation.

12.5.7 Patient Assessment for Initial Definitive Surgery Versus Temporizing Orthopedic Surgery

The initial patient assessment usually is performed using scoring systems such as the ISS or NISS. For life-threatening conditions, frequently due to penetrating trauma, the "triad of death" (blood loss, coagulopathy, and loss of temperature) approach has been used. In patients with blunt orthopedic injuries, it is important to account for soft tissue injuries as well and parameters of oxygenation to assess the clinical status of the patient [9].

Table 12.4 documents parameters and scoring systems that can be used to categorize a patient's condition. Three out of the four criteria should be present to qualify a patient for a specific category [10]. It is important to note that the combination of these parameters is a suggestion only and has a low level of evidence. Nevertheless, most of the components are scores that have been routinely used in the past and are widely accepted. For screening purposes, the following threshold levels have been used: pulmonary dysfunction (PaO_2/FiO_2 <250), platelet count (<95,000), hypotension unresponsive to therapy >10 blood units per 6 h, and requirement for vasopressors.

Inflammatory parameters have been identified to have predictive power for the development of systemic complications (multiple organ failure) [17]. An exaggerated posttraumatic systemic inflammatory may occur leading to immunosuppression [18]. Recent studies support the idea of simultaneous induction of the innate immune system ("genomic storm") and suppression of adaptive immune system with activation of more than 5,136 genes [19]. Trauma has promoted the expression of genes involved in innate immunity, microbial recognition, and inflammation. Additionally, the expression was decreased in genes for antigen presentation. Patients with uncomplicated recovery were associated with a down regulation of genes within 7–14 days after trauma [19].

The "inflammation and host response to injury collaborative research program" has investigated the activation and expression of genes in response to trauma and evaluated the prognostic value of this response. According to this prospective investigation, leukocyte gene expression scores obtained in the first hours after trauma can provide useful information that cannot be obtained by scoring anatomical or physiological parameters [20]. The genomic characterization appears to allow the stratification of patients with poor outcome (development of MOF) [21]. Whether this technique and scoring has application in the clinical setting remains an area of future study [20].

12.6 Special Situations

12.6.1 Surgical Priorities in the Presence of Additional Head Injuries

According to the pathophysiology of head injury, the brain loses the autoregulation of blood flow in zones of contusion. Also, an increase in the utilization of glucose occurs, adding to the susceptibility to ischemic injury [22]. Head trauma patients are at greatest risk for decreased cerebral blood flow during the first 12–24 h following injury [23]. Intraoperative hypotension is an important risk factor for secondary brain injury ("second hit" to the brain) [24]. The primary goal of management for traumatic brain injury is the avoidance of secondary insults (hypoperfusion) [25].

The management needs to be performed in close cooperation with the neurosurgical team and sudden changes in the strategy can occur according to the degree of cerebral swelling, imminent herniation, or increase in bleeding.

The orthopedic surgeon and the neurosurgeon need to reveal how much operative time, blood loss, and temperature loss can be accepted for each individual case. General rules are currently not available. If in doubt, monitoring of the intracranial pressure (ICP) is safer and should be performed. During fracture fixation, secondary insults should be avoided by maintaining adequate cerebral perfusion.

12.6.2 Surgical Priorities in the Presence of Additional Chest Injuries

The pathophysiology in chest trauma is well described. A lung contusion is a separate entity from rib fractures and has a higher association with ARDS than rib fractures [26]. In isolated rib fractures, a decrease in biomechanical (lack of rib cage motion) and pain-related hypoxemia is reversed by artificial ventilation. With lung contusion despite ventilation, intrapulmonary edema can develop. This is mediated by inflammatory cells and causes a local immunologic reaction [27]. The progressive nature of a pulmonary contusion can cause problems and is frequently underestimated. Early after injury, the blood gas parameters can still be within normal limits, and the chest X-ray may also present as a false negative. The immunologic mechanisms initiated by pulmonary contusions are comparable to those seen after severe injury [28, 29]. Thus, the host response to pulmonary contusion is similar to nonpulmonary injury, resulting in an increased risk of ARDS.

Patient evaluation focuses on the following clinical criteria: presence of a lung contusion on the initial chest x-ray or CT scan, worsening oxygenation (requirement of increased FiO$_2$ >40 % or PaO$_2$/FiO$_2$ <250), and increased airway pressures (e.g., >25–30 cm H$_2$O). The pulmonary function can change within hours after the injury and repeated blood gases should be obtained.

12.6.3 Surgical Priorities in the Presence of Additional Pelvic Ring Injuries

The pathophysiology of systemic effects in severe pelvic injuries is dictated by the degree of local blood loss from the pelvic floor, the presacral venous plexus, and any arterial damage. Unlike other injuries, autotamponade does not occur and retroperitoneal bleeding may mimic intra-abdominal injury. Soft tissue disruption can have more severe side effects than in the extremities since a higher degree of kinetic energy is required to cause substantial displacement. In open injuries with intestinal damage, a substantial increase in the risk of infection and late sepsis occurs [29, 30].

Timing of pelvic fixation is based on the hemodynamic status and the presence of associated abdominal injuries. The decision to attempt definitive fixation within 2448 h appears to be dependent upon the pelvic ring fracture pattern [31] and can be attempted in stable and borderline patients. In unstable patients, the use of sheets wrapped about the pelvis or a pelvic binder allows for rapid circumferential splinting of the pelvic ring most effectively at the level of the greater trochanter [32].

The paucity of studies in the literature seems to support early surgical management of such injuries. Favorable patterns may be treated by percutaneous fixation when several factors coincide: closed reduction can be achieved, the injury pattern is amenable to screw fixation alone, and the surgeon and operating team are available and experienced [33]. In cases of exsanguinations from a pelvic ring injury, direct packing of the true pelvic space has been described [34]. This technique is dependent upon achieving provisional stability of the pelvic ring with a binder, external fixation, or internal fixation.

Table 12.5 Mean duration until definitive treatment of major fractures in patients with multiple injuries, specified according to body regions [35]

Duration until definitive treatment	USA n=77	GER n=93	p-value
All fractures	5.5 days±4.2	6.6 days±8.7	n.s
Humerus fractures	5 days ± 3.7	6.6 days±6.1	n.s
Radius fractures	6 days ± 4.7	6.1 days±8.7	n.s.
Femur fractures	7.9 days±8.3	5.5 days±7.9	n.s
Tibia fractures	6.2 days±5.6	6.2 days±9.1	n.s
Pelvis fractures	5 days ± 2.8	7.1 days±9.6	n.s

Current recommendations are to identify the source of pelvic hemorrhage and to stop the bleeding, followed by stabilization of the pelvic ring. The use of a binder is often successful for achieving a physiologic state that allows surgery unless a single artery is damaged. This may be treated by coiling.

12.6.4 Surgical Priorities Depending on Trauma System

Some authors have argued that the trauma system dictates patient care. The early total care of all fractures was advocated by certain clinicians in the 1980s. However, a recent survey on the management of major fractures in multiply injured patients demonstrates that the timing of fracture fixation is similar in two groups of trauma centers in both the United States and Germany. Thereby, a staged approach toward fracture management appears to be the rule in both systems [35] (Table 12.5).

References

1. Gustilo RB, Mendoza RM, Williams DN. Problems in the management of type III (severe) open fractures: a new classification of type III open fractures. J Trauma. 1984;24(8):742–6.
2. Pape HC, Probst C, Lohse R, et al. Predictors of late clinical outcome following orthopedic injuries after multiple trauma. J Trauma. 2010;69(5):1243–51.
3. Johansen K, Daines M, Howey T, Helfet D, Hansen Jr ST. Objective criteria accurately predict amputation following lower extremity trauma. J Trauma. 1990; 30(5):568–72.

4. Meara JG, Guo L, Smith JD, Pribaz JJ, Breuing KH, Orgill DP. Vacuum-assisted closure in the treatment of degloving injuries. Ann Plast Surg. 1999;42(6):589–94.
5. DeFranzo AJ, Marks MW, Argenta LC, Genecov DG. Vacuum-assisted closure for the treatment of degloving injuries. Plast Reconstr Surg. 1999;104(7):2145–8.
6. Mullner T, Mrkonjic L, Kwasny O, Vecsei V. The use of negative pressure to promote the healing of tissue defects: a clinical trial using the vacuum sealing technique. Br J Plast Surg. 1997;50(3):194–9.
7. Banwell P, Withey S, Holten I. The use of negative pressure to promote healing. Br J Plast Surg. 1998;51(1):79.
8. Fleischmann W, Lang E, Russ M. [Treatment of infection by vacuum sealing]. Unfallchirurg. 1997;100(4)):301–4.
9. Pape HC, Tornetta III P, Tarkin I, Tzioupis C, Sabeson V, Olson SA. Timing of fracture fixation in multitrauma patients: the role of early total care and damage control surgery. J Am Acad Orthop Surg. 2009;17(9):541–9.
10. Pape HC, Giannoudis PV, Krettek C, Trentz O. Timing of fixation of major fractures in blunt polytrauma: role of conventional indicators in clinical decision making. J Orthop Trauma. 2005;19(8):551–62.
11. Kobbe P, Lichte P, Wellmann M, et al. [Impact of hypothermia on the severely injured patient]. Unfallchirurg. 2009;112(12):1055–61.
12. Sturm JA, Wisner DH, Oestern HJ, Kant CJ, Tscherne H, Creutzig H. Increased lung capillary permeability after trauma: a prospective clinical study. J Trauma. 1986;26(5):409–18.
13. Nuytinck JK, Goris JA, Redl H, Schlag G, van Munster PJ. Posttraumatic complications and inflammatory mediators. Arch Surg. 1986;121(8):886–90.
14. Pape HC, Schmidt RE, Rice J, et al. Biochemical changes after trauma and skeletal surgery of the lower extremity: quantification of the operative burden. Crit Care Med. 2000;28(10):3441–8.
15. Duwelius PJ, Huckfeldt R, Mullins RJ, et al. The effects of femoral intramedullary reaming on pulmonary function in a sheep lung model. J Bone Joint Surg Am. 1997;79(2):194–202.
16. Schemitsch EH, Jain R, Turchin DC, et al. Pulmonary effects of fixation of a fracture with a plate compared with intramedullary nailing. A canine model of fat embolism and fracture fixation. J Bone Joint Surg Am. 1997;79(7):984–96.
17. Pape HC, Remmers D, Grotz M, et al. Reticuloendothelial system activity and organ failure in patients with multiple injuries. Arch Surg. 1999;134(4):421–7.
18. Bone RC. Toward a theory regarding the pathogenesis of systemic inflammatory response syndrom: what we do and do not know about cytokine regulation. Crit Care Med. 2008;24:163–72.
19. Xiao W, Mindrinos MN, Seok J, et al. A genimic storm in critically injured humans. J Exp Med. 2011;208(13):2581–90.
20. Warren HS, Elson CM, Hayden DL, et al. A genomic score prognostic of outcome in trauma patients. Mol Med. 2009;15(7–8):220–7.
21. Desai KH, Tan CS, Leek JT, Maier RV, Tompkins RG, Storey JD. Dissecting inflammatory complications in critically injured patients by within-patient gene expression changes: a longitudinal clinical genomics study. PLoS Med. 2011;8(9), e1001093.
22. Reinert M, Hoelper B, Doppenberg E, Zauner A, Bullock R. Substrate delivery and ionic balance disturbance after severe human head injury. Acta Neurochir Suppl. 2000;76:439–44.
23. Miller JD. Head injury and brain ischaemia–implications for therapy. Br J Anaesth. 1985;57(1):120–30.
24. Chesnut RM, Marshall LF, Klauber MR, et al. The role of secondary brain injury in determining outcome from severe head injury. J Trauma. 1993;34(2):216–22.
25. Siegel JH, Gens DR, Mamantov T, Geisler FH, Goodarzi S, MacKenzie EJ. Effect of associated injuries and blood volume replacement on death, rehabilitation needs, and disability in blunt traumatic brain injury. Crit Care Med. 1991;19(10):1252–65.
26. Stellin G. Survival in trauma victims with pulmonary contusion. Am Surg. 1991;57(12):780–4.
27. Regel G, Dwenger A, Seidel J, Nerlich ML, Sturm JA, Tscherne H. [Significance of neutrophilic granulocytes in the development of post-traumatic lung failure]. Unfallchirurg. 1987;90(3):99–106.
28. Tate RM, Repine JE. Neutrophils and the adult respiratory distress syndrome. Am Rev Respir Dis. 1983;128(3): 552–9.
29. Weiland JE, Davis WB, Holter JF, Mohammed JR, Dorinsky PM, Gadek JE. Lung neutrophils in the adult respiratory distress syndrome. Clinical and pathophysiologic significance. Am Rev Respir Dis. 1986;133(2):218–25.
30. Keel M, Trentz O. Pathophysiology of polytrauma. Injury. 2005;36:691–709.
31. Olson SA, Burgess A. Classification and initial management of patients with unstable pelvic ring injuries. Instr Course Lect. 2005;54:383–93.
32. Bottlang M, Simpson T, Sigg J, Krieg JC, Madey SM, Long WB. Noninvasive reduction of open-book pelvic fractures by circumferential compression. J Orthop Trauma. 2002;16(6):367–73.
33. Routt Jr ML, Falicov A, Woodhouse E, Schildhauer TA. Circumferential pelvic antishock sheeting: a temporary resuscitation aid. J Orthop Trauma. 2006;20(1 Suppl):S3–6.
34. Cothren CC, Osborn PM, Moore EE, Morgan SJ, Johnson JL, Smith WR. Preperitonal pelvic packing for hemodynamically unstable pelvic fractures: a paradigm shift. J Trauma. 2007;62(4):834–9.
35. Schreiber V, Tarkin HI, Hildebrand F, et al. The timing of definitive fixation for major fractures in polytrauma – a matched pair comparison between a US and European level I centers. Injury. 2011;42(7):650–4.
36. Oestern HJ, Tscherne H. Pathophysiology and classification of soft tissue injuries associated with fractures. In fractures with soft tissue injuries. 1st ed. Berlin: Springer; 1984.
37. Kobbe P, Lichte P, Pape HC. Complex extremity fractures following high energy injuries: the limited value of existing classifications and a proposal for a treatment-guide. Injury. 2009;40 Suppl 4:S69–74.

Mangled Extremity: Management in Isolated Extremity Injuries and in Polytrauma

13

Mark L. Prasarn, Peter Kloen, and David L. Helfet

Contents

13.1	Introduction	169
13.2	Mechanism of Injury	170
13.3	Common Injury Patterns	170
13.4	Scoring Systems	170
13.5	Management	171
13.6	Complications	175
13.7	Predictive Ability of Scoring Systems to Predict Final Outcome	180
13.8	Outcomes Following Limb Salvage Versus Amputation	180
13.9	Cost of Care	182
13.10	The Mangled Upper Extremity	182
13.11	The Pediatric Mangled Extremity	183
13.12	The Mangled Extremity and Polytrauma	183
	Conclusions	184
	References	184

M.L. Prasarn, MD (✉)
Department of Orthopaedics and Rehabilitation,
University of Texas, 6400 Fanon Suite 1700,
Houston 77030, TX, USA
e-mail: markprasarn@yahoo.com

P. Kloen, MD, PhD
Orthopaedic Surgery, Academic Medical Center,
Meibergdreef 9, Amsterdam 1100 DD,
The Netherlands
e-mail: p.kloen@amc.uva.nl

D.L. Helfet, MD
Orthopedic Trauma Service, Hospital
for Special Surgery, 535 E. 70th Street,
New York 10021, NY, USA
e-mail: helfetd@hss.edu

13.1 Introduction

Clinical decision making for trauma patients with extremity injuries is typically straightforward with resulting maintenance of viability and function of the involved limb. Damage control orthopedics (DCO) has produced similar outcomes in the severely injured, unstable trauma victim with a relatively simple extremity injury. Numerous reports have described the beneficial effects of such temporizing measures that then allow the patient to be stabilized [26, 27, 40, 55, 68]. The decision process becomes much more clouded when dealing with trauma victims with severe extremity injuries, that is, mangled extremities. There has been much debate as to whether limb salvage or amputation results in the best clinical outcomes in such a patient.

The emergent management of severe extremity trauma poses a difficult clinical decision for the entire treating surgical team. Resuscitation and management of all life-threatening injuries always must take precedence over any extremity injury. In a small subset of patients with complete

© Springer-Verlag Berlin Heidelberg 2016
H.-C. Pape et al. (eds.), *The Poly-Traumatized Patient with Fractures:
A Multi-Disciplinary Approach*, DOI 10.1007/978-3-662-47212-5_13

traumatic disruption and clearly irreparable injuries an immediate complete amputation should be performed. Likewise, in the setting of prolonged limb ischemia, severe soft tissue loss that cannot be reconstructed or concurrent life-threatening injuries elsewhere in an unstable polytrauma patient, a primary amputation is likely indicated. Also, patients with severe ipsilateral foot and ankle crush injuries may be better served with immediate amputation.

There exists a significant population of trauma patients in whom such clear indications for amputation are absent. It has been questioned whether or not attempted preservation of the limb in such patients is appropriate, or whether the patient would be better served with primary amputation. In many circumstances, the patient undergoes prolonged unsuccessful attempts at limb salvage only to be subject to great physical, psychological, financial, and social suffering. Various scoring systems have been devised to attempt to identify patients who should have limb salvage attempted versus those who should undergo primary amputation. The reliability of such scoring systems has been questioned, and the outcomes of limb salvage versus amputation debated. It still remains unclear in the literature as to which modality results in the optimal outcome, and in whom each should be performed. The treating surgeon and patient therefore still have no objective simple criteria to assist in making such a monumental decision.

13.2 Mechanism of Injury

The vast majority of injuries that pose the possible risk of amputation are due to blunt trauma. Motor vehicle crashes and industrial/farm accidents are the leading causes of such injuries in both the upper and lower extremities [1, 3, 8, 11, 17, 18, 31, 33, 45, 60, 69]. Falls from a height, high-velocity gunshots, and explosion injuries constitute the remainder of mechanisms [2, 9, 33]. The most significant factor involved with the injury mechanism is the amount of energy transferred to the extremity rather than the actual mechanism. The relative amount of energy absorbed directly translates into the amount of destruction to the bone and soft tissues. The concept of the "zone of injury" has been coined to define the area of the extremity affected by the injuring force. This zone may be defined by the fracture type, the amount of comminution, the area of crush, laceration, or shearing of the soft tissues, or devascularization of the entire limb [18].

13.3 Common Injury Patterns

Most studies have defined severe extremity trauma as those with associated complex fractures, dysvascular limbs, significant soft-tissue loss, neurological injury, and severe injuries to the distal extremity (hand, foot, and ankle). In all instances, there is a high-energy transfer to the involved limb that results in some combination of injuries to bone, arteries, tendon, nerves, and soft tissue. Complicated fractures are typically Gustilo grade IIIB and IIIC, but sometimes include select grade IIIA open fractures. These injuries often times have significant bone loss that requires either later bone grafting or bone transport using Ilizarov techniques. Dysvascular limbs can result from knee dislocations, internal amputation of the upper extremity, vascular injury secondary to a closed fracture, or penetrating wounds. Patients that have concomitant vascular disruption of the involved limb often constitute a great number of these injuries and are more likely to result in amputation [8, 35, 41]. Significant soft-tissue injuries are those secondary to crush mechanisms, those with degloving wounds, or avulsion injuries. Distal extremity injuries that result in consideration of amputation include Gustilo grade III pilon fractures, severe hindfoot or midfoot injuries, and loss of multiple digits in the hand.

13.4 Scoring Systems

Multiple scoring systems have been proposed by various authors to help guide in the management of complex extremity trauma. Even so, there is still much debate regarding the criteria that should be utilized in predicting which limbs can be successfully reconstructed versus those that should undergo amputation [36, 42, 43, 46, 63]. Most of these predictive indices have been criticized as being too subjective, complex, difficult to universally apply, derived retrospectively from small patient series, and not validated with functional outcome data [5, 18]. Of note is that most of these scores were developed more than 15 years ago. Since then there have been enormous advances in plastic, orthopedic and vascular techniques that now permit limb salvage in the majority of these cases. The four most commonly used scoring systems are presented.

In 1987, Howe et al. proposed the Predictive Salvage Index (PSI) to be used in the setting of combined orthopedic and vascular injuries involving the lower extremity. In this system, points are assigned for the level of arterial injury, the degree of bone and muscle injury, and the amount of time elapsed from injury to arrival to the operating room. In a small, retrospective analysis of 21 patients, all 12 patients with successful limb salvage had a PSI <8, while 7 of the 9 who underwent amputation had a PSI of at least 8. They concluded that the PSI had a sensitivity of 78 % and specificity of 100 % for predicting amputation in this setting [42]. Other authors have reported much lower sensitivity and specificity of the PSI [5, 62].

In 1990, Johansen et al. introduced a system known as the Mangled Extremity Severity Score (MESS) after retrospectively reviewing 26 mangled lower limbs [43]. Under this system the patient receives a numerical score for four different factors: skeletal/soft-tissue injury, ischemia, shock, and patient age. The scores are summated, and a value of <7 has been shown to be predictive of salvage [36, 43] (Table 13.1). The proposed advantages of this predictive index are that the information is readily available upon presentation, its relative simplicity, and its reproducibility. Others have criticized its subjectivity, and review of larger series of patients has shown lower sensitivity of the index than initially reported [5, 53, 61].

In 1991, Russell et al. proposed the Limb Salvage Index (LSI) based on the review of 70 limb-threatening injuries. The index predicts the likelihood of limb salvage based on ischemia time and injury severity to six types of tissue that may be involved [63]. In order to specifically quantify each of these categories, extensive examination during an operation is necessary. The system is therefore very detailed and difficult to use in the acute decision-making process [5]. Another detailed scoring system, known as the NISSSA (nerve injury, ischemia, soft-tissue contamination, skeletal injury, shock, and age), was introduced by McNamara et al. in 1994. This system is a more complex modification of the MESS that separates the skeletal and soft-tissue injury, and adds a score for nerve injury. In a small retrospective series (24 patients), the authors concluded that the system is more sensitive and specific than the MESS [51].

13.5 Management

Initial management of the patient with a limb-threatening injury begins with ATLS protocol emphasizing a primary survey with immediate assessment of ABC's (Table 13.2). Following this, the field dressing should be removed and any significant bleeding immediately controlled. This should be done with direct pressure, tourniquet, a compressive dressing, or proximal clamping (in that order of preference). Once the resuscitative effort is underway, further assessment of other injuries should be undertaken as well as a thorough neurovascular examination. If there is disruption to the arterial flow to the extremity, and salvage is being considered, an intraluminal shunt may be used. Wound dressing, gross alignment and splinting should be performed. Following this, any radiographic studies may be obtained (including vascular studies if necessary), and intravenous antibiotic and tetanus prophylaxis administered. We always calculate a MESS for each patient at the onset of treatment.

If an early amputation is deemed necessary it is often advantageous to take medical record photographs to document the severity of the injury. Also, obtaining and documenting additional

Table 13.1 Criteria of Mangled Extremity Severity Score

Type	Characteristics	Injuries	Points
Skeletal/soft-tissue			
Group			
1	Low energy	Stab wounds, simple closed fractures, small-caliber gunshot wounds	1
2	Medium energy	Open or multiple-level fractures, dislocations, moderate crush injuries	2
3	High energy	Shotgun blast (close range) and high-velocity gunshot wounds	3
4	Massive crush	Logging, railroad, oil rig accidents	4
Shock			
Group			
1	Normotensive hemodynamics	BP stable in field and in OR	0
2	Transiently hypotensive	BP unstable in field but responsive to intravenous fluids	1
3	Prolonged hypotension	Systolic BP <90 mmHg in field and responsive to intravenous fluid only in OR	2
Ischemia			
Group			
1	None	A pulsatile limb without signs of ischemia	0[a]
2	Mild	Diminished pulses without signs of ischemia	1[a]
3	Moderate	No pulse by Doppler, sluggish capillary refill paresthesia, diminished motor activity	2[a]
4	Advanced	Pulseless, cool, paralyzed, and numb without capillary refill	3[a]
Ischemia			
Group			
1	<30 years		0
2	>30, <50 years		1
3	>50 years		2

Reprinted from Helfet et al. [36]
OR operating room, *BP* blood pressure
[a]Points X 2 if ischemic time exceeds 6 h

"attending" opinions on the need for the primary amputation is highly recommended. We also recommend keeping a photographic record throughout the course of treatment if reconstruction is performed, to document both progress and decline. Our indications for early amputation include unreconstructable osseous or soft-tissue injuries, irreparable vascular injuries, and severe loss of the plantar soft tissue (Fig. 13.1). Previous authors have recommended amputation if plantar sensation is absent. Recent evidence has suggested that initially absent plantar sensation does not predict a poor functional outcome, and that it may return in more than half of patients followed out to 24 months [8]. We therefore do not use absent plantar sensation as criteria for a primary amputation alone.

The amputation should be performed at the most distal level possible, but should not include clearly nonviable tissues. Examining color, consistency, contractility, and bleeding determine tissue viability. It has been shown that trans-tibial amputations have significantly better functional outcomes and lower energy expenditure than more proximal levels of amputation [18, 49]. A thorough irrigation and debridement should be

13 Mangled Extremity: Management in Isolated Extremity Injuries and in Polytrauma 173

Table 13.2 Algorithm for the management of the patient with severe extremity trauma

```
              Initial resuscitation as per ATLS protocol
                              │
         Presence of factors indicating limb unsalvageable
              (clinical or scoring eg. MESS)
         or patient with mortality risk secondary to involved limb
                      ╱                    ╲
                    No                     Yes
                    │                        │
            Hard signs of           Primary amputation with
           vascular injury           thorough irrigation and
            ╱         ╲                    debridement
          Yes          No
           │            │
      Irrigation and   Irrigation and
      debridement and  debridement and
      external         skeletal stabilization
      stabilization
           │                    │
        Angiogram                │
        ╱      ╲                 │
      (+)     (-)                │
       │        ╲                ↓
    Shunt/       →    Repeat irrigation  →  Definitive skeletal repair
    Definitive        and debridement(s)    and soft tissue coverage
    Vascular repair
```

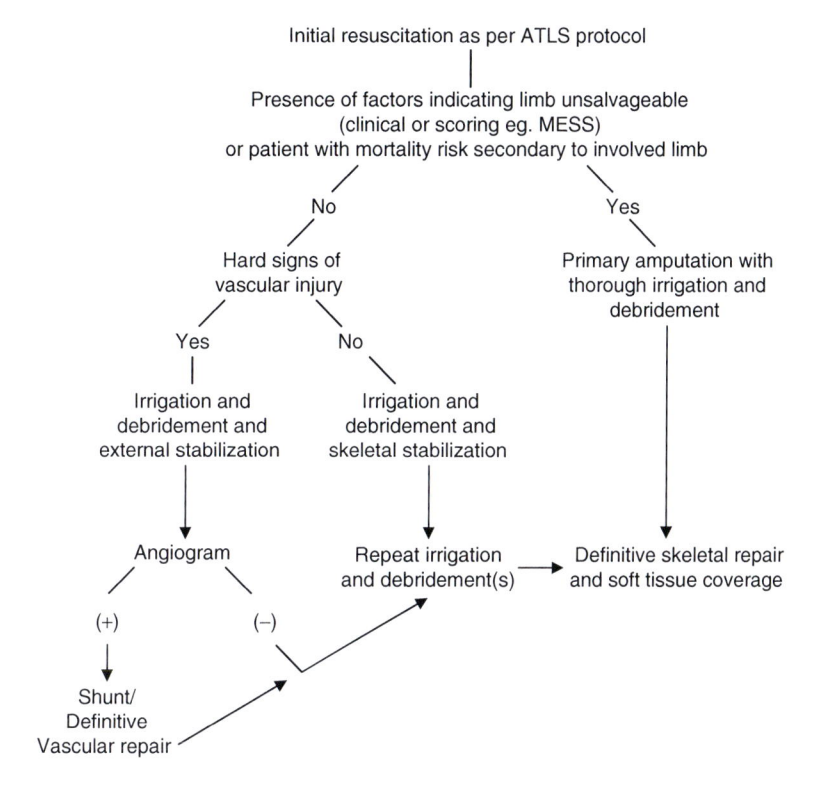

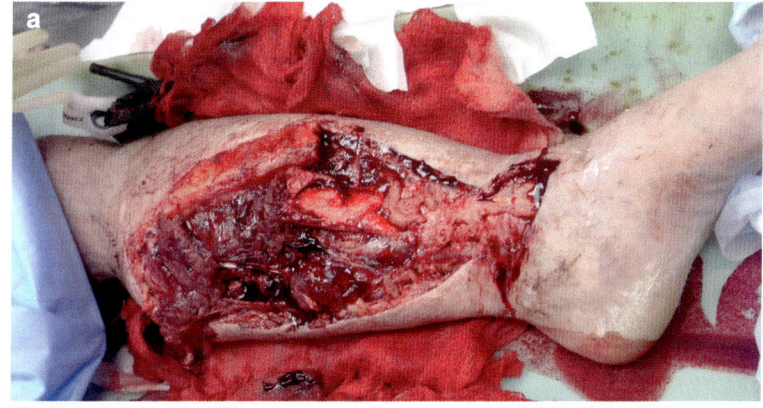

Fig. 13.1 A 25-year-old male sustained Gustillo and Anderson Type IIIB open tibia/fibula fractures following a crush injury in a motor cycle accident. (**a**) Clinical photo (*top image*) of the injury. (**b**) AP and lateral radiographs (*bottom images*) demonstrate a comminuted tibia and fibula fractures. The patient underwent early below-knee amputation secondary to unreconstructable soft tissue envelope

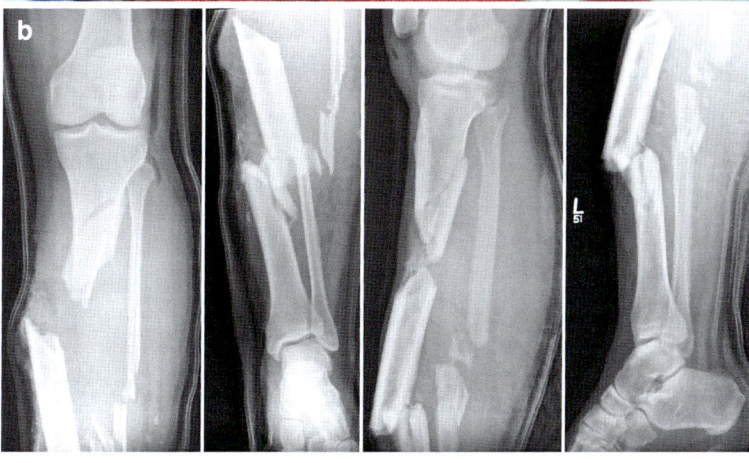

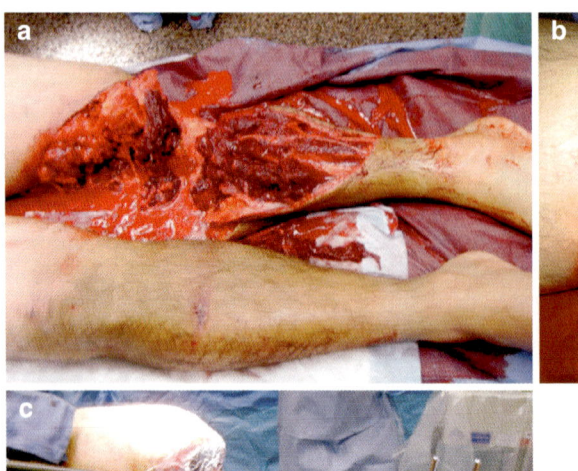

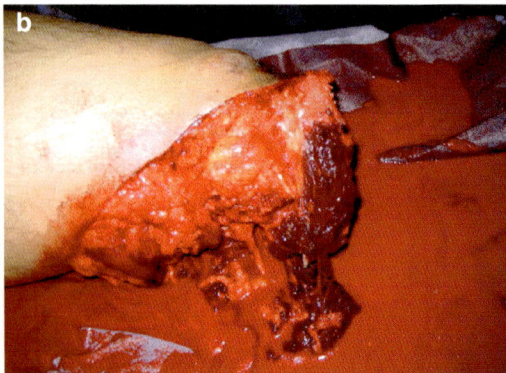

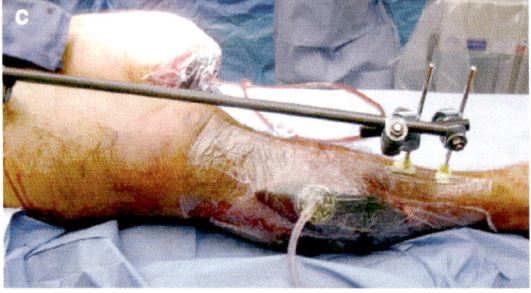

Fig. 13.2 A 21-year-old male presented to the emergency department following a motorcycle collision with bilateral lower extremity injuries. (**a**) Left-sided pulse-less (Grade IIIC) "mangled" knee/lower extremity injuries and a right-sided bicondylar closed tibial plateau fracture with compartment syndrome. (**b**) Left-sided completion of the above knee amputation retaining as much viable soft tissue as possible. (**c**) Application of negative pressure wound therapy dressing to left-sided amputation site, as well as external fixation of right bicondylar tibial plateau fracture and leg fasciotomies for compartment syndrome

performed without any attempt to close the wound at this time. A sterile dressing or wound VAC can be applied, and a splint placed if the amputation is below the level of the knee or elbow (Fig. 13.2). Return to the operating room with repeat surgical debridements should be performed as deemed necessary. In most instances several irrigation and debridements are undertaken prior to closure of the stump site to ensure adequate removal of nonviable tissue and a clean environment for wound healing.

If the need for amputation is not clear upon initial examination, then limb salvage should be attempted. Once again a thorough irrigation and debridement with removal of any contaminants and nonviable tissue performed emergently. This may be the most important operation the patient undergoes during the entire course of treatment. External fixation to gain stability of fractures and to aid in wound care is typically performed at this time. If necessary, a definitive vascular repair should be performed following skeletal stabilization. Ex-fix pins should be placed strategically away from the zone of injury and based on future incisions for definitive ORIF. Compromise of formal ORIF after DCO using external fixation is generally not an issue [68]. Fasciotomies should be performed as necessary. Antibiotic bead pouches and negative pressure wound therapy can be used to help decrease infection and assist with wound care [16, 37, 38, 54, 57]. The extremity is closely monitored over the next 24–72 h for soft tissue viability and sensorimotor function. Wounds should be regularly inspected, and repeat irrigation and debridements performed based on wound appearance (tissue viability, presence of contaminants, infection, etc.). VAC dressings are changed every 48–72 h.

If at any point the limb is deemed unsalvageable, or if the patient's life is in jeopardy secondary to the extremity, amputation should be performed. If the extremity remains viable for reconstruction and the patient condition permits then definitive skeletal stabilization and early soft tissue coverage should be performed [28, 29]. The use of BMP-2 has been approved in complex open tibia fractures. It was shown to accelerate fracture healing, reduce infection rate, and decrease the need for secondary procedures to obtain union in a randomized, prospective study involving 450 open tibia fractures [32]. Further research involving a larger cohort of patients with longer follow-up is necessary to confirm these results, and analyze the long-term complications and outcomes. Some recent studies have shown increased rates of heterotopic ossification in peri-articular injuries [7], possibly increased risk of malignancies [13], and unreliable reporting of complications in the literature [14]. Until more data is available, the utility and safety of BMP in the setting of open fractures is still uncertain.

Various modalities are available for surgical fixation including uniplanar external fixators, hyrid external fixators, thin-wire ring external fixators, plate and screw constructs, and intramedullary nails. There are pros and cons of each modality. It is beyond the scope of this chapter to recommend the type of fixation to use in the setting of complex extremity trauma. Many patients will require further surgery to achieve osseous union and this should be discussed along with possible complications with each patient thoroughly [8, 29, 35] (Figs. 13.3 and 13.4).

13.6 Complications

A major factor in the decision making in the treatment of the mangled extremity is the possible major complications associated with the treatment arm chosen. Harris et al. reported the nature and incidence of major complications for patients enrolled in the LEAP study group. Their cohort consisted of 545 patients with severe lower-extremity injuries followed prospectively for 24 months. A physician examined each patient at 3-, 6-, 12-, 24-month intervals and major complications recorded. The two most common complications were wound infection (28.3 %) and nonunion (23.7 %), and the majority of each of these required operative intervention and inpatient care. Approximately a quarter of each of these complications were considered severe enough to compromise long-term function. The overall incidence of wound dehiscence was 8.6 % and that of osteomyelitis 7.7 %. There was also a 5.3 % incidence of symptomatic hardware [35]. The complication data from the cohort was further examined based on treatment arm in the study. A total of 149 patients underwent amputations, and the revision amputation rate was 5.4 %. The most common complications in this group were wound infection (34.2 %), followed by stump revision (14.5 %), phantom limb pain and wound breakdown (13.4 % each), and stump complications (10.7 %). In the limb reconstruction group the most common complication was nonunion (31.5 %), followed by wound infection (23.2 %). Of these infections 8.6 % developed into osteomyelitis. There was an incidence of post-traumatic arthrosis of 9.4 % and wound necrosis or breakdown of 6.5 %. The late amputation group (patients amputated after initial discharge) experienced the highest rate of major complications (85 %) [35].

This fact clearly highlights the need for appropriate decision making in the patient with a mangled extremity at the onset of treatment. Although there were no late mortalities reported, an incidence of up to 21 % has been reported in the literature. Bondurant et al. undertook an investigation looking at the effects of delayed versus primary amputation. There was a significant increase in length of hospital stay (22 versus 53 days) and number of surgical interventions (1.6 versus 6.9). The cost was almost double ($28,964 versus $53,462), and there was a 21 % mortality rate in the delayed amputation group [6]. It is quite evident that every effort should be made to avoid a late amputation given such high costs for all involved.

In a prospective cohort study (using LEAP study patients), Castillo et al. examined the specific effect of smoking on complication rate in severe open tibia fractures. A total of 268 patients with unilateral injuries were followed prospectively. Nonunion rates were significantly higher in both the current and previous smoking groups

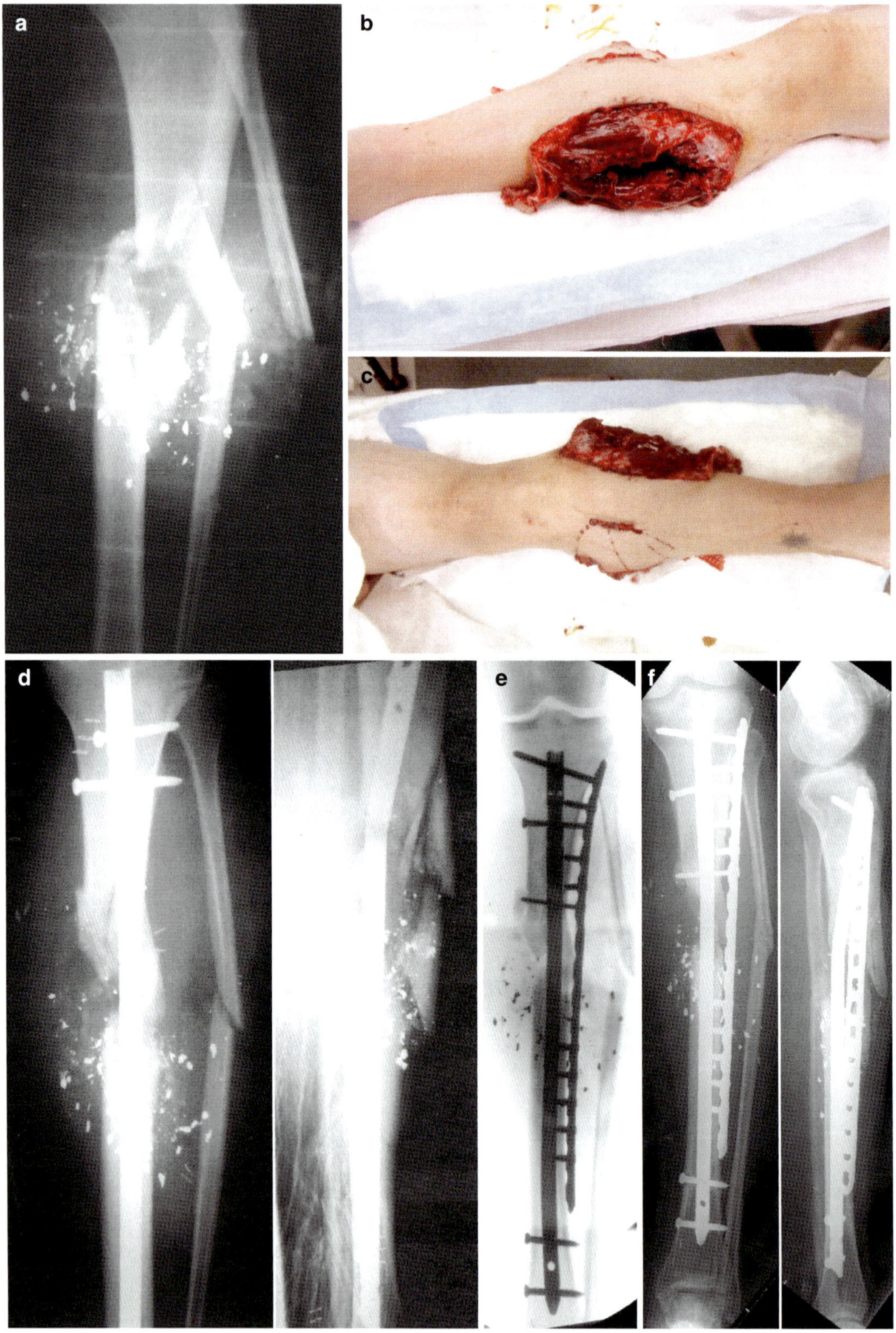
a
b
c
d
e
f

13 Mangled Extremity: Management in Isolated Extremity Injuries and in Polytrauma

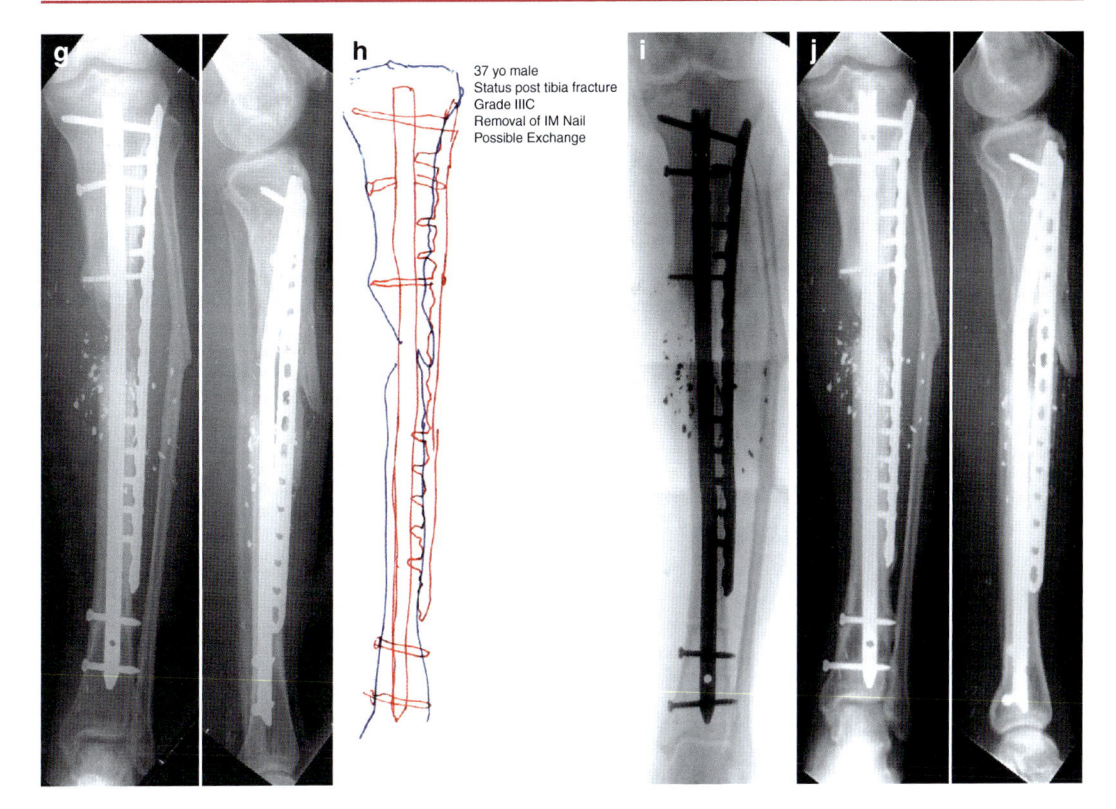

Fig. 13.3 (continued)

Fig. 13.3 A 36-year-old male was accidentally shot in the leg with a shotgun during a hunting trip. (**a–c**) He suffered an open, left-sided grade IIIC tibial shaft fracture with marked comminution. He also presented with complete functional deficit to his anterior compartment. He was taken to a local trauma center for irrigation and debridement (I&D), stabilization with and external fixation and a saphenous vein revascularization of the popliteal artery. Subsequent multiple I&D procedures were performed (including compromised bone). A negative pressure wound therapy dressing was placed over the wound sites. An inferior vena cava (IVC) filter was also inserted. (**d**) On day 3 a reamed, locked tibial intramedullary nail was inserted. (**e**) At 2 weeks following the injury, the patient was transferred to our institution for definitive management of his injuries. Repeat I&D was performed, the proximal interlocking screw was then removed to allow some correction of alignment and a percutaneous locking plate and screws was placed along the lateral surface of the tibia and a VAC dressing was applied. (**g**) Radiographs at 19 months illustrate some callus formation and a broken proximal interlocking screw. (**h, i**) Exchange IM nailing was planned and performed with placement of demineralized bone matrix (DBM) and a bone morphogenetic protein-7 (BMP-7) supplement. (**j**) At the latest follow-up visit at 29 months following revision surgery, he presented with good radiographic and clinical findings including increased callus formation and consolidation of the fracture, well-healed soft tissues, resolution of most pain symptoms, a return to activities of daily living and some recreational activities including weight training and skiing. A slight dorsiflexion lag was still present

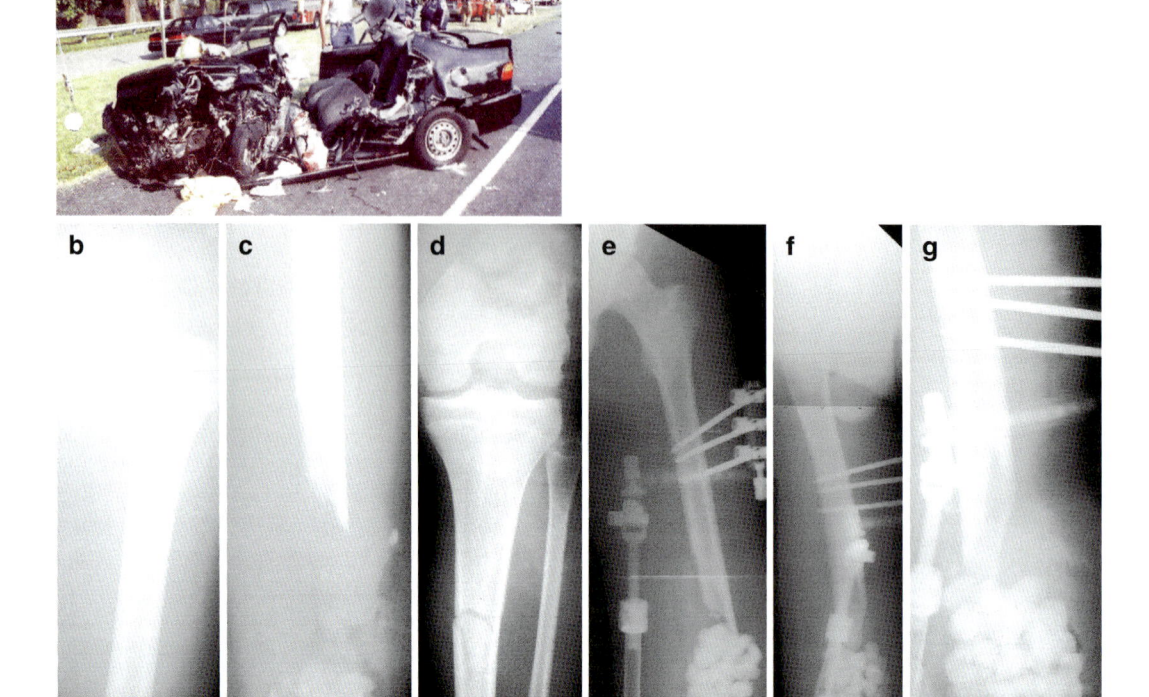

Fig. 13.4 A 17-year-old male was involved in a head-on collision with a tractor trailer. After being trapped inside the vehicle for approximately 1 h, he was extricated and flown to a local trauma center. He was diagnosed with an open, Grade IIIC left-sided AO/OTA Type C3.3 distal femur fracture with segmental defect and an ipsilateral tibial shaft fracture. External fixation was placed for initial stabilization and antibiotic beads were subsequently placed in the defect at 3 days following injury. Open reduction and internal fixation (ORIF) was performed with placement of an intramedullary (IM) locked nail for treatment of the tibial shaft fracture and then ORIF of the distal femur fracture with placement of a less invasive stabilization system (LISS) locking plate and screws. One week later the antibiotic beads were removed and the defect was prepared for bone graft placement. A second incision was made along the lateral border of the ipsilateral fibula and a free vascularized fibula bone graft was harvested for transplant to the femoral defect. It was docked in a double barrel fashion and stabilized using screw fixation. Following surgery he returned for regular follow-up visits. Three months after surgery all of the fractures were healing with incorporation of bone graft. The LISS plate was removed 4.5 years following the initial surgery. The clinical and radiographic follow-up illustrated excellent results with bony union, full range of motion, and complete resolution of pain and return to pre-injury activities. (**a**) Photograph of the vehicle and the scene following the accident. (**b–d**) Anteroposterior (AP) X-rays illustrating an AO/OTA Type C3.3 distal femur fracture with segmental bone defect and an ipsilateral tibial shaft fracture. (**e–g**) AP and lateral radiographs following placement of external fixation and antiobiotic beads at the site of the segmental bone defect. (**h**) Counterclockwise from top-left; preoperative plan, fluoroscopic images showing placement of intramedullary nail for the tibial shaft fracture and locking screws and open reduction and internal fixation (ORIF) of the distal femur fracture with placement of a LISS locking plate and screws. (**i–k**) Immediate postoperative radiographs demonstrating adequate fixation and alignment (**l**) AP radiographs illustrating preparation of distal femoral bone defect for placement of vascular bone graft. (**m**) AP X-radiograph following free vascularized fibular bone and placement of screw fixation. (**n–q**) AP and lateral X-rays 3.5 years following ORIF showing healed a distal femur fracture with incorporation of the fibular bone graft and a healed tibial shaft fracture. (**r, s**) AP and lateral X-rays 8 months following removal of LISS plate and screws and 4.5 years following fracture surgery

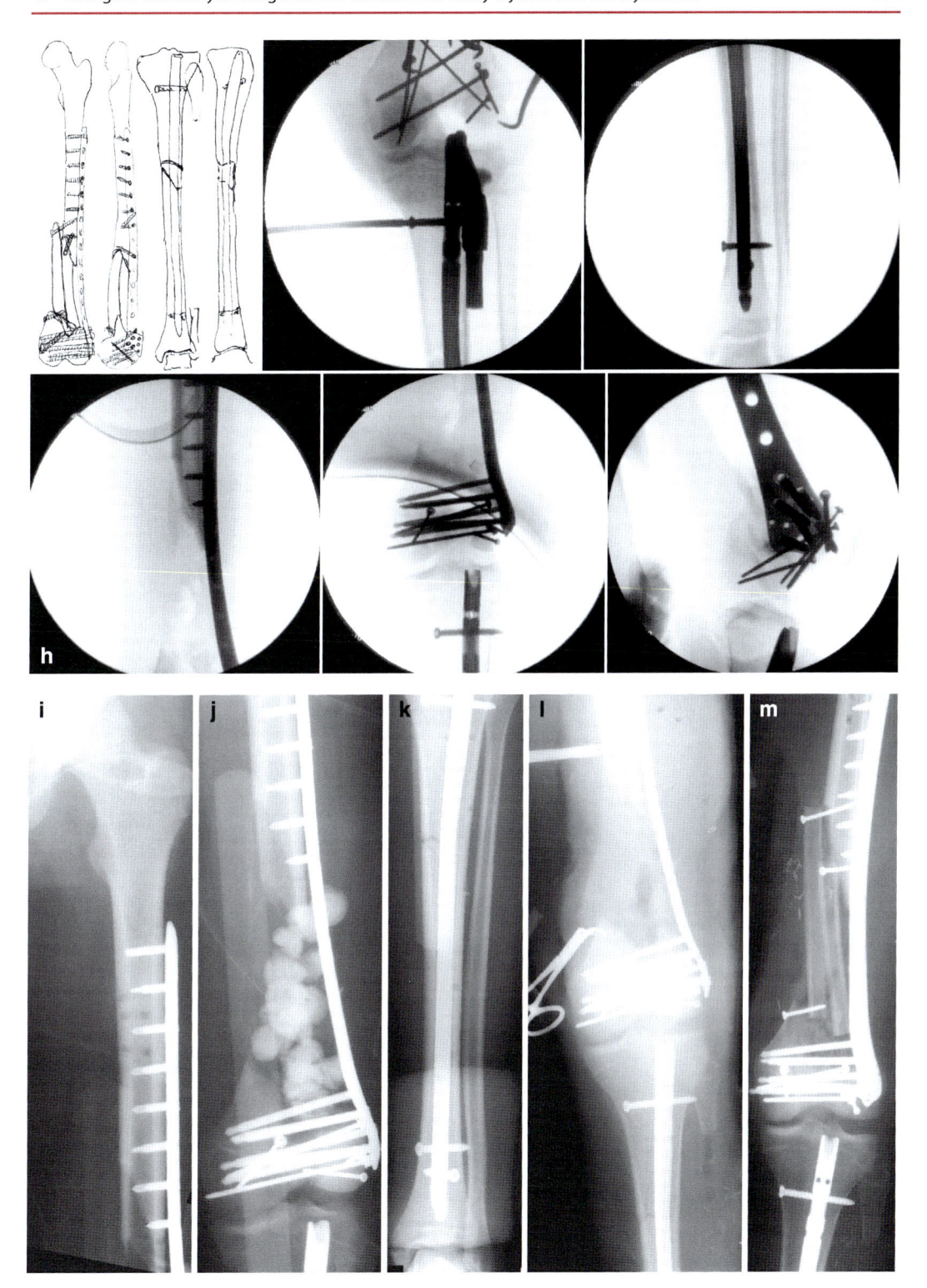

13 Mangled Extremity: Management in Isolated Extremity Injuries and in Polytrauma 179

Fig. 13.4 (continued)

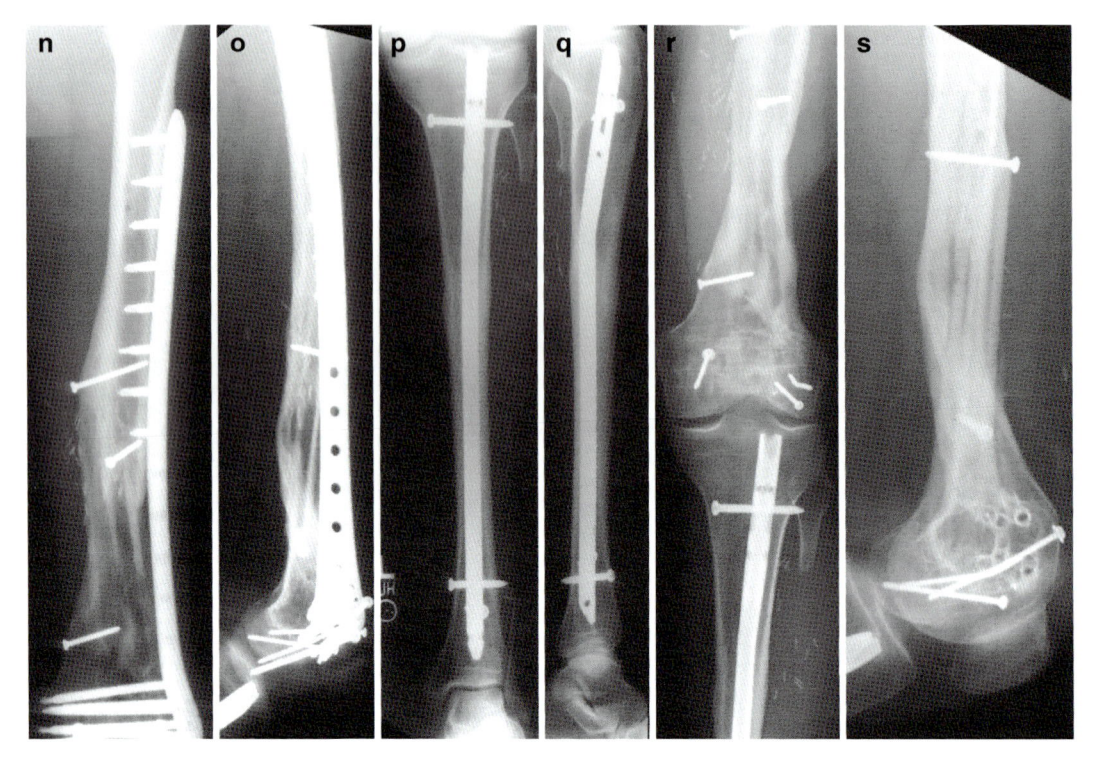

Fig. 13.4 (continued)

(37 % and 32 % respectively). The authors were able to demonstrate that current smokers were more than twice as likely to develop an infection, and 3.7 times more likely to have osteomyelitis. Previous smoking history was detrimental as well, and this group was 2.8 times more likely to develop osteomyelitis than nonsmokers. Their recommendation was that orthopedic surgeons should encourage patients to enter smoking cessation programs [15]. It has been shown that with the assistance of a surgeon up to 40 % of patients are able to quit smoking [58].

13.7 Predictive Ability of Scoring Systems to Predict Final Outcome

Some authors have examined the ability of the previously discussed scoring systems to predict functional outcome following treatment. Durham et al. performed a retrospective analysis of upper and lower severe extremity injuries to determine the validity and ability to predict outcome of the predictive indices discussed earlier. For each of the four systems analyzed, there were no significant differences between patients with good or poor functional outcomes [19]. Ly et al. reported on the ability of the five most commonly used predictive indices (above plus Hannover Fracture Scale-98) to determine functional recovery following limb salvage in a cohort of 507 patients (LEAP study group). The authors showed that none of the scoring systems analyzed were able to determine the outcome based on the SIP out to 24 months following injury [47]. One can conclude, based on these two studies, that the commonly applied predictive indices may be useful in early decision making, but are unable to predict functional recovery.

13.8 Outcomes Following Limb Salvage Versus Amputation

Recent medical and surgical technological advances have dramatically improved the surgeon's ability to salvage severely injured extremities. Limbs that historically would have been amputated can now be managed with complex reconstruction techniques. Although the limb remains viable, it is often questioned whether or not the patient would have been better served with an amputation. Limb salvage patients often still complain of edema, loss of motion, pain, decreased sensation, difficultly with footwear and ambulation [41]. The end result is often a physical, psychological, financial, and social cripple with a useless salvaged limb [34, 36].

Hoogendorn and van der Werken looked at the long-term outcome and quality of life of patients treated with reconstruction versus amputation following Grade III open tibia fractures. A total of 64 patients were assessed, including 43 with successful limb salvage and 21 who underwent amputations (including both primary and delayed). Lower extremity impairment was determined using "Guides to the Evaluation of Permanent Impairment" of the American Medical Association. Quality of life was measured using the Nottingham Health Profile (NHP), the SF-36, and a questionnaire the authors specifically designed for the study examining pain, daily function, psychological factors, and handicap with working. Patients who underwent amputations had more severe injuries, and had a higher number of vascular injuries (77 % versus 17 %). The limb salvage group underwent more operations and had more complications [41].

Delayed amputations were performed in eight patients, most commonly secondary to persistent infection and poor soft tissues. They were hospitalized twice as long as those who underwent primary amputation. Others have shown that delayed amputation results in poorer functional outcome versus primary amputation [6, 25]. From the reported health surveys the authors found low scores in both groups but no significant differences. In both groups over half the patients considered themselves disabled, with a slightly higher percentage of patients who had amputations reporting difficulty with practicing a profession (60 % versus 40 %). Of particular interest was that the mean lower extremity impairment score was significantly worse for amputees (73.5 %) as compared to the limb salvage group (17.6 %). These patients therefore perceived a higher level of function than those who were amputated [41].

The LEAP study group recently examined the functional outcome following limb salvage versus amputation. A total of 569 patients with severe leg-threatening injuries were studied in this multicenter, prospective, observational study. Eight level I trauma studies participated in this investigation. Functional outcome was measured using the sickness impact profile (SIP) and follow-up at 24 months was 84.4 %. Comparisons of outcomes for the SIP were adjusted for potential confounding variables of the patient characteristics as well as their specific injuries [8]. It was noted that patients who underwent amputation had more severe injuries, but otherwise did not differ from those who had reconstruction [8, 48].

Upon examining final functional outcome there were no significant differences in scores between either treatment group, although 42 % of the patients had scores greater than 10 indicating severe disability. Patients who underwent limb salvage were more likely to have been re-hospitalized than those who had amputation performed (47.6 % versus 33.9 %, $p=0.002$). Multivariate analysis reveal several factors that were significant factors for a poor outcome including: re-hospitalization for a major complication, having less than a high-school education, low household income, having no insurance or Medicaid, being nonwhite, smoking, having a poor social-support network, having a low-level of self-efficacy, and being involved with the legal system for injury compensation. At final follow-up, approximately 50 % of patients had returned to work and this rate did not differ between the two groups [8].

Patients with bilateral mangled extremities were excluded from the initial above analysis in the LEAP study but were followed prospectively and reported on separately. There were a total of 32 bilateral injuries, of which 14 had bilateral salvage, 10 had bilateral amputation, and 8 had unilateral salvage/amputation. Forty six percent of patients were severely disabled at 24 month follow-up as demonstrated by SIP scores >10. Once again, the groups where salvage procedures were performed had higher re-hospitalization rates for complications than the bilateral amputation group. The return to work rate was higher in the unilateral amputation/salvage group, and they had faster walking speeds. Examination of all three combinations of treatment of bilateral limb-threatening injuries demonstrated similar outcomes at 2 years. The evidence from this study suggested that the disability for bilateral limb-threatening injuries is high, but no more so than the unilateral group described above. The authors therefore concluded that treatment strategies for bilateral mangled extremities should be derived from the results from the larger cohort study of unilateral injuries [66].

MacKenzie et al. later reported on the long-term follow up of the original patients included in the LEAP study. The main goals of the study were to determine if the previously reported outcomes improved after 2 years, and whether there were any late differences between the treatment groups. Of the 569 patients from the original cohort, 397 were contacted by phone at an average of 84 months post-injury (range 70–90 months). On average, most of the patients reported physical and psychosocial functioning that had deteriorated since their 24-month follow-up ($p<0.05$). This increase in SIP scores was consistent across both treatment groups. It should be noted that patients who underwent knee amputations were at the highest risk for a poor outcome. More than a third of patients in both groups had been re-hospitalized between 2 and 7 years post-injury. At final follow-up almost half of the patients indicated severe disability, with SIP scores >10. Only 34.5 % of the cohort had a physical SIP subscore typical of the general population (<5) [49]. Recently, the LEAP study reported on a subset of patients with mangled foot and ankle injuries that required free tissue transfer or ankle arthrodesis, and they have shown to have SIP outcomes significantly worse than below-knee amputation [20].

13.9 Cost of Care

There have been conflicting reports in the literature over the long-term health-care costs of limb salvage versus amputation. Hertel et al. [40] calculated a 15 % higher hospital cost for the reconstruction patients over those who underwent amputation over the first 4 years post-injury [31].

Georgiadis also showed that patients who undergo reconstruction have higher hospital charges over those with primary amputations [25]. Bondurant et al. demonstrated a substantially higher hospital cost for patients who had delayed amputations over those who had primary amputations [6]. The LEAP study group found that the average 2-year costs for amputation versus reconstruction were very similar. When the cost of prosthetic devices was included, health-care costs were significantly higher for patients who had amputations. The projected lifetime health-care cost was three times higher for patients in the amputation group ($509,275 versus $163,282). The large number of patients in this particular study (545 patients), and the fact that this study is much more recent than the other mentioned reports, make this data more valuable [50].

13.10 The Mangled Upper Extremity

Of note are the differences between the mangled upper and mangled lower extremity, which must be carefully considered by the treating surgeon. Critical time for reperfusion is longer in the upper (8–10 h) versus the lower extremity (6 h) [69]. Bumbasirevic reviewed the literature on the mangled upper extremity and found upper limits of cold ischemia in the hand and digits up to 24 h [10]. A trans-tibial amputation carries a much better functional prognosis than a trans-radial amputation. This is due to the fact that upper extremity prostheses do not work as well as lower extremity prostheses, that is, both from a functional and an aesthetic perspective [10]. Loss of an upper extremity is probably emotionally much more difficult than loss of a lower extremity and might warrant more aggressive limb salvage attempts [64]. The mangled hand warrants special care and attention as was reiterated by Bumbasirevic who quoted the famous hand surgeon Bunnell "when you have nothing, a little is a lot" [10]. Shortening of the humerus to reduce soft tissue defects is tolerated well up to 5 cm, and the forearm up to 4 cm [10] in contrast to the lower extremity that does not tolerate shortening of more than 2 cm. Nerve reconstruction in the upper extremity is done with reasonable success, whereas in the lower extremity many consider major nerve injury an indication for primary amputation. The rehabilitation process is also more imperative when the upper extremity is involved with emphasis on the gliding of the tendons, preventing contractures and edema [10, 33]. One consistency to both is that the MESS has also been shown to be useful for predicting amputation following mangled upper extremities [65].

13.11 The Pediatric Mangled Extremity

The mangled extremity in the child poses its own challenges and has not been adequately investigated in the literature. Fortunately, these injuries are rare and this is likely the reason for a paucity of studies to help guide treatment. Vascular injuries in this population usually involve the femoral or brachial arteries, and most commonly result from penetrating injuries [52]. The MESS seems to have some utility in helping guide treatment, although the scoring system fails to be adequately specific in this patient population [4, 22, 52]. Stewart et al. studied the MESS, LSI, PSI, NISSSA, and the Hannover Fracture Scale 1998 (HFS-98), and none were recommended for use as an absolute indication for early amputation in children [67]. Of note is that the MESS uses age itself as a scoring criterion (no points below age 30) making comparison between pediatric patients difficult. The same holds true for the scoring of shock based on systolic blood pressure which is known to respond differently to hypovolemia in children. The MESS should therefore be used with some caution given the reports seem to demonstrate a lower need for amputation and limb salvage percentages especially in recent years [21, 67]. These complex injuries are rare in children and studies on this topic all contain fewer than 40 patients, and further investigation into this topic is clearly needed. We agree with others [52] that at this time the decision for limb salvage versus primary amputation still has to be made individually in pediatric patients.

13.12 The Mangled Extremity and Polytrauma

Severely injured patients that would not have survived their trauma in the past now survive because of improved resuscitation. Mangled limbs that used to be considered beyond reconstruction can now be salvaged. However, the decision of whether to reconstruct or amputate a mangled extremity in a polytrauma today still requires complex and careful decision making. An undisputed rule in polytrauma is "life before limb", meaning life-threatening issues are always addressed first. Orthopedic efforts in the initial resuscitation of the severely injured patient with extremity injury often involve damage control orthopedics (DCO) [55, 60, 69]. DCO polytrauma patients are typically categorized into stable, borderline, unstable, and *in extremis*. The goal of DCO is to minimize subsequent stresses after the first hit (i.e., injury) and its effectiveness in the context of major orthopedic fractures has been shown [55, 60, 68].

The question whether amputation of a mangled limb is advisable for a severely injured patient cannot be answered [44]. There are no clear guidelines with respect to the *isolated* mangled extremities, let alone the polytrauma patient. As an exception, utilizing DCO guidelines, salvage of the *stable* polytrauma patient's mangled limb is possibly the most relevant. For these, techniques involving early free tissue transfer and internal fixation as proposed by the "fix-and-flap" technique might be successful, but require a highly specialized trauma center [30]. There are not many centers than can provide soft tissue coverage with a free flap within 24 h. Recent evidence suggests that the use of VAC for up to 7 days does not compromise outcome. Others have found that the use of VAC has allowed them to "step down" the reconstructive ladder, needing fewer free vascular transfers [16, 56, 59]. Still, for these patients, the decision whether to salvage or amputate faces the same dilemmas as for the patient with the isolated mangled limb as described elsewhere in this chapter.

Borderline patients that stabilize after resuscitation can undergo early total care (ETC), but reconstructive efforts need to anticipate potential deterioration. Long procedures (e.g., "fix-and-flap") are not justified in these patients. Wound debridement, revascularization, and external fixation are all that can be done while a rapid turn for the worse should be anticipated. In the *unstable* or *in extremis* polytrauma patients there might be a role for primary amputation as prolonged revascularization and stabilization procedures add to the patient's catabolic state and will increase the second hit enormously. Any other reconstructive efforts for the extremities are not justified.

Next step in limb salvage should not be undertaken until the patient has stabilized and is beyond the systemic inflammatory response syndrome (SIRS) stage. As a rule, timing of second and subsequent major procedures (longer than 3 h) should be at least after 4 days [40]. If the limb develops evidence of sepsis, early amputation should still be considered. The use of fresh warm blood, plasma, and recombinant factor VII defined as damage control resuscitation before surgery help to optimize the physiologic parameters and theoretically allows for more prolonged surgical procedures such as revascularization [24].

Conclusions

The combination of osseous, vascular, soft-tissue, and nerve injury present following severe trauma to an extremity make such injuries a challenge to treat. Unfortunately, the data regarding the management of the mangled extremity are conflicting, and the literature is without Class I studies. It is therefore imperative that an experienced surgical team at a trauma center that cares for such patients with some regularity care for the patient with a complex extremity injury [50]. The treating team must always keep in mind the high prevalence of associated multisystem trauma and systemic problems related to these injuries. Even though the treatment goal is limb salvage, it must be kept in mind that in many instances a primary amputation might provide the best outcome, and can be an imperative life-saving procedure in the setting of polytrauma. New insights, therapies, and techniques will improve outcomes in even the most severely injured patients with complex extremity injuries. Good collaboration between plastic, orthopedic, and vascular surgeons makes an enormous difference in terms of limb salvage as well as secondary reconstruction [23].

As for the mangled limb in these patients it is unlikely a scoring system will allow a clear cut-off point for amputation versus salvage. What has become clear is that primary amputation should not be considered a treatment failure but rather a means of meeting goals of treatment [12]. As Hansen pointed out long ago, we should not let heroism triumph over reason [34].

References

1. Ball CG, Rozycki GS, Feliciano DV. Upper extremity amputations after motor vehicle rollovers. J Trauma. 2009;67(2):410–2.
2. Bartlett CS, Helfet DL, Hausman MR, Strauss E. Ballistics and gunshot wounds: effects on musculoskeletal tissues. J Am Acad Orthop Surg. 2000;8(1):21–36.
3. Beatty ME, Zook EG, Russell RC, Kinkead LR. Grain auger injuries: the replacement of the corn picker injury? Plast Reconstr Surg. 1982;69(1):96–102.
4. Behdad S, Rafiei MH, Taheri H, Behdad S, Mohammadzadeh M, Kiani G, et al. Evaluation of Mangled Extremity Severity Score (MESS) as a predictor of lower limb amputation in children with trauma. Eur J Pediatr Surg. 2012;22(6):465–9.
5. Bonanni F, Rhodes M, Lucke JF. The futility of predictive scoring of mangled lower extremities. J Trauma. 1993;34(1):99–104.
6. Bondurant FJ, Cotler HB, Buckle R, Miller-Crotchett P, Browner BD. The medical and economic impact of severely injured lower extremities. J Trauma. 1988; 28(8):1270–3.
7. Boraiah S, Paul O, Hawkes D, Wickham M, Lorich DG. Complications of recombinant human BMP-2 for treating complex tibial plateau fractures: a preliminary report. Clin Orthop Relat Res. 2009;467(12): 3257–62.
8. Bosse MJ, MacKenzie EJ, Kellam JF, Burgess AR, Webb LX, Swiontkowski MF, et al. An analysis of outcomes of reconstruction or amputation after leg-threatening injuries. N Engl J Med. 2002;347(24): 1924–31.
9. Brown KV, Ramasamy A, McLeod J, Stapley S, Clasper JC. Predicting the need for early amputation in ballistic mangled extremity injuries. J Trauma. 2009;66(4 Suppl):S93–7.

10. Bumbasirevic M, Stevanovic M, Lesic A, Atkinson HD. Current management of the mangled upper extremity. Int Orthop. 2012;36(11):2189–95.
11. Campbell DC, Bryan RS, Cooney III WP, Ilstrup D. Mechanical cornpicker hand injuries. J Trauma. 1979;19(9):678–81.
12. Cannada LK, Cooper C. The mangled extremity: limb salvage versus amputation. Curr Surg. 2005;62(6):563–76.
13. Carragee EJ. Re: BMP-2 augmented fusion in the low-risk, healthy subjects: a confirmation of effectiveness, harms highlights the need for study in high risk patients. Spine (Phila Pa 1976). 2014;39(4):341–2. http://www.ncbi.nlm.nih.gov/pubmed/24162132.
14. Carragee EJ, Baker RM, Benzel EC, Bigos SJ, Cheng I, Corbin TP, et al. A biologic without guidelines: the YODA project and the future of bone morphogenetic protein-2 research. Spine J. 2012;12(10):877–80.
15. Castillo RC, Bosse MJ, MacKenzie EJ, Patterson BM. Impact of smoking on fracture healing and risk of complications in limb-threatening open tibia fractures. J Orthop Trauma. 2005;19(3):151–7.
16. Dedmond BT, Kortesis B, Punger K, Simpson J, Argenta J, Kulp B, et al. The use of negative-pressure wound therapy (NPWT) in the temporary treatment of soft-tissue injuries associated with high-energy open tibial shaft fractures. J Orthop Trauma. 2007;21(1):11–7.
17. Dente CJ, Feliciano DV, Rozycki GS, Cava RA, Ingram WL, Salomone JP, et al. A review of upper extremity fasciotomies in a level I trauma center. Am Surg. 2004;70(12):1088–93.
18. Dirschl DR, Dahners LE. The mangled extremity: when should it be amputated? J Am Acad Orthop Surg. 1996;4(4):182–90.
19. Durham RM, Mistry BM, Mazuski JE, Shapiro M, Jacobs D. Outcome and utility of scoring systems in the management of the mangled extremity. Am J Surg. 1996;172(5):569–73.
20. Ellington JK, Bosse MJ, Castillo RC, MacKenzie EJ, LEAP Study Group. The mangled foot and ankle: results from a 2-year prospective study. J Orthop Trauma. 2013;27(1):43–8.
21. Elsharawy MA, Maher K, Elsaid AS. Limb salvage in a child with severely injured mangled lower extremity and muscle rigor. Vascular. 2012;20(6):321–4.
22. Fagelman MF, Epps HR, Rang M. Mangled extremity severity score in children. J Pediatr Orthop. 2002;22(2):182–4.
23. Fodor L, Sobec R, Sita-Alb L, Fodor M, Ciuce C. Mangled lower extremity: can we trust the amputation scores? Int J Burns Trauma. 2012;2(1):51–8.
24. Fox CJ, Gillespie DL, Cox ED, Kragh Jr JF, Mehta SG, Salinas J, et al. Damage control resuscitation for vascular surgery in a combat support hospital. J Trauma. 2008;65(1):1–9.
25. Georgiadis GM, Behrens FF, Joyce MJ, Earle AS, Simmons AL. Open tibial fractures with severe soft-tissue loss. Limb salvage compared with below-the-knee amputation. J Bone Joint Surg Am. 1993;75(10):1431–41.
26. Giannoudis PV, Dinopoulos H, Chalidis B, Hall GM. Surgical stress response. Injury. 2006;37 Suppl 5:S3–9.
27. Giannoudis PV, Giannoudi M, Stavlas P. Damage control orthopaedics: lessons learned. Injury. 2009;40 Suppl 4:S47–52. doi:10.1016/j.injury.2009.10.036.:S47-S52.
28. Godina M. Early microsurgical reconstruction of complex trauma of the extremities. Plast Reconstr Surg. 1986;78(3):285–92.
29. Gopal S, Giannoudis PV, Murray A, Matthews SJ, Smith RM. The functional outcome of severe, open tibial fractures managed with early fixation and flap coverage. J Bone Joint Surg Br. 2004;86(6):861–7.
30. Gopal S, Majumder S, Batchelor AG, Knight SL, De Boer P, Smith RM. Fix and flap: the radical orthopaedic and plastic treatment of severe open fractures of the tibia. J Bone Joint Surg Br. 2000;82(7):959–66.
31. Gorsche TS, Wood MB. Mutilating corn-picker injuries of the hand. J Hand Surg Am. 1988;13(3):423–7.
32. Govender S, Csimma C, Genant HK, Valentin-Opran A, Amit Y, Arbel R, et al. Recombinant human bone morphogenetic protein-2 for treatment of open tibial fractures: a prospective, controlled, randomized study of four hundred and fifty patients. J Bone Joint Surg Am. 2002;84-A(12):2123–34.
33. Gupta A, Wolff TW. Management of the mangled hand and forearm. J Am Acad Orthop Surg. 1995;3(4):226–36.
34. Hansen Jr ST. Overview of the severely traumatized lower limb. Reconstruction versus amputation. Clin Orthop Relat Res. 1989;243:17–9.
35. Harris AM, Althausen PL, Kellam J, Bosse MJ, Castillo R. Complications following limb-threatening lower extremity trauma. J Orthop Trauma. 2009;23(1):1–6.
36. Helfet DL, Howey T, Sanders R, Johansen K. Limb salvage versus amputation. Preliminary results of the mangled extremity severity score. Clin Orthop Relat Res. 1990;256:80–6.
37. Henry SL, Ostermann PA, Seligson D. The antibiotic bead pouch technique. The management of severe compound fractures. Clin Orthop Relat Res. 1993;295:54–62.
38. Herscovici Jr D, Sanders RW, Scaduto JM, Infante A, Dipasquale T. Vacuum-assisted wound closure (VAC therapy) for the management of patients with high-energy soft tissue injuries. J Orthop Trauma. 2003;17(10):683–8.
39. Hertel R, Strebel N, Ganz R. Amputation versus reconstruction in traumatic defects of the leg: outcome and costs. J Orthop Trauma. 1996;10(4):223–9.
40. Hildebrand F, Giannoudis P, Kretteck C, Pape HC. Damage control: extremities. Injury. 2004;35(7):678–89.
41. Hoogendoorn JM, van der Werken C. Grade III open tibial fractures: functional outcome and quality of life in amputees versus patients with successful reconstruction. Injury. 2001;32(4):329.
42. Howe Jr HR, Poole Jr GV, Hansen KJ, Clark T, Plonk GW, Koman LA, et al. Salvage of lower extremities

following combined orthopedic and vascular trauma. A predictive salvage index. Am Surg. 1987;53(4):205–8.
43. Johansen K, Daines M, Howey T, Helfet D, Hansen Jr ST. Objective criteria accurately predict amputation following lower extremity trauma. J Trauma. 1990;30(5):568–72.
44. Kobbe P, Lichte P, Pape HC. Complex extremity fractures following high energy injuries: the limited value of existing classifications and a proposal for a treatment-guide. Injury. 2009;40 Suppl 4:S69–74. doi:10.1016/j.injury.2009.10.039.:S69-S74.
45. Korompilias AV, Beris AE, Lykissas MG, Vekris MD, Kontogeorgakos VA, Soucacos PN. The mangled extremity and attempt for limb salvage. J Orthop Surg Res. 2009;4:4. doi:10.1186/1749-799X-4-4.:4.
46. Lange RH, Bach AW, Hansen Jr ST, Johansen KH. Open tibial fractures with associated vascular injuries: prognosis for limb salvage. J Trauma. 1985;25(3):203–8.
47. Ly TV, Travison TG, Castillo RC, Bosse MJ, MacKenzie EJ. Ability of lower-extremity injury severity scores to predict functional outcome after limb salvage. J Bone Joint Surg Am. 2008;90(8):1738–43.
48. MacKenzie EJ, Bosse MJ, Kellam JF, Burgess AR, Webb LX, Swiontkowski MF, et al. Characterization of patients with high-energy lower extremity trauma. J Orthop Trauma. 2000;14(7):455–66.
49. MacKenzie EJ, Bosse MJ, Pollak AN, Webb LX, Swiontkowski MF, Kellam JF, et al. Long-term persistence of disability following severe lower-limb trauma. Results of a seven-year follow-up. J Bone Joint Surg Am. 2005;87(8):1801–9.
50. MacKenzie EJ, Rivara FP, Jurkovich GJ, Nathens AB, Egleston BL, Salkever DS, et al. The impact of trauma-center care on functional outcomes following major lower-limb trauma. J Bone Joint Surg Am. 2008;90(1):101–9.
51. McNamara MG, Heckman JD, Corley FG. Severe open fractures of the lower extremity: a retrospective evaluation of the Mangled Extremity Severity Score (MESS). J Orthop Trauma. 1994;8(2):81–7.
52. Mommsen P, Zeckey C, Hildebrand F, Frink M, Khaladj N, Lange N, et al. Traumatic extremity arterial injury in children: epidemiology, diagnostics, treatment and prognostic value of Mangled Extremity Severity Score. J Orthop Surg Res. 2010;5:25. doi:10.1186/1749-799X-5-25.:25.
53. O'Sullivan ST, O'Sullivan M, Pasha N, O'Shaughnessy M, O'Connor TP. Is it possible to predict limb viability in complex Gustilo IIIB and IIIC tibial fractures? A comparison of two predictive indices. Injury. 1997;28(9–10):639–42.
54. Ostermann PA, Seligson D, Henry SL. Local antibiotic therapy for severe open fractures. A review of 1085 consecutive cases. J Bone Joint Surg Br. 1995;77(1):93–7.
55. Pape HC. Effects of changing strategies of fracture fixation on immunologic changes and systemic complications after multiple trauma: damage control orthopedic surgery. J Orthop Res. 2008;26(11):1478–84.
56. Parrett BM, Matros E, Pribaz JJ, Orgill DP. Lower extremity trauma: trends in the management of soft-tissue reconstruction of open tibia-fibula fractures. Plast Reconstr Surg. 2006;117(4):1315–22.
57. Prasarn ML, Zych G, Ostermann PA. Wound management for severe open fractures: use of antibiotic bead pouches and vacuum-assisted closure. Am J Orthop (Belle Mead NJ). 2009;38(11):559–63.
58. Rechtine GR, Frawley W, Castellvi A, Gowski A, Chrin AM. Effect of the spine practitioner on patient smoking status. Spine (Phila Pa 1976). 2000;25(17):2229–33.
59. Rinker B, Amspacher JC, Wilson PC, Vasconez HC. Subatmospheric pressure dressing as a bridge to free tissue transfer in the treatment of open tibia fractures. Plast Reconstr Surg. 2008;121(5):1664–73.
60. Roberts CS, Pape HC, Jones AL, Malkani AL, Rodriguez JL, Giannoudis PV. Damage control orthopaedics: evolving concepts in the treatment of patients who have sustained orthopaedic trauma. Instr Course Lect. 2005;54:447–62.
61. Robertson PA. Prediction of amputation after severe lower limb trauma. J Bone Joint Surg Br. 1991;73(5):816–8.
62. Roessler MS, Wisner DH, Holcroft JW. The mangled extremity. When to amputate? Arch Surg. 1991;126(10):1243–8.
63. Russell WL, Sailors DM, Whittle TB, Fisher Jr DF, Burns RP. Limb salvage versus traumatic amputation. A decision based on a seven-part predictive index. Ann Surg. 1991;213(5):473–80.
64. Scalea TM, DuBose J, Moore EE, West M, Moore FA, McIntyre R, et al. Western Trauma Association critical decisions in trauma: management of the mangled extremity. J Trauma Acute Care Surg. 2012;72(1):86–93.
65. Slauterbeck JR, Britton C, Moneim MS, Clevenger FW. Mangled extremity severity score: an accurate guide to treatment of the severely injured upper extremity. J Orthop Trauma. 1994;8(4):282–5.
66. Smith JJ, Agel J, Swiontkowski MF, Castillo R, MacKenzie E, Kellam JF. Functional outcome of bilateral limb threatening: lower extremity injuries at two years postinjury. J Orthop Trauma. 2005;19(4):249–53.
67. Stewart DA, Coombs CJ, Graham HK. Application of lower extremity injury severity scores in children. J Child Orthop. 2012;6(5):427–31.
68. Taeger G, Ruchholtz S, Waydhas C, Lewan U, Schmidt B, Nast-Kolb D. Damage control orthopedics in patients with multiple injuries is effective, time saving, and safe. J Trauma. 2005;59(2):409–16.
69. Togawa S, Yamami N, Nakayama H, Mano Y, Ikegami K, Ozeki S. The validity of the mangled extremity severity score in the assessment of upper limb injuries. J Bone Joint Surg Br. 2005;87(11):1516–9.

Management of Spinal Fractures

14

Keith L. Jackson, Michael Van Hal, Joon Y. Lee, and James D. Kang

Contents

14.1	**Incidence**...	187
14.1.1	Associated Injuries and Premorbid Factors..	187
14.2	**Mortality**..	188
14.3	**Prehospital Management**...................	188
14.3.1	Initial Management...............................	188
14.3.2	Clinical History and Examination........	188
14.3.3	Immobilization.....................................	188
14.3.4	Management During Transport............	189
14.4	**In-Hospital Management**...................	189
14.4.1	History...	189
14.4.2	Initial Management...............................	189
14.4.3	Spinal Imaging.....................................	193
14.4.4	Hospital Resuscitation: Workup...........	195
14.4.5	Cervical Traction..................................	195
14.5	**Treatment**..	196
14.5.1	Nonsurgical Treatment.........................	196
14.5.2	Surgical Treatment...............................	196
14.5.3	Special Situations.................................	197
14.6	**Clinical Outcomes**.............................	198
References..		199

K.L. Jackson, MD (✉)
Staff Spine Surgeon, Womack Army Medical Center,
2817 Reilly Road, Fort Bragg, NC 28310, USA
e-mail: lynn.jackson9@gmail.com

M. Van Hal, MD • J.D. Kang, MD
Department of Orthopaedic Surgery,
University of Pittsburgh Medical Center,
3471 Fifth Avenue Kaufmann Building Suite 1010,
15213 Pittsburgh, PA, USA
e-mail: VanHalMD@gmail.com; KangJD@UPMC.edu

J.Y. Lee, MD
Division of Spine Surgery, Department
of Orthopaedic Surgery, University of Pittsburgh
Medical Center, Pittsburgh, PA, USA
e-mail: LeeJY3@UPMC.edu

14.1 Incidence

Vertebral column fractures are a heterogeneous group of injuries, whose severity varies significantly by injury type. In total, fractures of the spine occur in 6 % of trauma patients with approximately half of this group having some degree of neurologic deficit [1]. The anatomic distribution of these injuries vary by region with approximately 55 % of cases involving the cervical spine, 15 % involving the thoracic region, 15 % affecting the thoracolumbar region, and the final 15 % of injuries occurring in the lumbosacral region [2]. Once a vertebral column fracture is detected, the clinician should carefully examine the entire spine since approximately 15–20 % of these patients have multiple injuries at noncontiguous levels [3].

14.1.1 Associated Injuries and Premorbid Factors

Patients with fractures of the spine frequently present with associated injuries to the head, visceral organs, or appendicular skeleton. In general, patients with cervical spine fractures are more likely to have associated head injuries while those with thoracic or lumbar fractures have a greater risk of sustaining associated chest or abdominal injury. The incidence of these associated injuries dramatically increases in the presence of neurologic injury. In fact, 80 % of patients presenting with paraplegia or quadriplegia have associated injuries and 41 % have documented head injuries [4]. The prevalence of these associated pathologies heightens the importance of performing a thorough examination and radiographic review in each patient, particularly those who are obtunded or intubated. These associated injuries directly impact healthcare cost, morbidity, and mortality in this population.

14.2 Mortality

The mortality rate of patients with vertebral column injury is influenced by the presence and severity of associated injuries and neurologic impairment. Among patients who die prior to hospitalization with an identifiable injury, only 4 % are due to spine injuries [5] and the reported 90-day mortality rate after thoracolumbar spine fracture surgery is 1.4 % [6]. This mortality rate increases dramatically in the presence of associated neurologic injury as the overall mortality rate for patients with spinal cord injuries has been reported as high as 17 % during the initial hospitalization [1]. While still high, the mortality rate of these injuries has decreased in recent years with modern advances such as more rapid forms of transport, enhanced imaging and diagnostic capabilities, and improved surgical and rehabilitative techniques.

14.3 Prehospital Management

14.3.1 Initial Management

The initial goals of prehospital care in the severely traumatized patient with a vertebral column injury are to provide immediate life saving medical interventions while maintaining spinal immobilization to prevent further neurologic injury. In keeping with these goals, every trauma patient should be considered at risk for spinal injury as the delayed or missed diagnosis of a spinal injury has been associated with tenfold risk of permanent neurologic deficit [7, 8]. First responders should exercise great care in patient extraction and transport to mitigate this risk while applying the fundamentals of Advanced Trauma Life Support (ATLS).

14.3.2 Clinical History and Examination

Elements of the initial history and patient evaluation are often helpful in predicting the risk of vertebral column injury. A history of a high-energy mechanism, high-speed motor vehicle accident (MVA), fall from heights (>4 m), physical signs of a head injury with or without unconsciousness, pain from the spine, and or neurological signs (weakness, radiculopathy, sensory loss) can help predict injury to the spine [9]. Findings such as hypotension, bradycardia, and warm dry skin are also important, as they may be indicative of neurogenic shock. Once recognized, first responders should relay this information to the receiving facility in order to help facilitate timely and appropriate care on arrival.

14.3.3 Immobilization

Patients with suspected spinal injury should be immobilized immediately at the scene of the accident. Emergency responders should maintain patients with a suspected cervical spine injury in a neutral position during both transport and potential intubation attempts. A cervical collar and spine board or a vacuum mattress should be applied. In cases of spinal deformity, the patient should be immobilized in a position of comfort without attempts at reduction in the field environment. Immobilization with a rigid cervical collar, firm spinal board, sandbags, and tape provides superior stability than a rigid cervical collar alone in patients with cervical injury [10]. In athletes, bicyclists, and motorcycle riders, helmets and shoulder pads should be left in place until individuals trained in their removal are available. Facemasks should be removed if access to the airway is needed.

14.3.4 Management During Transport

Patients with vertebral column injuries require strict adherence to immobilization and spinal precautions during transport in order to mitigate the risk of subsequent neurologic deterioration. Loading the patient onto the spinal board and subsequent examinations requires logrolling by a team of medical professionals or a specially designed scoop-stretcher in order to minimize spinal motion. Because of the relatively larger head size of pediatric patients, a spinal board with a head cutout or padding to elevate the trunk is necessary to maintain neutral head-neck alignment in this population. The development of pressure ulcers has been reported in up to 55 % of immobilized patients [11]. This risk is increased in patients who are unconscious, neurologically impaired, or immobilized for an extended period (>6 h). To mitigate this risk, a turning program should be initiated every 2 h while immobilized. Patients on a spine board are also often unable to secure their own airway and should be under supervision during transportation. Due to the possible untoward effects of immobilization patients should be evaluated and removed from spine boards as quickly as possible.

14.4 In-Hospital Management

The management in the hospital continues according to the principles of ATLS (Fig. 14.1).

14.4.1 History

The patient, eyewitnesses, paramedics, and emergency physicians should be questioned regarding the circumstances of the accident in order to determine the direction of force and mechanism of injury. Extrication from motor vehicle and traumatic brain injuries are associated with high risk of spinal injury [12]. In the ER setting, it is important to request information on the injury to continue the workup of spinal trauma. If stable and alert on admission, the patient should be asked about drug intake, pain from other injuries that could distract from the spine injury, as well as the type and mechanism of motor vehicle collision, and the onset of neck pain. A subset of fully awake and cooperative patients may have their cervical spine cleared based on careful and history and physical examination alone (Table 14.1).

14.4.2 Initial Management

The initial evaluation and management of a polytraumatized patient with a spinal injury should follow standard ATLS principles in a stepwise fashion (Fig. 14.1).

14.4.2.1 Ventilation

Establishing a patent airway is of utmost importance in the patient with vertebral column injuries as these patients may suffer from inadequate respiratory function due to concomitant rib fractures, facial injuries, paralysis of the diaphragm, or other injuries. In patients requiring intubation, maintenance of spinal alignment is critical. This procedure is best performed by nasotracheal tube or by fiber-optic procedures in conjunction with inline cervical traction. Regardless of the adjuvant tools used to establish an airway, avoiding extreme cervical extension is critical as this position narrows the spinal canal more than positions of flexion [12].

14.4.2.2 Circulation

Once an airway is secured, attention should turn to ensuring appropriate end organ perfusion and potentially avoid secondary ischemic injuries to the spinal cord. A spinal cord injury may cause vasospasm due to dysfunction of blood flow auto regulation. The hallmarks of neurogenic shock are hypotension, due to loss of sympathetic vascular tone; and bradycardia, due to loss of sympathetic innervations of the heart. If neurogenic shock is present, fluid resuscitation is a vital first intervention and follows the treatment principles used for brain injury. Central venous catheter and arterial lines are required for assessment of heart rate,

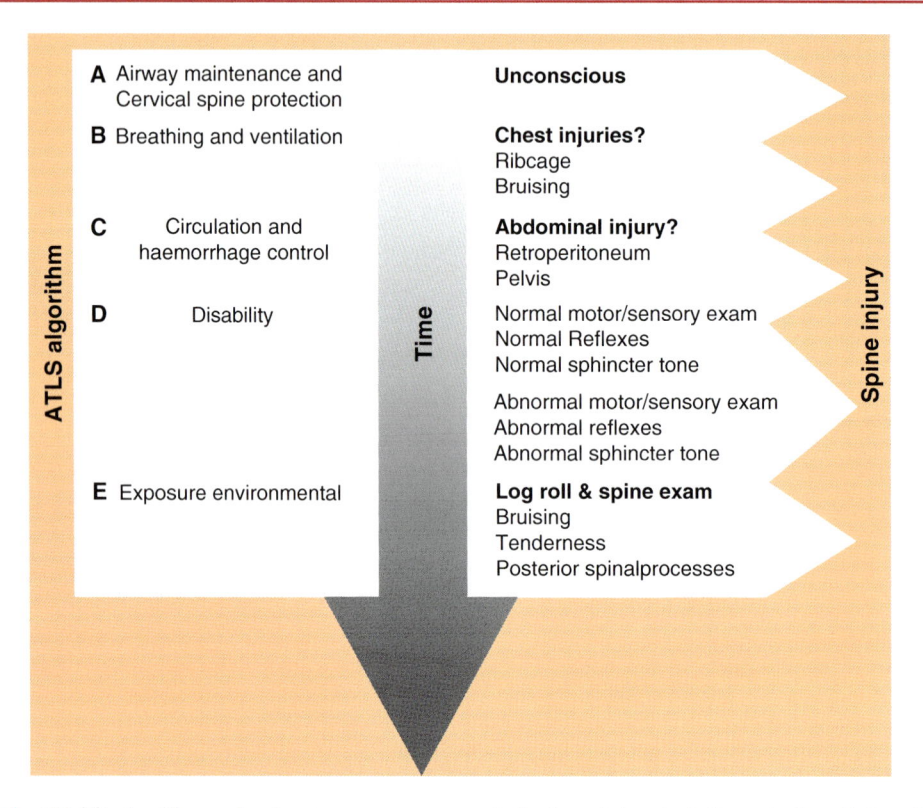

Fig. 14.1 ATLS® algorithm and spine trauma assessment. In step A, cervical spine (C-Spine) protection is essential. Every unconscious patient is stabilized by a stiff-neck orthosis. Patients with signs of chest injury in step B and abdominal injury in step C, especially retroperitoneal, are highly suspicious for thoracic and/or lumbar spine injury. Normal motor examination and reflexes do not rule out significant spine injury in the comatose patient. Abnormal neurologic examination is a sign for substantial spinal column injury including spinal cord injury. Log roll in step E is important to assess the posterior elements of the cervical to the sacral spine and looking for any signs of bruising, open wounds, tender points, and palpation of paravertebral tissue and posterior spinous processes in search for distraction injury. Spine precautions should only be discontinued when patients regain consciousness and are able to communicate sufficiently on spinal discomfort or neurologic sensations before the spine is cleared

Table 14.1 Criteria for cervical spine clearance based on history and physical examination alone

Fully alert
Not intoxicated
Involved in isolated blunt trauma
Neurologically normal
No midline tenderness to palpation

blood pressure, and perfusion, while urinary catheters monitor urine output. The early use of blood products is recommended in the multiple injured patients with an associated spinal cord injury to maximize the oxygen-carrying capacity and to minimize the secondary ischemic injury to the spinal cord. Early use of vasopressors such as dopamine or atropine is recommended to maintain spinal cord perfusion. While the goal in many trauma centers is to maintain the MAP >85 – 90 mmHg, the overall objective is to avoid hypotension in order to protect the cord from further neurologic injury and potentially improve neurologic recovery [13–15].

14.4.2.3 Physical and Neurological Examination

Once airway and circulatory concerns are stabilized, the provider should devote attention to performing a complete physical and neurologic examination. Often during the course of examination, inspection and palpation can provide important clues to help identify both associated and concomitant levels of spinal injury. For example, a transverse band of ecchymosis across the abdomen can suggest

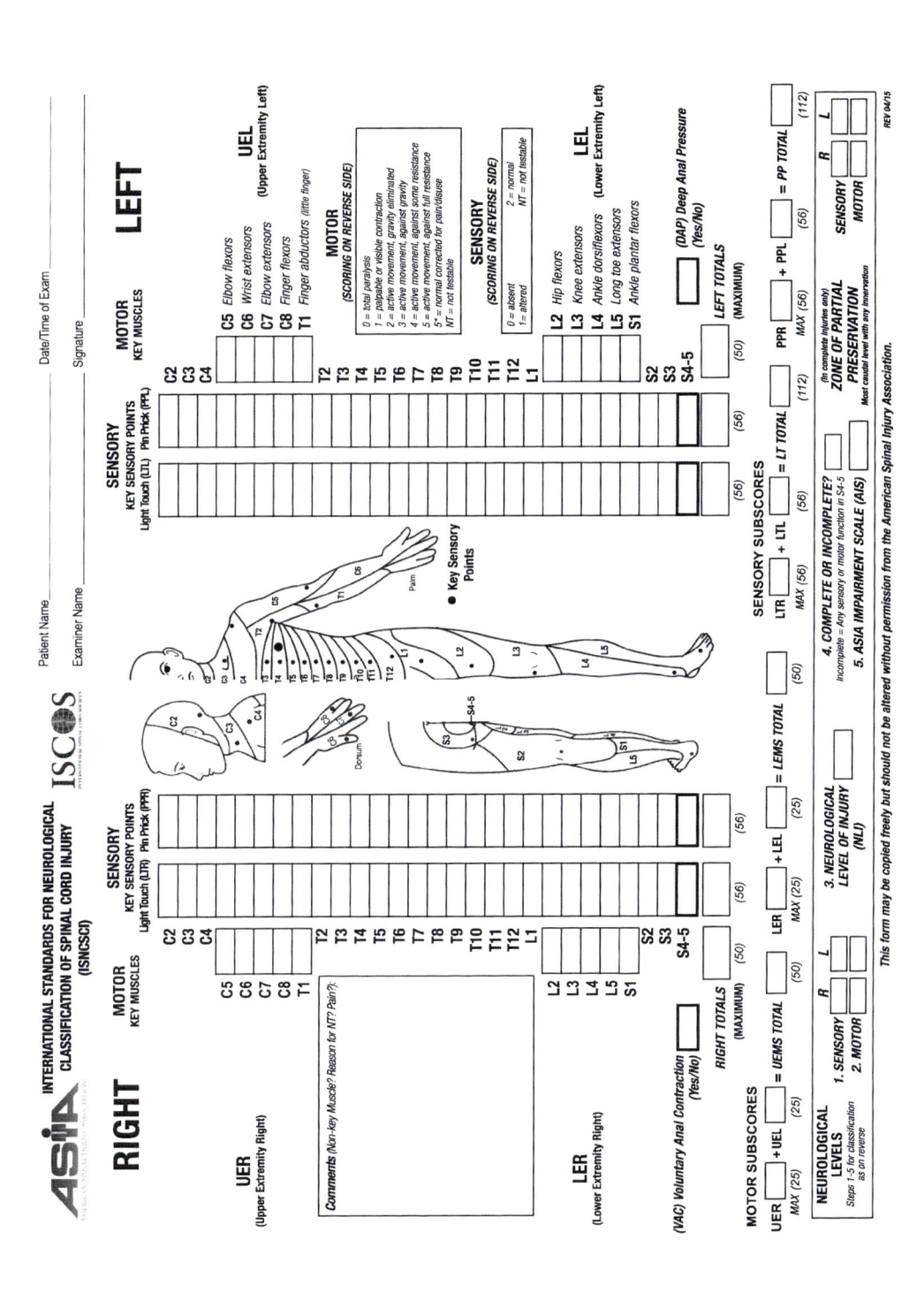

Fig. 14.2 American Spinal Injury Association (ASIA) exam worksheet detailing the essential portions of a complete neurologic examination in a trauma setting (Reprinted with permission from the American Spinal Injury Association: International Standards for Neurological Classification of Spinal Cord Injury, revised 2013; Atlanta, GA. Reprinted 2014.)

a flexion-distraction type of injury caused by a seat belt and may help predict an associated intra-abdominal injury. Similarly, bruising along the rib cage may suggest a thoracic fracture. When the patient is log-rolled in ATLS® step "E," any spontaneous pain from spine is noted as well as local hematomas. The spine must be palpated systematically for tenderness, step-off, or interspinous process gapping. During this portion of the examination it is also critical to perform a rectal exam. Elements of the rectal exam that should be noted include perianal sensation, rectal tone, voluntary contraction, and presence of the bulbocavernosus reflex. In awake and cooperative patients, the American Spinal Injury Association (ASIA) has provided a useful template for the essential elements of a complete neurologic examination (Fig. 14.2) [16]. In this system, the level of neurologic injury is determined by performing objective strength testing in ten myotomes and both light touch and pinprick sensation in 28 dermatomes. From this examination the motor, sensory, and neurologic levels of injury can be elucidated and an impairment classification can be assigned (Table 14.2). Throughout the course of the encounter, providers should remain mindful of associated injuries such as fractures, peripheral nerve injuries, and traumatic brain injuries due to their potential effects on the patient's examination.

The unconscious patient can often be difficult to examine. In these situations, questioning first responders on observations in the field and in transport may help preliminarily detect neurologic deficits. In patients without radiographic injury, serial reexaminations should be performed until a complete assessment is possible. In patients with radiographic evidence of a vertebral column injury, providers may also consider obtaining a magnetic resonance imaging (MRI) to directly assess for injury to the neurologic elements in addition to serial examinations.

14.4.2.4 Classification of Neurological Injury

Upon identification of a neurologic deficit, providers should attempt to classify the level of injury in an accurate and reproducible manner. The ASIA classification is a commonly employed

Table 14.2 ASIA Impairment Scale (AIS)

A = Complete. No sensory or motor function is preserved in the sacral segments S4-5.

B = Sensory Incomplete. Sensory but not motor function is preserved below the neurological level and includes the sacral segments S4-5 (light touch or pin prick at S4-5 or deep anal pressure) AND no motor function is preserved more than three levels below the motor level on either side of the body.

C = Motor Incomplete. Motor function is preserved below the neurological level[a], and more than half of key muscle functions below the neurological level of injury (NLI) have a muscle grade less than 3 (Grades 0–2).

D = Motor Incomplete. Motor function is preserved below the neurological level[a], and at least half (half or more) of key muscle functions below the NLI have a muscle grade > 3.

E = Normal. If sensation and motor function as tested with the ISNCSCI are graded as normal in all segments, and the patient had prior deficits, then the AIS grade is E. Someone without an initial SCI does not receive an AIS grade.

Reprinted with permission from the American Spinal Injury Association: International Standards for Neurological Classification of Spinal Cord Injury, revised 2013; Atlanta, GA. Reprinted 2014
NOTE: When assessing the extent of motor sparing below the level for distinguishing between AIS B and C, the motor level on each side is used, whereas to differentiate between AIS C and D (based on proportion of key muscle functions with strength grade 3 or greater) the neurological level of injury is used
[a]For an individual to receive a grade of C or D, i.e., motor incomplete status, they must have either (1) voluntary anal sphincter contraction or (2) sacral sensory sparing with sparing of motor function more than three levels below the motor level for that side of the body. The International Standards at this time allows even nonkey muscle function more than 3 levels below the motor level to be used in determining motor incomplete status (AIS B versus C)

and widely accepted modification of the Frankel grading system (Table 14.2). Within this system, the distinction of complete and incomplete injuries hinges on the presence or absence of the motor and sensory function of the lowest sacral segment (sacral sparing). Sacral sparing represents at least partial structural continuity of the white matter long tracts. Clinically, it is demonstrated by perianal sensation, rectal motor function, and great toe flexor activity. This distinction has important prognostic implications as sacral sparing has been shown to predict neurologic improvement and improved chances of regaining the ability to ambulate [17, 18].

Spinal Shock

In complete transections of the spinal cord, spinal areflexia occurs. This state is named spinal shock. It is clinically graded by testing the bulbocavernosus reflex, a spinal reflex mediated by the S3–S4 region of the medullary cone. If no evidence of spinal cord function is noted below the level of injury, and the bulbocavernosus reflex has not returned, no determination can be made regarding the lesion or the patient's prognosis. After 24 h, most patients emerge from spinal shock, as observed by the return of sacral reflexes [19]. If no sacral function exists at this point, the injury is considered complete and the probability of neurologic recovery is low. One exception is a direct injury to the conus medullaris where some functional recovery typically occurs. These patients may also have persistent absence of the bulbocavernosus reflex as a result of direct injury to the sacral portion of the spinal cord.

Incomplete Spinal Cord Injury Syndrome

Incomplete spinal cord injury can present as one of the following syndromes.

Anterior cord syndrome implies complete motor and sensory loss except retained trunk and lower extremity deep pressure sensation and proprioception. Only one out of ten patients has a chance of recovery.

Central cord syndrome represents central gray matter destruction with preservation of just the peripheral spinal cord structures. The patient usually has tetraplegia with preserved perianal sensation. Often, there is early return of bowel and bladder control. The neural axons nourishing the upper extremity pass more medial than the axons to the lower extremity. Therefore, the leg is stronger than the arm. The most common cause is cervical hyperextension injury in patients with narrow spinal canals. This injury can be mechanically stable. The syndrome has a good prognosis with recovery up to 75 %.

Brown-Séquard syndrome (lateral cord syndrome) is a unilateral cord injury, often caused by penetrating trauma. It is characterized by loss of motor deficit ipsilateral to the spinal cord injury and contralateral pain and temperature hypoesthesia. This syndrome usually has a good prognosis with most patients regaining bowel and bladder function and ability to walk.

Conus medullaris injury affects both the sacral most portion of the spinal cord and the lumbar nerve roots. As such, it presents as both an upper and lower motor neuron injury. These patients exhibit variable lower extremity weakness, paresthesia (particularly in the perianal region), bowel incontinence, poor rectal tone, and overflow urinary incontinence.

Cauda equina syndrome affects only the lower motor neurons in contrast to conus medullaris injuries. Symptoms of cauda equina include bowel and bladder dysfunction, saddle anesthesia, and variable amounts of radicular weakness and paresthesia. In these injuries, the prognosis for recovery depends largely on the time until decompression(<48 h) and the extent of the preoperative symptoms [20, 21].

14.4.3 Spinal Imaging

Recent advances in imaging techniques and availability have contributed to the delivery of more rapid and improved care to the multiply injured patient. Despite these advances, the rate of delayed or missed diagnosis of vertebral column injury has been reported as high as 16.5 % in North American trauma centers [22]. This alarming rate of delayed recognition or missed spinal fractures heightens the importance of both obtaining and carefully evaluating appropriate imaging studies in these cases.

14.4.3.1 Plain Film Radiography: Primary Assessment

Plain radiographs have long been mainstays in the radiographic evaluation of trauma patients due in part to their cost, wide availability, and portability. In modern trauma centers, these studies can often be obtained rapidly in the trauma bay without the need to transport a critically injured patient. Though these benefits are obvious, plain radiographs are often limited by the poor specificity and potential difficulty evaluating transitional areas of the spine such as the craniocervical and cervicothoracic junctions. In the cervical spine, the percentage of plain radiographs that provide adequate visualization ranges from 27.8 to 48 % [23, 24]. In regards to specificity, a meta-analysis comparing plain radiographs to computed tomography (CT) in the detection of cervical spine fractures demonstrated that plain radiographs were only able to identify only 52 % of injuries while 98 % were seen on CT [25]. This low specificity and difficulty obtaining adequate films coupled with increased availability of CT has limited the current use of x-ray in many modern trauma centers.

14.4.3.2 Computed Tomography: Secondary Assessment

In most trauma centers, CT has replaced plain radiographs as the primary tool for evaluation of the polytrauma patient with spinal injuries because of its established role in screening for concomitant brain, visceral, or bony injuries and improved accuracy. In a prospective study comparing orthogonal radiographs of the thoracolumbar spine to helical truncal CTs, Hauser and associates demonstrated superior sensitivity and specificity as well as superior positive and negative predictive values using CT in a traumatized population [26]. CT has also been shown to require less time with similar cost compared to using plain radiographs alone in an emergency room setting when comparing patients that have inadequate x-rays performed initially or require CT secondary to neck pain [24, 27].

The use of CT as an imaging modality also helps surgeons assess the stability of various spine fractures by demonstrating better bony detail while minimizing radiographic overlap. In a cervical sagittal CT scan, a displacement of more than 3.5 mm as well as segmental kyphosis of more than 11°may account for instability [28]. A widened intervertebral space and facet joint distraction of more than 50 % represent unstable discoligamentous injury [29]. Bony avulsion injuries of the anterior or posterior portions of both upper and lower vertebral endplate might indicate a rupture of the anterior or posterior longitudinal ligaments. At C1, this accounts for bony avulsion injuries of the transverse ligament. The frontal and axial CT reconstructions should rule out rotational offset of the vertebral segment, which indicates rotational instability with special attention to the C1-2 area. In the thoracolumbar region, a loss of more than 50 % of vertebral height, sagittal angulations of more than 25°, spinal canal encroachment more than 50 %, and increased interspinous distances are associated with unstable spine injuries [30, 31]. This enhanced ability to visualize specific injury patterns helps surgeons determine and predict the success of their treatment plans.

14.4.3.3 Computed Tomography Contrast Angiography

Blunt cerebral vascular injuries (BCVI), primarily arterial dissection, may occur in association with cervical spine trauma [32]. Early diagnosis and treatment reduces morbidity (stroke) and mortality in patients with vertebral artery injuries [33, 34]. While catheter angiography has been the historic gold standard, it is invasive. Currently, routine screening with MR angiography and 16-slice CT angiography can be performed in the initial radiologic workup [35]. Trauma victims with any of the following signs or symptoms should be considered to have BCVI until proven otherwise: coma unexplained by CT; neurologic deficit, including hemiparesis, transient ischemic attack, Horner's syndrome, oculosympathetic paresis or vertebrobasilar insufficiency, and evidence of cerebral infarction on CT; arterial hemorrhage from neck, mouth, nose, ears, large or expanding cervical hematoma, and cervical bruit in a patient younger than 50 years; fracture subluxation in cervical spine at any level, fractures from C1 to C3, and fractures into the transverse foramen at any level; and displaced mid-face fracture (LeFort II or III), basilar skull fracture with carotid canal involvement, closed head injury with consistent diffuse axonal injury, neck belt sign or significant swelling, and near hanging with anoxia [36, 37]. The timely recognition and treatment of these associated vertebral artery injuries can help prevent stroke and associated neurologic injuries.

14.4.3.4 Magnetic Resonance Imaging

The role of magnetic imaging in the evaluation of patients with vertebral column injuries is currently being defined. This imaging modality provides superior visualization of the disc complexes, ligaments, and the spinal cord. While the routine use of MRI may improve provider's ability to detect purely ligamentous injuries, the clinical significance of this increased sensitivity remains in question. Stassen and associates detected 13 ligamentous injuries in 44 obtunded patients (30 %) with no evidence of injury on helical CTs. None of the injuries detected on MRI alone required more than a rigid collar for treatment [38]. In a larger study examining 1577 patients with plain films and CTs, Diaz et al. obtained MRIs in 85 patients without fracture who could not be cleared and found 14 purely ligamentous injuries that were not detected by CT. However, none of these injuries required surgical treatment [39]. While the routine use of MRI in awake, neurologically, radiographically normal patients may not prove cost effective, it should be considered in cases of neurologic injury, cervical spine clearance in the obtunded patient population and evaluation of patients with persistent midline pain, and normal CTs.

14.4.4 Hospital Resuscitation: Workup

In the acute phase of recovery, patients with spinal cord injury benefit from admission into an intensive care setting. Management in this environment allows providers to closely monitor the patient's neurologic status, aggressively resuscitate any fluid deficits present, and optimally maintain hemodynamic parameters. A recent systematic review demonstrated that early ICU care was associated with decreased mortality and decreased cardiovascular and pulmonary morbidity [40]. These results underscore the need for both close monitoring and aggressive care in these injuries.

14.4.5 Cervical Traction

Cervical traction can improve cervical spine deformity, decompress nerves, and provide temporizing stability until more definitive means can be employed. In patients with unstable cervical spine fractures who cannot undergo immediate operative stabilization, halo ring application can often provide an efficient means of obtaining improved stability with minimal risk to the patient until definitive stabilization can be performed. Heary et al. found no cases of neurologic deterioration in 77 patients whose unstable cervical fractures were stabilized in a halo vest while further diagnostic studies and both neurosurgical and nonneurosurgical procedures were performed [41]. The principles for halo ring/cervical tong application have been well described [42–45]. Adherence to established application guidelines is critical to minimize morbidity such as pin penetration, pin site infection, peripheral nerve injury, and neurologic deterioration.

14.4.5.1 Spinal Cord Injury Units

Neuroprotective Drugs
In patients with spinal cord injury, neural damage occurs both as the result of the inciting trauma (primary injury) and the bodies initial response to injury (secondary injury). This secondary injury mechanism has been studied, and different neuroprotective substances have been tried to mitigate its consequences. Of these pharmacologic agents, corticosteroids are the most widely investigated. While early studies showed improved neurologic outcomes with their use, subsequent investigations have failed to document discernible benefits while documenting significant risks associated with the administration of high dose steroids [46–50]. These conflicting studies make the role of steroids in the routine treatment of patients extremely controversial. Currently, steroids are not routinely used in complete injuries, injuries from penetrating trauma, or in cases of delayed presentation. Their use in other situations is often dictated by the judgment of the treating physician and institutional norms.

Stroke Prevention
Blunt cerebral vascular injuries (BCVI), primarily arterial dissection, may occur in association with cervical spine trauma [32]. Early diagnosis and treatment reduces morbidity and mortality in patients with these injuries [51]. In patients with a known BCVI, antithrombotic therapy in form of antiplatelet (aspirin) and or anticoagulation (heparin) should be started, if there are no contraindications, and endovascular consultation should be obtained.

Thrombosis Prophylaxis

Due to the frequent requirements for early immobilization and their associated injuries, patients with vertebral column fractures are at an increased risk of developing a deep venous thrombosis (DVT). This risk is increased in the presence of neurologic deficit as these injuries cause low smooth muscle tone in peripheral vessels and promote venous stasis. As a result of these risks, providers should develop a comprehensive plan focused on DVT prevention that includes early mobilization, mechanical prophylaxis, and pharmacologic antithrombotic agents, if not contraindicated.

14.5 Treatment

The goals of treatment for all vertebral column injuries are to prevent or reverse neurologic injury, to augment spinal stability for early mobilization, and to promote fracture healing while maintaining acceptable alignment. While recent advances in surgical techniques and instrumentation have increased the safety and efficacy of operative treatment in select injury patterns, the majority of all vertebral column fractures are still effectively treated with nonoperative management.

14.5.1 Nonsurgical Treatment

14.5.1.1 External Orthoses

Spinal orthoses are a major component of the nonoperative treatment of spinal fractures. They restrict spinal motion by acting indirectly to reinforce the intervening soft tissue. Despite the heterogeneity of designs, the functions of all braces are analogous and include restriction of spinal movements, maintenance of spinal alignment, reduction of pain, and support of the trunk musculature. While many orthoses can prove helpful in achieving these goals, none are able to fully restrict motion, compliance is often difficult to ensure and all require careful fitting to maximize efficacy while limiting complications. Providers should understand the advantages and limitations of a number of devices to make informed decisions and maximize the effectiveness of nonoperative treatment.

14.5.1.2 Cervical Braces: Rigid Collars

Rigid collars are frequently used to treat cervical fractures. Examples of rigid collars include Miami J, Philadelphia, and Aspen. Inherent limitations of these devices are the relatively small contact area afforded between the mandible and clavicle, the lack of endpoint control of either the head or thorax, and restrictions on how much external pressure these devices can apply. Because of these limitations, rigid collars are effective in restricting sagittal motion but are less effective in limiting rotation and side bending [52]. These orthoses also tend to lose effectiveness in both the upper (occiput-C2) and lower levels (C6-7) of the cervical spine [53]. In situations where more cephalad or caudal immobilization is required, providers should consider the use of devices like a Halo vest or Minerva brace.

14.5.1.3 Thoracolumbar Orthoses

In the thoracic spine, the rib cage provides some natural support for fractures. The upper thoracic region (T5) and above is a very difficult region to immobilize with an external orthosis, often requiring immobilization with a halo vest or cervicothoracic orthosis (CTO). Spinal fractures from T8 to L2 are typically braced with a three-point fixation system (Jewett brace) that maintains extension of the thoracolumbar area or with a custom molded, hard shell orthosis (Body Jacket). Below L3, a lumbosacral orthosis is used for support. In order to increase the immobilization at the lumbosacral junction, a leg extension can be fitted to the orthosis to assist in limiting motion across the pelvis. Casting is another option for lumbar and thoracolumbar fractures and can provide better support and eliminate concerns of noncompliance.

14.5.2 Surgical Treatment

Only a small select group of unstable spine injuries with or without neurologic association merit surgical treatment. The primary goals for the surgical treatment are decompression of compromised/threatened neuronal elements and augmentation of spinal stability. The spinal fracture, patient, and associated injury/factors have to be interpreted before surgery is chosen as a treatment option. Controversy persists in the surgical community regarding the optimal treatment of many traumatic spinal injuries, especially regarding timing of surgical intervention and type of surgical approach employed in various situations.

14.5.2.1 Damage Control Spine Surgery

Frequently, the question arises as to which patient needs definitive surgery according to the principles of early total spine care and which patient is in need of a staged procedure after initial stabilization. Since no data exists for the multiple injured patients with spine trauma, one has to adopt information from general trauma [54–56]. Hemodynamically unstable patients with signs of shock, suffering from the lethal triad of hypothermia, coagulopathy, and acidosis, have high mortality rates and should undergo delayed or staged definitive fixation [57]. Since no definitive cutoff parameters are defined to help determine which patients may tolerate early definitive care, decision making should occur on an individual basis.

14.5.2.2 Surgical Timing

There are many factors that influence the optimal surgical timing in the multiply injured patient with spinal fractures. Variables such as neurologic injury, concomitant injuries, hemodynamic stability, and institutional ability to perform often complex and lengthy procedures should all be considered by the treating team in determining the optimal time for surgery. In general, early fixation of spinal fractures allows for quicker mobilization and decreases the risk of pulmonary, skin, and gastrointestinal complications often seen in this patient population.

The optimal timing of surgery in the face of neurologic injury remains a controversial and understudied topic. In animal trials, multiple investigators have shown improved neurologic recovery with early decompression of the spinal cord [58–60]. While these preclinical studies present convincing evidence that early decompression may lead to improved neurologic outcomes, the controlled nature of these spinal cord injury studies as well as the frequently used short time to decompression, 30–180 min from injury, limit the transferability of this information to human populations. Though the majority of data in human subjects has been derived from retrospective reviews, a recent large multicenter prospective cohort demonstrated significantly higher rates of neurologic recovery, defined as >2 AIS grades, in patients undergoing early surgical decompression (14.2 ± 5.4 h) compared to those treated in a delayed fashion (48.3 ± 29.3 h) with similar complication rates between the two groups [61]. While promising, the results of this investigation should be interpreted carefully as only 20 % of patients in the early surgical cohort achieved significant neurologic improvement at 6 months. Despite the limitations of the existing literature, early surgical decompression should be considered in hemodynamically stable patients with neurologic deficits when the appropriate resources and personnel are available.

Surgical Options

With recent advances in surgical techniques and instrumentation, modern surgeons have unprecedented options to effectively treat spinal fractures. Though many injuries can be safely addressed with a variety of surgical strategies and approaches, a few considerations may help guide the decision making process. In cases of burst fractures with spinal canal compromise, anterior approach surgeries allow more effective decompression and tend to provide more stability than posterior decompression and fusion techniques [62, 63]. Three column injuries, treated with anterior instrumentation, typically require posterior fixation or postoperative bracing to augment the injured posterior ligamentous structures [64]. These circumstances heighten the importance of close scrutiny of all preoperative imaging studies as well as the need for familiarity with a number of surgical options and approaches.

14.5.3 Special Situations

14.5.3.1 Gunshot/Open Injury

Gunshot wounds represent a distinct type of penetrating trauma with a unique treatment as decompression does not improve recovery if the bullet traverses the canal without any residual mass effect on neural elements [65]. For complete and incomplete neural deficits at the cervical and thoracic levels, operative decompression is of little benefit and can lead to higher complication rates than nonoperative management. Operative indications in these injuries include plumbism, an intracanal copper bullet, persistent cerebrospinal fluid leak, or new onset of neurologic deficit. With gunshots to the T12 to L5 levels, better motor recovery has been reported after intracanal bullet removal than with nonoperative treatment [66]. Debridement and removal of the bullet is also an option during laparotomy for abdominal injury. Finally, in cases where the projectile traverses the oropharynx or intestine, intravenous broad-spectrum antibiotics should be administered for 3 days as prophylaxis against infection.

14.5.3.2 Ankylosing Spondylitis (AS), Diffuse Idiopathic Skeletal Hyperostosis (DISH)

AS and DISH are related disorders where the individual spinal segments spontaneously fuse over time. As these segments ankylose, the spine begins to function similar to other long bones in the body. Because of this, fractures in these should be considered unstable until proven otherwise. These patients usually have a preinjury deformity, and their normal alignment should be maintained throughout their workup and eventual treatment. Strict logroll precautions are used until definitive management has been decided. These patients also have a propensity to develop epidural hematomas with even minor fractures that can lead to severe neurologic deficits. As a result, they should be monitored closely with frequent neurologic checks in the acute period after admission. Prompt decompression and fixation should be performed if a deficit does develop. Due to the altered spinal biomechanics, this patient population often requires long segment fixation.

14.5.3.3 Pediatric Patients

In children, ligamentous injuries are more frequent than bony injury. These pediatric injury patterns transition to adult types at 11 years of age. Most pediatric injuries occur in the upper cervical spine between the occiput and C3 because the ratio of mass between the head and the body is disproportionate at this location. Spinal cord injury without radiographic abnormality (SWICORA) commonly occurs in children younger than 11 years. The mechanism of these injuries is not fully understood but is likely to be a fracture of the cartilaginous vertebral endplate, which in turn leads to distraction of the cord and ischemic injury. Children with spinal tenderness or questionable radiographic findings should be treated with immobilization until their symptoms resolve.

14.5.3.4 Geriatric Patients

The high incidences of both spondylosis and osteoporosis present many challenges in treating geriatric patients with vertebral column fracture. Degenerative spinal stenosis results in a higher prevalence of associated spinal cord injury even in the face of stable spinal injuries in older patients. The optimal treatment of these typically central cord injuries is controversial at present. With the high incidence of osteoporosis in this population, these patients can also develop unstable fractures from even low energy mechanisms. Treatment of these injuries are often demanding, as older patients tolerate external bracing poorly, and surgical interventions often carry an additional risk of complications due to age-related medical conditions and the difficulties achieving stable fixation. In addition, Halo vest treatment is associated with a high mortality rate and should typically be avoided in patients over the age of 65.

14.6 Clinical Outcomes

The clinical outcome of patients with spinal injuries is difficult to assess as most of the available literature focuses on isolated spinal injuries. Logically, patients with spine injuries and multiple other severe injuries have a higher rate of disability, occupational handicap, and risk for ongoing incapacitating pain [67]. A 5-year follow-up study by McLain et al. assessed functional outcome in patients with thoracic, thoracolumbar, or lumbar fractures as a result of all high-energy trauma [68].

In this study, 38 % percent of those analyzed were considered polytrauma patients (ISS >26). Overall 44 % of patients had functional limitations at final follow-up. Of this group, neurologic injury had the largest impact on ultimate functional outcome.

References

1. Burney RE, et al. Incidence, characteristics, and outcome of spinal cord injury at trauma centers in North America. Arch Surg. 1993;128(5):596–9.
2. Advanced trauma life support for doctors ATLS: manuals for coordinators and faculty. Chicago: American College of Surgeons; 2008.
3. Vaccaro AR, et al. Noncontiguous injuries of the spine. J Spinal Disord. 1992;5(3):320–9.
4. Davidoff G, et al. Assessment of closed head injury in trauma-related spinal cord injury. Paraplegia. 1986;24(2):97–104.
5. Gomes E, Araújo R, Carniero A, Dias C, Lecky F, Costa-Pereira A. Mortality distribution in a trauma system: from data to health policy. Eur J Trauma Emerg Surg. 2008;34(6):561–9.
6. Jansson KA, et al. Thoracolumbar vertebral fractures in Sweden: an analysis of 13,496 patients admitted to hospital. Eur J Epidemiol. 2010;25(6):431–7.
7. Morris CG, McCoy EP, Lavery GG. Spinal immobilisation for unconscious patients with multiple injuries. BMJ. 2004;329(7464):495–9.
8. Reid DC, et al. Etiology and clinical course of missed spine fractures. J Trauma. 1987;27(9):980–6.
9. Determination of cervical spine stability in trauma patients. http://www.east.org/tpg/chap3u.pdf. EAST (Easatern Association for the Surgery of Trauma); 2009.
10. Green BA, Eismont FJ, O'Heir JT. Pre-hospital management of spinal cord injuries. Paraplegia. 1987;25(3):229–38.
11. Davis JW, et al. Clearing the cervical spine in obtunded patients: the use of dynamic fluoroscopy. J Trauma. 1995;39(3):435–8.
12. Ching RP, et al. The effect of post-injury spinal position on canal occlusion in a cervical spine burst fracture model. Spine (Phila Pa 1976). 1997;22(15):1710–5.
13. Amar AP, Levy ML. Pathogenesis and pharmacological strategies for mitigating secondary damage in acute spinal cord injury. Neurosurgery. 1999;44(5):1027–39;discussion 1039–40.
14. Dolan EJ, Tator CH. The effect of blood transfusion, dopamine, and gamma hydroxybutyrate on posttraumatic ischemia of the spinal cord. J Neurosurg. 1982;56(3):350–8.
15. Ducker TB, Kindt GW, Kempf LG. Pathological findings in acute experimental spinal cord trauma. J Neurosurg. 1971;35(6):700–8.
16. Waring 3rd WP, et al. 2009 review and revisions of the international standards for the neurological classification of spinal cord injury. J Spinal Cord Med. 2010;33(4):346–52.
17. Crozier KS, et al. Spinal cord injury: prognosis for ambulation based on sensory examination in patients who are initially motor complete. Arch Phys Med Rehabil. 1991;72(2):119–21.
18. Kiwerski J, Weiss M. Neurological improvement in traumatic injuries of cervical spinal cord. Paraplegia. 1981;19(1):31–7.
19. Kakulas BA. Pathology of spinal injuries. Cent Nerv Syst Trauma. 1984;1(2):117–29.
20. Ahn UM, et al. Cauda equina syndrome secondary to lumbar disc herniation: a meta-analysis of surgical outcomes. Spine (Phila Pa 1976). 2000;25(12):1515–22.
21. Shapiro S. Medical realities of cauda equina syndrome secondary to lumbar disc herniation. Spine (Phila Pa 1976). 2000;25(3):348–51;discussion 352.
22. Schunemann HJ, et al. An official ATS statement: grading the quality of evidence and strength of recommendations in ATS guidelines and recommendations. Am J Respir Crit Care Med. 2006;174(5):605–14.
23. Gale SC, et al. The inefficiency of plain radiography to evaluate the cervical spine after blunt trauma. J Trauma. 2005;59(5):1121–5.
24. McCulloch PT, et al. Helical computed tomography alone compared with plain radiographs with adjunct computed tomography to evaluate the cervical spine after high-energy trauma. J Bone Joint Surg Am. 2005;87(11):2388–94.
25. Holmes JF, Akkinepalli R. Computed tomography versus plain radiography to screen for cervical spine injury: a meta-analysis. J Trauma. 2005;58(5):902–5.
26. Hauser CJ, et al. Prospective validation of computed tomographic screening of the thoracolumbar spine in trauma. J Trauma. 2003;55(2):228–34;discussion 234–5.
27. Antevil JL, et al. Spiral computed tomography for the initial evaluation of spine trauma: a new standard of care? J Trauma. 2006;61(2):382–7.
28. White AA, Panjabi MM. Clinical biomechanics of the spine. 2nd ed. Philadelphia: Lippincott; 1990. xxiii, 722 p.
29. Blauth M, et al. Complex injuries of the spine. Orthopade. 1998;27(1):17–31.
30. McLain RF, Benson DR. Urgent surgical stabilization of spinal fractures in polytrauma patients. Spine (Phila Pa 1976). 1999;24(16):1646–54.
31. Rihn JA, et al. A review of the TLICS system: a novel, user-friendly thoracolumbar trauma classification system. Acta Orthop. 2008;79(4):461–6.
32. Fassett DR, Dailey AT, Vaccaro AR. Vertebral artery injuries associated with cervical spine injuries: a review of the literature. J Spinal Disord Tech. 2008;21(4):252–8.
33. Biffl WL, et al. Western trauma association critical decisions in trauma: screening for and treatment of blunt cerebrovascular injuries. J Trauma. 2009;67(6):1150–3.
34. Schmidt OI, et al. Closed head injury–an inflammatory disease? Brain Res Brain Res Rev. 2005;48(2):388–99.

35. Mueller CA, et al. Vertebral artery injuries following cervical spine trauma: a prospective observational study. Eur Spine J. 2011;20(12):2202–9.
36. Cothren CC, et al. Cervical spine fracture patterns mandating screening to rule out blunt cerebrovascular injury. Surgery. 2007;141(1):76–82.
37. Dunham CM, et al. Risks associated with magnetic resonance imaging and cervical collar in comatose, blunt trauma patients with negative comprehensive cervical spine computed tomography and no apparent spinal deficit. Crit Care. 2008;12(4):R89.
38. Stassen NA, et al. Magnetic resonance imaging in combination with helical computed tomography provides a safe and efficient method of cervical spine clearance in the obtunded trauma patient. J Trauma. 2006;60(1):171–7.
39. Diaz Jr JJ, et al. The early work-up for isolated ligamentous injury of the cervical spine: does computed tomography scan have a role? J Trauma. 2005; 59(4):897–903;discussion 903–4.
40. Management of acute spinal cord injuries in an intensive care unit or other monitored setting. Neurosurgery. 2002;50(3 Suppl):S51–7.
41. Heary RF, et al. Acute stabilization of the cervical spine by halo/vest application facilitates evaluation and treatment of multiple trauma patients. J Trauma. 1992;33(3):445–51.
42. Barnett GH, Hardy RW. Gardner tongs and cervical traction. Med Instrum. 1982;16(6):291–2.
43. Gardner WJ. The principle of spring-loaded points for cervical traction. Technical note. J Neurosurg. 1973; 39(4):543–4.
44. Kang M, Vives MJ, Vaccaro AR. The halo vest: principles of application and management of complications. J Spinal Cord Med. 2003;26(3):186–92.
45. Manthey DE. Halo traction device. Emerg Med Clin North Am. 1994;12(3):771–8.
46. Bracken MB, et al. Efficacy of methylprednisolone in acute spinal cord injury. JAMA. 1984;251(1):45–52.
47. Bracken MB, et al. Methylprednisolone or naloxone treatment after acute spinal cord injury: 1-year follow-up data. Results of the second National Acute Spinal Cord Injury Study. J Neurosurg. 1992;76(1):23–31.
48. George ER, et al. Failure of methylprednisolone to improve the outcome of spinal cord injuries. Am Surg. 1995;61(8):659–63;discussion 663–4.
49. Prendergast MR, et al. Massive steroids do not reduce the zone of injury after penetrating spinal cord injury. J Trauma. 1994;37(4):576–9;discussion 579–80.
50. Qian T, et al. High-dose methylprednisolone may cause myopathy in acute spinal cord injury patients. Spinal Cord. 2005;43(4):199–203.
51. Schneidereit NP, et al. Utility of screening for blunt vascular neck injuries with computed tomographic angiography. J Trauma. 2006;60(1):209–15;discussion 215–6.
52. Gavin TM, et al. Biomechanical analysis of cervical orthoses in flexion and extension: a comparison of cervical collars and cervical thoracic orthoses. J Rehabil Res Dev. 2003;40(6):527–37.
53. Agabegi SS, Asghar FA, Herkowitz HN. Spinal orthoses. J Am Acad Orthop Surg. 2010;18(11):657–67.
54. Pape HC, et al. Impact of intramedullary instrumentation versus damage control for femoral fractures on immunoinflammatory parameters: prospective randomized analysis by the EPOFF study group. J Trauma. 2003;55(1):7–13.
55. Rupp RE, et al. Thoracic and lumbar fractures associated with femoral shaft fractures in the multiple trauma patient. Occult presentations and implications for femoral fracture stabilization. Spine (Phila Pa 1976). 1994;19(5):556–60.
56. Scalea TM, et al. External fixation as a bridge to intramedullary nailing for patients with multiple injuries and with femur fractures: damage control orthopedics. J Trauma. 2000;48(4):613–21;discussion 621–3.
57. Pape HC, Giannoudis P, Krettek C. The timing of fracture treatment in polytrauma patients: relevance of damage control orthopedic surgery. Am J Surg. 2002;183(6):622–9.
58. Carlson GD, et al. Early time-dependent decompression for spinal cord injury: vascular mechanisms of recovery. J Neurotrauma. 1997;14(12):951–62.
59. Delamarter RB, Sherman J, Carr JB. Pathophysiology of spinal cord injury. Recovery after immediate and delayed decompression. J Bone Joint Surg Am. 1995;77(7):1042–9.
60. Guha A, et al. Decompression of the spinal cord improves recovery after acute experimental spinal cord compression injury. Paraplegia. 1987;25(4):324–39.
61. Fehlings MG, et al. Early versus delayed decompression for traumatic cervical spinal cord injury: results of the Surgical Timing in Acute Spinal Cord Injury Study (STASCIS). PLoS One. 2012;7(2), e32037.
62. Kaneda K, et al. Anterior decompression and stabilization with the Kaneda device for thoracolumbar burst fractures associated with neurological deficits. J Bone Joint Surg Am. 1997;79(1):69–83.
63. Shono Y, McAfee PC, Cunningham BW. Experimental study of thoracolumbar burst fractures. A radiographic and biomechanical analysis of anterior and posterior instrumentation systems. Spine (Phila Pa 1976). 1994;19(15):1711–22.
64. Mann KA, et al. A biomechanical investigation of short segment spinal fixation for burst fractures with varying degrees of posterior disruption. Spine (Phila Pa 1976). 1990;15(6):470–8.
65. Stauffer ES, Wood RW, Kelly EG. Gunshot wounds of the spine: the effects of laminectomy. J Bone Joint Surg Am. 1979;61(3):389–92.
66. Bono CM, Heary RF. Gunshot wounds to the spine. Spine J. 2004;4(2):230–40.
67. Hebert JS, Burnham RS. The effect of polytrauma in persons with traumatic spine injury. A prospective database of spine fractures. Spine (Phila Pa 1976). 2000;25(1):55–60.
68. McLain RF, Burkus JK, Benson DR. Segmental instrumentation for thoracic and thoracolumbar fractures: prospective analysis of construct survival and five-year follow-up. Spine J. 2001;1(5):310–23.

The Management of the Multiply Injured Elderly Patient

Charles M. Court-Brown and N. Clement

Contents

15.1	Introduction	201
15.2	**Multiple Injuries**	203
15.2.1	Epidemiology	203
15.2.2	Injury Patterns	204
15.2.3	Motor Vehicle Accidents	205
15.2.4	Falls from a Height	206
15.2.5	Suicide	206
15.2.6	Falls	207
15.2.7	Treatment	207
15.2.8	Predictors of Mortality	207
15.2.9	Outcome	208
References		216

C.M. Court-Brown, MD, FRCS Ed (Orth) (✉)
N. Clement, MRCS
Department of Orthopaedic Surgery,
Royal Infirmary of Edinburgh,
Old Dalkeith Road, Edinburgh EH16 4SU, UK
e-mail: ccb@courtbrown.com; courtbrown@aol.com;
nickclement@doctors.org.uk

15.1 Introduction

The management of the multiply injured patient improved significantly in the 1960s and 1970s in a number of countries. Specialist trauma centres were established and the importance of early resuscitation and surgical treatment was appreciated. There was however very little interest in the management of the severely injured elderly patient until the 1980s when a number of papers on this topic were published. In 1984, Oreskovich et al. [1] published the results of the treatment of 100 consecutive patients who were older than 70 years of age. They documented a 15 % mortality but noted that while 85 % of their patients survived, 88 % of them did not return to their previous level of independence. They also observed that the Injury Severity Score (ISS) [2] was not predictive of survival in this elderly group.

DeMaria and his colleagues [3] took a somewhat more optimistic view of the benefits of aggressive trauma care in the multiply injured elderly. In 1987, they published the results of 63 survivors of blunt trauma who were over 65 years of age. They pointed out that the overall level of injury was moderate with a mean ISS of 15.8 and that only 62 % of their patients had injuries in two or more body regions but that 71 % of the patients had pre-existing cardiovascular disease. Prior to injury 97 % of the patients were independent but after treatment and rehabilitation, 89 %

of patients returned to an independent existence although they pointed out that these patients tended to be younger and to have had a shorter hospital stay and fewer complications. Of the 12 patients in their study who were aged 80 years or more 8 (66.6 %) returned home. Their conclusion was that aggressive support of the elderly was justified as few required permanent nursing home care and the majority returned to independent living.

This study also examined the factors related to failure to survive trauma in older patients. The authors showed that non-survivors were older and had more severe overall injury. They also had more serious head and neck trauma but there was no difference in the severity of non-head and neck trauma, the mechanism of injury or the requirement for surgery. Non-survivors had more frequent complications including a higher prevalence of cardiovascular complications and a greater requirement for ventilation for 5 or more days. They took the view that a number of complications were potentially avoidable and therefore aggressive treatment of geriatric trauma was indicated.

Champion et al. [4], in 1989, analysed data from 3833 patients aged 65 years or more in the Major Trauma Outcome Study (MTOS) and showed that 20.7 % of older patients injured in motor vehicle accidents died. In this analysis they pointed out that 28.2 % of the elderly patient group had been injured in motor vehicle accidents compared with 40.6 % who had been injured in falls and that 11.7 % of this latter group had died. They concluded that the perception of injury as a disease of the young resulted in people failing to recognise the importance of trauma in the elderly. They suggested that trauma systems and trauma centres might be put in place to treat elderly patients. Champion and his colleagues [5] also analysed a group of 180 elderly trauma patients aged 65 years or more and compared their results with a similarly injured group of younger patients. They also used a nationally collected database to analyse mortality at different ages. They showed that mortality increased with age and that this increase occurred at all ISS scores, in all mechanisms of injury and in all body regions. Older patients had higher complication rates and this was particularly true for pulmonary and infectious complications. They theorised that triaging elderly trauma patients to trauma centres at a lower threshold of injury to similarly injured younger patients would be beneficial.

Since these papers were published there has been an increasing awareness of the importance of trauma in the elderly population, this usually being defined as patients aged at least 65 years of age. However there are difficulties in defining what constitutes severe trauma in the elderly population. Superficially the concept of severe injury is straightforward and one can specify that the ISS should be at least 16 or that there should be injuries in multiple body systems. However Champion et al. [4] pointed out that there was a significant mortality following simple falls and that in the elderly population an ISS of 0–8 was associated with a mortality of 2.9 % and a complication rate of 16.2 %. An ISS of 9–15 was associated with a mortality of 6.9 % and a complication rate of 31.1 %. It is now generally accepted that in the elderly population the mortality of low-energy injury is relatively high and the common fragility fractures, particularly those of the proximal femur, are associated with significant mortality. It is also accepted that minor head injuries in the elderly may prove fatal.

Despite the excellent studies published in the 1980s and 1990s it is salutary to observe that there are a number of recent papers showing that elderly patients are not infrequently undertriaged compared with their younger counterparts. This problem was highlighted by Grant et al. [6] in a study of elderly patients injured by blunt trauma in Scotland. They pointed out that elderly patients injured in motor vehicle accidents, falls from a height or standing falls were less likely to be managed in a resuscitation room and they were more likely to be treated by less senior staff at an early stage. As a result of this they had a less interventional approach to treatment. This was despite contemporaneous studies showing that elderly patients who survived severe injury fared as well as younger patients with both groups having the same functional outcome 2 years after injury [7].

More recent studies from the United States and Canada have also highlighted the problems of undertriage. Chang et al. [8] examined 26,565 patients in Maryland, USA, and showed that the overall undertriage rate was 49.9 % in the ≥65 year group compared with 17.8 % in the <65 year group. They also showed that the undertriage started at 50 years of age with another decrease being seen at 70 years of age. Moore et al. [9] carried out a similar study in Quebec, Canada and showed that geriatric patients were less likely to be admitted to a Level I or II Trauma Centre, even if they had a severe head injury.

In a recent study Staudenmayer et al. [10] examined the triage of elderly trauma patients in California and Utah, USA. They showed that patients who were taken to non-trauma centres were on average older, more often female and less often had an ISS >15. They documented a 60-day mortality of 17 % in patients who had an ISS >15 but they did not find any difference in mortality between trauma and non-trauma centres although the costs of patient treatment in trauma centres was significantly higher than in non-trauma centres. They theorised that mortality in the elderly is influenced by more factors than simply the severity of the initial injury. This view is supported by a study from Clement et al. [11] who examined the role of pre-existing medical conditions in the elderly. They showed a higher mortality following minor trauma in elderly patients with pre-existing medical conditions.

It seems likely that the ageism referred to by Grant et al. [6] remains a problem in the management of the injured elderly patient. It is accepted that there is a higher mortality in elderly patients compared with younger patients but the fact that elderly survivors fare as well as young survivors means that great care must be taken not to undertriage elderly patients.

One of the consequences of admitting patients with different degrees of injury to different centres is that there is some confusion in papers discussing the problem of trauma in the elderly population. Some studies have specifically looked at polytraumatised older patients with injuries in more than one body system or an ISS of at least 16 whereas other studies have examined all patients admitted to certain types of hospital. Obviously the results from these two types of study will be different. In this chapter we have accepted that it is difficult to define what constitutes multiple trauma or severe injury in the elderly population and we have examined both patients with multiple body system injures and those with multiple fractures. We have also examined a group of elderly patients who presented with open fractures and compared them with a group of younger patients who presented with open fractures during the same period.

15.2 Multiple Injuries

15.2.1 Epidemiology

It is generally assumed that the incidence of polytrauma in the elderly is increasing and indeed this seems to be the case. There is no doubt that the incidence of the elderly in the population is increasing quickly. In 2000, 12 % of the population of the United States was at least 65 years of age with 5.9 % being 75 years or older and 1.5 % being 85 years or older. It has been postulated that by 2030, 20 % of the population will be aged 65 years or more and 2.5 % will be aged 85 years or more [12]. In the United Kingdom it has become clear that the fastest growing group in the population are the nonogenarians (≥90 years) who made up 0.58 % of the population in 2001 but will probably comprise 1.2 % of the population in 2025 [13]. United Kingdom statistics have also shown that the increase in the population of the elderly is not being matched by improved health. In 2008 the National Office of Statistics stated that while the population of the United Kingdom had been living longer over the previous 23 years the time that both sexes could be expected to be in poor health or have a limiting illness or disability had risen between 1988 and 2004 [14]. There were some minor improvements after 2004 but it is clear that increased longevity will be matched by poorer health and an increasing incidence of medical comorbidities. This is particularly important in severe or multiple trauma as medical comorbidities help to dictate the prognosis in the elderly. The fact is that the problem is already occurring in orthopaedic trauma. Figures from Edinburgh, Scotland in 2000 show that while nonogenarians made up 0.58 % of the population they accounted for 3.02 % of the fractures in the community, 8.7 % of the in-patient admissions and 7.6 % of the acute orthopaedic trauma surgery [15].

However the increase in the elderly population has to be balanced against a presumed decrease in motor vehicle accidents in many countries. In the United Kingdom in 2001, 9.7 % of motor vehicle accident casualties were 60 years or more but this represents a decline of 2.1 % since 1994–1998 [16]. As there was no formalised trauma system in the United Kingdom at that time, this improvement shows the value of accident prevention. It seems reasonable to assume that accident prevention will improve in other countries and the incidence of motor vehicle accident casualties will decline. However a contrary view has been put forward by the World Health Organisation who listed motor vehicle trauma as the 11th most common cause of death in 2002 but forecast that it would become the third most common cause of death by 2020 [17].

It is difficult to be precise about the future epidemiology of multiple trauma in the elderly but there is no doubt that low-energy multiple fractures will be an increasing problem because of the increasing number of falls in a progressively older, less fit population. It has been estimated that about 10 % of falls cause severe injury [18] and a recent Swedish study has shown that 7 % of falls in the elderly result in fracture [19]. It is likely that fall-related fractures will increase in frequency in the future and the Center of Disease Control and Prevention in the United States has suggested that in 2020 the cost of falls may reach $54.9 billion [20].

As has already been pointed out it is difficult to estimate the prevalence of severe injury in the elderly population as the published data comes from hospitals that admit different categories of patients and different severities of injury. However a review of the Trauma Audit and Research Network (TARN) database in the United Kingdom, which reviews all injured patients who arrive alive at hospitals and who are admitted for more than 72 h or who die within the 72 h period, shows that only 1.8 % of patients have an ISS ≥ 16 and are 65 or more years of age [21]. Forty-two per cent of the injuries followed motor vehicle accidents.

15.2.2 Injury Patterns

The patterns of injury seen in the elderly patient will, to an extent, vary with the mode of injury. Patterns of injury for specific modes of injury are discussed later in the chapter. However a study by Gowing and Jain [22] looked at 125 trauma victims aged >65 years who presented with an ISS >12. Falls accounted for 64 % of admissions with motor vehicle accidents, machine injuries, natural accidents and suicides accounting for 27 %, 3 %, 2 % and 3 % of admissions, respectively. The average age was 77 years and the average ISS was 23.

Head injuries accounted for 64 % of the principal diagnoses with thoracic and orthopaedic injuries making up 14 % and 12 % of the principal diagnoses. Subdural haematoma was the commonest injury after a standing fall but the authors emphasised that a number of the principal diagnoses were in fact pre-existing medical conditions. The mortality was 31.2 % and the ISS correlated with mortality. As in other studies Gowing and Jain [22] pointed out that a large number of the patients were discharged home.

If one simply examines fractures in the ≥65 year group prospectively collected inpatient and outpatient data from the Royal Infirmary of Edinburgh in a 1-year period in 2010/2011 [23] shows that 34.0 % of non-spinal fractures occur in patients who are aged at least 65 years. Spinal fractures were not assessed because it is likely that the majority of spinal fractures in the elderly are not admitted to hospital. However 91.9 % of fractures in the ≥65 year group followed a simple fall with only 1.5 % of fractures occurring as a result of a motor vehicle accident and 0.4 % occurring as a result of a fall from a height these being the two common causes of high-energy injury. It is therefore apparent that severe injury in the elderly population is relatively rare whether this be polytrauma or high-energy fracture.

15.2.3 Motor Vehicle Accidents

Motor vehicle accidents are the cause of most high-energy injuries in the elderly although as has previously been discussed their overall prevalence is relatively low. If one excludes the studies that have included fall-related accidents it becomes clear that the other causes of high-energy injury are relatively rare. Tornetta et al. [24] showed that 73.9 % of high-energy polytrauma in the elderly was caused by motor vehicle accidents compared with 18.1 % which were caused by falls from a height and 8 % by crush injury and other causes. In this study only 31.1 % of the motor vehicle accident polytrauma cases were in elderly pedestrians. These figures are similar to European figures but generally speaking there are more pedestrian injuries in Europe. Broos et al. [25] studied 126 multiply injured elderly patients in Belgium. If the 30 fall-related injuries are excluded 75 % of their injuries followed motor vehicle accidents with 44 % being pedestrians. A further 28 % were car occupants, 21 % were bicyclists and 7 % were motorcyclists. In a large study from Germany, Kuhne et al. [26] showed that 53.2 % of patients aged 56–75 years and 44.9 % aged 76–95 years sustained multiple injuries as a result of motor vehicle accidents.

In view of the relatively high numbers of pedestrian injuries and fatalities in the elderly population it is worth examining these injuries in more detail. A study from Los Angeles of 5000 pedestrian versus motor vehicle accidents between 1994 and 1996 [27] showed that only 8 % of the victims were aged 65 years or more. The average ISS of the elderly group was 12.3 which was higher than the paediatric and adult groups. The highest prevalence of injuries was musculoskeletal (40 %) followed by head and neck injuries (31 %) and external injuries (13.9 %). There were very few spinal (5.4 %) or chest injuries (3.4 %). An analysis of the musculoskeletal injuries showed that in the elderly group there were twice as many upper limb as lower limb fractures. The overall mortality for the 5000 patients was 7.7 % but it varied greatly with age with 3.1 % mortality in the paediatric group, 8.1 % in the adult group and 27.8 % in the elderly group.

Another analysis of a trauma registry in Los Angeles between 1993 and 2003 [28] involving 5838 patients showed that 9.3 % of pedestrians injured in motor vehicle accidents were older than 65 years of age. The authors analysed two groups of patients, those with an ISS >15 and those with an ISS >30. In both groups patients over 65 years of age had the highest prevalence of injury. The elderly showed a high prevalence of severe head injury with an AIS >3 (23.7 %) but lower prevalences of severe chest injury (8.8 %), spinal injury (8.5 %), abdominal injury (8.3 %) and extremity injury (1.3 %). The main head injuries were subarachnoid haematomas and brain contusions. The main extremity injuries were fractures of the pelvis and tibia. There was a similar distribution of fractures of the cervical, thoracic and lumbar spines. The overall mortality for all age groups was 7.7 % but in the 65+ year group the mortality was 25.1 %.

Siram et al. [29] examined the pattern of injury in elderly pedestrians and compared it to younger pedestrians. They examined data on 79,307 patients and showed that pedestrians aged ≥65 years sustained more fractures of the skull, pelvis, upper limbs and lower limbs and more intracranial injuries than younger pedestrians. They found no difference in the rates of pancreatic, splenic and genitourinary injuries but the elderly had higher rates of rib fractures, pneumothoraces and haemothoraces. The authors defined the odds of mortality more precisely than other studies had done. They stated that the odds of mortality in patients aged 25–34 years were 1.08. They rose with increasing age such that they were 3.67 in the 65–74 year group and 8.27 in the ≥85 year group.

Demetriades et al. [28] also reviewed the relationship of age to injury type and injury severity in pedestrians. They defined the associated fractures carefully and showed that in patients ≤14 years the prevalence of pelvic fractures, femoral fractures, tibial fractures and spinal fractures were 6.3 %, 15.5 %, 15 % and 0.4 % respectively. In the >65 year group the prevalence of pelvic, tibial and spinal fractures had risen to 22.6 %,

32 % and 8.5 % but the prevalence of femoral fractures had fallen to 9.8 %. In their >14 year group 11.2 % had an ISS >15 compared with 36.8 % of the >65 year group. The mortality was 3.2 % in the <14 year group and 25.1 % in the >65 year group.

Similar figures were seen in an Australian study [30] where pedestrians aged 17–39 years had an average ISS of 14.1 and a mortality of 3.7 %. The 40–64 year group had an average ISS of 13.4 and a mortality of 5.5 % but the ≥65 year group had an average ISS of 14.9 and a mortality of 22.7 %. The authors highlighted intoxication in young males and injuries in the elderly population as being the two most important problems in pedestrian injuries.

In a study from Ireland [31] the authors analysed 3232 accidents involving adult pedestrians. They documented that older adults represent 36 % of adult pedestrian fatalities and 23 % of serious injuries although they only accounted for 19 % of adult pedestrian motor vehicle accidents. In this study they attempted to analyse which conditions were associated with a higher rate of elderly pedestrian injuries and deaths. They showed that most accidents involving elderly pedestrians occurred in daylight with good visibility (56 %) and in good weather conditions (77 %). Older adults were less likely to be injured at night than younger adults but they were more likely to be struck by trucks or heavy goods vehicles than younger patients. Accidents involving older pedestrians occurred at every type of road crossing but the elderly were less likely to be injured at traffic lights or roundabouts. The authors emphasised the need for specialised accident prevention schemes for the elderly.

Another potential problem is increasing cognitive dysfunction in the elderly population. There is evidence that elderly patients may have an impaired ability to judge automobile speed [32] and may show poorer attention at road crossings [33]. A recent study has shown evidence that more elderly patients killed in pedestrian accidents had evidence of dementia than age-matched controls [34]. This may well prove to be a significant problem in an increasingly aging population.

15.2.4 Falls from a Height

The other cause of high-energy injury is falls from a height. The extent of injury depends on the height of the fall and the elderly tend to fall from lower heights than younger patients. However it is likely that injuries caused by falls from a height are more common than they previously were. An analysis of the distribution of fractures between the 1950s and 2010/2011 in the United Kingdom [23] shows that a number of fractures which used to be seen in the young now often occur in older patients. A good example of this is the calcaneal fracture which is often caused by a fall from a height. This fracture is now relatively common in older patients.

An analysis of 1613 patients who had fallen more than 15 feet [35] showed that in the 65+ year group severe head and spinal injuries were most common with a prevalence of 18.9 % and 16.2 % respectively. The frequency of pelvic, femoral and tibial fractures increased with age such that the prevalence of these fractures in the 65+ year group was 18.7 %, 18.9 % and 8.1 %, respectively.

In a study of unintentional fatal injuries arising from unpaid work at home Driscoll et al. [36] showed that falls from a height were the commonest cause of fatal accidents at home in older people. In people aged 15–54 years it was contact with electricity which caused most domestic fatalities. However in the 55–74 year group 41.8 % of domestic fatalities occurred as a result of a fall from a height. The equivalent figure in the ≥75 year group was 34.3 % and overall a fall from a height was the commonest cause of domestic fatalities.

15.2.5 Suicide

A cause of multiple injuries which has been underestimated, particularly in the elderly, is attempted suicide. Gowing and Jain [22] pointed out that 3 % of injured patients admitted to a Canadian Trauma Centre were as a result of attempted suicides. A recent study from Germany [37] has also emphasised the frequency of suicide attempts by stating that the number of suicides and suicide attempts in Germany is more than twice the number of traffic deaths. An analysis of the Trauma Registry of the German Trauma society of all suicide attempts in adults ≥18 years with an ISS ≥9 showed that there were 1894 attempts of which 274 (14.5 %) were in people ≥65 years. In females jumping from a height was the commonest method of attempting suicide whereas in males it was the use of firearms. Psychiatric dysfunction was more common in females.

15.2.6 Falls

In recent years there has been increased interest in the epidemiology and outcome of falls. As has already been pointed out there is good evidence that in older people falls from a standing height may cause considerable injury and may be responsible for significant mortality. In an analysis of the changing epidemiology of injuries and mortality following falls in patients aged 50 years and over in Finland Kannus et al. [38] showed significant changes between 1970 and 1995. They demonstrated that there had been a 284 % increase in the number of older persons with a fall-induced injury during this period and they showed that the annual increase in fall-induced injury was 9.9 % for males and 12.1 % for females. They did note that there had been a slight decline between 1970 and 1977 but there has been a rapid sharp increase in the incidence of fall-induced injury after 1977. They recorded that the mean age of older persons with a fall-induced injury rose from 67.3 years in 1970 to 73.0 years in 1995. The figures for males were 63.6 and 68.0 years respectively and for females they were 69.2 and 75.3 years. Analysis of the injuries caused by falls showed that the prevalence of long bone fractures had stayed constant in the study period but that soft tissue injuries and dislocations had increased although head injury, other than fracture, had apparently decreased. They thought that the incidence of fall-induced injury would continue to rise.

15.2.7 Treatment

The treatment of polytrauma in the elderly is essentially the same as for young patients although there are two important caveats. Firstly, as Champion et al. [4] pointed out elderly patients may well require more aggressive resuscitation and treatment than younger patients with equivalent injuries. This particularly applies to apparently less severe injuries. Secondly the frequency of medical comorbidities is usually higher than may be seen in younger patients and a good history of associated medical conditions must be obtained. However the principles of assessment, resuscitation and treatment are similar to younger patients and are discussed elsewhere in this book. A good analysis of the principles of management of the multiply injured patients is contained in the chapter dealing with the 'Management of the multiply injured patient' by Pape and Giannoudis in the 8th edition of Rockwood and Green [39].

15.2.8 Predictors of Mortality

An analysis of medical comorbidities in the New York State Registry between 1994 and 1998 shows that, not unexpectedly, the frequency of medical comorbidities increases with age. In an analysis of 76,466 patients, Hannan et al. [40] showed that in their 13–39 year age group only 3.5 % of patients had associated comorbidities compared with 29.4 % in the 65–74 year group, 34.7 % in the 75–84 year group and 37.3 % in the 85+ year group. Their possible comorbidities included chronic obstructive pulmonary disease, congestive heart failure, acute myocardial infarction, other ischaemic heart disease, cerebrovascular disease and peripheral vascular disease. When combined with factors such as intubation status, low systolic blood pressure, low motor response, male gender and lower ICISS, the presence of comorbidities was associated with increased mortality. The adjusted odds ratios for mortality relative to the 13–39 year group were 2.67 for 40–64 year old patients, 8.41 for 65–74 year old patients, 17.4 for 75–84 year old patients and 34.98 for the 85+ year group.

McGwin et al. [41] analysed the relationship between mortality and chronic medical comorbidities together with the severity of the injury in both younger and older patients. In older patients they showed that in less severely injured patients the presence of medical comorbidities increased mortality whereas the same effect was not noted in more severely injured patients with an ISS >26. They concluded that older patients with medical comorbidities should be considered to have an increased risk of death compared with their non-chronically ill counterparts. Older patients with minor injuries (ISS 1–15) had a significantly increased risk of death if they had coexisting haematological disease, diabetes, cardiac disease, renal disease, hepatic disease, neurological disease, respiratory disease or spinal injury. In moderately injured elderly patients (ISS 16–25) cardiac, respiratory and cardiac disease influenced mortality but hypertension was protective. In severely injured patients (ISS ≥26) hypertension and spinal injury appeared to be protective. However the authors pointed out that there may well have been under-reporting of associated medical comorbidities but they suggested that their results showed that elderly patients with minor injuries and associated medical comorbidities should be treated aggressively.

Similar results were reported by Broos et al. [25] who found that early survivors of multiple injuries had a significantly lower prevalence of diabetes and cardiopulmonary, neuropsychiatric and renal disease. Tornetta et al. [24] looked at other predictors of outcome in the multiply injured elderly. They showed that the requirement for transfusion and fluid replacement predicted outcome as did the type of surgery that the patient required. They found that patients who only underwent a general surgical procedure were 2.5 times more likely to die and patients who required both general surgery and orthopaedic surgery were 1.5 times more likely to die. Those who underwent an orthopaedic procedure were less likely to die than those who had no surgery. They could not demonstrate a positive correlation between mortality and early or late surgery.

The Injury Severity Score (ISS) [2] is the most widely used determinant of injury. Early studies suggested that it was less predictive of outcome in

Table 15.1 Factors which increase mortality in elderly patients

Injury severity score <25
GCS <9
Systolic blood pressure <90 mmHg
Pulse >90/min
Increased transfusion requirement
Increased volume replacement
Associated injuries (AIS >3)
Head
Chest
Abdomen
Spine
Comorbidities
Haematological disease
Diabetes
Cardiovascular disease
Renal disease
Hepatic disease
Neurological disease
Respiratory disease

Data from Giannoudis et al [21], Tornetta et al [24] and McGwin et al [41]

the elderly than in younger patients. However Tornetta et al. [24] theorised that this was because minor fall-related injuries were included in these studies, and when they were excluded that in fact elderly patients who died had a higher ISS than those who survived (33.1 and 16.4). They found that the Glasgow Coma Scale (GCS) [42] was also predictive of survival in the elderly. Giannoudis et al. [21] also showed that the ISS, GCS and blood pressure (BP) on admission were predictive of survival in elderly patients. They showed that a pulse rate of >90 on admission and severe (Abbreviated injury scale (AIS) ≥3) head, chest, abdominal and spinal injury were associated with higher mortality in elderly patients. In the elderly group cardiac arrest on admission was associated with 100 % mortality. A list of predictors of mortality in elderly patients is given in Table. 15.1.

15.2.9 Outcome

15.2.9.1 Polytrauma

There is very little information about outcome, other than mortality, in the elderly admitted with

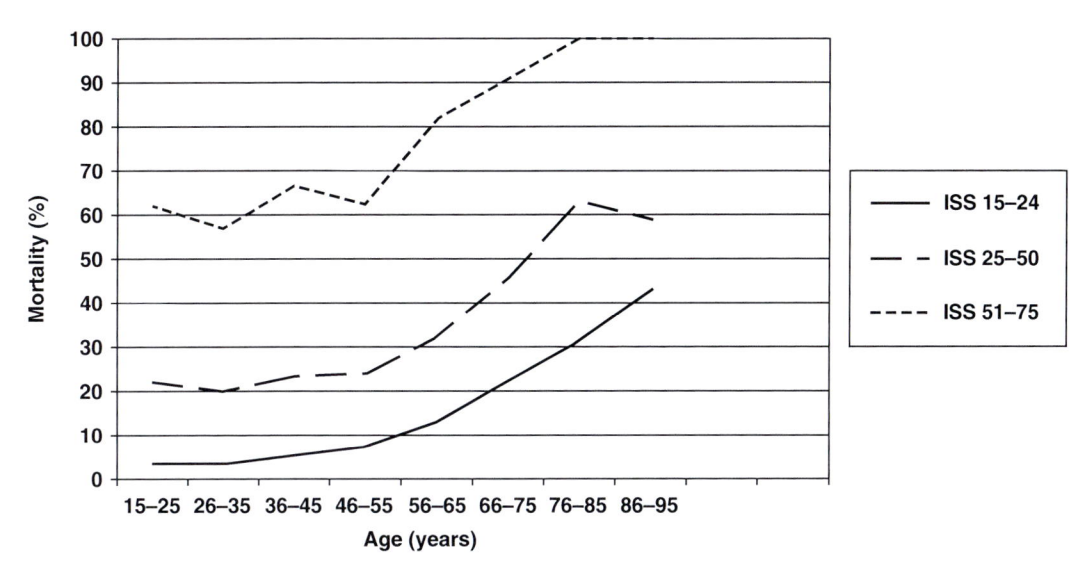

Fig. 15.1 Mortality following multiple injury for different ages and different severity of injury. Note the increase in mortality in the 6th decade (Data taken, with permission, from Kuhne et al [26])

severe injury. There is some evidence that older patients have fewer long-term psychological problems than younger patients [43] but these results came from patients who were not polytraumatised patients but were patients admitted with severe fractures. Studies on the outcome of elderly polytrauma survivors are required.

Mortality following polytrauma clearly varies with the degree of injury and in the large multicentre studies where a wide spectrum of injury has been included the mortality is less than in studies that concentrate on polytrauma victims. There is also considerable variation between mortality in different countries. In countries such as the United States and Germany where there are formal trauma systems, the results are better than in the United Kingdom where such a system has only recently been instituted. The literature suggests that the average mortality for elderly polytraumatised patients in countries with a formal trauma system is 15–25 % [24–26] but of course it depends on the age of the patients and the severity of injury. Kuhne et al. [26] analysed mortality in 5375 patients in Germany who had an ISS ≥16 and were aged between 15 and 95 years. The overall mortality was 23 % but it was 8.1 % if the ISS was 16–24, 27.2 % if the ISS was 25–50 and 66.1 % if the ISS was 51–75.

Their results are shown in Fig. 15.1. They stated that mortality rose from 56 years onwards. These overall mortality figures are not dissimilar from those reported from other trauma centres but higher figures have been reported. Aldrian et al. [44] reported a mortality of 53.3 % in the elderly with 31.1 % dying within 24 h. Their average ISS was 32.1.

The statement by Kuhne et al. [26] that mortality in polytraumatised patients rose after the age of 56 once again highlights the polarisation of much of the literature dealing with severely injured patients. Their assessment of a group of polytraumatised patients admitted to trauma centres in Germany should be compared with the study of Caterino et al. [45] in the United States. They examined the Ohio State Registry which records a wider range of admissions from both trauma and non-trauma centres. They found that 70 years was the equivalent age at which mortality increased. They recommended that 70 years should be taken as the cut-off age for considering a patient to be elderly in trauma studies but it is vital that the type of injury be accurately recorded given the differences between these two papers.

In the United Kingdom, which lacked a formal trauma system until recently, Giannoudis et al. [21] reported 42 % mortality in elderly polytraumatised patients. As with other studies the mortality was age dependent and it reached almost 50 % in patients aged over 75 years. In their earlier study DeMaria et al. [3] had reported 80 % mortality in patients with an ISS ≥25 who were at least 80 years of age. More recently it has been shown that elderly patients with an ISS >30 require less ICU facilities than younger patients because of their higher mortality [46]. It is also interesting to note that in the United States mortality following injury in the very elderly (>80 years) is less in trauma centres than in acute care hospitals [47]. Mortality obviously increases with age and degree of injury but it is also influenced significantly by the type of hospital and the trauma system within the country.

15.2.9.2 Falls

The mortality from falls has increased in the last few decades. As with the incidence of fall-induced injury Kannus and his co-workers have used the Finnish Cause-of-Death register to assess the incidence of fall-induced mortality between 1971 and 2002 [48]. They pointed out that in 2002 falls were responsible for 285 % more deaths than motor vehicle accidents and that there had been an overall 136 % increase in fall-induced deaths in the study period. The relevant figures for males and females were 201 % and 97 % respectively. They also showed that while the incidence of fall-induced deaths had been relatively steady in females between 1975 and 2002 it had continued to increase in females. They theorised that there would be a 108 % increase in mortality by 2030.

15.2.9.3 Types of Injury

In the elderly there are two main types of serious injury that frequently occur with both low-energy and high-energy injuries and may be associated with significant mortality. These are head injuries and fractures. Obviously injuries may occur in other body systems but they are usually caused by high-energy trauma and their characterisation and management is discussed elsewhere in the book.

Head Injury

In a recent study of head injury in the elderly Mitra et al. [49] analysed 96 patients and showed that 31.2 % of head injuries followed a low fall, 30.2 % occurred because the patient was struck by a motor vehicle and 17.7 % were caused by a high fall. All patients presented with an initial GCS <8 which had not been caused by sedation or paralysis. They reported that 62.2 % of patients aged 65–74 years died compared with 68.2 % aged 75–84 years and 100 % of patients aged at least 85 years. Increasing age and brainstem injury were identified as predictors of mortality. Frankel et al. [50] analysed the outcome of traumatic brain injury in the elderly and showed that elderly patients were significantly less likely to be discharged home. However they felt that the results of treatment were encouraging and they stated that older patients exhibited the potential to achieve functional goals.

Multiple Fractures

Multiple fractures in the elderly may occur as a result of high-energy or low-energy injuries. The assumption is often made that they are mainly caused by motor vehicle accidents or falls from a height but this is simply not the case. In a review of 6872 inpatient and outpatient fractures in the Royal Infirmary of Edinburgh in 2007/8 [51] there were 2293 patients aged at least 65 years of age. Of these 117 (5.1 %) presented with multiple fractures. One hundred and ten (94.0 %) had two fractures, 6 (5.1 %) had three fractures and one (0.9 %) 75 year old pedestrian presented with four fractures after a motor vehicle accident. Table 15.2 shows the causes of multiple fractures in the elderly population. It can be seen that the highest prevalence is indeed related to motor vehicle accidents with 36.4 % of patients presenting with multiple fractures. Predictably the next most common cause of multiple fractures in the elderly was falls from a height followed by falls down stairs. However although the prevalence of multiple fractures following simple falls was only 4.4 %, the frequency of fall-related fractures in the elderly population means that 92 patients presented with multiple fractures following a fall during the year, this constituting 78.6 % of all the multiple fractures. Table 15.2 shows that the average of the multiple fracture group was 71.3 years and about 80 % were female.

Table 15.2 The epidemiology of multiple fractures in patients aged at least 65 years of age presenting to the Royal Infirmary of Edinburgh in a 1-year period in 2007/2008

	Patients (n)	Multiple fractures	%	Average age (year)	Gender ratio
Simple fall	2111	96	4.5	79.0	16/84
Fall from height	11	3	27.3	72.0	67/33
Fall down stairs	80	10	12.5	77.0	30/70
Motor vehicle accident	22	8	36.4	80.2	75/25
Direct blow/assault	45	2	4.4	77.5	0/100
Sport	17	0	–	–	–
Spontaneous	24	0	–	–	–
Others	25	0	–	–	–
	2335	119	5.1	78.7	22/78

A review of the 32 fractures that resulted from motor vehicle accidents shows that they occurred in 22 patients with 7 patients presenting with two fractures and one patient with four fractures. The average age was 80.2 years and 75 % of the patients were male. Five (22.7 %) of the 22 patients were bicyclists all of whom presented with a single fracture. A further 4 (18.2 %) were vehicle occupants and one vehicle passenger presented with two fractures. The remaining 13 (59.1 %) elderly patients were pedestrians struck by a vehicle of which 7 (53.8 %) presented with multiple fractures. The average age of this group was 78.9 years and 14 (63.6 %) of the fractures were in the lower limb or pelvis and 8 (36.4 %) were in the upper limb. Three (13.6 %) of the fractures were open.

Table 15.2 shows that the prevalence of multiple fractures following a fall from a height approaches that of motor vehicle accidents but all the fractures were closed suggesting that either falls from a height in the elderly are not as severe as in younger patients or possibly that many falls are fatal. Table 15.2 also shows that falls down stairs are associated with a high prevalence of multiple fractures. The results indicate that the highest frequency of multiple fractures in the 65+ year group follows motor vehicle accidents where the elderly patient is a pedestrian struck by a vehicle.

Table 15.3 shows the fractures associated with all index fractures in a consecutive group of patients aged ≥65 years treated in the Royal Infirmary of Edinburgh in a 1-year period in 2010/2011. All modes of injury are included. Overall 8.5 % of fractures were associated with other fractures during the study year but 12 fracture types were associated with at least 10 % other fractures. Obviously the location of some of the associated fractures depended on the circumstances of the injury but generally speaking the associated fractures are often near the location of the index fracture. The greatest number of multiple fractures in the elderly that will present to orthopaedic surgeons will follow a simple fall and these will be examined in more detail.

Fall-Related Multiple Fractures

A review of all patients aged at least 16 years presenting to the Orthopaedic Trauma Unit of the Edinburgh Royal Infirmary in a 1-year period during 2007/2008 shows that 3843 fractures were caused by simple falls, this being 55.9 % of all the fractures. Analysis of the patients of at least 65 years of age shows that 2213 fractures were caused by simple falls. These fractures occurred in 2111 patients with 2015 patients presenting with a single fracture, 90 presenting with two fractures and six patients presenting with three fractures. Table 15.2 shows that the average age of patients presenting with multiple fractures after a fall was 79 years. The average age of males was 76.6 years with 79.5 years being recorded for females. This compares with 79.2 years and 80.0 years for males and females who presented with single fractures. The gender ration for single fractures was 20/80 indicating that multiple fractures are more common in

Table 15.3 The prevalence of multiple fractures in patients aged ≥65 years

Fractures	No	Multiple fractures No	%	Other fractures (%)
Scapula	12	2	16.7	Clavicle 66.6 %, proximal humerus 33.3 %
Clavicle	54	4	7.4	Scapula 50 %, proximal humerus 25 %
Proximal humerus	267	27	10.1	Proximal femur 59.3 %, distal radius 14.8 %
Humeral diaphysis	30	1	3.3	Distal radius 100 %
Distal humerus	27	5	18.5	Metacarpal 40 %, proximal radius 40 %
Proximal radius/ulna	2	0	0	
Proximal radius	31	5	16.1	Distal humerus 40 %, pelvis 20 %
Proximal ulna	32	6	18.7	Proximal femur 33.3 %, distal radius 33.3 %
Radial/ulna diaphyses	2	0	0	
Radial diaphysis	7	0	0	
Ulnar diaphysis	6	1	16.7	Proximal tibia 100 %
Distal radius/ulna	510	42	8.2	Proximal femur 31.0 %, bilateral 26.2 %
Carpus	15	1	6.7	Metatarsal 100 %
Metacarpus	64	14	21.9	Metacarpal 71.4 %, distal radius 28.6 %
Finger phalanges	90	13	14.4	Phalanx 53.8 %, pelvis 15.4 %
Pelvis	89	8	9.0	Phalanx 25 %, metacarpal 25 %
Proximal femur	683	38	5.6	Distal radius 44.7 %, proximal humerus 39.5 %
Femoral diaphysis	55	2	3.6	Ankle 50 %, proximal humerus 50 %
Distal femur	19	2	10.5	Ankle 50 %, proximal humerus 50 %
Patella	27	0	0	
Proximal tibia	18	5	27.8	Metacarpal 40 %, pelvis 20 %
Tibia/fibular diaphyses	12	1	8.3	Clavicle 8.3 %
Distal tibia	8	0	0	
Ankle	170	9	5.3	Metatarsal 33.3 %, calcaneus 11.1 %
Talus	0	0	0	
Calcaneus	6	2	33.3	Metatarsal 50 %, ankle 50 %
Midfoot	2	0	0	
Metatarsus	79	13	12.7	Metatarsal 61.5 %, ankle 23.1 %
Toes	4	0	0	
Total	2355	201	8.5	

The fractures represent all inpatient and outpatient fractures treated in the Royal Infirmary of Edinburgh in a 1-year period in 2010/2011. The two commonest other fractures associated with each index fracture are shown

elderly females but the average ages of males and females are not dissimilar.

An analysis of multiple fractures of all ages in 2007/2008 shows that they are much more common in older patients. Figure 15.2 shows the age-related incidence of multiple fall-related fractures in the whole population. There were none in the 15–19 years group but Fig. 15.2 shows that the incidence starts to rise in the 6th decade of life and continues to rise until the 10th decade presumably mainly because of increased osteopenia and other medical comorbidities predisposing the patients to falls.

Only six (6.25 %) of the elderly patients who presented with fall-related multiple fractures had three fractures. It was not possible to define any relationship between different fracture combinations. Two involved the upper limb only and four involved both upper and lower limbs. Five (83.3 %) of these fractures occurred in females with an average age of 78.6 years with only one 71 year old male presenting with three fractures after a fall.

A review of the 90 patients who presented with double fracture combinations showed there were three groups. Group 1 consisted of 29 (32.2 %)

Fig. 15.2 The incidence of multiple fall-related fractures in different age groups (Data from the Royal Infirmary of Edinburgh, Scotland)

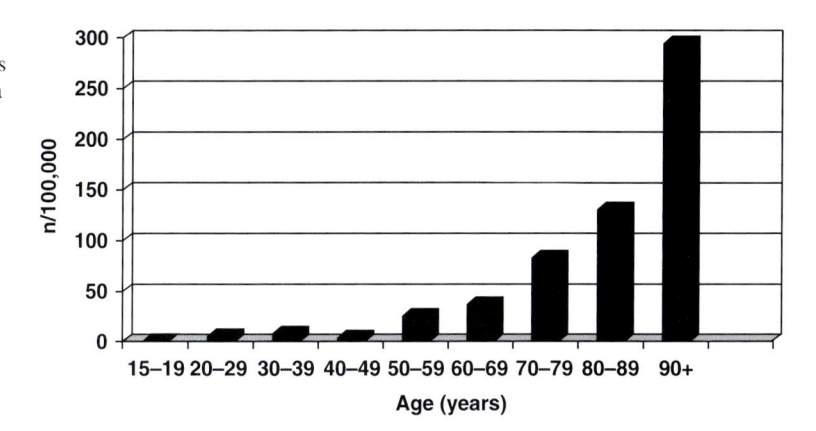

patients who presented with two upper limb fractures. Group 2 comprised 11 (12.2 %) patients who presented with two lower limb fractures and Group 3 consisted of the remaining 50 (55.6 %) patients who presented with fracture combinations involving both upper and lower limbs. Pelvic fractures were included with the lower limb fractures. Group 1 had an average age of 75 years and a male/female gender ratio of 17/83. Group 2 had an average age of 83.4 years and a gender ratio of 18/82 and Group 3 had an average age of 80.6 years and a gender ratio of 14/86.

Analysis of the Group I patients showed that combinations of fractures involving the distal radius and proximal humerus were most commonly seen. Of the 29 upper limb double fracture combinations 19 (65.5 %) involved the distal radius and 11 (37.9 %) the proximal humerus with 4 (13.8 %) patients presenting with fractures of the distal radius and proximal humerus. There were in fact only three double upper limb fracture combinations that did not involve the distal radius or proximal humerus. The commonest Group 1 combinations were bilateral distal radial fractures (27.5 %), the distal radius/proximal humerus combination (13.8 %) and the combination of distal radius and finger phalanx (10.3 %).

Of the 11 Group 2 patients five (45.4 %) involved the proximal femur, four (36.4 %) the pelvis and four (36.4 %) involved the ankle. In fact there was only one combination of midfoot and metatarsal fractures that did not involve the proximal femur, pelvis or ankle. The commonest lower limb combinations were fractures of the proximal femur and pelvis and fractures of the ankle and metatarsal which both occurred in 27.3 % of Group 2 fractures.

Group 3 fractures were most commonly seen. Of the 50 Group 3 fractures 34 (68 %) involved the proximal femur and 17 (34 %) presented with a combination of proximal femoral and proximal humeral fractures, this being the commonest double fracture combination. A further 11 (22 %) patients presented with proximal femoral and distal radial fractures. Of the fracture combinations that did not involve the proximal femur, the common combination was that of the proximal humerus and pelvis which presented in 8 % of Group 3 cases followed by that of the distal radius and pelvis which occurred in 6 % of the patients.

The results show that the four commonest fractures in double fracture combinations involve fractures of the proximal femur, distal radius, proximal humerus and pelvis. Proximal femoral fractures occurred in 39 (43.3 %) of the double fracture combinations with distal radius fractures being involved in 38 (42.2 %). The average ages of these fractures groups were 81.4 and 77.6 years respectively and the gender ratios were 20/80 and 13/87. Proximal humeral fractures occurred in 34 (37.8 %) of the double fracture combinations. These patients had an average age of 79.7 years and a gender ratio of 15/85. Pelvic fractures occurred in 11 (12.2 %) patients with an average age of 87.7 years and a gender ratio of 8/92.

Table 15.4 shows the basic epidemiological data of the nine most common double fracture

Table 15.4 Epidemiological criteria of the nine double fracture configurations that occurred at least three times in a 1-year period

Fracture combination	n	%	Age (year)	Gender ratio (%)
Proximal humerus/proximal femur	17	18.9	80.9	18/82
Distal radius/proximal femur	11	12.2	80.2	18/82
Distal radius/distal radius	8	8.9	74.2	20/80
Distal radius/proximal humerus	4	4.4	79.5	0/100
Proximal humerus/pelvis	4	4.4	87.2	25/75
Distal radius/finger phalanx	3	3.3	74.7	0/100
Distal radius/pelvis	3	3.3	85.0	0/100
Proximal femur/pelvis	3	3.3	92.3	0/100
Ankle/metatarsal	3	3.3	75.3	0/100

combinations, these being the fracture combinations that presented at least three times during the year. It is evident that fractures of the proximal femur, proximal humerus and distal radius are involved in all the common combinations except for the ankle metatarsal combination. It is also worth noting the extreme age of patients who present with a combination of a fall-related pelvic fracture and a fracture of the proximal femur, distal radius and proximal humerus.

With increased longevity it seems likely that multiple fall-related fractures will become more common and that they will present in patients who have multiple medical comorbidities and who require aggressive medical management to increase the chance of survival from these apparently straightforward injuries.

Open Fractures

The common perception is that open fractures are not infrequently associated with more severe injury and are commoner in high-energy modes of injury such as motor vehicle accidents or falls from a height. There is very little information about the epidemiology of open fractures in the elderly population which is perhaps surprising as they are often difficult to treat and they can cause considerable morbidity. We have undertaken a review of open fractures in ≥65 year patients over a 15-year period and we have compared the open fractures with those seen in patients aged <65 years.

All open fractures presenting to the Royal Infirmary of Edinburgh between 1995 and 2009 were analysed. Only patients from the hospital's catchment area were included in the study so that an accurate epidemiological analysis could be obtained. All adults ≥15 years were included and the population figures of 1995–2009 were averaged to allow fracture incidences to be calculated.

In the 15-year period 20.3 % of all open fractures presented in patients aged ≥65 years. They were more commonly seen in females with a gender ratio of 35/65. Open fractures in the <65 year group were more commonly seen in males and the gender ratio was 78/22. The incidence of all fractures in the two groups is shown in Table 15.5. It may be surprising to observe that the incidence of open fractures is slightly higher in older patients than in younger patients. Indeed if the incidence is calculated for the ≥80 year group it rises to $446.7/10^5$/year indicating that the frequency of open fractures in the adult population correlates with increasing age. Table 15.5 shows that the spectrum of open fractures is remarkably similar between the two patient groups. Exceptions are open distal radial and ankle fractures which have a much higher incidence in the ≥65 year group and open hand fractures which have a higher incidence in the <65 year group.

Table 15.5 also shows the prevalence of Gustilo Type III [52] open fractures in both groups of patients. There is a higher prevalence of Gustilo Type III fractures in the <65 year group but 20.4 % of the open fractures in the ≥65 year group were Type III fractures. With the exception of femoral diaphyseal fractures the prevalence of Gustilo Type III fractures in the femur, patella and tibia was remarkably similar in both groups of patients despite the fact that 64.8 % of open femoral, patellar and tibial fractures in the ≥65 year group followed a fall compared with 13 % in the <65 year group. This emphasises the importance of falls in causing serious injuries in frail elderly patients. Another interesting observation is that if open fractures of

15 The Management of the Multiply Injured Elderly Patient

Table 15.5 The numbers and incidence of open fractures treated in the Royal Infirmary of Edinburgh in a 15-year period

Fractures	≥65 years			<65 years		
	No	×/10⁶/year	GIII (%)	No	×/10⁶/year	GIII (%)
Scapula	0	0	–	2	0.3	0
Clavicle	1	0.7	0	8	1.2	0
Proximal humerus	3	2.1	0	9	1.4	11.1
Humeral diaphysis	6	4.1	33.3	10	1.6	10.0
Distal humerus	6	4.1	16.7	12	1.9	58.3
Proximal radius/ulna	1	0.7	0	1	0.2	0
Proximal radius	0	0	–	0	0	–
Proximal ulna	15	10.3	6.7	36	5.6	16.7
Radial/ulna diaphyses	9	6.2	0	35	5.5	5.7
Radial diaphysis	1	0.7	0	4	0.6	25.0
Ulnar diaphysis	4	2.7	50.0	21	3.3	9.5
Distal radius/ulna	124	85.1	2.4	60	9.4	1.7
Carpus	0	0	–	1	0.2	100.0
Metacarpus	8	5.5	12.5	96	15.0	10.4
Finger phalanges	146	100.2	8.6	944	147.2	23.8
Pelvis	1	0.7	0	6	0.9	0
Proximal femur	0	0	–	1	0.2	0
Femoral diaphysis	2	1.4	0	40	6.2	67.5
Distal femur	5	3.4	60.0	21	3.3	76.2
Patella	5	3.4	20.0	41	6.4	31.7
Proximal tibia	7	4.8	57.1	22	3.4	59.1
Tibial diaphysis	48	32.9	37.5	219	34.1	46.1
Distal tibia	7	4.8	42.8	24	3.7	45.8
Ankle	54	37.1	51.9	72	11.2	44.4
Talus	0	0	–	6	0.9	50.0
Calcaneus	4	2.7	75.0	14	2.2	71.4
Midfoot	0	0	–	5	0.8	80.0
Metatarsus	7	4.8	42.9	96	15.0	26.0
Toes	20	13.7	10.0	150	23.4	18.0
Total	484	332.3	20.4	1902	296.6	28.4

The prevalence of Gustilo Type III fractures [52] is also shown

the fingers and toes are excluded the average ISS for the ≥65 year patients who presented with an open fracture was 12 with 11 being recorded for the younger group. Again this emphasises that the spectrum of injuries after lower energy trauma in the elderly is not dissimilar from the injuries seen in younger patients after higher energy trauma.

Analysis of the high-energy injuries in the ≥65 year patients shows that 13.2 % of open fractures followed a motor vehicle accident or a fall from a height. The average ISS was 14 and 43.7 % of the fractures were Gustilo Type III in severity. In the younger group of patients 24.7 % of the open fractures followed a motor vehicle accident or a fall from a height. The average ISS was also 14 and 48.5 % of the open fractures were Gustilo Type III in severity. The only real difference between the two groups was that in the ≥65 year group 79.2 % of the motor vehicle accident open fractures occurred in pedestrians compared with 29.4 % of the <65 year group.

It is apparent that the spectrum of open fractures is very similar in both groups of patients.

There are a few exceptions such as the open femoral diaphyseal fracture which are very rare in older patients but Table 15.5 highlights the fact that frailer older patients have a similar distribution of open fractures to younger patients.

References

1. Oreskovich MR, Howard JD, Copass MK, et al. Geriatric trauma: injury patterns and outcomes. J Trauma. 1984;24:565–72.
2. Baker SP, O'Neill B, Haddon W, et al. The injury severity score: a method for describing patients with multiple injuries and evaluating emergency care. J Trauma. 1974;14:187–96.
3. DeMaria EJ, Kenney PR, Merriam MA, et al. Survival after trauma in geriatric patients. Ann Surg. 1987;206:738–43.
4. Champion HR, Copes WS, Buyer D, et al. Major trauma in geriatric patients. Am J Public Health. 1989;79:1278–82.
5. Finelli FC, Jonsson J, Champion HR, et al. A case control study for major trauma in geriatric patients. J Trauma. 1989;29:541–8.
6. Grant PT, Henry JM, McNaughton GW. The management of elderly blunt trauma victims in Scotland: evidence of ageism? Injury. 2000;31:519–28.
7. Van der Sluis CK, Klasen HJ, Eisma WH, et al. Major trauma in young and old: what is the difference? J Trauma. 1996;40:78–82.
8. Chang DC, Bass RR, Cornwell EE, et al. Undertriage of elderly trauma patients to state-designated trauma centers. Arch Surg. 2008;143:776–81.
9. Moore L, Turgeon AF, Sirois M-J, et al. Trauma Centre outcome performance: a comparison of young adults and geriatric patients in an inclusive trauma system. Injury. 2012;43:1580–5.
10. Staudenmayer KL, Hsia RY, Mann NC, et al. Triage of elderly trauma patients: a population-based perspective. J Am Coll Surg. 2013;217:569–76.
11. Clement ND, Tennant C, Muwanga C. Polytrauma in the elderly: predictors of the cause and time of death. Scand J Trauma Resusc Emerg Med. 2010;18:26.
12. US Census Bureau. International data base (IDB). Available at http://www.census.gov/ipc/www/idb.
13. UK Office of National Statistics. http://www.statistics.gov.uk/cci/nugget_print.asp?ID=1875.
14. UK Office of National Statistics. http://www.statistics.gov.uk/cci/nugget_print.asp?ID=934.
15. Court-Brown CM, Clement N. Four score years and ten. An analysis of the epidemiology of fractures in the very elderly. Injury. 2009;40:1111–4.
16. UK Office of National Statistics. http://www.dft.gov.uk/pgr/statistics.
17. Safety on roads. What's the vision? OECD Publishing; 2002.
18. Tinetti ME, Speechley M, Ginter SF. Risk factors for falls among elderly persons living in the community. N Engl J Med. 1988;319:1701–7.
19. Von Wågert Heideken P, Gustafson Y, Kallin K, et al. Falls in very old people: the population based Umeå study in Sweden. Arch Gerontol Geriatr. 2009;49:390–6.
20. Center of Disease Control and Prevention. Cost of falls among older adults. http://www.cdc.gov/homeandrecreationalsafety/falls/fallscost.html
21. Giannoudis PV, Harwood PJ, Court-Brown C, et al. Severe and multiple trauma in older patients: incidence and mortality. Injury. 2009;40:362–7.
22. Gowing R, Jain MK. Injury patterns and outcomes associated with elderly trauma victims in Kingston, Ontario. Can J Surg. 2007;50:437–44.
23. Court-Brown CM. The epidemiology of fractures and dislocations. In Tornetta P, Heckman JD, Court-Brown CM, McQueen MM, Ricci W, McKee M, editors. Rockwood and Green's fractures in adults. 8th ed. Philadelphia: Lippincott, Williams & Wilkins. 2015.
24. Tornetta P, Mostavi H, Riina J, et al. Morbidity and mortality in elderly trauma patients. J Trauma. 1999;46:702–6.
25. Broos PLO, D'Hoore A, Vanderschot P, et al. Multiple trauma in elderly patients. Factors influencing outcome: importance of aggressive care. Injury. 1993;24:365–8.
26. Kuhne CA, Ruchholtz S, Kaise GN, et al. Mortality in severely injured elderly trauma patients-when does age become a risk factor. World J Surg. 2005;29:1476–82.
27. Peng RY, Bongard FS. Pedestrian versus motor vehicle accidents: an analysis of 5000 patients. J Am Coll Surg. 1999;189:343–8.
28. Demetriades D, Murray J, Martin M, et al. Pedestrians injured by automobiles: relationship of age to injury type and severity. J Am Coll Surg. 2004;199:382–7.
29. Siram SM, Sonaike V, Bolorunduro OB, et al. Does the pattern of injury in elderly pedestrian trauma mirror that of the younger pedestrian? J Surg Res. 2011;167:14–8.
30. Small TJ, Sheedy JM, Grabs AJ. Cost, demographics and injury profile of adult pedestrian trauma in inner Sydney. ANZ J Surg. 2006;76:43–7.
31. Martin AJ, Hand EB, Trace F, et al. Pedestrian fatalities and injuries involving Irish older people. Gerontology. 2009;56:266–71.
32. Oxley J, Fildes B, Ihsen E, et al. Differences in traffic judgements between young and old adult pedestrians. Accid Anal Prev. 1997;29:839–47.
33. Sparrow WA, Bradshaw EJ, Lamoureux E, et al. Ageing effects on the attention demands of walking. Hum Mov Sci. 2002;21:961–72.
34. Gorrie CA, Rodriguez M, Sachdev P, et al. Increased neurofibrillary tangles in the brains of older pedestrians killed in traffic accidents. Dement Geriatr Cogn Disord. 2006;22:20–6.
35. Demetriades D, Murray J, Brown C, et al. High-level falls: type and severity of injuries and survival outcome according to age. J Trauma. 2005;58:342–5.

36. Driscoll TR, Mitchell RJ, Hendrie AL, et al. Unintentional fatal injuries arising from unpaid work at home. Inj Prev. 2003;9:15–9.
37. Topp T, Lefering R, Mueller T, et al. Suicide in old age: the underestimated risk. An analysis of 1,894 patients in the Trauma Registry of the German Trauma Society. Unfallchirurg. 2013;116:332–7.
38. Kannus P, Parkkari J, Koskinen S, et al. Fall-induced injuries and deaths among older adults. JAMA. 1999;281:1895–9.
39. Pape HC, Giannoudis PV. Management of the multiply injured patient. In: Tornetta P, Heckman JD, Court-Brown CM, McQueen MM, Ricci W, McKee M, editors. Rockwood and Green's fractures in adults. 8th ed. Philadelphia: Lippincott, Williams & Wilkins. 2015.
40. Hannan EL, Hicks Waller C, Szypulski Farrell L, et al. Elderly trauma inpatients in New York state: 1994–1998. J Trauma. 2004;56:1297–304.
41. McGwin G, MacLennan PA, Bailey Fife J, et al. Preexisting conditions and mortality in older trauma patients. J Trauma. 2004;56:1291–6.
42. Teasdale G, Jennett B. Assessment of coma and impaired consciousness. A practical scale. Lancet. 1974;2(7872):81–4.
43. Ponsford J, Hill B, Karamitsios M, Bahar-Fuchs A. Factors influencing outcome after orthopaedic trauma. J Trauma. 2008;64:1001–9.
44. Aldrian S, Nau T, Koenig F, et al. Geriatric polytrauma. Wien Klin Wochenschr. 2005;117:145–9.
45. Caterino JM, Valasek T, Werman HA. Identification of an age cutoff for increased mortality in patients with elderly trauma. Am J Emerg Med. 2010;28:151–8.
46. Taylor MD, Tracy JK, Meyer W, et al. Trauma in the elderly: intensive care unit resource use and outcome. J Trauma. 2002;53:407–14.
47. Meldon SW, Reilly M, Drew BL, et al. Trauma in the very elderly: a community-based study of outcomes at trauma and nontrauma centers. J Trauma. 2002;52:79–84.
48. Kannus P, Parkkari J, Niemi S, et al. Fall-induced deaths among elderly people. Am J Public Health. 2005;95:422–4.
49. Mitra B, Cameron PA, Gabbe BJ, et al. Management and hospital outcome of the severely head injured elderly patient. ANZ J Surg. 2008;78:588–92.
50. Frankel JE, Marwitz JH, Cifu DX, et al. A follow-up study of older adults with traumatic brain injury: taking into account decreasing length of stay. Arch Phys Med Rehabil. 2006;87:57–62.
51. Court-Brown CM, Aitken SA, Forward D, O'Toole D. The epidemiology of fractures. In: Bucholz RW, Heckman JD, Court-Brown CM, Tornetta P, editors. Rockwood and Green's fractures in adults. 7th ed. Philadelphia: Lippincott, Williams & Wilkins; 2010.
52. Gustilo RB, Mendoza RM, Williams DN. Problems in the management of type III (severe) open fractures: a new classification of type III open fractures. J Trauma. 1984;24:742–6.

General Management in the Elderly: Preoperative and ICU

16

Alain Corcos and Andrew B. Peitzman

Contents

16.1	**Introduction**	219
16.2	**The Physiology of Aging and How It Impacts Trauma Care**	220
16.2.1	Cardiovascular	220
16.2.2	Pulmonary	221
16.2.3	Neurologic	221
16.2.4	Renal	222
16.3	**Optimizing Perioperative Management**	222
16.3.1	Comorbidity	222
16.3.2	Medications and Supplements	223
16.3.3	Nutrition	223
16.3.4	Cognitive Impairment	224
16.3.5	Functional Capacity	225
16.4	**Frailty**	225
16.5	**Reversal of Therapeutic Anticoagulation**	226
16.6	**Advanced Directives**	227
References		227

A. Corcos, MD, FACS
Trauma Services, Division of Multisystem Trauma,
UPMC Mercy Hospital, 1400 Locust St.,
Suite 6538, Pittsburgh, PA 15219, USA
e-mail: corcosac@upmc.edu

A.B. Peitzman, MD (✉)
Department of Surgery, UPMC-Presbyterian,
F-1281, UPMC-Presbyterian, Pittsburgh,
PA 15213, USA
e-mail: peitzmanab@upmc.edu

16.1 Introduction

Our population is aging rapidly. Currently, individuals over the age of 65 are nearly one tenth of the world's population and the oldest old, those over 85, are now the fastest growing segment of many nations' population [1]. Not surprisingly, these trends are reflected in incidence rates for the polytraumatized patient. In the United States, patients over 65 comprise one third or more of trauma service hospital admissions and account for up to half of all injury-related fatalities. Falls and traffic-related motor vehicle accidents remain the leading causes of injury among the aged. One sixth of all traffic fatalities and one fifth of all automobile-pedestrian fatalities in the United States are 65 and older [2]. Ground-level falls have become the leading cause of hospitalization for the elderly [3].

The elderly are an ill-defined and nonhomogenous demographic in general, and what exactly constitutes severe injury in this age group further obscures not only statistical analysis but in some cases triage and treatment. Although Champion et al. [4] showed as early as 1989 that minor, single-system injury can carry a relatively high mortality rate in patients over age 65, evidence since suggests that the injured elderly remain less likely to receive care at trauma centers than younger trauma patients. A comprehensive literature review by the practice management guideline committee of The Eastern Association for the

Surgery of Trauma (EAST) concluded that elderly patients with at least one body system with an abbreviated injury scale (AIS) severity score of three or higher should be triaged to a trauma center and cared for in an intensive care unit staffed by a critical care specialist [5].

Preexisting illness, commonly associated with advanced age, further complicates the study and treatment of patients in this age group, particularly as it pertains to multisystem trauma. Cardiovascular, neurological, pulmonary, and renal system comorbidities compound normal physiologic changes associated with aging, rendering the elderly polytrauma patient particularly vulnerable to inhospital and postoperative complications, which, in turn, are associated with a significantly higher mortality rate.

A practical approach to the management of the polytraumatized elderly patient necessitates an understanding of the clinical impact of the physiologic changes that occur with aging and involve nearly every organ system. Additionally, an appreciation of the comorbid conditions, medications, nutritional status, and functional capacity specific to each patient can help guide preoperative management, avoid postoperative complications, and optimize outcomes. Finally, chronological age alone is neither a good predictor of performance nor an adequate contraindication to aggressive intervention. An understanding of what is increasingly considered a distinct entity from aging, frailty, is extremely helpful. Although few truly evidenced-based recommendations apply uniquely to an elderly trauma patient cohort, there are some important age-related considerations with peri- and postoperative management with substantial consensus agreement.

16.2 The Physiology of Aging and How It Impacts Trauma Care

16.2.1 Cardiovascular

Cardiac output is the product of heart rate and stroke volume, while stroke volume is a function of factors affecting ventricular volume (preload), intrinsic myocardial contractility, and vascular resistance (after-load). Normal age-related changes alter every element of this equation. As the heart ages, myocytes are progressively lost while those that remain increase in volume. This myocardial hypertrophy leads to a stiff and less compliant ventricle, impairing end-diastolic filling and diminishing the preload contribution to stroke volume. Predictably, the heart's ability to compensate adequately with increases in cardiac output during periods of hypovolemia or hemorrhage is decreased. Additionally, aged myocytes do not respond as well to adrenergic stimulation, affecting the chronotropic (lower maximum heart rate) and inotropic (impaired contractility) contributions to the epinephrine-mediated compensatory response critical to maintaining tissue oxygen delivery. Finally, normal aging also affects the arterial vascular system and the afterload component of cardiac output. Smooth muscle cells of the arterial wall undergo intimal hyperplasia with age, leading to a stiff, less compliant peripheral vasculature, while stiffness of the great vessels is the main contributor to the baseline elevation in systolic blood pressure that is associated with normal aging and is independent of atherosclerosis. Compounding this constellation of changes is an overall decrease in myocardial muscle oxygen delivery, as 50 % of those over the age of 65 and 80 % of those over 80 have some degree of coronary artery stenosis.

As a result of these cardiovascular age-related changes, early evaluation (primary survey) and resuscitation of the geriatric trauma patient can be challenging. While shock can be difficult to appreciate, early identification and intervention is essential to avoid prolonged periods of hypoperfusion. "Normal" blood pressure and heart rate during the primary survey should not reassure against ongoing, occult hemorrhage. Evidence from a survey of the National Trauma Data Bank (NTDB) suggests that mortality begins to increase in older trauma patients (age >43) at initial presenting systolic blood pressures of 117 mmHg, considerably higher than the traditional standard of 90 mmHg [6].

Early in the evaluation, an arterial blood gas analysis should be obtained. Metabolic acidosis, reflecting elevated serum lactate levels, is the most sensitive indicator of occult hypoperfusion.

A thorough search for cavitary hemorrhage should be performed, including a chest radiograph and the focused assessment sonography in trauma (FAST). A pelvis radiograph that includes both hip joints is an essential adjunct to the primary survey in this age group, as relatively minor pelvic or hip fractures can be associated with significant, often occult, retroperitoneal hemorrhage. Early monitoring of the cardiovascular system should be considered. Although the ubiquitous use of pulmonary artery catheters in moderately to severely injured elderly patients to optimize cardiac output to supratherapeutic values is no longer recommended, invasive hemodynamic monitoring in selected cases clearly has potential benefits. Echocardiography to determine and follow ventricular volume and cardiac performance is helpful and increasingly available at the critical care bedside. Although nonoperative management of blunt abdominal solid organ injury should be considered in any patient who is hemodynamically stable regardless of age, the risk of nonoperative management in an elderly individual with a splenic laceration may actually be greater than the risk of an early operation.

16.2.2 Pulmonary

Both oxygenation and ventilation are altered in the elderly in such a way that function and reserve are compromised. Decompensation and failure can occur rapidly and with little warning in the face of injury. Inspiratory and expiratory forces, both functional (vital capacity) and forced (FEV_1), decrease with age as a result of chest wall stiffness and impaired pulmonary elastic recoil. Musculoskeletal changes also contribute to poor chest wall compliance in the form of kyphosis, osteoporosis, calcification of the intercostal cartilages, arthritis of the costovertebral joints, and atrophy of the intercostal muscles. The diaphragm's contribution to respiratory function increases with age. Conditions that limit optimal diaphragm function affect this age group disproportionately. Pulmonary parenchymal tissue properties also change, leading to impaired gas exchange and a lower baseline arterial oxygen partial pressure. Alveolar wall thickening and inelastic small airways predispose to poor diffusion and collapse-induced ventilation-perfusion mismatches. Finally, the compensatory response to both hypoxia and hypercarbia are attenuated, mucocilliary clearance mechanisms are impaired, and narcotic-induced respiratory depression is hypersensitive.

The loss of respiratory reserve that accompanies aging is further complicated by the effects of chronic disease in the geriatric polytrauma patient. Smoking, chronic obstructive pulmonary disease, emphysema, and chronic bronchitis often conspire with age-related changes to undermine respiratory function, even with minor injury. During the primary survey, careful monitoring (continuous pulse oximetry and respiratory rate assessment) and early supplemental oxygen are mandatory; an arterial blood gas analysis is important. Although oxygen administration may depress a "hypoxemic drive" in patients whose ventilation depends in part on a relatively low blood oxygen saturation level, hypercarbia in the acute trauma situation is an acceptable risk in order to maximize oxygenation delivery. Mechanical ventilator support for early, rapid decompensation, persistent hypoxemia, or ventilatory failure should not be delayed. Simple pneumothorax and hemothorax are poorly tolerated and should be diagnosed and treated early. Multiple rib fractures and pulmonary contusions are associated with a high complication rates and require adequate pain control to mobilize secretions and optimize respiratory mechanics. Regional analgesia supplied by thoracic epidural, thoracic paravertebral, or intercostal blocks, supplemented with nonsteroidal anti-inflammatory drugs and low-dose opioids, is superior to systemic analgesia alone [7].

16.2.3 Neurologic

The two major age-related central nervous system changes are cortical atrophy and decreased cerebral blood flow. Both of these processes increase the risk of traumatic injury in the elderly individual. Age-related cortical atrophy can begin as early as age 60 and is ubiquitous to a varying degree above age 80. From a purely mechanical perspective, cortical atrophy renders the geriatric patient vulnerable to subdural hemorrhage, as the shrunken brain, prone to more movement within the calvarium, stretches parasagittal veins, priming them for rupture upon impact or rotation. A progressive, low-level decrease in cerebral blood flow can impair cortical oxygen delivery and consumption, leading to sub-clinical changes in mental status, visual and auditory function, tactile and proprioception sensation, orthostatic tolerance, and reflex time. Add to these normal age-related changes, which account for the relatively high incidence of falls in this age group, the increasingly frequent use of anticoagulants and platelet inhibitors as part of the routine management of chronic conditions in this age group, which exacerbate the traumatic brain injury seen in the elderly [8].

Extradural hemorrhage is more common in the elderly polytrauma patient, and due to the accommodating pericortical space afforded by atrophy, mass effect symptoms may be absent or delayed. The expeditious and liberal use of computed tomography (CT) is encouraged in this age group and should be obtained as early in the evaluation as is safe and reasonable. Neurological assessment beyond the Glasgow Coma Scale is important but can be limited by preexisting cognitive impairment or the sequelae of previous cerebral vascular events. Injuries to the bony spine can be difficult to diagnosis on physical exam or plain radiographs, particularly in the face of osteoporosis or osteoarthritis. CT has become the standard radiologic modality for assessing the spinal column for fracture. Incomplete cord injury (central or anterior cord syndrome) is also relatively more common in this age group and is often associated with preexisting canal stenosis. They can occur with even mild cervical hyperextension, typically after a fall or motor vehicle crash. These injuries usually require MRI for diagnosis. A thorough history from family or caregivers is important. Cognitive dysfunction and dementia are not only common but predictive of outcome.

16.2.4 Renal

Kidney function, particularly as measured by the glomerular filtration rate (GFR), decreases with age while renal reserve function is even more markedly impaired. This occurs primarily as nephrons become sclerotic, tubules lengthen and become fibrotic, and blood flow attenuates. The most common laboratory marker for renal function, serum creatinine may remain in the normal range despite a significant decrease in GFR, as protein production drops off secondary to lost muscle mass. It is important to keep in mind that the elderly patient tolerates both dehydration and volume overload poorly. Electrolyte and acid-base regulation are also at risk. Acute renal failure in the face of polytrauma is known to be associated with increased mortality, particularly in the elderly. Fluid and electrolyte balance should be carefully monitored, and exposure to nephrotoxic drugs should be minimized. Oliguria requires prompt attention and aggressive treatment. Creatinine clearance should be used when dosing medications that undergo renal elimination.

16.3 Optimizing Perioperative Management

Fractures are the most common injury in the elderly, and the majority require operative fixation for optimal outcome. Nearly one third of patients aged 75 and older who suffer a ground-level fall will sustain a fracture [9]. In the United States, according to The National Center for Health Statistics, 5/1000 people over the age of 65 required open reduction with internal fixation of a fracture in 2010, more than double the rate for all those under the age of 65 [10]. In the elderly polytraumatized patient, important preexisting conditions that require special consideration during the perioperative period include comorbidity, medications, nutritional status, cognitive impairment, and functional capacity.

16.3.1 Comorbidity

Two-thirds of older individuals have multiple chronic conditions. As a contributor to death, heart disease and cancer top the list, followed by chronic lower respiratory disease, stroke, Alzheimer's disease, and diabetes [11]. Obtaining an accurate and complete medical history is important and should not be deferred or omitted when assessing the elderly polytrauma patient. If the patient cannot supply the information, it should be sought from a family member, care-provider, or the medical record. The American Society of Anesthesiologists (ASA) physical status classification, widely used to risk-stratify operative patients, has been shown to accurately reflect severity of preexisting illness and has even been validated as an independent predictor of postoperative morbidity and mortality in older surgical patients [12].

16.3.2 Medications and Supplements

Over 80 % of elderly adults take at least one medication and one third take at least five. Nearly half of all older individuals use at least one over the counter medication and the same amount use some sort of dietary supplement [13]. Polypharmacy, either in the form of an excessive absolute number of drugs taken, use of medications without appropriate indication, or the use of duplicate medications, is a problem among the elderly. A thorough review of the polytrauma patient's current medication list is essential and unnecessary medications should be discontinued in the perioperative or postinjury period. Keep in mind that the abrupt withdrawal of serotonin reuptake inhibitors (SSRIs), beta-blockers, clonidine, statins, and corticosteroids can lead to significant withdrawal symptoms and complications. ACE inhibitors and angiotensin receptor blockers should be continued.

Dietary supplements are no longer simple vitamins and minerals. Today's supplements include herbals, amino acids, enzymes, botanicals, and animal extracts. Unless prompted, many patients who use supplements fail to report taking them, as there is a perception that since they are "natural," they are inherently safe. Many of the more commonly used herbal supplements can complicate perioperative care through interactions with anesthetic agents (Valerian, Kava), inhibition of platelet function (ginseng, garlic, ginkgo biloba), alterations in the catecholamine response (Ephedra), suppression of immune function (Echinacea), and interaction with important drug classes (St John's wort and CYP450 enzymes, Ephedra and MAO inhibitors) [14]. The pharmacokinetic and pharmacodynamics properties of these alternative medicines are not well established. Consequently, they should be discontinued at least 1 week before elective surgery. If surgery cannot wait, as is often the case in the polytraumatized patient with fractures, knowledge of their use and appreciation of associated side effects is important.

Recent enthusiasm for the use of perioperative beta-blockade has been tempered by the results of the POISE trial, which demonstrated that despite a significant reduction in cardiac events and myocardial infarction, an increase in both stroke and death in the patients treated with metoprolol was observed [15]. According to the American College of Cardiology Foundation/American Heart Association Task Force, evidence-based class I recommendations for perioperative beta-blocker therapy are limited to continuation in patients undergoing surgery who are already receiving beta blockers for approved indications. Evidence-based class II recommendations are limited to patients who are at high cardiac risk from coronary artery disease and are undergoing vascular surgery. They should have beta-blockers titrated to heart rate and blood pressure [16]. Excepting these recommendations, nine credible randomized trials indicate that the routine use of beta-blockade for perioperative protection increases mortality [17].

The Beers criteria for potentially inappropriate medication use in adults 65 and older was revised in 2003 by a consensus panel of experts and identified 48 individual medications or classes of medications to avoid in older individuals, many of which are felt to have adverse outcomes of high severity [18]. Of importance to the perioperative polytrauma patients, these include long-acting benzodiazepines, anticholinergics, antihistamines, Diphenhydramine, Digoxin, barbiturates, Meperidine, Ketorolac, and Amioderone.

16.3.3 Nutrition

Prevalent among the elderly, malnutrition will impact recovery from injury and surgery profoundly. Nearly one quarter of all individuals over the age of 65 are undernourished, ranging from less than ten percent of community-dwelling individuals to over half of those in rehabilitation centers. Another 46 % are at risk [19]. A serum albumin of less than 3.5 g/deciliter accurately reflects nutritional status and should be a part of the routine laboratory evaluation in elderly polytrauma patients. In 1999 the Geriatric Hip Fracture Research Group at the Hospital for Joint Diseases in New York described that patients with abnormal albumin levels were 2.9 times more likely to have a length of stay greater than 2 weeks, 3.9 times more likely to die within 1 year of surgery, and 4.6 times less likely to recover their prefracture level of independence [20]. Also, a nation-wide Veteran's Administration Surgical Risk Study published in 1999 established a strong association between decreasing serum albumin levels and increasing mortality and morbidity in patients undergoing major noncardiac operations [21]. Other important indicators of severe malnutrition include a body mass index less than 20 kg/m^2, an unintentional weight loss of more than ten percent in 6 months, and a total lymphocyte count of less than 1500 cells/ml. Although the urgent nature of an acute fracture in a polytrauma patient does not allow for the preoperative optimization of a patient's nutritional status, intervention with nutritional supplementation containing calorie-rich and protein-rich preparations in the postoperative period may help lower the rates of infection and pressure ulcer formation, improve wound healing, shorten hospitalization, and improve mortality rates. Serum prealbumin levels biweekly should be obtained to monitor catabolic status.

16.3.4 Cognitive Impairment

Cognitive dysfunction disorders in the elderly, although prevalent, are not well understood. Included are simple cognitive impairment, dementia, Alzheimer's disease, cerebrovascular disease, stroke, and idiopathic medical conditions. Who is at risk, how quickly a condition progresses, and how different diagnoses interact or are defined, are all areas of active investigation. Cognitive impairment without dementia is estimated to be present in up to one in five septuagenarians, and the progression to clinical dementia in these individuals is roughly 10 % per year. Comprehensive assessment tools such as the Mini-Mental State Examination, Memory Impairment Screen, and Cognitive Abilities Screening Instrument are available and have been validated in the literature. Their use is increasing among primary care physicians and geriatricians, and their results can supply a valuable baseline objective assessment of cognitive functioning in an elderly individual who is not known to the acute care practitioner or trauma physician. The Mini-Cog is a brief screen that employs three-item recall and a clock drawing task. It has sensitivity and specificity of 99 and 96 %, respectively, for dementia; can be administered rapidly; and strikes the best balance between accuracy and ease of administration in the acute setting [22].

Dementia, as a preexisting condition, is a strong predictor for developing delirium, an acute confusion state, in polytrauma patients and postoperative elderly patients, particularly those being cared for in the intensive care unit. Delirium is associated with longer hospital stays, functional decline, and higher mortality rates [23]. Prevention, detection, and treatment of delirium are important goals in the management of the geriatric polytrauma patient. Typically, onset occurs 1–2 days following surgery or injury and can persist for several days. Prevention should begin with a medication assessment and a reduction or elimination of psychoactive drugs whenever possible. Nonpharmacologic approaches to managing sleep, anxiety, and agitation are preferable to medication. Benzodiazepines should be avoided whenever possible, and haloperidol, or olanzapine, should be reserved for those patients with agitation severe enough to risk interruption of essential medical therapies or self-injury, or for those with distressing psychotic symptoms such as hallucinations or delusions. Involving family members in care is crucial, particularly for reorientation and prevention of self-harm. Encourage mobility, ensure that if needed, patients have glasses, hearing aids, and dentures, and communicate regularly with them and their families [24].

16.3.5 Functional Capacity

Functional status in the geriatric population is most often described by an individual's ability to perform activities of daily living (ADL), which typically include bathing, dressing, transferring, feeding, continence, and toileting. An easy screening method, with utility in the acute care situation, involves asking four simple questions: "Can you get out of bed yourself?" "Can you dress yourself?" "Can you make your own meals?" and "Can you do your own shopping?" [25] An appreciation for a patient's preexisting functional status is important as poor status and impaired mobility are associated with postoperative delirium, discharge institutionalization, mortality, and even surgical site infections due to methicillin-resistant *Staphylococcus aureus* [26]. Lawrence and colleagues from the Veterans Evidence-based Research Dissemination and Implementation Center followed an elderly cohort of patients for 6 months following abdominal surgery and concluded that better preoperative physical performance status independently predicted better recovery and shorter time to recovery across multiple measures including ADL and Mini-Mental State Exam [27].

16.4 Frailty

Geriatricians have long recognized frailty as an entity distinct from chronological age, comorbidity, or disability. Although not well defined, frailty has been characterized as "a biologic syndrome of decreased reserve and resistance to stressors, resulting from cumulative declines across multiple physiologic systems, and causing vulnerability to adverse outcomes" [28]. In 2001, Linda Fried and colleagues from the Cardiovascular Health Study Collaborative Research Group operationalized and validated a phenotype that included the following five criteria: unintentional weight loss (10 lbs in past year), self-reported exhaustion, weakness (grip strength), slow walking speed, and low physical activity. Using data from community-dwelling older adults that participated in the Cardiovascular Health Study, they

Table 16.1 Frailty phenotype [30]

Frailty criteria	Measurement
Shrinkage	Loss of 10 or more pounds in the last year
Weakness	Decreased grip strength (<20 % by gender and BMI)
Exhaustion	Self reported "exhaustion," poor energy or endurance
Slowness	Slow walking (lowest 20 % by age and gender)
Low activity	Low weekly energy expenditure: lowest 20 % Males: <383 kcals/week Females: <270 kcals/week

1 point for each criterion met
0–1 = not frail
2–3 = intermediate frail or prefrail
4–5 = frail

defined frailty as a clinical syndrome in which three or more criteria were present. Frailty, as described in this way, was found to be an independent predictor of incident falls, hospitalization, disability, and death within 3 years. Intermediate status, in which one or two of the criteria were present, placed individuals at significant risk of progression to frailty over 3–4 years. Woods et al. corroborated these findings in 2005 using data from 40,657 women aged 65–70 who participated in the Women's Health Initiative Observational Study [29]. In 2010, Martin Makary and his surgical colleagues prospectively measured frailty using the Fried criteria (also referred to as the Hopkins Frailty Score) in 594 patients, age 65 and older, undergoing elective surgery, and found that frailty independently predicted postoperative complications, length of stay, and discharge to a skilled or assisted-living facility [30] (Table 16.1).

Other researchers have attempted to quantify the presences of frailty by measuring abnormalities across the various elements of the Comprehensive Geriatric Assessment, a nonstandardized, multidisciplinary evaluation introduced in 1987 by the American Geriatric Society. Relevant domains include cognition, function, nutrition, chronic disease burden, and geriatric syndromes such as frequent falls. This method of defining frailty has come to be known as the accumulation of deficits model and has also

Table 16.2 Modified Frailty Index criteria [32]

History of diabetes mellitus	Functional status 2 (not independent in 30 days prior to surgery)	History of COPD or pneumonia
History of congestive heart failure	History of hypertension requiring medication	History of transient ischemic attack or stroke
History of myocardial infarction	History of peripheral vascular disease or rest pain	History of Stroke with neurologic deficit
History of prior cardiac surgery or percutaneous intervention	History of impaired sensorium	

been studied and validated as a predictor of outcome in elderly surgical patients. Robinson et al. described a simple frailty score that employs seven characteristics and takes an average of 5 min per patient to collect; when used within 30 days of elective operation (colorectal or cardiac), a score of four or more was found to be associated with increased postoperative complications [31]. Perhaps a more useful measure in the acute care setting is the Modified Frailty Index. (Table 16.2) This 11-item survey can be obtained mostly by history and is scored as a ratio; one point is given for each feature, and the total is divided by the number of variables for which the patient has data, resulting in an ordinal variable with stepwise increases from zero to one. In a retrospective analysis of emergency general surgery cases in patients over the age of 60 using the National Surgical Quality Improvement Program database, the Modified Frailty Index showed a strong correlation with infection complications and mortality and was 11 times more predictive of death than age alone [32].

16.5 Reversal of Therapeutic Anticoagulation

Oral anticoagulation for stroke prevention in patients with chronic cardiac arrhythmias is well established, and warfarin remains the most common medication in use for this purpose. Newer oral direct thrombin inhibitors such as Dabigatran are approved in the United States, Europe, Australia, and Japan, and their use is growing. Platelet inhibition for a number of cardiovascular and vascular indications is also well established with widespread use of aspirin and Clopidogrel, often in combination. Clearly, polytrauma patients with medication-induced coagulopathy are at an increased risk of hemorrhage, and a large retrospective survey of the California Office of Statewide Planning and Development database over a 14 year period revealed a two fold increase in all-cause mortality following a ground-level fall in elderly patients (age >65) who take oral anticoagulants [33]. Patients with traumatic brain injury are most at risk for the rapid development of life-threatening complications and in the same survey, 31.6 % of those taking oral anticoagulants died with head injury as compared to 23.8 % of patients not anticoagulated. Even patients with injuries limited to fractures, however, can be impacted by delays to surgical reduction, longer hospital stays, and higher rates of disposition to rehabilitation or nursing facilities [34]. Based primarily on work done by Ivascu and colleagues [35, 36], the EAST practice management guidelines committee was able to offer recommendations based on Class III scientific evidence (retrospectively collected data): [37]

1. All elderly patients who were taking medications for systemic anticoagulation before their injury should have appropriate assessment of their coagulation profile as soon as possible after admission.
2. All elderly patients with suspected head injury (e.g., those with altered GCS, headache, nausea, external trauma, or high-energy mechanism) who were taking medications for systemic anticoagulation before their injury should be evaluated with head computed tomography as soon as possible after admission.
3. Patients receiving warfarin with a posttraumatic intracranial hemorrhage should receive initiation of therapy to correct their international normalized ratio (INR) toward a normal range (e.g., <1.6× normal) within 2 h of admission.

Transfusion of thawed fresh-frozen plasma (FFP) and the administration of parenteral vitamin K remain first line therapies for the reversal of the effects of warfarin. The use of FFP in the case of bleeding complications or emergency surgery, however, can be time-consuming and poorly tolerated in elderly patients with limited cardiopulmonary reserve. An appealing alternative is the prothrombin complex concentrate which contains human plasma-derived prothrombin and coagulation factors VII, IX, and X (vitamin K-dependent). These lyophilized products (Kcentra/Beriplex and Octaplex) are standardized to potency, unlike FFP, and can be rapidly administered at low volumes. Several other plasma-derived coagulation factor concentrates exist (e.g., fibrinogen, antithrombin), and their use, particularly when combined with the rapid diagnostic accuracy of rotational thromboelastometry point-of-care testing, allows a more specific targeting of coagulation deficiencies. As a result, newer treatment algorithms are emerging that could considerably reduce the use of allogeneic blood products [38].

16.6 Advanced Directives

Nearly one third of individuals over the age of 65 will undergo an inpatient surgical procedure during the year before their death [39]. Many of these procedures fall within the scope of acute care surgery, polytrauma, or fracture fixation. Increasingly, these patients present to the hospital with advanced directives, such as a Living Will or Healthcare Power of Attorney, that include "Do Not Resuscitate" (DNR) orders. According to data from the Health and Retirement Study, over two-thirds of people aged 60 and above had an advanced directive at the time of their death in 2010, an increase of more than 50 % over the previous decade [40]. In the emergency care setting, interpreting advanced directives and DNR orders can be complicated. Although they may be specific for listing limitations (e.g., no cardiac compressions, endotracheal intubation, advanced airway management, or defibrillation), they are often vague as to application (e.g., in the event of a terminal illness, or no reasonable hope of functional recovery). The administration of anesthesia alone will expose patients to practices and procedures that might be viewed as "resuscitation." Additionally, operative interventions and anesthesia may subject patients to new and potentially reversible risks of cardiopulmonary arrest. Policies that automatically suspend DNR orders or other treatment limitations prior to surgery are no longer recommended by the American College of Surgeons or American Society of Anesthesiologists [41, 42]. Rather, a policy of "required reconsideration" is felt to be more appropriate. This process should involve a candid discussion with the patient or the patient's representative that outlines intraoperative and perioperative risks associated with the surgical procedure as well as an approach for potentially life-threatening problems. Any clarifications of or modifications to the patient's directives should be documented in the medical record and communicated to other members of the health care team.

References

1. http://www.nia.nih.gov/health/publicaztion/why-population-aging-matters-global-perspective/tren-1-aging-population. Page last updated 7 Oct 2011.
2. National Highway Traffic Safety Administration. Traffic Safety Facts 2012 data: older population. Available at: http://www-nrd.nhtsa.dot.gov/Pubs/812005.pdf. Mar 2014.
3. Samaras N, Chevalley T, Samaras D, et al. Older patients in the emergency department: a review. Ann Emerg Med. 2010;56:261–9.
4. Champion HR, Copes WS, Buyer D, et al. Major trauma in geriatric patients. Am J Public Health. 1989;79:1278–82.
5. Calland JF, Ingraham AM, Martin N, et al. Evaluation and management of geritrictrauma: an Eastern Association for the Surgery of Trauma practice management guideline. J Trauma Acute Care Surg. 2012; 73:S345–50.
6. Eastridge BJ, Salinas J, McManus JG. Hypotension begins at 110: redifining "hypotension" with data. J Trauma Acute Care Surg. 2007;63:291–9.
7. Ho AM, Karmaker MK, Critchley LAH. Acute pain management of patients with multiple fractured ribs: a focus on regional techniques. Curr Opin Crit Care. 2011;17:323–7.
8. Thompson HJ, McCormick WC, Kagan SH. Traumatic brain injury in older adults: epidemiology, outcomes,

8. and future implications. J Am Geriatr Soc. 2006;54:1590–5.
9. Pfortmueller CA, Kunz M, Lindner G, et al. Fall-related emergency department admission: fall environment and settings and related injury patterns in 6357 patients with special emphasis on the elderly. ScientificWorldJournal. 2014;2014:6 Article ID 256519.
10. Centers of Disease Control. CDC/NCHS national hospital discharge survey. 2010. Available at: http://www.cdc.gov/nchs/data/nhds/4procedures/2010pro4_procedurecategoryage.pdf.
11. Centers for Disease Control and Prevention. The state of aging & health in America 2013. Atlanta: Centers for Disease Control and Prevent, US Department of Health and Human Services; 2013. Available at: http://www.cdc.gov/aging/pdf/state-aging-health-in-america-2013.pdf.
12. Bo M, Cacello E, Ghiggia F, et al. Predictive factors of clinical outcome in older surgical patients. Arch Gerontol Geriatr. 2007;44:215–24.
13. Qato DM, Alexander GC, Conti RM, et al. Use of prescription and over-the-counter medications and dietary supplements among older adults in the United States. JAMA. 2008;300:2867–78.
14. Rispler DT, Sara J. The impact of complementary and alternative treatment modalities on the care of orthopaedic patients. J Am Acad Orthop Surg. 2011;19:634–43.
15. POISE Study Group. Effects of extended-release metoprolol succinate in patients undergoing non-cardiac surgery (POISE trial): a randomized controlled trial. Lancet. 2008;371:1839–47.
16. Fleischmann KE, Beckman JA, Buller CE, et al. 2009 ACCF/AHA focused update on perioperative beta blockade. Circulation. 2009;120:2123–51.
17. Nowbar AN, Cole GD, Shun-Shin MJ, et al. International RGT-based guidelines for use of perioperative stress testing and perioperative beta-blockers and statins in non-cardiac surgery. Int J Cardiol. 2014;172:138–43.
18. Fick DM, Cooper JW, Wade WE, et al. Updating the beers criteria for potentially inappropriate medication use in older adults. Arch Intern Med. 2003;163:2716–24.
19. Kaiser MJ, Bauer JM, Ramsch C, et al. Frequency of malnutrition in older adults: a multinational perspective using the mini nutritional assessment. J Am Geriatr Soc. 2010;58:1734–8.
20. Koval KJ, Maurer SG, Su ET, et al. The effects of nutritional status on outcome after hip fracture. J Orthop Trauma. 1999;13:164–9.
21. Gibbs J, Cull W, Henderson W, et al. Preoperative serum albumin level as a predictor of operative mortality: results from the National VA Surgical Risk Study. Arch Surg. 1999;134:36–42.
22. Borson S, Scanlan J, Brush M, et al. The mini-cog: a cognitive 'vital signs' measure for dementia screening in multi-lingual elderly. Int J Geriatr Psychiatry. 2000;15:1021–7.
23. Robinson TN, Raeburn CD, Tran ZV, et al. Postoperative delirium in the elderly: risk factors and outcomes. Ann Surg. 2009;249:173–8.
24. Inouye SK, Westendrop RGJ, Saczynski JS. Delirium in elderly people. Lancet. 2014;383:911–22.
25. Lachs MS, Feinstein AR, Cooney LM, et al. A simple procedure for general screening for functional disability in elderly patients. Ann Intern Med. 1990;112:699–706.
26. Chen TY, Anderson DJ, Chopra T, et al. Poor functional status is an independent predictor of surgical site infections due to methicillin-resistant Staphylococcus aureus in older adults. J Am Geriatr Soc. 2010;58:527–32.
27. Lawrence VA, Hazuda HP, Cornell JE, et al. Functional independence after major abdominal surgery in the elderly. J Am Coll Surg. 2004;199:762–72.
28. Fried LP, Tangen CM, Walston J, et al. Frailty in older adults: evidence for a phenotype. J Gerontol A Biol Sci Med Sci. 2001;56:M146–56.
29. Woods NF, LaCroix AZ, Gray SL, et al. Frailty: emergence and consequences in women aged 65 and older in the women's health initiative observational study. J Am Geriatr Soc. 2005;53:1321–30.
30. Makary MA, Segev DL, Pronovost PJ, et al. Frailty as a predictor of surgical outcomes in older patients. J Am Coll Surg. 2010;210:901–8.
31. Robinson TN, Wu DS, Pointer L, et al. Simple frailty score predicts postoperative complications across surgical specialties. Am J Surg. 2013;206:544–50.
32. Farhat JS, Velanovich V, Falvo AJ, et al. Are the frail destined to fail: frailty index as predictor of surgical morbidity and mortality in the elderly. J Trauma Acute Care Surg. 2012;72:1526–31.
33. Inui TS, Parina R, Chang DC, et al. Mortality after ground-level fall in the elderly patient taking oral anticoagulation for atrial fibrillation/flutter: a long-term analysis of risk versus benefit. J Trauma Acute Care Surg. 2014;76:642–50.
34. Kirsch MJ, Vrabec GA, Marley RA, et al. Preinjury warfarin and geriatric orthopedic trauma patients: a case-matched study. J Trauma. 2004;57:1230–3.
35. Ivascu FA, Howells GA, Junn FS, et al. Rapid warfarin reversal in anticoagulated patients with traumatic intracranial hemorrhage reduces hemorrhage progression and mortality. J Trauma. 2005;59:1131–9.
36. Ivascu FA, Janczyk RJ, Junn FS, et al. Treatment of trauma patients with intracranial hemorrhage on preinjury warfarin. J Trauma. 2006;61:318–21.
37. Calland JF, Ingraham AM, Martin N, Marshall GT, Schulman CI, Stapleton T, et al. Geriatric trauma, evaluation and management of. Available at: https://www.east.org/resources/treatment-guidelines/geriatric-trauma,-evaluation-and-management-of. Accessed 30 Sep 2014.
38. Tanaka KA, Esper S, Bolliger D. Peroperative factor concentrate therapy. Br J Anaesth. 2013;111:i35–49.
39. Kwok AC, Semel ME, Lipsitz SR, et al. The intensity and variation of surgical care at the end of life: a retrospective cohort study. Lancet. 2011;378:1408–13.

40. Silveira MJ, Wiitala W, Piette J. Advance directive completion by elderly Americans: a decade of change. J Am Geriatr Soc. 2014;62:706–10.
41. American College of Surgeons. Statement on advance directives by patients: "do not resuscitate" in the Operating Room (2014). Available at: http://www.facs.org/fellows_info/statements/st-19.html. Accessed 30 Sept 2014.
42. American Society of Anesthesiologists. Ethical guidelines for the anesthesia care of patients with do-not-resuscitate orders or other directives that limit treatment (2013). Available at: https://www.asahq.org/For-Members/Standards-Guidelines-and-Statements.aspx. Accessed 30 Sept 2014.

Polytrauma in Young Children

17

Achim Braunbeck and Ingo Marzi

Contents

17.1	Introduction	231
17.2	Definition and Epidemiology	231
17.3	Anatomic and Physiologic Specifics in Children and Adolescents	232
17.4	Patterns of Injury	232
17.5	Scoring	233
17.6	First-Line Treatment in Polytraumatized Patients	233
17.7	Specific Injuries	235
17.7.1	Head Trauma	235
17.7.2	Thoracic Trauma	236
17.7.3	Abdominal Trauma	236
17.7.4	Bones	237
References		243

A. Braunbeck, MD (✉) • I. Marzi
Department of Trauma Surgery,
Johann Wolfgang Goethe-University,
Theodor-Stern-Kai 7, 60590 Frankfurt, Germany
e-mail: achim.braunbeck@kgu.de

17.1 Introduction

The treatment of severely injured, polytraumatized children is not only, due to its social context, one of the most feared challenges in medicine and recent data showed that even specialized trauma centers only take care of up to 1–2 polytraumatized children per month [1], making it difficult to gain profound experience with the various different physiological and anatomical conditions and reactions of children to polytraumat. A lack of large, evidence-based studies goes along with it. But not only traumatic injuries remain the main causes of death and impairment in children above 1 year [2], data further suggests that up to 30 % of the early deaths in polytraumatized children would have been avoidable with adequate trauma therapy [3].

17.2 Definition and Epidemiology

Similar to adults, polytrauma in children is defined as a combination of injuries involving two or more organ systems as a result of a single incident that accounts for life-threatening condition [4].

The incidence of pediatric patients in emergency services in general is estimated about 5–10 % of all patients admitted to an emergency department [5]. Similarly, the incidence of polytraumatized children is estimated to be 360 per

100,000, which accounts for 6 % of all polytrauma patients. The overall mortality is about 12 % [6].

According to the German Trauma Register, a nationwide database established in 1993 and collecting data from multiply injured trauma patients, between January 1993 and December 2007, 1333 children below 14 suffered from an Injury Severity Score (ISS) >9. Age breakdown was as follows:

Up to 1 year ~ 9 %, 1–5 years ~ 23 %, 6–9 years ~ 30 %, 10–14 years ~ 38 % [7].

17.3 Anatomic and Physiologic Specifics in Children and Adolescents

Children are not small adults, especially not in traumatology. Starting with a lower body mass, which has to absorb the energy of the trauma, a thinner soft tissue cover and exposed organs predispose children to severe injuries. Beneath the specific anatomically related injury preconditions that will be more deeply addressed in the sections below, general physiological preconditions and changes during aging have to be known. Blood pressure, heart rate, and breathing frequency are just some of the basic changes between adults and children (Table 17.1) [8]. In relation, children exhibit a larger body surface compared to their total body volume. That puts especially small children in danger for hypothermia. Although the relative blood loss is tolerated better by children than adults, the total amount can easily be underestimated and deterioration of the child can happen very fast when critical borders are crossed (Table 17.2). A physiological blood volume of 80 ml/kg is usually calculated, more than 40 % blood loss will trigger severe hemorrhagic shock. Consequently, data from the German Trauma Register shows that polytraumatized children also need higher rates of basic CPR than adults with comparable injuries (16.2 vs- 3.1 %) [9]. Therefore, focused and continuous careful reassessment is essential.

17.4 Patterns of Injury

In young children, fall from a high place with resulting head and neck injuries leads statistics. The physiological conditions, especially the disproportionately large head and weak neck musculature in children less than 3 years of age (Fig. 17.1), puts them at a high risk for brain and neck injuries, even at low velocities [10].

With growing age, traffic accidents account for the majority of trauma and incidence of abdominal, thoracic, and extremity injuries gain numbers. Thereby, trauma by bicycle, pedestrian, and motor vehicle accidents each account a third of trauma cause, showing a higher proportion of pedestrians and bicyclists among children compared to adults.

Common trauma-injury conditions like the dash board injury in adults are the combination of lumbar spine fractures with abdominal lesions of liver or small bowels due to an insufficient safety belt [6] or the combination of thoracic injury and femur fracture on one side and head injuries on the opposite side due to hit facing the car (so-called Waddell's triad).

Table 17.2

Age (years)	Blood loss (ml)	% total blood volume
4	500	40
8	500	25
Adult	500	10

Table 17.1

Age group	Weight range (kg)	Heart rate (beats/min)	Blood pressure (mmHg)	Respiratory rate (breaths/min)	Urinary output (ml/kg/h)
0–12 months	3.5–10	<160	>60	<60	2.0
1–2 years	10–14	<150	>70	<40	1.5
3–5 years	14–18	<140	>75	<35	1.0
6–12 years	18–36	<120	>80	<30	1.0
13–18 years	36–70	<100	>90	<30	0.5–1.0

Modified from ATLS [8]

17 Polytrauma in Young Children

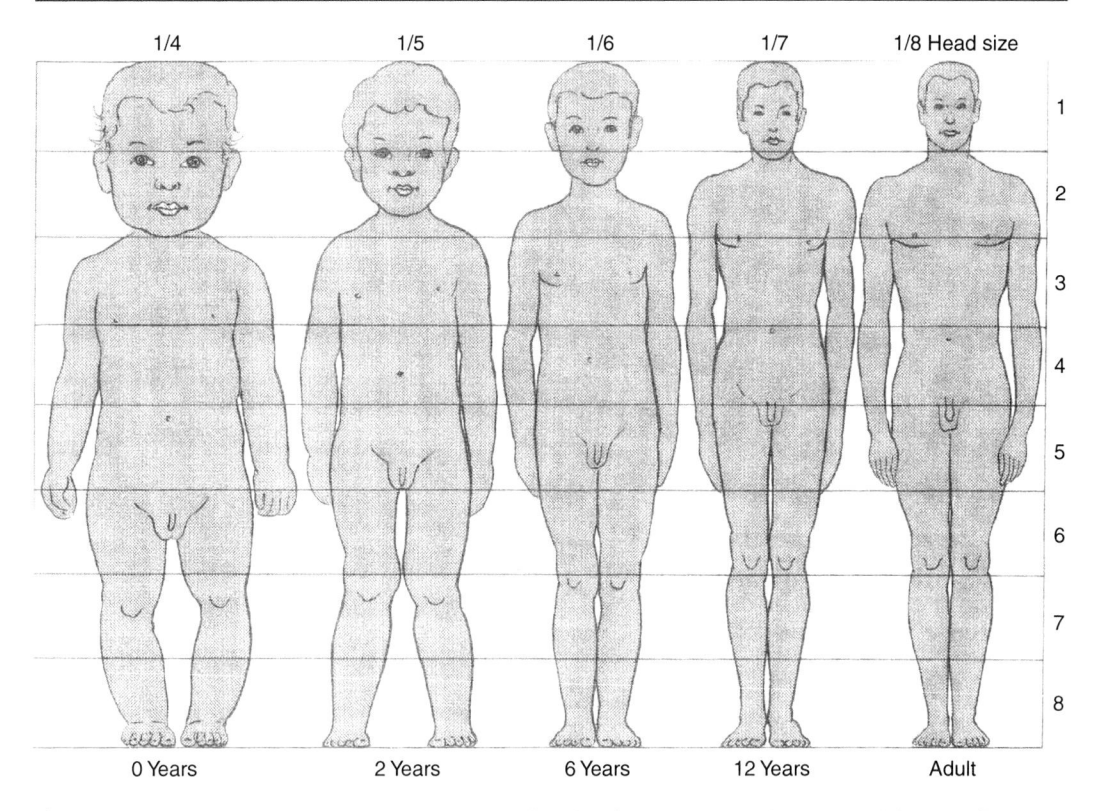

1/4	1/5	1/6	1/7	1/8 Head size
0 Years	2 Years	6 Years	12 Years	Adult

Fig. 17.1 Relationship between head, trunk, and extremities in children vs. adults (Bernbeck and Dahmen [31])

17.5 Scoring

Although the Injury Severity Score (ISS) is usable for children, other scores like the PTS, the Pediatric Trauma Score, have been developed to provide a more specific tool. The PTS consists of a combination of anatomical and physiological values and can reach scores between −6 and +12. It has been shown that the PTS correlates with overall survival rates in polytraumatized children (Table 17.3) [11].

17.6 First-Line Treatment in Polytraumatized Patients

Training as well as strategic and attentive treatment algorithms are the answer to optimal first-line care in polytraumatized children [1].

Under optimal circumstances, an experienced trauma team, consisting of the basic disciplines of anesthesia, radiology, and trauma surgery, should be supported by a pediatric competence, ideally a pediatric surgeon, pediatric intensive care provider, or abdominal surgeon with pediatric experience. Similar to adult trauma therapy, additional disciplines like thoracic surgery, neuro surgery, etc., may be useful if available. Besides this ideal setting, in prehospital settings or small trauma centers, there might be initially only one medical professional. In any way, the major goal of the first-line treatment consists of three parts:

- Protection of vital functions
- Rapid diagnostic
- Initiation of differentiated therapy

To reach these goals, working according to common and daily used algorithms is useful. The ATLS (advanced trauma life support), well established in adult polytrauma treatment, applies to most situations and is often recommended. The primary survey follows the maxim "treat first

Table 17.3

I Scoring
A. Weight
  1. Weight > 20 kg: score +2
  2. Weight 10–20 kg: score +1
  3. Weight < 10 kg: score −1
B. Airway
  1. Normal airway: score +2
  2. Maintained airway: score +1
  3. Invasive airway (e.g., intubated): −1
C. Systolic blood pressure
  1. SBP > 90 mmHg: score +2
  2. SBP 50–90 mmHg: score +1
  3. SBP < 50 mmHg: score −1
D. Central nervous system
  1. Awake: score +2
  2. Obtunded: score +1
  3. Coma: score −1
E. Open wound
  1. No open wound: score +2
  2. Minor open wound: score +1
  3. Major open wound: score −1
F. Skeletal trauma
  1. No skeletal trauma: score +2
  2. Closed fracture: score +1
  3. Open fracture or multiple fractures: score −1

II. Interpretation
A. Score range: +12 to −6
B. Trauma score ≤8 indicates significant mortality risk

III. References
A. Tapas (1987) J Pediatr Surg 22:14

what kills first" and consists of an easy A-B-C-D-E pattern.

A. Airway: To maintain or reestablish supplementation of oxygen, neck protection with a rigid cervical collar is necessary until definitive exclusion of instability. As young children often have a prominent occiput, a pad placed under the thoracic spine should be used to provide neutral alignment of the spine.

Oxygen is delivered in all cases initially with high flow rates via mask. Foreign bodies, occluding the upper airways, have to be removed. If internal airways have collapsed, naso-pharyngeal or oro-pharyngeal tubes are available and easy to insert (Guedel or Wendel tube). If the neurological status puts the patient at risk for aspiration, endotracheal intubation should be performed. The anatomical preconditions, like a bigger tongue or cephaled larynx, have to be anticipated, as well as vagal reflexes due to pharyngeal stimulation like bradycardia or hypersecretion.

B. Breathing: Hypoxia is the major cause of cardiac arrest in children. Thereby, due to the increased elasticity of the pediatric thorax, also without clinical signs of thorax trauma, severe lung contusion can exist. Sufficient ventilation, also with high PEEP (15 mmHg), can be necessary, for example, in drowning accidents. In case of asymmetric thorax movement, thoracic emphysema, missing breathing murmurs, or other signs of a hemato-/pneumothorax, a chest tube must be inserted to evacuate hematoma and/or air.

C. Circulation: It is essential to know age-related heart frequencies and blood pressure to validly assess pediatric status. In hemodynamic shock and centralization, even for experienced pediatrics or emergency doctors it can be difficult and time consuming to establish sufficient peripheral vein access (that means 2 big and working intravenous accesses). In these cases, it is essential not to waste time (especially not in a prehospital setting with a central line), but to establish temporary intraosseous access through the proximal tibia. This can be easily done by modern drilling machine systems. If the proximal tibia is fractured on both sides, other body parts like the malleolus or proximal humerus may be chosen. The administered volume must be carefully reevaluated especially in young children, as hypervolemia can easily be reached. A bolus of 20 ml/kg saline is recommended.

D. The assessment of neurological status must be related to the age. In all case, the (p)GCS (Pediatric Glasgow Coma Scale) should be assessed. Neurological impairments due to injuries of the extremities or spine have to be assessed according to the overall situation and age-related compliance of the child.

E. Exposure: Hypothermia must be avoided and addressed as soon as possible; the younger the child, the more it is important. Due to the proportional larger surface compared to their body volume, loss of warmth can be enormous, especially when loss of warm body liquids accompany.

First-line treatment in the emergency room is completed by a body-to-toe examination and ultrasound examination of the abdomen. Native x-rays of the thorax and pelvis are taken synchronous to the clinical evaluation. In some cases, the diagnostic via MSCT (Multi Slice Computer Tomography) can be avoided, as sensitivity of repeated ultrasound examinations in combination with clinical assessment can reach sensitivity for abdominal injuries of up to 100 %. In most cases, a CT scan must be performed, to avoid missing life-threatening injuries in difficult accessible regions and incompliant patients. In this respect, it is of utmost importance that a dedicated CT protocol is performed with lower dosis and protection of the eyes.

The combination of volume refractory hypovolemia and abdominal or thoracic stab injury can lead to direct operative treatment without further diagnostic, especially if sonography detects intra-abdominal or thoracic fluids with a severe hemorrhagic shock.

In all other cases, a secondary survey with initiation of initial injury-specific therapy completes the emergency room treatment [8].

management. In all cases of higher trauma or significant deficits of the GCS, a CT scan should be performed. Kupperman et al. showed that the age-related combination of different clinical signs can help to decide whether in some cases a CT scan can be avoided in mild traumatic head injuries [14].

Hypoxia and hypotension exacerbate the direct brain injury and must be avoided at any price. The indication for intubation should be made generously, especially if the neurological state preconditions aspiration (GCS < 8). A GCS < 8 is also often seen as indication for the operative insertion of a ventricular or intraparenchymal pressure monitoring device.

Diffuse traumatic brain injury (TBI) is the most common type of injury and results in a range of injury severity from concussion to diffuse axonal injury (DAI). If CT scan detects a shift of the midline, a hemicraniectomy is recommended, and, as outcome even in severe brain injuries is better in children than in adults, should be considered even in worst injuries.

Large hematoma should be urgently evacuated via trepanation or hemicraniectomy, as both subdural hematoma as well as epidural hematoma have shown to worsen outcome after 3 h (Fig. 17.2). In some cases, also the evacuation of intraparenchymal bleedings can be useful.

17.7 Specific Injuries

17.7.1 Head Trauma

Approximately 80 % of all pediatric polytrauma patients suffer from head injuries, with fall from a height being the main cause of trauma up to the age of 5. Thereby, brain injuries are the leading cause of trauma-related death in children [12]. Due to the disproportionately large head and weak neck musculature, even fall from low height may lead to severe brain injuries. Additional physiological preconditions of the immature brain like higher metabolic rate and vasoreactivity lead to significantly higher rates of posthypoxic edema [13].

Therefore, the fast assessing and addressing of brain injuries is crucial in pediatric polytrauma

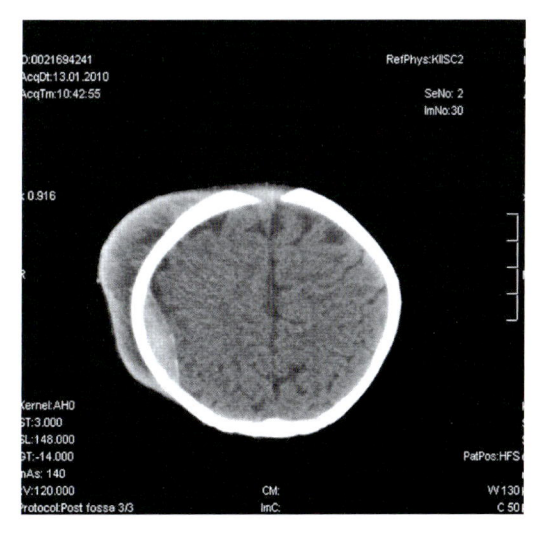

Fig. 17.2 Epidural hematoma, patient 8 months, fall from chair

In all cases, intensive care therapy and intense neuromonitoring is mandatory. Therapy principles are similar and often relay on guidelines for adult head and brain injuries, giving special interest to the cerebral perfusion pressure (CCP). The CCP is the difference between the intracranial pressure (ICP) and the mean arterial pressure (MAP). As the MAP physiologically is much lower in children, the recommended CCP values differ.

Intensive therapy further requires the maintaining of norm glycemia, norm hypoxia, and an elevated position of the head (30°) for optimal venous blood flow. The effects of controlled hypothermia in TBI to reduce ICP and improve outcome are still discussed as side effects like cardiac arrhythmias, impaired immune function, and coagulation cascade seem to eradicate previously reported benefits [13].

17.7.2 Thoracic Trauma

Approximately 25–50 % of all pediatric polytrauma patients suffer from thoracic injuries. Thereby, thoracic trauma worsens outcome and increases risk of death in polytraumatized children. Due to the increased compliance of the children's chest wall, severe lung injuries can occur without signs of external instability in the primary survey [15]. Special interest has to be put on specific signs of lung injuries like cutaneous emphysema, paradoxical breathing, or missing breathing murmurs. Critical edema and relevant pneumothorax should be assessed and addressed before CT scan via insertion of a chest tube by minithoracotomy.

Pericardial effusion and injuries of the central vessels are rare, but may need direct intervention like pericardiocentesis. In cases of instable, volume refractory circulation and sharp chest trauma, emergency (lateral) thoracotomy may be the only choice.

Injuries of the tracheobronchial structures can be treated up to a quarter of the circumference conservative; otherwise, operative reconstruction is necessary.

Lung contusion and ARDS (acute respiratory distress syndrome) with resulting respiratory insufficiency are in focus of further therapy. Due to the trauma mechanism, contusion should also be anticipated without signs in the initial CT scan. Oxygenation indices like the Horrowitz Index (PaO_2/FIO_2) can help to further classify the respiratory insufficiency and control therapy.

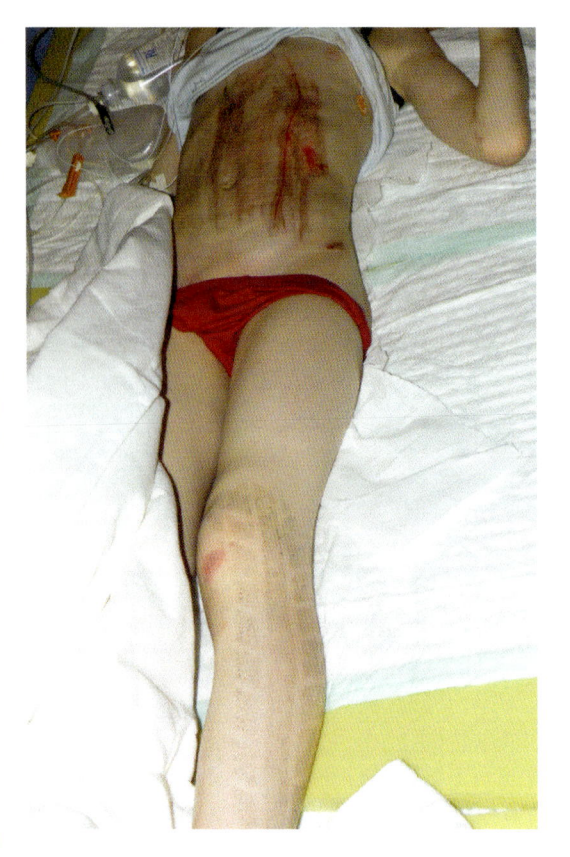

Fig. 17.3 Thoracic and abdominal marks in a patient run over by a tractor

Therapy includes kinematic therapy and different oxygenation protocol, and use of antibiotics in case of postcontusion pneumonia, but also ECMO (extracorporeal membrane oxygenation), may be required in exceptional situations [16, 17].

17.7.3 Abdominal Trauma

Approximately 15–45 % of all pediatric polytrauma patients suffer from abdominal injuries. The in relation larger liver and spleen in combination with a lower diaphragm and less developed abdominal wall musculature preconditions organ lesions, for example, due to an insufficient lumbar safety belt (Fig. 17.3). Different classification systems exist to further specify liver, spleen, and kidney injuries, like the classification of the American

Association for the Surgery of Trauma (AAST). Most of the organ lesions today are treated conservative [18], but close monitoring is mandatory. In case of deterioration, a secondary organ rupture has to be considered [19]. Although it could be shown that the risk of death from intra-abdominal bleeding by a spleen rupture is considerably low, hematothorax, surprisingly much more significant than hematoperitoneum, had a 45 times higher risk of death [20].

In cases of hemodynamic instability and need for surgical bleeding control, operation is required. Depending of the overall situation of the patient, like in volume refractory thoracic sharp trauma, even emergency laparotomy without further diagnostic evaluation can be mandatory. Operation includes local blood control by direct suturing, ligation, and coagulation. In cases of uncontrollable organ bleeding, an abdominal (most often perihepatical) packing can be performed to gain time for resuscitation or neurosurgery operations. After stabilization, definitive treatment, for example, closing of biliary leakages, follows.

Ruptures of the bowel and intestines are addressed by direct suturing or segmental resection. In case of massive contamination (e.g., anal impalement trauma), a temporary artificial anus can be established. Intramural hematomas usually are treated conservative.

Injuries to the urogenital system are addressed primarily by different drainage systems, a secondary reconstruction may follow.

Injuries to the pancreas are also usually treated conservative; in case of larger laceration, early partial resection seems to have benefits [21].

Diaphragm injuries have to be excluded and tears closed by direct suture.

17.7.4 Bones

Most pediatric polytrauma patients suffer from fractures. The risk of sustaining a fracture increases with age until approximately the age of 13. Orthopedic injuries are rarely life threatening. However, they can be the major cause for long-term morbidity. Depending on the age, up to 90 % of all fractures occur at the extremities whereas fractures of the trunk only constitute 10 %. In young children, fractures of the lower extremities are common. In older children, fractures of the long bones of the upper extremity are observed two to three times more often than fractures of the lower extremities.

The main localization of the fracture is the bone metaphysis with 65 % followed by the diaphysis with 25 %. Injuries of the epiphysis are less common and approximately 10 %. Depending on the age of the child, the corrective potential can be used in pediatric fracture therapy. Initial management of extremity injuries should include the following:

- Covering all wounds with sterile dressing
- Realigning the deformed extremities
- Splinting all potentially injured extremities
- Neurovascular examination before and after the reduction and splinting with assessment of the tightness [8, 22]

17.7.4.1 Pelvis

Similar to thoracic injuries, severe injuries of the pelvic bones or organs can occur without suspicious external signs of trauma. If a pelvis fracture is detected, high-energy trauma must be considered and organ lesion excluded. Most of the pelvic fractures belong to the A/B type and can be treated conservative; in C-type, external fixation with a fixateur externe must be performed and, most of the time, is also definitive treatment. If secondary reconstruction is necessary, disturbance of the growth zones/physis should be avoided [23–25]. Although most of the pediatric pelvic fractures are not hemodynamically relevant, in some cases, especially "open book" fractures, indirect blood control may be essential. As age-related devices are not available everywhere, classical bondage of in-rotated legs is still performed. If arterial bleeding is detected via CT scan, interventional radiology and coiling of bleeding vessels may be useful [25].

17.7.4.2 Spine

Due to a disproportionately large head and weak neck musculature in children (especially in those less than 3 years of age), horizontally adjusted facet joints, and other preconditions, fractures of the upper cervical spine are much more common up to the age of 12 [10]. Atlanto-occipital dislocation often combined with severe brain injuries most often results in ultimate resuscitation need and has a bad prognosis. Fractures of the dens are usually Salter-Harris-Type-I-fractures and can be treated with a Halofixateur. Ligamentous instabilities, often located segment C2/3, are addressed by posterior fusion and immobilization with a Halofixateur.

Injuries to the lower cervical spine are commonly seen in older children. Here, as well as in thoracic and lumbar spine fractures, type A fractures dominate and can most of the time be treated conservative. Indications for operative treatment (internal fixation) are similar to adult treatment, with emergency indication for hemilaminectomy if neurological impairment is diagnosed [9].

17.7.4.3 Extremities

Although fractures of the extremities are the second most abundantly seen injuries in polytraumatized children, their diagnosis and therapy is often of secondary relevance, with exception of femur fractures and their associated blood loss [26]. Therefore, femur fractures need to be stabilized operatively during day one surgery. Due to a larger compliance of the young bone, a thicker periost, etc., severe soft tissue damage due to bone fragmentation is less common. Stabilization in casts can be performed in most cases to facilitate first-line treatment of abdominal or neurosurgical injuries, but in each and every case, a preferably ultimate stabilization should be achieved early (day 1–2), to avoid additional production of proinflammatory signals [27]. If open fractures are apparent, a compartment syndrome develops or neurovascular injuries accompany the fracture, operative stabilization and soft tissue care is needed (Fig. 17.4). In general, surgical debridement should be more conservative in children because of the more regenerative potential [28].

17.7.4.4 Corrective Potential

Growing bones possess the ability to correct deformities like angulations during growing. The potential is dependent on the age of the child, the fracture type, the direction of dislocation, degree of angulation, and other factors. In general, the younger the child, the better angulations are correctable, and varus deformities are much better to correct than rotational deformities.

The contribution of each physis to the longitudinal growth is different. The highest growth potential at the upper extremity is located at the proximal humerus and the distal forearm, whereas the physis located around the elbow only contributes a small extent to the longitudinal growth (Fig. 17.5). The physis with the greatest part of the longitudinal growth at the lower extremity is located centrically around the knee joint [29].

17.7.4.5 Diaphyseal Fractures

In polytraumatized children, first-line operation has to be performed in diaphyseal fractures of the large diaphysis or large soft tissue damage. Fixateur externe and intramedullary nailing (ESIN) is performed to allow intensive care (Fig. 17.6). Elastic stable intramedullary nailing (ESIN) is a minimally invasive, movement- and partial load-stable procedure in the treatment of diaphyseal and metaphyseal shaft fractures. The principle is based on a three-point support with two pretwisted flexible titanium nails inserted into the bone shaft. The ideal fracture is a diaphyseal transverse fracture, but also diagonal and spiral fractures can be treated. Due to the high correction potential, depending on the exact location, sometimes also large varus displacements can be tolerated. ESIN is – if possible – the treatment of choice in shaft fractures besides the external fixateur. Plate osteosynthesis is reserved for seldom instable fractures of adolescents close to the proximal femur, in order to achieve a rapid stabilization in the polytraumatized child.

17.7.4.6 Articular and Methapyseal Fractures

Articular fractures require exact anatomical reconstruction and should be addressed after the proinflammatory phase of polytrauma. Depending on the exact localizations and fracture type, operative treatment often includes K-wire stabilization to avoid additional disturbance to

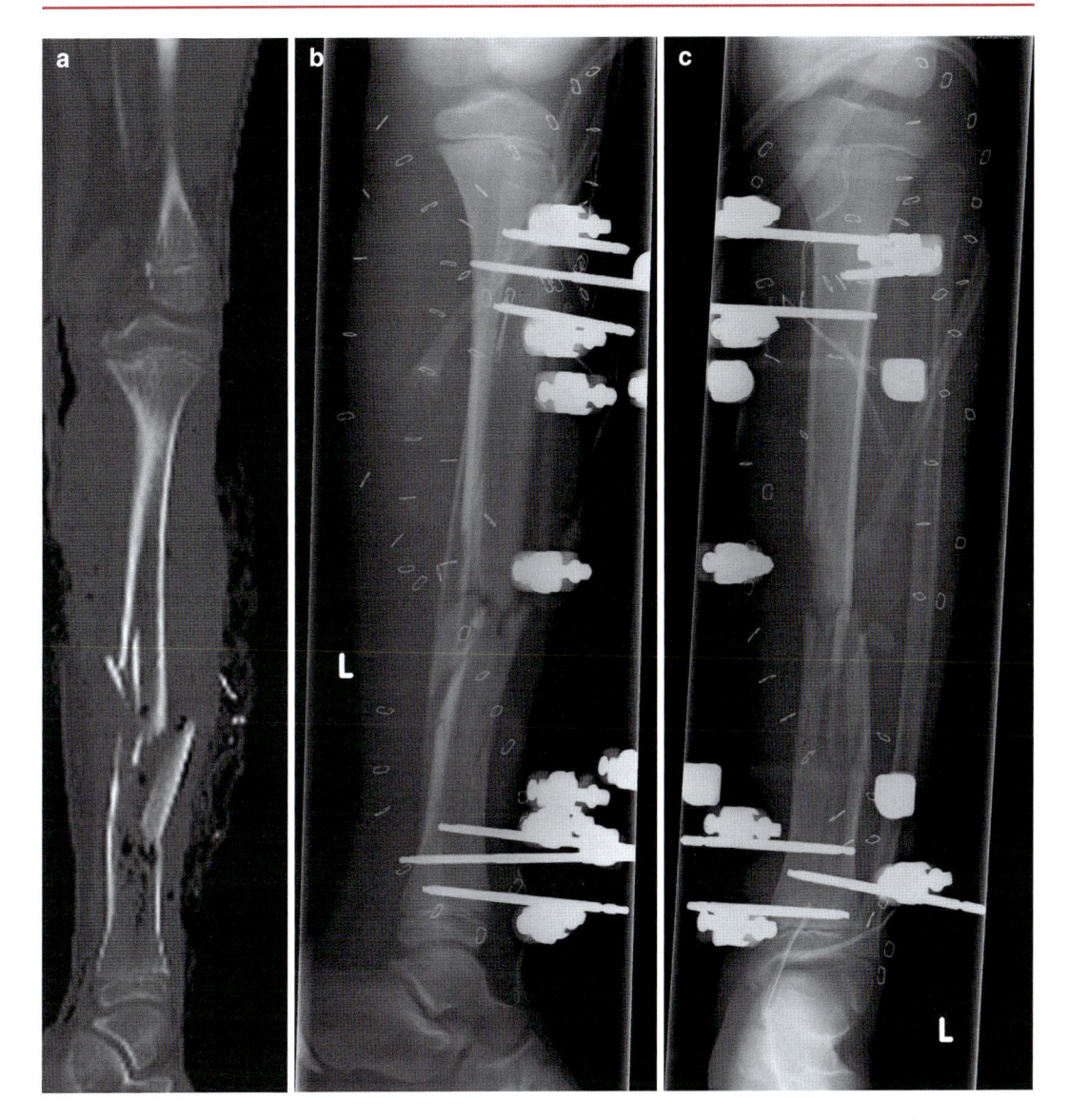

Fig. 17.4 Preoperative CT scan of the left tibia of a 12-year-old boy after a severe train accident (**a**) and postoperative x-rays (**b**, **c**) after primary osteosynthesis with fixateur externe

the physis. With growing age, operative implants become more and more similar to adult implants. The classical indication for percutaneous placed K-wires using a minimal invasive is in metaphyseal fractures. It can also be used in epiphyseal fractures in small children or in patients with small fracture fragments (Fig. 17.7). Removal of the percutaneous placed and above the skin left wires can usually be performed without anesthesia. However, immobilization with a cast is necessary to maintain stability [29].

17.7.4.7 Growth Plate
Injuries to the physis may disturb future bone growth and lead to angular or length discrepancies. The classification of Salter and Harris and Aitken is most often used to asses fractures (Fig. 17.8). Type 1 is characterized by physeal separation, type II by a fracture that traverses the physis and exits through the metaphysis. Type III is characterized by a fracture that traverses the physis before exiting through the epiphysis and type IV by a fracture that traverses the epiphysis,

physis, and metaphysis. Type V describes a crush injury of the physis.

17.7.4.8 Transitional Fractures

Transitional fractures are fractures of the partially closed growth plate during adolescence, mostly in girls around 11–13 years, in boys 12–14 years. Transitional fractures most often occur in the distal tibia. The affecting force deflects to the joint by means of the already ossified part of the physis, so that a ventrolateral epiphyseal fragment is produced analogous to an osseous syndesmosis rupture, the so-called two-plane fracture. It is related to the size of the closure of the physis that has already taken place. Additional torsional forces can lead to further dorsal fragments, corresponding to a Volkmann triangle, the so-called triplane fracture.

Because of the low remaining growth potential of these children, reconstruction of the joint surface is the primary aim in these injuries; growth disorders are not usually expected in adolescents [29, 30].

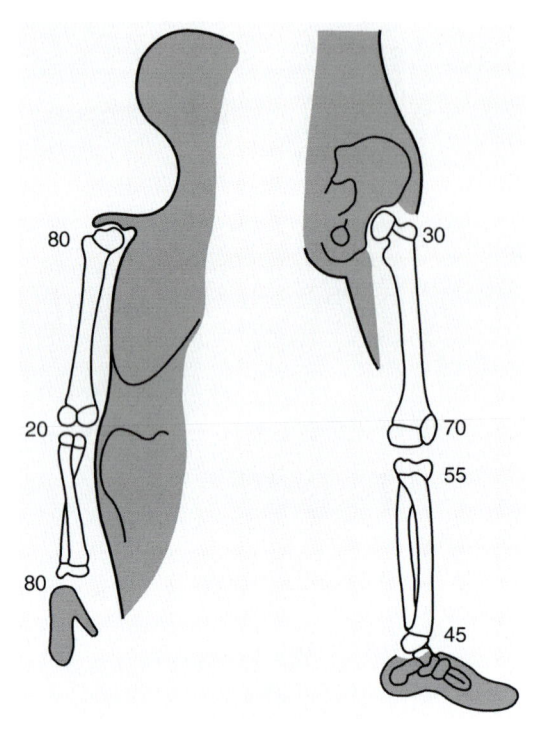

Fig. 17.5 Corrective potential (in %) in extremity injuries [32]

Fig. 17.6 Pre- (**a**) and postoperative (**b**) x-rays of a diaphyseal femur fracture in an 2.5-year-old boy after primary osteosynthesis with ESIN

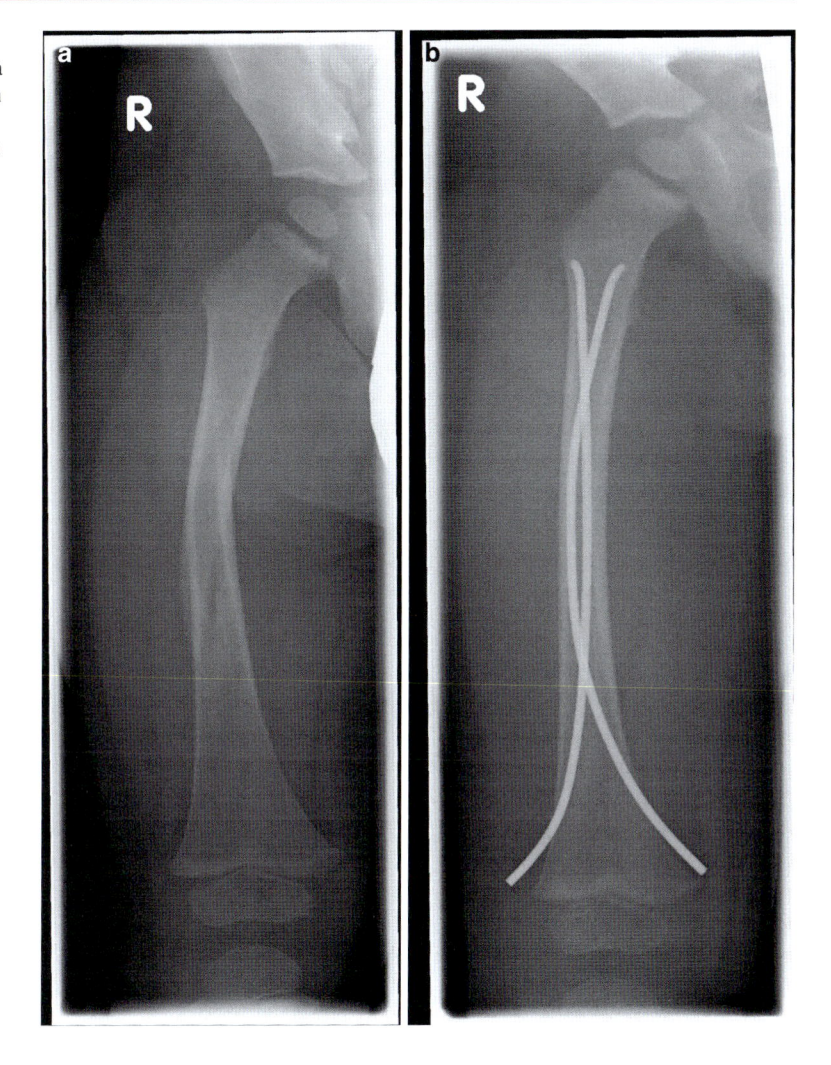

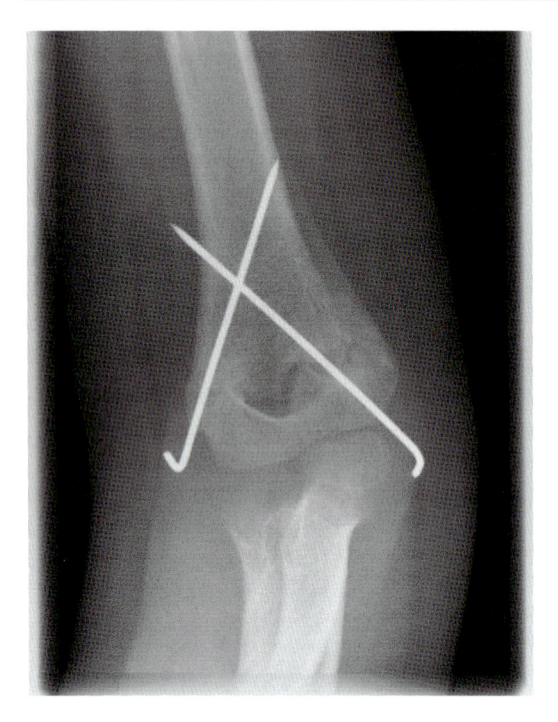

Fig. 17.7 X-ray of a supracondylar humerus fracture after stabilization with crossed K-wires in a 5-year-old girl

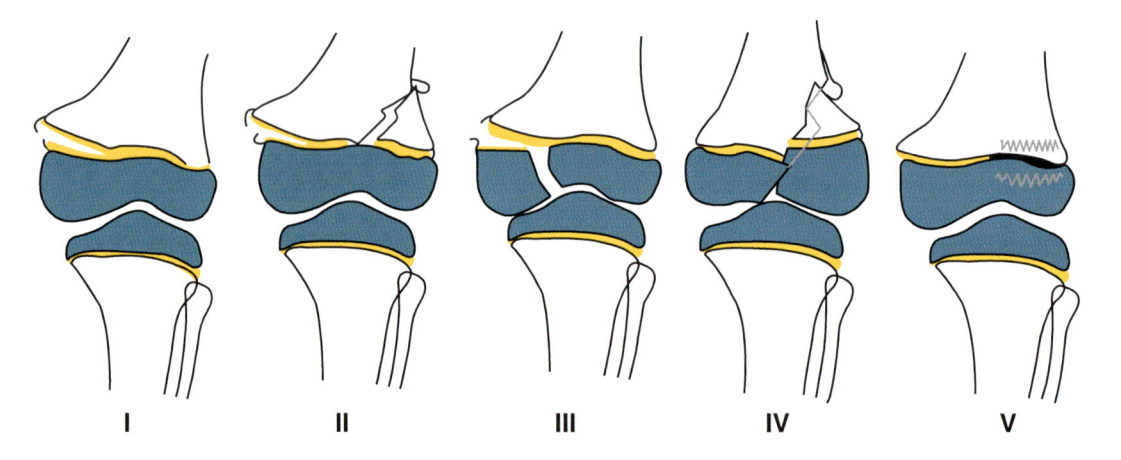

Fig. 17.8 Anatomical classification of Salter and Harris [33]

References

1. Osler TM, Vane DW, Tepas JJ, Rogers FB, Shackford SR, Badger GJ. Do pediatric trauma centers have better survival rates than adult trauma centers? An examination of the National Pediatric Trauma Registry. J Trauma. 2001;50:96–101.
2. Meyer AA. Death and disability from injury: a global challenge. J Trauma. 1998;44:1–12.
3. Laurer H, Wutzler S, Wyen H, et al. Quality of prehospital and early clinical care of pediatric trauma patients of school age compared to an adult cohort. A matched-pair analysis of 624 patients from the DGU trauma registry. Unfallchirurg. 2009;112:771–7.
4. Kay RM, Skaggs DL. Pediatric polytrauma management. J Pediatr Orthop. 2006;26:268–77.
5. Schlechtriemen T, Masson R, Burghofer K, Lackner CK, Altemeyer KH. Pediatric cases in preclinical emergency medicine: critical aspects in the range of missions covered by ground ambulance and air rescue services. Anaesthesist. 2006;55:255–62.
6. Meier R, Krettek C, Grimme K, et al. The multiply injured child. Clin Orthop Relat Res. 2005;44(432):127–31.
7. Wyen H, Jakob H, Wutzler S, Lefering R, Laurer H, Marzi I, Lehnert M. The Trauma Registry of DGU. Prehospital and early clinical care of infants, children, and teenagers compared to an adult cohort. Eur J Trauma Emerg Surg. 2010;36:300–7.
8. American College of Surgeons Committee on Trauma. Advanced trauma life support for doctors ATLS student course manual. 8th ed. Chicago: WB Saunders, Ed. American College of Surgeine; 2008.
9. Jakob H. Trauma Berufskrankheit. 2013;15:67–74.
10. Dykes EH. Paediatric trauma. Br J Anaesth. 1999;83:130–8.
11. Letts M, Davidson D, Lapner P. Multiple trauma in children: predicting outcome and long-term results. Can J Surg. 2002;45:126–31.
12. Remmers D, Regel G, Neumann C, Pape HC, Post-Stanke A, Tscherne H. Pediatric polytrauma. A retrospective comparison between pediatric, adolescent and adult polytrauma. Unfallchirurg. 1998;101:388–94.
13. Sookplung P, Vavilala MS. What is new in pediatric traumatic brain injury? Curr Opin Anaesthesiol. 2009;22:572–8.
14. Kuppermann N, Holmes JF, Dayan PS, et al. Identification of children at very low risk of clinically-important brain injuries after head trauma: a prospective cohort study. Lancet. 2009;374:1160–70.
15. Ceran S, Sunam GS, Aribas OK, Gormus N, Solak H. Chest trauma in children. Eur J Cardiothorac Surg. 2002;21:57–9.
16. Tovar JA, Vazquez JJ. Management of chest trauma in children. Paediatr Respir Rev. 2013;14:86–91.
17. Pauze DR, Pauze DK. Emergency management of blunt chest trauma in children: an evidence-based approach. Pediatr Emerg Med Pract. 2013;10:1–22 ; quiz –3.
18. Landau A, van As AB, Numanoglu A, Millar AJ, Rode H. Liver injuries in children: the role of selective non-operative management. Injury. 2006;37:66–71.
19. Nouira F, Kerkeni Y, Ben Amor A, et al. Liver injuries in children: the role of selective non-operative management. Tunis Med. 2012;90:144–7.
20. Podkamenev VV, Pikalo IA, Zaitsev AP. The risk of death from splenic ruptures by polytrauma among children. Khirurgiia. 2013;44(3):61–5.
21. Meier DE, Coln CD, Hicks BA, Guzzetta PC. Early operation in children with pancreas transection. J Pediatr Surg. 2001;36:341–4.
22. Abdelgawad AA, Kanlic EM. Orthopedic management of children with multiple injuries. J Trauma. 2011;70:1568–74.
23. Smith WR, Oakley M, Morgan SJ. Pediatric pelvic fractures. J Pediatr Orthop. 2004;24:130–5.
24. Niedzielski KR, Guzikiewicz N, Malecki K, Golanski G. Pelvic fractures in children and adolescents in polytrauma and high-energy injuries. Ortop Traumatol Rehabil. 2013;15:41–8.
25. Schneidmueller D, Wutzler S, Kelm A, Wyen H, Walcher F, Marzi I. Pelvic injuries in childhood and adolescence: retrospective analysis of 5-year data from a national trauma centre. Unfallchirurg. 2011;114:510–6.
26. Zwingmann J, Schmal H, Sudkamp NP, Strohm PC. Injury severity and localisations seen in polytraumatised children compared to adults and the relevance for emergency room management. Zentralbl Chir. 2008;133:68–75.
27. Gatzka C, Begemann PG, Wolff A, Zorb J, Rueger JM, Windolf J. Injury pattern and clinical course of children with multiple injuries in comparison to adults, Ab 11-year analysis at a clinic of maximum utilization. Unfallchirurg. 2005;108:470–80.
28. Strohm PC, Bannasch H, Helwig P, Momeni A, Stark GB, Sudkamp NP. Open fracture and soft tissue injury. Z Orthop Unfall. 2010;148:95–111; quiz 2.
29. Schneidmueller D, Marzi I. Grundprinzipien der Kindertraumatologie. In: Scharf H-PR A, editor. Orthopädie und Unfallchirurgie. München: Elsevier; 2009. pp. 345 ff.
30. Jakob H, Sander AL, Marzi I. Das Polytrauma im Kindesalter. In: Marzi IR, editor. Praxisbuch Polytrauma. Koeln: Deutscher Aerzte-Verlag GmbH; 2012. p. 247.
31. Bernbeck R, Dahmen G. Kinder-Orthopädie. 3rd ed. Stuttgart/New York: Thieme; 1983. p. 8.
32. von Lutz L. Frakturen und Luxationen im Wachstumsalter. 4th ed. Stuttgart/New York: Thieme Verlag; 2001. p. 12.
33. Salter RB, Harris WR. Injuries involving the epiphyseal plate. J Bone Joint Surg Am. 1963;45:587–622.

Fracture Management in the Pregnant Patient

18

Erich Sorantin, Nima Heidari, Karin Pichler, and Annelie-Martina Weinberg

Contents

18.1	**Epidemiology**	245
18.2	**Anatomic and Physiologic Changes in Pregnancy**	246
18.3	**Assessment of the Injured Pregnant Patient**	248
18.3.1	General Assessment	248
18.3.2	Radiological Assessment	249
18.4	**Surgical Intervention**	253
18.4.1	Anaesthesia	253
18.4.2	Intraoperative Radiology	254
18.5	**Outcomes**	256
	References	256

E. Sorantin (✉)
Department of Radiology,
Medical University of Graz,
Auenbruggerplatz 34,
Graz 8036, Austria
e-mail: erich.sorantin@medunigraz.at

N. Heidari, MBBS, MSc, FRCS
Centre for Orthopaedics, Royal London Hospital,
Barts Health NHS Trust, Whitechapel Road, London,
England E1 1BB, UK
e-mail: n.heidari@gmail.com

K. Pichler, MD, PhD Student
Department of Orthopedic Surgery,
Medical University of Graz,
Auenbruggperplatz 5, Graz 8036, Austria

Department of Orthopedic Surgery,
Medical University of Innsbruck,
Anichstr. 35, Innsbruck A-6020, Austria
e-mail: karin_pichler@hotmail.com

A.-M. Weinberg
Department of Orthopedic Surgery,
Medical University of Graz,
Auenbruggerplatz 5, Graz 8036, Austria
e-mail: Annelie.Weinberg@t-online.de

18.1 Epidemiology

Trauma in pregnancy is a relatively uncommon problem but it is complicated due to the alterations of the maternal anatomy and physiology as well as the presence of the foetus in the gravid uterus. Between 4 and 8 % of all pregnant women have an accident resulting in an injury [1–4] but only 0.3–0.4 % require admission to hospital [5]. Trauma is the leading non-obstetric cause of maternal mortality accounting for 46 % of maternal deaths [6]. This translates to approximately one million deaths per year worldwide. Pregnancy itself is not a risk factor for mortality following trauma, which has been shown to be a function of the severity of the injury [7, 8]. The risk of trauma to both the foetus and the mother increases as the pregnancy progresses with approximately 15 % of injuries occurring in the first trimester and up to 55 % in the third trimester. The pregnant patient seems to be more vulnerable to abdominal trauma and less prone to head or thoracic injury. It is not clear however whether the severity of the head injury is less or the potential for

© Springer-Verlag Berlin Heidelberg 2016
H.-C. Pape et al. (eds.), *The Poly-Traumatized Patient with Fractures:
A Multi-Disciplinary Approach*, DOI 10.1007/978-3-662-47212-5_18

recovery is greater [8]. The increase in the relative incidence of abdominal trauma with increasing gestation is most likely due to change in the shape of the patient as well as inappropriate positioning of seat belts in motor vehicles. The leading cause of trauma is road traffic accidents, followed by falls [6]. Other important causes such as domestic violence should not be overlooked, and some studies suggest this to be the leading cause of maternal mortality [9]. These injury patterns are described in reports from western countries.

The leading cause of foetal death is road traffic accidents with the main aetiologies being maternal death and placental abruption. A combination of a non-viable pregnancy (less than 23 weeks of gestation) and an injury severity score of greater than 8 has been shown to increase foetal mortality fivefold [10].

Several risk factors have been identified for the occurrence of injuries and trauma in the pregnant patient including young age, history of domestic violence and drug abuse [11]. It is interesting that some racial risk factors have been identified in the occurrence of trauma in pregnancy in the USA. It has been shown that African-American and Hispanic pregnant women are at higher risk for trauma in pregnancy [12]. This is more likely to be a function of the patient's socioeconomic status. After the high energy injuries, described above, pregnant women sustain low energy fractures associated with falls. Osteoporosis of pregnancy has been implicated in these injuries [13, 14].

18.2 Anatomic and Physiologic Changes in Pregnancy
(Table 18.1)

The most obvious and dramatic change during pregnancy is the enlargement of the uterus brought about by the growth of the foetus. The uterus becomes an intra-abdominal organ at approximately 12 weeks of gestation. At 20 weeks, the vertex of the uterus can be palpated at the level of the umbilicus, and by the 36th week, the uterus reaches the costal margin. In the last few weeks of pregnancy, fundal height decreases as the foetal head engages into the pelvis in preparation for the birth.

Anatomical changes during pregnancy should be borne in mind when interpreting initial radiological assessment of the patient. The elevation of the diaphragm by approximately 4 cm and its widening by 2 cm during late pregnancy should be appreciated on the chest radiograph. This may give the appearance of widened mediastinum and an enlarged heart. Increased levels of circulating progesterone lead to the softening of the sacroiliac ligaments, hence widening the joint space. The pubic symphysis may also be widened by 4–8 mm [16].

The changes of the cardiovascular system are numerous and begin from the 8th week of gestation. Progesterone induces relaxation of the smooth muscle in the walls of the peripheral vasculature. There is a gradual decline in blood pressure from week 10 reaching its lowest point by week 28 of gestation. In the third trimester, the blood pressure gradually returns to pre-pregnancy levels. The heart rate also shows an increase by 10–15 beats per minute driving an increase in the cardiac output of 30–50 %. This gradually returns to normal over the first 2 post-partum weeks. There is a 50 % increase in the blood volume which is mostly due to an expansion of the plasma volume with only 30 % increase in the volume of red cells. This brings about a dilutional anaemia referred to as physiological anaemia of pregnancy. The hypervolaemic and hyperdynamic circulation allows the mother to tolerate blood loss of 500–1000 mL with little change in blood pressure and pulse rate. This however is achieved to the detriment of the foetus following trauma. Vasoconstriction of uterine and splanchnic blood vessels and diversion of circulatory volume masks maternal blood loss although signs of foetal distress will be apparent prior to the mother showing the expected signs of shock [17].

Almost all of the coagulation factors increase in pregnancy. This along with the expansion of blood volume and cardiac output are important adaptations for the expected blood loss at the time of delivery [11]. This hypercoagulable state predisposes the mother to thromboembolic disease.

18 Fracture Management in the Pregnant Patient

Table 18.1 Changes in maternal anatomy and physiology in pregnancy [11, 15]

Conditions	Change during pregnancy	Normal pregnancy values
Cardiovascular		
Heart rate	Increases 15–20 bpm	75–95 bpm
Cardiac output	Increases 30–50 %	6–8 l/min
Mean arterial blood pressure	Decreases 10 mmHg in mid trimester	80 mmHg
Systemic vascular resistance	Decreases 10–15 %	1200–1500 dyn/s/cm^{-5}
ECG	Flat or inverted T waves in leads III, V1 and V2	
	Q waves in leads III and aVF	
Hematologic		
Blood volume	Increases 30–50 %	4500 mL
Erythrocyte volume	Increases 10–15 %	
Hematocrit	Decreased	
White blood cell count	Increased	5000–15,000/mm^3
Factors I, II, V, VII, VIII, IX, X and XII	Increased	
Fibrinogen	Increased	>400 mg/dL
Prothrombin time	Decreased by 20 %	
Partial thromboplastin time	Decreased by 20 %	
Respiratory		
Tidal volume	Increased 40 %	700 mL
Minute ventilation	Increased 40 %	10.5 mL
Expiratory reserve volume	Decreased 15–20 %	550 mL
Functional residual capacity	Decreased 20–25 %	1350 mL
Upper airway	Increased oedema; capillary engorgement	
Diaphragm	Displaced 4 cm cephalad	
Thoracic anteroposterior diameter	Increased	
Risk of aspiration	Increased	
Respiratory rate	Slightly increases in the first trimester	
Oxygen consumption	Increased 15–20 % at rest	
Blood gas		
pH	Unchanged	7.4–7.45
PCO$_2$	Decreased	27–32 mmHg
PO$_2$	Increased	100–108 mmHg
HCO$_3$	Decreased	18–21 mEq/L
Abdomen and genitourinary system		
Intra-abdominal organs	Compartmentalization and cephalad displacement	
Gastrointestinal tract	Decreased gastric emptying; decreased motility; increased risk of aspiration	
Peritoneum	Small amounts of intra-peritoneal fluid normally present; desensitized to stretching	
Musculoskeletal system	Widened symphysis pubis and sacroiliac joints	
Kidneys	Mild hydronephrosis (right > left)	
Renal blood flow	Increased 50–60 %	700 mL/min
Glomerular filtration rate	Increased 60 %	140 mL/min
Serum creatinine	Decreased	<0.8 mg/dL
Serum urea nitrogen	Deceased	<13 mg/dL

The respiratory system also undergoes some changes. There is engorgement of the respiratory mucosa that leads to difficulties in intubation and mucosal bleeding [18, 19]. This may result in severe airway compromise. There are also adaptations related to the increased metabolic demands. The presence of the foetus necessitates an increase in oxygen consumption of 15–20 %. Progesterone stimulates the respiratory centre leading to hyperventilation, which brings about a compensated respiratory alkalosis with a concomitant drop in the PCO_2. There is a 4 cm elevation of the diaphragm with a 2 cm increase in the thoracic anteroposterior diameter. This results in a 20–25 % decrease in the functional residual capacity [15]. The pregnant patient is therefore much less tolerant of hypoxia and the associated acidosis. Foetal oxygenation remains constant if maternal PaO_2 is kept above 60 mmHg, because below this level there is a profound drop in foetal oxygenation [11].

Progesterone reduces gastrointestinal motility and the gravid uterus displaces the stomach cephalad. This results in the incompetence of the gastroesophageal pinchcock mechanism placing the pregnant patient at greater risk of regurgitation and aspiration [20]. Therefore, all pregnant patients should be assumed to have a full stomach and the threshold for insertion of a gastric tube lowered.

In the genitourinary system, there is gradual ascent of the uterus from the pelvis where it is well protected into the abdomen from the 12th week of gestation. Once the uterus becomes intra-abdominal, it is at greater risk of injury from blunt and penetrating trauma. The bladder is displaced anteriorly and superiorly. The renal pelvis and the ureters become dilated due to the compressive effect of the uterus as well as the effect of circulating progesterone. The increased cardiac output and blood volume increases renal perfusion by up to 60 % with a concomitant increase in the glomerular filtration rate. This leads to a significant reduction in the serum urea and creatinine levels [15].

18.3 Assessment of the Injured Pregnant Patient

18.3.1 General Assessment

The initial assessment and management of the injured pregnant patient follows the well-established routine of Advanced Trauma Life Support. The best initial treatment of the foetus is the provision of optimum resuscitation for the mother accompanied by foetal monitoring particularly when the foetus is viable. The safe and judicious assessment of the pregnant patient should be a multidisciplinary exercise with the early involvement of an obstetrician, neonatologist, radiologist and trauma surgeon [11, 15, 16, 21, 22].

Pregnant trauma patients can be divided into four groups. The first group are women, who are not aware that they are pregnant. Therefore, all female trauma patients in the reproductive age group should have a pregnancy test performed [23]. Identification of these patients is especially important because routine radiographic studies, performed in the trauma assessment, have the greatest teratogenic potential in early pregnancy. But this consideration should not interfere with life-saving investigations or interventions for the patient. Patients belonging to the second group are injured women of less than 26 weeks of gestation. In these patients, resuscitation is aimed primarily at the mother since the foetus is not yet independently viable. The third and perhaps the most challenging group consists of women with pregnancies more than 26 weeks of gestation. At this stage, there are two patients to consider during the assessment and resuscitation. Finally there are those patients, who present in the perimortem stage. In these patients, early caesarean section may facilitate maternal resuscitation and preserve the life of the foetus [16].

After 20 weeks of gestation, nursing the pregnant patient supine will induce supine hypotension syndrome as the gravid uterus compresses the vena cava, reducing the venous return and embarrassing maternal cardiac output by 30 %.

18 Fracture Management in the Pregnant Patient

This can be alleviated by either displacing the uterus to the left side or, if possible, to nursing the patient tilted left side down by 15°. Due to reduction in the mother's respiratory reserve, supplemental oxygen should be provided. Loss of up to 2000 mL of blood is well tolerated, but this is at the expense of uterine blood supply. The use of vasopressors further compromises uterine blood flow and their use should be avoided unless it is a life-saving intervention. Monitoring of uterine activity and the assessment of the foetus is imperative and should continue for 2–6 h after an injury, even with relatively minor trauma [24, 25]. Signs of foetal distress may be the first signs of maternal hypovolaemia and haemodynamic compromise. The use of vasopressors should be avoided as they further embarrass uteroplacental perfusion. It is preferable to manage cardiac output and blood pressure by replacing volume.

In case of a positive Kleihauer-Betke test, indicating foetal blood in the maternal circulation, the rhesus-negative patients should receive anti-D antibody to prevent isoimmunization [26–28].

As part of the secondary survey, a complete medical and obstetric history should be obtained, particularly details relating to pre-existing hypertension, eclampsia and diabetes. Information about the mechanism of injury, use of drugs and alcohol should be sought. Otherwise, all limbs and body system should be examined in the usual manner. Radiological examination of all suspected fractures should be carried out with the involvement of a radiologist, as a close check needs to be kept on the cumulative dose of radiation received by the patient [22, 29–32].

Early vaginal examination is important. Ideally, this should be performed with an obstetrician in attendance to assess cervical effacement and dilation, foetal position and the presence of amniotic fluid or blood. In the presence of vaginal bleeding, it is prudent to rule out a placenta previa prior to the formal examination of the cervix [31]. The bleeding may be due to placental abruption, labour or placenta previa. Other more traumatic causes such as uterine rupture and an open pelvic fracture must also be considered.

Focused Assessment with Sonography (FAST) scan is important to assess the presence of intra-abdominal haemorrhage. An ultrasound examination of the foetus and placenta can be performed after the FAST scan or incorporated as part of the trauma scan. If a chest tube thoracostomy is needed, it has to be placed one or two intercostal spaces higher than usual to avoid diaphragmatic injury.

Tetanus prophylaxis is not contraindicated and should be administered according to standard protocols.

18.3.2 Radiological Assessment

18.3.2.1 General Considerations

Trauma in pregnancy represents a special situation as two patients are involved – the mother and the child. Radiographic and CT examinations of the pregnant patient irradiate the unborn and can cause severe harm. Intrauterine development consists of three phases and radiation sensitivity is related to gestational age.

As a general guideline, the "ALARA Principle" should be mentioned. It entails that radiation should be used "*a*s *l*ow *a*s *r*easonably *a*chievable" [33].

18.3.2.2 Basics of Radiation Protection

The following types of radiation have to be differentiated: α-, β-, γ- and x-rays. For medical imaging, only γ-radiation (nuclear medicine) and x-rays are used.

Important Units for Radiation Benchmarking

Ion dose: measures radiation by the amount of the induced ionization – the SI unit is R.

Absorbed dose: defines the absorbed dose per kg mass, the SI unit is gray (Gy) = 1 J/kg.

Dose output: is dose/time, the SI unit is Gy/s.

Due to the inherent different properties of α-, β-, γ- and x-rays, they are converted into units that are representative of their varying biologic activity. This is achieved by multiplying the absorbed dose by a dimensionless radiation weighting factor (*WR*, prior *Q* – relative biological

Table 18.2 Weighting factor by radiation type [34]

Radiation type		Radiation weighting factor
Photons		1
Electrons, muons		1
Neutrons	<10 keV	5
	10 to 100 keV	10
	>100 keV to 2 MeV	20
	>2 to 20 MeV	10
	>20 MeV	5
Protons (energy >2 MeV)		5
α–Radiation		20

Table 18.3 Tissue/organ weighting factor with due consideration of the different sensitivity of tissues/organs to radiation [35]

Tissue/organ	Weighting factor (W_t)
Gonads	0.08
Red bone marrow	0.12
Colon	0.12
Lung	0.12
Stomach	0.12
Urinary bladder	0.04
Mamma	0.12
Liver	0.04
Oesophagus	0.04
Thyroid	0.04
Skin	0.01
Bone surface	0.01
Brain	0.01
Salivary glands	0.01
Others	0.12

Table 18.4 Typical effective doses in imaging – can vary due to technical factors (e.g. additional filtration) as well as adjustment of the exposure settings to body mass/size, age and several other factors [37]

Examination	Typical effective dose (mSv)	Number of chest x-rays leading to the comparable exposure
Chest (p.a.)	0.02	1.0
Extremities/joints	0.01	0.5
Skull	0.07	3.5
Thoracic vertebra	0.70	35.0
Hip	0.30	15.0
Pelvis	0.70	35.0
Mammography (bilateral, two planes)	0.50	25.0
Intravenous urography	2.50	125.0
Head CT	2.30	115.0
Chest CT	8.00	400.0
Abdomen/pelvis CT	10.00	500.0
Renal function scintigraphy	0.80	40.0
Thyroid scintigraphy	0.90	45.0
Lung perfusion scintigraphy	1.10	55.0
Skeletal scintigraphy	4.40	220.0
Myocardial perfusion scintigraphy	6.80	340.0
Positron emission tomography	7.20	360.0
Myocardial scintigraphy	17.00	865.0

effectiveness). The result is the *dose equivalent*, which is measured in sievert (Sv):

Sievert (Sv) = Gy × WR – the corresponding values can be found in Table 18.2.

Organ dose: represents the absorbed dose output of an organ, tissue or body part, which is multiplied by the radiation weighting factor – the SI unit is again Sv.

Effective dose equivalent: considers the different radiation sensitivity for various human tissues by the so-called tissue/organ weighting factor (W_t – Table 18.3). The effective dose equivalent is calculated by first multiplying the organ dose with the tissue/organ weighting factor, followed by adding all individual doses.

Natural Background Radiation

The source of natural background radiation falls into two broad categories – natural (from ground and space) and artificial (medicine, radioactive fallout, nuclear waste, consumer products, etc.). The cumulative dose is approximately 4 mSv. It is interesting to note that medical diagnostic imaging and nuclear medicine are responsible for about 79 % of man-made radiation [36]. Typical radiation doses for medical imaging can be found in Table 18.4.

Deterministic Versus Stochastic Radiation Effects

In *deterministic effects*, there is a classic dose–effect relationship such as the LD50/30 (the dose of whole-body irradiation where 50 % of subjects die within 30 days) [38] of ~4.0 Sv, or after a

Table 18.5 Effects of irradiation during intrauterine life* [36]

Effect	Gestational age	Lower threshold (mSv)	Risk-coefficient
Death during pre-implantation phase	0–10 days	100	0.1 %/mSv*
Malformation	10 days–8 weeks	100	0.05 %/mSv*
Severe mental retardation	8–15 weeks	300	0.04 %/mSv*
	16–25 weeks	300	0.01 %/mSv*
IQ-reduction	8–15 weeks		0.03 IQ/mSv
	16–25 weeks		0.01 IQ/mSv
Cancer/leukaemia			0.006 %/mSv
Genetic defects			0.0003 %/mSv male
			0.0001 %/mSv female

3.0 Sv there are severe skin burns, after 3.0–4.0 Sv cataracts occur – just to name some examples.

Stochastic effects are those that occur in a random manner, including cancer and genetic defects. These events cannot be related to a single dose but the cumulative effect of multiple exposures may result in damage and for this reason, the concept of the excess lifetime risk was introduced. The risk is higher for younger people, which can be partly explained by the higher sensitivity of dividing cells to radiation. The "International Commission on Radiation Protection (ICRP)" suggests an excess rate of 5 % per Sv for lower doses and 10 % for higher ones.

An excess lifetime risk factor of 10 % means after exposing 10,000 individuals to 10 mSv dose of radiation, there will be about ten additional deaths due to leukaemia or cancer, but it is important to note that even without this radiation there would be 2,500 cancer-related deaths [37].

Radiation Effects During Intrauterine Life

The following facts are based on the report of "German Society for Medical Physics" and the "German Roentgen Society" [39]. A summary of all effects can be found in Table 18.5.

The period of intrauterine life can be divided into three phases. These are the pre-implantation phase (until 10 days post-conception), the phase of organogenesis (10 days to 8 weeks of gestation) and the foetal period (from the 3 months of gestation to term). Exposure to radiation in each phase has characteristic effects.

Pre-implantation phase: high doses (>100.0 mSv) result in spontaneous abortion which is often clinically silent, since pregnancy is not known yet. Birth defects are possible with a risk coefficient of 0.1 % per mSv.

Organogenesis: high doses (>100.0 mSv) cause organ malformations as well as growth retardation and functional disorders. The risk coefficient for organ malformations is 0.05 % per mSv which doubles at 200 mSv.

Foetal period: the central nervous system is the most susceptible organ during this phase and radiation exposure has been linked to severe neuromotor development disorders with risk coefficient of 0.04 % per mSv from the 8th to 15th week of gestation and 0.01 % per mSv from the 16th to 25th week of gestation. Reduction in the "intelligent quotient" (IQ) represents another known radiation effect, being more severe during early pregnancy: 30 IQ points for the 8th to 15th week of gestation and 10 IQ points for the 16th to 25th week of gestation.

Cancer Risk After Intrauterine Irradiation

A linear dose–effect relationship is presumed; however, there is no known threshold. It is assumed that doses of less than 100 mSv may pose a significant risk for the development of leukaemia and cancer. The risk coefficient is about 0.006 % per mSv. Typical foetal doses delivered by imaging are listed in Table 18.6.

Genetic Effects After Irradiation

A linear dose–effect relationship is also assumed. There are no data available from human studies; we have however extrapolated from some animal studies, as shown in Table 18.5.

Table 18.6 Estimated doses to the foetus during imaging (from [40])

Examination	Typical foetal dose (mGy)
Cervical spine (AP, lat)	0.001
Extremities	0.001
Chest (PA, lat)	2
Thoracic spine (AP, lat)	3
Abdomen (AP)	0
21-cm patient thickness	1
33-cm patient thickness	3
Lumbar spine (AP, lat)	1
Limited IVP[a]	6
Small-bowel study[b]	7
Double-contrast barium enema study[c]	7

AP anteroposterior projection, *lat* lateral projection, *PA* posteroanterior projection
[a]Limited IVP is assumed to include four abdominopelvic images. A patient thickness of 21 cm is assumed
[b]A small-bowel study is assumed to include a 6-min fluoroscopic examination with the acquisition of 20 digital spot images
[c]A double-contrast barium enema study is assumed to include a 4-min fluoroscopic examination with the acquisition of 12 digital spot images

18.3.2.3 Imaging of the Pregnant Patient

Radiographs of the extremities can be safely performed during all stages of pregnancy, but adequate shielding is a MUST and can reduce the radiation dose to the unborn by up to 30 %. The generator settings should be on the lowest possible values where diagnostic information can still be gleaned. This necessitates discussion and close collaboration with both radiologists and radiographers.

In stable patients with suspected ligamentous injuries (e.g. ankle), MRI is preferable over repeated stress radiographs.

In abdominal trauma or poly-trauma patients, ultrasound is the preferred first-line imaging modality – e.g. "focus assessment sonographic trauma scan" (FAST) in order to detect free, intraperitoneal fluid. It is imperative to include the foetus as well as the placenta in every sonographic evaluation of the abdomen and pelvis – in the optimal case together with an obstetrician [41].

CT is the preferred modality of choice in unstable patients or in patients with clinical/sonographic signs of injuries to chest, mediastinum, aorta, spine, retroperitoneum, bowel, bladder and pelvis. Intravenous iodine contrast may be administered as indicated clinically, but this may induce hypothyroidism in the unborn in addition to causing renal anomalies. Therefore, after delivery, follow-up investigations of thyroid and renal function are needed.

Modern CT scanners offer several features for dose reduction. Usually for scout views, the tube is above the patient ("position up"), but if the tube is below the CT couch ("position down"), then a reduction of about two-third the anterior parts of the body (thyroid, mamma and foetus) can be achieved [42]. This is especially important for the childbearing mother, since there are considerable concerns about the radiation sensitivity of the female breast during pregnancy [43–45]. The tube current of the scout view should be reduced and the extension as tight as possible as well as the scan range.

CT should be performed with adapted tube current values for the mother with considerations of her body habitus. It is important to note that a 20–30 % reduction of the ideal adjusted dose will lead to more image noise; however, the image quality will be sufficient for diagnosing traumatic lesions. In case of using "automated exposure control," the choice of an appropriate reconstruction kernel is a crucial point (usual "soft tissue kernel"), since "bone sharp kernels" will lead to an unnecessary high degree of radiation exposure [46]. For CT image reconstruction, "filtered back projection" was used for more than 40 years. The dramatic increase in today's computational power facilitates to use sophisticated algorithms for image reconstruction such as the so-called "iterative image reconstruction" (sometimes referred as "statistical reconstruction") – those techniques are now offered by all vendors. Iterative image reconstruction enables considerable dose saving of more than 50 % [47–55].

In CT examinations, a total radiation dose of more than 100 mSv should not be exceeded using standard trauma protocols. In case of abdominal CT, the effective dose will be between 10 and 40 mSv. Another point is represented by the differences between various CT scanner generations.

All multidetector CT scanners suffer from "overbeaming" where the x-ray beam extends beyond the edge of detector rows, exposing the patient to a greater radiation dose within those areas. This effect is more pronounced in CT scanners with less rows (e.g. 4–16 rows). Multidetector-row CT systems with more than 64 rows suffer from "overranging" as the reconstruction algorithm requires additional raw data on the beginning and end of the planned scan. Therefore extra rotations outside the planned length are needed for image reconstruction. This can be reduced by adequate tailoring of scan length [56].

Nevertheless, calculations of the International Committee on Radiation Protection (IRCP) estimate that a foetal dose of 10 mGy will increase the risk of leukaemia or cancer considerably [57].

MRI is not an option in unstable pregnant patients, since the examinations are time consuming. Moreover for vascular imaging, intravenous injection of Gadolinium-based contrast medium is necessary, which should be avoided in regard to the embryo/foetus (see below). Furthermore, not at all MRI suites offer sufficient monitoring facilities for the pregnant mother.

Field strengths of up to 1.5 T are preferable, as there are concerns about the heating effects of radio-frequency pulses as well as the effect of acoustic noise on the unborn. Gadolinium-based MRI contrast media have been shown to be teratogenic in animal studies if administered in doses of two to seven times greater than normal. Gadolinium crosses the placenta and is excreted by the foetal kidney into the amniotic fluid. In the light of new insights in Gadolinium side effects including Nephrogenic Systemic Fibrosis (NSF), guidelines from the "European Society of Urogenital Radiology" (ESUR – http://www.esur.org/guidelines/en/index.php) state that in pregnancy those contrast media should be used with caution and at the lowest possible dose [58]. NSF occurs in people with severely impaired renal function, but as foetal kidneys are immature, the potential harm to the unborn is unquantifiable and extreme caution should be exercised.

It is the authors belief that if intravenous contrast is essential for clinical decision making, then CT should be considered, as the side effects of radiation and iodine contrast are known, whereas this is not the case for Gadolinium and MRI.

In general, imaging departments, dealing with pregnant trauma patients, should elaborate an imaging strategy in advance. This algorithm should address the use of imaging – both for initial assessment of the patient (including dose settings for plain radiography and CT) as well as the subsequent clinical treatment (including intraoperative use of imaging or further imaging for follow-up). For CT, the scanner with most rows and iterative image reconstruction should be used.

Polytraumatized pregnant patients should only referred to centres, where a specialized team of trauma surgeons, anaesthetists, obstetricians as well as radiologists, medical physicists and radiographers is available on a 24/7 basis.

18.4 Surgical Intervention

It is logical to postpone all elective procedures until after delivery [59, 60]. However, provision of optimum emergency surgical care should not be compromised. Surgical management of fractures as dictated by the bony and soft tissue injury and it may not be feasible to postpone these procedures [29]. Most can be safely carried out in the pregnant patient. Consideration specific to anaesthesia, intraoperative radiology and orthopaedics should be taken into account.

18.4.1 Anaesthesia

Pregnancy is not a contraindication to anaesthesia. No increase in stillbirths, birth defects [61] or neural tube defects [62] has been demonstrated as a result of pregnant women receiving anaesthesia.

The management of the airway can be a challenge in pregnant patients. The incidence of difficult intubations is 17-fold higher in advanced pregnancy. There is an increased risk of aspiration, and the risk of hypoxia higher due to reduced functional reserve and increased oxygen consumption [63]. The combination of limited maternal reserve and a foetus sensitive to changes in maternal metabolism requires close monitoring and expedient action on the part of the anaesthetist. The goals of ventilation include a high PaO$_2$ and a PaCO$_2$ normal for the gestation [64]. Frequent measurements of blood gases may be invaluable in these circumstances.

Uterine and foetal monitoring are useful as foetal distress maybe the first sign of maternal hypovolaemia. Monitoring volume status in pregnancy may be difficult as some data show poor correlation between central venous and left ventricular filling pressures. Some authors suggest inserting a Sawn-Ganz catheter if accurate haemodynamic monitoring is required [65, 66].

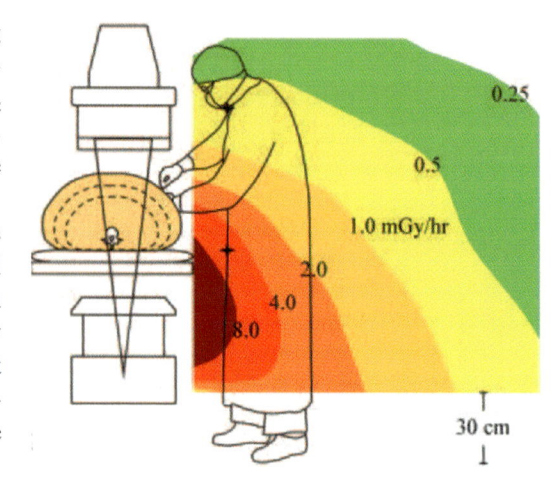

Fig. 18.1 Isodose lines during fluoroscopy – colours represent areas of almost same dose (Modified after [67])

18.4.2 Intraoperative Radiology

18.4.2.1 General Considerations

The following *hardware features* should be available:

Pulsed fluoroscopy: in cases where live imaging is required, 25 frames per second are used for fluoroscopy, but this temporal resolution is rarely needed in trauma surgery. Many machines allow a rate of two frames per second, which is often adequate in most circumstances.

Last image hold: the last fluoroscopy image stays on the screen and can be referred to, without further radiation.

Leaf shutter: allows the operator to control the size of the radiation field by coning onto the region of interest. An example of dose distribution in dependence of the field of view can be seen in Fig. 18.1.

Field of view (FOV) (magnification): using the system's zoom functions increases the dose, e.g. changing the FOV from 28 to 20 cm (usually one magnification step) doubles the dose [67].

Powerful generator: primary rapid and steep increase in kV should be possible.

18.4.2.2 Intraoperative Imaging

During any operative procedure, the fluoroscopy unit should be handled by the radiographer. Exact placement of the primary beam, tight use of the shutter and lead shielding are mandatory – especially the uterus should be as far as possible from the primary beam. Lead shielding reduces the scatter from the units itself and other outside sources, whereas scatter from the irradiated tissues cannot be reduced. Of course, irradiation time should be as short as possible and extensive use of the "last image hold" technique is mandatory. The same guidelines apply for intraoperative radiography; typical doses to the foetus can be found in Table 18.5. The dose output of C-arm systems can differ considerably between manufacturers. This makes it difficult to estimate an absolute tolerable time period for irradiation of the pregnant uterus. Using the data published by Schueler et al. [67], and assuming that the gravid uterus is directly in the x-ray beam, the threshold dose of 100 mSv will be reached in about 3 min at a FOV of 28 cm, but in only 1.5 min at a FOV of 20 cm.

Take-Home Points for Imaging

- Design of clinical pathways for pregnant trauma patients requires involvement of radiologists and medical physicists. Detailed knowledge of the cumulative dose received by the patient is essential for ongoing management decisions.
- Ultrasound is the first modality of choice – this should be carried out in the presence of an obstetrician.

- If a CT scan is necessary, the region scanned should be kept as small as possible. Utilize all inherent possibilities to reduce dose of ionizing radiation including rigorous mAs lowering.
- If administration of intravenous iodine contrast is necessary, close monitoring of thyroid and renal function and referral to a paediatrician are essential for the child after birth.
- In non-acute imaging, detailed counselling of the mother is necessary if the foetal dose is likely to go beyond 1 mGy.
- Intraoperative Imaging – avoid direct irradiation the uterus.

18.4.2.3 Orthopaedic Surgical Management

There is a paucity of literature on the outcomes of orthopaedic injuries in pregnancy. In a study from New Orleans, only 4 % of pregnant trauma patients had orthopaedic injuries [68]; however, this may not be representative in other populations. Extremity fractures should be treated much in the same way as they would be in the non-pregnant patient. The pregnant patient tends to be young, and suboptimal surgical management of their fracture have profound long-term consequences. As long as direct irradiation of the uterus is avoided and adequate shielding is employed, there are no contraindications to intraoperative imaging. This is of course not the case with pelvic and proximal femoral fracture fixation. Modifications of surgical technique may reduce the need of intraoperative imaging. Most minimally invasive techniques are highly dependent on intraoperative imaging and are not advocated in this situation. An open technique of fracture reduction and fixation reduces the need for imaging.

Pregnancy is a prothrombotic state and prolonged immobilization, and bed rest should be avoided. The increased risk of venous thromboembolism (VTE) begins in the first trimester and has a tendency to occur in the left lower limb [69]. There is little data regarding VTE in pregnant women and recommendations are based on expert opinion derived from evidence in non-pregnant populations [70]. Some specific risk factors that may relate to the traumatized pregnant patient include immobility [71], blood loss and transfusion [72] as well as having any surgical procedures. It appears that low-molecular-weight heparins are safe to use in these patients [73]. The decision to prescribe anticoagulation should be based on the assessment of individual patients and with consideration of risk factors. Surgical treatment of an injury to get the patient mobile is clearly desirable and the benefits outweigh the risks of the procedure.

Positioning the patient in the left lateral decubitus position (lying left side down) moves the gravid uterus away from the vena cava and avoids the development of supine hypotension syndrome. If it is not possible to position the patient in this way, the uterus should be manually displaced. Any blood loss should be directly communicated with the anaesthetist. Although the patients' haemodynamic parameters may remain within normal limits, this is at the expense of the blood flow to the uterus and foetus.

Fractures of the pelvis and acetabulum present a particular challenge in this patient cohort. The literature on this subject is restricted to mostly case reports [74–77] and there is a general trend for conservative management of these fractures. A retrospective review from a major trauma centre of 24 years only reported seven pregnant patients with a pelvic fracture [21]. Of these patients, five mothers and three foetuses survived. This group represents severely traumatized patients, and their care need to be undertaken in specialist units. Up to 9 % of the women and 35 % of foetuses die following these injuries [75]. The surgical management of these fractures can also be hazardous to the patient and the foetus, with maternal blood loss and risk of direct injury to the uterus or the foetus [76]. In these circumstances, a caesarean section could save the life of the mother and her unborn child [75, 76, 78]. The cumulative dose of ionizing radiation to the foetus may be prohibitive in employing minimally invasive techniques for the fixation of pelvic and acetabular fractures. The issue of emergent external fixation of the pelvis in the pregnant patient has not been address in the literature. The gestational age is important as the third trimester the gravid uterus may interfere with the placement of both high and low anterior external fixator half pins.

There are a few logical issues that must be considered to aid decision making in situations where operative intervention is required. These include foetal gestational age and viability, level of maternal and foetal compromise, the cumulative dose of ionizing radiation and the necessities of fracture fixation. There simply are no easy answers and the treatment needs to be individually tailored to the patient.

18.5 Outcomes

Trauma represents a risk for both - mother and foetus. Quantification of the risks to the foetus and the mother is substantiated by only few reports. This is a reflection of the unusual nature of injuries in the pregnant patient and the difficulties in collecting data on their outcome. Most of the data concentrates on the severely injured patient, but it should be borne in mind that even relatively minor trauma can lead to preterm labour and foetal loss. It has been estimated that between 4 and 61 % of injured pregnant patients lose their foetuses [2].

Weiss et al. [79] in a study reported on the causes of foetal death related to maternal injury. The data was collected from 16 states in the USA over a 3-year period. Motor vehicle accidents were by far the most common cause of foetal death (82 %) with firearms (6 %) and falls (3 %) far behind. The physiological diagnoses associated with foetal loss were placental abruption (42 %) and maternal death (11 %). They noted an association between placental abruption accompanied by uterine rupture and advancing gestational age.

Information on maternal haemodynamic parameters is sparse and does not provide a reliable indication of the foetal status [80]. Some risk factors have been identified that herald the possibility of acute termination of pregnancy. Theodorou and colleagues [10] showed that an ISS ≥ 9 and gestational age of ≤ 23 weeks are strong predictors of foetal loss. Other authors have demonstrated adverse foetal outcomes with increasing injury severity [81, 82], but it is interesting that even moderate maternal trauma can result in foetal death. The issue of gestational age is also contentious as some authors have not made this link [80, 82, 83]. The rates of preterm labour are increased in the presence of head injuries with patients who have a GCS ≤ 12 being three times more likely to go into labour. This has not been related to increased foetal death [83].

In general, it is difficult to truly predict the outcome. On one hand, devastating foetal outcomes are seen in relatively minor trauma; on the other hand, patients with pelvic fractures often have been reported to have an uneventful child birth. It is wise to be cautious and plan every step on a case-to-case basis. All pregnant patients with a viable foetus need to be closely monitored and the early involvement of obstetricians is essential for the correct and judicious interpretation of foetal monitoring data.

References

1. El-Kady D, Gilbert WM, Anderson J, Danielsen B, Towner D, Smith LH. Trauma during pregnancy: an analysis of maternal and fetal outcomes in a large population. Am J Obstet Gynecol. 2004;190(6):1661–8.
2. Esposito TJ. Trauma during pregnancy. Emerg Med Clin North Am. 1994;12(1):167–99.
3. Rosenfeld JA. Abdominal trauma in pregnancy. When is fetal monitoring necessary? Postgrad Med. 1990;88(6):89–91, 94.
4. Vaizey CJ, Jacobson MJ, Cross FW. Trauma in pregnancy. Br J Surg. 1994;81(10):1406–15.
5. Lavin Jr JP, Polsky SS. Abdominal trauma during pregnancy. Clin Perinatol. 1983;10(2):423–38.
6. Connolly AM, Katz VL, Bash KL, McMahon MJ, Hansen WF. Trauma and pregnancy. Am J Perinatol. 1997;14(6):331–6.
7. Shah AJ, Kilcline BA. Trauma in pregnancy. Emerg Med Clin North Am. 2003;21(3):615–29.
8. Shah KH, Simons RK, Holbrook T, Fortlage D, Winchell RJ, Hoyt DB. Trauma in pregnancy: maternal and fetal outcomes. J Trauma. 1998;45(1):83–6.
9. Chang J, Berg CJ, Saltzman LE, Herndon J. Homicide: a leading cause of injury deaths among pregnant and postpartum women in the United States, 1991–1999. Am J Public Health. 2005;95(3):471–7.
10. Theodorou DA, Velmahos GC, Souter I, Chan LS, Vassiliu P, Tatevossian R, Murray JA, Demetriades D. Fetal death after trauma in pregnancy. Am Surg. 2000;66(9):809–12.

11. Hill CC. Trauma in the obstetrical patient. Womens Health (Lond Engl). 2009;5(3):269–83.
12. Ikossi DG, Lazar AA, Morabito D, Fildes J, Knudson MM. Profile of mothers at risk: an analysis of injury and pregnancy loss in 1,195 trauma patients. J Am Coll Surg. 2005;200(1):49–56.
13. Aynaci O, Kerimoglu S, Ozturk C, Saracoglu M. Bilateral non-traumatic acetabular and femoral neck fractures due to pregnancy-associated osteoporosis. Arch Orthop Trauma Surg. 2008;128(3):313–6.
14. Willis-Owen CA, Daurka JS, Chen A, Lewis A. Bilateral femoral neck fractures due to transient osteoporosis of pregnancy: a case report. Cases J. 2008;1(1):120.
15. Muench MV, Canterino JC. Trauma in pregnancy. Obstet Gynecol Clin North Am. 2007;34(3):555–83, xiii.
16. Tsuei BJ. Assessment of the pregnant trauma patient. Injury. 2006;37(5):367–73.
17. Greiss Jr FC. Uterine vascular response to hemorrhage during pregnancy, with observations on therapy. Obstet Gynecol. 1966;27(4):549–54.
18. Jouppila R, Jouppila P, Hollmen A. Laryngeal oedema as an obstetric anaesthesia complication: case reports. Acta Anaesthesiol Scand. 1980;24(2):97–8.
19. Kuczkowski KM, Reisner LS, Benumof JL. Airway problems and new solutions for the obstetric patient. J Clin Anesth. 2003;15(7):552–63.
20. Vanner RG. Mechanisms of regurgitation and its prevention with cricoid pressure. Int J Obstet Anesth. 1993;2(4):207–15.
21. Pape HC, Pohlemann T, Gansslen A, Simon R, Koch C, Tscherne H. Pelvic fractures in pregnant multiple trauma patients. J Orthop Trauma. 2000;14(4):238–44.
22. Petrone P, Asensio JA. Trauma in pregnancy: assessment and treatment. Scand J Surg. 2006;95(1):4–10.
23. Hirsh HL. Routine pregnancy testing: is it a standard of care? South Med J. 1980;73(10):1365–6.
24. Chames MC, Pearlman MD. Trauma during pregnancy: outcomes and clinical management. Clin Obstet Gynecol. 2008;51(2):398–408.
25. Sperry JL, Casey BM, McIntire DD, Minei JP, Gentilello LM, Shafi S. Long-term fetal outcomes in pregnant trauma patients. Am J Surg. 2006;192(6):715–21.
26. Eager R, Sutton J, Spedding R, Wallis R. Use of anti-D immunoglobulin in maternal trauma. Emerg Med J. 2003;20(5):498.
27. Huggon AM, Watson DP. Use of anti-D in an accident and emergency department. Arch Emerg Med. 1993;10(4):306–9.
28. Weinberg L. Use of anti-D immunoglobulin in the treatment of threatened miscarriage in the accident and emergency department. Emerg Med J. 2001;18(6):444–7.
29. Flik K, Kloen P, Toro JB, Urmey W, Nijhuis JG, Helfet DL. Orthopaedic trauma in the pregnant patient. J Am Acad Orthop Surg. 2006;14(3):175–82.
30. Lowe SA. Diagnostic radiography in pregnancy: risks and reality. Aust N Z J Obstet Gynaecol. 2004;44(3):191–6.
31. Mattox KL, Goetzl L. Trauma in pregnancy. Crit Care Med. 2005;33(10 Suppl):S385–9.
32. Patel SJ, Reede DL, Katz DS, Subramaniam R, Amorosa JK. Imaging the pregnant patient for nonobstetric conditions: algorithms and radiation dose considerations. Radiographics. 2007;27(6):1705–22.
33. Sardanelli F, Hunink MG, Gilbert FJ, Di LG, Krestin GP. Evidence-based radiology: why and how? Eur Radiol. 2010;20(1):1–15.
34. Wikipedia. Strahlungswichtungsfaktor. http://de.wikipedia.org/wiki/Strahlenwichtungsfaktor. 2010. 14-3-2010. Ref Type: Online Source
35. Wikipedia. Effektive Dosis. http://de.wikipedia.org/wiki/Effektive_Dosis. 2010. 13-4-2010. Ref Type: Online Source
36. Committee to Assess Health Risks from Exposure to Low Levels of Ionizing Radiation N R C. Health risks from exposure to low levels of ionizing radiation: BEIR VII Phase 2. 2006. Online resource - http://www.cirms.org/pdf/NAS%20BEIR%20VII%20Low%20Dose%20Exposure%20-%202006.pdf.
37. Shannoun F, Blettner M, Schmidberger H, Zeeb H. Radiation protection in diagnostic radiology. Dtsch Arztebl Int. 2008;105(3):41–6.
38. The United States Nuclear Regulatory Commission. Lethal dose (LD). http://www.nrc.gov/reading-rm/basic-ref/glossary/lethal-dose-ld html [serial online] 2010; [cited 2010 Apr 27].
39. Dierker J, Eschner W, Gosch D, et al. Pränatale Strahlenexposition aus medizinischer Indikation. Deutsche Gesellschaft für Medizinische Physik; 2002. Online Resource: http://www.dgmp.de/media/document/205/bericht7-neuauflage2002.pdf.
40. Wikipedia. Strahlenbelastung. http://de.wikipedia.org/wiki/Strahlenbelastung. 2010. 14-3-2010. Ref Type: Online Source
41. Ma OJ, Mateer JR, Ogata M, Kefer MP, Wittmann D, Aprahamian C. Prospective analysis of a rapid trauma ultrasound examination performed by emergency physicians. J Trauma. 1995;38(6):879–85.
42. Sorantin E, Weissensteiner S, Hasenburger G, Riccabona M. CT in children – dose protection and general considerations when planning a CT in a child. Eur J Radiol. [Internet]. 2012 [cited 16 Apr 2012]. Available from: http://www.ncbi.nlm.nih.gov/pubmed/22227258.
43. Matthews S. Short communication: imaging pulmonary embolism in pregnancy: what is the most appropriate imaging protocol? Br J Radiol. [Internet]. 2006;79(941):441–4 [cited 30 Dec 2013]. Available from: http://www.ncbi.nlm.nih.gov/pubmed/16632627.
44. Baysinger CL. Imaging during pregnancy. Anesth Analg. [Internet]. 2010;110(3):863–7 [cited 30 Dec 2013]. Available from: http://www.ncbi.nlm.nih.gov/pubmed/20185662.
45. Topatan B, Basaran A. Imaging during pregnancy: computed tomography pulmonary angiography versus ventilation perfusion scintigraphy. Anesth Analg. [Internet]. 2011;112(2):483–4 [cited 14 Dec 2013]. author reply 484–5. Available from: http://www.ncbi.nlm.nih.gov/pubmed/21257704.

46. Lee CH, Goo JM, Ye HJ, Ye S-J, Park CM, Chun EJ, et al. Radiation dose modulation techniques in the multidetector CT era: from basics to practice. Radiographics [Internet]. 2008;28(5):1451–9. Available from: http://www.ncbi.nlm.nih.gov/pubmed/18794318.
47. Funama Y, Taguchi K, Utsunomiya D, Oda S, Yanaga Y, Yamashita Y, et al. Combination of a low-tube-voltage technique with hybrid iterative reconstruction (iDose) algorithm at coronary computed tomographic angiography. J Comput Assist Tomogr. [Internet]. 2011;35(4):480–5 [cited 28 Dec 2013]. Available from: http://www.pubmedcentral.nih.gov/articlerender.fcgi?artid=3151159&tool=pmcentrez&rendertype=abstract.
48. Kihara S, Murazaki H, Hatemura M, Sakumura H, Morisaki T, Funama Y. Radiation reduction and image quality improvement with iterative reconstruction at multidetector-row computed tomography. Nihon Hoshasen Gijutsu Gakkai Zasshi [Internet]. 2011;67(11):1426–32 [cited 28 Dec 2013]. Available from: http://www.ncbi.nlm.nih.gov/pubmed/22104234.
49. Kazakauskaite E, Husmann L, Stehli J, Fuchs T, Fiechter M, Klaeser B, et al. Image quality in low-dose coronary computed tomography angiography with a new high-definition CT scanner. Int J Cardiovasc Imaging [Internet]. 2013;29(2):471–7 [cited 28 Dec 2013]. Available from: http://www.ncbi.nlm.nih.gov/pubmed/22825255.
50. Hou Y, Liu X, Xv S, Guo W, Guo Q. Comparisons of image quality and radiation dose between iterative reconstruction and filtered back projection reconstruction algorithms in 256-MDCT coronary angiography. AJR Am J Roentgenol. [Internet]. 2012;199(3):588–94. [cited 28 Dec 2013]. Available from: http://www.ncbi.nlm.nih.gov/pubmed/22915398.
51. Fuchs TA, Fiechter M, Gebhard C, Stehli J, Ghadri JR, Kazakauskaite E, et al. CT coronary angiography: impact of adapted statistical iterative reconstruction (ASIR) on coronary stenosis and plaque composition analysis. Int J Cardiovasc Imaging [Internet]. 2013;29(3):719–24 [cited 28 Dec 2013]. Available from: http://www.ncbi.nlm.nih.gov/pubmed/23053859.
52. Sinitsyn VE, Glazkova MA, Mershina EA, Arkhipova IM. Possibilities of decreasing radiation load during MSRT coronarography: using adaptive statistic iterative reconstruction. Angiol Sosud Khir. [Internet]. 2012;18(3):44–9 [cited 28 Dec 2013]. Available from: http://www.ncbi.nlm.nih.gov/pubmed/23059606
53. Nakaura T, Kidoh M, Sakaino N, Utsunomiya D, Oda S, Kawahara T, et al. Low contrast- and low radiation dose protocol for cardiac CT of thin adults at 256-row CT: usefulness of low tube voltage scans and the hybrid iterative reconstruction algorithm. Int J Cardiovasc Imaging [Internet]. 2013;29(4):913–23 [cited 28 Dec 2013]. Available from: http://www.ncbi.nlm.nih.gov/pubmed/23160977.
54. Geyer LL, Körner M, Hempel R, Deak Z, Mueck FG, Linsenmaier U, et al. Evaluation of a dedicated MDCT protocol using iterative image reconstruction after cervical spine trauma. Clin Radiol. [Internet]. 2013;68(7):e391–6 [cited 28 Dec 2013]. Available from: http://www.ncbi.nlm.nih.gov/pubmed/23537577.
55. Laqmani A, Buhk JH, Henes FO, Klink T, Sehner S, von Schultzendorff HC, et al. Impact of a 4th generation iterative reconstruction technique on image quality in low-dose computed tomography of the chest in immunocompromised patients. Rofo [Internet]. 2013;185(8):749–57 [cited 28 Dec 2013]. Available from: http://www.ncbi.nlm.nih.gov/pubmed/23749649.
56. Sorantin E, Riccabona M, Stücklschweiger G, Guss H, Fotter R. Experience with volumetric (320 rows) pediatric CT. Eur J Radiol. [Internet]. 2012 [cited 16 Apr 2012]. Available from: http://www.ncbi.nlm.nih.gov/pubmed/22227261.
57. Brenner D, Elliston C, Hall E, Berdon W. Estimated risks of radiation-induced fatal cancer from pediatric CT. AJR Am J Roentgenol. 2001;176(2):289–96.
58. European Society of Urogenital Radiology. Guidelines of the European Society of Urogenital Radiology. Guidelines of the European Society of Urogenital Radiology [serial online] 2010; [cited 2010 Apr 27]. Online resource: http://www.esur.org/guidelines/.
59. Brodsky JB, Cohen EN, Brown Jr BW, Wu ML, Whitcher C. Surgery during pregnancy and fetal outcome. Am J Obstet Gynecol. 1980;138(8):1165–7.
60. Steinberg ES, Santos AC. Surgical anesthesia during pregnancy. Int Anesthesiol Clin. 1990;28(1):58–66.
61. Mazze RI, Kallen B. Reproductive outcome after anesthesia and operation during pregnancy: a registry study of 5405 cases. Am J Obstet Gynecol. 1989;161(5):1178–85.
62. Kallen B, Mazze RI. Neural tube defects and first trimester operations. Teratology. 1990;41(6):717–20.
63. Meroz Y, Elchalal U, Ginosar Y. Initial trauma management in advanced pregnancy. Anesthesiol Clin. 2007;25(1):117–29, x.
64. Cook PT. The influence on foetal outcome of maternal carbon dioxide tension at caesarean section under general anaesthesia. Anaesth Intensive Care. 1984;12(4):296–302.
65. Visser W, Wallenburg HC. Central hemodynamic observations in untreated preeclamptic patients. Hypertension. 1991;17(6 Pt 2):1072–7.
66. Wallenburg HC. Invasive hemodynamic monitoring in pregnancy. Eur J Obstet Gynecol Reprod Biol. 1991;42(Suppl):S45–51.
67. Schueler BA, Vrieze TJ, Bjarnason H, Stanson AW. An investigation of operator exposure in interventional radiology. Radiographics. 2006;26(5):1533–41.
68. Timberlake GA, McSwain Jr NE. Trauma in pregnancy. A 10-year perspective. Am Surg. 1989;55(3):151–3.
69. James AH, Tapson VF, Goldhaber SZ. Thrombosis during pregnancy and the postpartum period. Am J Obstet Gynecol. 2005;193(1):216–9.
70. Gates S, Brocklehurst P, Davis L J. Prophylaxis for venous thromboembolic disease in pregnancy and the early postnatal period. Cochrane Database Syst Rev. 2002;(2):CD001689.
71. Jacobsen AF, Skjeldestad FE, Sandset PM. Ante- and postnatal risk factors of venous thrombosis: a hospital-

based case–control study. J Thromb Haemost. 2008;6(6): 905–12.
72. James AH, Jamison MG, Brancazio LR, Myers ER. Venous thromboembolism during pregnancy and the postpartum period: incidence, risk factors, and mortality. Am J Obstet Gynecol. 2006;194(5): 1311–5.
73. Andersen AS, Berthelsen JG, Bergholt T. Venous thromboembolism in pregnancy: prophylaxis and treatment with low molecular weight heparin. Acta Obstet Gynecol Scand. 2010;89(1):15–21.
74. Dunlop DJ, McCahill JP, Blakemore ME. Internal fixation of an acetabular fracture during pregnancy. Injury. 1997;28(7):481–2.
75. Leggon RE, Wood GC, Indeck MC. Pelvic fractures in pregnancy: factors influencing maternal and fetal outcomes. J Trauma. 2002;53(4):796–804.
76. Loegters T, Briem D, Gatzka C, Linhart W, Begemann PG, Rueger JM, Windolf J. Treatment of unstable fractures of the pelvic ring in pregnancy. Arch Orthop Trauma Surg. 2005;125(3):204–8.
77. Yosipovitch Z, Goldberg I, Ventura E, Neri A. Open reduction of acetabular fracture in pregnancy. A case report. Clin Orthop Relat Res. 1992;282:229–32.
78. Prokop A, Swol-Ben J, Helling HJ, Neuhaus W, Rehm KE. Trauma in the last trimester of pregnancy. Unfallchirurg. 1996;99(6):450–3.
79. Weiss HB, Songer TJ, Fabio A. Fetal deaths related to maternal injury. JAMA. 2001;286(15):1863–8.
80. Esposito TJ, Gens DR, Smith LG, Scorpio R, Buchman T. Trauma during pregnancy. A review of 79 cases. Arch Surg. 1991;126(9):1073–8.
81. Rogers FB, Rozycki GS, Osler TM, Shackford SR, Jalbert J, Kirton O, Scalea T, Morris J, Ross S, Cipolle M, Fildes J, Cogbill T, Bergstein J, Clark D, Frankel H, Bell R, Gens D, Cullinane D, Kauder D, Bynoe RP. A multi-institutional study of factors associated with fetal death in injured pregnant patients. Arch Surg. 1999;134(11):1274–7.
82. Scorpio RJ, Esposito TJ, Smith LG, Gens DR. Blunt trauma during pregnancy: factors affecting fetal outcome. J Trauma. 1992;32(2):213–6.
83. Kissinger DP, Rozycki GS, Morris Jr JA, Knudson MM, Copes WS, Bass SM, Yates HK, Champion HR. Trauma in pregnancy. Predicting pregnancy outcome. Arch Surg. 1991;126(9):1079–86.

Open Fractures: Initial Management

19

Michael Frink and Steffen Ruchholtz

Contents

19.1	Introduction and Historical Background	261
19.2	Patient Evaluation and Diagnostics	262
19.3	Surgical Treatment	263
19.3.1	Hair Removal	263
19.3.2	Debridement and Irrigation	264
19.3.3	Time to Debridement	265
19.4	Antibiotic Treatment	265
19.5	Fracture Stabilization	266
19.6	Timing of Wound Closure	268
19.7	Limb Salvage Versus Amputation: Do Scores Help?	271
19.8	Future Perspectives	271
Conclusion		272
References		272

M. Frink, MD, MHBA (✉) • S. Ruchholtz, MD
Department for Trauma, Hand and Reconstructive Surgery, University Medical Center,
Baldingerstraße, Marburg 35043, Germany
e-mail: frink@med.uni-marburg.de;
michaelfrink@web.de;
ruchholt@med.uni-marburg.de

19.1 Introduction and Historical Background

More than 300 years ago, an open fracture commonly resulted in a local infection and in consecutive sepsis and death when no early amputation was performed [1, 2]. This treatment strategy continued back to the end of the eighteenth century and beginning of the nineteenth century [3]. Until World War I mortality rate after open fractures remained greater than 80 %. Improvement in surgical techniques, wound management, and antimicrobial therapy dramatically reduced mortality associated with open wounds [2, 4].

Open limb fractures mostly affect the lower extremity especially the lower leg due to minor soft tissue coverage. The incidence of tibial fracture is described as 17–21 per 100,000 persons while the incidence for open fractures is 11.5 per 100,000 persons per year [5, 6]. Open tibial fractures represent 44.5 % of all open fractures [5] and are mostly caused by high-energy mechanisms. Therefore, additional severe injuries are common in those patients. Treatment costs for the most severe open injuries are calculated up to half a million US$ per patient [7]. Until 30 years ago, treatment of open fractures resulted in an amputation rate of 15 % [2]. Thus, treatment of open fractures remains a medical as well as socioeconomic challenge and may be even complicated in the context of additional injuries.

© Springer-Verlag Berlin Heidelberg 2016
H.-C. Pape et al. (eds.), *The Poly-Traumatized Patient with Fractures:
A Multi-Disciplinary Approach*, DOI 10.1007/978-3-662-47212-5_19

Initial treatment consists of early appropriate antibiotic and surgical treatment. Surgical management includes thorough debridement of contaminated and nonviable soft tissue as well as bone, internal or external fracture stabilization, and timed wound coverage.

While initial systemic treatment with antibiotics is "almost universal in all high income countries" [8] controversies regarding local treatment persists. This chapter focuses on local wound treatment and highlights current controversies in open fracture management.

19.2 Patient Evaluation and Diagnostics

While preclinical treatment primarily focuses on preventing further contamination and tissue damage in the early clinical period careful examination and diagnostics represent the first steps in open fracture management.

A detailed history of mechanism and setting of injury should be obtained as well as the tetanus immunization status should be evaluated.

It remains controversial at which time point the preclinical applied wound cover should be removed. In a study investigating the value of initial wound cultures in 5 of 7 wound infections, primary wound cultures were negative suggesting a potential further contamination during the clinical course. In contrast, an open fracture should be examined by an experienced surgeon to plan further diagnostic and therapeutic steps as early as possible which may require removal of the preclinical applied wound dressing in the emergency department.

A physical examination should be performed particularly focusing on the neurovascular status of the affected anatomic region. The vascular status should be documented and in absence of palpable pulses Doppler ultrasonography or apparative imaging to detect vascular injuries may be required. Most patients with severe injuries are admitted after analgosedation on scene and therefore reliable information regarding sensory and motor impairment will be limited.

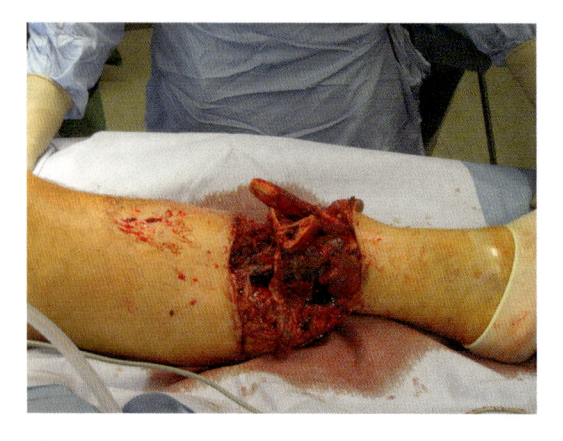

Fig. 19.1 Open fracture of the right lower limb following a motorcycle accident showing extensive loss of soft tissue injury, periosteal stripping an major wound contamination (Type IIIb following the Gustilo classification [13])

Since acute compartment syndromes requiring a fasciotomie are associated with open fractures [9] an exclusion of this complication is mandatory. Clinical diagnosis of an acute compartment syndrome is limited in severely injured patients; thus, measurement of intracompartmental pressure should be considered [10].

In the literature, various classification systems of open fractures are described. In 1959, a grading system was published on the basis of possible wound closure [11]. The currently most common is the classification of Gustilo and Anderson [12] which was amended in 1984 with a further differentiation of type III open fractures (Fig. 19.1) (Table 19.1) [13]. However, there remain certain problems regarding the interobserver reliability of this classification system [14, 15] which may raise problems in comparing therapeutic and outcome studies.

In 2010, a new system was introduced by the Orthopedic Trauma Association based on an extensive literature search as well as weighting by an expert panel. The Orthopedic Trauma Association Open Fracture Classification (OTA-OFC) considers skin injury, muscle injury, arterial injury, contamination and bone loss, and each category is subdivided into three grades of severity (Table 19.2) [16]. This new grading system showed moderate to excellent results regarding the interobserver reliability in a first evaluation [17] and therefore may serve as a new classification system in future investigations. Moreover, the OTA-OFC showed a predictive value regarding treatment strategies [18].

Table 19.1 Classification of open fractures [12, 13]

Degree	Characteristics
Type I	Wound less than one centimeter long Clean
Type II	Laceration more than one centimeter long Without extensive soft-tissue damage, flaps, or avulsions

Type III	Open segmental fracture or
	Open fracture with extensive soft-tissue damage or
	Traumatic amputation
	Special categories:
	  Gunshot injuries
	  Any open fracture caused by a farm injury
	  Any open fracture with accompanying vascular injury requiring repair

Type IIIa	Open fractures with adequate soft tissue coverage of a fractured bone despite Extensive soft tissue laceration or flaps, or high-energy trauma regardless of the size of the wound
Type IIIb	Extensive soft tissue injury with periosteal stripping and bone exposure Usually associated with massive contamination
Type IIIc	Arterial injury requiring repair

Per definition, an open fracture requires communication of the fracture site with a skin discontinuity. If exclusion is not possible or there remains any doubt, the fracture should be treated as an open fracture.

Regardless of the soft tissue damage X-rays in two views (ap. and lateral) with adjacent joints should be performed. Dependent on the patient's status articular fractures should be visualized with CT imaging before definitive treatment is performed.

19.3 Surgical Treatment

19.3.1 Hair Removal

Removal of gross contamination should be performed before disinfection of the surgical site. Historically, preparation for surgery included routine hair removal of the surgical site. Although no studies are specifically focusing on open fracture treatment, a recent Cochrane review was not able to recommend a strategy for or against hair removal before surgery. When hair removal is necessary, a clipper should be used since shaving

Table 19.2 Orthopedic trauma association open fracture classification [18]

Parameter	Skin	Muscle	Arterial	Contamination	Bone loss
1	Laceration with edges that approximate	None or minor muscle tissue necrosis; some muscle tissue injury with intact muscle unit function	No major vessel disruption	No grossly visible contamination	None or insignificant bone loss
2	Laceration with edges that does not approximate	Loss of significant muscle tissue but the involved muscle unit remains in longitudinal continuity	Vessel injury without distal ischemia	Surface contamination, visible	Bone missing or devascularized bone fragments, but some contact between proximal and distal fragments
3	Laceration associated with extensive degloving	Extensive muscle necrosis with loss of muscle tendon unit continuity, (partial or complete compartment excision, muscle defect that does not reapproximate)	Vessel injury with distal ischemia	Contaminant embedded in bone or deep soft tissues or high risk environmental conditions (barnyard, fecal, dirty water, etc.)	Segmental bone loss

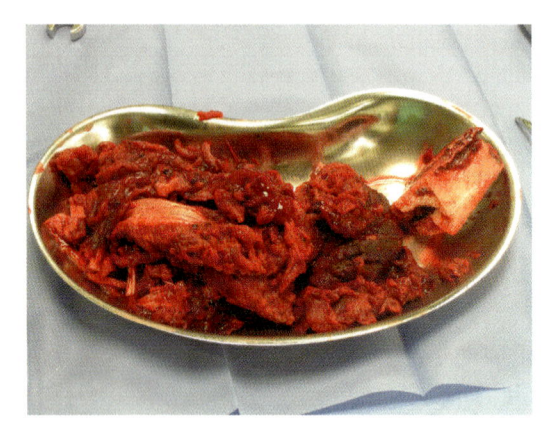

Fig. 19.2 Discarded contaminated and devitalized soft tissue and isolated bone fragments after thorough debridement of an open fracture of the lower leg

was associated with an increased incidence of surgical site infections [19].

19.3.2 Debridement and Irrigation

Debridement and irrigation remain the key steps in surgical treatment of open fractures. Small puncture wounds or lacerations need to be extended until complete soft tissue damage can be evaluated. Debridement should include grossly contaminated and/or devitalized soft tissue. Since perfusion is impaired in this tissue the host defense system is not able to eliminate foreign organisms. Moreover, if dead or contaminated tissue remains it serves as a medium for bacterial growth. Thus, debridement of subcutaneous tissue, muscle, and fascia is necessary until viability of marginal tissue is preserved and a clean and stable wound is achieved.

Fragments of devitalized bone or bone unattached from soft tissue should be discarded (Fig. 19.2). A most recent study comparing more or less aggressive bone debridement in type II and III open supracondylar femur fractures suggests bone debridement restricted to grossly contaminated bone "with retention of other bone fragment" and without use of antibiotic cement spacers. The less aggressive protocol leads to more frequent healing after the index procedure without increasing the incidence of infections [20] suggesting a cautious debridement.

In fractures with extensive soft tissue injury, initial evaluation of tissue viability is difficult. In those patients, a repeat debridement after 48–72 h "to eliminate devitalized tissue that subsequently develops" should be performed [21].

Multiple irrigation protocols in open fracture treatment including distilled water, boiled water, soap, antibiotic solutions as well as antiseptic solutions are described [22–24]. Regarding antiseptic solutions, povidone iodine wound irrigations [25, 26] and chlorhexidine gluconate [27] were evaluated. These antiseptic solutions show a broad spectrum activity versus antimicrobials including bacteria, viruses, and fungi [28]. However, to date, all clinical trials failed to show beneficial results regarding the incidence of infections. These observations are supported from experimental studies in which various antiseptic irrigation solutions were tested [29–31]. Moreover, toxicity towards intrinsic cells may cause problems in wound healing.

In a recent prospective randomized trial comparing soap solution with normal saline, no differences regarding infection, wound healing problems, and nonunion were observed [32].

In experimental studies, use of antibiotic solutions with an antimicrobial effect was able to show beneficial effects in infection prevention as compared to normal saline which could not be confirmed in human studies [33]. In an international survey, the majority of responding surgeons favored normal saline solution for irrigation in open fractures [34].

To date, there is no convincing data showing superior effects of solutions other than normal saline for irrigation of wounds associated with open fractures.

Pulsatile lavages for wound irrigation using different pressures are commercially available (Fig. 19.3). Under experimental settings, high pressure pulsatile lavage caused penetration of particulates up to 15.6 mm; cellular damage was detected 1.3 mm under the surface while low pressure pulsatile lavage showed penetration of particulates and cell death in 0.7 mm below tissue

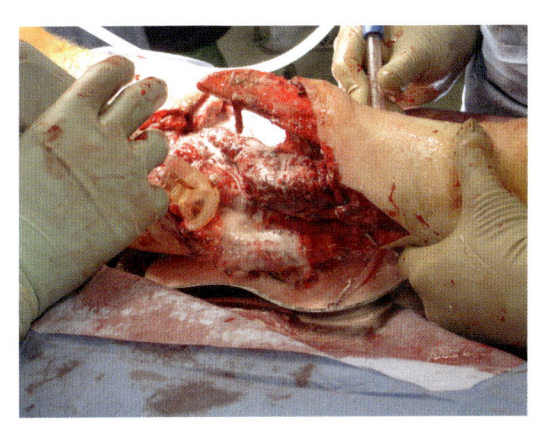

Fig. 19.3 Low pressure pulsatile lavage during the initial surgical debridement

surface [35]. In parallel, bacterial penetration was twofold increased by high pressure pulsatile lavage and number of retained bacteria was two- to fourfold lower after low pressure pulsatile lavage [36]. In another study, propagation of bacteria into the intramedullary canal due to high pressure irrigation was detected [37]. High pressure pulsatile lavage showed no beneficial effects regarding removal of adherent bacteria after 3 h delay as compared to low pressure pulsatile lavage. In contrast after 6 h delay, only high pressure pulsatile lavage removed adherent bacteria from bone [38].

Experimental studies investigating the effect of high pressure pulsatile lavage consistently showed detrimental effects on fracture healing. In a rabbit model, high pressure pulsatile lavage resulted in diminished bone density during the early phase (2 weeks) of fracture healing [39]. Mechanical testing of fractured femora in a rat model revealed a decreased mechanical stability 3 weeks after high-pressure irrigation [40].

In a clinical study, the use of low pressure pulsatile lavage showed superior results regarding the infection-associated reoperation rate when compared to high pressure devices [32].

Low pressure pulsatile lavage seems sufficient for bacterial clearance and is associated with less detrimental effects regarding propagation of bacteria and bone healing.

19.3.3 Time to Debridement

Three decades ago, initial treatment of all open fractures was stated as an emergency procedure [41]. In accordance, clinical guidelines recommended initial debridement within 6 h following injury, commonly known as the "six-hour rule" [42]. Circumstances of infrastructural or human resources as well as the patient's condition may delay the period between injury and surgical treatment raising the question of the consequences. A recent systematic review including six prospective and ten retrospective cohort studies with 3539 open fractures provides an overview about this issue. For type I and II injuries, the infection rate was 12 % following early (within 6 h) and 5 % following late debridement without reaching statistical significance [42]. In type III fractures, early debridement resulted in 15 % and late debridement in an 11 % infection rate without reaching statistical significance [42]. A beneficial effect of early surgical debridement could not be shown in dependence of anatomic location or when only deep infections were considered. In children, delay in surgical debridement (>6 h) did not influence the rates of acute infection as well [43]. Additionally, early debridement did not lead to improved bone healing [44].

Surgical debridement should be performed within 12 h after injury. There seems no evidence for the historical six-hour rule in the current literature.

19.4 Antibiotic Treatment

Superficial and deep infections including bony structures and joints remain a major problem in patients suffering from open fractures. In polytraumatized patients, reduced perfusion of the fracture may additionally impair the body's response to contamination. Besides surgical debridement, administration of antibiotics is recommended in various guidelines [45, 46]. The administration of antibiotics before or at the time of primary treatment reduced the number of early infections independently of agents used or duration [8]. There is consensus that administration should be initiated within 1 hour after detection of an open fracture [47]. Prehospital treatment of open fractures with antibiotics was associated with a shortened time interval between injury and antibiotic administration. However, it did not show an additional benefit regarding incidence of infections or impaired fracture healing [48].

Multiple studies investigated the effectiveness of different antibiotics for infection prevention [8, 46]. Although antibiotic treatment is routinely used for therapeutic purposes in open fractures, most studies suffer from "methodologic problems, including commingling of prospective and retrospective data sets, absence of or inappropriate statistical analysis, lack of blinding, or failure of randomization" [46]. However, the most current Cochrane review states that "further placebo controlled randomized trials are unlikely to be justified in middle and high income countries" [8].

For Type I and type II fractures, it is recommended to cover gram-positive germs, especially against *Staphylococcus aureus* [46]. Antibiotic treatment of type III fractures should be effective against gram-negative bacteria and therefore additional treatment with an aminoglycoside is recommended [8].

Duration of treatment with antibiotics is mostly dependent on wound closure. In most studies, antibiotic coverage for 24 h after primary closure was sufficient. Thus, antibiotic coverage for 2–3 days should be sufficient for type I and II fractures. More severe open fractures may benefit from prolonged treatment.

Systemic antibiotic treatment should be initiated as early as possible after injury. Cephalosporines covering a gram positive spectrum can be used for type I and II injuries, for type III fractures an aminoglycoside should be added to cover gram-negative bacterias.

Local antibiotic administration using the bead pouch technique produce high local levels of antibiotics in the wound. This should inhibit local bacterial growth without side effects induced by systemic administration. Especially in complication wounds that cannot be initially closed local antibiotic treatment offers a tool in wound management. In an animal study, deleterious effects of gentamycine on bone healing were excluded [49]. The advantage of high concentrations of locally administered antibiotics seems questionable after introducing negative-pressure wound therapy for temporary closure [50].

Only few retrospective clinical trials investigated the benefit of local antibiotic administration. In a prophylactic approach, local antibiotic treatment with tobramycin-impregnated bead chains decreased incidences of wound infections and osteomyelitis when compared to a systemic antibiotic triple therapy with cefazolin, tobramycin, and penicillin [51]. The beneficial effect of tobramycine beads seems pronounced in higher degree open fractures [52, 53]. The same authors confirmed the results in a study including more than 1000 patients in which local aminoglycoside treatment resulted in a decrease of the infection rate from 12 to 3.7 % [54]. The only prospective study comparing local and systemic antibiotic treatment included 67 patients with 75 open fractures. No difference regarding incidence of infection was detected. In addition, local treatment in conjunction with intravenous antibiotics did not show an additional benefit [55].

According to the actual literature, a beneficial effect of local antibiotic administration in open fractures is limited based on studies with low evidence.

19.5 Fracture Stabilization

Mechanical stabilization of open fractures is a key step in the treatment strategy. Surgical stabilization of the fracture prevents secondary damage to adjacent soft tissue, and decreases the incidence of infection. Various factors including patient's parameter (e.g., age, comorbidities, etc.), additional injuries, and localization of the fracture influence the type of fracture stabilization. Common techniques include external fixation, intramedullary stabilization, conventional and locking plate osteosynthesis as well as screw and wire fixation. During the last two decades the concept of damage control orthopedics changed treatment strategies in polytraumatized patients [56, 57]. Considering the patient's status with regard to hemodynamics as well as head and thoracic injuries temporary stabilization of long bone fractures with external fixation devices (Fig. 19.4a, b) will reduce further surgical harm and therefore prevent an exaggerated posttraumatic immune response [58, 59].

Insofar, complex reconstruction of articular surfaces is not part of the initial management. In these cases, mechanical stabilization is achieved with a temporary external fixation. Placement of pins should consider future surgical approaches, implant placement or local flaps.

Fixation with external devices is fast and easy to handle but a high incidence of pin tract infections, difficulties in soft tissue management and malunions are described when definitive stabilization was attempted [60]. When external fixation is used as a bridging device until the patient is stabilized and definitive internal fixation is performed during the clinical course no increased risk of infectious complications was evident [61]. Moreover, results regarding osseous healing showed a 97 % union rate at 6 months in open and closed femoral fractures following temporary stabilization [62]. Consistently, the interval between temporary external stabilization and definitive intramedullary nailing influences incidence of complications. The delay before conversion was identified as a risk factor for infections [61, 63, 64].

External fixation as definitive treatment has shown an increased risk for reoperation, nonunion, and infection as compared with unreamed nailing [65].

Definitive treatment of diaphyseal fractures with intramedullary nailing (Fig. 19.4c, d) should be considered in patients in type I and II fractures in whom adequate soft tissue coverage is possible and no severe contamination is present. Although only minimal incisions are required preventing further damage to the soft tissue in the fracture region, nail locking must be possible.

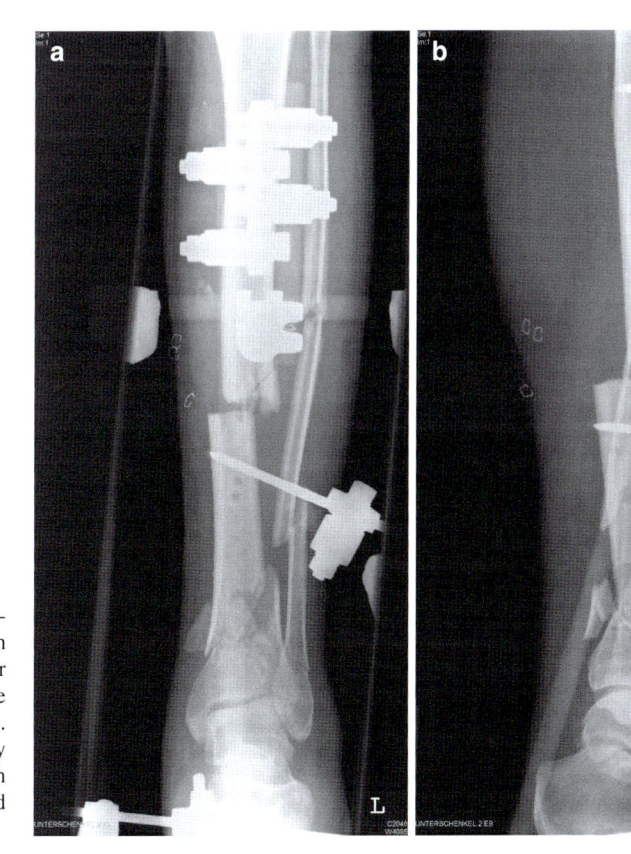

Fig. 19.4 X-ray of temporary external fixation in an open fracture of the lower limb (Type IIIb following the Gustilo classification) in ap. (**a**) and lateral (**b**) view. X-ray of definitive stabilization with a locking nail in ap. (**c**) and lateral (**d**) view

Fig. 19.4 (continued)

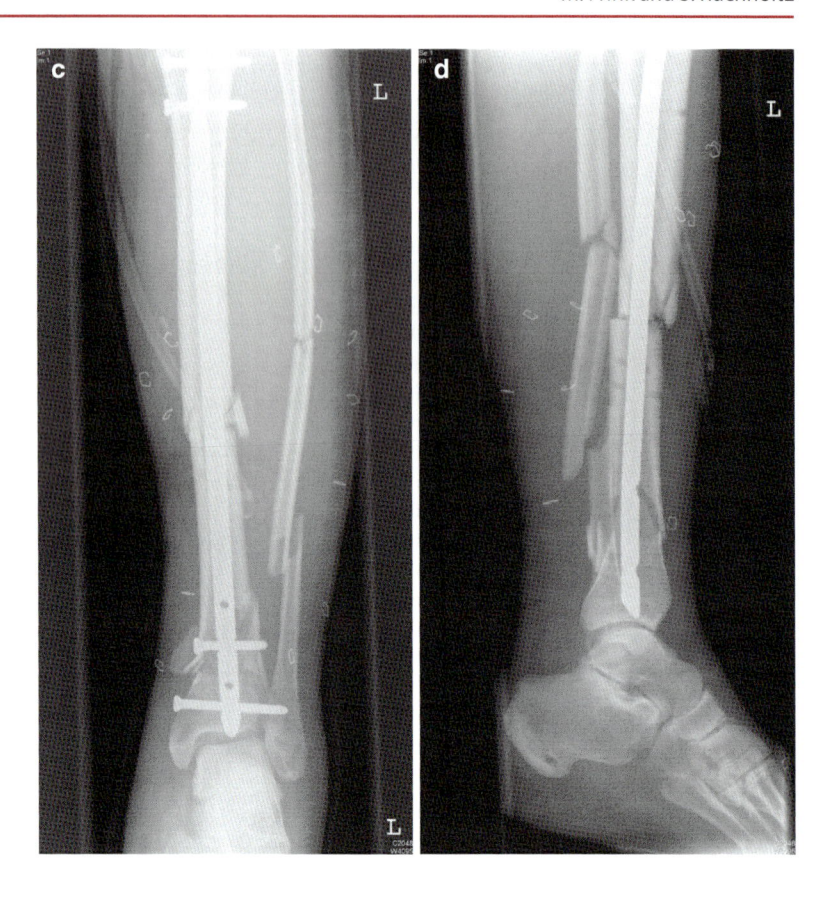

The debate of reamed versus unreamed nailing remains in the current literature. In a prospective multi-center study (SPRINT) including over 1300 patients with open and closed tibial fractures, no differences regarding required revision surgery between reamed and unreamed nailing were revealed. However, patients with open fractures treated with reamed nailing had an increased risk for revision surgery as compared to unreamed nailing [66]. In contrast, reamed nailing showed superior results in closed fractures [66]. An earlier meta-analysis including 2 randomized trials with 132 patients did not show any effects regarding revision surgery, infections or mal- and nonunions [65]. Nonetheless, an insignificant advantage for reamed nailing was detected which is in contrast to the SPRINT study [65].

Primary plate osteosynthesis for fixation in open fracture has been shown to be associated with high infection rates [67, 68]. Although introduction of locking plates allows minimal invasive surgical techniques, infection rates remain high [69]. In contrast, staged protocols including temporary external fixation were shown to decrease infection rates [70].

An algorithm for fracture stabilization is provided in Fig. 19.5.

19.6 Timing of Wound Closure

The question of wound closure timing remains a controversial debate since more than 25 years ago a landmark paper about emergency free flaps was published in 1988 [71]. Approximately 10 years before, open wound management was still propagated to prevent fatal infections such as gas gangrene [72]. The advantage of emergency free tissue transfer was seen in reduced infection rates occurring in longstanding open wounds [73]. Moreover delayed wound closure prolongs in-hospital time as well as treatment costs. A reduced infection rate and time to union due to microsurgical reconstruction within 72 h following injury was

Fig. 19.5 Algorithm for fracture stabilization considering patient's status and local injury characteristics

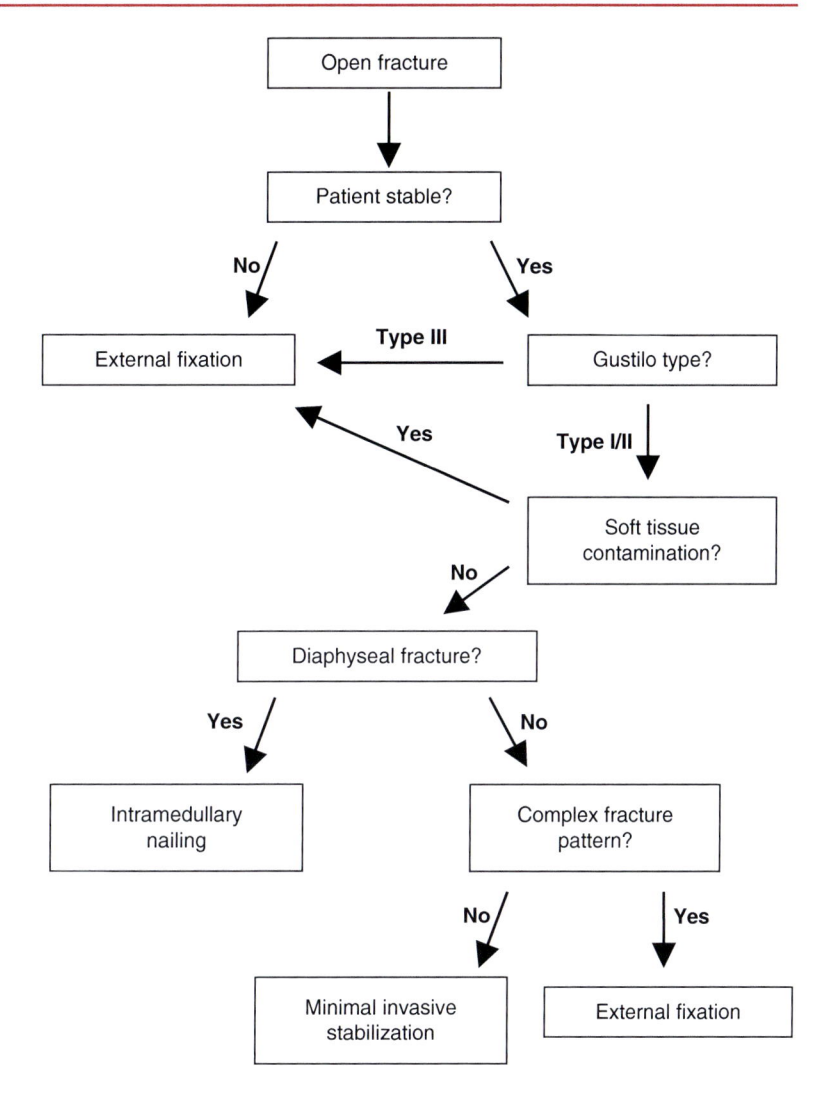

published 2 years earlier by Godina with inclusion of 532 patients [74]. The concept of early wound closure was supported by microbiologic analyses showing that initial microbial contamination before surgical debridement was not predicting occurrence of infections. In a prospective study investigating 117 open fractures, in only two of seven infections initial wound cultures were positive [75]. In another study, 39.3 % of predebridement swabs were positive while none of them developed an infection [76]. To date, the concept of fix and flap is still used and excellent results regarding infection rate, limb salvage as well as function were reported in a retrospective study [77]. However, the evidence for immediate closure remains low with mostly retrospective studies conducted before vacuum-assisted wound closure (VAC) was introduced. Additionally, most of the studies investigating early versus delayed wound closure were criticized since a selection bias with less severe injuries in the early closure group was evident [78, 79].

Nowadays, the concept of immediate or early wound closure requiring microsurgical techniques comes into conflict with the damage control approach. Free tissue transfer may take several hours and require repositioning of the patient during the surgical procedure [80] which may not be tolerated from patients suffering from severe head, pulmonary, and abdominal trauma.

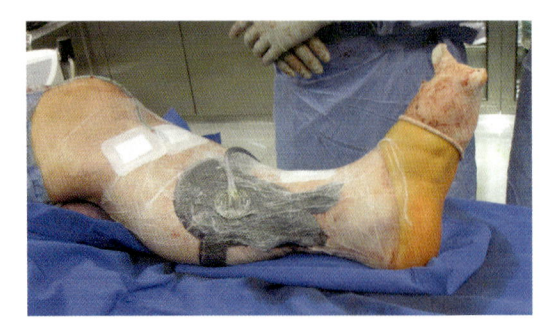

Fig. 19.6 Temporary vacuum-assisted wound closure after shortening and definitive stabilization before definitive wound coverage

Delayed wound closure was already suggested in 1959 when problems with more severe open fractures occurred [11]. The well-known study by Gustilo and Anderson revealed an increased infection rate up to 44 % in open segmental tibial fractures with extensive tissue injury [12]. Hence, they concluded that type I and type II injuries could be primarily closed while type III open fractures should benefit from a delayed closure. Following this treatment protocol a reduced infection rate from 14 to 4.5 % was observed [81].

In general, studies favoring delayed closure or not showing significant differences show a higher level of evidence with mostly prospective studies. Another problem remains with the different definitions of time periods referring to early and late wound closure [82] impeding the comparability of published studies.

The introduction of VAC-systems (Fig. 19.6) provided a bridging tool until definitive wound coverage can be performed. Negative-pressure wound therapy prevents further contamination and may increase perfusion close to the wound edge [83]. The VAC therapy stimulates granulation tissue, supports skin grafts and therefore prevents requirement of more complex reconstruction procedures [84, 85]. Treatment with VAC resulted in decreased edema improving local tissue perfusion which may have positive effects on wound healing [85]. Although a diminished number of bacteria due to VAC therapy were advocated, recent studies question this hypothesis [86, 87]. For type III open

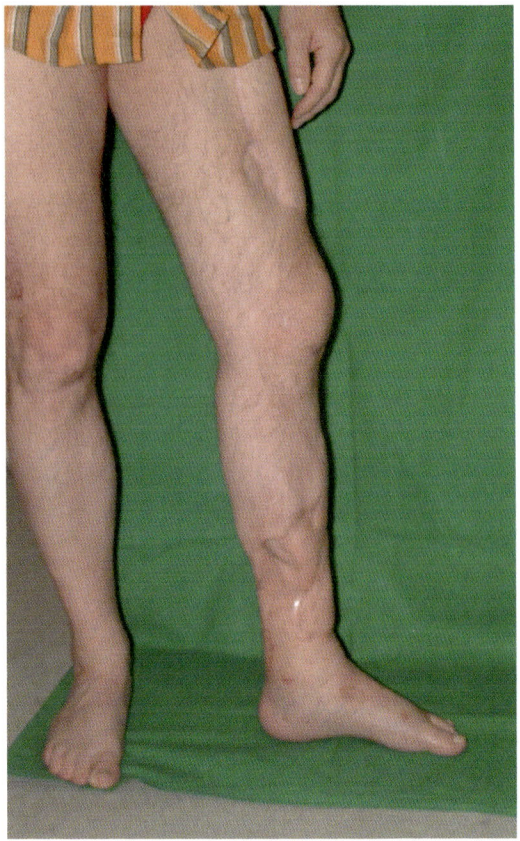

Fig. 19.7 Clinical result of patient shown in Figs. 19.4 and 19.6 one year after definitive wound closure

diaphyseal tibial fractures a reduction of required free tissue transfer due to application of negative-pressure wound therapy was shown (Fig. 19.7) [88].

In a recent communication, the benefit of using negative-pressure wound therapy in Gustilo Grade IIIB/IIIC open tibial fracture was questioned. In a pooled analysis including three retrospective case control studies with 119 patients a tendency towards an increased infection rate in patients treated VAC for more than 7 days was detected [89].

Following current recommendations, type I and II injuries should be primarily closed following a thorough debridement as described earlier. Type III injuries can be treated using negative-pressure wound therapy and staged coverage of soft tissue defects can be performed after transfer to a specialized center.

19.7 Limb Salvage Versus Amputation: Do Scores Help?

Soft tissue injury severity has the greatest impact on decision making regarding limb salvage versus amputation. Multiple scores predictive of the functional recovery of patients with open fractures are described. They should help in decision making whether limb salvage or amputation should be performed. The most commonly used scores are as follows:

- Mangled Extremity Severity Score (MESS) [90]
- Limb Salvage Index (LSI) [91]
- Predicting Salvage Index (PSI) [92]
- Nerve Injury, Ischemia, Soft Tissue Injury, Skeletal Injury, Shock and Age of Patients Score (NISSSA) [93]

The initial decision has an impact on the further course since results after delayed amputation are poor. Increased rate of infections are reported, when amputation was performed later than 5 days after injury [94]. In contrast, other reports from combat zones describe a high degree of patient satisfaction following elective late amputation [95].

During the Lower Extremity Assessment Project (LEAP), 601 patients with high-energy lower-extremity trauma were evaluated. In 407 patients, limb salvage and reconstruction was successful 6 months after discharge. None of the evaluated scores mentioned above was able to predict functional recovery in patients with successful limb reconstruction [96]. Most scores showed a high specificity and therefore low scores reliably predict limb salvage potential [97]. In contrast, a low sensitivity lead to a failure in predicting amputation [97]. Vascular reconstruction represents a time consuming procedure with more than 3 h at thigh or upper arm [98]. In more distal vascular injuries, a more complicated surgical procedure should be expected. Most polytraumatized patients will not sustain long and demanding operations without the risk of developing complications during the further clinical course.

To date, the assessment for or against limb salvage remains an individual decision. Scores predicting functional outcome are not reliable. Before complex surgeries for limb salvage are planned the status of the polytraumatized patients should be carefully examined.

19.8 Future Perspectives

Although advances in diagnostics and therapy of open fractures improved outcome, there still remain problems that need to be addressed in future research. Few aspects of latest innovations are described.

New debridement techniques were recently investigated in an experimental study [99]. A plasma-mediated bipolar radiofrequency ablation was able to reduce bacterial load as compared to a cold steel curette.

The most recent development in treatment of bone defects offers new therapeutic options. In an animal model, osteoinductive and osteoconductive bone-graft substitutes with tobramycin-impregnated calcium sulfate pellets placed in a bone defect resulted in prevention of *Staphylococcus aureus* infection [100]. The combination of bone grafts with antibiotic delivery may reduce number of required surgical procedures in the future.

The problem of deep infections leads to the development of new implants implication antibiotic-coated surfaces. Incorporated metal implants are known to promote infections and systemic antibiotic administration does not deliver antibiotics to the interface between tissue and implant. Antibiotic-coated implants may overcome this problem. Promising results from experimental animal studies [101] lead to a clinical trial using a gentamycine-coated tibia nail. In a preliminary study including closed as well as open fractures no implant-related infections occurred. No adverse effects were observed [102]. Other antibiotics (daptomycin [103], vancomycin [103], rifampicin [104], fusidic acid [104]) were tested in experimental settings.

To date, no trials with convincing data are available supporting routine clinical use of these implants.

Conclusion

Decision making in polytraumatized patients with open fractures still represents a major challenge. Adequate treatment requires an individual assessment of patient's comobidities, local damage as well as injury pattern and severity. The surgeon in charge should be experienced in polytrauma management and soft tissue injury treatment.

If there remains any doubt regarding local soft tissue injury and fracture pattern, a fracture should be treated as an open fracture and temporary negative-pressure wound therapy should be considered.

References

1. Keeman JN. Treatment of open fractures before Lister and the management of the fatal leg fracture of Admiral Michiel Adriaensz de Ruyter, 1676. Ned Tijdschr Geneeskd. 2004;148(52):2607–15.
2. Pape HC, Webb LX. History of open wound and fracture treatment. J Orthop Trauma. 2008;22(10 Suppl):S133–4.
3. Stansbury LG, Branstetter JG, Lalliss SJ. Amputation in military trauma surgery. J Trauma. 2007;63(4):940–4.
4. Pailler JL, Labeeu F. Gas gangrene: a military disease? Acta Chir Belg. 1986;86(2):63–71.
5. Court-Brown CM, Rimmer S, Prakash U, McQueen MM. The epidemiology of open long bone fractures. Injury. 1998;29(7):529–34.
6. Howard M, Court-Brown CM. Epidemiology and management of open fractures of the lower limb. Br J Hosp Med. 1997;57(11):582–7.
7. Chung KC, Saddawi-Konefka D, Haase SC, Kaul G. A cost-utility analysis of amputation versus salvage for Gustilo type IIIB and IIIC open tibial fractures. Plast Reconstr Surg. 2009;124(6):1965–73.
8. Gosselin RA, Roberts I, Gillespie WJ. Antibiotics for preventing infection in open limb fractures. Cochrane Database Syst Rev. 2004;1, CD003764.
9. Branco BC, Inaba K, Barmparas G, Schnuriger B, Lustenberger T, Talving P, et al. Incidence and predictors for the need for fasciotomy after extremity trauma: a 10-year review in a mature level I trauma centre. Injury. 2011;42(10):1157–63.
10. Frink M, Hildebrand F, Krettek C, Brand J, Hankemeier S. Compartment syndrome of the lower leg and foot. Clin Orthop Relat Res. 2010;468(4):940–50.
11. VELISKAKIS KP. Primary internal fixation in open fractures of the tibal shaft; the problem of wound healing. J Bone Joint Surg Br. 1959;41-B(2):342–54.
12. Gustilo RB, Anderson JT. Prevention of infection in the treatment of one thousand and twenty-five open fractures of long bones: retrospective and prospective analyses. J Bone Joint Surg Am. 1976;58(4):453–8.
13. Gustilo RB, Mendoza RM, Williams DN. Problems in the management of type III (severe) open fractures: a new classification of type III open fractures. J Trauma. 1984;24(8):742–6.
14. Horn BD, Rettig ME. Interobserver reliability in the Gustilo and Anderson classification of open fractures. J Orthop Trauma. 1993;7(4):357–60.
15. Brumback RJ, Jones AL. Interobserver agreement in the classification of open fractures of the tibia. The results of a survey of two hundred and forty-five orthopaedic surgeons. J Bone Joint Surg Am. 1994;76(8):1162–6.
16. Orthopaedic Trauma Association: Open Fracture Study Group. A new classification scheme for open fractures. J Orthop Trauma. 2010;24(8):457–64.
17. Agel J, Evans AR, Marsh JL, Decoster TA, Lundy DW, Kellam JF, et al. The OTA open fracture classification: a study of reliability and agreement. J Orthop Trauma. 2013;27(7):379–84.
18. Agel J, Rockwood T, Barber R, Marsh JL. Potential predictive ability of the orthopedic trauma association open fracture classification. J Orthop Trauma. 2014;28(5):300–6.
19. Tanner J, Norrie P, Melen K. Preoperative hair removal to reduce surgical site infection. Cochrane Database Syst Rev. 2011;11, CD004122.
20. Ricci WM, Collinge C, Streubel PN, McAndrew CM, Gardner MJ. A comparison of more and less aggressive bone debridement protocols for the treatment of open supracondylar femur fractures. J Orthop Trauma. 2013;27(12):722–5.
21. Melvin JS, Dombroski DG, Torbert JT, Kovach SJ, Esterhai JL, Mehta S. Open tibial shaft fractures: I. Evaluation and initial wound management. J Am Acad Orthop Surg. 2010;18(1):10–9.
22. Museru LM, Kumar A, Ickler P. Comparison of isotonic saline, distilled water and boiled water in irrigation of open fractures. Int Orthop. 1989;13(3):179–80.
23. Anglen JO. Comparison of soap and antibiotic solutions for irrigation of lower-limb open fracture wounds. A prospective, randomized study. J Bone Joint Surg Am. 2005;87(7):1415–22.
24. Rosenstein BD, Wilson FC, Funderburk CH. The use of bacitracin irrigation to prevent infection in postoperative skeletal wounds. An experimental study. J Bone Joint Surg Am. 1989;71(3):427–30.
25. Rogers DM, Blouin GS, O'Leary JP. Povidone-iodine wound irrigation and wound sepsis. Surg Gynecol Obstet. 1983;157(5):426–30.
26. Gilmore OJ, Sanderson PJ. Prophylactic interparietal povidone-iodine in abdominal surgery. Br J Surg. 1975;62(10):792–9.
27. Vallance S, Waldron R. Antiseptic vs. saline lavage in purulent and faecal peritonitis. J Hosp Infect. 1985;6(Suppl A):87–91.

28. Edmiston Jr CE, Bruden B, Rucinski MC, Henen C, Graham MB, Lewis BL. Reducing the risk of surgical site infections: does chlorhexidine gluconate provide a risk reduction benefit? Am J Infect Control. 2013;41(5 Suppl):S49–55.
29. Penn-Barwell JG, Murray CK, Wenke JC. Comparison of the antimicrobial effect of chlorhexidine and saline for irrigating a contaminated open fracture model. J Orthop Trauma. 2012;26(12):728–32.
30. Gaines RJ, DeMaio M, Peters D, Hasty J, Blanks J. Management of contaminated open fractures: a comparison of two types of irrigation in a porcine model. J Trauma Acute Care Surg. 2012;72(3):733–6.
31. Owens BD, White DW, Wenke JC. Comparison of irrigation solutions and devices in a contaminated musculoskeletal wound survival model. J Bone Joint Surg Am. 2009;91(1):92–8.
32. Petrisor B, Sun X, Bhandari M, Guyatt G, Jeray KJ, Sprague S, et al. Fluid lavage of open wounds (FLOW): a multicenter, blinded, factorial pilot trial comparing alternative irrigating solutions and pressures in patients with open fractures. J Trauma. 2011;71(3):596–606.
33. Anglen JO. Wound irrigation in musculoskeletal injury. J Am Acad Orthop Surg. 2001;9(4):219–26.
34. Petrisor B, Jeray K, Schemitsch E, Hanson B, Sprague S, Sanders D, et al. Fluid lavage in patients with open fracture wounds (FLOW): an international survey of 984 surgeons. BMC Musculoskelet Disord. 2008;9:7.
35. Boyd III JI, Wongworawat MD. High-pressure pulsatile lavage causes soft tissue damage. Clin Orthop Relat Res. 2004;427:13–7.
36. Hassinger SM, Harding G, Wongworawat MD. High-pressure pulsatile lavage propagates bacteria into soft tissue. Clin Orthop Relat Res. 2005;439:27–31.
37. Bhandari M, Adili A, Lachowski RJ. High pressure pulsatile lavage of contaminated human tibiae: an in vitro study. J Orthop Trauma. 1998;12(7):479–84.
38. Bhandari M, Schemitsch EH, Adili A, Lachowski RJ, Shaughnessy SG. High and low pressure pulsatile lavage of contaminated tibial fractures: an in vitro study of bacterial adherence and bone damage. J Orthop Trauma. 1999;13(8):526–33.
39. Caprise Jr PA, Miclau T, Dahners LE, Dirschl DR. High-pressure pulsatile lavage irrigation of contaminated fractures: effects on fracture healing. J Orthop Res. 2002;20(6):1205–9.
40. Adili A, Bhandari M, Schemitsch EH. The biomechanical effect of high-pressure irrigation on diaphyseal fracture healing in vivo. J Orthop Trauma. 2002;16(6):413–7.
41. Gustilo RB, Merkow RL, Templeman D. The management of open fractures. J Bone Joint Surg Am. 1990;72(2):299–304.
42. Schenker ML, Yannascoli S, Baldwin KD, Ahn J, Mehta S. Does timing to operative debridement affect infectious complications in open long-bone fractures? a systematic review. J Bone Joint Surg Am. 2012;94(12):1057–64.
43. Skaggs DL, Friend L, Alman B, Chambers HG, Schmitz M, Leake B, et al. The effect of surgical delay on acute infection following 554 open fractures in children. J Bone Joint Surg Am. 2005;87(1):8–12.
44. Singh J, Rambani R, Hashim Z, Raman R, Sharma HK. The relationship between time to surgical debridement and incidence of infection in grade III open fractures. Strategies Trauma Limb Reconstr. 2012;7(1):33–7.
45. Hoff WS, Bonadies JA, Cachecho R, Dorlac WC. East Practice Management Guidelines Work Group: update to practice management guidelines for prophylactic antibiotic use in open fractures. J Trauma. 2011;70(3):751–4.
46. Hauser CJ, Adams Jr CA, Eachempati SR. Surgical Infection Society guideline: prophylactic antibiotic use in open fractures: an evidence-based guideline. Surg Infect (Larchmt). 2006;7(4):379–405.
47. Obremskey WT, Molina CS, Collinge C, Tornetta III P, Sagi C, Schmidt A, et al. Current practice in the initial management of open fractures among orthopaedic trauma surgeons. J Orthop Trauma. 2013. [Epub ahead of print].
48. Thomas SH, Arthur AO, Howard Z, Shear ML, Kadzielski JL, Vrahas MS. Helicopter emergency medical services crew administration of antibiotics for open fractures. Air Med J. 2013;32(2):74–9.
49. Fassbender M, Minkwitz S, Kronbach Z, Strobel C, Kadow-Romacker A, Schmidmaier G, et al. Local gentamicin application does not interfere with bone healing in a rat model. Bone. 2013;55(2):298–304.
50. Large TM, Douglas G, Erickson G, Grayson JK. Effect of negative pressure wound therapy on the elution of antibiotics from polymethylmethacrylate beads in a porcine simulated open femur fracture model. J Orthop Trauma. 2012;26(9):506–11.
51. Henry SL, Ostermann PA, Seligson D. The prophylactic use of antibiotic impregnated beads in open fractures. J Trauma. 1990;30(10):1231–8.
52. Ostermann PA, Henry SL, Seligson D. Value of adjuvant local antibiotic administration in therapy of open fractures. A comparative analysis of 704 consecutive cases. Langenbecks Arch Chir. 1993;378(1):32–6.
53. Ostermann PA, Henry SL, Seligson D. The role of local antibiotic therapy in the management of compound fractures. Clin Orthop Relat Res. 1993;295:102–11.
54. Ostermann PA, Seligson D, Henry SL. Local antibiotic therapy for severe open fractures. A review of 1085 consecutive cases. J Bone Joint Surg Br. 1995;77(1):93–7.
55. Moehring HD, Gravel C, Chapman MW, Olson SA. Comparison of antibiotic beads and intravenous antibiotics in open fractures. Clin Orthop Relat Res. 2000;372:254–61.
56. Scalea TM, Boswell SA, Scott JD, Mitchell KA, Kramer ME, Pollak AN. External fixation as a bridge to intramedullary nailing for patients with multiple injuries and with femur fractures: damage control orthopedics. J Trauma. 2000;48(4):613–21.

57. D'Alleyrand JC, O'Toole RV. The evolution of damage control orthopedics: current evidence and practical applications of early appropriate care. Orthop Clin North Am. 2013;44(4):499–507.
58. Flierl MA, Stoneback JW, Beauchamp KM, Hak DJ, Morgan SJ, Smith WR, et al. Femur shaft fracture fixation in head-injured patients: when is the right time? J Orthop Trauma. 2010;24(2):107–14.
59. Taeger G, Ruchholtz S, Waydhas C, Lewan U, Schmidt B, Nast-Kolb D. Damage control orthopedics in patients with multiple injuries is effective, time saving, and safe. J Trauma. 2005;59(2):409–16.
60. Velazco A, Fleming LL. Open fractures of the tibia treated by the Hoffmann external fixator. Clin Orthop Relat Res. 1983;180:125–32.
61. Harwood PJ, Giannoudis PV, Probst C, Krettek C, Pape HC. The risk of local infective complications after damage control procedures for femoral shaft fracture. J Orthop Trauma. 2006;20(3):181–9.
62. Nowotarski PJ, Turen CH, Brumback RJ, Scarboro JM. Conversion of external fixation to intramedullary nailing for fractures of the shaft of the femur in multiply injured patients. J Bone Joint Surg Am. 2000;82(6):781–8.
63. McGraw JM, Lim EV. Treatment of open tibial-shaft fractures. External fixation and secondary intramedullary nailing. J Bone Joint Surg Am. 1988;70(6):900–11.
64. Blachut PA, Meek RN, O'Brien PJ. External fixation and delayed intramedullary nailing of open fractures of the tibial shaft. A sequential protocol. J Bone Joint Surg Am. 1990;72(5):729–35.
65. Bhandari M, Guyatt GH, Swiontkowski MF, Schemitsch EH. Treatment of open fractures of the shaft of the tibia. J Bone Joint Su rg Br. 2001;83(1):62–8.
66. Schemitsch EH, Bhandari M, Guyatt G, Sanders DW, Swiontkowski M, Tornetta P, et al. Prognostic factors for predicting outcomes after intramedullary nailing of the tibia. J Bone Joint Surg Am. 2012;94(19):1786–93.
67. Bach AW, Hansen Jr ST. Plates versus external fixation in severe open tibial shaft fractures. A randomized trial. Clin Orthop Relat Res. 1989;241:89–94.
68. Clifford RP, Beauchamp CG, Kellam JF, Webb JK, Tile M. Plate fixation of open fractures of the tibia. J Bone Joint Surg Br. 1988;70(4):644–8.
69. Kim JW, Oh CW, Jung WJ, Kim JS. Minimally invasive plate osteosynthesis for open fractures of the proximal tibia. Clin Orthop Surg. 2012;4(4):313–20.
70. Sohn OJ, Kang DH. Staged protocol in treatment of open distal tibia fracture: using lateral MIPO. Clin Orthop Surg. 2011;3(1):69–76.
71. Lister G, Scheker L. Emergency free flaps to the upper extremity. J Hand Surg Am. 1988;13(1):22–8.
72. Brown PW, Kinman PB. Gas gangrene in a metropolitan community. J Bone Joint Surg Am. 1974;56(7):1445–51.
73. Levin LS, Erdmann D. Primary and secondary microvascular reconstruction of the upper extremity. Hand Clin. 2001;17(3):447–55, ix.
74. Godina M. Early microsurgical reconstruction of complex trauma of the extremities. Plast Reconstr Surg. 1986;78(3):285–92.
75. Valenziano CP, Chattar-Cora D, O'Neill A, Hubli EH, Cudjoe EA. Efficacy of primary wound cultures in long bone open extremity fractures: are they of any value? Arch Orthop Trauma Surg. 2002;122(5):259–61.
76. Faisham WI, Nordin S, Aidura M. Bacteriological study and its role in the management of open tibial fracture. Med J Malaysia. 2001;56(2):201–6.
77. Gopal S, Majumder S, Batchelor AG, Knight SL, De Boer P, Smith RM. Fix and flap: the radical orthopaedic and plastic treatment of severe open fractures of the tibia. J Bone Joint Surg Br. 2000;82(7):959–66.
78. Ostermann PA, Henry SL, Seligson D. Timing of wound closure in severe compound fractures. Orthopedics. 1994;17(5):397–9.
79. Henley MB, Chapman JR, Agel J, Harvey EJ, Whorton AM, Swiontkowski MF. Treatment of type II, IIIA, and IIIB open fractures of the tibial shaft: a prospective comparison of unreamed interlocking intramedullary nails and half-pin external fixators. J Orthop Trauma. 1998;12(1):1–7.
80. Khan MA, Jose RM, Taylor C, Ahmed W, Prinsloo D. Free radial forearm fasciocutaneous flap in the treatment of distal third tibial osteomyelitis. Ann Plast Surg. 2012;68(1):58–61.
81. Patzakis MJ, Wilkins J, Moore TM. Considerations in reducing the infection rate in open tibial fractures. Clin Orthop Relat Res. 1983;178:36–41.
82. Levin LS. Early versus delayed closure of open fractures. Injury. 2007;38(8):896–9.
83. Borgquist O, Anesater E, Hedstrom E, Lee CK, Ingemansson R, Malmsjo M. Measurements of wound edge microvascular blood flow during negative pressure wound therapy using thermodiffusion and transcutaneous and invasive laser Doppler velocimetry. Wound Repair Regen. 2011;19(6):727–33.
84. Horch RE, Gerngross H, Lang W, Mauckner P, Nord D, Peter RU, et al. Indications and safety aspects of vacuum-assisted wound closure. MMW Fortschr Med. 2005;147 Suppl 1:1–5.
85. DeFranzo AJ, Argenta LC, Marks MW, Molnar JA, David LR, Webb LX, et al. The use of vacuum-assisted closure therapy for the treatment of lower-extremity wounds with exposed bone. Plast Reconstr Surg. 2001;108(5):1184–91.
86. Assadian O, Assadian A, Stadler M, Diab-Elschahawi M, Kramer A. Bacterial growth kinetic without the influence of the immune system using vacuum-assisted closure dressing with and without negative pressure in an in vitro wound model. Int Wound J. 2010;7(4):283–9.
87. Moues CM, van den Bemd GJ, Heule F, Hovius SE. Comparing conventional gauze therapy to vacuum-assisted closure wound therapy: a prospective

randomised trial. J Plast Reconstr Aesthet Surg. 2007;60(6):672–81.
88. Dedmond BT, Kortesis B, Punger K, Simpson J, Argenta J, Kulp B, et al. The use of negative-pressure wound therapy (NPWT) in the temporary treatment of soft-tissue injuries associated with high-energy open tibial shaft fractures. J Orthop Trauma. 2007;21(1):11–7.
89. Cheng HT, Hsu YC, Wu CI. Risk of infection with delayed wound coverage by using negative-pressure wound therapy in Gustilo Grade IIIB/IIIC open tibial fracture: an evidence-based review. J Plast Reconstr Aesthet Surg. 2013;66(6):876–8.
90. Helfet DL, Howey T, Sanders R, Johansen K. Limb salvage versus amputation. Preliminary results of the Mangled Extremity Severity Score. Clin Orthop Relat Res. 1990;256:80–6.
91. Russell WL, Sailors DM, Whittle TB, Fisher Jr DF, Burns RP. Limb salvage versus traumatic amputation. A decision based on a seven-part predictive index. Ann Surg. 1991;213(5):473–80.
92. Howe Jr HR, Poole Jr GV, Hansen KJ, Clark T, Plonk GW, Koman LA, et al. Salvage of lower extremities following combined orthopedic and vascular trauma. A predictive salvage index. Am Surg. 1987;53(4):205–8.
93. McNamara MG, Heckman JD, Corley FG. Severe open fractures of the lower extremity: a retrospective evaluation of the Mangled Extremity Severity Score (MESS). J Orthop Trauma. 1994;8(2):81–7.
94. Jain A, Glass GE, Ahmadi H, Mackey S, Simmons J, Hettiaratchy S, et al. Delayed amputation following trauma increases residual lower limb infection. J Plast Reconstr Aesthet Surg. 2013;66(4):531–7.
95. Helgeson MD, Potter BK, Burns TC, Hayda RA, Gajewski DA. Risk factors for and results of late or delayed amputation following combat-related extremity injuries. Orthopedics. 2010;33(9):669.
96. Ly TV, Travison TG, Castillo RC, Bosse MJ, MacKenzie EJ. Ability of lower-extremity injury severity scores to predict functional outcome after limb salvage. J Bone Joint Surg Am. 2008;90(8): 1738–43.
97. Bosse MJ, MacKenzie EJ, Kellam JF, Burgess AR, Webb LX, Swiontkowski MF, et al. A prospective evaluation of the clinical utility of the lower-extremity injury-severity scores. J Bone Joint Surg Am. 2001;83-A(1):3–14.
98. McHenry TP, Holcomb JB, Aoki N, Lindsey RW. Fractures with major vascular injuries from gunshot wounds: implications of surgical sequence. J Trauma. 2002;53(4):717–21.
99. Sonnergren HH, Strombeck L, Aldenborg F, Faergemann J. Aerosolized spread of bacteria and reduction of bacterial wound contamination with three different methods of surgical wound debridement: a pilot study. J Hosp Infect. 2013;85(2): 112–7.
100. Beardmore AA, Brooks DE, Wenke JC, Thomas DB. Effectiveness of local antibiotic delivery with an osteoinductive and osteoconductive bone-graft substitute. J Bone Joint Surg Am. 2005;87(1):107–12.
101. Lucke M, Schmidmaier G, Sadoni S, Wildemann B, Schiller R, Haas NP, et al. Gentamicin coating of metallic implants reduces implant-related osteomyelitis in rats. Bone. 2003;32(5):521–31.
102. Fuchs T, Stange R, Schmidmaier G, Raschke MJ. The use of gentamicin-coated nails in the tibia: preliminary results of a prospective study. Arch Orthop Trauma Surg. 2011;131(10):1419–25.
103. Smith JK, Bumgardner JD, Courtney HS, Smeltzer MS, Haggard WO. Antibiotic-loaded chitosan film for infection prevention: a preliminary in vitro characterization. J Biomed Mater Res B Appl Biomater. 2010;94(1):203–11.
104. Kalicke T, Schierholz J, Schlegel U, Frangen TM, Koller M, Printzen G, et al. Effect on infection resistance of a local antiseptic and antibiotic coating on osteosynthesis implants: an in vitro and in vivo study. J Orthop Res. 2006;24(8):1622–40.

Vascular Injuries: Indications for Stents, Timing for Vascular and Orthopedic Injuries

20

Luke P.H. Leenen

Contents

20.1	Introduction	277
20.2	**Signs and Symptoms**	277
20.2.1	Clinical Evaluation	277
20.2.2	Doppler Evaluation	278
20.2.3	Angiography	278
20.2.4	CT Angiogram	279
20.2.5	Digital Subtraction Angiography	279
20.2.6	MR Angiography	279
20.3	**Treatment**	279
20.3.1	General Tactics	279
20.3.2	Severe Torso Bleeding	280
20.3.3	Extremity Bleeding	280
20.3.4	Specific Anatomic Considerations	282
Conclusion		287
References		287

L.P.H. Leenen, MD, PhD, FACS
Department of Trauma, UMC Utrecht,
Heidelberglaan 100, Utrecht 3584 CX,
The Netherlands
e-mail: lleenen@umcutrecht.nl

20.1 Introduction

Although rare, every orthopedic trauma has the chance of accompanying vascular injury. Nevertheless, delay in recognition can lead to loss of the limb [1]. The combination of fracture and arterial injury can go with amputations rates as high as 10–40 % [2]! Therefore, every effort should be made to exclude synchronous injury of the vascular system. Simple diagnostic methods can reveal early on the compromise of the vascular system. The real challenge in these combined injuries however is the timing and logistics throughout initial management and definitive care. Irreversible tissue damage may occur if more than 6 h passes before blood flow to the leg is restored [3]. The ultimate problem however is the patient with the multisystem injuries, where the preservation of life should prevail over the preservation of the limb. Quick diagnosis and targeted, temporary treatment modalities however can make the best of both worlds. Specifics of management of the mangled extremity are dealt with elsewhere in this book.

20.2 Signs and Symptoms

20.2.1 Clinical Evaluation

Immediate clinical evaluation is of utmost importance in the evaluation of a patient with fracture or dislocation of the musculoskeletal system. Hard

Table 20.1 Hard signs of vascular injury

Absent distal pulses
Expanding hematoma
Pulsatile bleeding
Palpable thrill
Audible bruit

and soft signs of vascular injury are presented in Table 20.1. Paleness of the extremity distal of the supposed lesion is a warning sign of vascular compromise if the patient is hemodynamically normal. Palpation of the peripheral pulses guides further need for evaluation. If there is no palpable pulse in a patient without further hemodynamic problems, then further evaluation is needed. Reduction of fractures and dislocations should be performed, where after renewed evaluation should take place. If there is still no palpable pulse, urgent further evaluation is warranted, preferably by angiography. Palpable thrill or an audible bruit also indicates serious injury to the vascular system.

In rare cases, with an expanding hematoma in an extremity, no further evaluation should take place, and the patient should be taken to the operating room to stop the bleeding, preferably by proximal control or direct exploration. If free pulsatile bleeding is obvious from the open fracture, tamponade is done as quickly as possible and a tourniquet should be considered. Recent experiences in the Iraq and Afghanistan conflicts show good results in these devastating events [4, 5].

20.2.2 Doppler Evaluation

In many occasions, Doppler evaluation is done for further evaluation of the vascular system. However, Doppler evaluation is only a valuable tool, if it is accompanied with a Doppler-guided pressure reading and an ankle-brachial index evaluation. The ankle-brachial index should be above 90 % to exclude vascular injury [6]. Picking up a Doppler-positive signal does not exclude a major vascular compromise, as in most times some signal is picked up from collaterals, which are however insignificant for the survival of the extremity. Only very experienced vascular surgeons or vascular technicians can evaluate the spectrum of the Doppler signal in such cases, although they mostly rely on a formal spectral analysis.

20.2.3 Angiography

Angiography is the gold standard in the evaluation of vascular injuries (Fig. 20.1) [7]. Depending on the urgency or complexity of the case, this can be done either in the angio suite, which is in many cases preferable because of the extensive and high quality radiological possibilities, or in the OR, most times with less sophisticated equipment. The OR environment, however, is favorable in patients with multisystem injuries or for instance in damage control situations [8]. A sim-

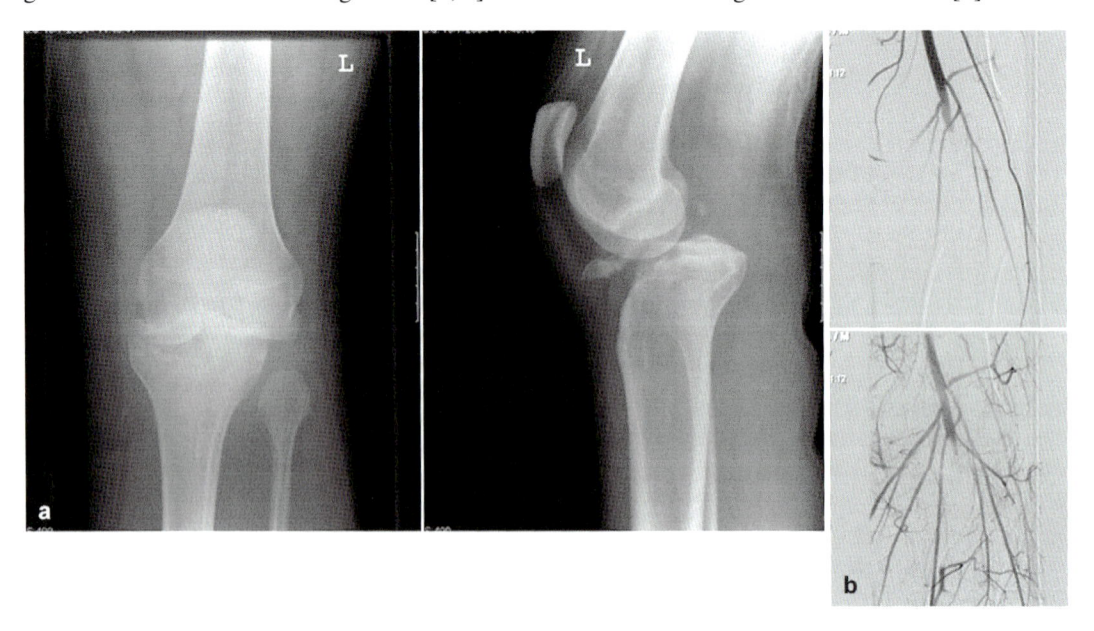

Fig. 20.1 (a) Patient with a knee dislocation. (b) Subsequent arteriography demonstrated a complete stop at the popliteal artery

ple one-shot angiogram through a proximal arterial puncture gives most times a very adequate overview of the vascular system and the level of the problem.

Advantage of the arteriography is the possibility of angio embolization. In cases of severe arterial bleeding, e.g., in pelvic fractures, angio embolization can be an important adjunct in the treatment of these severely injured patients, after initial mechanical stabilization and packing. Intraluminal manipulation, when performing an angiogram, gives also the possibility of using intraluminal stents. These stents can be used for bridging defects, occluded trajectories and coverage of traumatic pseudoaneurysms [9]. The use of large amounts of intravenous contrast has the disadvantage of a chance of contrast nephropathy and allergic reaction. In emergency cases also, the chance of local vessel injury is relevant [10].

20.2.4 CT Angiogram

As in the current practice, CT is very often used for the evaluation of the trauma patient. CT angiography is an option for further evaluation of the vascular status of the trauma patient. A specific protocol and timing should be used for an optimal result. This modality is less invasive compared to the classic angiogram; however, contrast related problems can occur as well with this technique. It has largely replaced the invasive angiography for initial diagnostics in the trauma setting as it is readily available [11].

20.2.5 Digital Subtraction Angiography

Intravenous digital subtraction angiography (DSA) can be used in selected cases; however, it results in inferior image quality and requires a trip to the radiology department. In children, however, this can be an option, as the vascular system is less easy to catheterize (Fig. 20.2). The relative high dose of contrast that is given is the disadvantage of this technique.

Fig. 20.2 Digital intravenous subtraction angiography in a child with a supracondylar humeral fracture. Disruption of the brachial artery in the area of the fracture fixed with two K-wires

20.2.6 MR Angiography

Increasing popular in vascular surgery is the use of the MR angiography. Because of the very specific circumstances where the multiply injured patient is many times on the ventilator, this modality is, infrequently used in the early evaluation of the trauma patient [7].

20.3 Treatment

20.3.1 General Tactics

20.3.1.1 Logistics

In the last years, major changes in the logistics for the optimal care of the trauma patient have been imminent. Not only the further advancement of the

CT, even into the emergency room and even used in hemodynamically unstable patient, but also the further development of the endovascular techniques in vascular surgery led to the development of hybrid operation suites in which classic operative and endovascular techniques could be used in unison. This development is of major importance in patients with multisystem injuries with skeletal, visceral and vascular lesions. In such a suite, all modalities can be used for optimal care of the trauma patient without the necessity to transport the patient, with all its dangers and intricacies. For instance, patients with major pelvic injuries after initial mechanical stabilization and packing can undergo catheter-guided embolization in the same instance and place. Moreover, patients with severe hepatic injuries can undergo after initial packing embolization of the remaining arterial intrahepatic lesions, without dangerous transport to the angio suite.

20.3.2 Severe Torso Bleeding

Non-compressible torso bleeding still remains a major cause of death in the trauma patient. Morrison and coworkers [12] reported the use of the resuscitative endovascular balloon occlusion of the aorta in severe torso trauma with impressive experimental results, reducing mortality to 25 % with continuous balloon occlusion. In a clinical series of both blunt and penetrating injury, Brenner and coworkers [13] reached hemorrhage control by aortic balloon occlusion through percutaneous or direct cut-down of the femoral artery. There was no hemorrhage-related death in this small series, with descending aortic and infrarenal aortic lesions. New aortic balloon systems can be used fluoroscopy free; therefore, it is suitable for use in the mergence room or trauma bay.

20.3.3 Extremity Bleeding

Several tactics can be chosen after the diagnosis is obvious. For an overview, see the treatment algorithm. In severe open wounds with heavy bleeding, tamponade is the treatment of choice. This can be manually done (Fig. 20.8). Recent experiences in the Iraq and Afghanistan conflicts showed a renewed interest and good results of the application of tourniquets, as mentioned above.

After the prehospital and initial resuscitation phase, gaining proximal control is of utmost importance. Thereafter, revascularization is done as soon as possible. In case of complex combined vascular and musculoskeletal injuries, regaining perfusion of the distal part of the extremity is very important. Nevertheless, it should not compromise the possibilities for orthopedic intervention neither should the orthopedic intervention make an adequate vascular procedure impossible. Although 6 h of ischemia time is tolerated in an injured leg, as little ischemia time as possible should be allowed. The longer the ischemia time in an injured leg, the higher the coagulation disposition will be. An adequate option is to use a shunt (Fig. 20.3) to bridge the time to definitive care with a well-perfused distal part of the extremity.

From the orthopedic standpoint, temporary stabilization of the fracture with an external fixator is a good option. It shortens the time to vascular reconstruction as well as reperfusion and leaves the opportunity for extensive reconstruction after the vascular continuity is restored. Care should be taken to restore adequate length, so the definitive reconstruction of the bone can be done without major shortening or lengthening of the extremity, as this can compromise the vascular conduit later on.

For repair, mostly an interposition vein graft is used [14], because of the immunological properties (Fig. 20.8). A PTFE conduit can be used; however in case of open fractures and contaminated wounds, this is the less preferable option. Direct repair can be used in selected cases; however in order to prevent a relevant stenosis after direct repair, a vein patch is often used instead [15].

In case of an incomplete occlusion of an artery, most times based on a stretching mechanism, with an intimal tear as the result, several options are available [16]. Antiplatelet therapy

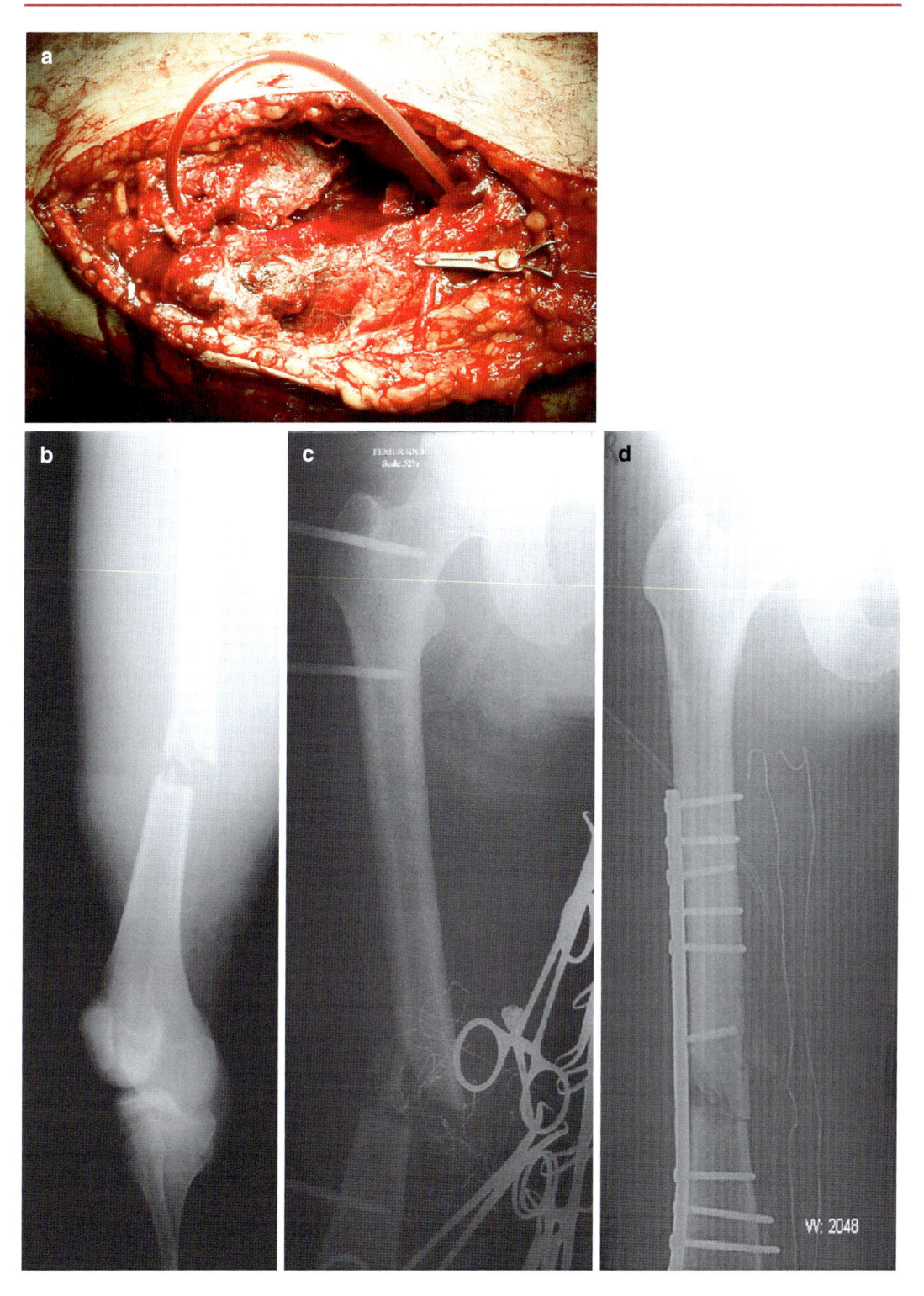

Fig. 20.3 (**a**) Shunt in situ in the superficial femoral artery in a patient with a femoral fracture and severe head injury. After hemorrhage control, restoration of flow by (**b**) a shunt, (**c**) temporary external fixation, and (**d**) ultimate plate fixation of the femur

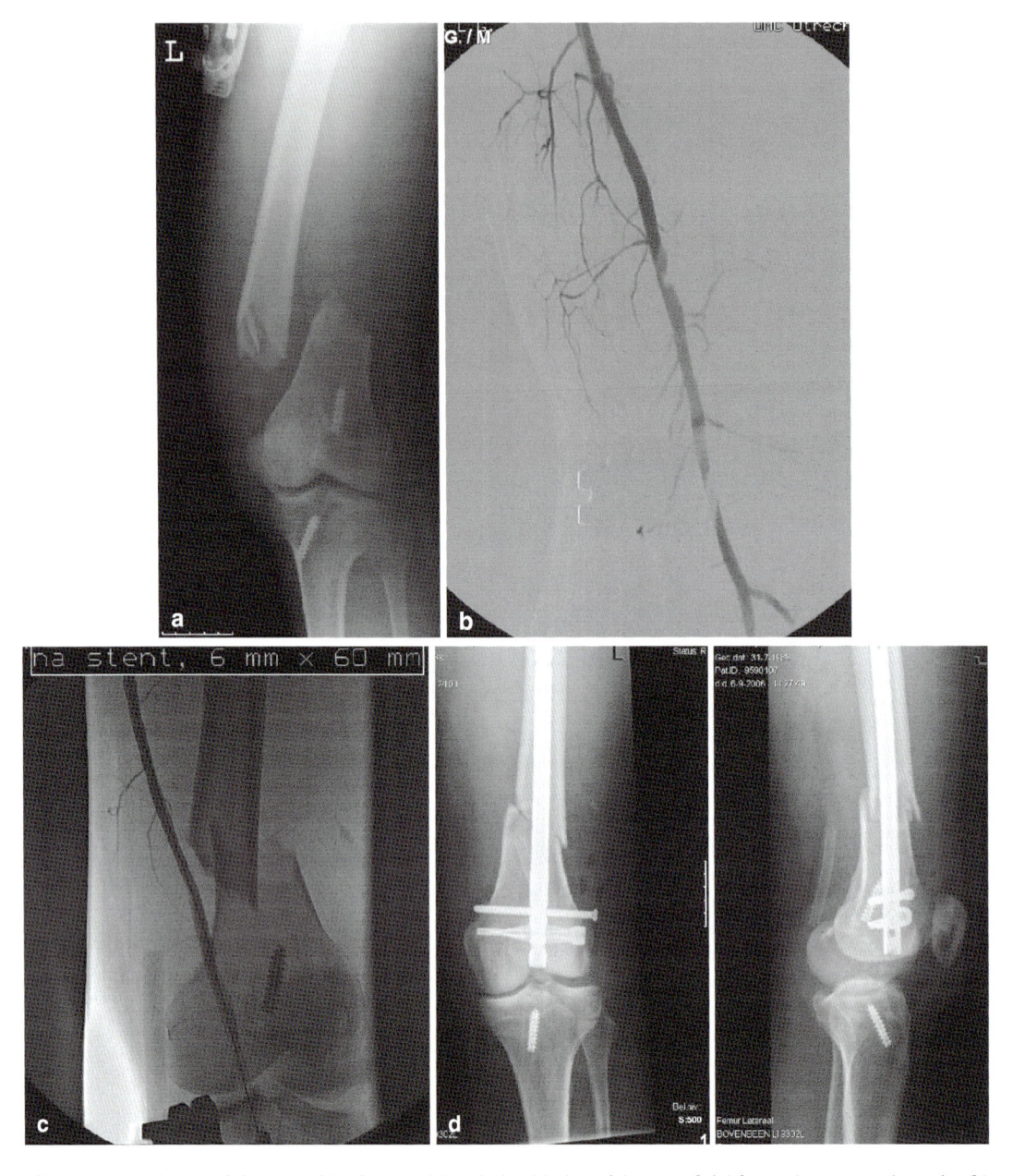

Fig. 20.4 (**a**) Distal shaft fracture of the femur, with an intimal lesion of the superficial femoral artery, as shown by (**b**) arteriography, treated with (**c**) a wall stent after initial external fixator, with (**d**) a distal femoral nail

has been advised in such cases, e.g., in case of carotid artery lesions after cervical fractures [17]. Other authors advocate the use of a wall stent, placed through radiological intervention (Fig. 20.4). Stents have been used for a variety of vascular problems like aneurysms, dissections and hematoma [18].

20.3.4 Specific Anatomic Considerations

20.3.4.1 Neck

As mentioned earlier, patients with stretch to the neck, signified for instance by fractures of the cervical spine [19] have to be evaluated by plain angiography or CT angiography [20]. A 16-slice CT scan can do the job [20]. This has to be a separate sequence/run after evaluating the neck for other traumatic injuries. In case of an intimal lesion (Fig. 20.5), currently, antiplatelet therapy is the treatment of choice [17].

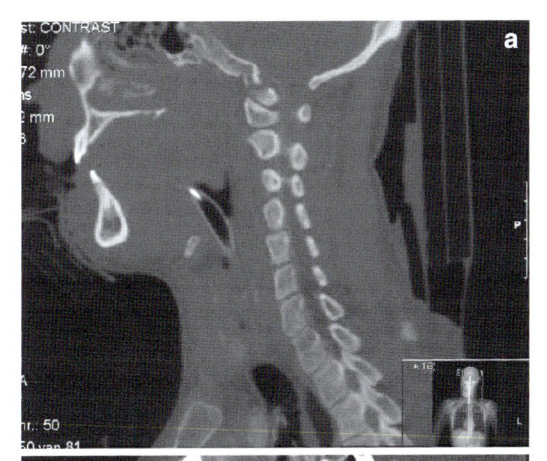

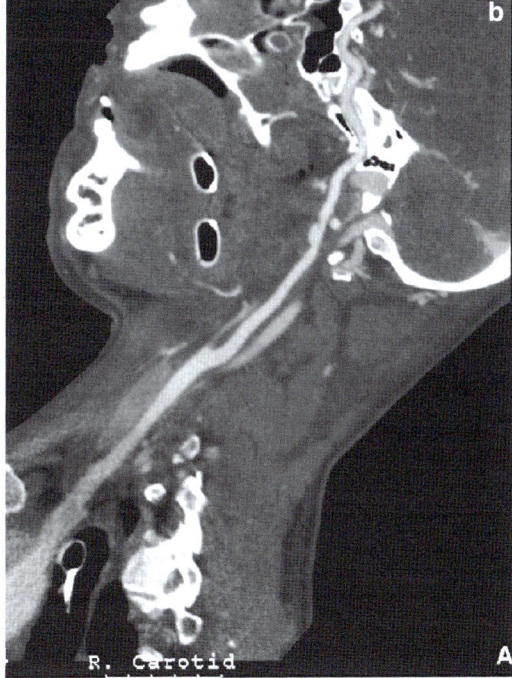

Fig. 20.5 (a) Fracture of the foraminal condyel after a motor vehicle accident with head-on collision. (b) Routine evaluation with CT Angio demonstrated an intimal flap. The patient was treated with anti platelet medication with good outcome

20.3.4.2 Upper Extremity

Fractures of the proximal humerus also are known for vascular compromise, as shown in Fig. 20.6. This area is not easily approachable surgically and can be managed with recanalization and stents as shown here. Castelli and coworkers [21] used stents in this area successfully without major complications.

Gaining proximal control in case of severe bleeding in the area of the subclavian artery is very problematic. With catheterization and subsequent use of intraluminal detachable balloons, control can be obtained, as described by Scalea and Sclafani [22].

The highest incidence of vascular compromise in upper extremity injuries is with the distal humeral supracondylar fractures common in young children. Mainly, the extension type is related with vascular injuries (Fig. 20.2). Vascular problems in the area of the elbow should be repaired, as the brachial artery is the principal end artery for the lower part of the arm. Mostly, a short bypass is the treatment of choice in this area.

Below the level of the elbow, because of the duplicate pursuance of the vasculature, mostly no major problems occur. In case of severe bleeding, the vessel can be tied off, with the premise that the counterpart is open. Through the arc of the hand, mostly sufficient flow is available for the downstream area of the lower arm.

20.3.4.3 Pelvic Bleeding

Exsanguination after pelvic fractures remains a major challenge. After initial stabilization and packing, angio embolization should be contemplated [23]. Local circumstances however dictate whether this is safe and can be done in timely fashion. A vascular interventional radiology team should be readily available around the clock. Intricacies of pelvic trauma are dealt with elsewhere in this book.

Also after the initial resuscitation, pelvic bleeding can remain a challenge as smaller vessels can demonstrate ongoing bleeding, as shown

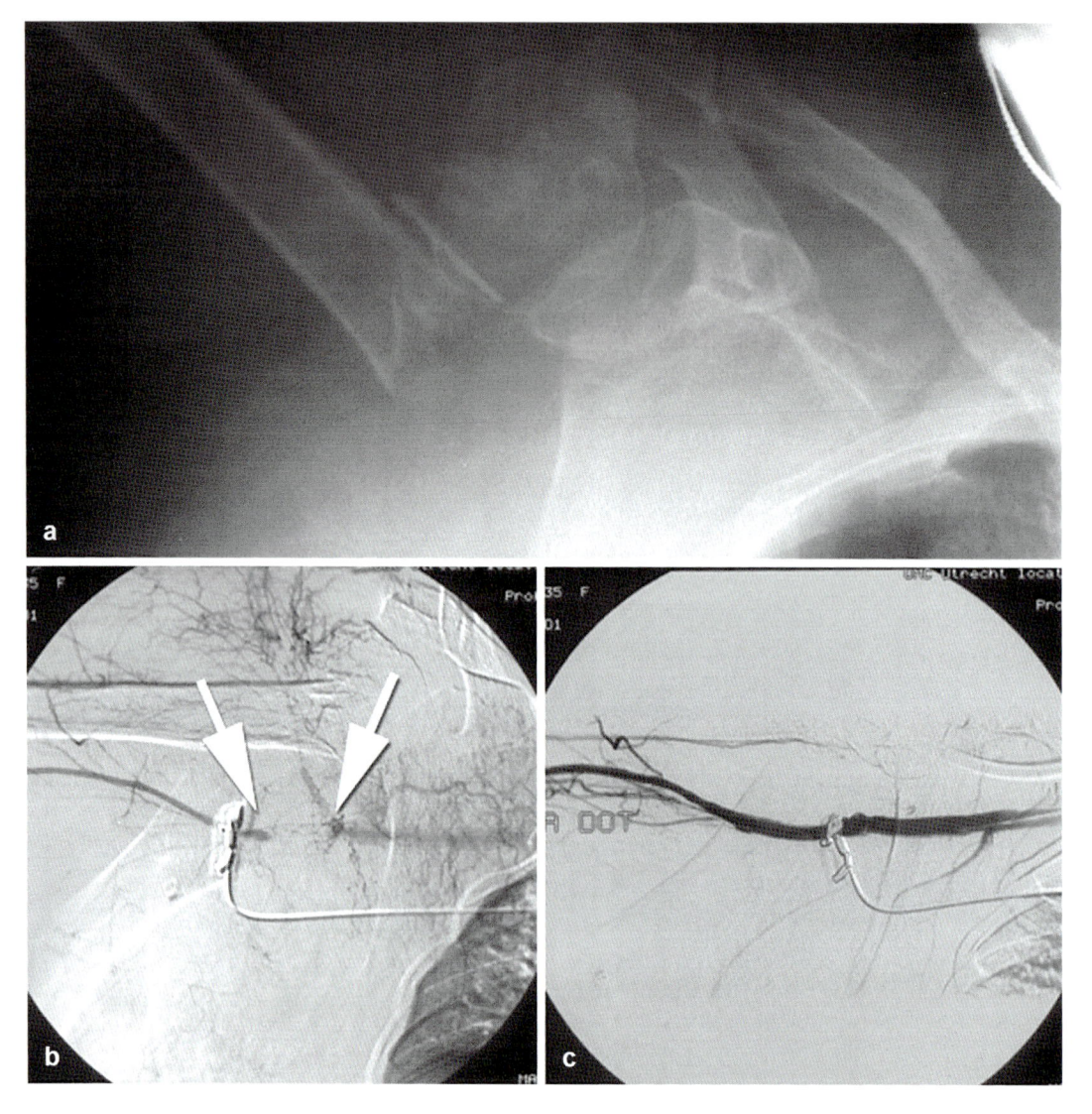

Fig. 20.6 (a) Proximal humeral fracture, with (b) vascular compromise of the axillary artery. (c) Good patency after a Dotter procedure and stent placement

in Fig. 20.7. Interventional radiology is an elegant way of tackling this problem.

In pelvic cases, the evaluation of bleeding vascular injury precedes the evaluation of vascular compromise. Thereafter exact evaluation of the integrity of the iliac arteries should be done. In case where vascular and nervous injuries coexist, together with disruption of the SI joint and symphysis, an internal hemipelvectomy should be suspected. In these lesions, a crossover bypass is one of the possibilities; however, care should be taken to shut down the proximal side to preclude bleeding after revascularization.

20.3.4.4 Lower Extremity

Currently, based on the experiences from the Iraq war, tourniquets are gaining popularity in case of severe open exsanguinating extremity wounds. Revascularization should be done as early as possible, however taking into account the general condition of the patient and the vital functions.

For acute revascularization, temporary stents have been of value (Fig. 20.3), as has been dis-

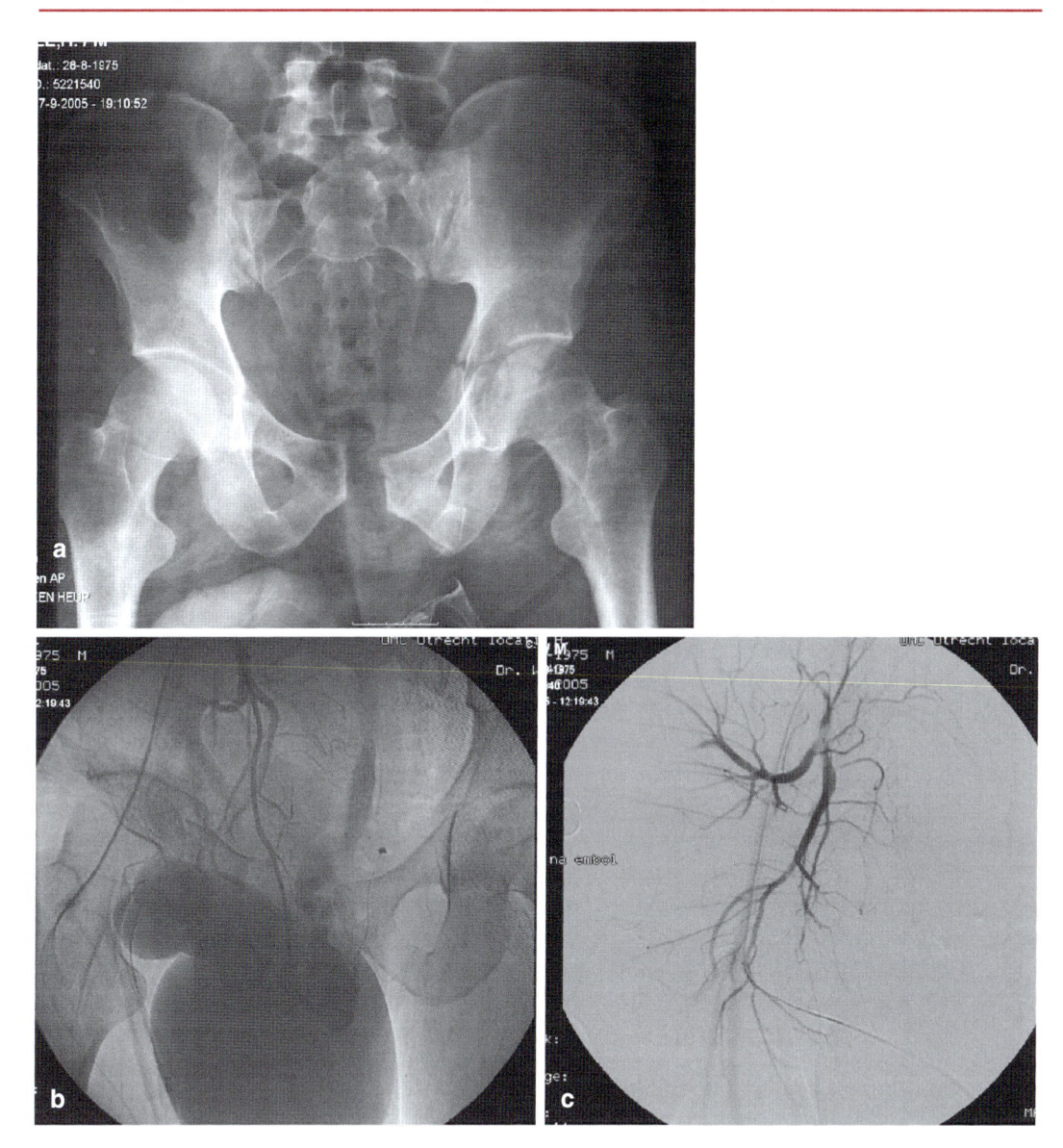

Fig. 20.7 (**a**) Pelvic fracture, with pelvic ring and acetabular involvement. (**b**) Further evaluation of persistent blood loss showed arterial bleeding. (**c**) Treatment with embolization with good result

cussed above, followed by a venous bypass, preferably with the great saphenous vein from the contralateral side (Fig. 20.8). The use of the homolateral saphenous vein is contraindicated as it might be damaged and with concomitant injury of the deep veins, swelling of the homolateral leg can compromise venous return altogether [7]. For intimal tears, both, stenting (Fig. 20.4), but also follow up and platelet inhibiting is used.

Huynh evaluated the skeletal injuries of the lower extremity associated with arterial injury and found that tibia and fibula fractures are related the most with arterial injury [15], followed by knee dislocations (Fig. 20.1). The below-knee popliteal artery and the distal superficial femoral artery are involved most. They advise to reconstruct the vascular injury where after the bone should be repaired, as they did in 63 % of cases. In general, they do not use shunting in this area. Their proto-

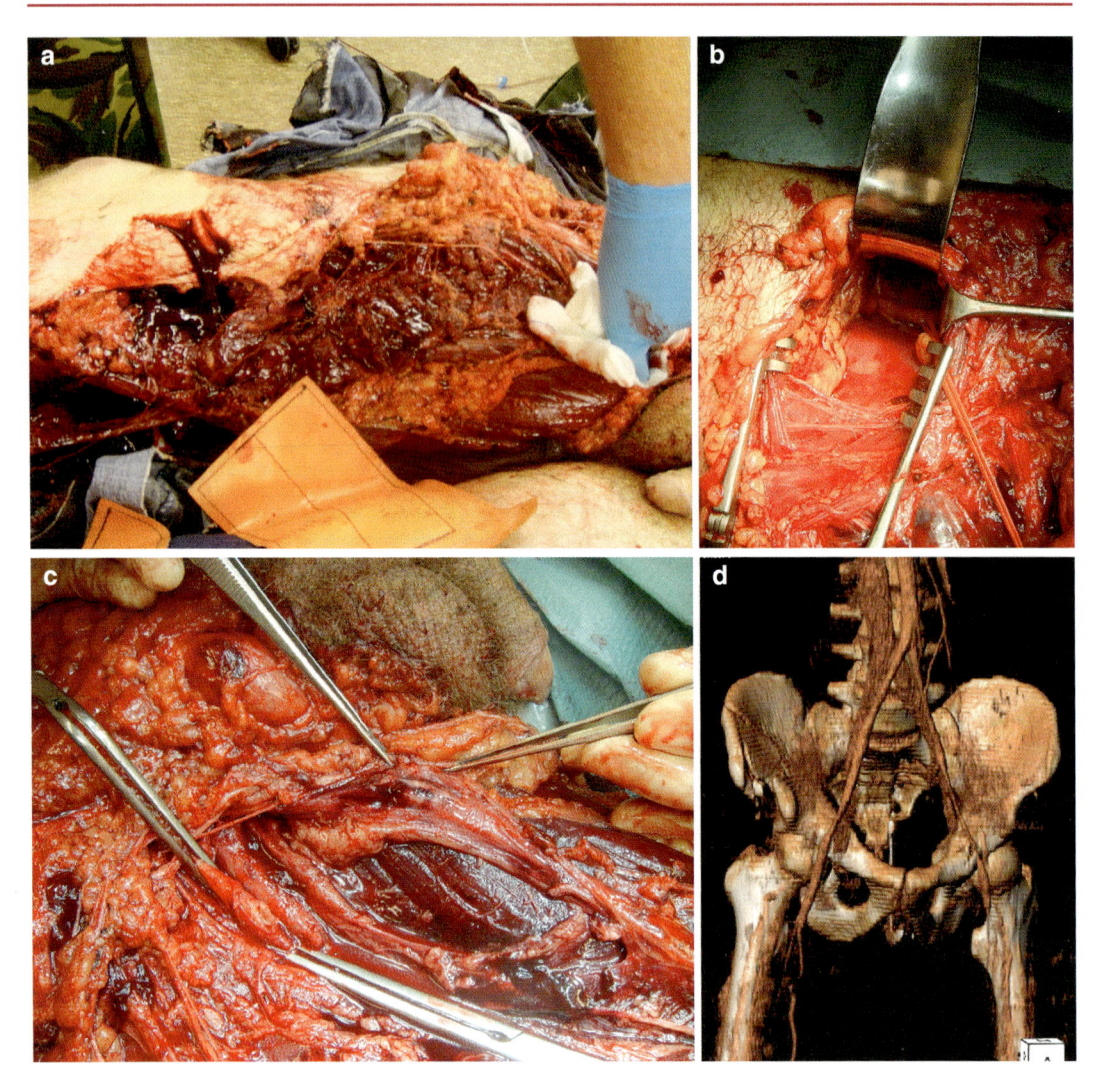

Fig. 20.8 (**a**) Severe open injury of right leg and pelvic region. Direct manual tamponade of the arterial bleeding. Head of the patient is to the right. (**b**) Proximal control of external iliac artery through an incision above the iliac crest and retroperitoneal approach. (**c**) Vascular lesion of the femoral artery. (**d**) Postoperative CT with volume rendering technique of pelvic region with pelvic fracture after vascular repair with interposition vein graft

col comprises of a medial approach to the vessel, debridement of the injured segment, heparinization, embolectomy, if needed, and reconstruction with graft, venous patch and in the minority direct repair. They advise a low threshold for fasciotomy to prevent compartment syndrome as they did in 60 % of their cases. With this algorithm, they achieved 92 % salvage.

Lesions below the trifurcation are mostly not amenable for repair. Most times, one artery will suffice for adequate perfusion [24]. Brinker et al. [25] evaluated the opinion of 200 vascular surgeons on lesions in this area; however, no consensus was reached on the treatment of the various lesions. Hafez et al. [26] evaluated in a total series of 550 vascular injuries of which in the majority were penetrating injury reported fairly good results for repair of crural arteries. Segal evaluated 18 patients with lower limb injuries and vascular repair. Hafez et al, as well as al-Salman [14] also used contralateral vein graft with fairly good results. Nevertheless, the last authors report that these lower limb injuries carry a high incidence of amputation of up to 30 %.

The development of a compartment syndrome is generally seen as a major compromise of orthopedic injury with concomitant vascular injury. Therefore, after revascularization, there is a general idea that a fasciotomy should be added.

Conclusion

Vascular injury accompanying skeletal trauma is relatively rare. However, prompt diagnosis and expeditious repair are the prerequisites for prevention of amputation. A wealth of new techniques like CT and intraluminal catheterization has become available for diagnostics and repair. When treated early, the general prognosis is good.

References

1. Andrikopoulos V, Antoniou I, Panoussis P. Arterial injuries associated with lower-extremity fractures. Cardiovasc Surg. 1995;3:15–8.
2. Bishara RA, Pasch AR, Douglas DD, Schuler JJ, Lim LT, Flanigan DP. The necessity of mandatory exploration of penetrating zone II neck injuries. Surgery. 1986;100:655–60.
3. Persad IJ, Reddy RS, Saunders MA, Patel J. Gunshot injuries to the extremities: experience of a U.K. trauma centre. Injury. 2005;36:407–11.
4. Kragh Jr JF, Walters TJ, Baer DG, et al. Practical use of emergency tourniquets to stop bleeding in major limb trauma. J Trauma. 2008;64:S38–49.
5. Kragh Jr JF, Littrel ML, Jones JA, et al. Battle casualty survival with emergency tourniquet use to stop limb bleeding. J Emerg Med. 2011;41(6):590–7.
6. Lynch K, Johansen K. Can Doppler pressure measurement replace "exclusion" arteriography in the diagnosis of occult extremity arterial trauma? Ann Surg. 1991;214:737–41.
7. Doody O, Given MF, Lyon SM. Extremities–indications and techniques for treatment of extremity vascular injuries. Injury. 2008;39:1295–303.
8. Leenen L, Moll FL. Vascular injuries in polytrauma patients. In: Pape H-C, Peitzman AB, Schwab CW, editors. Damage control management in the polytrauma patient. New York: Springer; 2009. p. 315–30.
9. Onal B, Ilgit ET, Kosar S, Akkan K, Gumus T, Akpek S. Endovascular treatment of peripheral vascular lesions with stent-grafts. Diagn Interv Radiol. 2005;11:170–4.
10. Peck MA, Rasmussen TE. Management of blunt peripheral arterial injury. Perspect Vasc Surg Endovasc Ther. 2006;18:159–73.
11. Fleiter TR, Mervis S. The role of 3D-CTA in the assessment of peripheral vascular lesions in trauma patients. Eur J Radiol. 2007;64:92–102.
12. Morrison JJ, Ross JD, Rasmussen TE, Midwinter MJ, Jansen JO. Resuscitative endovascular balloon occlusion of the aorta: a gap analysis of severely injured UK combat casualties. Shock. 2014;41:388–93.
13. Brenner, Megan L, Moore, et al. A clinical series of resuscitative endovascular balloon occlusion of the aorta for hemorrhage control and resuscitation. J Trauma Acute Care Surg. 2013;75(3):506–11.
14. Al-Salman MM, Al-Khawashki H, Sindigki A, Rabee H, Al-Saif A, Al-Salman FF. Vascular injuries associated with limb fractures. Injury. 1997;28:103–7.
15. Huynh TT, Pham M, Griffin LW, et al. Management of distal femoral and popliteal arterial injuries: an update. Am J Surg. 2006;192:773–8.
16. Frykberg ER, Vines FS, Alexander RH. The natural history of clinically occult arterial injuries: a prospective evaluation. J Trauma. 1989;29:577–83.
17. Cothren CC, Moore EE, Biffl WL, et al. Anticoagulation is the gold standard therapy for blunt carotid injuries to reduce stroke rate. Arch Surg. 2004;139:540–5.
18. Piffaretti G, Tozzi M, Lomazzi C, et al. Endovascular treatment for traumatic injuries of the peripheral arteries following blunt trauma. Injury. 2007;38:1091–7.
19. Cothren CC, Moore EE, Biffl WL, et al. Cervical spine fracture patterns predictive of blunt vertebral artery injury. J Trauma. 2003;55:811–3.
20. Biffl WL, Egglin T, Benedetto B, Gibbs F, Cioffi WG. Sixteen-slice computed tomographic angiography is a reliable noninvasive screening test for clinically significant blunt cerebrovascular injuries. J Trauma. 2006;60:745–51.
21. Castelli P, Caronno R, Piffaretti G, et al. Endovascular repair of traumatic injuries of the subclavian and axillary arteries. Injury. 2005;36:778–82.
22. Scalea TM, Sclafani SJ. Angiographically placed balloons for arterial control: a description of a technique. J Trauma. 1991;31:1671–7.
23. Leenen LPH. Pelvic fractures: soft tissue trauma. Eur J Trauma Emerg Surg. 2010;35:117–23.
24. Segal D, Brenner M, Gorczyca J. Tibial fractures with infrapopliteal arterial injuries. J Orthop Trauma. 1987;1:160–9.
25. Brinker MR, Caines MA, Kerstein MD, Elliott MN. Tibial shaft fractures with an associated infrapopliteal arterial injury: a survey of vascular surgeons opinions on the need for vascular repair. J Orthop Trauma. 2000;14:194–8.
26. Hafez HM, Woolgar J, Robbs JV. Lower extremity arterial injury: results of 550 cases and review of risk factors associated with limb loss. J Vasc Surg. 2001;33:1212–9.

Management of Articular Fractures

21

Tak-Wing Lau and Frankie Leung

Contents

21.1	Types of Articular Injuries...............	289
21.2	Assessment..................................	290
21.3	Strategy of Management of Articular Fractures in Polytrauma Patients...........	290
21.4	Principles of Managing Open Articular Fractures.....................	291
21.5	Floating Joint Injuries....................	292
21.5.1	Floating Knee Injury......................	292
21.5.2	Floating Shoulder Injuries................	297
21.5.3	Floating Elbow............................	299
21.6	Traumatic Knee Dislocation.............	300
21.6.1	Classification..............................	300
21.6.2	Associated Injuries........................	301
21.6.3	Evaluation and Assessment..............	301
21.6.4	Management...............................	303
21.6.5	Rehabilitation.............................	303
21.6.6	Outcome and Complications............	304
Conclusions...		304
References...		304

T.-W. Lau (✉) • F. Leung
Department of Orthopaedics and Traumatology,
Queen Mary Hospital, The University of Hong Kong,
5/F, Professorial Block, 102 Pokfulam Road,
Hong Kong, China
e-mail: catcherlau@yahoo.com.hk; klleunga@hku.hk

21.1 Types of Articular Injuries

Articular injuries are common in polytraumatized patients. While these are unlikely to be life threatening in the acute setting, they would cause significant disability if not appropriately treated. The joint can be affected in one of the following ways.

First, the high-energy trauma causes a fracture that involves the articular surface. Such intra-articular fractures can cause severe disability to the patient if they are not treated appropriately. Accurate joint reconstruction with stable fixation allowing early mobilization of the joint is important for good cartilage healing and good joint motion recovery. The surgical reconstruction will require careful preoperative planning and should be done later as definitive fixation.

Second, the term 'floating joint' injury refers to the fractures occurring both proximal and distal to the joint, resulting in a total lack of bony support of the affected joint. The fractures may not extend to the articular surface. Since nerves and blood vessels are commonly in close vicinity to the joint, the risks of neurovascular complications are usually much higher in the presence of a floating joint.

Third, dislocations of major joints or fracture-dislocations occur and represent are orthopaedic emergencies. These conditions must be recognized promptly in the emergency room during secondary survey. Reduction should be achieved with appropriate analgesics or anaesthesia as soon as possible. If the dislocations are left unattended, there will be a high chance of vascular or neurological complications.

21.2 Assessment

In the emergency room, resuscitation should follow the ATLS protocol. After the primary survey, a thorough secondary survey should be performed and the whole body should be examined for other injuries. The presence of any open fractures or compartment syndrome should not be missed. Floating joint injuries or major joint dislocations have to be recognized based on the deformity of the limbs. However, the treating doctor should not be distracted by the obvious deformity and overlook other associated complications. The distal circulation of the limb must be checked and if the patient is conscious, a quick motor and sensory examination should be recorded as a baseline for further reference. Afterwards, appropriate splints must be applied. Radiographs in two planes, including the full length of the long bones, must be obtained to confirm the diagnosis.

21.3 Strategy of Management of Articular Fractures in Polytrauma Patients

Complex fractures around joints remain challenges in the management of polytraumatized patients and they are associated with an increased risk of complications. During decision making to formulate the plan of management, the surgeon must take into account any associated injuries to other major internal organs and body parts (Table 21.1). At the same time, the local soft tissue condition around the joint must be carefully assessed. These two factors will affect the timing of the fracture fixation and the method of fracture fixation [1].

In general, complex fractures around the joints are better managed with a staged strategy [2, 3]. First, the soft tissue condition around the injured joint, especially the knee and the ankle, is usually in an unfavourable condition. There are usually severe oedema and blisters, thus rendering primary fracture fixation very risky with high complication rates. Second, intra-articular fractures are complex injuries. In order to achieve a good outcome, the articular surface should be reconstructed anatomically, the limb axis be restored correctly and a stable fixation connecting the articular block to the metaphysis and diaphysis should be obtained to allow for early joint motion. This often necessitates a good preoperative assessment of the fracture including good quality radiographs, CT scans with reconstruction and in indicated cases, MRI. Good and accurate surgical planning and meticulous surgical skills are crucial in achieving a good fixation. Hence, these

Table 21.1 Surgical priorities in the treatment of complex articular fractures in polytrauma

A. Primary surgical procedures in the emergency setting
1. Limb-saving procedures:
Reduction of large joints, such as hip, knee, by close or open means with temporary stabilization by splint or traction
Bony stabilization with urgent vascular surgery for acute damage to vascular supply
Debridement and spanning external fixation for open articular fractures together with appropriate intravenous antibiotics
Fasciotomy and spanning external fixation for articular fractures complicated by compartment syndrome
2. Spanning transarticular external fixation as a damage control procedure
To stabilize floating joint injuries or unstable joint dislocation after reduction in unstable patients
To stabilize peri-articular fractures with unfavourable local soft tissue conditions
B. Secondary surgical procedures that should be done when the general condition of the patient is stabilized or the soft tissue condition has improved
1. Definitive fixation of intra-articular fractures with initially unfavourable soft tissue conditions
2. Definitive fixation of unstable fracture dislocations, e.g. shoulder, acetabulum
3. Definitive fixation of floating joint injuries that are initially treated with spanning external fixation
4. Soft tissue coverage and definitive fixation of open intra-articular fractures

Fig. 21.1 Temporary knee-spanning external fixation in a 34-year-old polytrauma victim with comminuted proximal tibial fracture complicated by compartment syndrome. Emergency fasciotomy was performed. Vacuum-assisted closure was applied and wound closure was performed on day 10 after injury. The definitive fixation was then carried out on day 14

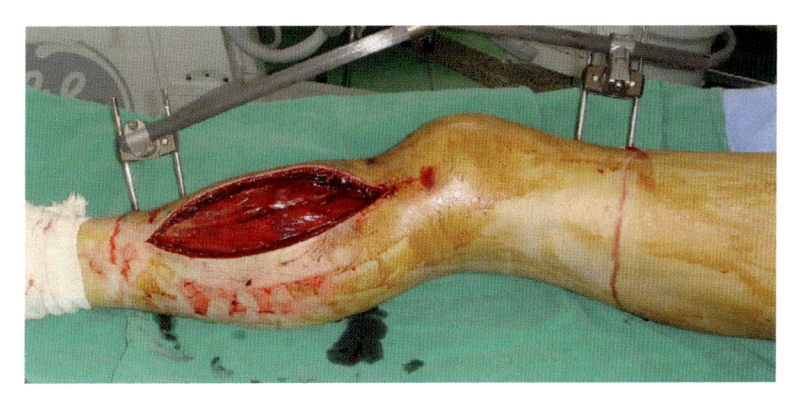

difficult definitive reconstructions should not be performed in the setting of emergency surgery in a polytraumatized patient.

Generally speaking, the management of articular fractures in polytrauma patients should include a primary spanning external fixation applied in the emergency setting (Fig. 21.1). The configuration should be simple and allows easy access to the soft tissues during subsequent surgeries. The surgeon applying the external fixation should preferably be the surgeon who fixes the fracture definitively. Definitive fixation should be carried out when both the general condition of the patient and the local soft tissue conditions are optimized. Vacuum-assisted closure (VAC) therapy has been shown to be effective in managing large soft tissue defects and in assisting wound closure [4–6].

Sometimes in high-energy articular fractures, the stability of the joint is affected, resulting in a fracture-dislocation. In principle, a major joint dislocation that causes significant deformity should be reduced as soon as possible as the distal circulation will be affected. In the case of posterior hip dislocation that commonly occurs with posterior wall fracture of the acetabulum, reduction can usually be done quickly with closed manipulation once the patient is anaesthetized. Fixation of the posterior wall fracture should be done at a later stage after thorough assessment with CT scan. Similarly, fracture dislocations involving the ankle should be reduced urgently to avoid compromises of the soft tissue envelope and the distal circulation.

21.4 Principles of Managing Open Articular Fractures

Open articular fractures are often the result of high-energy trauma and are often associated with severe fracture comminution and bone loss. The common causes include road traffic accidents, industrial accidents or fall from heights. In lower extremity trauma, open injuries are more common especially in the knee and ankle regions [7].

Initial care of open joint injuries includes a good assessment of the patient's general condition and urgent management of all life threatening events. Repeated debridement should be preformed. The size and the degree of contamination of the open wound are assessed. The wound should be first irrigated with copious amount of normal saline and the debris is removed. Broad-spectrum antibiotics should be started. If wound is grossly contaminated with dirt or soil, antibiotics covering anaerobes should also be started. The devitalized soft tissue including skin, fascia, fat and muscle should be debrided. Exposed cartilage should be covered with viable soft tissue if possible. Important peri-articular structures including tendons, stabilizing ligaments, neurovascular bundle should be debrided with caution. In case of suspected vascular damage associated with open joint injury, vascular surgeon should be brought in to revitalize the distal circulation as soon as possible. The return of blood circulation in the limb is one of the important factors to fight against future complications including infection. Amputation rate can be as high as 86 % if revascularization is delayed [8].

Appropriate imaging studies will always include a CT scan of the injured site that will allow surgeons to formulate an operative plan. One of the challenges in treating open joint injuries is the preservation of free osteochondral fragments. In general, these should be preserved by all means. Small osteochondral fragment can be removed. However, large unstable osteochondral fragment should be stabilized to anatomical position by minimal implants [9]. This can usually be achieved by appropriately sized lag screws or multiple K-wires. The cartilage portion should be cleaned well with copious amount of saline. The bony part should be debrided and contamination removed. Intra-articular drains should be placed to drain all the fluid collection in post-operative period.

Bony fragments which are contaminated or are without any soft tissue attachments should be excised. We use the technique of antibiotic cement spacer placed within the osseous void followed by staged bone grafting [10]. The induced biomembrane formed around the spacer prevents graft resorption, improves vascularity and later corticalization. Unpublished data from our institution between 2009 and 2012 showed four patients with open articular fractures (including two in distal tibia, one in distal femur and one in olecranon) undergoing bone reconstruction by the induced membrane technique. The antibiotic used for cement spacer is either gentamicin or vancomycin. Cancellous bone grafting was done after an average of 44 days (Fig. 21.2). All patients demonstrated radiological consolidation over the defect after treatment.

For most of the open articular injuries, joint spanning external fixator is almost a must for the temporary stabilization and immobilization of the damaged joint. The spanning external fixator should be rigid but versatile enough to allow daily observation of the open wound area.

In general, definitive internal fixation of the joint and metaphyseal area is best done as early as possible to minimize joint stiffness and improve cartilage healing. However, this can only be done when the joint has no sign of infection and there is adequate soft tissue coverage. Timing of definitive fixation depends on the general status of the patient as well as local soft tissue and bony conditions. This needs a lot of experience and careful planning. Therefore, open articular fractures are best managed by experienced trauma surgeons in trauma centres where other related experts such as plastic surgeons or vascular surgeons are readily available.

21.5 Floating Joint Injuries

21.5.1 Floating Knee Injury

A floating knee refers to the injury when the ipsilateral femur and tibia are both fractured. A significant force must be needed in order to break these two bones and therefore this injury frequently implies a more substantial mechanism of injury. The patients are commonly haemodynamically unstable and may have significant injuries of other organs and the other extremities. This injury is also associated with complications that carry an increased risk of morbidity and mortality.

Fraser et al. classified floating knee injuries by whether there is joint involvement [11] (Fig. 21.3).

Type I is the injury with extra-articular fractures of both bones.
Type II is subdivided into three groups, as follows:
- Type IIa involves femoral shaft and tibial plateau fractures.
- Type IIb includes fractures of the distal femur and the shaft of the tibia.
- Type IIc indicates fractures of the distal femur and tibial plateau.

This is the commonest classification system for floating knee injury and is of prognostic value since type I fractures have better functional outcome than type II with various extent of intra-articular involvement. This classification was recently modified by Ran et al. to include patellar fractures as a type III fracture [12]. They reported worse outcome with complex articular fractures and type III fractures.

21 Management of Articular Fractures

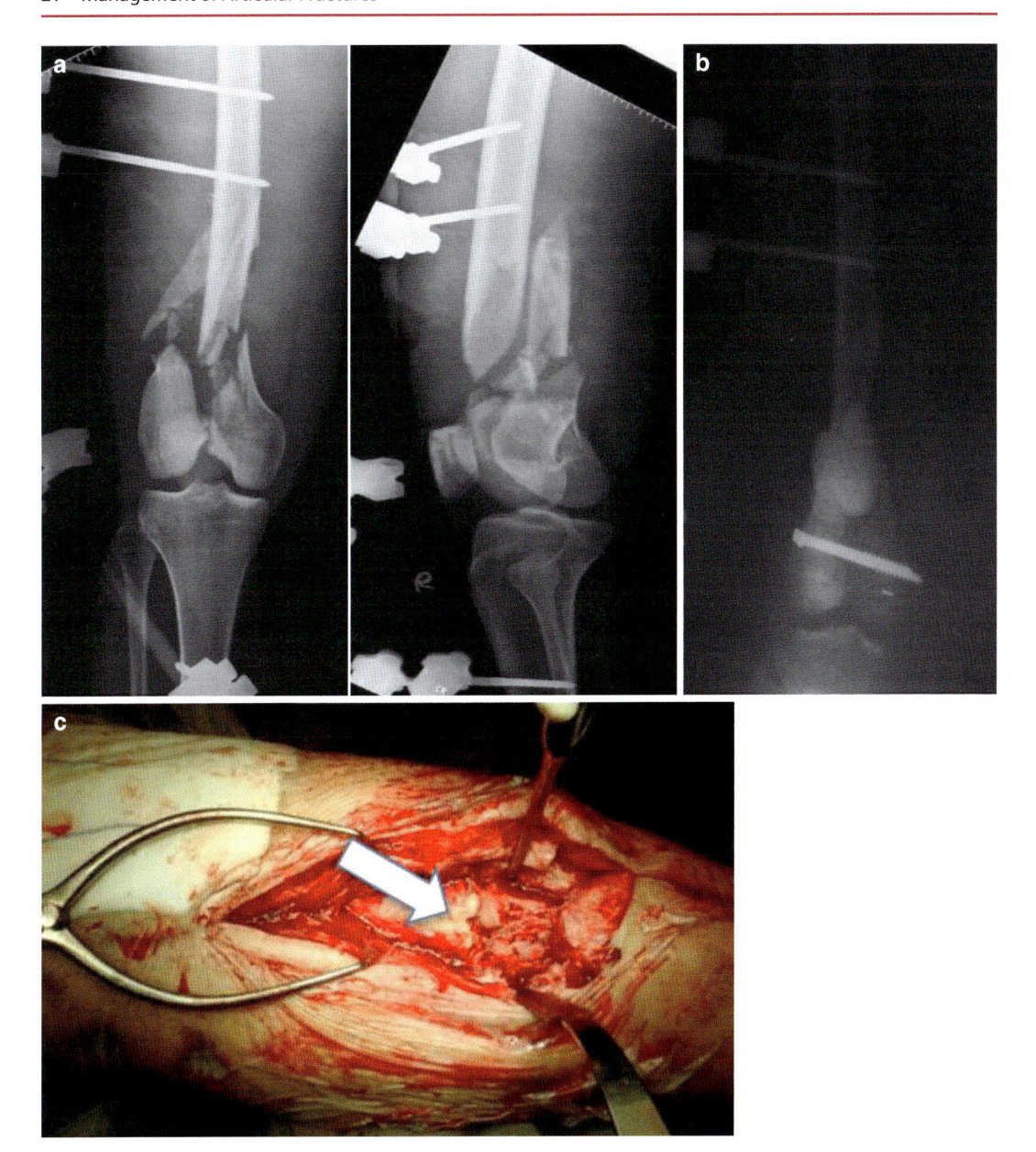

Fig 21.2 (a) A type IIA open distal femoral fracture in a 25-year-old male sustaining multiple injuries after a motor cycle crash. Debridement and spanning external fixator was applied. (b) Placement of antibiotic cement spacer into the defect after the wound bed had been adequately debrided. (c) The cement spacer was removed and the defect grafted with autograft. Six weeks later, internal fixation was performed. Notice the rapid incorporation of the bone graft into richly vascularized bone (*white arrow*)

21.5.1.1 Management of Fractures in Floating Knee Injury

Historically, floating knee injuries were totally treated or partially treated non-operatively. However, the results were unsatisfactory [11]. The current recommended treatment of the bony injuries is surgical fixation of both the femoral and the tibial fracture. There is no single ideal method of fixation [13]. The surgeon should take into consideration the extent of soft tissue injury, the location and pattern of the fractures and the associated injuries.

Isolated floating knee injury without significant articular involvement should be treated

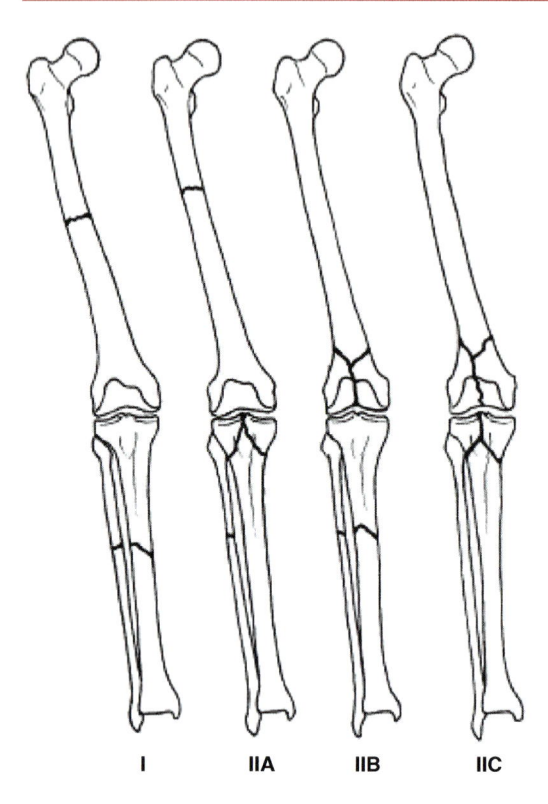

Fig. 21.3 Floating knee classification of Fraser et al. [11]

acutely if the patient is haemodynamically stable. If both fractures occur in the diaphysis, then both the femoral shaft and tibial shaft should be treated with intramedullary nailing. There is still a controversy as to whether antegrade or retrograde femoral nailing should be used. Rethnam suggested that antegrade nailing should be done [14]. Advocates for retrograde femoral nailing suggested that the quickest surgical procedure is to perform a retrograde intramedullary nailing of the femur with an intramedullary nailing of the tibia using a single incision over the knee (Fig. 21.4).

Alternatively, the tibia fracture is temporarily splinted with a cast and an antegrade femoral nailing is done first, followed by the tibial nailing. If either one or both fractures involve the epi-metaphyseal region, then the appropriate peri-articular plate fixation should be performed according to the location (Fig. 21.5). In case of severe soft tissue swelling as in tibial plateau or plafond fractures, the definitive fixation may be delayed until the soft tissue condition improves and there is a lower chance of soft tissue complications. In case of complex articular involvement with significant fracture comminution, such as tibial plateau fracture, then one can also elect to apply an external fixator temporarily and the definitive fixation done at a later stage when the required surgical expertise is available.

On the other hand, in unstable patients or those in extremis, life-threatening injuries such as haemothorax, pneumothorax, intraabdominal haemorrhage, intracranial haematoma must be managed as the first priority. Under these circumstances, a temporary stabilization with a spanning external fixator should be performed, following the principles of damage control orthopaedic surgery. Once the patient's physiological status is stabilized, conversion to internal fixation and definitive surgery can then be performed.

In the post-operative period, range of motion of the knee joint should be started early. Continuous passive motion can be used until satisfactory knee motion has been achieved. The patient should do partial-weight-bearing walking if both fractures are extra-articular. If one or both fractures involve the knee joint articular surface, then weight bearing should be delayed for 6–8 weeks.

21.5.1.2 Associated Injuries in Floating Knee Injuries

Vascular injuries of the affected limb can occur in a floating knee injury. The reported incidence ranges from 21 to 29 % [15, 16]. Limb ischaemia may occur if the popliteal or posterior tibial arteries are injured. As a result, a thorough vascular assessment is crucial in early detection of this injury. Preoperatively, the peripheral pulses should be assessed with palpation and hand-held Doppler in all floating knee injuries. If arterial injury is suspected, an intra-operative arteriogram should be performed and vascular repair should be performed together with the bony stabilization.

The incidence of open fractures in a floating knee injury can be as high as 50–70 % [16]. The commonest pattern is a closed femoral fracture with an open tibial fracture. Paul et al. reported

Fig. 21.4 A 48-year-old gentleman was hit by a moving car from front. He sustained a Type I floating knee injury. Distal neurovascular status was normal. He also had a concomitant stable pelvis injury. There was no other major internal organ injury. (**a**) X-ray of his left femur showed a transverse fracture left mid-shaft of femur. (**b**) X-ray of his left tibia showed a spiral fracture left mid-shaft of tibia and proximal fibula. Surgical fixation was performed the next day. (**c, d**) Single medial parapatellar approach was used for the retrograde femur nail and the antegrade tibial nail. The patient was allowed to freely mobilize the hip and knee after surgery with protected-weight-bearing walking

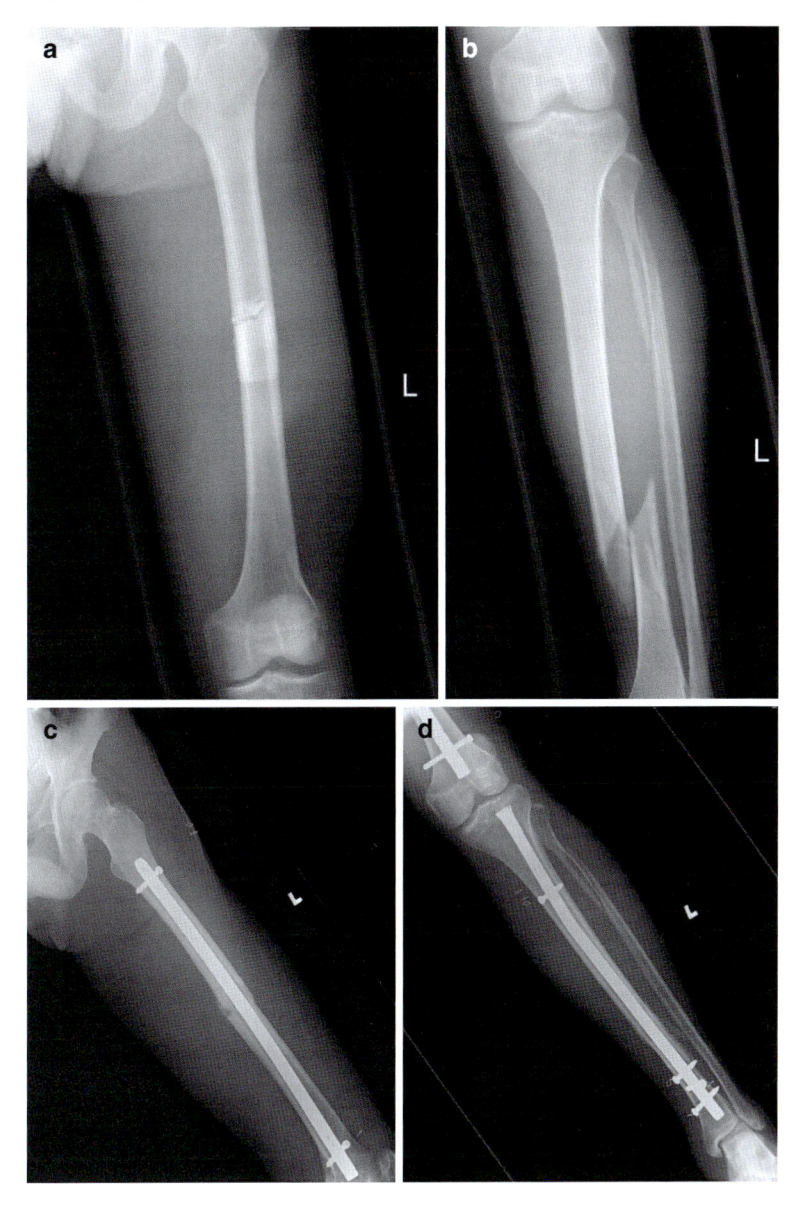

that 17 of 21 patients had open fractures of one or more bones and 76 % of these were either grade II or grade III [16]. In general, the management of open fractures associated with floating knee injuries should follow the principles of open fracture management. This should include adequate debridement and stabilization of the fractures with either external fixation or intramedullary nailing depending on the grading of the open fractures. It is expected that multiple surgical procedures are usually required and in patients with severe mangled limbs and unstable general conditions, amputation should be considered [16].

Associated ipsilateral knee ligaments injuries are common in the floating knee injury [17]. Anterolateral rotatory instability is the commonest instability pattern. However, there is a diagnostic difficulty as the floating joint cannot be tested for ligamentous injuries. Hence after stabilization of the fractures, stress testing of the knee ligaments must be performed. If a ligamentous injury is suspected, then an acute arthroscopy can be performed and the injured ligaments can be repaired acutely or at a later stage.

21.5.1.3 Complications

The management of the fractures in floating knee injuries is challenging to orthopaedic surgeons. Fraser et al. reported 35 % of patients with floating knee injuries required late surgery for delayed union or non-union, osteomyelitis, refracture and malunion [11]. There are several explanations to this high rate of complications. The first reason is that most of the fracture fixation surgeries are performed in the emergency setting. The level of surgical expertise available is a crucial factor to the success of the surgery since sometimes a good fixation can be difficult for the average surgeon. Moreover, the floating knee segment presents great difficulty in achieving an accurate reduction of either fracture. Hence, floating knee injuries are prone to delayed union or non-union. Rotational mal-alignment can also be difficult to diagnose intra-operatively (Fig. 21.2). The overall leg length should be checked at the end of the surgery and in the early post-operative period. If the patient's general condition allows, any malreduction should be corrected within the first few weeks before hard bone is formed, necessitating an osteotomy surgery.

Fat embolism can occur in a floating knee injury. Karlstrom and Olerud reported 6 out of 31 patients with fat embolism syndrome [18]. Veith et al. reported 13 % incidence of fat embolism syndrome in 54 patients of floating knee injuries [19]. The diagnosis is made if the patient has pyrexia, tachycardia, tachypnoea and altered sensorium within 48 h of admission. To confirm the diagnosis, an arterial blood gas test should be done and will reveal hypoxia. The patient should be managed in an intensive care unit with mechanical ventilation. The fractures should also be provisionally stabilized to minimize further haemorrhage and the chance of the fatty bone marrow entering the circulation. Hence, a spanning external fixator should be applied in the emergency surgery. Definitive fixation of the fractures should be delayed until the patient's conditions improve, which usually take place after 1 week of supportive care.

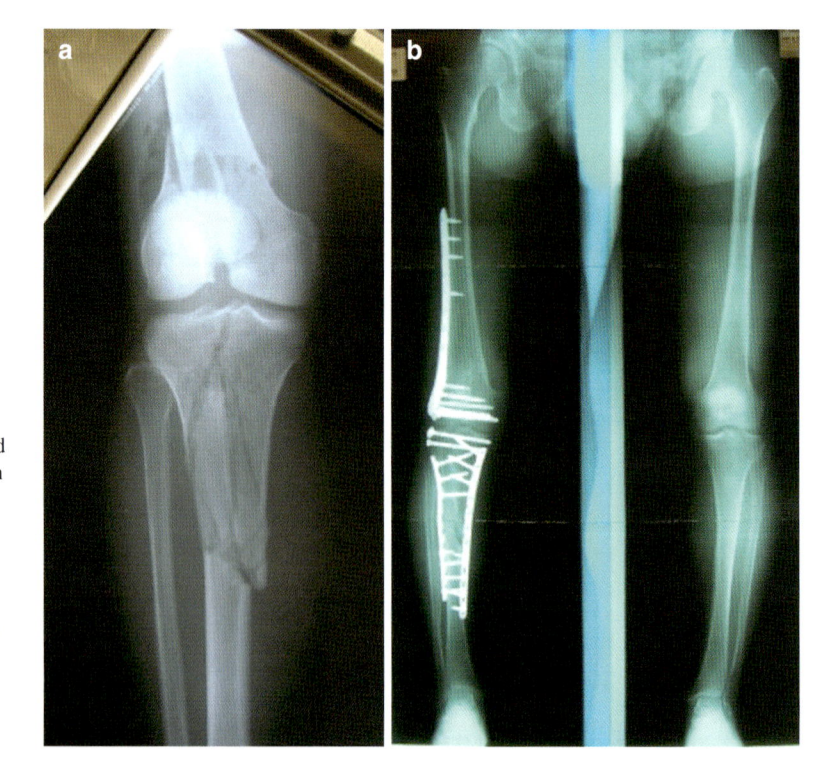

Fig. 21.5 (**a**) A 42-year-old male was injured by a fallen heavy object and sustained a type IIc floating knee injury. (**b**) Early fixation of both fractures were done after initial stabilization. Good lower limb alignment was obtained. (**c**, **d**) Good knee range of motion was obtained at 6 months after injury

Fig. 21.5 (continued)

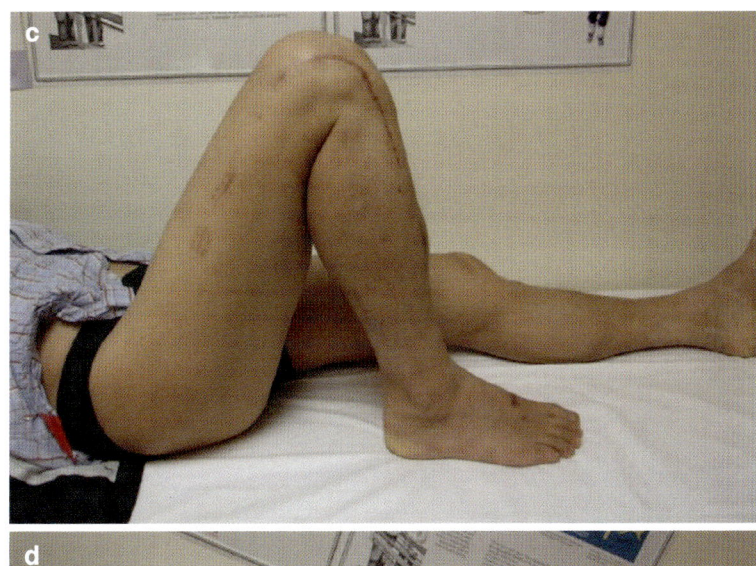

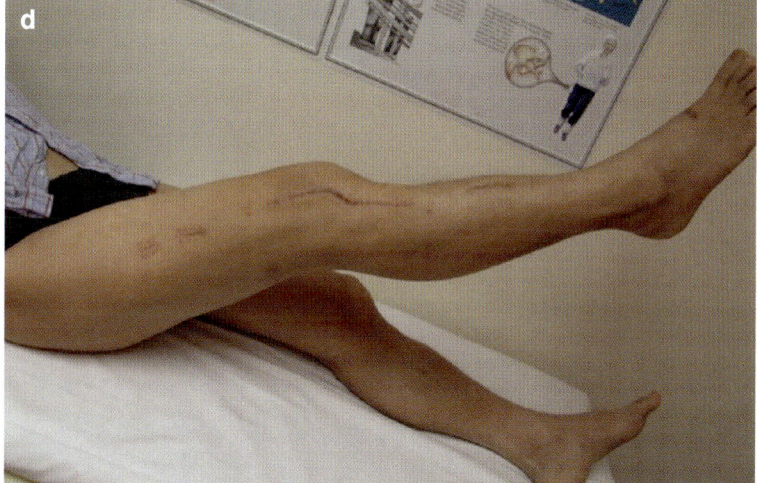

21.5.2 Floating Shoulder Injuries

Floating shoulder is an uncommon injury with both clavicle and scapular neck fractured, resulting in gross instability and severe displacement of the shoulder girdle. The term floating shoulder is initially describing the inherent bony instability as described similarly in elbow and knee joints. Later Goss introduced the important concept of superior shoulder suspensory complex [20, 21]. It is a ring of complex soft tissue structures existing between two struts. The middle third of the clavicle acts as the superior strut while the scapular body and spine serves as the inferior strut. The complex maintains a normal relationship between the upper extremity and axial skeleton. The scapula is suspended to the clavicle by ligaments and acromioclavicular joint. It can be further sub-classified into three components [21]:

1. The clavicle-acromioclavicular joint-acromial strut
2. The clavicle-coracoclavicular ligamentous-coracoid linkage
3. The three-process-scapular body junction

A single disruption of the ring is a stable injury. A double disruption will result in unstable injury [21, 22] (Fig. 21.6).

Fig. 21.6 The superior shoulder suspensory complex has three components: *1* – the acromioclavicular joint-aromial strut, *2* – the clavicular-coracoclavicular ligamentous-coracoid linkage, *3* – the three-process-scapular body junction

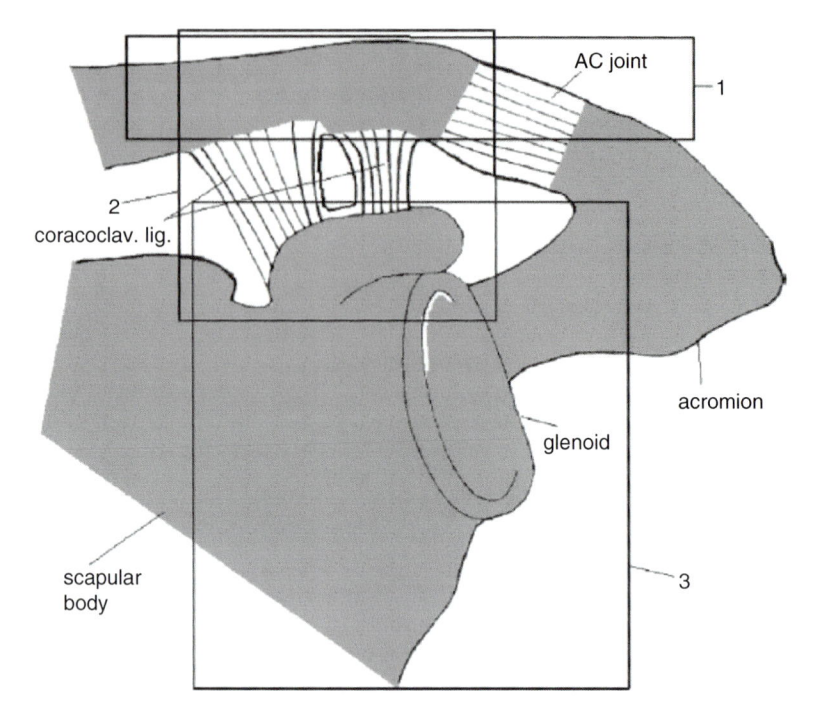

21.5.2.1 Clinical Presentation and Diagnosis

The clinical presentation of floating shoulder injuries varies with the associated injuries. When there are other serious injuries, the condition is often overlooked. During secondary survey, one can notice that the shoulder is usually grossly swollen and tender. A displaced clavicle fracture or the prominent lateral clavicular end in an acromioclavicular joint dislocation may be visible. Movements in all directions will be severely limited. Ribs fractures are not uncommon. Shortly after the injury, a detailed neurovascular examination around the shoulder may be difficult. Nevertheless, the distal neurovascular status should still be checked as the nearby brachial plexus and axillary vessels may be injured. This is one of the most important prognostic factors with regard to final clinical outcome [23]. Open injuries occur commonly in the area of the clavicle.

Radiological examination includes the anteroposterior view of the scapula and the transcapular lateral view is usually most informative. Important factors to assess include the amount of clavicular displacement, glenoid angulation and medialization, the extent of intra-articular involvement and the extent of communication [24]. If the patient is physiologically stable, further evaluation with CT scan and three-dimensional reconstructions can help to better delineate the fracture pattern.

21.5.2.2 Management

Floating shoulder injuries normally do not require emergency management, unless there is an associated open clavicula fracture, thus requiring urgent debridement. Once diagnosed, the shoulder should be supported with a broad arm sling and additional evaluation along with a CT scan should be performed when the patient's general condition is stable.

To date there is no consensus regarding the most adequate treatment of floating shoulder injuries because of the small patient number and heterogeneity of all the studies. Based on current literature review, the treatment options are now evenly divided into nonsurgical treatment, as well as open reduction and internal fixation [24]. The degree of displacement of both clavicle and scapular neck fractures plays an important role in deciding the stability of the fractures.

Non-surgical management is popular because of its non-invasiveness and low morbidity [25, 26].

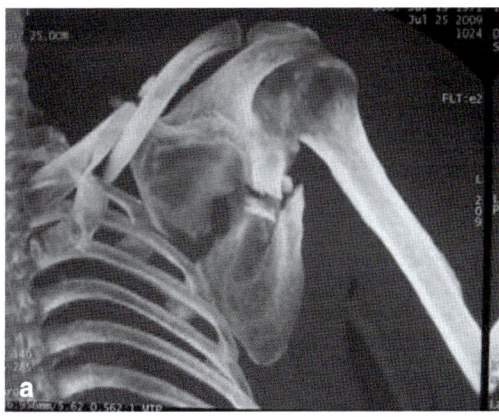

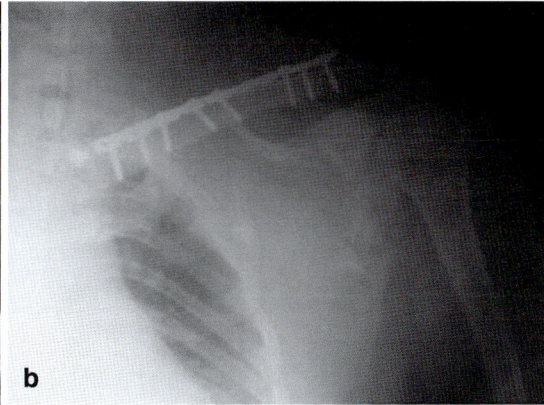

Fig. 21.7 (**a**) A 38 year-old man fell from 20 feet during work and sustained head concussion, fractures of right fourth to sixth ribs and left second and sixth ribs, fracture left clavicle and left scapula fracture with comminution over the scapular body and an undisplaced glenoid neck fracture. CT thorax revealed bilateral small apical pneumothorax. (**b**) In order to improve the ventilatory effort and to facilitate nursing processes in intensive care unit, plate fixation of left clavicle was performed

It includes a period of immobilization and pain management, followed by gradual mobilization exercise and strengthening exercise in 4–6 weeks time. Minimally displaced fractures with no sign of significant ligament disruptions can be successfully treated by conservative means [24, 27]. It is also indicated when the multiply injured patient is in a haemodynamically unstable condition or in extremis.

In a multiply injured patient with a floating shoulder injury, surgical intervention should be considered because the unstable shoulder girdle is difficult in terms of nursing. This is true especially in the intensive care stage when they require breathing exercises and chest physiotherapy. Hence, once the patient is stable haemodynamically, one should consider at least fixing the clavicular fracture, which can indirectly reduce and stabilize the glenoid fracture. The patient is then allowed to perform early mobilization exercises. This has the benefit of reducing pain and minimizes the chance of frozen shoulder. Hashiguchi and Ito reported successful treatment in patients with floating shoulder injuries by clavicular fixation alone [28] (Fig. 21.7).

If significant displacement of glenoid remains after clavicle fixation, reduction and fixation of the glenoid may be indicated because of the theoretical restoration of the rotator cuff lever arm [24, 27]. However, surgical fixation of the scapular neck needs surgical expertise. It usually involves a posterior skin incision in a prone position which is not good especially for a chest injured patient. This will also lead to an inevitable increase in surgical trauma with more intraoperative blood loss and more post-operative pain. Hence the scapular fixation may be performed later at a second stage procedure.

21.5.2.3 Complications

In the setting of untreated or neglected floating shoulder, the weight of the arm and the contraction of the biceps, triceps and coracobrachial muscles results in downward pull of the distal fragment, with resultant change of the shoulder contour, the 'drooping shoulder'. This shortening will cause loss of mechanical advantage of the rotator cuff muscles [23, 29]. The increase in displacement of the fracture will result in complications such as malunion, non-union, post-traumatic arthritis, subacromial impingement or chronic brachial plexopathy [30–33].

21.5.3 Floating Elbow

Ipsilateral diaphyseal fractures of the humerus and the forearm are termed as 'floating elbow' injuries. These injuries are rare and they can happen in both adults and children. Usually, these injuries are the results of high-energy trauma, such as road traffic accidents, industrial accidents or falls from a height. As a result, open injuries are common. Nevertheless, with the advance of modern plating and nailing, debridement and antibiotics, there is a major improvement in the outcome of this severe injury compared with two decades ago [34–38].

There is no special classification for floating elbow injury. The fracture pattern of the humerus and forearm are classified individually using the traditional ways, e.g. AO/OTA classification. In addition, this injury is usually associated with conditions such as open fractures, nerve injuries, vascular injuries, compartment syndrome and multisystem injuries [34, 37, 39, 40]. In the literature, the incidence of open fracture is above 50 % [37]. In many cases, the soft tissue injury is so severe that multiple staged operations are required for soft tissue coverage before the fracture fixation. Uncommonly, the elbow joint itself can also be dislocated [41].

21.5.3.1 Management
Grossly contaminated wound should be thoroughly debrided. The use of external fixator in open humeral fractures is applied in case of grossly contaminated wounds, or when rapid skeletal stabilization is required for urgent revascularization.

Following soft tissue coverage, the humeral fractures can be stabilized by plating or nailing. The use of different implants and techniques depends on the local soft tissue condition and individual surgeon experience. At present, there is no clear advantage of whether plate or nail fixation is better than the other in the setting of floating elbow [36]. The forearm fracture is treated like isolated one. Stable plate fixation is the standard with attention paid to the alignment, rotation and the interosseous distance.

21.5.3.2 Outcome
Although excellent and good functions can be achieved after surgical treatment in up to 67 % of patients, the presence of brachial plexus injury and peripheral nerve injury seems to have an adverse effect in functional outcome [37, 38].

The timing of surgery, the existence of open fractures, multisystem injuries and presence of neurovascular injuries are all not significantly related to poor functional outcome. These patients' functional outcome falls into a bimodal distribution. One group of patients recovers at around 1 year time and behaves similar to isolated fracture. However, another group of patients have significant problem afterwards and remains disable for long period of time.

21.5.3.3 Complications
Despite all the improvement of management, floating elbow is a complex injury and prone to have complications. The incidence of non-union, malunion and infection, myositis ossificans are exceptionally high [34, 40, 41]. Another common problem is loss of elbow flexion and extension range of movement. Supination and pronation problem is less frequent but it is usually associated with high-energy trauma to the forearm [36].

21.6 Traumatic Knee Dislocation

Traumatic knee dislocation is an uncommon problem. It accounts <0.02 % of all orthopaedic problems [42, 43]. However, this may be an underestimation of the real situation because a high percentage of the knee is spontaneously reduced at the scene [44]. Besides fall from height and motor vehicle accidents, people involved in high-speed sport activities also have a chance of getting knee dislocation. They are present with multiple ligamentous disruption, but vascular and nerve injuries are common as well. The historical way of conservative treatment using simple immobilization with inconsistent outcomes [45, 46] has evolved to the present principles of early surgical intervention with ligamentous repair and reconstruction, to be followed by early mobilization [44].

21.6.1 Classification

Classification can be done according to the time of presentation after injury. An interval of 3 weeks is used to differentiate acute and chronic injury [47]. On the other hand, the anatomical classification proposed by Kennedy, is more commonly used [43]. The classification is based on the direction of tibia displacement in relation to the femur, i.e. anterior, posterior, medial or lateral. The fifth type, rotatory dislocation is the combination of the multidirectional displacement. Among these, anterior dislocation is the commonest type as a result of hyperextension injury. It comprises 40 % of all knee dislocations. The second commonest one is posterior dislocation, which is usually due to 'dash-board' type injury in motor vehicle. It comprises another 1/3 of cases [8]. Rotatory dislocation is the least common type, roughly about 5 %. It is further subdivided into anteromedial, posteromedial, anterolateral and posterolateral, in which posterolateral is the commonest with a high incidence of irreducibility [48]. However, the major drawback of this classification is the difficulty of application when the knee is spontaneously reduced. Another more recent classification, proposed by Schenck [49], is based on the status of the ligamentous disruptions and any associated intra-articular fractures. This can provide more information on the nature and severity of the problem which guides to specific management.

21.6.2 Associated Injuries

Traumatic knee dislocation is often associated with other concomitant injuries. Vascular injury, mainly involving popliteal artery, may result in disastrous consequence. The reported incidence can be up to 65 % [50]. The variable incidence is due to different degrees of damage to vessels, ranging from minor intimal damage to complete transaction. Besides, there may be a lot of occult injury not being diagnosed. The degree of suspicion and the use of arteriography greatly affect the pick-up rate of any vascular compromise.

Another commonly associated injury is the common peroneal nerve damage in about 20 % of cases [51]. The incidence is much higher in posterolateral dislocation or involvement of the posterolateral complex. The reported incidence can be up to 45 % [52]. Tibial nerve injury can also occur but it is much less common.

Fractures, especially avulsion fractures, are often encountered. The usual sites are origins of PCL or lateral tibial plateau in the form of Segond fracture. Fractures of the distal femur or proximal tibia are not uncommon as well.

21.6.3 Evaluation and Assessment

In emergency setting, the initial examination should be directed to neurovascular examination since the consequence of missing the vascular injury is disastrous. The dislocated knee usually presents with significant pain and swelling. A pitfall in diagnosis would be those spontaneously reduced knee dislocations which may look benign on presentation (Fig. 21.8).

Since most of the time, the joint capsules and ligaments are severely disrupted, a spontaneously reduced knee is presented with severe and extensive bruising on medial and lateral side of the leg because of the uncontained haemarthrosis. In addition, the presence of multiple ligamentous laxity is another clue to spontaneously reduced dislocated knee.

The current trend of vascular assessment is now based on both clinical assessment and imaging, with clinical evaluation as the more important aspect. Selective arteriography in patients with abnormal physical abnormalities is practised nowadays. The manual palpation of the pulses of dorsalis pedis and posterior tibialis is sufficient to detect any clinically significant vascular injury. Although minor intimal injuries are not detected by clinical examination, these non-flow-limiting intimal injuries rarely progressed to occlusive lesion [53]. Nevertheless, repeated serial careful vascular examination within the first 48 h is important. Whenever there is an abnormal clinical finding, one should proceed to urgent arteriography without delay. Ankle-brachial index (ABI) is a useful and non-invasive adjunct to detect vascular compromise. It is the ratio of Doppler systolic pressure in injured limb (ankle) to the Doppler systolic pressure in uninjured limb (brachial). The presence of ABI <0.9 indicates immediate further investigation of the arterial status, usually an arteriography [54]. However, the result can be inaccurate in patients with peripheral vascular disease.

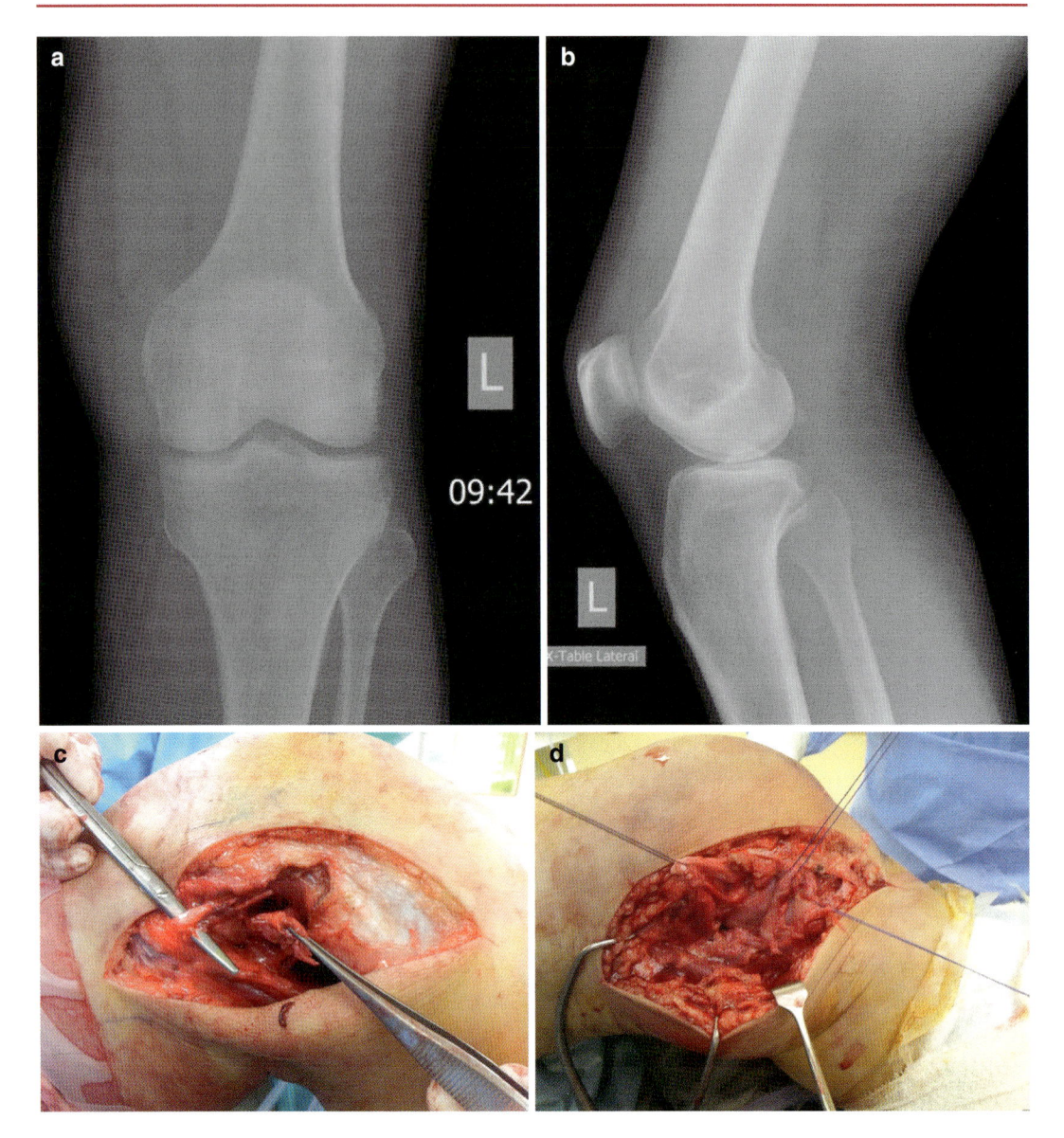

Fig. 21.8 (**a**, **b**) A 45-year-old man crushed by machinery in construction site. He had a closed left knee injury. AP and lateral X-rays of the knee showed no gross abnormality. Distal neurovascular status was intact. Knee was swollen and painful. Clinical examination showed multidirectional instability. (**c**) Lateral side of the knee showed torn medial collateral ligament and rupture Popliteus tendon insertion. (**d**) Medial side of the knee showed complete rupture of the medial collateral ligament complex. Medial meniscus was tagged with suture

A complete neurological examination should be obtained. The degree of damage can be as minor as neuropraxia to complete neuronotmesis. Like vascular assessment, serial neurological reassessment should also be done, as the development of deteriorate neurological deficit can be a sign of developing compartment syndrome or ischemia.

The evaluation of the knee stability should be done after the lower limb is cleared of any impending vascular damage. The examination is usually difficult because of intense pain, muscle spasm and gross swelling. It should be done as gently as possible to minimize the chance of iatrogenic damage. The ACL is best tested by

Lachman test and the PCL by posterior drawer test. The presence of valgus and varus instability signifies medial and lateral collateral ligaments disruptions [44].

The radiological assessment must include plain radiographs during injury and after the reduction. Besides confirmation of the reduction of joint, they also give details on any associated fractures and avulsions. Nevertheless, these investigations should not delay the vascular assessment and intervention. Angiography should be done when there is suspicion of vascular compromise. Magnetic resonance imaging is useful in evaluation of the type and extent of ligamentous injuries as well as cartilage and meniscal damage.

21.6.4 Management

21.6.4.1 Acute

The joint should be reduced gently by gentle traction and manipulation under conscious sedation. The direction of reduction should be guided by the direction of dislocation. The reduced knee joint is then temporarily held with a long leg splint.

Once the reduction is done, the vascular status should be reassessed clinically immediately. If pulses are absent or ABI is <0.9, urgent angiography should be obtained and vascular surgeon opinion is sought. When the site of vascular injury is confirmed, urgent revascularization, using bypass grafting of the popliteal artery or repair using a reverse saphenous vein graft, is required [51]. Fasciotomy is usually performed after revascularization. The knee is preferably immobilized by a knee-spanning external fixator to protect the vascular repair and the knee from re-dislocation. The use of the joint spanning external fixator is also indicated in open injury and joint that failed to maintain reduction in a splint.

In case of knee dislocation necessitates vascular repair, concomitant repair of the torn medial or lateral collateral ligaments can be attempted but the use of sutures and magnitude of the procedure should be kept to minimal. On the other hand, a late repair of these ligaments in a few days time is also a good option [44]. The delay in repair can help the surgeon to monitor the vascular status of the limb in the next 48 h after the repair. It also allows further imaging study for better preoperative planning and delineation of the extent of ligamentous injuries. In open injuries, all ligamentous procedure should be delayed until the wound is well covered and clean.

21.6.4.2 Definitive

The definitive management of multi-ligamentous knee injuries is controversial. However, there are more and more well-designed studies, which provide guidelines for the management of this difficult problem [44, 55–57].

Surgical treatment is the treatment of choice. Absolute indications for surgical treatment include irreducible knees, dysvascular limbs and open injuries. Studies have shown that the surgically treated dislocated knees usually have better range of movement, higher level of activities and better knee scores [46, 55–58].

Another important issue is the timing of ligamentous repair and reconstruction. Meanwhile, there is no consensus on the right timing of surgery. Although many studies showed that the range of movement, knee stability, knee scores (Lysholm score and International Knee Documentation Committee (IKDC) score) and level of activities are better in patients managed within 3 weeks of injury [55–57], there is evidence showing no significant difference between the early and late management groups [59]. The delay of surgery in 3–6 weeks time may allow the healing of the capsule to facilitate the use of arthroscopic repair. In fact, the timing of the definite ligamentous repair is affected by many other factors, especially the vascular status, swelling of the knee, soft tissue coverage and the presence of concomitant fractures.

21.6.5 Rehabilitation

The goal of rehabilitation is to restore the knee range of movement followed by progressive strengthening exercise. The reconstructed knee should be protected by a hinged knee brace or a mobile hinged external fixator. The knee was immobilized for first 3 weeks followed by passive mobilization exercise in brace in the next 3 weeks. Starting from 7th week, the patient is allowed to start gradual weight-bearing training till full-weight-bearing walking. Range of movement and strengthening exercises should continue up to 3 months and then followed by further training to allow patients to reintegrate into his/her previous activities of daily living [44, 51].

21.6.6 Outcome and Complications

Acute traumatic knee dislocation is a severe injury with multiple ligamentous disruption and a high incidence of neurovascular damage. The most disastrous local consequence is probably amputation. The chance of it in failed revascularization within first 8 h can be up to 86 % [8]. Return to normal function is rare. Using the IKDC score, about 39 % of patients are nearly normal, 40 % are abnormal and the remaining 21 % are severely abnormal [55, 56, 60]. The most common complications are joint stiffness and failure of some of the component of ligamentous reconstruction. Post-traumatic osteoarthritis can be up to 50 % [61]. Common peroneal nerve injuries are common. Many of them are neuropraxia and they are managed by observation. Unfortunately, spontaneous full recovery is only about 20 % [55].

> **Conclusions**
> The timing of surgical treatment of articular fractures in polytrauma patients must be based on priorities and be integrated into the optimal management of the overall patient. Open fractures and associated neurovascular injuries are common and often require urgent treatment in the emergency setting. On the other hand, the complex fractures will require careful preoperative planning and preparation. Although primary definitive fracture fixation can be performed in selected patients, a spanning transarticular external fixation should be used most of the time as an initial immobilization method while the patient's physiological status is being stabilized or the soft tissue injury is improving. Large metaphyseal defects can be managed by staged bone grafting with the use of antibiotics cement spacers. In general, the overall injury severity and the extent of soft tissue injury will dictate the timing of definitive fracture fixation.

References

1. Kobbe P, Lichte P, Pape HC. Complex extremity fractures following high energy injuries: the limited value of existing classifications and a proposal for a treatment-guide. Injury. 2009;40 Suppl 4: S69–74.
2. Sirkin M, Sanders R, DiPasquale T, Herscovici Jr D. A staged protocol for soft tissue management in the treatment of complex pilon fractures. J Orthop Trauma. 1999;13(2):78–84.
3. Mills WJ, Nork SE. Open reduction and internal fixation of high-energy tibial plateau fractures. Orthop Clin North Am. 2002;33(1):177–98. ix.
4. Argenta LC, Morykwas MJ. Vacuum-assisted closure: a new method for wound control and treatment: clinical experience. Ann Plast Surg. 1997;38(6):563–76; discussion 577.
5. Webb LX. New techniques in wound management: vacuum-assisted wound closure. J Am Acad Orthop Surg. 2002;10(5):303–11.
6. Lee HJ, Kim JW, Oh CW, Min WK, Shon OJ, Oh JK, et al. Negative pressure wound therapy for soft tissue injuries around the foot and ankle. J Orthop Surg Res. 2009;4:14.
7. Collins DN, Temple SD. Open joint injuries. Classification and treatment. Clin Orthop Relat Res. 1989;243:48–56.
8. Green NE, Allen BL. Vascular injuries associated with dislocation of the knee. J Bone Joint Surg Am. 1977;59(2):236–9.
9. Olson SA, Wills MD. Initial management of open fractures. In: Rockwood CA Jr, Green DP, editors. Fractures in adults 6th edition. Philadelphia: Lippincott Williams & Wilkins; 2006.
10. Masquelet AC, Fitoussi F, Begue T, Muller GP. Reconstruction of the long bones by the induced membrane and spongy autograft. Ann Chir Plast Esthet. 2000;45(3):346–53.
11. Fraser RD, Hunter GA, Waddell JP. Ipsilateral fracture of the femur and tibia. J Bone Joint Surg Br. 1978; 60-B(4):510–5.
12. Ran T, Hua X, Zhenyu Z, Yue L, Youhua W, Yi C, Fan L. Floating knee: a modified Fraser's classification and the results of a series of 28 cases. Injury. 2013; 44(8):1033–42.

13. Lundy DW, Johnson KD. "Floating knee" injuries: ipsilateral fractures of the femur and tibia. J Am Acad Orthop Surg. 2001;9(4):238–45.
14. Rethnam U, Yesupalan RS, Nair R. Impact of associated injuries in the floating knee: a retrospective study. BMC Musculoskelet Disord. 2009;10:7.
15. Adamson GJ, Wiss DA, Lowery GL, Peters CL. Type II floating knee: ipsilateral femoral and tibial fractures with intraarticular extension into the knee joint. J Orthop Trauma. 1992;6(3):333–9.
16. Paul GR, Sawka MW, Whitelaw GP. Fractures of the ipsilateral femur and tibia: emphasis on intra-articular and soft tissue injury. J Orthop Trauma. 1990;4(3):309–14.
17. Szalay MJ, Hosking OR, Annear P. Injury of knee ligament associated with ipsilateral femoral shaft fractures and with ipsilateral femoral and tibial shaft fractures. Injury. 1990;21(6):398–400.
18. Karlstrom G, Olerud S. Ipsilateral fracture of the femur and tibia. J Bone Joint Surg Am. 1977;59(2):240–3.
19. Veith RG, Winquist RA, Hansen Jr ST. Ipsilateral fractures of the femur and tibia. A report of fifty-seven consecutive cases. J Bone Joint Surg Am. 1984;66(7):991–1002.
20. Goss TP. Double disruptions of the superior shoulder suspensory complex. J Orthop Trauma. 1993;7(2):99–106.
21. Goss TP. Scapular fractures and dislocations: diagnosis and treatment. J Am Acad Orthop Surg. 1995;3(1):22–33.
22. Williams Jr GR, Narania J, Klimkiewicz J, Karduna A, Iannotti JP, Ramsey M. The floating shoulder: a biomechanical basis for classification and management. J Bone Joint Surg Am. 2001;83-A(8):1182–7.
23. van Noort A, van der Werken C. The floating shoulder. Injury. 2006;37(3):218–27.
24. DeFranco MJ, Patterson BM. The floating shoulder. J Am Acad Orthop Surg. 2006;14(8):499–509.
25. Edwards SG, Whittle AP, Wood II GW, et al. Nonoperative treatment of ipsilateral fractures of the scapula and clavicle. J Bone Joint Surg Am. 2000;82(6):774–80.
26. Ramos L, Mencia R, Alonso A, Fernandez L. Conservative treatment of ipsilateral fractures of the scapula and clavicle. J Trauma. 1997;42(2):239–42.
27. Owens BD, Goss TP. The floating shoulder. J Bone Joint Surg Br. 2006;88(11):1419–24.
28. Hashiguchi H, Ito H. Clinical outcome of the treatment of floating shoulder by osteosynthesis for clavicular fracture alone. J Shoulder Elbow Surg. 2003;12(6):589–91.
29. Obremskey WT, Lyman JR. A modified judet approach to the scapula. J Orthop Trauma. 2004;18(10):696–9.
30. Hardegger FH, Simpson LA, Weber BG. The operative treatment of scapular fractures. J Bone Joint Surg Br. 1984;66(5):725–31.
31. van Noort A, te Slaa RL, Marti RK, van der Werken C. The floating shoulder. A multicentre study. J Bone Joint Surg Br. 2001;83(6):795–8.
32. Rikli D, Regazzoni P, Renner N. The unstable shoulder girdle: early functional treatment utilizing open reduction and internal fixation. J Orthop Trauma. 1995;9(2):93–7.
33. Egol KA, Connor PM, Karunakar MA, Sims SH, Bosses MJ, Kellam JF. The floating shoulder: clinical and functional results. J Bone Joint Surg Am. 2001;83-A(8):1188–94.
34. Rogers JF, Bennett JB, Tullos HS. Management of concomitant ipsilateral fractures of the humerus and forearm. J Bone Joint Surg Am. 1984;66(4):552–6.
35. Lange RH, Foster RJ. Skeletal management of humeral shaft fractures associated with forearm fractures. Clin Orthop Relat Res. 1985;195:173–7.
36. Simpson NS, Jupiter JB. Complex fracture patterns of the upper extremity. Clin Orthop Relat Res. 1995;318:43–53.
37. Yokoyama K, Itoman M, Kobayashi A, Shindo M, Futami T. Functional outcomes of "floating elbow" injuries in adult patients. J Orthop Trauma. 1998;12(4):284–90.
38. Solomon HB, Zadnik M, Eglseder WA. A review of outcomes in 18 patients with floating elbow. J Orthop Trauma. 2003;17(8):563–70.
39. Levin LS, Goldner RD, Urbaniak JR, et al. Management of severe musculoskeletal injuries of the upper extremity. J Orthop Trauma. 1990;4(4):432–40.
40. Pierce RO, Hodorski DF. Fractures of the humerus, radius, and ulna in the same extremity. J Trauma. 1979;19(3):182–5.
41. Viegas SF, Gogan W, Riley S. Floating dislocated elbow: case report and review of the literature. J Trauma. 1989;29(6):886–8.
42. Hoover N. Injuries of the poplitealartery associated with dislocation of the knee. Surg Clin North Am. 1961;41:1099–112.
43. Kennedy JC. Complete dislocation of the knee joint. J Bone Joint Surg Am. 1963;45:889–904.
44. Rihn JA, Groff YJ, Harner CD, Cha PS. The acutely dislocated knee: evaluation and management. J Am Acad Orthop Surg. 2004;12(5):334–46.
45. Taylor AR, Arden GP, Rainey HA. Traumatic dislocation of the knee. A report of forty-three cases with special reference to conservative treatment. J Bone Joint Surg Br. 1972;54(1):96–102.
46. Richter M, Bosch U, Wippermann B, Hofmann A, Krettek C. Comparison of surgical repair or reconstruction of the cruciate ligaments versus nonsurgical treatment in patients with traumatic knee dislocations. Am J Sports Med. 2002;30(5):718–27.
47. Palmer I. On the injuries of the ligaments of the knee joint: a clinical study. Acta Chir Scand. 1938;53:1–28.
48. Quinlan AG. Irreducible posterolateral dislocation of the knee with button-holing of the medial femoral condyle. J Bone Joint Surg Am. 1966;48(8):1619–21.
49. Schenck RC. The dislocated knee. Instr Course Lect. 1994;43:127–36.
50. Meyers MH, Harvey Jr JP. Traumatic dislocation of the knee joint. A study of eighteen cases. J Bone Joint Surg Am. 1971;53(1):16–29.

51. Robertson A, Nutton RW, Keating JF. Dislocation of the knee. J Bone Joint Surg Br. 2006;88(6):706–11.
52. Niall DM, Nutton RW, Keating JF. Palsy of the common peroneal nerve after traumatic dislocation of the knee. J Bone Joint Surg Br. 2005;87(5):664–7.
53. Stain SC, Yellin AE, Weaver FA, Pentecost KN. Selective management of nonocclusive arterial injuries. Arch Surg. 1989;124(10):1136–40; discussion 1140–1.
54. Mills WJ, Barei DP, McNair P. The value of the ankle-brachial index for diagnosing arterial injury after knee dislocation: a prospective study. J Trauma. 2004;56(6):1261–5.
55. Liow RY, McNicholas MJ, Keating JF, Nutton RW. Ligament repair and reconstruction in traumatic dislocation of the knee. J Bone Joint Surg Br. 2003;85(6):845–51.
56. Harner CD, Waltrip RIL, Bennett CH, Francis KA, Cole B, Irrgang JJ. Surgical management of knee dislocations. J Bone Joint Surg Am. 2004;86-A(2):262–73.
57. Tzurbakis M, Diamantopoulos A, Xenakis T, Gergoulis A. Surgical treatment of multiple knee ligament injuries in 44 patients: 2–8 years follow-up results. Knee Surg Sports Traumatol Arthrosc. 2006;14(8):739–49.
58. Dedmond BT, Almekinders LC. Operative versus nonoperative treatment of knee dislocations: a meta-analysis. Am J Knee Surg. 2001;14(1):33–8.
59. Fanelli GC, Giannotti BF, Edson CJ. Arthroscopically assisted combined posterior cruciate ligament/posterior lateral complex reconstruction. Arthroscopy. 1996;12(5):521–30.
60. Wascher DC, Dvirnak PC, DeCoster TA. Knee dislocation: initial assessment and implications for treatment. J Orthop Trauma. 1997;11(7):525–9.
61. Werier J, Keating JF, Meek RN. Complete dislocation of the knee: the long-term results of ligamentous reconstruction. Knee. 1998;5:255–60. 1996.

Outcome and Management of Primary Amputations, Subtotal Amputation Injuries, and Severe Open Fractures with Nerve Injuries

William W. Cross III and Marc F. Swiontkowski

Contents

22.1	Introduction	307
22.2	**Traumatic Primary Amputations: Considerations and Completions**	308
22.2.1	Outcome of Traumatic Primary Amputations	310
22.3	**The Subtotal Amputation Injury: Limb Salvage or Amputation**	311
22.3.1	Factors Influencing Initial Salvage Decisions	311
22.3.2	Lower-Extremity Injury-Severity Scales and Scores: Tools for Assisting Surgeons with Salvage or Amputation Decisions	312
22.3.3	Lower-Extremity Injury-Severity Scales and Scores: Predicting Functional Outcomes of Salvaged Limbs After Limb-Threatening Trauma	317
22.3.4	Lower-Extremity Injury-Severity Scales and Scores: Summary	318
22.3.5	Outcomes in Patients Undergoing Limb Salvage or Amputation for Limb-Threatening Injuries	318
22.3.6	Outcomes of the Mangled Foot and Ankle	320
22.3.7	Complications in the Treatment of Severe Lower-Extremity Trauma	321
22.3.8	Psychological Distress in Patients with Severely Injured Lower Extremities	322
22.3.9	Societal Costs Associated with Limb Salvage and Amputation	323
22.4	**The Open Fracture with Severe Nerve Injury**	323
22.5	**Summary**	324
References		325

W.W. Cross III (✉)
Division of Orthopedic Trauma,
Department of Orthopedic Surgery, Mayo Clinic,
200 1 St. SW, Rochester, MN 55902, USA
e-mail: cross.william@mayo.edu

M.F. Swiontkowski
Department of Orthopaedic Surgery,
University of Minnesota,
University of Minnesota Medical School,
2450 Riverside Ave. S. Suite R200, Minneapolis,
MN 55454, USA
e-mail: swion0001@umn.edu

22.1 Introduction

More than three out of five accidental injuries in the USA are to the musculoskeletal system. Costs associated with the care of these injuries have been estimated to be $849 billion or 7.7 % of the US gross domestic product (GDP) in the year 2004. Musculoskeletal disease and injury continue to account for the majority of both lost wages and hospital bed days in the USA [1]. We must improve the care of these injuries so that we may help patients rehabilitate from injury and prevent future morbidity.

A small but resource-heavy subset is the high-energy trauma patient with a mangled extremity [2]. The evaluation and subsequent management of this patient group can be a great source of stress for both the patient and the treating surgical team. The decision-making processes are difficult and can be controversial, and the clinical evidence for these decisions has been largely based upon small case series and historical Level V evidence [3]. These data have influenced the treatment of limb-threatening trauma and have potentially led to large numbers of limb amputations with severe lower-extremity trauma where limb salvage may have been technically possible but not recommended [4, 5]. As medical and surgical technology, skills, procedures, and concepts have evolved, so has our ability to salvage limbs previously thought to be unsalvageable. Particular areas of advancement include soft-tissue handling, less invasive fracture management, microvascular repair, and soft-tissue coverage [6–13]. Limb-salvage protocols have been evaluated, and many of them have influenced our current treatment strategies [14, 15]. These studies and others reviewing complicated limb trauma have suggested that early amputation may be preferable due to the mental and physical toll limb salvage can levy on patients [16–18]. Most studies have included small numbers of patients, and their results have correspondingly not yielded definitive results [6, 7, 18, 19].

In an effort to provide evidence for clinicians to rely upon when making amputation versus salvage decisions, a large multicenter, prospective, observational study was undertaken entitled the Lower Extremity Assessment Project (LEAP) [20–22]. Utilizing data from this project and more recent data from military services involved with combat-related injuries, several areas of the amputation – limb-salvage debate – have been explored. Evidence from this trial and others is presented in the following chapter to assist treatment teams in these difficult and complex situations. The goals of this chapter are to present the data from this study and provide a framework for surgical treatment teams to employ when evaluating the high-energy trauma patient with a mangled extremity.

22.2 Traumatic Primary Amputations: Considerations and Completions

The patient presenting with a complete or near-complete traumatic amputation as the result of high-energy trauma requires an evaluation consistent with the latest recommendations of the American College of Surgeons and the principles of Advanced Trauma Life Support [23–25]. Once the patient's life-threatening issues have been stabilized, attention can then be focused on the injured extremity. It is perhaps best to have the orthopedic surgeon present prior to any surgical intervention. It is typically this surgeon who will follow the patient through subsequent recovery and functional gain with the affected extremity. In addition, any further surgical interventions are likely to be performed by an orthopedic surgeon.

Standard open wound protocols should be followed in accordance with open fracture principles surrounding the acute zone of injury (see Chap. 20). Once the patient is physiologically stable, the zone of injury on the affected limb is defined in the surgical suite, and the limb is deemed appropriate for definitive amputation, and appropriate surgical steps are taken according to the desired amputation level and planned technique (i.e., bone cut lengths, muscle flap coverage, myodesis planning).

In the orthopedic trauma setting, there are three primary lower-extremity amputations we consider appropriate: below-the-knee, above-the-knee, and, in some select cases, through-the-knee. In the high-energy trauma patient, more often than not, the heel pad has been traumatized over the hind foot making the Syme amputation less optimal and rarely used option (Fig. 22.1). The hip disarticulation is also rarely used except for the most severe proximal injuries. This usually includes those with massive soft-tissue injury and/or an obvious vascular and complete sciatic nerve transection. The indications and techniques for the above three primary amputations have been well described [26] and are not the focus of this chapter. However, when contemplating an amputation through-the-knee, the surgeon must

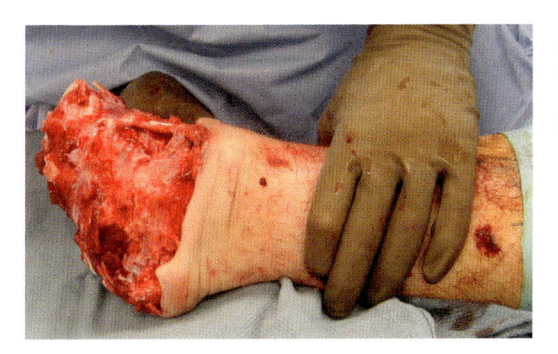

Fig. 22.1 This 28-year-old male was involved in a high-speed motorcycle crash and sustained significant forefoot and midfoot trauma. The heel pad was severely damaged in this case which happens commonly in these injury patterns. This makes subsequent reconstrutive efforts difficult with amputation levels below the midsection of the tibia (i.e., Syme amputations)

critically evaluate the soft-tissue envelope around this tenuous area. If there is any evidence that the zone of injury includes this area, most especially the proximal gastroc-soleus musculature, then there should be strong consideration to proceed with an amputation level above-the-knee. Data from the LEAP study [20, 22, 27] has suggested that through-the-knee amputations do not perform as well as above-the-knee amputations in the mangled extremity patient. This finding was most likely attributed to the condition of the soft-tissue envelope in their patient cohort and to difficulties with prosthetic fitting. In the absence of compromised soft-tissues in this area and in the properly selected patient with experienced prosthetics support, a through-the-knee amputation has been shown to provide good muscular balance and has a low risk for the late development of joint contractures [28].

Severe upper extremity injuries, which present as complete or near-complete amputations, warrant special consideration and evaluation by a surgeon who is familiar with reconstruction procedures in this area. The decision-making process in the mangled upper extremity can be challenging, especially when limb salvage becomes an option [29]. Primary amputation may not be in the best interest of some patients as it has been suggested that a sensate hand with minimal prehensile function can outperform a prosthesis [30]. Standard principles of wound care

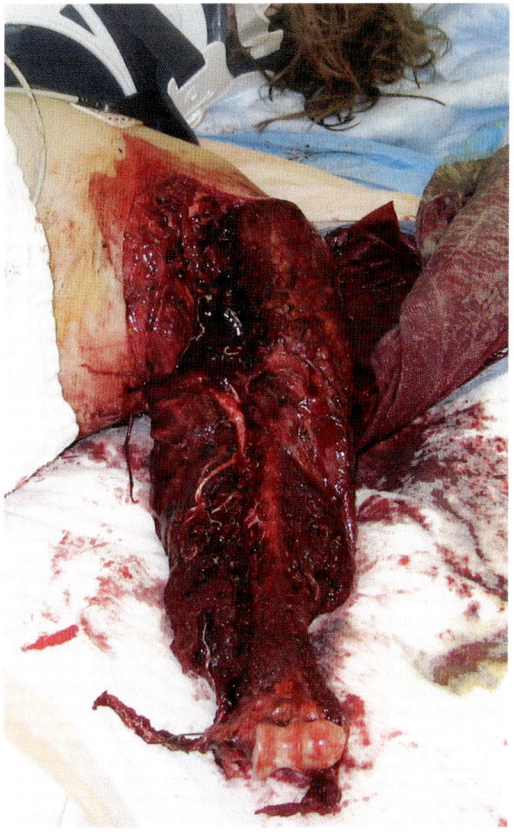

Fig. 22.2 This 16-year-old female was involved in a high-speed motor vehicle crash in which the vehicle rolled multiple times. She sustained a traumatic amputation of the forearm including the entire radius and ulna. The proximal soft-tissue involvement was extensive, and she underwent a proximal amputation leaving 14 cm of residual humerus. She was ultimately fit with a myoelectric hand

should be employed until appropriate consultation can be obtained. When definitive surgical intervention is required, preservation of length is critical and can decrease the energy needed for the patient to suspend their prosthesis (Fig. 22.2). Furthermore, the increased surface area of the limb can help with load distribution, prosthesis propulsion in space, and counterpressure with task performance [26].

Absolute indications for primary limb amputation have been suggested in the literature with varying algorithms. Generally, these indications have included a patient presenting with a total or near-total leg amputation or complete tibial or

Table 22.1 Primary amputation guidelines

Absolute indications	1. Presentation with complete or near-complete limb amputation
	2. Complete sciatic or tibial nerve transection in an adult
Relative indications	1. Concurrent ipsilateral severe foot injury
	2. Large intercalary soft-tissue or bone loss
	3. Warm ischemia time of >6 h
	4. Severe concurrent multiple injuries

sciatic nerve transection in an adult [14, 31, 32]. Relative indications have included two or more of the following: concurrent severe ipsilateral foot injury, large intercalary soft-tissue or bone loss, warm ischemia time of greater than 6 h, and severe concurrent multiple injuries (Table 22.1) [8, 15, 31, 33–35]. Uniformly, however, these studies indicate that the clinician's judgment at the time of initial evaluation is critical; amputation decision-making should employ a multitude of factors. We also advise seeking multispecialty input with this difficult decision (i.e., orthopedics, plastic surgery, general surgery). In one study, a combined approach led to 89 % of patients achieving a successful viable limb, and only 11 % went on to secondary amputation [31].

22.2.1 Outcome of Traumatic Primary Amputations

There is little in the literature reporting the long-term outcome of traumatic amputations. Recently, Dougherty published a study evaluating the outcomes of 123 transtibial amputees from the Vietnam War – 65 % of which were victims of land mines and booby traps. He found that with isolated amputations, these patients led relatively normal lives. However, when concomitant injuries were sustained by these patients, their SF-36 scores lowered, and their incidence of psychological illness increased [36]. Smith et al. [37] published a descriptive study describing outcomes of 20 patients with unilateral transtibial amputations. They found that SF-36 scores were lower than normal age-matched scores in the categories of physical function and role limitations because of physical health problems and pain. Aside from those two sections, scores from the normal population were not significantly different. Lerner et al. [38, 39] evaluated three groups of patients: posttraumatic fracture nonunion, chronic refractory osteomyelitis, and lower-extremity amputation. In their group of 109 patients, they found that the chronic osteomyelitis patients were the most adversely affected among the three groups. Interestingly, 85 % of the amputee patients believed they had been "mentally scarred" by their orthopedic problem, but despite that complaint, they had minimal restriction in lifestyle and activity – a direct contrast to the poorer functioning osteomyelitis group.

In 2004, a study was published which reviewed 161 trauma-related amputation patients that were participants in the LEAP study [27]. This study found no differences in outcomes between the above-the-knee amputees and the below-the-knee amputees. The exception to this finding was with walking speeds in which the below-the-knee group performed better. A key finding in this study was the significantly poorer outcomes of patients that had undergone a through-the-knee amputation. The poorer outcome was associated with worse walking speeds and also less physician-measured satisfaction in terms of clinical, functional, and cosmetic recoveries of their patients. As we noted earlier, we believe the surgeon must critically evaluate the zone of injury prior to proceeding with a through-the-knee amputation. Furthermore, when faced with the decision to proceed with an above-the-knee amputation, surgeons should take whatever steps are necessary to preserve femoral length [40]. It was recently shown that retained length of the femur significantly improves temporospatial and kinematic gait outcomes. Careful attention to the adductors, either with preservation or reconstruction, can benefit this group of patients and improve their mobility.

The outcome of isolated traumatic lower-extremity amputations is mixed but can generally be associated with residual disability and lower outcome scores than the general population. While Dougherty's [41] study of transtibial amputations demonstrated relatively normal scores with a select population with an isolated lower-extremity injury, other studies indicate substantially poorer outcomes. In another study by Dougherty examining more proximal transfemoral amputations, substantial disability was found in patient follow-up [36]. Smith et al. [37] and the LEAP study [27] also identified significant disability with traumatic amputations in follow-up. These studies indicate that when lower-extremity injuries are among a constellation of traumatic injuries, which they often are, outcomes demonstrate increased disability. An extensive rehabilitation program offered at the treating US Army hospital may have influenced the better outcomes identified in Dougherty's transtibial amputation study. This finding and those of the LEAP study underscore the need to have high-energy traumatic amputation patients closely followed and managed by a multidisciplinary team including surgeons, rehabilitation physicians, nurses, prosthetists, and therapists. It is also the surgeon's responsibility to inform patients of expected outcomes and ensure that unrealistic expectations are not confusing patients during their recovery. These discussions can allay patient fears and allow the patient, their families, and support networks to adjust to the trauma and plan ahead for expected changes.

Table 22.2 Inclusion criteria of the LEAP study [22]

1. Traumatic amputations below the distal femur
2. Gustilo Type IIIA fracture *with*
   (a) Length of hospital stay >4 days
   (b) Two or more surgical limb procedures
   (c) Two or more of the following: (a) severe muscle damage (>50 % loss of one or more major muscle groups or associated compartment syndrome with myonecrosis); (b) associated nerve injury (posterior tibial or peroneal deficit); (c) major bone loss or bone injury (associated fibula fracture; >50 % displacement, comminution, and segmental-type fracture; and >75 % probability of requiring bone graft/transport)
3. Gustilo Type IIIB tibia fracture
4. Gustilo Type IIIC tibia fracture
5. Dysvascular injuries below the distal femur excluding the foot include knee dislocations, closed tibia fractures, and penetrating wounds with vascular injury documented from arteriogram, surgery, or ultrasound
6. Major soft-tissue injuries below the distal femur excluding the foot include:
   (a) AO[a] type IC3–IC5 degloving injuries
   (b) Severe soft-tissue crush/avulsion injuries with muscle disruption or compartment syndrome
   (c) Compartment syndrome resulting in myonecrosis and requiring partial or full muscle unit resection
7. Severe foot injuries including:
   (a) Type IIIB open ankle fractures
   (b) Severe open hindfoot or midfoot injury (i.e., either insensate plantar surfaces, devascularization, major degloving injury, or open soft-tissue injury requiring coverage)
   (c) Open Type III pilon fractures

Lower Extremity Assessment Project
[a]Arbeitsgemeinschaft für Osteosynthesefragen

22.3 The Subtotal Amputation Injury: Limb Salvage or Amputation

The high-energy trauma patient with a subtotal amputation to an extremity presents immediate challenges to the trauma team. The Lower Extremity Assessment Project (LEAP) was a prospective cohort study of 601 patients who had been admitted to eight Level I trauma centers for the treatment of severe lower-extremity injuries below the distal part of the femur [21]. This study sought to provide evidence for clinicians to use when faced with this dilemma and has recently published 7-year follow-up data [20]. The singular study has produced multiple projects investigating various facets of the lower-extremity injured patient, and many are discussed in the ensuing sections. Inclusion criteria for the LEAP study are listed in Table 22.2 and highlight the severity of trauma evaluated in this study as well as the breadth of injuries included. Please refer to Case 1 in Figs. 22.3a, 22.4, and 22.5c and Case 2 in Figs. 22.6a and 22.7c for limb-salvage and amputation examples.

22.3.1 Factors Influencing Initial Salvage Decisions

Initial decisions for the acute trauma patient with a severely injured lower extremity include

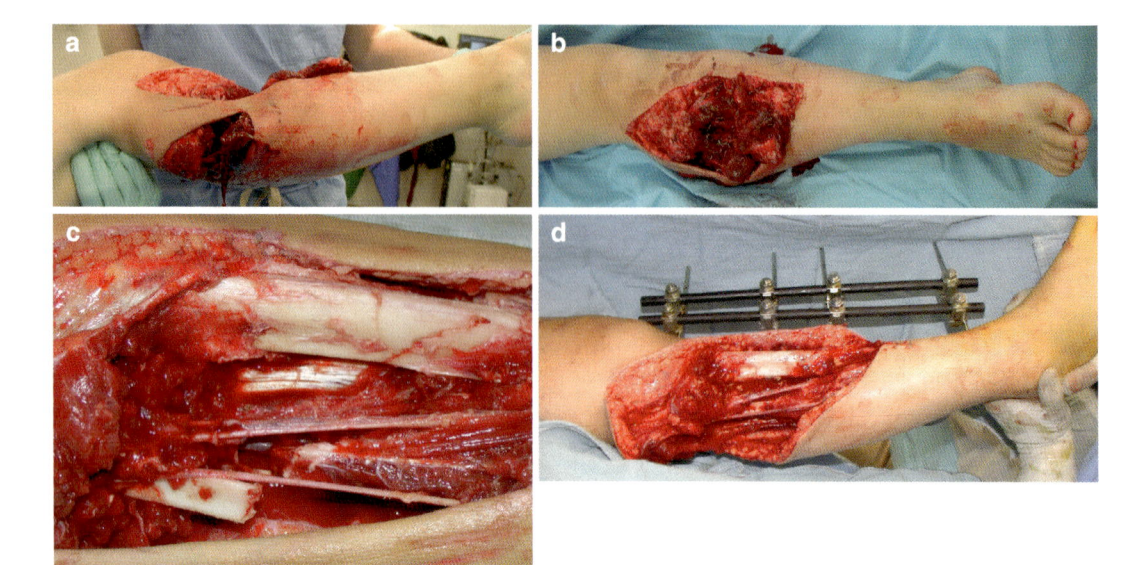

Fig. 22.3 This 20-year-old female sustained severe right lower leg trauma after being run over by a personal watercraft. (**a–d**) Initial surgical evaluation and debridement with subsequent external fixation. (**c**) Extensive soft-tissue loss and intact neurovascular bundle posterior to the tibia fracture. At this time, we confirmed our decision to salvage the limb. This wound had a vacuum-assisted closure device until the plastic surgery team could evaluate and ultimately place a tissue flap over the wound (Case and photographs courtesy of David P. Barei, MD)

immediate amputation (i.e., within the first 24 h) or delayed (i.e., secondary procedure with the first hospitalization) [8, 14, 15, 17, 42, 43]. There are a multitude of factors influencing this decision: those related directly to the leg injury itself, the extent and severity of associated injuries, the physiologic reserve of the patient, and their social support network. The training and experience of the attending surgeon may also play a role in the decision-making process [44].

Mackenzie et al. published the results of a survey pertaining to surgeons and their decision to amputate or reconstruct traumatized lower extremities. This study highlighted various factors that different specialties (general surgeons and orthopedic surgeons) deemed most important to consider in the critical decision of amputation versus salvage (Table 22.3). Interesting perspectives representative of specialty-specific training and goals were identified. Namely, the general surgeons tended to emphasize the overall physiologic condition and reserve of the patient as a whole (the injury-severity scale, limb ischemia), whereas the orthopedic surgeon emphasized functional outcome prognosis (nerve integrity, soft-tissue coverage, limb ischemia). The study conclusions suggest that the main factor influencing surgeons on the question of salvageable limbs is apparent soft-tissue damage: muscle injury, absence of sensation, arterial injury, and vein injury. Patient factors were found to play much less of a role, although alcohol consumption and socioeconomic status were noted to be of some influence [44].

22.3.2 Lower-Extremity Injury-Severity Scales and Scores: Tools for Assisting Surgeons with Salvage or Amputation Decisions

Lower-extremity injury-severity scores were developed by clinicians to assist surgical teams in making the often difficult initial decision of whether to attempt limb salvage or amputate a severely traumatized extremity. Surgeons have hypothesized that patients who undergo initial salvage attempts but subsequently require later amputation have worse outcomes than those who

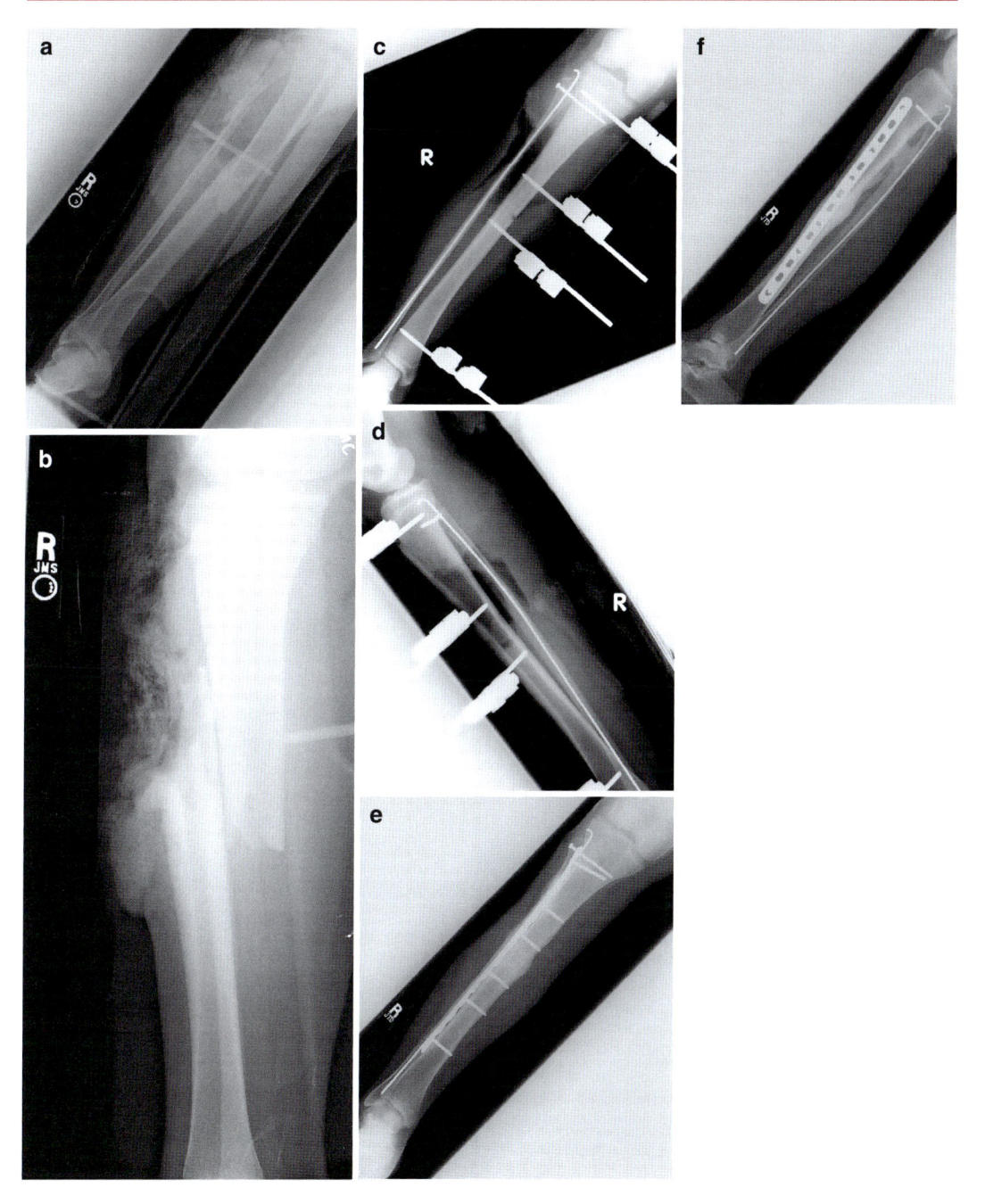

Fig. 22.4 (**a**, **b**) Anterior-posterior and lateral radiographic views of the injured lower extremity. Note significant soft-tissue shadow highlighting the extensive damage. This patient was fortunate and did not sustain substantial bone loss. (**c**, **d**) Provisional external fixation was employed to restore length, alignment, and rotation to the injured limb. (**e**, **f**) One year post-injury radiographs demonstrating complete union of both the tibia and fibula (Case and photographs courtesy of David P. Barei, MD)

have early amputation. This makes intuitive sense and was shown to be correct in the LEAP study [16] and highlights the importance of early and accurate selection on which patients should proceed with a limb amputation during their first hospitalization.

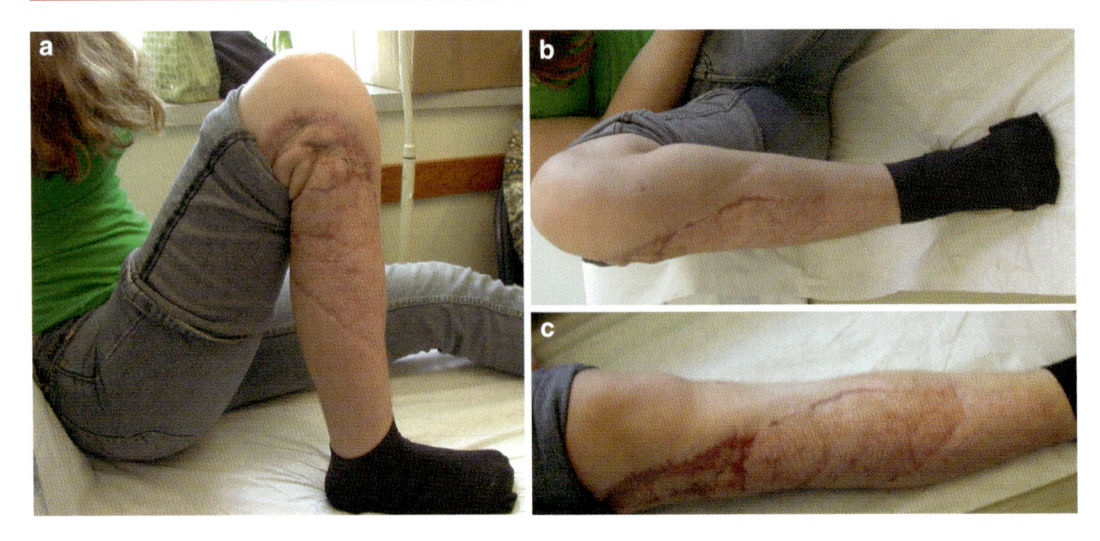

Fig. 22.5 (**a–c**) Clinical follow-up demonstrating good result of limb salvage with this patient. She was able to gain excellent range of motion and had an outstanding support network aiding her in the recovery process (Case and photographs courtesy of David P. Barei, MD)

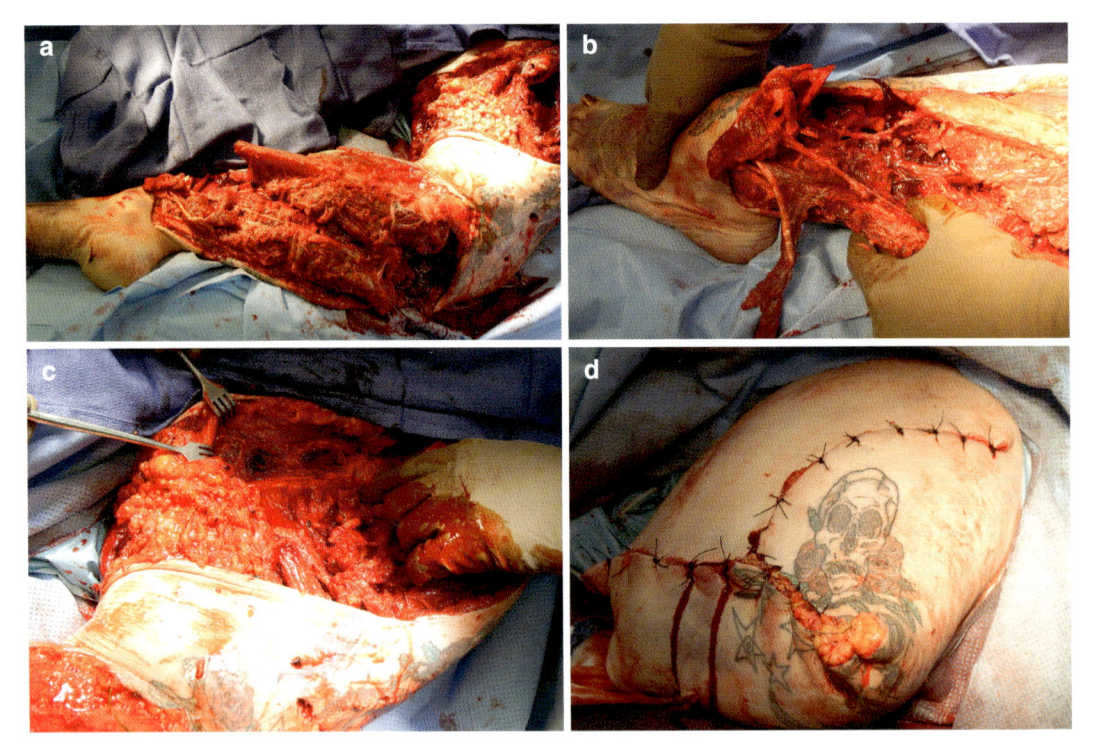

Fig. 22.6 This 40-year-old female was involved in a severe motorcycle crash. In Figures (**a–c**) profound soft-tissue and osseous damage was sustained. Emergency department evaluation demonstrated the foot be avascular. The patient underwent emergent operative intervention and underwent an acute above-the-knee amputation (**d**). She returned to the operating suite several times over the ensuing days for further debridement and, ultimately, a disarticulation of the hip joint. Radiographs for this patient are shown in Fig. 22.7

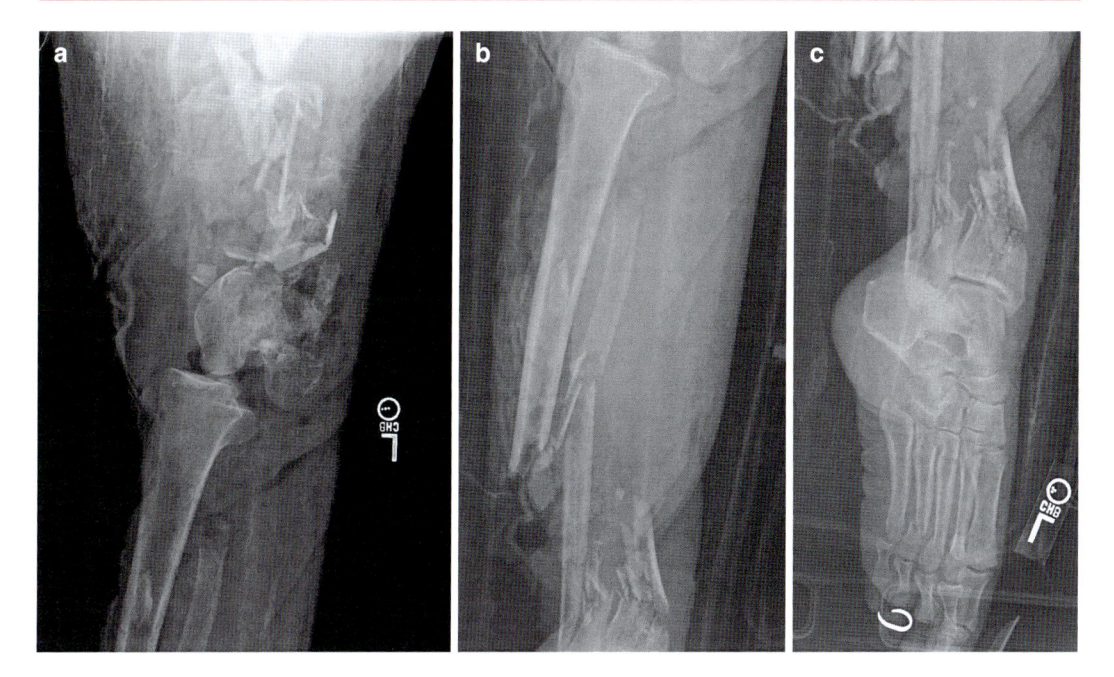

Fig. 22.7 (a–c) Radiographs of the patient pictured in Fig. 22.6a

Several studies [31, 33, 45–47] have examined the application of high-energy lower-extremity trauma scoring systems to patients with severe lower-extremity trauma. The LEAP study [21] contained the largest patient cohort of 565 prospectively evaluated high-energy lower-extremity injured patients. Each patient in this study had five well-known injury-severity scoring systems applied to their case in an effort to determine the clinical utility of each system [45]. The five systems evaluated were the Mangled Extremity Severity Score (MESS) [29, 48], the Limb Salvage Index (LSI) [32], the Predictive Salvage Index (PSI) [34], the Nerve Injury, Ischemia, Soft-Tissue Injury, Skeletal, Shock, and Age of Patient Score (NISSSA) [49], and the Hannover Fracture Scale (HFS) [50]. Table 22.4 represents the components of each injury-severity scale with the addition of a newer scale that was developed in India to predict hospital days required, flap requirements, rate of infection, and the number of secondary procedures required. This scale also incorporates patient comorbidities but emphasized primarily the evaluation of type IIIB open tibia fractures [51]. It was not assessed in the LEAP trial but is included for the sake of completeness. See Tables 22.5, 22.6, 22.7, 22.8, and 22.9 for details on each extremity trauma scale.

Table 22.3 Percent distribution of most important factor typically considered in the decision to amputate vs. reconstruct by specialty

Factor	Total (%)	General surgeons (%)	Orthopedic surgeons (%)
Nerve integrity/plantar sensation	32	21	38
Limb ischemia	20	27	15
Soft-tissue coverage	14	9	17
Muscle damage	7	6	8
Neurovascular damage	3	0	6
Fracture pattern/bone loss	4	0	6
High Injury Severity Scale (ISS)	12	31	0
Patient characteristics	2	0	4
Others	6	6	6

Adapted from MacKenzie et al. [44]

When reviewing the initial studies for each of these instruments, reports indicated both high sensitivity and specificity for their respective scores [29, 32, 34, 48, 49]. However, when

Table 22.4 Components of lower-extremity injury-severity scoring systems

Severity scale factors	Lower-extremity injury-severity scales					
	MESS[a]	LSI[b]	PSI[c]	NISSSA[d]	HFS[e]	GHOISS[f]
Age	X			X		X
Shock	X			X	X	X
Warm ischemia time	X	X	X	X	X	X
Bone injury		X	X		X	
Muscle injury		X	X			X
Skin injury		X			X	X
Nerve injury		X		X	X	X
Deep-vein injury		X				
Skeletal/soft-tissue injury	X			X		
Contamination				X	X	X
Time to treatment			X			
Comorbidities						X
Score predicting amputation	≥7	≥6	≥8	≥11	≥9	≥17 (14–17 gray zone)

Adapted from Bosse et al. [45] and Rajasekaran et al. [51]
[a]Mangled Extremity Severity Score (MESS) [29, 48]
[b]Limb Salvage Index (LSI) [32]
[c]Predictive Salvage Index (PSI) [34]
[d]Nerve Injury, Ischemia, Soft-Tissue Injury, Skeletal, Shock, and Age of Patient Score (NISSSA) [49]
[e]Hannover Fracture Scale (HFS) [50, 87]
[f]Ganga Hospital Open Injury Severity Score (GHOISS) [51]

Table 22.5 The Mangled Extremity Severity Scale (MESS) [29]

A. Skeletal/soft-tissue injury	
Low energy (stab; simple fracture; civilian GSW)	1
Medium energy (open or multiple Fxs, dislocation)	2
High energy (close-range shotgun or "military" GSW, crush injury)	3
Very high contamination, soft-tissue avulsion	4
B. Limb ischemia	
Pulse reduced or absent but perfusion normal	1[a]
Pulseless, paresthesias, diminished capillary refill	2[a]
Cool, paralyzed, insensate limb	3[a]
C. Shock	
Systolic BP always >90 mmHg	0
Hypotensive transiently	1
Persistent hypotension	2
D. Age (years)	
<30	0
30–50	1
>50	2

[a]Score doubled for ischemia >6 h

these scoring instruments have been evaluated subsequently by other clinicians, the initial results have been unable to reproduce (Table 22.10) with widely varying sensitivity and specificity values. The differences among these instruments (typically a higher specificity) demonstrate that they may be more helpful to treatment teams in determining which injuries may support entry of the injured extremity into a limb-salvage pathway [45] and not to which extremities should undergo immediate amputation. The sensitivities were generally low in the LEAP study demonstrating that their accuracy at predicting which extremities may eventually require amputation is poor and certainly should not be relied upon to make acute treatment decisions. Furthermore, in the face of low test sensitivity, placing too much emphasis upon these scores may delay an inevitable amputation risking complications in patient care potentially resulting in sepsis and even death [42].

Bosse et al. and Bonanni et al. [33, 45] were unable to recommend any scale for independent use in determining the fate of an injured limb. With the initial presentation of a trauma patient, they concluded that lower-extremity injury-severity scales have limited usefulness and that scores at or above respective amputation thresholds should be used cautiously in decision-making with high-energy trauma patients. Their utility is

Table 22.6 The Limb Salvage Index [32]

Artery	0	Contusion, intimal tear, partial laceration or avulsion (pseudoaneurysm) with no distal thrombosis and palpable pedal pulses; complete occlusion of one of three shank vessels or profunda
	1	Occlusion of two or more shank vessels, complete laceration, avulsion or thrombosis of femoral or popliteal vessels without palpable pedal pulses
	2	Complete occlusion of femoral or popliteal or three of three shank vessels with no distal runoff available
Nerve	0	Contusion or stretch injury, minimal clean laceration of femoral, peroneal, or tibial nerve
	1	Parietal transection or avulsion of sciatic nerve; complete transection or partial transection of femoral, peroneal, or tibial nerve
	2	Complete transection or avulsion of sciatic nerve; complete transection or avulsion of both peroneal and tibial nerves
Bone	0	Closed fracture of one or two sites; open fracture without comminution or with minimal displacement; closed dislocation without fracture; open joint without foreign body; fibula fracture
	1	Closed fracture at three or more sites on the same extremity; open fracture with comminution or moderate to large displacement; segmental fracture; fracture dislocation; open joint with foreign body; bone loss <3 cm
	2	Bone loss >3 cm; Type IIIB or IIIC fracture (open fracture with periosteal stripping, gross contamination, extensive soft-tissue injury loss)
Skin	0	Clean laceration, single or multiple, or small avulsion injuries, all with primary repair; first-degree burns
	1	Delayed closure due to contamination; large avulsion requiring STSG or flap closure. Second- or third-degree burns
Muscle	0	Laceration or avulsion involving a single compartment or single tendon
	1	Laceration or avulsion involving two or more compartments; complete laceration or avulsion of two or more tendons
	2	Crush injury
Deep vein	0	Contusion, partial transection, or avulsion; complete laceration or avulsion if alternate route of venous return is intact; superficial vein injury
	1	Complete laceration, avulsion, or thrombosis with no alternate route of venous return
Warm ischemia time	0	<6 h
	1	6–9 h
	2	9–12 h
	3	12–15 h
	4	>15 h

in providing a list of the factors to consider when making the clinical decision.

22.3.3 Lower-Extremity Injury-Severity Scales and Scores: Predicting Functional Outcomes of Salvaged Limbs After Limb-Threatening Trauma

It has been hypothesized that lower-extremity injury-severity scores may have utility in the accurate prediction of functional outcome in the limbs that underwent salvage after severe trauma. This important and useful question has been studied recently in a number of studies [33, 46, 52, 53]. Ly et al. [53] evaluated the clinical and functional outcomes of the patient cohort in the LEAP study as determined by the Sickness Impact Profile [54, 55] and the patients' scores on the MESS, PSI, and LSI lower-extremity injury-severity scores. They found no correlation among these instruments with patient clinical or functional outcomes. A unique point that this study investigated was the specific evaluation of functional scores on patients in whom the injury-severity threshold scores had recommended an amputation, but the patients had undergone limb-salvage instead. Very interestingly, these "amputation-recommended" patients had outcome scores that were *no worse* than those

Table 22.7 The Predictive Salvage Index [34]

Level of arterial injury	
Suprapopliteal	1
Popliteal	2
Infrapopliteal	3
Degree of bone injury	
Mild	1
Moderate	2
Severe	3
Degree of muscle injury	
Mild	1
Moderate	2
Severe	3
Interval from injury to operating room (hr)	
<6	0
6–12	2
>12	4

Table 22.8 The Hannover Fracture Scale [87, 88]

Bone loss		Deperiostation	
No	0	No	0
<2 cm	1	Yes	1
>2 cm	2	*Local circulation*	
Skin injury		Normal pulse	0
No	0	Capillary pulse only	1
<¼ circumference	1	Ischemia <4 h	2
¼–½ circumference	2	Ischemia 4–8 h	3
½–¾ circumference	3	Ischemia >8 h	4
>¾ circumference	4	*Systemic circulation (syst. BP mm Hg)*	
Muscle injury		Constantly >100	0
No	0	Until admission <100	1
<¼ circumference	1	Until operation <100	2
¼–½ circumference	2	Constantly <100	3
½–¾ circumference	3	*Neurology*	
>¾ circumference	4	Palmarly-plantarly: yes	0
Wound contamination		Sensibility: no	1
No	0	Finger – toe yes	0
Partly	1	Active motion: no	1
Massive	2		
Score range 0–22		Cutoff point (COP)	≥11

patients who had salvaged limbs and had injury-severity scores indicating that amputation was not recommended. Durham et al. [46] studied 30 limbs that had undergone limb salvage and had similar findings as Ly et al. Based upon phone interviews and clinic visits where return to work, impairment, and disability were assessed, they also concluded that none of the extremity injury scales could predict functional outcome.

22.3.4 Lower-Extremity Injury-Severity Scales and Scores: Summary

Whenever evaluating patients and deciding upon optimal care for their injured limb, due caution should be exercised when interpreting the lower-extremity injury-severity scales. This holds true with both initial management and extrapolating ultimate functional outcomes with patients. It is the author's opinion that these lower-extremity scoring systems should still play a role in the management decisions for some patients but should simply be used as one data point among many in the complex processes surrounding the care of the high-energy trauma patient.

22.3.5 Outcomes in Patients Undergoing Limb Salvage or Amputation for Limb-Threatening Injuries

In 2002, Bosse et al. [21] and LEAP study group published their initial report on a prospective cohort of 569 patients that had sustained high-energy lower-extremity trauma from March 1994 to June 1997. The patients in this study had either undergone limb salvage or amputation and were followed prospectively for 24 months and then reported on again at 7-year follow-up [20].

The initial report demonstrated that patients had similar functional outcomes regardless of whether they underwent limb reconstruction/salvage or amputation. The results also indicated that although the outcomes were similar, both groups had substantial levels of disability, and only half had returned to work at 2 years post-injury. Indeed, patients in both groups were able to show significant improvement over the study period, but an important overreaching finding of the study was the profound disability and persistently low psychosocial-functioning subscale [54, 56].

Table 22.9 The Nerve Injury, Ischemia, Soft-Tissue Injury, Skeletal, Shock, and Age of Patient Score [49]

Type of injury	Degree of injury	Points	Description
Nerve injury (N)	Sensate	0	No major nerve injury
	Dorsal	1	Deep or superficial peroneal nerve femoral nerve[a]
	Plantar partial	2	Tibial nerve injury[a]
	Plantar complete	3	Sciatic nerve injury[a]
Ischemia (I)	None	0	Good to fair pulses, no ischemia
	Mild	1[b]	Reduced pulses, perfusion normal
	Moderate	2[b]	No pulse(s), prolonged capillary refill, Doppler pulses present
	Severe	3[b]	Pulseless, cool, ischemic, no Doppler pulses
Soft tissue/contamination (S)	Low	0	Minimal to no ST contusion, no contamination [Gustilo Type I] [89]
	Medium	1	Moderate ST injury, low-velocity GSW, moderate contamination, minimal crush [Gustilo Type II] [89]
	High	2	Moderate crush, deglove, high-velocity GSW, moderate ST injury may require soft-tissue flap, considerable contamination [Gustilo Type IIIA] [90]
	Severe	3	Massive crush, farm injury, severe deglove, severe contamination, requires soft-tissue flap [Gustilo Type IIIB] [90]
Skeletal (S)	Low energy	0	Spiral fractures, oblique fracture, no or minimal displacement [Winquist and Hansen Type I, Johner and Wruhs A_1, A_2] [91, 92]
	Medium energy	1	Transverse fracture, minimal comminution, small-caliber GSW [Winquist and Hansen Type II, Johner and Wruhs A_3, B_1] [91, 92]
	High energy	2	Moderate displacement, moderate comminution, high-velocity GSW, butterfly fragment(s) [Winquist and Hansen Types III–IV, Johner and Wruhs B_1, B_2, B_3] [91, 92]
	Severe energy	3	Segmental, severe comminution, bony loss [Winquist and Hansen Type IV, Johner and Wruhs C_1, C_2, C_3] [91, 92]
Shock (S)	Normotensive	0	Blood pressure normal, always >90 mmHg systolic
	Transient hypotension	1	Transient hypotension in field or emergency center
	Persistent hypotension	2	Persistent hypotension despite fluids
Age (A)	Young	0	<30 years
	Middle	1	30–50 years
	Old	2	>50 years
Total score (N+I+S+S+S+A)			

ST Soft Tissue, *GSW* Gunshot wound
[a]Nerve injury as assessed primarily in the emergency room
[b]Score doubles with ischemia > 6 hours

This study was also able to enlighten surgeons on particular factors not related to the injury itself that may predispose some trauma patients to a poorer or less than optimal outcome. These included a lower level of education, poverty, lack of private health insurance, smoking, and involvement with disability-compensation litigation [21]. The elucidation of these factors provides areas for treatment teams to intervene and assist patients in achieving a better outcome. We advocate for the early involvement and intervention by psychosocial and vocational rehabilitation specialists. Their function in the patient's recovery we believe is imperative and a key component for a better functional outcome. With their expertise, they can directly address the variables listed above and change or even prevent adverse outcomes.

In addition to the listed factors above, self-efficacy and an involved social support network are important determinants of outcome and should be emphasized in rehabilitation [57–59].

Table 22.10 Independent analyses of lower-extremity injury-severity scales[a]

	MESS	PSI	LSI	NISSSA	HFS-97
Bosse et al. [45]					
Sensitivity	0.45	0.47	0.51	0.33	0.37
Specificity	0.93	0.84	0.97	0.98	0.98
Bonanni et al. [33]					
Sensitivity	0.22	0.33	0.61		
Specificity	0.53	0.70	0.43		
Durham et al. [46]					
Sensitivity	0.79	0.96	0.83		
Specificity	0.83	0.50	0.83		
Dagum et al. [31]					
Sensitivity	0.40	0.60	0.60		
Specificity	0.89	0.94	0.83		

[a]Evaluating Gustilo-Anderson type III fractures including immediate amputations

The orthopedic surgeon evaluating this patient in the outpatient setting can be instrumental in this area and help empower the social support network to assist the patient through both the difficult physical and mental recoveries. The orthopedist is also likely the only clinician who can help determine the activity level of the patient in the postoperative time frame and, with this knowledge and assistance from the social workers and disability specialists, can help make vocational retraining possible. Both of the above functions should help facilitate the patient's return to work as excessive delay in this area could potentially lead to poorer outcomes [60, 61].

Longer-term follow-up on the LEAP patient cohort was published at 7 years post-injury [20]. Perhaps unexpectedly, one-half of the patients in the LEAP study remained "severely" disabled and one-quarter were "very severely" disabled [54, 55]. Only one-third of the patients had outcome scores similar to the general population. As found in the initial LEAP 2-year results, there were no significant differences identified among limb-salvage and amputation groups. This follow-up study confirmed and added other factors that were found to be predictive of poor outcomes in the LEAP patient cohort: older age, female gender, nonwhite race, lower education level, living in a poor household, current or previous smoking history [62], low self-efficacy, poor self-reported health status before the injury, and involvement with the legal system in an effort to obtain disability payments. Conclusions drawn from this study warrant attention from treatment teams and do not necessarily involve the acute surgical management of this traumatized population. The optimization of recovery in these patients should emphasize the involvement of professionals who can address certain areas of recovery beyond the operating theater, namely, job retraining, intensive rehabilitative therapy, and education [63–65]. Furthermore, educating patients and their families on realistic and typical expected outcomes is important, as many patients will foster unrealistic expectations. The presence and mental fixation on these unrealistic expectations may predispose patients to poorer outcomes and generalized dissatisfaction with their condition and care [20, 60, 61].

22.3.6 Outcomes of the Mangled Foot and Ankle

A specific subset of patients within the LEAP study that underwent limb salvage with mangled foot and ankle trauma was recently reported upon [66]. This cohort included 174 patients with severely injured foot or hindfoot injuries. The spectrum of injuries included mostly complex foot trauma and tibial pilon fractures. Salvage was undertaken in 116 patients and 58 had an immediate BKA. Assessed outcomes included primarily the Sickness Impact Profile, walking speed, rehospitalizations related to injury complications, time to full weightbearing, visual analog pain scale, and return to work. At 2-year follow-up, the authors found that the limb-salvage group, those that had *free tissue transfers* and/or ankle fusions, had significantly poorer outcomes than the standard BKA group with standard skin flap design closure. This relationship was not found with standard soft-tissue coverage in the salvage group, which highlights the priority of careful soft-tissue management, specifically that around the vulnerable heel pad [67]. The greatest deficit identified in these study groups revolved around the psychosocial aspect of the limb-salvage group. This demonstrates, as shown in the LEAP study as well, the immense psychological toll these injuries exhibit upon patients during their recovery and onwards.

Another recent study reviewed the outcome of 63 military service members with 89 mangled lower limbs resulting from blast injuries sustained in a combat environment [68]. This study, along with that of Ellington et al. [66], showed that open fractures of the hindfoot were associated with higher rates of amputation, 29 % in this study with six of those conducted for chronic pain 18 months following the injury. The authors also noted higher rates of amputation when the trauma was associated with a vascular injury. At final follow-up, 74 % of the injured limbs still had persisting pain and disability related to injury. Only 14 % of the service members were ultimately fit to return to their preinjury duties.

Adding to the mangled lower-extremity data set from a combat theater, 90 % of patients in another study (91 of 102 patients) sustained open calcaneal fractures [69] from a blast-type mechanism. With an average of 4 years follow-up, 42 % of this cohort went on to amputation. Fifteen percent of these were done in a delayed fashion. This study highlighted several factors predictive of eventual amputation: blast-type mechanism, plantar wound location and size, and escalating Gustilo-Anderson classification type. It is also quite interesting to note that the authors reported statistically significant lower visual analog scores (2.1 compared to 4.0) in the amputation group than the limb-salvage group.

22.3.7 Complications in the Treatment of Severe Lower-Extremity Trauma

The management of limb-threatening trauma is challenging and complications can be significant. Harris et al. [70] reported that among the 149 amputations performed among the LEAP patients, there was a 5.4 % amputation revision rate. There was an overall 24 % complication rate with most of these being reported at 3 months post-injury. The most common complications were wound infection (34 %) followed by wound dehiscence (13 %). In the 371 limb-salvage patients, 3.9 % required a late amputation, which was defined as a limb undergoing amputation after the initial hospitalization. Most complications were noted at 6 months post-injury and included a total of 37.7 % of this group. Again, the most common complication noted was wound infection (23.2 %). The complications of osteomyelitis and nonunions were, not surprisingly, seen predominantly in the salvage group and entailed 8.6 % and 31 %, respectively.

Soft-tissue coverage associated with limb salvage and reconstruction is also associated with significant complications and has been reported to occur in 53 % of flap procedures within the LEAP patient cohort. Operative intervention was required in 87 % of these patients [71]. Rehospitalization, often a setback in recovery, occurred in one-third of LEAP study patients and involved the limb-salvage/reconstruction group more than the amputation group.

When complications become unsalvageable or limb-salvage techniques fail for various reasons, some patients may opt for an elective amputation rather than proceed with further efforts. Choosing an elective amputation in this situation is a particularly sensitive issue and certainly one of the most difficult decisions to make for the patient. The time already invested in recovery and the lure of anticipated functional gain can make this decision all the more challenging. Quon et al. [72] reviewed a small cohort of patients undergoing elective amputations for a functionally impaired lower limb that limited those patients' ability to do their everyday activities. They identified three key factors in their patients' decisions: pain, function, and participation. While the study subjects voiced differing reasons within these categories, basic tenets of the study related to patients feeling the leg was potentially holding them back and that amputating the leg may afford them decreased pain, improved function with daily activities, and future participation in hobbies or activities they were previously forced to give up due to the trauma.

Complications in the management of this severely injured group of patients are sadly unavoidable. It is in our and our patients best interest to understand the nature of the complications and how then to best avoid them. From the initial evaluation and subsequent follow-up of these patients, treatment teams should not underestimate the difficult nature of the recovery process and the potential for complications and secondary procedures. Further, a future area of research may be warranted with investigation into when salvage efforts have stalled and patients may be better suited with an elective amputation over continued salvage techniques. As clinicians we have a duty to inform patients on all treatment options, and perhaps early involvement of an amputation team may help some patients opt for an earlier amputation rather than struggle with the ostensibly successful limb salvage with an unpredictable recovery.

22.3.8 Psychological Distress in Patients with Severely Injured Lower Extremities

Accompanying the significant challenges with physical recovery and impairment is an often underappreciated source of morbidity with orthopedic trauma patients – psychological distress and mental illness [73, 74]. This is especially evident in the high-energy lower-extremity trauma patient where limb salvage and amputations are being debated and subsequent recoveries managed. During the course of the LEAP study, patients were evaluated for psychological distress [75] utilizing the Brief Symptom Inventory [76, 77]. At 2 years post-injury, 42 % of the patients screened positive for a psychological disorder, yet only 22 % had reported receiving any mental health services. Almost 20 % of the study group reported severe phobic anxiety and/or depression. The authors of the study were able to identify factors that were likely to be associated with patients that had psychological distress. These included poorer physical function, younger age, nonwhite race, poverty, a likely drinking problem, neuroticism, a poor sense of self-efficacy, and limited social support. Interestingly, some of these same factors have been attributed to chronic pain syndromes which could certainly exacerbate any coexisting psychological distress these patients may be suffering from [78].

Another study utilizing the LEAP study participants worked to characterize the relationships between pain, psychological distress, and physical function in the early and later stages of recovery [79]. They reported that the presence of depression and anxiety, at any detectable level, led to decreased levels of function during recovery after injury. Complimenting this data set, a study by Castillo et al. [80] showed that during the early phases of recovery, levels of pain were able to predict corresponding levels of anxiety and depression symptoms. Stronger relationships were seen with anxiety and pain throughout the recovery stages. Based upon these and other studies, it is quite clear that the patient with severe lower-extremity trauma would benefit significantly from interventions specifically aimed at decreasing negative emotions, especially anxiety, in the recovery period.

As emphasized previously, the orthopedic surgeon is most likely going to be the primary coordinator of care with these patients in the postoperative period during their lengthy functional recoveries. Along with recognizing the physical dysfunction and instituting appropriate referrals for therapy and job retraining, the treating surgeon must also be astute enough to evaluate and screen these traumatized patients for psychological distress. If mental distress is suspected or identified, appropriate consultation or referral should be initiated to a provider trained in this area. Furthermore, by understanding and recognizing potential risk factors for psychological distress and thus poorer outcomes with this patent population (i.e., drinking problems, poor social support network, or poor self-efficacy), prophylactic referrals can be made early in the patient's recovery. Ultimately, for patients to be given the best chance for the most favorable outcome, the physical and psychological needs of this population should to be addressed simultaneously [75]. Adding directed therapy toward these areas could prove to decrease acute pain associated with recovery and improve overall functional gains.

22.3.9 Societal Costs Associated with Limb Salvage and Amputation

An argument we have heard and understand is that of the cost of limb salvage and its toll on society in comparison to a "quick amputation and be done with it" attitude… "let the patient get on with their life." The cost burden of the limb-salvage and amputation debate was recently reported [2], and the results directly counter what many have argued in the past. At 2 years of follow-up, both groups had essentially the same healthcare costs. However, projected lifetime costs were $509,000 for amputees and $163,000 for limb-salvage patients (2002 US dollar figures) – over a threefold difference. The difference was mainly attributed to the repair and replacement costs associated with prostheses for the amputation population, which had an estimated 40–45 years of life remaining. In regard to complications, they found a 46 % increase in costs if patients had required a rehospitalization – a finding that underscores the importance of clinicians having a solid understanding of risk factors for both complications and poorer outcomes.

22.4 The Open Fracture with Severe Nerve Injury

The management of severe limb-threatening injuries is challenging and often requires difficult decisions to be made acutely. Predicting the outcome of patients with this type of trauma (Table 22.2) has proved challenging, and the utility of limb-salvage predictive scores has been shown to be limited. A repetitive and concerning theme in the scientific literature surrounding limb salvage and amputation is the severe open fracture with associated nerve injury and purported poor results of 60–100 % disability with this type of injury [81–83]. This scenario represents a unique conundrum in the decision-making process.

The loss of foot plantar sensation has been ingrained into the trauma surgeon's psyche as a major, if not sometimes the primary predictor of acute amputation. In fact, MacKenzie et al. [44] showed that nearly 40 % of orthopedic surgeons place nerve integrity and plantar sensation as the primary determinant in the decision to amputate or reconstruct (Table 22.3). Often, this decision is made based on initial emergency room evaluation even though this sometimes rudimentary exam has been shown to be unpredictable [35]. The influence of nerve integrity on the trauma community has been borne out by its direct and independent inclusion into three of the major limb-salvage prediction scales: the LSI, NISSSA, and HFS (Tables 22.4, 22.6, 22.8, and 22.9).

The insensate foot was evaluated among 55 patient cohort of the LEAP study [84]. This group presented to the emergency department with an insensate foot and underwent either amputation (26 patients) or limb salvage (29 patients). The insensate-salvage group was also matched and compared with a sensate-salvage group as a control group in the study. The authors identified some interesting and important findings directly impacting commonly held beliefs pertaining to limb-salvage versus amputation debates and predicted outcomes. First and foremost, patients that had absent plantar sensation demonstrated substantial impairment at final follow-up. However, their outcomes were similar and appeared to be unaffected whether undergoing amputation or limb salvage. Second and perhaps most interesting, the patients with the insensate foot on presentation that underwent limb salvage did not have worse outcomes than the matched cohort with intact sensation that underwent limb salvage. This included no differences in final plantar sensation or the need for late amputation. In fact, 67 % of the patients in the insensate foot group regained normal foot sensation over the study period – a highlight that supports increased diligence in treatment decisions utilizing emergency department nerve exams. Ultimately, the 2-year outcome of patients that had undergone limb salvage with an insensate foot did not appear to be influenced or adversely affected by the presence or absence of plantar sensation [84].

More recently, Beltran et al. [85] reviewed 32 open type III tibia fractures with a total of 43 peripheral nerve injuries (peroneal or tibial) sustained in a combat environment. Complimenting the LEAP data, this study specifically investigated nerve injuries sustained with high-energy mechanisms such as seen in military combat. With nearly 2-year follow-up, 89 % of injured motor nerves were functional, and 93 % of sensory nerve injuries were functional as well. Full return of function was seen in 37 % of the motor nerve injuries and in 25 % of sensory nerves. The authors conclude that improvement can be expected in 50 % of motor nerve injuries and in 27 % of sensory nerve injuries.

The decisions in these analyses and others are often based upon emergency department evaluation and not upon direct surgical observation. The initial evaluation demonstrating a loss of plantar sensation can easily be attributed to a transient neurapraxia from compression or stretch and/or temporary ischemia, which can be reversible. Furthermore, in the combat situation, both blast injuries and high-velocity gunshot wounds can cause local tissue cavitation leading to nerve dysfunction. The intraoperative finding of complete nerve transection or segmental neural element loss could be suggestive of an absolute indication for primary limb amputation, especially in light of associated vascular injuries or other severe traumas. However, it is important to note that often clinicians treat patients with insensate feet in the clinical setting, namely, in the diabetic and spinal cord injury patient populations [84]. In the surgical suite, we do not advocate invasive surgical exploration of nerve structures in the lower extremity when they are not already exposed secondary to the trauma itself. This practice is associated with unwarranted tissue damage and should be avoided. With evidence to support return of both motor and sensory functions including plantar sensation during recovery, the reliance specifically upon plantar sensation and nerve function in general in the lower extremity during the initial physical exam finding should be avoided in the amputation decision-making process.

22.5 Summary

The high-energy lower-extremity trauma patient presents many challenges to treatment teams. Past literature has not been overly supportive of limb salvage and often makes the point that early amputation is advantageous to save patients from lengthy suffering [15, 17]. However, as technology and surgical concepts have evolved, so have our abilities to salvage limbs previously thought to be candidates only for amputation. These salvaged limbs, although demonstrating generally poor outcomes, have been shown to have equivalent results to limbs treated with primary amputation [20–22] and entail an equivalent of 2-year healthcare costs and substantial savings over the long term.

Often, given the option of limb salvage or amputation, most patients opt to save their extremity rather than undergo an amputation. While data presented here and in the LEAP data show equivalent results among the salvage/amputation groups, it should be noted that most of the data were derived from care patients had received at Level I trauma centers. It has been argued that these centers, with their experienced trauma staff, may impart different outcomes than patients treated elsewhere [86].

We believe that limb salvage is a reasonable goal for clinicians and patients at experienced Level I trauma centers. The LEAP data and other studies present sufficient evidence to support this conclusion. The early involvement of post-acute-care services, such as therapists, rehabilitation specialists, psychologists, and many others, is imperative for the optimization of patient outcomes and potentially holds the highest value in recovery efforts. Diligence, thoughtful care, and presenting realistic expectations will allow these traumatized patients to achieve their best recovery and functional outcomes.

References

1. Andersson G, American Academy of Orthopaedic Surgeons. United States Bone and Joint Initiative: The Burden of Musculoskeletal Diseases in the United States (BMUS), Third Edition, 2014. Rosemont, IL. Available at http://www.boneandjointburden.org. Accessed on 15 July 2015.
2. MacKenzie EJ, Jones AS, Bosse MJ, Castillo RC, Pollak AN, Webb LX, et al. Health-care costs associated with amputation or reconstruction of a limb-threatening injury. J Bone Joint Surg Am. 2007;89(8):1685–92.
3. Wright JG. A practical guide to assigning levels of evidence. J Bone Joint Surg Am. 2007;89(5):1128–30.
4. Dillingham TR, Pezzin LE, MacKenzie EJ. Incidence, acute care length of stay, and discharge to rehabilitation of traumatic amputee patients: an epidemiologic study. Arch Phys Med Rehabil. 1998;79(3):279–87.
5. Dillingham TR, Pezzin LE, MacKenzie EJ. Limb amputation and limb deficiency: epidemiology and recent trends in the United States. South Med J. 2002;95(8):875–83.
6. Godina M. Early microsurgical reconstruction of complex trauma of the extremities. Plast Reconstr Surg. 1986;78(3):285–92.
7. Francel TJ, Vander Kolk CA, Hoopes JE, Manson PN, Yaremchuk MJ. Microvascular soft-tissue transplantation for reconstruction of acute open tibial fractures: timing of coverage and long-term functional results. Plast Reconstr Surg. 1992;89(3):478–87; discussion 88–9.
8. Caudle RJ, Stern PJ. Severe open fractures of the tibia. J Bone Joint Surg Am. 1987;69(6):801–7.
9. Anglen J, Kyle RF, Marsh JL, Virkus WW, Watters III WC, Keith MW, et al. Locking plates for extremity fractures. J Am Acad Orthop Surg. 2009;17(7):465–72.
10. Collinge CA, Sanders RW. Percutaneous plating in the lower extremity. J Am Acad Orthop Surg. 2000;8(4):211–6.
11. Haidukewych GJ. Innovations in locking plate technology. J Am Acad Orthop Surg. 2004;12(4):205–12.
12. Haidukewych GJ, Ricci W. Locked plating in orthopaedic trauma: a clinical update. J Am Acad Orthop Surg. 2008;16(6):347–55.
13. Gorman PW, Barnes CL, Fischer TJ, McAndrew MP, Moore MM. Soft-tissue reconstruction in severe lower extremity trauma. A review. Clin Orthop Relat Res. 1989;243:57–64.
14. Lange RH. Limb reconstruction versus amputation decision making in massive lower extremity trauma. Clin Orthop Relat Res. 1989;243:92–9.
15. Hansen Jr ST. The type-IIIC tibial fracture. Salvage or amputation. J Bone Joint Surg Am. 1987;69(6):799–800.
16. Smith DG, Castillo R, MacKenzie E, Bosse MJ, Group TLS. Functional outcomes of patients who have late amputation after trauma is significantly worse than for those who have early amputation. Orthopaedic trauma association 2003 annual meeting; 10/9/2003; Salt Lake City, UT2003.
17. Hansen Jr ST. Overview of the severely traumatized lower limb. Reconstruction versus amputation. Clin Orthop Relat Res. 1989;243:17–9.
18. Georgiadis GM, Behrens FF, Joyce MJ, Earle AS, Simmons AL. Open tibial fractures with severe soft-tissue loss. Limb salvage compared with below-the-knee amputation. J Bone Joint Surg Am. 1993;75(10):1431–41.
19. Francel TJ. Improving reemployment rates after limb salvage of acute severe tibial fractures by microvascular soft-tissue reconstruction. Plast Reconstr Surg. 1994;93(5):1028–34.
20. MacKenzie EJ, Bosse MJ, Pollak AN, Webb LX, Swiontkowski MF, Kellam JF, et al. Long-term persistence of disability following severe lower-limb trauma. Results of a seven-year follow-up. J Bone Joint Surg Am. 2005;87(8):1801–9.
21. Bosse MJ, MacKenzie EJ, Kellam JF, Burgess AR, Webb LX, Swiontkowski MF, et al. An analysis of outcomes of reconstruction or amputation after leg-threatening injuries. N Engl J Med. 2002;347(24):1924–31.
22. MacKenzie EJ, Bosse MJ, Kellam JF, Burgess AR, Webb LX, Swiontkowski MF, et al. Characterization of patients with high-energy lower extremity trauma. J Orthop Trauma. 2000;14(7):455–66.
23. Alexander RH, Proctor HJ, American College of Surgeons, Committee on Trauma. Advanced trauma life support program for physicians: ATLS. 5th ed. Chicago: American College of Surgeons; 1993.
24. Trauma ACoSCo. Advanced trauma life support for doctors ATLS: manuals for coordinators and faculty. 8th ed. Chicago: American College of Surgeons; 2008.
25. Trauma ACoSCo. Advanced trauma life support for doctors: student course manual. 6th ed. Chicago: American College of Surgeons; 1997.
26. Pinzur M. Amputations in trauma. In: Browner BD, Jupiter J, Levine A, Trafton P, Krettek C, editors. Skeletal trauma: basic science, management, and reconstruction, vol. 2. 4th ed. Philadelphia: Saunders/Elsevier; 2009. p. 2 v. xxv, 2882, I56 p.
27. MacKenzie EJ, Bosse MJ, Castillo RC, Smith DG, Webb LX, Kellam JF, et al. Functional outcomes following trauma-related lower-extremity amputation. J Bone Joint Surg Am. 2004;86-A(8):1636–45.
28. Pinzur MS, Smith DG, Daluga DJ, Osterman H. Selection of patients for through-the-knee amputation. J Bone Joint Surg Am. 1988;70(5):746–50.
29. Johansen K, Daines M, Howey T, Helfet D, Hansen Jr ST. Objective criteria accurately predict amputation following lower extremity trauma. J Trauma. 1990;30(5):568–72; discussion 72–3.
30. Pinzur MS, Angelats J, Light TR, Izuierdo R, Pluth T. Functional outcome following traumatic upper limb amputation and prosthetic limb fitting. J Hand Surg Am. 1994;19(5):836–9.

31. Dagum AB, Best AK, Schemitsch EH, Mahoney JL, Mahomed MN, Blight KR. Salvage after severe lower-extremity trauma: are the outcomes worth the means? Plast Reconstr Surg. 1999;103(4):1212–20.
32. Russell WL, Sailors DM, Whittle TB, Fisher Jr DF, Burns RP. Limb salvage versus traumatic amputation. A decision based on a seven-part predictive index. Ann Surg. 1991;213(5):473–80; discussion 80–1.
33. Bonanni F, Rhodes M, Lucke JF. The futility of predictive scoring of mangled lower extremities. J Trauma. 1993;34(1):99–104.
34. Howe Jr HR, Poole Jr GV, Hansen KJ, Clark T, Plonk GW, Koman LA, et al. Salvage of lower extremities following combined orthopedic and vascular trauma. A predictive salvage index. Am Surg. 1987;53(4):205–8.
35. Lange RH, Bach AW, Hansen Jr ST, Johansen KH. Open tibial fractures with associated vascular injuries: prognosis for limb salvage. J Trauma. 1985;25(3):203–8.
36. Dougherty PJ. Long-term follow-up of unilateral transfemoral amputees from the Vietnam war. J Trauma. 2003;54(4):718–23.
37. Smith DG, Horn P, Malchow D, Boone DA, Reiber GE, Hansen Jr ST. Prosthetic history, prosthetic charges, and functional outcome of the isolated, traumatic below-knee amputee. J Trauma. 1995;38(1):44–7.
38. Lerner RK, Esterhai Jr JL, Polomano RC, Cheatle MD, Heppenstall RB. Quality of life assessment of patients with posttraumatic fracture nonunion, chronic refractory osteomyelitis, and lower-extremity amputation. Clin Orthop Relat Res. 1993;295:28–36.
39. Lerner RK, Esterhai Jr JL, Polomono RC, Cheatle MC, Heppenstall RB, Brighton CT. Psychosocial, functional, and quality of life assessment of patients with posttraumatic fracture nonunion, chronic refractory osteomyelitis, and lower extremity amputation. Arch Phys Med Rehabil. 1991;72(2):122–6.
40. Bell JC, Wolf EJ, Schnall BL, Tis JE, Tis LL, Potter BK. Transfemoral amputations: the effect of residual limb length and orientation on gait analysis outcome measures. J Bone Joint Surg Am. 2013;95(5):408–14.
41. Dougherty PJ. Transtibial amputees from the Vietnam War. Twenty-eight-year follow-up. J Bone Joint Surg Am. 2001;83-A(3):383–9.
42. Bondurant FJ, Cotler HB, Buckle R, Miller-Crotchett P, Browner BD. The medical and economic impact of severely injured lower extremities. J Trauma. 1988;28(8):1270–3.
43. Dirschl DR, Dahners LE. The mangled extremity: when should it be amputated? J Am Acad Orthop Surg. 1996;4(4):182–90.
44. MacKenzie EJ, Bosse MJ, Kellam JF, Burgess AR, Webb LX, Swiontkowski MF, et al. Factors influencing the decision to amputate or reconstruct after high-energy lower extremity trauma. J Trauma. 2002;52(4):641–9.
45. Bosse MJ, MacKenzie EJ, Kellam JF, Burgess AR, Webb LX, Swiontkowski MF, et al. A prospective evaluation of the clinical utility of the lower-extremity injury-severity scores. J Bone Joint Surg Am. 2001;83-A(1):3–14.
46. Durham RM, Mistry BM, Mazuski JE, Shapiro M, Jacobs D. Outcome and utility of scoring systems in the management of the mangled extremity. Am J Surg. 1996;172(5):569–73; discussion 73–4.
47. O'Sullivan ST, O'Sullivan M, Pasha N, O'Shaughnessy M, O'Connor TP. Is it possible to predict limb viability in complex Gustilo IIIB and IIIC tibial fractures? A comparison of two predictive indices. Injury. 1997;28(9–10):639–42.
48. Helfet DL, Howey T, Sanders R, Johansen K. Limb salvage versus amputation. Preliminary results of the Mangled Extremity Severity Score. Clin Orthop Relat Res. 1990;256:80–6.
49. McNamara MG, Heckman JD, Corley FG. Severe open fractures of the lower extremity: a retrospective evaluation of the Mangled Extremity Severity Score (MESS). J Orthop Trauma. 1994;8(2):81–7.
50. Tscherne H, Gotzen L. Fractures with soft tissue injuries. Berlin/New York: Springer; 1984. vi, 164 p. p.
51. Rajasekaran S, Naresh Babu J, Dheenadhayalan J, Shetty AP, Sundararajan SR, Kumar M, et al. A score for predicting salvage and outcome in Gustilo type-IIIA and type-IIIB open tibial fractures. J Bone Joint Surg Br. 2006;88(10):1351–60.
52. Lin CH, Wei FC, Levin LS, Su JI, Yeh WL. The functional outcome of lower-extremity fractures with vascular injury. J Trauma. 1997;43(3):480–5.
53. Ly TV, Travison TG, Castillo RC, Bosse MJ, MacKenzie EJ. Ability of lower-extremity injury severity scores to predict functional outcome after limb salvage. J Bone Joint Surg Am. 2008;90(8):1738–43.
54. Bergner M, Bobbitt RA, Carter WB, Gilson BS. The sickness impact profile: development and final revision of a health status measure. Med Care. 1981;19(8):787–805.
55. Bergner M, Bobbitt RA, Kressel S, Pollard WE, Gilson BS, Morris JR. The sickness impact profile: conceptual formulation and methodology for the development of a health status measure. Int J Health Serv. 1976;6(3):393–415.
56. de Bruin AF, de Witte LP, Stevens F, Diederiks JP. Sickness impact profile: the state of the art of a generic functional status measure. Soc Sci Med. 1992;35(8):1003–14.
57. Berkman L, Glass T. Social integration, social networks, social support, and health. In: Berkman L, Kawachi I, editors. Social epidemiology. New York: Oxford University Press; 2000.
58. MacKenzie EJ, Morris Jr JA, Jurkovich GJ, Yasui Y, Cushing BM, Burgess AR, et al. Return to work following injury: the role of economic, social, and job-related factors. Am J Public Health. 1998;88(11):1630–7.
59. MacKenzie EJ, Bosse MJ. Factors influencing outcome following limb-threatening lower limb trauma: lessons learned from the Lower Extremity Assessment Project (LEAP). J Am Acad Orthop Surg. 2006;14(10 Spec No):S205–10.
60. O'Toole RV, Castillo RC, Pollak AN, MacKenzie EJ, Bosse MJ. Surgeons and their patients disagree regarding cosmetic and overall outcomes after surgery

for high-energy lower extremity trauma. J Orthop Trauma. 2009;23(10):716–23.
61. O'Toole RV, Castillo RC, Pollak AN, MacKenzie EJ, Bosse MJ. Determinants of patient satisfaction after severe lower-extremity injuries. J Bone Joint Surg Am. 2008;90(6):1206–11.
62. Castillo RC, Bosse MJ, MacKenzie EJ, Patterson BM. Impact of smoking on fracture healing and risk of complications in limb-threatening open tibia fractures. J Orthop Trauma. 2005;19(3):151–7.
63. Archer KR, MacKenzie EJ, Bosse MJ, Pollak AN, Riley 3rd LH. Factors associated with surgeon referral for physical therapy in patients with traumatic lower-extremity injury: results of a national survey of orthopedic trauma surgeons. Phys Ther. 2009;89(9): 893–905.
64. Archer KR, Castillo RC, Mackenzie EJ, Bosse MJ. Gait symmetry and walking speed analysis following lower-extremity trauma. Phys Ther. 2006;86(12): 1630–40.
65. Castillo RC, MacKenzie EJ, Archer KR, Bosse MJ, Webb LX. Evidence of beneficial effect of physical therapy after lower-extremity trauma. Arch Phys Med Rehabil. 2008;89(10):1873–9.
66. Ellington JK, Bosse MJ, Castillo RC, MacKenzie EJ, Group LS. The mangled foot and ankle: results from a 2-year prospective study. J Orthop Trauma. 2013;27(1):43–8.
67. Lawrence SJ, Singhal M. Open hindfoot injuries. J Am Acad Orthop Surg. 2007;15(6):367–76.
68. Ramasamy A, Hill AM, Masouros S, Gibb I, Phillip R, Bull AM, et al. Outcomes of IED foot and ankle blast injuries. J Bone Joint Surg Am. 2013;95(5), e25.
69. Dickens JF, Kilcoyne KG, Kluk MW, Gordon WT, Shawen SB, Potter BK. Risk factors for infection and amputation following open, combat-related calcaneal fractures. J Bone Joint Surg Am. 2013;95(5), e24.
70. Harris AM, Althausen PL, Kellam J, Bosse MJ, Castillo R. Complications following limb-threatening lower extremity trauma. J Orthop Trauma. 2009;23(1): 1–6.
71. Pollak AN, McCarthy ML, Burgess AR. Short-term wound complications after application of flaps for coverage of traumatic soft-tissue defects about the tibia. The Lower Extremity Assessment Project (LEAP) Study Group. J Bone Joint Surg Am. 2000;82-A(12): 1681–91.
72. Quon DL, Dudek NL, Marks M, Boutet M, Varpio L. A qualitative study of factors influencing the decision to have an elective amputation. J Bone Joint Surg Am. 2011;93(22):2087–92.
73. Crichlow RJ, Andres PL, Morrison SM, Haley SM, Vrahas MS. Depression in orthopaedic trauma patients. Prevalence and severity. J Bone Joint Surg Am. 2006;88(9):1927–33.
74. Singh G, Harkema JM, Mayberry AJ, Chaudry IH. Severe depression of gut absorptive capacity in patients following trauma or sepsis. J Trauma. 1994;36(6):803–8; discussion 8–9.
75. McCarthy ML, MacKenzie EJ, Edwin D, Bosse MJ, Castillo RC, Starr A. Psychological distress associated with severe lower-limb injury. J Bone Joint Surg Am. 2003;85-A(9):1689–97.
76. Derogatis LR, Melisaratos N. The Brief Symptom Inventory: an introductory report. Psychol Med. 1983;13(3):595–605.
77. Derogatis LP. BSI: brief symptom inventory. 3rd ed. Minneapolis: National Computer Systems; 1993.
78. Castillo RC, MacKenzie EJ, Wegener ST, Bosse MJ. Prevalence of chronic pain seven years following limb threatening lower extremity trauma. Pain. 2006;124(3):321–9.
79. Wegener ST, Castillo RC, Haythornthwaite J, Mackenzie EJ, Bosse MJ, Group LS. Psychological distress mediates the effect of pain on function. Pain. 2011;152(6):1349–57.
80. Castillo RC, Wegener ST, Heins SE, Haythornthwaite JA, Mackenzie EJ, Bosse MJ, et al. Longitudinal relationships between anxiety, depression, and pain: results from a two-year cohort study of lower extremity trauma patients. Pain. 2013;154(12):2860–6.
81. Lusskin R, Battista A. Evaluation and therapy after injury to peripheral nerves. Foot Ankle. 1986;7(2): 71–81.
82. Aldea PA, Shaw WW. Management of acute lower extremity nerve injuries. Foot Ankle. 1986;7(2):82–94.
83. Bateman JE. Trauma to nerves in limbs. Philadelphia: Saunders; 1962. p. 443.
84. Bosse MJ, McCarthy ML, Jones AL, Webb LX, Sims SH, Sanders RW, et al. The insensate foot following severe lower extremity trauma: an indication for amputation? J Bone Joint Surg Am. 2005;87(12):2601–8.
85. Beltran MJ, Burns TC, Eckel TT, Potter BK, Wenke JC, Hsu JR, et al. Fate of combat nerve injury. J Orthop Trauma. 2012;26(11):e198–203.
86. MacKenzie EJ, Rivara FP, Jurkovich GJ, Nathens AB, Egleston BL, Salkever DS, et al. The impact of trauma-center care on functional outcomes following major lower-limb trauma. J Bone Joint Surg Am. 2008;90(1):101–9.
87. Krettek C, Seekamp A, Kontopp H, Tscherne H. Hannover Fracture Scale '98 – re-evaluation and new perspectives of an established extremity salvage score. Injury. 2001;32(4):317–28.
88. Tscherne H, Oestern HJ. A new classification of soft-tissue damage in open and closed fractures (author's transl). Unfallheilkunde. 1982;85(3):111–5.
89. Gustilo RB, Anderson JT. Prevention of infection in the treatment of one thousand and twenty-five open fractures of long bones: retrospective and prospective analyses. J Bone Joint Surg Am. 1976;58(4):453–8.
90. Gustilo RB, Mendoza RM, Williams DN. Problems in the management of type III (severe) open fractures: a new classification of type III open fractures. J Trauma. 1984;24(8):742–6.
91. Winquist RA, Hansen Jr ST, Clawson DK. Closed intramedullary nailing of femoral fractures. A report of five hundred and twenty cases. J Bone Joint Surg Am. 1984;66(4):529–39.
92. Johner R, Wruhs O. Classification of tibial shaft fractures and correlation with results after rigid internal fixation. Clin Orthop Relat Res. 1983;178:7–25.

High-Energy Injuries Caused by Penetrating Trauma

23

Yoram A. Weil and Rami Mosheiff

Contents

23.1	**Gunshot Ballistics and Injuries**	330
23.1.1	Mechanism of Injury	330
23.1.2	Vascular Injuries	330
23.1.3	Principles of Treatment	331
23.1.4	Fracture Care	331
23.2	**Blast Injuries**	332
23.2.1	Mechanisms of Blast Injury	334
23.2.2	Triage and Primary Resuscitation	337
23.2.3	Treatment of Specific Injuries	338
Conclusions		339
References		339

While blunt trauma care involving orthopedic injuries evolved over the years, penetrating orthopedic trauma is still vastly under-represented in current literature. This is despite the recent rise in firearms-related causality rate, especially in North America and other industrialized countries [1], as well as the surge in global terrorism [2]. Therefore, the practicing orthopedic trauma surgeon is in a dire need of more information regarding the recognition, management, and preparation needed to cope with single/mass casualties with high energy penetrating musculoskeletal injuries.

Recent conflicts, especially in Iraq and Afghanistan, have taught us lessons regarding newer and deadlier limb threatening injuries [3, 4].

The topic of penetrating limb injuries is wide but can be divided mainly into two main subgroups – those inflicted by firearms of both high and low velocity, and those caused by blast or explosions. Both injuries had been long described in the military setting but nowadays are encountered in an alarmingly rising rate in the civilian setting.

The purpose of this chapter will be to characterize both gunshot- and blast-related extremity injuries, discuss their initial and definite management and review some of the recent experience with emphasis on the civilian setting.

Y.A. Weil, MD (✉) • R. Mosheiff
Orthopaedic Trauma Service,
Hadassah Medical Center, Jerusalem 911200, Israel
e-mail: weily@hadassah.org.il;
ramim@hadassah.org.il; ramim@cc.huji.ac.il

23.1 Gunshot Ballistics and Injuries

Many authors and surgeons attempted to classify the injury pattern of gunshots based on the type of weapon, bullet, energy transfer, and velocity of the projectile [1, 5, 6]. Traditionally, projectiles were classified into "high-velocity" vs. "low-velocity", the latter being mainly shot from handguns while the former produced by assault rifles. The cut-off point is controversial but it is still widely accepted that most high-velocity injuries refer to >2000 ft/s [1]. Since the kinetic energy is proportional to the square of the velocity, most modern guns have switched to smaller but faster ammunition. Both in vivo evidence in dogs [7] and in vitro simulation in gelatin blocks [1, 8] demonstrate a wider wound tract and tissue damage inflicted by faster ammunition. Early clinical reports from the Vietnam War describe the unusual injuries at that time inflicted by the M-16 assault rifle [9] represent this trend as well. Despite the above, the distinction between high velocity and low velocity can at times be artificial. A 0.45 caliber pistol bullet can transmit 900 joules of energy, equivalent to a 5-kg weight being dropped from a height of 20 m, which can cause severe musculoskeletal injury [10], while a military assault rifle, classified as high velocity, can hit a thigh with a clean in-and-out wound without significant damage [1].

23.1.1 Mechanism of Injury

After the bullet leaves the firearm muzzle, it propagates in a complex movement pattern consisting of yaw and rotation, creating a complex form of motion called nutation, although overall speaking, a well designed bullet will yaw less than 3° and will usually hit the target straight [11]. Upon striking its target, the bullet creates a temporary cavity due to the stretching forces and vacuum created by its passing. This temporary cavity is reported to be of an increased magnitude with missile velocities greater than 2000 ft per second [1, 12]. The cavitation process lasts only a few milliseconds and the amount of tissue damage is dictated by the tissue elasticity and tolerance to stretch. For example, near-liquid organs such as the brain, liver, or spleen might be violently disrupted during the temporary cavity formation [1, 5]. In the limbs, however, muscle is damaged mainly within the close vicinity of the passing bullet but can tolerate stretching quite well [13]. Major vessels are rarely injured by stretch. Nerves rarely tear due to cavitation, and most of the time nerve damage results only in neuropraxia. Therefore nerve exploration should not be routinely performed in gunshot wounds [14].

Bone can be damaged incompletely or completely. Various fracture morphological patterns have been described for both complete and incomplete fractures, but the clinical value of these classifications is questionable. However, it should be mentioned that bone fragments can be propelled to the area of temporary cavity and can cause damage to adjacent structures. This is not always obvious when looking at injury films since most bone fragments usually retract back to the original bone [1].

Virtually all gunshot wounds are contaminated [1, 5]. In an experimental model, the number of organisms in a gunshot wound tends to multiply 10–100 times from 6 to 24 h, while basically all cultures from devitalized muscle are positive at the time of the initial injury [15]. Therefore the time of wound excision is believed to be of a greater value if done earlier in the course of treatment.

23.1.2 Vascular Injuries

Vascular injuries are not uncommon in high-energy gunshot wounds and should be sought after [1, 16]. They are predictive of a more guarded outcome, especially with regard to infection [17]. Although physical examination is often sufficient to rule out major vascular injury [18], some minor vascular injuries have been reported to appear later on in the course of events. On the one hand, angiography, although very accurate, is not routinely warranted if the physical examination is normal [19]. On the other, with a hypo perfused limb with a localized lesion and "hard" signs of vascular injury, the location of injury becomes obvious and immediate exploration is indicated without further studies [1]. When angiography is performed in absence of "hard physical findings (ischemic, cold, pulseless limb, bruit or expanding hematoma), it is either negative in many cases or demonstrate benign lesions in other cases [20]. Therefore, arteriography should be judiciously used only for borderline cases when physical examination is hard to perform or is unreliable. Recently, high resolution multi-slice CT angiography has been used in subsets of patients with a high accuracy rate approaching 94 % [21]. This modality seems to be promising in the initial diagnosis of vascular extremity trauma.

23.1.3 Principles of Treatment

As with all high-grade open fractures, treatment goals should be: stabilization of bone, adequate care of soft tissue, wound coverage, and restoration of limb function. This is true both for diaphyseal as well as for articular fractures [22].

Perhaps the most commonly known term of wound excision coined has been "debridemént", defined by Larrey (1812) and widely adopted in World War I, which implied incision and decompression of the wound [23], more likely to be a kind of fasciotomy [5] than the current mode of excisional surgery. It is now well accepted that the single most important factor in reducing the risk of infection is the timely administration of intravenous antibiotics [1, 5]. While it is accepted that in many cases of low-velocity gunshots non-operative treatment can succeed, most high-velocity/energy wounds require some degree of surgical wound care.

The degree of wound excision necessary for healing is highly controversial and remains one of the greatest challenges for the trauma surgeon. Although ideally, devitalized tissue should be removed in order to decrease the infection burden and to promote angiogenesis, it is clinically hard to judge which tissue warrants removal, especially in regard to muscle. In an experimental high-energy gunshot produced in pigs, extensive tract excision did not result in a better outcome than simple wound drainage and antibiotic treatment [24]. On the other hand, retaining devitalized tissue can result in necrosis and sepsis [25]. The four "Cs" – color, consistency, contractility, and circulation, known for more than 40 years – have served as a rough guideline for identification of dead muscle for generations of surgeons and are still valid today [26]. Attempts to correlate between gross findings associated with these "Cs" to microscopic findings yielded some inaccuracies. This is related to temporary ischemia around the injury zone which resolves after a few hours [1]. Therefore, it is advised not to apply the *"when in doubt, cut it out"* regime but rather to repeat wound exploration within 48–72 h, thus sparing more viable tissue [5]. This is of course, by no means an advice to retain detached, devascularized, or contaminated tissue that obviously needs to be excised during the first session. Wounds traditionally are best left open initially, but closure should be considered early in order to improve joint motion and reduce stiffness, ideally not longer than 5–10 days after injury [26]. In case tension occurs, primary skin graft should be used. Recently, vacuum assisted closure serves as a powerful tool in reducing infection, promoting granulation, and expediting closure in war wounds and is continuing to evolve rapidly [27]. This is especially helpful in situations where prolonged transport is expected, such as in remote war zones.

An increased rate of vascular injuries occurs in extremity fractures caused by gunshots [10]. Despite controversies about the sequence of fixation and vascular supply restoration, recent war experience demonstrates the advantage of immediate vascular temporary shunting, especially when prolonged and remote evacuation is expected [28].

23.1.4 Fracture Care

Most modern texts nowadays refer to gunshot-related fractures as open fractures caused by other mechanisms [1]. However, there are special considerations, which are unique to these injuries and require different management strategies. Due to the limited scope of this chapter we will review only some of the important ones.

23.1.4.1 Long Bones

Traditionally, gunshot injuries were considered grossly contaminated and therefore external fixation was the mainstay of their treatment for many years. However, reports from the 1990s demonstrated the efficacy of immediate intramedullary nailing of femoral shaft fracture caused by low- to mid-energy gunshot with acceptable clinical results [29], and also with higher energy firearm trauma [30].

With femoral fractures, care should be taken to avoid rotational mal-alignment since comminution will often distort the anatomical landmarks such as the cortical step sign [31] used to judge femoral rotation. An example is provided in Fig. 23.1. Recently, computer navigation has been shown to be effective in restoring rotational alignment. One of the cases in a recently published series dealt with a comminuted proximal femur fracture caused by a gunshot successfully restored for length and rotation by using computer navigation [32].

Bone loss and tissue coverage are a major challenge for tibial fractures especially involving the distal third, when soft tissue availability is scant. Due to these difficulties, our policy in high-energy gunshot wounds is to minimize bone debridement to minimum necessary and preserve as much bone as possible. Many times, these will be osteoconductive and would not result in massive infection. Recent experience in open distal femur fractures demonstrated that more conservative bone debridement does not increase infection rate, but decreases nonunion [33]. Several solutions adopted from the general orthopedic trauma are acceptable. These include rotational flaps, free tissue transfer, and immediate or late bone grafting [34, 35]. Each of these techniques can be used in conjunction with a circular frame for distraction osteogenesis [36], but not exclusively. A case of an IIIB distal tibial gunshot fracture with significant bone loss is depicted in Fig. 23.2.

It should be noted, however, that even with high success rates of these hard-to-treat fractures, one can look at a prolonged recovery period with multiple reconstructive attempts [37]. As with other mangled extremity injuries, the option for primary amputation should be considered. Increased bulk of literature, both civilian and military [38, 39], denounce the use of "scoring systems" such as the MESS score for determination of the fate of a mangled extremity, and prefer to rely on the surgeon's experience and on the general condition of the gunshot victim as predictors for early amputation.

23.1.4.2 Joints

As with other articular fractures, joint reconstruction and stable fixation should become the primary goal of treatment. However, some unique considerations should be given. First, due to the contaminated nature of gunshot injury, either arthroscopic or open irrigation should be strongly considered even in cases where a joint space violation is suspected [22]. A rare but a relevant example is the case where abdominal penetration occurred concomitant with pelvic or hip involvement. In these cases, contamination and joint sepsis would ultimately result in a catastrophic joint destruction.

Restoring metaphyseal comminution and building the articular block back to the shaft might be challenging in cases of severe comminution (Fig. 23.3), but effort should be made, like with any articular fracture, to restore joint congruency in order to maintain function and early motion [22]. Finally, chronic retention of metallic foreign bodies can result either in a local reaction [40, 41], or in rare cases, in systemic toxicity, such as lead poisoning [42, 43]. These will mandate early removal even in asymptomatic patients.

23.2 Blast Injuries

Another penetrating trauma type to be discussed here is related to explosion or blast. These are by far less common injuries than gunshot wounds, but their number is unfortunately growing in both civilian and military setting due to geopolitical reasons [44]. Despite the daunting threat of chemical and biological warfare, conventional terrorism is still the most common form of attacks resulting in high causality events [45]. Examples from recent years include the

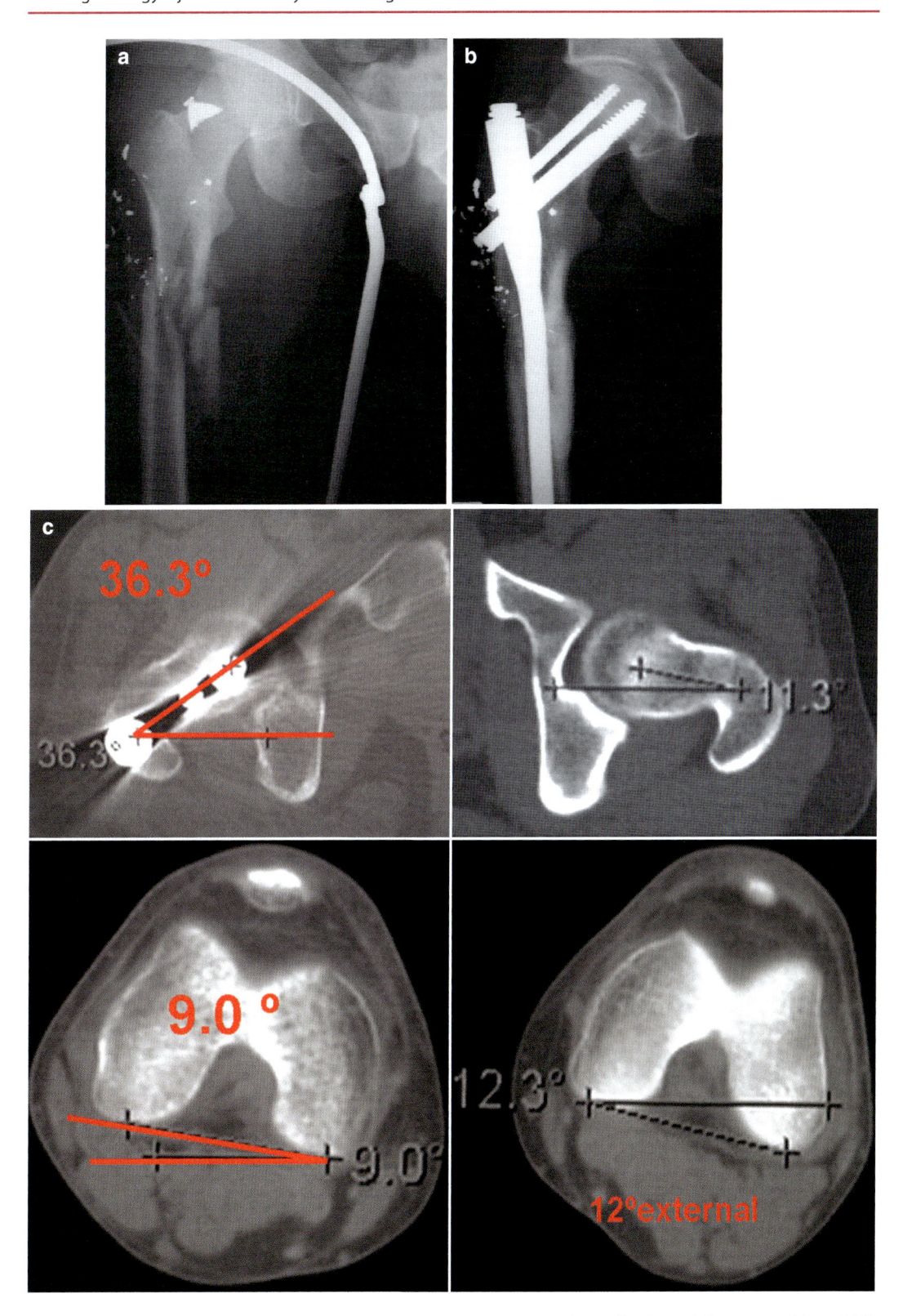

Fig. 23.1 (a) Gunshot fracture to a 25-year-old counter-terrorist fighter, caused by an AK-47. (b) Three months following irrigation, debridement, and fixation with a proximal femoral nail (Synthes, Battlach, Switzerland) with no apparent wound complications. (c) Due to complaints of difficulty running and intoeing, a CT scanogram was performed which revealed almost 40° internal rotation deformity. (d) Fracture revised by a derotation osteotomy and a reamed nail

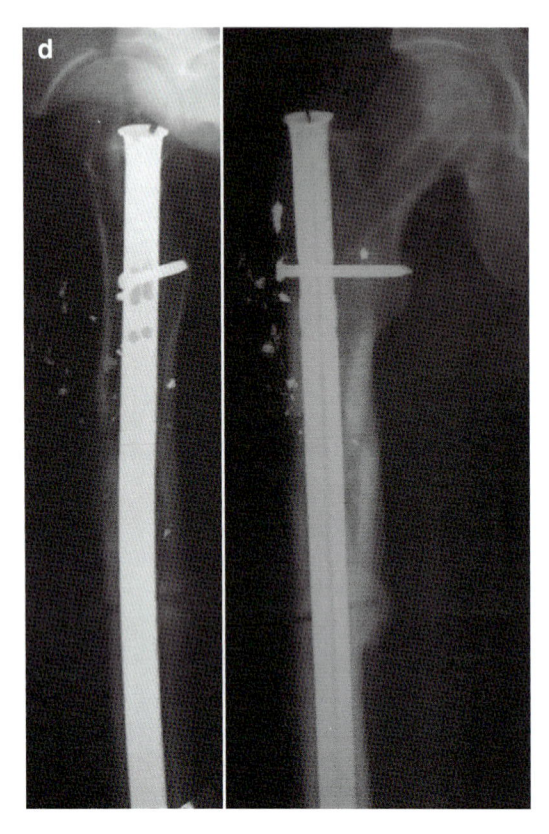

Fig. 23.1 (continued)

London attacks of 2005 [46], Madrid train attacks of 2004 [47], and the succession of suicide bombing during the Palestinian uprising between 2000 and 2005 [48, 49]

Blast injuries are different from gunshots mainly because of their multiple mechanisms of injury [16, 50, 51]. They tend to involve more body regions; and generally tend to be of higher severity scores, with increased overall potential for prolonged ICU stay and mortality [16, 52]. Although the surgical management of individual injuries may be similar to that of other types of trauma, the overall management of these patients as individuals as well as in the context of mass casualty event is worthy of consideration.

In this section we will overview the mechanisms of injury, principles of triage, and team approach as well as damage control strategies and definite orthopedic treatment of complex injuries with a special emphasis on the civilian setting.

23.2.1 Mechanisms of Blast Injury

The primary blast effect is related to the rapid pressure wave created during the detonation of an explosive [53]. The scene location and type of explosive used have a direct effect on the severity of injuries. Blast wave energy tends to decrease rapidly in space and dissipate [54]. However, when blast occurs in a closed or confined space, such in a bus or a room, the blast waves are reverberated from the walls instead of dissipating [54–56], thus inflicting more damage on human victims. In a series of suicide bombing in Israel occurring in buses during the years 1995–1996, a threefold increase in primary blast injuries was observed when compared to those of open-space explosions, exemplifying this phenomenon [56]. When the pressure wave created by detonation encounters certain air-fluid interfaces, unique tissue damage may occur. The most common and perhaps the most life-threatening injury involves the lung. Pressure differentials across the alveolar-capillary interface cause disruption, hemorrhage, pulmonary contusion, pneumothorax, hemothorax, pneumomediastinum, and subcutaneous emphysema [57].

The second most common type of primary blast injury is that to hollow viscera. The intestines, most usually the colon, are affected by the detonation wave. Mesenteric ischemia or infarct can cause delayed rupture of the large or the small intestine; these injuries are difficult to detect initially. Rupture, infarction, ischemia, and hemorrhage of solid organs such as the liver, spleen, and kidney are generally associated with very high blast forces or proximity of the patient to the blast center [58].

Tympanic membrane injury was extensively discussed in the literature discussing terror attacks. It is the most common non-lethal injury caused by relatively low-pressure blast waves.

23 High-Energy Injuries Caused by Penetrating Trauma 335

Fig. 23.2 (**a**) A 50-year-old schizophrenic smoker was injured by an M-16 military assault rifle sustaining a grade III-B open distal tibial fracture. (**b**) After irrigation and debridement with an attempt to preserve as much bone as possible, and immediate fixation with an unreamed tibial nail. (**c, d**) A year after a definite treatment that included further wound irrigation, latissimus dorsi free-flap, and iliac crest bone graft. Despite the imperfect ankle alignment, the patient was doing clinically well and returned to function

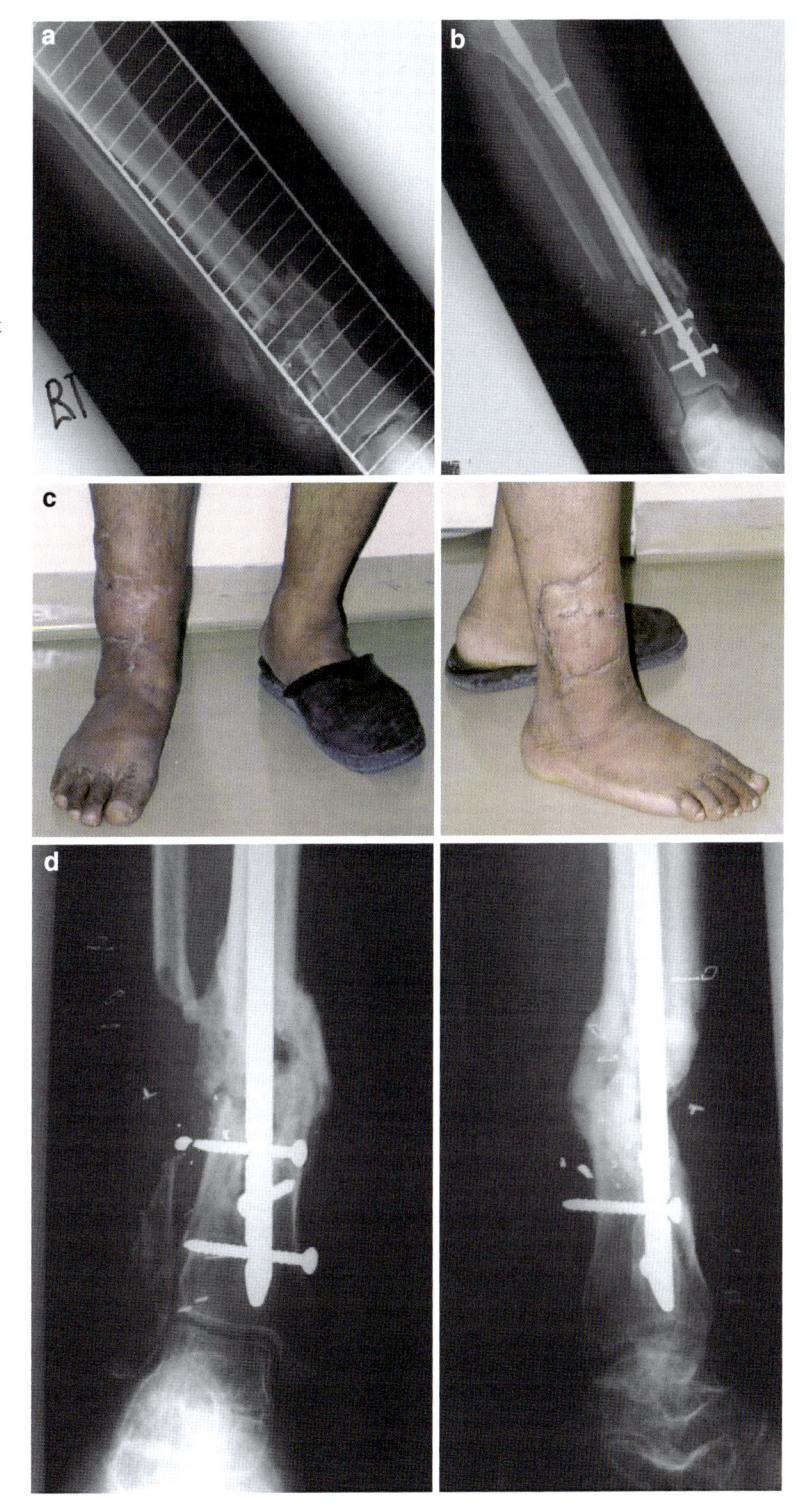

Fig. 23.3 (**a**) A 40-year-old patient was shot in his left lower arm and treated elsewhere with an external fixator with attempted internal fixation and brought to us for evaluation. (**b**) A staged protocol was used – removal of external fixator and wound debridement, definite treatment using parallel locked 3.5 reconstruction plates and iliac crest bone graft. Eighteen months after injury, the fracture is solidly healed and the patient has a reasonable range of painless motion

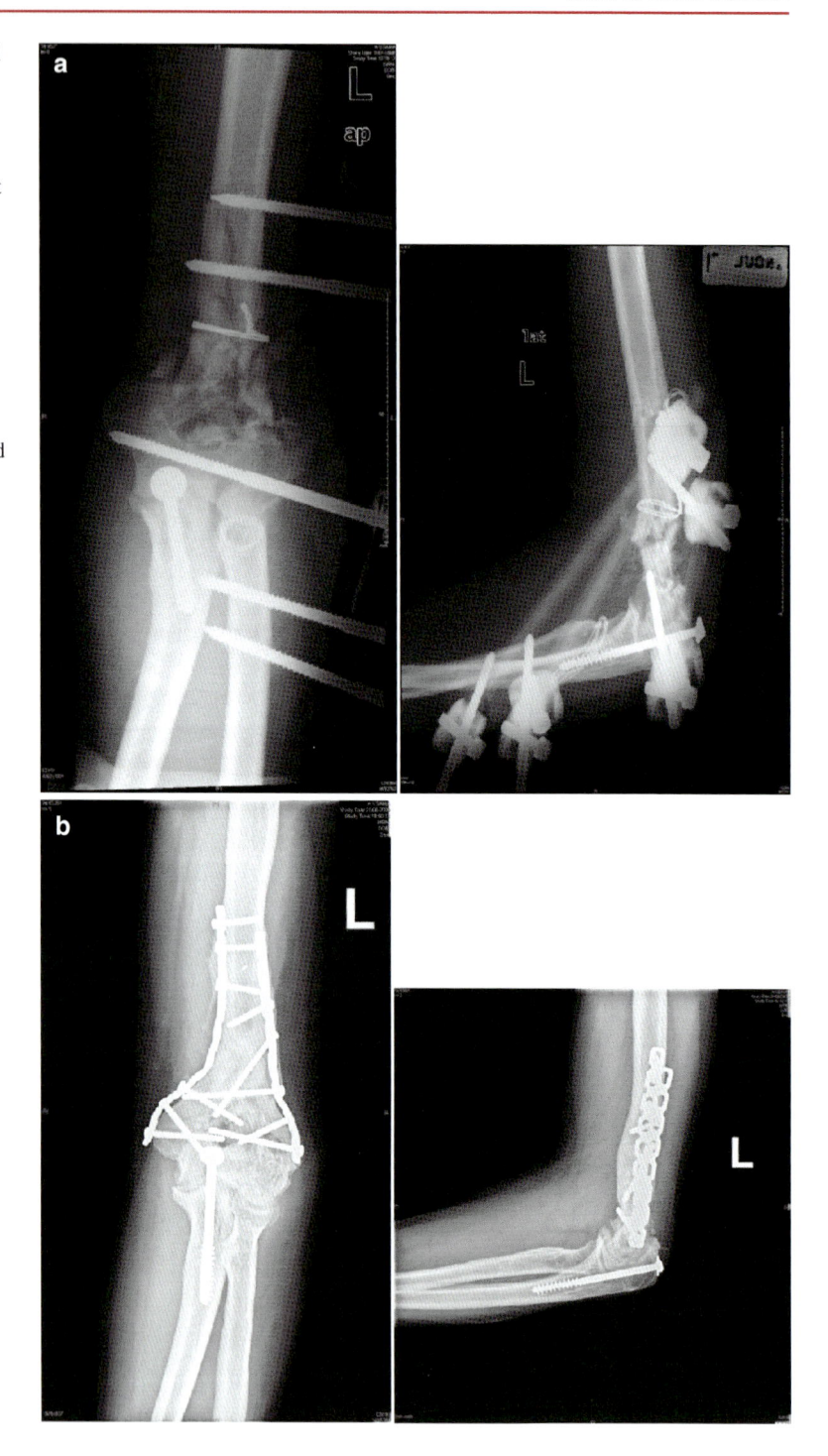

Traditionally, its presence was used to predict severe primary blast injuries (such as the lung or bowel), yet it is now referred to as questionable and unreliable [59], as will be further discussed in the triage section.

Traditionally, limb injury due to primary blast was considered a rarity. Hull and Cooper studied primary blast effects on the extremities resulting in traumatic amputations in Northern Ireland [60]. Only 9 out of 52 victims with traumatic amputations due to primary blast survived, demonstrating the high level of energy needed to avulse a limb. However, in the military setting, with the use of body armor becoming a common practice, devastating primary blast amputation has become increasingly common. These can include multiple and higher limb avulsions as well as pelvic and perineal injuries – areas not well protected by armor [4, 61].

Traditionally, the secondary blast effects comprise the core of the orthopedic injuries observed in warfare [62, 63] and also in civilian terror attacks such as the Middle Eastern experience [50, 64, 65]. Secondary blast effects are related to penetrating injuries caused by fragments ejected from the explosives and/or by foreign bodies impregnated within it. The extent of this effect depends on the subject's distance from the detonation center, the shape and size of the fragments, and the number of foreign bodies implanted or created by the explosive. In contrast to most warfare injuries, the improvised explosives used by terrorists have multiple added fragments including screws, bolts, nails, and other objects that may increase the damage caused by penetrating injuries [66]. Open fractures, severe soft tissue injuries, and multi-organ penetrating injuries are the more common pattern seen in the severely injured victim [66, 67].

Tertiary blast injury refers to the blunt trauma component of the explosion. Flying or falling objects can cause additional traumatic elements to those described above. When structural collapse takes place, a high casualty and mortality event occurs [48]. Our experience in Israel did not demonstrate a significant proportion of additional blunt trauma, but reports from other parts of the world, such as those following the Oklahoma City explosion, state this as the primary mechanism of the injury, as well as the [68] cause of usually devastating results.

The quaternary blast effect is a recently added one, and includes the thermal and chemical damage caused by fire and noxious substances occurring at the vicinity of the explosion. Confined-space explosions significantly increase these types of injuries [56].

23.2.2 Triage and Primary Resuscitation

Perhaps the most significant difference between gunshots and the blast wounded, besides the individual injury pattern, is the "mass-casualty" effect caused by multiple military and civilian attacks. Instead of treating a single patient brought to the treating facility, the surgeon faces a scenario of mass casualty, and is required to simultaneously deal with multiple patients having multiple injuries. Hence, the initial effort should be to establish an orderly triage system, and to allocate medical team and hospital resources even before the first patient arrives to the hospital [69, 70].

In the military setting, front medical teams utilizing "damage control" strategies and performing only emergency surgeries have recently developed, especially in the global war against terrorism [28]. Further procedures are then performed in secondary and tertiary centers after further triage and usually prolonged transportation. However, this is not the case in the civilian setting, which is in the primary focus of this section.

At recent attacks – such in Jerusalem, Madrid, and London and recently the Boston Marathon – evacuation time to a definite care facility ranged between 18 min to 1–2 h [51, 71–73]. Some events, especially those on a large scale, described only a few severely injured patients in a majority of "walking wounded". The Middle East experience, such as in Jerusalem, paradoxically demonstrated that smaller bombing scenes resulted in overall less causalities but with more critically ill patients (4–8 per event) arriving in very short notice to treating facilities [48].

Logistics of Emergency Department management have a major impact on triage. We recommend evacuation of non-critical non-terror-related patients temporarily to the hospital floors while the seriously ill patients can be treated in designated areas. The trauma bays are thus devoted solely to resuscitative efforts done on critically ill patients, while the rest of the ED serves as an admitting area to the rest of the patients. Each area is staffed with surgeon-in-charge and other members of the treating team (surgical and orthopedic residents, nurses, medical students, etc.). A surgeon-in-charge should be designated beforehand and should serve in critical junctions as suggested by Almogy et al. [71]: Triage at the initial admitting phase and in various treating cycles as well as diagnostics and directions towards the operating room or admitting floors are constantly done, until the general chaos is reduced. Every hospital should explore and identify the logistics mechanism required to provide the best and most efficient setting for disaster management under its capacity, should an actual disaster occur.

The next important principle in managing an event of this nature is to direct the flow of patients in an orderly fashion in a "one way" system. Potential bottlenecks, such as in the CT scanner, ICUs, and limited number of available operating rooms, should be identified and the patients should be directed to their proper destination only after the available resources of the hospital are mapped and identified [48]. Since as many as 50 % of causalities would require a surgical intervention – ICU stay or both [74] – hospital management should be prepared to allocate these facilities in a timely fashion.

23.2.3 Treatment of Specific Injuries

Blast extremity injuries tend to be more varied and less predictable than gunshots [16, 75]. The energy of the penetrating foreign body is extremely variable and greatly depends on the distance from the detonation center [62]. The existence of an extremity fracture, therefore, indicates a high-energy mechanism and has proven many times to implicate a polytrauma situation [76]. This is in contrast to a gunshot patient who can present with an extremity injury as a sole manifestation [16, 77]. In fact, one of the studies performed at our center indicates that fractures caused by blast are highly associated with potentially lethal blast lung injuries [78]. Therefore, an "isolated" fracture caused by blast mechanism should alert the surgeon to aggressively seek and diagnose associated injuries. Although high-energy gunshots wounds may involve a higher rate of vascular injury, compartment syndrome, and higher grade open fractures [16], blast extremity injury can involve more multiple fracture sites than gunshots, as well as a higher ISS and more associated life threatening injuries [16, 79, 80]. In this context, the treatment plans and strategies of each patient should be meticulously defined. As many patients are expedited to the operating room, more orthopedic teams should be available to undertake emergent procedures.

Damage control orthopedic and soft tissue strategies should generally be the rule in these cases since 70 % of bone-injured blast patients have an ISS of above 20 [75, 77] and are highly prone to prolonged ICU stay, respiratory failure, and coagulopathy [52, 79]. An orthopedic surgeon-in-charge should direct the teams in decision making, and the scope of first stage treatment plans should be limited as reconstruction can be planned in subsequent phases.

As in blunt polytrauma situations, tertiary surveys are extremely important in order to identify missed injuries, most of them of musculoskeletal nature [81]. We established a routine of a "morning after" rounds using the records from the ED to allocate the patients in the entire hospital, performing vigilant physical examination and documenting the penetrating injury both in writing and in a graphic form using a burn-unit type chart. A significant number of missed fractures and retained foreign bodies were identified with these surveys.

23.2.3.1 Specific Considerations
Most treatment principles of gunshot injury in regards to bone, soft tissue and vascular management should be applied to blast skeletal injuries. However, specific considerations unique to blast injury in the civilian setting should also be applied. First, as mentioned above, a polytrauma situation dictates decision making in regard to staging of bone and soft tissue treatment that is slightly different than that applied to the typical gunshot wounded victim. An illustrative case is demonstrated on Fig. 23.4.

Also, the metal load and the amount of foreign bodies warrant removal, otherwise unnecessary

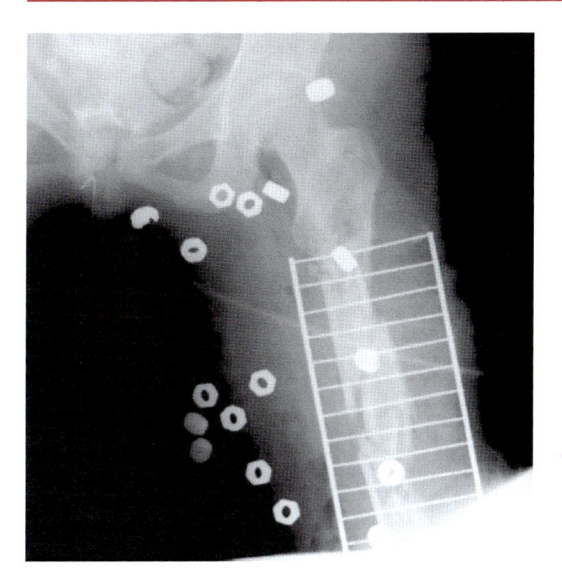

Fig. 23.4 A comminuted femoral fracture caused by secondary blast: note the numerous bolts implanted in the explosive

in gunshot wounds. The mere removal can cause further soft tissue damage, thus mandating minimal invasive techniques. We reported the use of computerized navigation as well as metal detectors to attempt and minimize dissection involved in these removals [40, 82].

Lastly, the fact that more and more suicide bombers are involved in modern terrorism may increase the risk of biological contamination of the victims with tissues originating from the terrorists themselves, such as bone fragments [83]. Concerns of blood borne infections such as Hepatitis B/C and HIV should be taken into consideration when dealing with suicide bombers [84].

Conclusions

At the turn of the twenty-first century, firearm and terror related violence has not shown signs of decline, and it seems the world is still facing a rise in casualties related to these mechanisms. Lots have been learnt during recent years about mechanisms of injury, scenarios of mass casualty events and treatment strategies. Despite this reality, principles of treating an isolated penetrating injury, such as gunshot and multiple penetrating limb injuries, such as blast, are not yet part of the standard medical education and upbringing of the average orthopedic trauma surgeon. Keeping in mind that no part of the world is immune at this point to these devastating injuries, research and investigation of outcome and treatment strategies are in strong need, as well as internalization of the current knowledge and principles among the orthopedic and trauma surgeons community.

References

1. Bartlett CS. Clinical update: gunshot wound ballistics. Clin Orthop Relat Res. 2003;408:28–57.
2. Almogy G, Rivkind AI. Terror in the 21st century: milestones and prospects–part I. Curr Probl Surg. 2007;44:496–554.
3. Bhatti JA, Mehmood A, Shahid M, Bhatti SA, Akhtar U, Razzak JA. Epidemiological patterns of suicide terrorism in the civilian Pakistani population. Int J Inj Contr Saf Promot. 2011;18:205–11.
4. Mamczak CN, Elster EA. Complex dismounted IED blast injuries: the initial management of bilateral lower extremity amputations with and without pelvic and perineal involvement. J Surg Orthop Adv. 2012;21:8–14.
5. Fackler ML. Gunshot wound review. Ann Emerg Med. 1996;28:194–203.
6. Fackler ML. Civilian gunshot wounds and ballistics: dispelling the myths. Emerg Med Clin North Am. 1998;16:17–28.
7. Liu YQ, Chen XY, Li SG, et al. Wounding effects of small fragments of different shapes at different velocities on soft tissues of dogs. J Trauma. 1988;28:S95–8.
8. Fackler ML, Malinowski JA. Ordnance gelatin for ballistic studies. Detrimental effect of excess heat used in gelatin preparation. Am J Forensic Med Pathol. 1988;9:218–9.
9. Dimond Jr FC, Rich NM. M-16 rifle wounds in Vietnam. J Trauma. 1967;7:619–25.
10. Volgas DA, Stannard JP, Alonso JE. Current orthopaedic treatment of ballistic injuries. Injury. 2005;36:380–6.
11. Hopkinson DA, Marshall TK. Firearm injuries. Br J Surg. 1967;54:344–53.
12. Hennessy MJ, Banks HH, Leach RB, Quigley TB. Extremity gunshot wound and gunshot fracture in civilian practice. Clin Orthop Relat Res. 1976;114:296–303.
13. Hollerman JJ, Fackler ML, Coldwell DM, Ben-Menachem Y. Gunshot wounds: 1. Bullets, ballistics, and mechanisms of injury. AJR Am J Roentgenol. 1990;155:685–90.

14. Omer Jr GE. Injuries to nerves of the upper extremity. J Bone Joint Surg Am. 1974;56:1615–24.
15. Tian HM, Deng GG, Huang MJ, Tian FG, Suang GY, Liu YG. Quantitative bacteriological study of the wound track. J Trauma. 1988;28:S215–6.
16. Weil YA, Petrov K, Liebergall M, Mintz Y, Mosheiff R. Long bone fractures caused by penetrating injuries in terrorists attacks. J Trauma. 2007;62:909–12.
17. Berg RJ, Okoye O, Inaba K, et al. Extremity firearm trauma: the impact of injury pattern on clinical outcomes. Am Surg. 2012;78:1383–7.
18. Dennis JW, Frykberg ER, Veldenz HC, Huffman S, Menawat SS. Validation of nonoperative management of occult vascular injuries and accuracy of physical examination alone in penetrating extremity trauma: 5- to 10-year follow-up. J Trauma. 1998;44:243–52; discussion 2–3.
19. McCorkell SJ, Harley JD, Morishima MS, Cummings DK. Indications for angiography in extremity trauma. AJR Am J Roentgenol. 1985;145:1245–7.
20. Frykberg ER, Crump JM, Vines FS, et al. A reassessment of the role of arteriography in penetrating proximity extremity trauma: a prospective study. J Trauma. 1989;29:1041–50; discussion 50–2.
21. White PW, Gillespie DL, Feurstein I, et al. Sixty-four slice multidetector computed tomographic angiography in the evaluation of vascular trauma. J Trauma. 2010;68:96–102.
22. Dougherty PJ, Vaidya R, Silverton CD, Bartlett C, Najibi S. Joint and long-bone gunshot injuries. J Bone Joint Surg Am. 2009;91:980–97.
23. Cleveland M, Manning JG, Stewart WJ. Care of battle casualties and injuries involving bones and joints. J Bone Joint Surg Am. 1951;33-A:517–27.
24. Fackler ML. Wound ballistics: the management of assault rifle injuries. Mil Med. 1990;155:222–5.
25. Mendelson JA. The relationship between mechanisms of wounding and principles of treatment of missile wounds. J Trauma. 1991;31:1181–202.
26. Bartlett CS, Helfet DL, Hausman MR, Strauss E. Ballistics and gunshot wounds: effects on musculoskeletal tissues. J Am Acad Orthop Surg. 2000;8:21–36.
27. Powell ET. The role of negative pressure wound therapy with reticulated open cell foam in the treatment of war wounds. J Orthop Trauma. 2008;22:S138–41.
28. Gifford SM, Aidinian G, Clouse WD, et al. Effect of temporary shunting on extremity vascular injury: an outcome analysis from the Global War on Terror vascular injury initiative. J Vasc Surg. 2009;50:549–55; discussion 55–6.
29. Nowotarski P, Brumback RJ. Immediate interlocking nailing of fractures of the femur caused by low- to mid-velocity gunshots. J Orthop Trauma. 1994;8:134–41.
30. Nicholas RM, McCoy GF. Immediate intramedullary nailing of femoral shaft fractures due to gunshots. Injury. 1995;26:257–9.
31. Langer JS, Gardner MJ, Ricci WM. The cortical step sign as a tool for assessing and correcting rotational deformity in femoral shaft fractures. J Orthop Trauma. 2010;24:82–8.
32. Weil YA, Greenberg A, Khoury A, Mosheiff R, Liebergall M. Computerized navigation for length and rotation control in femoral fractures: a preliminary clinical study. J Orthop Trauma. 2014;28(2):e27–33.
33. Ricci WM, Collinge C, Streubel PN, McAndrew CM, Gardner MJ. A comparison of more and less aggressive bone debridement protocols for the treatment of open supracondylar femur fractures. J Orthop Trauma. 2013;27:722–5.
34. Atesalp AS, Yildiz C, Basbozkurt M, Gur E. Treatment of type IIIa open fractures with Ilizarov fixation and delayed primary closure in high-velocity gunshot wounds. Mil Med. 2002;167:56–62.
35. Tropet Y, Garbuio P, Obert L, Jeunet L, Elias B. One-stage emergency treatment of open grade IIIB tibial shaft fractures with bone loss. Ann Plast Surg. 2001;46:113–9.
36. Hutson Jr JJ, Dayicioglu D, Oeltjen JC, Panthaki ZJ, Armstrong MB. The treatment of gustilo grade IIIB tibia fractures with application of antibiotic spacer, flap, and sequential distraction osteogenesis. Ann Plast Surg. 2010;64:541–52.
37. Celikoz B, Sengezer M, Isik S, et al. Subacute reconstruction of lower leg and foot defects due to high velocity-high energy injuries caused by gunshots, missiles, and land mines. Microsurgery. 2005;25:3–14; discussion 5.
38. Bosse MJ, MacKenzie EJ, Kellam JF, et al. A prospective evaluation of the clinical utility of the lower-extremity injury-severity scores. J Bone Joint Surg Am. 2001;83-A:3–14.
39. Brown KV, Ramasamy A, McLeod J, Stapley S, Clasper JC. Predicting the need for early amputation in ballistic mangled extremity injuries. J Trauma. 2009;66:S93–7; discussion S7–8.
40. Peyser A, Khoury A, Liebergall M. Shrapnel management. J Am Acad Orthop Surg. 2006;14:S66–70.
41. Eylon S, Mosheiff R, Liebergall M, Wolf E, Brocke L, Peyser A. Delayed reaction to shrapnel retained in soft tissue. Injury. 2005;36:275–81.
42. Peh WC, Reinus WR. Lead arthropathy: a cause of delayed onset lead poisoning. Skeletal Radiol. 1995;24:357–60.
43. Linden MA, Manton WI, Stewart RM, Thal ER, Feit H. Lead poisoning from retained bullets. Pathogenesis, diagnosis, and management. Ann Surg. 1982;195:305–13.
44. Stein M, Hirshberg A. Medical consequences of terrorism. The conventional weapon threat. Surg Clin North Am. 1999;79:1537–52.
45. Lerner EB, O'Connor RE, Schwartz R, et al. Blast-related injuries from terrorism: an international perspective. Prehosp Emerg Care. 2007;11:137–53.

46. Aylwin CJ, Konig TC, Brennan NW, et al. Reduction in critical mortality in urban mass casualty incidents: analysis of triage, surge, and resource use after the London bombings on July 7, 2005. Lancet. 2006;368:2219–25.
47. de Gutierrez Ceballos JP, Turegano Fuentes F, Perez Diaz D, Sanz Sanchez M, Martin Llorente C, Guerrero Sanz JE. Casualties treated at the closest hospital in the Madrid, March 11, terrorist bombings. Crit Care Med. 2005;33:S107–12.
48. Shamir MY, Rivkind A, Weissman C, Sprung CL, Weiss YG. Conventional terrorist bomb incidents and the intensive care unit. Curr Opin Crit Care. 2005;11:580–4.
49. Shapira SC, Adatto-Levi R, Avitzour M, Rivkind AI, Gertsenshtein I, Mintz Y. Mortality in terrorist attacks: a unique modal of temporal death distribution. World J Surg. 2006;30:2071–7; discussion 8–9.
50. Peleg K, Aharonson-Daniel L. Blast injuries. N Engl J Med. 2005;352:2651–3; author reply –3.
51. Peleg K, Aharonson-Daniel L, Michael M, Shapira SC. Patterns of injury in hospitalized terrorist victims. Am J Emerg Med. 2003;21:258–62.
52. Kluger Y, Peleg K, Daniel-Aharonson L, Mayo A. The special injury pattern in terrorist bombings. J Am Coll Surg. 2004;199:875–9.
53. Wightman JM, Gladish SL. Explosions and blast injuries. Ann Emerg Med. 2001;37:664–78.
54. Arnold JL, Halpern P, Tsai MC, Smithline H. Mass casualty terrorist bombings: a comparison of outcomes by bombing type. Ann Emerg Med. 2004;43:263–73.
55. DePalma RG, Burris DG, Champion HR, Hodgson MJ. Blast injuries. N Engl J Med. 2005;352:1335–42.
56. Leibovici D, Gofrit ON, Stein M, et al. Blast injuries: bus versus open-air bombings–a comparative study of injuries in survivors of open-air versus confined-space explosions. J Trauma. 1996;41:1030–5.
57. Mellor SG, Cooper GJ. Analysis of 828 servicemen killed or injured by explosion in Northern Ireland 1970–84: the Hostile Action Casualty System. Br J Surg. 1989;76:1006–10.
58. Almogy G, Mintz Y, Zamir G, et al. Suicide bombing attacks: can external signs predict internal injuries? Ann Surg. 2006;243:541–6.
59. Leibovici D, Gofrit ON, Shapira SC. Eardrum perforation in explosion survivors: is it a marker of pulmonary blast injury? Ann Emerg Med. 1999;34:168–72.
60. Hull JB, Cooper GJ. Pattern and mechanism of traumatic amputation by explosive blast. J Trauma. 1996;40:S198–205.
61. Jacobs N, Taylor DM, Parker PJ. Changes in surgical workload at the JF Med Gp Role 3 Hospital, Camp Bastion, Afghanistan, November 2008-November 2010. Injury. 2012;43:1037–40.
62. Covey DC. Blast and fragment injuries of the musculoskeletal system. J Bone Joint Surg Am. 2002;84-A:1221–34.
63. Covey DC. Combat orthopaedics: a view from the trenches. J Am Acad Orthop Surg. 2006;14:S10–7.
64. Barham M. Blast injuries. N Engl J Med. 2005;352:2651–3; author reply –3.
65. Ashkenazi I, Olsha O, Alfici R. Blast injuries. N Engl J Med. 2005;352:2651–3; author reply –3.
66. Ad-El DD, Eldad A, Mintz Y, et al. Suicide bombing injuries: the Jerusalem experience of exceptional tissue damage posing a new challenge for the reconstructive surgeon. Plast Reconstr Surg. 2006;118:383–7; discussion 8–9.
67. Aharonson-Daniel L, Klein Y, Peleg K. Suicide bombers form a new injury profile. Ann Surg. 2006;244:1018–23.
68. Teague DC. Mass casualties in the Oklahoma City bombing. Clin Orthop Relat Res. 2004;422:77–81
69. Hirshberg A, Scott BG, Granchi T, Wall Jr MJ, Mattox KL, Stein M. How does casualty load affect trauma care in urban bombing incidents? a quantitative analysis. J Trauma. 2005;58:686–93; discussion 94–5.
70. Hirshberg A, Stein M, Walden R. Surgical resource utilization in urban terrorist bombing: a computer simulation. J Trauma. 1999;47:545–50.
71. Almogy G, Belzberg H, Mintz Y, Pikarsky AK, Zamir G, Rivkind AI. Suicide bombing attacks: update and modifications to the protocol. Ann Surg. 2004;239:295–303.
72. Einav S, Feigenberg Z, Weissman C, et al. Evacuation priorities in mass casualty terror-related events: implications for contingency planning. Ann Surg. 2004;239:304–10.
73. Walls RM, Zinner MJ. The Boston Marathon response: why did it work so well? JAMA. 2013;309:2441–2.
74. de Ceballos JP, Turegano-Fuentes F, Perez-Diaz D, Sanz-Sanchez M, Martin-Llorente C, Guerrero-Sanz JE. 11 March 2004: The terrorist bomb explosions in Madrid, Spain–an analysis of the logistics, injuries sustained and clinical management of casualties treated at the closest hospital. Crit Care. 2005;9:104–11.
75. Weil YA, Mosheiff R, Liebergall M. Blast and penetrating fragment injuries to the extremities. J Am Acad Orthop Surg. 2006;14:S136–9.
76. Weil YA, Peleg K, Givon A, Mosheiff R, Israeli Trauma G. Penetrating and orthopaedic trauma from blast versus gunshots caused by terrorism: Israel's National Experience. J Orthop Trauma. 2011;25:145–9.
77. Weil YA, Peleg K, Givon A, Mosheiff R. Musculoskeletal injuries in terrorist attacks–a comparison between the injuries sustained and those related to motor vehicle accidents, based on a national registry database. Injury. 2008;39:1359–64.
78. Almogy G, Luria T, Richter E, et al. Can external signs of trauma guide management?: lessons learned from suicide bombing attacks in Israel. Arch Surg. 2005;140:390–3.

79. Peleg K, Aharonson-Daniel L, Stein M, et al. Gunshot and explosion injuries: characteristics, outcomes, and implications for care of terror-related injuries in Israel. Ann Surg. 2004;239:311–8.
80. Weil Y, Peleg K, Givon A, Mosheiff R, ITG. Penetrating and orthopaedic trauma from blast vs. Gunshots caused by terrorism – Israel's National Experience. J Orthop Trauma. 2011;25: 145–9.
81. Buduhan G, McRitchie DI. Missed injuries in patients with multiple trauma. J Trauma. 2000;49:600–5.
82. Mosheiff R, Weil Y, Khoury A, Liebergall M. The use of computerized navigation in the treatment of gunshot and shrapnel injury. Comput Aided Surg. 2004;9:39–43.
83. Leibner ED, Weil Y, Gross E, Liebergall M, Mosheiff R. A broken bone without a fracture: traumatic foreign bone implantation resulting from a mass casualty bombing. J Trauma. 2005;58:388–90.
84. Wong JM, Marsh D, Abu-Sitta G, et al. Biological foreign body implantation in victims of the London July 7th suicide bombings. J Trauma. 2006;60:402–4.

Management of Traumatic Bone Defects

24

Richard P. Meinig and Hans-Christoph Pape

Contents

24.1 Introduction.. 343
24.2 **Initial Management of Traumatic Bone Defects**... 344
24.3 **Skeletal Fixation and Soft Tissue Coverage**... 345
24.4 **Bone Defect Reconstitution**........................ 346
References... 349

24.1 Introduction

Traumatic bone loss has long been a challenging clinical problem. Contemporary techniques in the management of acute bone stabilization, revascularization, and soft tissue reconstruction have lead to an increase in limb salvage. [1] The reconstruction of bone defects has numerous options. Generally the management of bone defects can be divided into two approaches. The first approach involves reconstitution of a bone defect that has been stabilized in situ by autologous bone grafting or one of its variations. The second approach involves distraction osteogenesis. The two approaches are not mutually exclusive but have their relative indications and difficulties. Distraction osteogenesis therapy is generally more protracted, technically very challenging, and accompanied by high complication rates [2]. However, distraction osteogenesis can be spectacularly successful in the simultaneous management of soft tissue coverage, bone defect, and spatial deformity. Because of the complexity of frame construction, pin site management, patient compliance, and duration of treatment, distraction osteogenesis procedures are perhaps best reserved for specialty clinics. Management of bone defects by skeletal stabilization, early soft tissue coverage and by autologous reconstruction utilizes implants, techniques, and resources that are widely available. This paper will present a summary of contemporary techniques that will

R.P. Meinig, MD (✉)
Orthopaedic Services, Penrose-St Francis
Health Care System, 2222 North Nevada Ave.,
Colorado Springs, CO 80907, USA
e-mail: rickmeinig@gmail.com

H.-C. Pape
Department of Orthopaedic Trauma,
University of Aachen Medicine Center,
30 Pauwels Street, Aachen 52074, Germany
e-mail: hpape@ukaachen.de

© Springer-Verlag Berlin Heidelberg 2016
H.-C. Pape et al. (eds.), *The Poly-Traumatized Patient with Fractures:
A Multi-Disciplinary Approach*, DOI 10.1007/978-3-662-47212-5_24

allow for the primary therapy of complex traumatic bone loss.

24.2 Initial Management of Traumatic Bone Defects

Bone defects caused by direct trauma usually are the result of a high energy injury and are therefore associated with mortality and morbidity of visceral or traumatic brain injury. Assessment of the long bone injury will determine whether the limb is viable and should be amputated versus limb salvage. Considerations for limb salvage are obviously complex and efforts to quantitate the injury in regards to amputation such as the Mangled Extremity Severity Scale (MESS) or the Orthopaedic Trauma Association Limb Evaluation and Assessment Protocol (LEAP) are often helpful but not definitive [3, 4]. Limb salvage requires a limb in which vascularity can be reestablished, adequate neurologic function in terms of sensation and motor, viable muscle – tendon groups, and soft tissues – bone injury in which sepsis can be ultimately ablated. The actual extent of bone loss that limits limb salvage has yet to be defined regardless of reconstruction technique. In addition to the biological factors, the patient's psychosocial systems need to be evaluated as reconstruction and limb salvage is a relatively long process possibly requiring multiple surgical interventions, medical therapies, rehabilitation, and patient compliance.

The initial management of a limb deemed suitable for limb salvage will consist of emergent resuscitation of the patient. Priority for reestablishing hemodynamic stability and managing the closed head injury component will frequently preclude definitive skeletal stabilization. The concept of "damage control orthopaedics" has recently emerged in which spanning external fixation or unreamed intramedullary nailing is expediently performed to limit the anesthetic time and reduce pulmonary exposure to medullary canal contents in severely injured patients [5]. When performed under these circumstances, pin sites and implants should be chosen to allow for subsequent definitive fixation and stabilization of the bone defect. Conversion to definitive fixation should be performed as soon as feasible to minimize potential septic seeding from external fixation pin sites – generally less than 10 days. Stabilization of open fractures with intramedullary nails has been validated to be acceptable in terms of infection risk [6]. In general, locked intramedullary stabilization of a diaphyseal and some metaphyseal defects is preferred as length, rotation, and axial alignment can be reestablished and maintained in a single procedure. The IM nail allows for immediate rehabilitation of the limb in near anatomic position. In addition, the IM nail has the biomechanical advantages of strength and symmetric load sharing in comparison to plates. The locked plate is a relatively recent development which allows for improved mechanical stability in situations of poor bone quality, bone defect or comminution, and articular fracture patterns associated with metaphyseal or diaphyseal extension. Plate fixation can be performed with minimal exposure to provide stable bridging constructs for the management of bone defects. If a limb requires vascular repair, plate or IM nail fixation needs to be coordinated with the vascular reconstruction to provide a stable environment for the repair as well as utilize the surgical exposure if indicated. An essential early step in the management of a bone defect is the initial debridements of bone and soft tissue. The initial debridements are likely to reduce the septic burden and re-establish soft tissue viability in the shortest time. The principles of debridement are well established and consist of excision and removal of nonviable osseous and soft tissues. Serial debridements are frequently required to discern borderline tissues on a clinical basis. The decision of implant and technique is therefore dictated by patient hemodynamic and neurologic status, concomitant vascular repair, tissue and bone debridement, and soft tissue coverage Initial management of the bone defect is directed at managing the dead space of the defect in preparation for soft tissue coverage and bone reconstitution. During debridements, the defect can be provisionally managed with commercially available PMMA-antibiotic beads or surgeon fabricated PMMA-antibiotic spacers. The PMMA-antibiotic beads or PMMA-spacers can be serially exchanged during debridements to aide in reducing deep sepsis and has been well described [7].

In summary, the initial phase consists of patient resuscitation with provisional or possibly definitive fixation of the skeletal defect. This is combined with establishment of a sterile bone defect and clean wound by surgical debridement and soft tissue wound care. The time line would be days 1–3.

24.3 Skeletal Fixation and Soft Tissue Coverage

If the limb has been treated with spanning external fixation, conversion to IM nail or plate implant can be performed – ideally within 7 days. Soft tissue coverage should be obtained either by wound closure, wound V.A.C therapy, or local/free flap coverage. Early soft tissue reconstruction aids in the prevention of deep sepsis as well as preparing an environment advantageous for bone grafting. The bone defect can be managed either primarily with early bone grafting or vascularized bone transfer. However, the cultivation of an "induced membrane" has clinical and basic science advantages for delaying definitive autologous bone transfer into segmental defects for a period of 4–6 weeks [8, 9].

Conversion from spanning external fixation or provisional stabilization to definitive implant fixation should restore the limb to near anatomic length, axial alignment, and rotation. The definitive implant should have sufficient mechanical properties to function during the duration of bone reconstitution. With early restoration and maintenance of the limb in anatomic position, patient comfort, rehabilitation, and function is greatly enhanced – a distinct advantage over distraction osteogenesis.

The keystone step during the Interim Management Phase is perhaps the reestablishment of an environment amenable to successful bone grafting. Animal studies and clinical studies indicate that a biologically active membrane that facilitates bone regeneration can be induced by the temporary implantation of a PMMA cement spacer. Histological, immunohistochemical, and biochemical assay in animal models demonstrate that by 4–6 weeks, a fibrous, highly vascularized, growth factor rich encapsulating membrane has encapsulated about the PMMA spacer. At 4–6 weeks postimplantation, vascular endothelial growth factor (VEGF), transforming growth factor (TGF-beta), and bone morphogenetic protein-2 (BMP-2) are at peaking levels within the membrane [10]. Autologous bone techniques may therefore optimally performed at 4–6 weeks post-PMMA spacer implantation. The technique is easily performed. PMMA cement is prepared and a tubular or appropriately shaped spacer is fabricated to span the defect and overlap the native bone ends. Antibiotic cement can be utilized as an adjunct to around the bone defect to prevent deep sepsis. Commercial antibiotic-PMMA mixtures that are available for primary total joint arthroplasty can be utilized (Biomet, Warsaw IN, Zimmer, Warsaw, IN, and Stryker, Mahwah, NJ). The surgeon, however, can prepare PMMA with higher amounts of added heat stable antibiotic to produce a bactericidal spacer [11].

In summary, the interim phase consists of obtaining: (1) early definitive internal fixation to stabilize the bone defect and limb in near anatomic alignment, and (2) preparation of a sterile osteogenic defect for osseous regeneration. A stable soft tissue environment is reestablished by wound closure or flap coverage if needed. In many cases, an induced membrane is formed by the temporary implantation of bulk PMMA with planned autologous grafting at approximately 4 weeks.

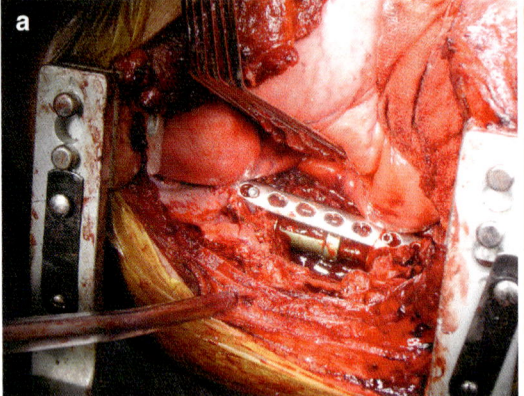

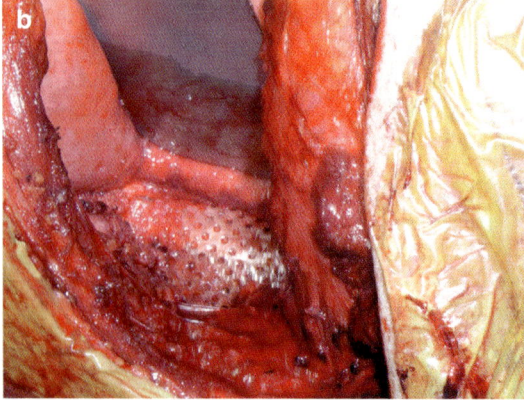

Fig. 24.1 (**a**) Thoracic spine burst fracture of T6 with corpectomy and stabilization of body with titanium mesh gauge and small fragment plate. Spinal cord is exposed. (**b**) PLA membrane is fabricated to form posterior wall following corpectomy of T6. Mixture of autologous bone from vertebral body fracture and DBM putty is grafted in cage as well as anterior to the membrane which protects cord from bone graft spillage into spinal canal

24.4 Bone Defect Reconstitution

Autologous bone grafting remains the gold standard in the reconstitution of bone defects. Autograft is the only material that provides osteogenic cells (osteocytes, osteoblasts, marrow stem cells), osteoconductive matrix (inorganic mineral), and osteoinductive molecules (BMPs, transforming growth factor-beta, vascular endothelial growth factor, and others) [12]. There are many techniques described for bone graft harvest including iliac graft harvest, local cancellous bone harvest, bone marrow aspirations and concentration, vascularized fibula, and most recently intramedullary canal harvest (Reamer Irrigator Aspirator-Synthes, Inc., West Chester, PA) [13–17]. In addition to autologous bone harvest, there are commercially available sources for recombinant osteoinductive bone morphogenetic proteins (BMP-7/OP01 and BMP-2). Alternatively, a spacer can be used in special indications, e.g., spine fractures (Fig. 24.1).

The primary limiting factor in autologous bone transplant has been reported morbidity and complications associated with the harvest site as well as adequate volume for large defects. With defects of 2 cm or less, traditional anterior iliac crest bone graft is usually sufficient as 5–72 ml can be harvested [13]. Larger defects can still be grafted with iliac crest by multiple harvest sites such as the contra lateral site or use of the posterior iliac crests with amounts of 25–90 ml be obtained [13]. In addition, the use of a small acetabular reamer may result in less donor site pain and larger volume of graft [14].

The most recent development in autologous harvest techniques is the intramedullary canal harvest. A recent review confirms that the use of the Reamer Irrigator Aspirator (RIA) in a single pass reaming of the femur produces significant amounts of bone graft (25–90 ml) with low rates of complications and postoperative pain [13]. While the rate of complication is lower than that described in conventional iliac harvest, iatrogenic femur fracture has occurred. In addition, studies of RIA harvest material suggest that it is rich in growth factors, viable cells, and morselized trabecular bone [15]. The RIA harvest can thus be considered biologically equivalent to iliac graft. The bone marrow harvest, however, lacks any structural properties that can be achieved with tricortical iliac harvest.

In addition to autologous bone graft, bone graft substitutes can be utilized to augment the autograft harvest. Bone graft substitutes include osteoconductive materials such as synthetic tricalcium phosphates, calcium sulfates, and coral. These materials are fabricated as granules, blocks, strips, putties, and pastes. However, the efficacy of these materials as stand-alone graft in segmental defects is unknown [18]. Similarly, there are currently at least 40 commercial preparations of Demineralized Bone Matrix (DBM). DBM is an acid extract of human cadaveric bone consisting largely of type I collagen and other acid stable proteins including bone morphogenetic proteins. The osteoinductive content of the DBM is low and subject to the variables of donor biological activity, processing, and carrier [16]. The osteoconductive properties of the various commercial DBMs relate to carrier chemistry, adjunctive inorganic additives such as cadaveric cancellous bone or synthetic mineral. At present there are no prospective studies proving the benefits of DBM for the reconstruction of segmental bone defect. The primary use of DBM may be as an extender for autologous bone harvests such as intramedullary reaming harvest, cancellous bone, or marrow aspirates and concentrates [16]. The role of recombinant bone morphogenetic proteins in bone defect reconstruction continues to evolve [19]. The high cost, carrier characteristics, biological activity and mechanical qualities of available commercial BMP preparations limit its use at present mainly to small cortical defects and acute open tibia fractures.

There are numerous options for the application of the autologous bone graft. Defects up to 29 cm have been successfully grafted using the induced membrane technique as described recently by Masquelet [9]. At 4–6 weeks post-PMMA block implantation, the block is removed by longitudinally incising the encapsulating membrane. Autologous bone in the form of iliac graft or RIA bone marrow harvest, or autologous bone-bone substitute or autologous bone-allograft mixture is then used to fill the resulting cavity. A defect stabilized with an intramedullary nail will require less bone graft volume than defects stabilized with external fixation or plate constructs. A resorbable polylactide membrane can also be used to shape and contain the graft for applications such as the distal tibia and femur. In addition, resorbable membranes can be used to contain the graft in applications near the spinal cord, interosseous membrane of the forearm, or other applications where the reconstruction needs to be precisely configured [20]. The polymeric membrane may be used where bone grafting is done primarily such as the reconstruction of an unstable thoracic burst fracture where cancellous bone graft is combined with a titanium vertebral reconstructions cage and a posterior vertebral body wall is fabricated by molding a polymer membrane (Fig. 24.1). Another technique for applying autograft is the use of cylindrical titanium cages to form a weight bearing diaphysis. In this technique, titanium mesh cages that are typically used in spinal vertebral reconstructions are fashioned to bridge the defect which has been stabilized with an intramedullary nail. The cage is packed with cancellous bone and the cage–host bone margins are autografted to create a construct which has considerable immediate mechanical stability [21].

In summary, Phase III consists of bone reconstitution of the defect with an autologous bone graft. The autologous bone graft can consist of harvested iliac crest, intramedullary reaming harvests, or combinations of autogenous materials with synthetic bone substitutes or allograft materials. The stabilized defect can be prepared with the formation of an induced membrane, or bridged with a resorbable polylactide membrane or titanium mesh cage. Alternatively, Ilizarov distraction osteogenesis can be used, as shown in the case example (Fig. 24.2).

Because of adequate mechanical stability from the internal fixation construct, functional rehabilitation can be instituted very early in the clinical course of limb salvage and bone defect reconstruction. The three phase algorithm incorporates surgical techniques and implants that are widely available.

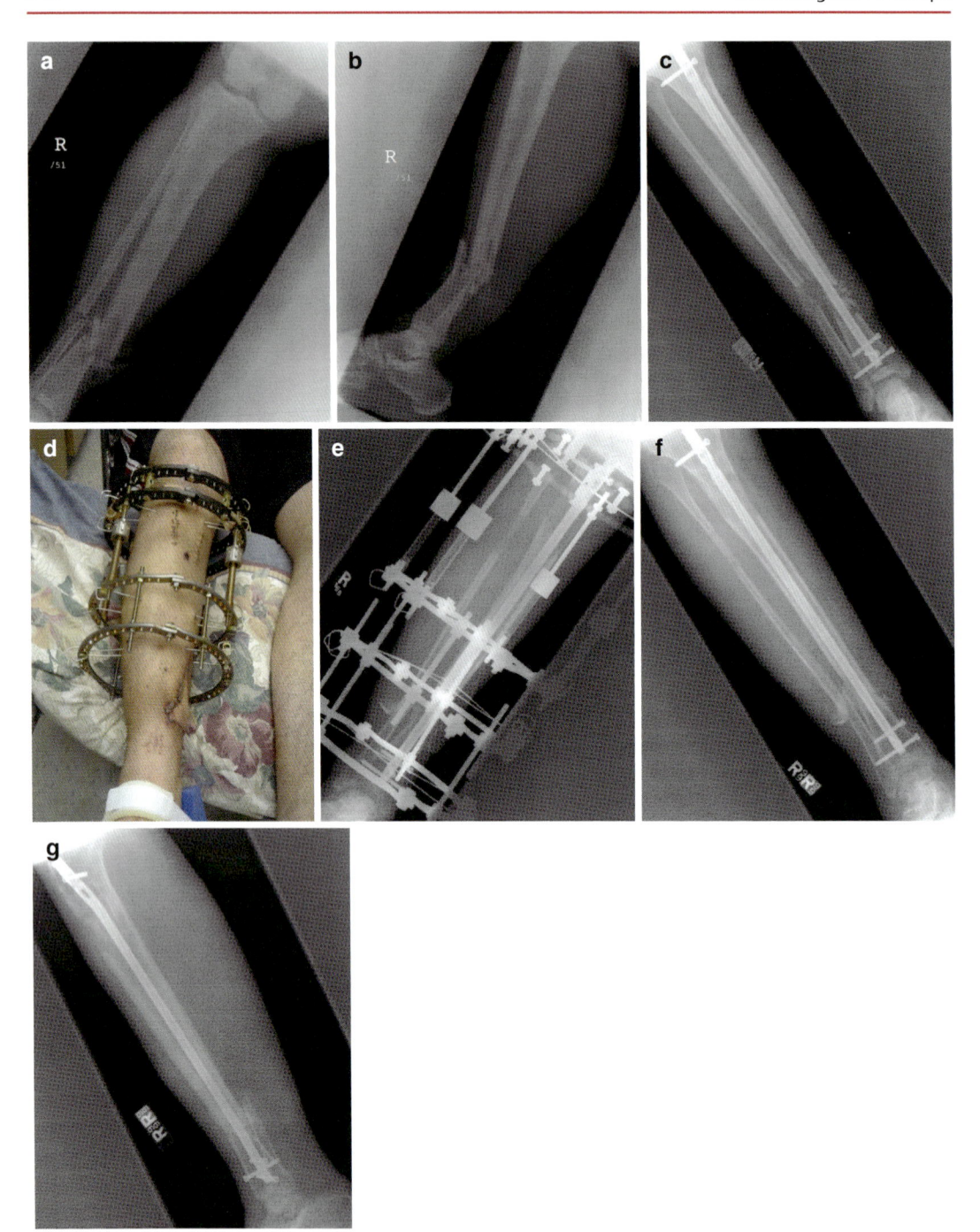

Fig. 24.2 (**a**, **b**) Injury film of open grade IIIb tibia and fibula fracture. (**c**) AP radiograph post op following debridement of diaphyseal bone segment, IM nailing, and wound closure by local flap. (**d**) Soft tissue status after placement of the Ilizarov distraction device. (**e**) Radiographic evidence of regeneration following Ilizarov transport. (**f**, **g**) Radiograph at 4 months after bone transport

References

1. Wiese A, Pape HC. Bone defects caused by high-energy injures, bone loss, infected nonunions, and nonunions. Orthop Clin North Am. 2010;41:1–4.
2. Abdel-Aal AM. Ilizarov bone transport for massive tibial bone defects. Orthopedics. 2006;29(1):70–4.
3. Webb LW, Bosse MJ, Castillo RC, Mackenzie EJ. Analysis of surgeon controlled variable in the type III open tibial diaphysis fracture. J Bone Joint Surg Am. 2007;89:923–8.
4. McNamara MG, Heckman JD, Corley EG. Severe open fractures of the lower extremity: a retrospective evaluation of the mangled extremity severity score. J Orthop Trauma. 1994;8:81–4.
5. Roberts CS, Pape HC, Jones AL, Malkini AL, Rodriquez JL, Giannoulis PV. Damage control orthopaedics: evolving concepts in the treatment of patients who have sustained orthopaedic trauma. J Bone Joint Surg Am. 2005;87:434–49.
6. Parekh AA, Smith W, Silva S, et al. Treatment of distal femur and proximal tibia fractures with external fixation followed by planned conversion to internal fixation. J Trauma. 2008;64(3):736–9. 5. Ristiniemi J, Lakovaara M, Flinikkila T, et al. Staged method using antibiotic beads and subsequent autografting for large traumatic tibial bone loss: 22 of 23 fractures healed after 5–20 months. Acta Orthop. 2007;78(4):520–7.
7. Adams K, Crouch L, Cierney G, Calhoun J. In vitro and in vivo evaluation of antibiotic diffusion from antibiotic impregnated polymethylmethracylate beads. Clin Orthop Relat Res. 1992;278:244–52. 15.
8. Klaue K, Anton C, Knothe U, et al. Biological implementation of in situ induced autologous foreign body membranes in consolidation of massive cancellous bone grafts. J Bone Joint Surg Br. 1993;79:236.
9. Masquelet AC, Begue T. The concept of induced membrane for reconstruction of long bone defects. Orthop Clin North Am. 2010;41:27–37.
10. Viateau V, Bensidhoum M, Guilleman G, et al. Use of the induced membrane technique for bone tissue engineering purposes: animal studies. Orthop Clin North Am. 2010;41:49–56.
11. Ristiniemi J, Lakovaara M, Flinikkila T, et al. Staged method using antibiotic beads and subsequent autografting for large traumatic tibial bone loss: 22 of 23 fractures healed after 5–20 months. Acta Orthop. 2007;78(4):520–7.
12. Marino JT, Ziran B. Use of solid and cancellous autologous bone graft for fractures and nonunions. Orthop Clin North Am. 2010;41:15–6.
13. Conway JD. Autograft and nonunions: morbidity with intramedullary bone graft versus iliac crest bone graft. Orthop Clin North Am. 2010;41:75–84.
14. Dick W. Use of the acetabular reamer to harvest autogenic bone graft material: a simple method for producing bone paste. Arch Orthop Trauma Surg. 1986;185:225–8.
15. McCall TA, Brokaw DS, Jelen BA, et al. Treatment of large segmental bone defects with reamer-irrigator-aspirator bone graft: technique and case series. Orthop Clin North Am. 2010;41:63–73.
16. Blum B, Moseley J, Miller L, et al. Measurement of bone morphogenetic proteins and other growth factors in demineralized bone matrix. Orthopedics. 2004;27(1):161–5.
17. Hernigou P, Poignard A, Buejean F, et al. Percutaneous autologous bone-marrow grafting for nonunions. Influence of number and concentration of progenitor cells. J Bone Joint Surg Am. 2005;87(7):1490–7.
18. Janhangir AA, Nunley RM, Mehta S, et al. Bone graft substitutes in orthopaedic surgery. J Am Acad Orthop Surg. 2008;2(1):35–7.
19. Jones AL, Bucholz RW, Bosse MJ, et al. Recombinant human BMP-2 and autogenous bone graft for reconstruction of diaphyseal tibial fractures with cortical defects. A randomized, controlled trial. J Bone Joint Surg Am. 2006;88(7):1431–41.
20. Meinig RP. Clinical use of resorbable polymeric membranes in the treatment of bone defects. Orthop Clin North Am. 2010;41:39–47.
21. Gugala Z, Lindsey RW, Gogolewski S. New approaches in the treatment of critical-size segmental defects in long bones. Macromol Symp. 2007;253:147–61.

Acute Soft Tissue and Bone Infections

25

Lena M. Napolitano

Contents

25.1	Introduction	351
25.2	Classification of SSTIs	351
25.3	Specific Types of SSTIs	353
25.3.1	Traumatic Wound Infections	353
25.3.2	Surgical Site Infections (SSIs)	353
25.3.3	Closed Long Bone Fractures	355
25.3.4	Open Fractures	356
25.3.5	Necrotizing Soft Tissue Infections (NSTIs)	357
25.3.6	Pyomyositis	359
25.3.7	Osteomyelitis	359
25.3.8	Microbiology of Osteomyelitis	361
25.4	Four Important Steps in SSTI Treatment	361
25.4.1	Epidemiology and Microbiology of SSTIs	363
Conclusion		366
References		367

L.M. Napolitano, MD, FACS, FCCP, FCCM
Acute Care Surgery [Trauma, Burns, Critical Care, Emergency Surgery], Trauma and Surgical Critical Care, Department of Surgery, University of Michigan Health System, Room 1C421, University Hospital, 1500 East Medical Drive, Ann Arbor, MI 48109-0033, USA
e-mail: lenan@umich.edu

25.1 Introduction

Skin and soft tissue infections (SSTIs) span a broad spectrum of clinical entities from limited cellulitis or small abscess to rapidly progressive necrotizing fasciitis, which may be associated with septic shock or toxic shock syndrome [1–5]. Severe and complicated SSTIs may result in critical illness and require management in the intensive care unit [6]. The complex interplay of environment, host, and pathogen are important to consider when evaluating SSTIs and planning therapy. The key to a successful outcome in caring for patients with severe SSTIs is (1) early diagnosis and differentiation of necrotizing vs. non-necrotizing SSTI, (2) early initiation of appropriate empiric broad-spectrum antimicrobial therapy with consideration of risk factors for specific pathogens, (3) "source control," that is, early aggressive surgical intervention for drainage of abscesses and debridement of necrotizing soft tissue infections, and (4) pathogen identification and appropriate de-escalation of antimicrobial therapy (Table 25.1).

25.2 Classification of SSTIs

The US Food and Drug Administration (FDA) previously classified SSTIs into two broad categories for the purpose of clinical trials evaluating new antimicrobials for the treatment of SSTIs:

Table 25.1 Steps in optimal management of patients with severe SSTIs

1. Early diagnosis and differentiation of necrotizing vs. non-necrotizing SSTI
2. Early initiation of appropriate empiric broad-spectrum antimicrobial therapy with anti-MRSA coverage and consideration of risk factors for specific pathogens
3. "Source control" of SSTI (i.e., early aggressive surgical intervention for drainage of abscesses and debridement of necrotizing soft tissue infections)
4. Pathogen identification and appropriate de-escalation of antimicrobial therapy

Table 25.2 Comparison of old and new classification of SSTIs by FDA

Uncomplicated	Complicated

Superficial infections, such as:
  Simple abscesses
  Impetiginous lesions
  Furuncles
  Cellulitis
Can be treated by antibiotics or surgical incision alone
	Deep soft tissue, such as:
	  Infected ulcers
	  Infected burns
	  Major abscesses
	Significant underlying disease state which complicates response to treatment
	Requires significant surgical intervention and antimicrobials

New FDA definition (October 2013):

Acute bacterial skin and skin structure infections (ABSSSI) defined as bacterial infection of the skin with a lesion size area of at least 75 cm^2 (lesion size measured by the area of redness, edema, or induration), including the following:

Cellulitis/erysipelas: A diffuse skin infection characterized by spreading areas of redness, edema, and/or induration

Wound infection: An infection characterized by purulent drainage from a wound with surrounding redness, edema, and/or induration

Major cutaneous abscess: An infection characterized by a collection of pus within the dermis or deeper that is accompanied by redness, edema, and/or induration

From: http://www.fda.gov/ohrms/dockets/98fr/2566dft.pdf

uncomplicated and *complicated* (Table 25.2). *Uncomplicated* SSTIs include superficial infections such as cellulitis, simple abscesses, impetigo, and furuncles. These infections can be treated by antibiotics and/or surgical incision for drainage of abscess alone. In contrast, *complicated* SSTIs include deep soft tissue infections that require significant surgical intervention, such as infected ulcers, infected burns, and major abscesses, and these patients also have significant underlying comorbidities, that is, disease states which complicate (and usually delay) response to treatment. Complicated SSTIs are a significant clinical problem, in part related to the increasing resistance of infecting bacteria to our current antibiotic therapies.

- *Uncomplicated SSTIs* are associated with *low* risk for life- or limb-threatening infection. These patients can be treated with empiric antibiotic therapy according to likely pathogen and local resistance patterns.
- *Complicated SSTIs* are associated with *high* risk for life- or limb-threatening infection. In these patients, it is of paramount importance to initiate appropriate and adequate broad-spectrum initial empiric antimicrobial therapy with coverage for MRSA and to consider the need for surgical intervention for abscess drainage or debridement.

Patients with complicated SSTIs require hospitalization for treatment. Specific circumstances that warrant hospitalization include the presence of tissue necrosis, sepsis, severe pain, altered mental status, immunocompromised state, and organ failure (respiratory, renal, and hepatic). SSTIs can lead to serious potentially life-threatening local and systemic complications. The infections can progress rapidly and early recognition and proper medical and surgical management is the cornerstone of therapy. A recent prospective observational US study of complicated SSTI patients (*n* = 1033) determined that the mean length of hospital stay was 7.1 days, 41.2 % underwent surgical procedures related to the study infection, the most common class of initial intravenous antibiotic prescribed was vancomycin, and the hospital mortality rate was 0.4 % [7]. In contrast, a similar study in Europe reported mean hospital length of stay of 18.5 days with a mortality rate of 3.4 % [8].

In October 2013, FDA changed the SSTI terminology and issued final guidance for the treatment of acute bacterial skin and skin structure infections (ABSSSI) [9]. This guidance defined

ABSSSI as cellulitis, erysipelas, wound infection, and major cutaneous abscess. An ABSSSI is defined as a bacterial infection of the skin with a lesion size area of at least 75 cm^2 (lesion size measured by the area of redness, edema, or induration). The minimum area of involvement of 75 cm^2 is chosen to select patients with acute bacterial skin infections for which a reliable control drug treatment effect can be estimated for the conduct of new antimicrobial treatment trials. While the FDA generally requires two Phase III trials to support approval of drugs to treat acute bacterial skin and skin structure infections (ABSSSI), this guidance stated that a single Phase III study that is supported by additional independent evidence may suffice.

Patients with the following infection types can be enrolled in ABSSSI clinical trials:

- *Cellulitis/erysipelas*: A diffuse skin infection characterized by spreading areas of redness, edema, and/or induration
- *Wound infection*: An infection characterized by purulent drainage from a wound with surrounding redness, edema, and/or induration
- *Major cutaneous abscess*: An infection characterized by a collection of pus within the dermis or deeper that is accompanied by redness, edema, and/or induration

Unfortunately, this new guidance does not address less serious skin infections, such as impetigo and minor cutaneous abscess, or more serious infections needing more complex treatment regimens, such as infections resulting from animal or human bites, necrotizing soft tissue infections, diabetic foot infection, decubitus ulcer infection, myonecrosis, osteomyelitis, and ecthyma gangrenosum.

25.3 Specific Types of SSTIs

25.3.1 Traumatic Wound Infections

A recent report from the Lower Extremity Assessment Project (LEAP), a multi-institutional prospective observational study of 545 patients with limb-threatening lower extremity trauma with 2-year follow-up at eight Level-1 trauma centers, documented that wound infection (34 %) was the most common complication in the primary amputation group, and that nonunion (31.5 %) and wound infection (23.2 %) were the most common complications in the limb salvage group. Furthermore, the late amputation group had the highest complication rate (68 %), mostly due to wound infection [10]. When traumatic wound infections occur, it is recommended to initiate early empiric broad-spectrum antibiotic therapy to cover methicillin-resistant *S. aureus* (MRSA) and all other potential pathogens, obtain wound cultures, and then tailor definitive antimicrobial therapy once the culture results return. In addition, the wound may require surgical debridement to provide adequate source control.

25.3.2 Surgical Site Infections (SSIs)

SSIs are one of the most common SSTIs that occur in orthopedic and trauma care. SSIs are defined as "superficial incisional" or "deep incisional" SSI based on the depth of the infection as defined by the Centers for Disease Control (CDC) and the National Healthcare Safety Network (NHSN) (Table 25.3).

It has recently been determined that scores commonly used to predict SSI in other types of surgery [National Nosocomial Infection Surveillance System (NNIS) and Study on the Efficacy of Nosocomial Infection Control (SENIC) scores] are not predictive of SSI after orthopedic fracture surgery [11]. A new score [RIOTS Composite Score includes fractures classification AO type C3 or Sanders type 4, 2 points; BMI < 30 kg/m^2, 1 point; ASA class ≥ 3, 1 point] was proposed for SSI prediction in orthopedic fracture surgery that incorporates fracture classification, American Society of Anesthesiologists classification, and body mass index with area under the ROC curve of 0.75, significantly higher than NNIS and SENIC scores.

25.3.2.1 SSI Prevention

A number of SSI prevention strategies have significantly decreased the rate of SSIs following orthopedic surgery and fracture repair in the past

Table 25.3 CDC/NHSN classification of surgical site infections (SSIs)

Type of SSI	Definition
Superficial incisional	Infection occurs within 30 days after the operative procedure and involves only skin and subcutaneous tissue of the incision and patient has at least one of the following: (a) Purulent drainage from the superficial incision (b) Organisms isolated from an aseptically obtained culture of fluid or tissue from the superficial incision (c) At least one of the following signs or symptoms of infection: pain or tenderness, localized swelling, redness, or heat, and superficial incision is deliberately opened by surgeon and is culture positive or not cultured. A culture-negative finding does not meet this criterion (d) Diagnosis of superficial incisional SSI by the surgeon or attending physician
Deep incisional	Infection occurs within 30 days after the operative procedure if no implant[a] is left in place or within 1 year if implant is in place and the infection appears to be related to the operative procedure and involves deep soft tissues (e.g., fascial and muscle layers) of the incision and patient has at least one of the following: (a) Purulent drainage from the deep incision but not from the organ/space component of the surgical site (b) A deep incision spontaneously dehisces or is deliberately opened by a surgeon and is culture positive or not cultured when the patient has at least one of the following signs or symptoms: fever (.388 C), or localized pain or tenderness. A culture-negative finding does not meet this criterion (c) An abscess or other evidence of infection involving the deep incision is found on direct examination, during reoperation, or by histopathologic or radiologic examination (d) Diagnosis of a deep incisional SSI by a surgeon or attending physician

From Horan et al. [102]
[a]Implant: A nonhuman-derived object, material, or tissue (e.g., prosthetic heart valve, nonhuman vascular graft, mechanical heart, or hip prosthesis) that is permanently placed in a patient during an operative procedure and is not routinely manipulated for diagnostic or therapeutic purposes

decade [12]. The Surgical Care Improvement Project (SCIP) has implemented three measures for antibiotic prophylaxis for SSI prevention: (1) antibiotic received within 1 h prior to surgical incision, (2) appropriate antibiotic selection based on surgical procedure performed, and (3) antibiotic discontinued within 24 h after surgery completed (Table 25.4). Additional evidence-based strategies for SSI prevention include the following: (1) appropriate hair removal (clipping, no shaving); (2) maintenance of normothermia intraoperatively and perioperatively; (3) glycemic control; (4) appropriate skin preparation; (5) supplemental oxygen administration.

25.3.2.2 Microbiology of SSIs

Staphylococcus aureus (*S. aureus*) is the most common causative pathogen for all SSIs in the United States data reported by the NHSN (Table 25.5), and an increasing percentage of these *S. aureus* isolates are methicillin-resistant (MRSA). Comparison of the causative pathogens for SSI in US hospitals documents that *S. aureus* increased from 22.5 % (1986–2003) to 30 %

Table 25.4 Antibiotics for SSI prevention in orthopedic surgery

Choice of antimicrobial agent
Cefazolin
If ß-lactam allergy, use clindamycin or vancomycin
Consider preoperative screening for MRSA colonization
If infected or colonized with MRSA, use vancomycin
Timing of administration
Start up to 60 min before incision: Cefazolin, Clindamycin
Start up to 120 min before incision: Vancomycin
Infusion completed 10 min before tourniquet inflation
Dosing
Cefazolin, 1–2 g (2 g for patient weighing >80 kg)
Vancomycin (15 mg/kg) and Clindamycin (600–900 mg) dosing based on patient mass
Pediatric dosing based on patient mass
Duration of antimicrobial use
Single preoperative dose
Redose antimicrobial intraoperatively for prolonged procedure or significant blood loss
Mupirocin should be given intranasally to all patients with documented colonization with *S. aureus*

Adapted from: Bratzler et al. [21]

Table 25.5 Causative pathogens for surgical site infections (SSI) in US hospitals 2006–2007, National Healthcare Safety Network

Organism	SSIs from all types of surgeries No. (%) of SSIs Total $n=7025$	SSIs from orthopedic surgeries No. (%) of SSIs Total $n=963$
Staphylococcus aureus	2108 (30.0 %)	548 (48.6 %)
Methicillin-sensitive (MSSA)	*1102 (50.8 %)*	
Methicillin-resistant (MRSA)	*1006 (49.2 %)*	
Coagulase-negative staphylococci	965 (13.7 %)	173 (15.3 %)
Enterococcus spp.	788 (11.2 %)	104 (10.8 %)
Escherichia coli	671 (9.6 %)	34 (3.0 %)
Pseudomonas aeruginosa	390 (5.6 %)	38 (3.4 %)
Enterobacter spp.	293 (4.2 %)	37 (3.3 %)
Klebsiella spp.	213 (3.0 %)	19 (2.0 %)

Adapted from Hidron et al. [14]

(2006–2007), with MRSA now the leading causative pathogen, comprising 49.2 % of all isolates [13, 14]. The advent of community-associated MRSA (CA-MRSA) has impacted SSI significantly. Recent studies document that CA-MRSA is replacing traditional health care-associated or nosocomial MRSA strains in SSI among inpatients [15]. CA-MRSA has emerged as a leading cause of health care-associated infections among patients with prosthetic joint SSIs [16].

In a study of 8302 patients readmitted to US hospitals from 2003 to 2007 with culture-confirmed SSI, the proportion of infections caused by MRSA increased significantly, from 16.1 to 20.6 %, and these infections were associated with higher mortality rates, longer stays, and higher hospital costs [17]. In view of this important finding, some surgeons have advocated strongly that patients be screened for nasal carriage of MRSA prior to elective surgery, with consideration of decolonization prior to surgery, and modification of antimicrobial agents for SSI prevention on the basis of the results.

Interestingly, when evaluating the microbiology of SSIs related to orthopedic surgical cases, S. aureus comprised an even greater percentage of isolates (48.6 %) when compared to isolates reported for SSIs from all surgical cases (30 %) (Table 25.5). Although knowledge of national microbiology of SSIs related to specific surgical procedures is important, it is of even greater importance to know the microbiology of SSIs within your own institution, and this should help to guide empiric antimicrobial management for treatment of SSIs in your local setting. Reports of resistant gram-negative isolates, particularly multi-drug-resistant *Enterobacter* isolates producing extended spectrum beta-lactamases (ESBLs), as the etiology of SSIs in orthopedic and trauma surgery is worrisome [18, 19]. This highlights the importance of pathogen identification, that is, obtaining material for gram stain and culture, in the management of all SSIs.

25.3.3 Closed Long Bone Fractures

A Cochrane Database systematic review of patients undergoing surgery for proximal femoral and other closed long bone fractures (data from 8447 participants in 23 studies) documented that single dose antibiotic prophylaxis significantly reduced deep incisional SSI (risk ratio 0.40, 95 % CI 0.24–0.67), superficial incisional SSI, urinary infections, and respiratory tract infections. Multiple dose antibiotic prophylaxis had an effect of similar size on deep incisional SSI. Therefore, appropriate antibiotic prophylaxis should be used in all patients undergoing surgical management of hip or other closed long bone fractures [20].

The recent publication of the "Clinical practice guidelines for antimicrobial prophylaxis in surgery" by the American Society of Health-System Pharmacists (ASHP), Infectious Diseases Society of America (IDSA), Surgical Infection Society (SIS), and Society for Healthcare

Epidemiology of America (SHEA) provides evidence-based national recommendations [21].

The recommended regimen in hip fracture repair or other orthopedic procedures involving internal fixation is cefazolin. Clindamycin and vancomycin should be reserved as alternative agents. If there are surveillance data showing that gram-negative organisms are a cause of SSIs for the procedure, practitioners may consider combining clindamycin or vancomycin with another agent (cefazolin if the patient is not β-lactam-allergic; aztreonam, gentamicin, or single-dose fluoroquinolone if the patient is β-lactam-allergic). Mupirocin should be given intranasally to all patients with documented colonization with *S. aureus*. (Strength of evidence for prophylaxis = A.)

25.3.4 Open Fractures

Antibiotics reduce the incidence of early infections in open fractures of the limbs, confirmed by a Cochrane Database systematic review of 913 participants in 7 studies. The use of antibiotics had a protective effect against early infection compared with no antibiotics or placebo (relative risk 0.41 [95 % confidence interval (CI) 0.27–0.63]; absolute risk reduction 0.08 (95 % CI 0.04–0.12); number needed to treat (NNT) 13 (95 % CI 8–25). There were insufficient data in the included studies to evaluate other outcomes [22]. The Surgical Infection Society evidence-based guidelines for prophylactic antibiotic use in open fractures recommend the use of a short course of first-generation cephalosporins, begun as soon as possible after injury, in addition to modern orthopedic fracture wound management (Table 25.6) [23]. Open fracture grade (Gustilo) and the degree of associated soft-tissue injury are independent determinants of infection risk. A recent single-institution review of patients with Gustilo IIIB tibial fractures ($n=52$) determined that nosocomial bacterial pathogens (Enterococci, Pseudomonas, Enterobacter, and MRSA) were responsible for deep tissue infections, and advocated for tailoring antimicrobial prophylaxis against nosocomial organisms at the time of definitive wound closure [24].

Table 25.6 Risk of SSTI in adult trauma patients with open extremity fractures and antimicrobial prophylaxis recommendations

Grade of open fracture	Characteristics of Gustilo grade open fracture	Infection rate	Amputation rate
Grade I	Clean wound smaller than 1 cm in diameter, simple fracture pattern, no skin crushing	0–2 %	0 %
Grade II	A laceration larger than 1 cm but without significant soft tissue crushing, including no flaps, degloving, or contusion. Fracture pattern may be more complex	2–7 %	0 %
Grade III	An open segmental fracture or a single fracture with extensive soft tissue injury. Also included are injuries older than 8 h. Type III injuries are subdivided into three types:		
Grade III A	Adequate soft tissue coverage of the fracture despite high energy trauma or extensive laceration or skin flaps	5–10 %	2.5 %
Grade III B	Inadequate soft tissue coverage with periosteal stripping. Soft tissue reconstruction is necessary	10–50 %	5.6 %
Grade III C	Any open fracture that is associated with an arterial injury that requires repair	25–50 %	25 %

Grade of open fracture	Recommended antibiotic	Alternate if PCN allergy
Grade I or II	Kefzol 1–2 g load then 1 g IV q8h for 48 h	Clindamycin 900 mg IV q8h for 48 h
Grade III	Ceftriaxone 1 g IV q24h for 48 h	Clindamycin 900 mg IV q8h and Aztreonam 1 g IV q8h for 48 h

Hauser et al. [23]; Luchette et al. [103]; Okike and Bhattachyaryya [104]; Holtom [105]; Gustilo and Anderson [106]

25.3.5 Necrotizing Soft Tissue Infections (NSTIs)

NSTIs are aggressive soft tissue infections that cause widespread necrosis, and can include necrotizing cellulitis, fasciitis, and myositis/myonecrosis [25, 26]. Establishing the diagnosis of NSTI can be the main challenge in treating patients with NSTI, and knowledge of all available tools is the key for early and accurate diagnosis (Table 25.7) [27]. There have been a number of recent advances in the definition, pathogenesis, diagnostic criteria, and treatment of necrotizing soft tissue infections [28, 29].

Patients with NSTIs require prompt aggressive surgical debridement, appropriate intravenous antibiotics, and intensive support. Despite aggressive treatment, their mortality and morbidity rates remain high, with some series reporting mortality rates of 25–35 % [30]. A high index of suspicion should be used in conjunction with laboratory and imaging studies to establish the diagnosis as rapidly as possible. Successful treatment requires early, aggressive surgical debridement of all necrotic tissue, appropriate broad-spectrum systemic antibiotic therapy, and supportive care (fluid resuscitation, organ and critical care support) to maintain oxygenation and tissue perfusion. Delayed definitive debridement remains the single most important risk factor for death.

A recent single-institution series of 166 patients documented that the overall mortality rate was 16.9 % and limb loss occurred in 26 % of patients with extremity involvement [31]. Independent predictors of mortality included white blood cell count greater than $30,000 \times 10^3$/µL, creatinine level greater than 2 mg/dL (176.8 µmol/L), and heart disease at hospital admission. Independent predictors of limb loss included heart disease and shock (systolic blood pressure <90 mmHg) at hospital admission. Clostridial infection was an independent predictor for both limb loss (odds ratio, 3.9 [95 % confidence interval, 1.1–12.8]) and mortality (odds ratio, 4.1 [95 % confidence interval, 1.3–12.3]) and was highly associated with intravenous drug use and a high rate of leukocytosis on hospital admission.

A 30-day postoperative mortality risk calculator for NSTI was recently developed using the National Surgery Quality Improvement Project (NSQIP) which identified seven independent variables that correlated with mortality: age older than 60 years (odds ratio [OR]=2.5; 95 % CI 1.7–3.6), functional status (partially dependent: OR=1.6; 95 % CI 1.0–2.7; totally dependent: OR=2.3; 95 % CI 1.4–3.8), requiring dialysis (OR=1.9; 95 % CI 1.2–3.1), American Society of Anesthesiologists class 4 or higher (OR=3.6; 95 % CI 2.3–5.6), emergent surgery (OR=1.6; 95 % CI 1.0–2.3), septic shock (OR=2.4; 95 % CI 1.6–3.6), and low platelet count (<50 K/µL: OR=3.5; 95 % CI 1.6–7.4; <150 K/µL but >50 K/µL: OR=1.9; 95 % CI 1.2–2.9). The receiver operating characteristic area was 0.85 (95 % CI 0.82–0.87), which indicated a strong predictive model that can aid physicians in the decision-making process [32].

25.3.5.1 Aids to Diagnosis of NSTIs

Early operative debridement is a major determinant of outcome in NSTIs. However, early recognition of NSTIs is difficult clinically. A novel diagnostic scoring system for distinguishing NSTIs from other severe soft tissue infections based on laboratory tests routinely performed for

Table 25.7 Clinical clues to the diagnosis of necrotizing soft tissue infections

Skin findings:	Erythema
	Tense edema
	Gray or discolored wound drainage
	Vesicles or bullae
	Skin necrosis
	Ulcers
	Crepitus
Systemic features:	Severe pain out of proportion to physical findings
	Pain that extends past margin of apparent skin infection
	Fever
	Tachycardia, tachypnea
	Diaphoresis
	Delirium

the evaluation of severe SSTIs is called the Laboratory Risk Indicator for Necrotizing Fasciitis (LRINEC) score (Table 25.8) [33].

The LRINEC score was initially developed in a retrospective observational study including 145 patients with necrotizing fasciitis and 309 patients with severe cellulitis or abscesses admitted to the two tertiary care hospitals. The cutoff value for the LRINEC score was 6 points with a positive predictive value of 92.0 % and negative predictive value of 96.0 %. The LRINEC score is a robust score capable of detecting even clinically early cases of necrotizing fasciitis. The variables used are routinely measured to assess severe soft tissue infections. Patients with a LRINEC score of ≥6 should be carefully evaluated for the presence of necrotizing fasciitis.

Since the initial development of the LRINEC score, a number of other cohort studies have validated its utility in the diagnosis of NSTIs [34]. A multicenter study in 229 patients with NSTIs from 2002 to 2005 reported an overall mortality rate of 15.8 % and amputation rate of 26.3 %. This study also documented that a LRINEC score ≥6 was associated with a higher rate of both mortality and amputation [34].

25.3.5.2 Diagnostic Imaging in NSTIs

A high clinical index of suspicion is required if the diagnosis is to be made sufficiently early for successful treatment. NSTIs necessitate prompt aggressive surgical debridement for satisfactory treatment in addition to antimicrobial therapy. It is critical to remember that because of the rapidly progressive and potentially fatal outcome of this condition, if imaging cannot be performed expeditiously, delaying treatment is not justified. Plain film findings may reveal extensive soft tissue gas. CT examination can reveal asymmetric thickening of deep fascia in association with gas, and associated abscesses may also be present. MR imaging can also assist in the diagnosis of NSTIs [35]. MR imaging has been documented to effectively differentiate between necrotizing and non-necrotizing infection of the lower extremity and other areas of the body, but should not delay prompt surgical intervention in NSTIs management [36–38].

25.3.5.3 Microbiology of NSTIs

Necrotizing fasciitis and myonecrosis are typically caused by infection with Group A streptococcus, *Clostridium perfringens*, or, most commonly, aerobic and anaerobic organisms as part of a polymicrobial infection that may include *S. aureus*. In case series, CA-MRSA has recently been described as a predominantly monomicrobial cause of necrotizing fasciitis [39, 40]. A retrospective review of patients presenting with necrotizing fasciitis between 2000 and 2006 indicated that MRSA was the most common pathogen, accounting for one-third of the organisms isolated [41].

NSTIs have been classified into two types, either polymicrobial (Type I) or monomicrobial (Type II). Polymicrobial infections are more common, due to both aerobic and anaerobic organisms, and commonly occur in the trunk and

Table 25.8 Laboratory risk indicator for necrotizing fasciitis (LRINEC) score

Variable, units	Score
C-reactive protein, mg/L	
<150	0
≥150	4
Total white cell count, per mm³	
<15	0
15–25	1
>25	2
Hemoglobin, g/dL	
>13.5	0
11–13.5	1
<11	2
Sodium, mmol/L	
≥135	0
<135	2
Creatinine, μmol/L	
≤141	0
>141	2
Glucose, mmol/L	
≤10	0
>10	1

The maximum score is 13; a score ≥6 should raise the suspicion of necrotizing fasciitis and a score of ≥8 is strongly predictive of this disease

perineum. NSTIs that are monomicrobial in origin commonly occur in the limbs and are typically caused by infection with Group A streptococcus, *Clostridium perfringens*, or *S. aureus*. NSTIs are categorized into these two specific types based on the microbiologic etiology of the infection, and this classification does impact on the specific antimicrobial agents required for treatment of these NSTIs.

- *Type 1, or polymicrobial*
- *Type 2, or monomicrobial*

Increasingly, MRSA has been identified as the causative microbe in NSTIs, but a separate category for this NSTI does not currently exist [42–46]. Given this finding, anti-MRSA empiric antimicrobial therapy should be initiated in all patients with NSTIs and pathogen-directed antimicrobial therapy considered once tissue culture results are available.

Uncommon microbiologic causes of NSTIs and primary sepsis include *Vibrio* and *Aeromonas* species, virulent gram-negative bacteria, and members of the Vibrionaceae family that thrive in aquatic environments [47]. These NSTIs are likely to occur in patients with hepatic disease, diabetes, and immunocompromised conditions [48]. These organisms are found in warm sea waters and are often present in raw oysters, shellfish, and other seafood. The diagnosis of *Vibrio* NSTIs should be suspected when a patient has the appropriate clinical findings and a history of contact with seawater or raw seafood [49]. Early fasciotomy and culture-directed antimicrobial therapy should be aggressively performed in those patients with hypotensive shock, leukopenia, severe hypoalbuminemia, and underlying chronic illness, especially a combination of hepatic dysfunction and diabetes mellitus. The rate of amputation and mortality is very high in these patients, and early definitive management is of paramount importance [50–52].

A recent study of 125 patients identified that a LRINEC score of 2 or greater and the presence of hemorrhagic bullous/blistering lesions in patients with *Vibrio vulnificus* SSTI are associated with an 11.9-fold increased risk for the presence of NSTI and necrotizing fasciitis [53].

25.3.6 Pyomyositis

Myositis is a rare infection that may lead to serious and potentially life-threatening local and systemic complications [54]. The infection can progress rapidly, and early recognition and proper medical and surgical management is therefore the cornerstone of therapy. With the increasing prevalence of community-associated MRSA as a pathogen in severe SSTIs, pyomyositis is more common than in past years. Myositis often occurs in muscle sites that have been compromised by injury, ischemia, malignancy, or surgery. The predominant pathogens are *S. aureus*, Group A streptococci (GAS), gram-negative aerobic and facultative bacilli, and the indigenous aerobic and anaerobic cutaneous and mucous membranes local microflora.

CT scan imaging is a rapid and sensitive diagnostic test and commonly demonstrates diffuse enlargement of the involved muscle and may demonstrate the presence of fluid or gas collections within the muscle suggesting the presence of abscesses. MRI is more sensitive in showing early inflammatory changes prior to development of abscesses in myositis [55]. Emergency surgical exploration is warranted in order to define the nature of the infective process which is accomplished by direct examination of the involved muscles. Surgical intervention is required to perform appropriate abscess drainage and debridement and to also evaluate for necrotizing myositis. Fasciotomies and extremity amputation are sometimes necessary.

25.3.7 Osteomyelitis

Bone and joint infections are challenging to diagnose and treat [56]. The key to successful management is early diagnosis. The recommended algorithm for diagnosis and treatment of acute osteomyelitis is presented in Fig. 25.1. Bone

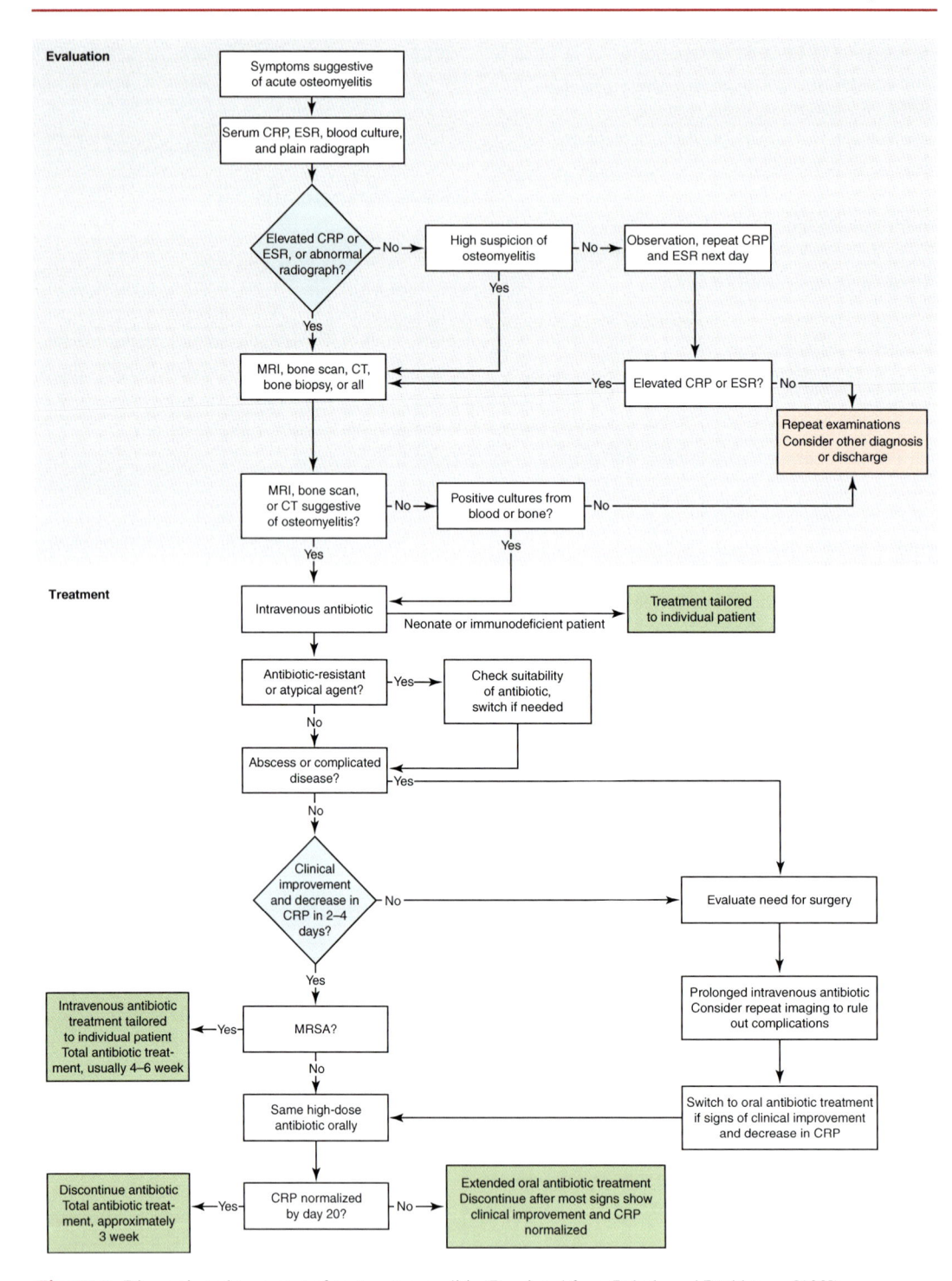

Fig. 25.1 Diagnosis and treatment of acute osteomyelitis (Reprinted from Peltola and Pääkkönen [100])

sampling may be required for microbiological and pathological examination to allow targeted appropriate antimicrobial therapy. There are three types of *acute* osteomyelitis (in order of decreasing frequency):

1. Osteomyelitis secondary to a contiguous focus of infection (after trauma, surgery, or insertion of a joint prosthesis)
2. Osteomyelitis secondary to vascular insufficiency (in diabetic foot infections or peripheral vascular disease)
3. Osteomyelitis secondary to hematogenous origin

The rate of osteomyelitis following severe limb-threatening lower extremity trauma reported in the LEAP study was 9.4 % in the total study cohort of 330 patients. The rates of osteomyelitis ranged from 3.1 % in the primary amputation group to the highest rate of 27.3 % in patients with Grade IIIC tibia fracture [57].

Acute osteomyelitis is treated with antibiotics and careful assessment of any associated wound to determine if the soft tissue and wound require infection source control by surgical debridement. In contrast, *chronic* osteomyelitis is associated with avascular necrosis of bone and formation of sequestrum (dead bone), and surgical debridement is necessary for cure in addition to antimicrobial therapy, with 6 weeks of parenteral antibiotic therapy recommended. However, oral antibiotics that achieve adequate levels in bone are now available, and similar cure rates have been achieved with oral and parenteral antimicrobial therapies [58].

25.3.8 Microbiology of Osteomyelitis

The most common pathogenic microorganism in any type of osteomyelitis is *S. aureus*, either susceptible (MSSA) or resistant (MRSA) to methicillin and coagulase-negative staphylococci are common in foreign-body-associated osteomyelitis. The ability of *S. aureus* to adhere is thought to be crucial for the early colonization of host tissues and implanted biomaterials.

25.4 Four Important Steps in SSTI Treatment

1. Early diagnosis and differentiation of necrotizing vs. non-necrotizing SSTI

A classification for SSTIs that is commonly used is the differentiation of *necrotizing soft tissue infections* (*NSTIs*) from *non-necrotizing infections*. This differentiation is critical since necrotizing infections warrant prompt aggressive surgical debridement. Clinical clues to the diagnosis of NSTIs are listed in Table 25.7. The differentiation of necrotizing infections from non-necrotizing infections is critical to achieving adequate surgical therapy [59]. A clear approach to these infections must allow rapid identification and treatment of NSTIs because they are limb-threatening and life-threatening.

When clinical "hard clinical signs" (bullae, crepitus, gas on x-ray, hypotension with SBP <90 mmHg, or skin necrosis) of NSTI are present, establishing the diagnosis of NSTI is not difficult. However, hard signs of NSTIs are often absent on presentation, thus potentially delaying diagnosis and surgical intervention. Studies have documented that less than 50 % of patients with a definitive diagnosis of NSTI presented with "hard clinical signs" of NSTI [60]. Admission white blood cell count >15,400×10^9/L and/or serum sodium <135 mEq/L was documented to help differentiate NSTI from non-NSTI and aided in early diagnosis [61, 62]. The Laboratory Risk Indicator for Necrotizing Fasciitis (LRINEC) score is also helpful as a laboratory aid in distinguishing necrotizing from non-necrotizing SSTIs (see NSTI section above).

If there is any question regarding the possible diagnosis of a NSTI, it is imperative to proceed with surgical intervention and to be certain that the surgical incision is continued down to the fascial and muscle level to make a definitive diagnosis.

2. Early initiation of appropriate empiric broad-spectrum antimicrobial therapy with anti-MRSA coverage and consideration of risk factors for specific pathogens

Antimicrobial therapy is an essential element in the management of severe SSTIs. As in all serious life-threatening infections, it is important to initiate *early* and *appropriate* empiric antimicrobial therapy. It is well established that prompt appropriate treatment of hospitalized infections reduces mortality [63]. Similar findings were reported in studies of patients with ventilator-associated pneumonia [64] and sepsis [65]. A study of ICU patients found that the higher mortality rate associated with inappropriate initial therapy is still observed when antibiotics are switched from an inappropriate to an appropriate treatment [66].

Furthermore, appropriate and timely antibiotic therapy improves treatment outcomes for SSTIs caused by methicillin-resistant *Staphylococcus aureus* (MRSA) [67]. In a study of 492 patients with community-onset MRSA SSTIs, 95 % of episodes treated with an active antibiotic within 48 h were treated successfully, compared with an 87 % rate of successful treatment in patients who did not receive an active antibiotic ($P=0.001$). In logistic regression analysis, failure to initiate active antimicrobial therapy within 48 h of presentation was the only independent predictor of treatment failure (adjusted OR, 2.80; 95 % CI, 1.26–6.22; $P=0.011$). Similarly, in a study of patients admitted to the hospital with MRSA sterile-site infection, multivariate analysis found inappropriate antimicrobial treatment to be an independent risk factor for hospital mortality (adjusted OR, 1.92; 95 % CI, 1.48–2.50; $P=0.013$) [68].

An empiric treatment algorithm for SSTI directed against CA-MRSA in the emergency department that promotes both the use of antibiotics likely active against CA-MRSA and early incision and drainage of abscesses was examined. Clinical failure occurred in only 3 % of cases treated according to the algorithm, compared with 62 % of those not treated according to the algorithm ($p<0.001$). Furthermore, among cases that underwent immediate incision and drainage, initial treatment with antibiotics active in vitro against the MRSA isolate was associated with a decreased clinical failure rate when compared to those treated with inactive antibiotics 0 % vs. 67 %, $p<0.001$) [69].

Empiric antibiotic therapy should be initiated in *all* patients with cSSTIs. Intravenous broad-spectrum antimicrobial therapy should be initiated when an infection is severe or progresses rapidly, when there are signs of systemic illness, when the patient has comorbidities or is immunosuppressed, for very old or young patients, when an abscess cannot be completely drained, and when the infection does not respond to incision and drainage [70].

Timely initiation of antimicrobial therapy is also important in the treatment of severe SSTIs, particularly if associated with septic shock [71, 72]. In a study of 2731 adult patients with septic shock, a strong relationship between the delay in effective antimicrobial initiation and in-hospital mortality was noted (adjusted odds ratio 1.119 [per hour delay], 95 % confidence interval 1.103–1.136, $p<.0001$) [73]. Administration of an antimicrobial effective for isolated or suspected pathogens within the first hour of documented hypotension was associated with a survival rate of 79.9 %. Each hour of delay in antimicrobial administration over the ensuing 6 h was associated with an average decrease in survival of 7.6 %. By the second hour after onset of persistent/recurrent hypotension, in-hospital mortality rate was significantly increased relative to receiving therapy within the first hour (odds ratio 1.67; 95 % confidence interval, 1.12–2.48). In multivariate analysis (including Acute Physiology and Chronic Health Evaluation II score and therapeutic variables), time to initiation of effective antimicrobial therapy was the single strongest predictor of outcome. Interestingly, only 50 % of septic shock patients received effective antimicrobial therapy within 6 h of documented hypotension.

25.4.1 Epidemiology and Microbiology of SSTIs

An understanding of the changing epidemiology and microbiology of all SSTIs is required for diagnosis and selection of appropriate empiric antibiotic therapy. Staphylococci and streptococci have long been the leading microbiologic causes of cSSTIs [74]. In recent years, however, *S. aureus* has emerged as the most common cause of SSTIs. In addition to Group A streptococci and *S. aureus*, the indigenous aerobic and anaerobic cutaneous and mucous membranes local microflora usually is responsible for polymicrobial infections, such as NSTIs and diabetic foot infections. Severe SSTIs can also be due to *Clostridium* spp., microorganisms associated with water sources (*Vibrio* spp., *Aeromonas*), and polymicrobial/mixed infections.

Community-associated MRSA (CA-MRSA) infections have risen rapidly in the last decade, and SSTIs are the predominant site of infection, accounting for 74 % of all CA-MRSA infections in one study [75]. A 15-year study of the changing epidemiology of MRSA infections from military medical facilities in San Diego from 1990 to 2004 documented that 65 % of MRSA infections were community-acquired, with SSTIs as the major site of infection in 95 % of cases [76].

MRSA was the most commonly identifiable cause of SSTI presenting to EDs in a prospective multicenter US study. *S. aureus* was isolated from 320 (76 %) of 422 patients with SSTI. The prevalence of MRSA was 59 % overall and ranged from 15 to 74 % by ED. Pulsed-field type USA300 accounted for 97 % of MRSA isolates; 72 % of these were a single indistinguishable strain (USA300-0114). SCC*mec* type IV and the Panton-Valentine leukocidin toxin gene were each detected in 98 % of MRSA isolates. Among methicillin-susceptible *S. aureus* isolates, 31 % were USA300 and 42 % contained PVL genes [77]. The spectrum of skin infections caused by CA-MRSA is wide and can range from simple cutaneous abscesses to large abscesses, severe pyomyositis, and fulminant necrotizing soft tissue infections [43, 78–81].

Importantly, since its emergence in 2000, epidemic spread of the MRSA clone USA300 has led to a high burden of SSTIs in the United States and is strongly correlated with MRSA bloodstream infection. It has been concluded that USA300 SSTIs serve as a source for bloodstream infection given that isolates from concurrent SSTI were the same genotypically as the USA300 isolates that caused bloodstream infections [82]. Given this important findings, it is imperative to provide prompt and definitive source control and antimicrobial therapy for CA-MRSA SSTIs in all patients.

MRSA has also been identified as the most common cause of severe SSTIs requiring surgical drainage and debridement in a single-center 7-year study from Houston [83]. From 2000 to 2006, 288 patients with SSTIs that required operative debridement were identified. The most common microorganism retrieved from intraoperative cultures was *S. aureus*, 70 % of which were MRSA. Streptococcus species accounted only for 15 % of microbes isolated. Monomicrobial etiology was identified in 67 % of patients and MRSA was also the predominant microbe isolated from such cultures (68 %). The frequency of MRSA isolates increased significantly during the study from 34 % in the year 2000 to 77 % in the year 2006, $p<0.001$). Interestingly, the examination of vancomycin MIC demonstrated a shift for MRSA isolates over this time period, with 38 % of the isolates having an MIC ≥ 1 μg/mL, with 31 % of isolates with MIC = 2 μg/mL. This is concerning given recent reports documenting high treatment failure rates for MRSA infections with increased MIC [84, 85].

In a study of 12,506 patients with culture-proven skin, soft tissue, and bone or joint infection in hospitalized patients, *S. aureus* caused infection in 54.6 % of patients and 28.0 % of the *S. aureus* isolates recovered were methicillin-resistant. Health care–associated infections and complicated SSTIs were associated with significantly higher mortality rates, longer and more costly length of hospital stay [86].

Based on this change in microbiologic etiology of SSTIs, all patients who present with or develop severe cSSTIs should be treated with broad-spectrum antimicrobial therapy, including mandatory coverage for MRSA. Patients who present to the hospital with severe infection or infection progressing despite antibiotic therapy should be treated aggressively. In these cases, if *S aureus* is cultured, the clinician should assume the organism may be resistant and should treat with agents effective against MRSA, such as vancomycin, linezolid, or daptomycin [87]. Although risk factors for MRSA SSTIs have been identified, in patients with severe SSTIs one should not rely solely on the use of risk factors for MRSA in the decision making regarding whether empiric anti-MRSA antimicrobials should be used.

Choice of empiric antimicrobial therapy for SSTIs is guided by a number of factors. For patients with severe SSTIs that are surgical site infections, it is important to choose an empiric antimicrobial agent that is different than the class of antibiotics that was used for surgical site infection prophylaxis as the time of the initial surgery. In the case of surgical site infection (SSI), the type and site of operation dictate which pathogens are suspected. Infections following operations in the gastrointestinal or genitourinary tract may be monomicrobial or mixed, and may be caused by gram-positive or gram-negative bacteria. In contrast, infections following clean operations in other parts of the body are typically caused by gram-positive pathogens. Immunocompromised or neutropenic patients are, of course, at increased risk of infection and are less able to control local infection and therefore should be treated with empiric, broad-spectrum antibiotics at the first clinical signs of infection, including fever.

It is important to provide anti-MRSA coverage in the empiric regimen of all patients with severe SSTIs. The MRSA carriage status of the patient should not be used as a guide to treatment for SSTIs, as it poorly predicts the need for anti-MRSA coverage in hospitalized orthopedic patients [88].

A number of intravenous anti-MRSA antimicrobials are approved by the FDA (vancomycin, linezolid, daptomycin, tigecycline, televancin, ceftaroline) and a number of new anti-MRSA antimicrobials are in development. Guideline-based recommendations for treatment of MRSA bone/joint infections vs. SSTIs and ABSSSIs are shown in Tables 25.9 and 25.10. A comprehensive review of SSTI antimicrobial studies has recently been published [89]. Options for oral treatment of MRSA SSTIs include clindamycin, trimethoprim-sulfamethoxazole, doxycycline, minocycline, linezolid, and tedizolid (a new oral oxazolidinone) [90]. Oral tedizolid at a dose of 200 mg once daily for 6 days was noninferior to 10 days of 600 mg twice daily linezolid for the treatment of ABSSSIs, including those of MRSA etiology [91]. There is currently no evidence to recommend any specific antibiotic in the treatment of MRSA SSIs [92] or MRSA infections in nonsurgical wounds [93].

When selecting empiric antimicrobials for treatment of severe cSSTIs, selection of specific antimicrobials that inhibit toxin production may

Table 25.9 Antimicrobial treatment of MRSA bone/joint infections compared with SSTIs or ABSSSIs

Antibiotic treatment of MRSA bone and joint infections	Dose
Vancomycin	15–20 mg/kg IV Q8–12 h
Daptomycin	6 mg/kg IV QD
Linezolid	600 mg PO/IV BID
TMP-SMX + rifampin	4 mg/kg/dose (TMP) Q8–12 h 600 mg QD (rifampin)
Clindamycin	600 mg PO/IV Q8h

Perlroth et al. [107]; Dombrowski and Winston [108]; Livorsi et al. [109]
Surgical debridement is the mainstay of therapy
Some experts recommend adding rifampin (300–450 mg BID)

Table 25.10 Antimicrobial treatment of MRSA complicated SSTIs or ABSSSIs

Antibiotic treatment of complicated SSTIs or ABSSSIs	Dose
Vancomycin	15–20 mg/kg IV Q8–12 h
Daptomycin	4 mg/kg IV QD
Linezolid	600 mg PO/IV BID
Tedizolid	200 mg PO QD
Telavancin	10 mg/kg/dose IV QD
Ceftaroline	600 mg IV Q12h
Clindamycin	600 mg PO/IV Q8h

be helpful, particularly in those patients with evidence of toxic shock syndrome. This is commonly present in patients with streptococcal and staphylococcal infections. Protein cytotoxins play an important role in the pathogenesis of a variety of staphylococcal infections, and toxin production should be considered when selecting an antimicrobial agent for gram-positive pathogens [94]. The recent identification of a class of secreted staphylococcal peptides [phenol-soluble modulin (PSM) peptides] document that they have a remarkable ability to recruit, activate and lyse human neutrophils, thus eliminating the main cellular defense against MRSA infection [95]. The β-lactams actually enhance toxin production. In contrast, both clindamycin and linezolid have the ability to inhibit toxin production by suppression of translation, but not transcription, of toxin genes for *S. aureus* and by direct inhibition of synthesis of group A streptococcal toxins. Particularly when patients exhibit signs and symptoms of streptococcal toxic shock syndrome (shock, coagulopathy, organ failure, and NSTI), anti-toxin antimicrobials (clindamycin or linezolid) should be promptly initiated [96].

3. "Source Control," that is, early aggressive surgical intervention for drainage of abscesses and debridement of necrotizing soft tissue infections

"Source control" includes drainage of infected fluids, debridement of infected soft tissues, removal of infected devices or foreign bodies, and finally, definite measures to correct anatomic derangement resulting in ongoing microbial contamination and to restore optimal function [97]. Source control represents a key component of success in the therapy of sepsis, since it is the best method of prompt reduction of the bacterial inoculum at the site of infection. Source control has been best identified as an important therapeutic strategy in the treatment of complicated abdominal infections, [98] but is of paramount importance in the treatment of cSSTIs as well. Appropriate and timely source control is mandatory in the treatment of severe SSTIs, particularly in the case of NSTIs. This is depicted as the main pillar of the "Treatment Triangle" of SSTIs in Fig. 25.2.

4. Pathogen identification and appropriate de-escalation of antimicrobial therapy

Given the increasing prevalence of multidrug-resistant pathogens as the etiology of severe SSTIs, pathogen identification is of paramount importance. All patients with severe SSTIs should have blood cultures obtained on admission, prior to initiation of empiric antimicrobial therapy if possible. In addition, cultures should be obtained directly from the SSTI site, either abscess fluid when incision and drainage is performed or tissue sample in the case of NSTIs when surgical debridement is performed.

Initial management of cSSTIs should include collection of specimens for culture and antimicrobial susceptibility testing from *all* patients with abscesses or purulent lesions. Culture and susceptibility findings are useful both for individual patient management and in monitoring local patterns of antimicrobial resistance. It has been documented that physicians and other health care workers cannot accurately predict if a SSTI is due to MRSA. A prospective observational study conducted in an urban tertiary academic center in emergency department patients presenting with purulent wounds and abscesses that received wound culture ($n=176$) documented that physician suspicion of MRSA had a sensitivity of 80 % (95 % CI 71–87 %) and a specificity of 23.6 % (95 % CI 14–37 %) for the presence of MRSA on wound culture with a positive likelihood ratio (LR) of 1.0 (95 % CI 0.9–1.3) and a negative LR of 0.8 (95 % CI 0.5–1.3). Prevalence was 64 %. Emergency physician's suspicion of MRSA infection was a poor predictor of MRSA infection [99].

It is important to de-escalate antimicrobial therapy in the treatment of severe SSTIs once culture results return. Pathogen-directed antimicrobial therapy is then initiated, with de-escalation from the initial broad-spectrum empiric antimicrobial regimen, with an attempt to decrease to monotherapy if at all possible. De-escalation of antimicrobial therapy should

Surgical drainage and debulking
- Incision and drainage of abscesses
- Removal of prosthetic material (if possible)

Prevention of transmission
- Improved hand hygiene
- Cleaning of shared equipment between uses
- Separation of infected patients; avoidance of overcrowding
- Selective decolonization

Wound culture
- Community-associated MRSA: consider TMP-SMX, tetracycline, erythromycin, clindamycin, vancomycin
- Health care-associated MRSA: consider vancomycin, rifampin, linezolid, (and possibly daptomycin, quinupristin-dalfopristin, fusidic acid)

Antibiotic therapy
- MSSA: antistaphylococcal penicillin, 1-CEF
- Community-associated MRSA: TMP-SMX, clindamycin, doxycycline
- Health care-associated MRSA: vancomycin, linezolid, daptomycin, rifampin plus fusidic acid

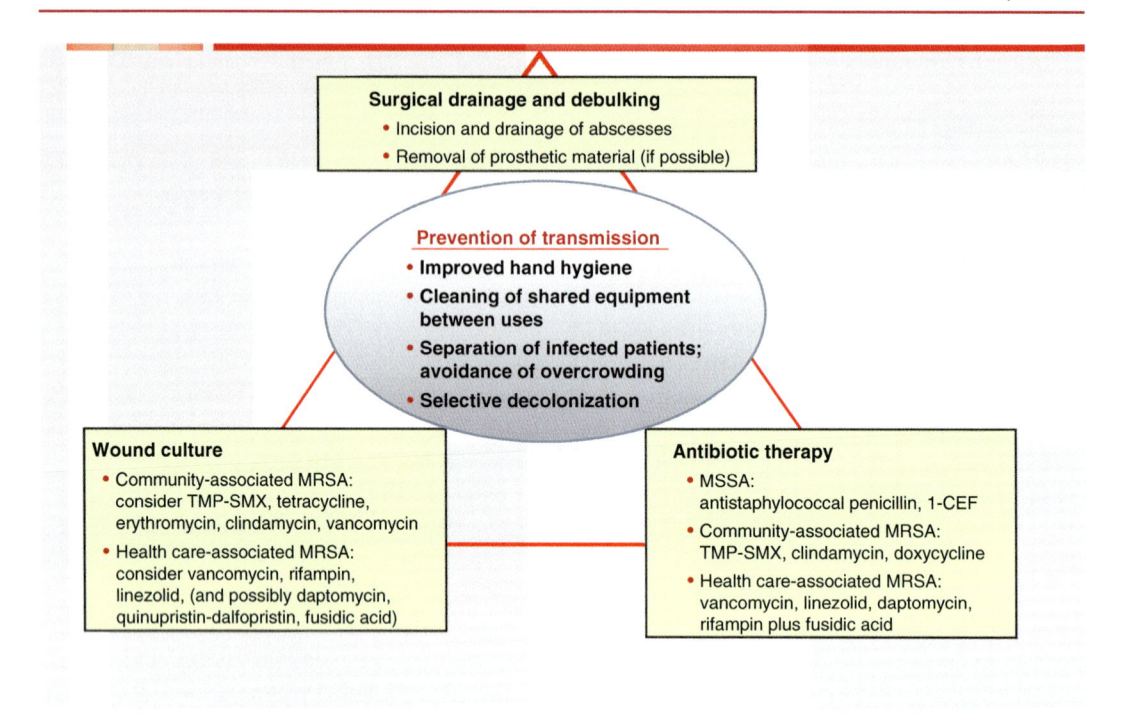

Fig. 25.2 Treatment triangle for *S. aureus* infection. The 3 components of the treatment of presumed *S. aureus* infection include surgical drainage and debridement, obtaining a wound culture, and initiation of appropriate empiric antimicrobial therapy. If MRSA SSTI is confirmed, it is critically important to utilize all methods to prevent microbial transmission, including hand hygiene. For wound cultures that are positive for community-associated MRSA (usually not a multidrug-resistant phenotype), in vitro susceptibility to trimethoprim-sulfamethoxazole (TMP-SMX), tetracycline, erythromycin, clindamycin, and vancomycin should be assessed. If the isolate is resistant to erythromycin but susceptible to clindamycin, the clindamycin D-zone test should be performed if clindamycin therapy is being considered. For wound cultures that are positive for health care–associated MRSA (usually a multidrug-resistant phenotype), in vitro susceptibility to vancomycin, rifampin, and linezolid should be assessed. Assessment of susceptibility to daptomycin and quinupristin-dalfopristin is not necessary unless therapy with these agents is being considered. Susceptibility to fusidic acid may be assessed in countries where this agent is available. Empirical antibiotic therapy should be reviewed once susceptibility data are known. For methicillin-susceptible *S. aureus* (MSSA), antistaphylococcal penicillin or a first-generation cephalosporin (1-CEF) may be suitable. For community-associated MRSA, TMP-SMX, clindamycin, or tetracycline may be suitable. For health care–associated MRSA, vancomycin, linezolid, daptomycin, or rifampin plus fusidic acid may be suitable (Adapted from Grayson [101])

occur as early as possible, but is only possible if appropriate microbiologic specimens are obtained at the time of SSTI source control. De-escalation is founded on identification of the pathogen and its antibiotic susceptibilities.

Conclusion

SSTIs are associated with significant morbidity and mortality, and it is important to differentiate necrotizing vs. non-necrotizing SSTIs early in the course of treatment. MRSA is the most common cause of purulent cSSTIs. All patients who present with complicated SSTIs should be treated with broad-spectrum antimicrobial therapy, including mandatory coverage for MRSA. Source control, including abscess drainage and surgical debridement, is the mainstay of therapy in severe cSSTIs. It is of paramount importance to obtain specimens for culture and antimicrobial susceptibilities given the high prevalence of MRSA as a causative pathogen in cSSTIs. Empiric broad-spectrum antimicrobial therapy should be de-escalated to narrower-spectrum agents based on culture pathogen identification and the patient's clinical response.

References

1. Napolitano LM. The diagnosis and treatment of skin and soft tissue infections (SSTIs). Surg Infect (Larchmt). 2008;9 Suppl 1:1.
2. Napolitano LM. Severe soft tissue infections. Infect Dis Clin North Am. 2009;23(3):571–9.
3. Napolitano LM. Early appropriate parenteral antimicrobial treatment of complicated skin and soft tissue infections caused by methicillin-resistant Staphylococcus aureus. Surg Infect (Larchmt). 2008;9(Suppl1):s17–27.
4. Napolitano LM. Perspectives in surgical infections: what does the future hold? Surgical Infection Society, North America, Presidential Address. Surg Infect (Larchmt). 2010;11(2):111–23.
5. DiNubile MJ, Lipsky BA. Complicated infections of skin and skin structures: when the infection is more than skin deep. J Antimicrob Chemother. 2004;53 Suppl 2:ii37–50.
6. Vinh DC, Embil JM. Severe skin and soft tissue infections and associated critical illness. Curr Infect Dis Rep. 2007;9(5):415–21.
7. Lipsky BA, Moran GJ, Napolitano LM, Vo L, Nicholson S, Kim M. A prospective, multicenter, observational study of complicated skin and soft tissue infections in hospitalized patients: clinical characteristics, medical treatment, and outcomes. BMC Infect Dis. 2012;12:227.
8. Garau J, Ostermann H, Medina J, Avila M, McBride K, Blasi F, REACH study group. Current management of patients hospitalized with complicated skin and soft tissue infections across Europe (2010–2011): assessment of clinical practice patterns and real-life effectiveness of antibiotics from the REACH study. Clin Microbiol Infect. 2013;19(9): E377–85.
9. http://www.fda.gov/downloads/Drugs/GuidanceComplianceRegulatoryInformation/Guidances/ucm071185.pdf.
10. Harris AM, Althausen PL, Kellam J, Bosse MJ, Castillo R, Lower Extremity Assessment Project (LEAP) Study Group. Complications following limb-threatening lower extremity trauma. J Orthop Trauma. 2009;23:1–6.
11. Paryavi E, Stall A, Gupta R, Scharfstein DO, Castillo RC, Zadnik M, Hui E, O'Toole RV. Predictive model for surgical site infection risk after surgery for high-energy lower-extremity fractures: development of the risk of infection in orthopedic trauma surgery score. J Trauma Acute Care Surg. 2013;74(6):1521–7.
12. Uçkay I, Hoffmeyer P, Lew D, Pittet D. Prevention of surgical site infections in orthopaedic surgery and bone trauma: state-of-the-art update. J Hosp Infect. 2013;84(1):5–12.
13. Gaynes R, Edwards JR. National nosocomial infections surveillance system: overview of nosocomial infections caused by gram-negative bacilli. Clin Infect Dis. 2005;41:848–54.
14. Hidron AI, Edwards JR, Patel J, National Healthcare Safety Network Team; Participating National Healthcare Safety Network Facilities, et al. NHSN annual update: antimicrobial- resistant pathogens associated with healthcare-associated infections: annual summary of data reported to the National Healthcare Safety Network at the Centers for Disease Control and Prevention, 2006–2007. Infect Control Hosp Epidemiol. 2008;29:996–1011.
15. Manian FA, Griesnauer S. Community-associated MRSA is replacing traditional healthcare-associated strains in surgical site infections among inpatients. Clin Infect Dis. 2008;47:434–5.
16. Kourbatova EV, Halvosa JS, King MD, et al. Emergence of community-associated MRSA USA 300 clone as a cause of healthcare-associated infections among patients with prosthetic joint infections. Am J Infect Control. 2005;33:385–91.
17. Weigelt JA, Lipsky BA, Tabak YP, et al. Surgical site infections: causative pathogens and associated outcomes. Am J Infect Control. 2010;38:112–20.
18. Haenie M, Podbielski A, Mittelmeier W, et al. Infections after primary and revision total hip replacement caused by enterobacteria producing extended spectrum beta-lactamases (ESBL): a case series. Hip Int. 2010;20(2):248–54.
19. Martinez-Pastor JC, Vilchez F, Pitart C, et al. Antibiotic resistance in orthopaedic surgery: acute knee prosthetic joint infections due to ESBL-producing Enterobacteriaceae. Eur J Clin Microbiol Infect Dis. 2010;29(8):1039–41.
20. Gillespie WJ, Walenkamp GH. Antibiotic prophylaxis for surgery for proximal femoral and other closed long bone fractures. Cochrane Database Syst Rev. 2010;3:CD000244.
21. Bratzler DW, Dellinger EP, Olsen KM, Perl TM, Auwaerter PG, Bolon MK, Fish DN, Napolitano LM, Sawyer RG, Slain D, Steinberg JP, Weinstein RA, American Society of Health-System Pharmacists (ASHP); Infectious Diseases Society of America (IDSA); Surgical Infection Society (SIS); Society for Healthcare Epidemiology of America (SHEA). Clinical practice guidelines for antimicrobial prophylaxis in surgery. Surg Infect (Larchmt). 2013;14(1):73–156. doi:10.1089/sur.2013.9999.
22. Gosselin RA, Roberts I, Gillespie WJ. Antibiotics for preventing infections in open limb fractures. Cochrane Database Syst Rev. 2004;(1):CD003764.
23. Hauser CJ, Adams Jr CA, Eachempati SR, Council of the Surgical Infection Society. Surgical Infection Society guideline: prophylactic antibiotic use in open fractures: An evidence-based guideline. Surg Infect (Larchmt). 2006;7(4):379–405.
24. Glass GE, Barrett SP, Sanderson F, et al. The microbiological basis for a revised antibiotic regimen in high-energy tibial fractures: preventing deep infections by nosocomial organisms. J Plast Reconstr Aesthet Surg. 2011;64(3):375–80.

25. Sarkar B, Napolitano LM. Necrotizing soft tissue infections. Minerva Chir. 2010;65(3):347–62.
26. Sarani B, Strong M, Pascual J, Schwab CW. Necrotizing fasciitis: current concepts and review of the literature. J Am Coll Surg. 2009;208(2):279–88.
27. Anaya DA, Dellinger EP. Necrotizing soft-tissue infection: diagnosis and management. Clin Infect Dis. 2007;44(5):705–10.
28. Cainzos M, Gonzalez-Rodriguez FJ. Necrotizing soft tissue infections. Curr Opin Crit Care. 2007;13(4):433–9.
29. Yilmazlar T, Ozturk E, Alsoy A, Ozquc H. Necrotizing soft tissue infections: APACHE II score, dissemination, and survival. World J Surg. 2007;31(9):1858–62.
30. Cuschieri J. Necrotizing soft tissue infection. Surg Infect (Larchmt). 2008;9(6):559–62.
31. Anaya DA, McMahon K, Nathens AB, et al. Predictors of mortality and limb loss in necrotizing soft tissue infections. Arch Surg. 2005;140:151–7.
32. Faraklas I, Stoddard GJ, Neumayer LA, Cochran A. Development and validation of a necrotizing soft-tissue infection mortality risk calculator using NSQIP. J Am Coll Surg. 2013;217(1):153–160.e3; discussion 160–1.
33. Wong CH, Khin LW, Heng KS, et al. The LRINEC (Laboratory Risk Indicator for Necrotizing Fasciitis) score: a tool for distinguishing necrotizing fasciitis from other soft tissue infections. Crit Care Med. 2004;32(7):1535–41.
34. Su YC, Chen HW, Hong YC, Chen CT, Hsiao CT, Chen IC. Laboratory risk indicator for necrotizing fasciitis score and the outcomes. ANZ J Surg. 2008;78(11):968–72.
35. Struk DW, Munk PL, Lee MJ, et al. Imaging of soft tissue infections. Radiol Clin North Am. 2001;39(2):277–303.
36. Brothers TE, Tagge DU, Stutley JE, et al. Magnetic resonance imaging differentiates between necrotizing and non-necrotizing fasciitis of the lower extremity. J Am Coll Surg. 1998;187:416–21.
37. Malghem J, Lecouvet FE, Omoumi P, Maldague BE, Vande Berg BC. Necrotizing fasciitis: contribution and limitations of diagnostic imaging. Joint Bone Spine. 2013;80(2):146–54.
38. Kim KT, Kim YJ, Won Lee J, Kim YJ, Park SW, Lim MK, Suh CH. Can necrotizing infectious fasciitis be differentiated from nonnecrotizing infectious fasciitis with MR imaging? Radiology. 2011;259(3):816–24.
39. Wong CH, Chang HC, Pasupathy S, et al. Necrotizing fasciitis: clinical presentation, microbiology, and determinants of mortality. J Bone Joint Surg Am. 2003;85-A(8):1454–60.
40. McHenry CR, Piotrowski JJ, Petrinic D, et al. Determinants of mortality for necrotizing soft-tissue infections. Ann Surg. 1995;221(5):558–65.
41. Elhabash S, Lee L, Farrow B, et al. Characteristics and microbiology of patients presenting with necrotizing fasciitis. Presented at the Association of VA Surgeons 31st Annual Meeting. Little Rock: Arkansas; 10–12 May 2007.
42. Lee TC, Carrick MM, Scott BG, et al. Incidence and clinical characteristics of MRSA necrotizing fasciitis in a large urban hospital. Am J Surg. 2007;194(6):809–12.
43. Miller LG, Perdreau-Remington F, Rieg G, et al. Necrotizing fasciitis caused by community-associated MRSA in Los Angeles. N Engl J Med. 2005;352(14):1445–53.
44. Young LM, Price CS. Community-acquired MRSA emerging as an important cause of necrotizing fasciitis. Surg Infect (Larchmt). 2008;9(4):469–74.
45. Olsen RJ, Burns KM, Chen L, et al. Severe necrotizing fasciitis in a human immunodeficiency virus-positive patient caused by MRSA. J Clin Microbiol. 2008;46(3):1144–7.
46. Dehority W, Wang E, Vernon PS, et al. Community-associated MRSA necrotizing fasciitis in a neonate. Pediatr Infect Dis J. 2006;25(11):1080–1.
47. Tsai YH, Hsu RW, Huang TJ, et al. Necrotizing soft tissue infections and sepsis caused by Vibrio vulnificus compared with those caused by Aeromonas species. J Bone Joint Surg Am. 2007;89(3):631–6.
48. Tsai YH, Hsu RW, Huang KC, et al. Systemic Vibrio infection presenting as necrotizing fasciitis and sepsis. A series of 13 cases. J Bone Joint Surg Am. 2004;86(11):2497–502.
49. Minnaganti VR, Patel PJ, Iancu D, et al. Necrotizing fasciitis caused by Aeromonas hydrophila. Heart Lung. 2000;29(4):306–8.
50. Bross MH, Soch K, Morales R, Mitchell RB. Vibrio vulnificus infection: diagnosis and treatment. Am Fam Physician. 2007;76(4):539–44.
51. Tsai YH, Huang TJ, Hsu RW, et al. Necrotizing soft tissue infections and primary sepsis caused by Vibrio vulnificus and Vibrio cholerae non-O1. J Trauma. 2009;66(3):899–905.
52. Kuo YL, Shieh SJ, Chiu HY, Lee JW. Necrotizing fasciitis caused by Vibrio vulnificus: epidemiology, clinical findings, treatment and prevention. Eur J Clin Microbiol Infect Dis. 2007;26(11):785–92.
53. Chao WN, Tsai SJ, Tsai CF, Su CH, Chan KS, Lee YT, Ueng KC, Lin DB, Chen CC, Chen SC. The Laboratory Risk Indicator for Necrotizing Fasciitis score for discernment of necrotizing fasciitis originated from Vibrio vulnificus infections. J Trauma Acute Care Surg. 2012;73(6):1576–82.
54. Brook I. Microbiology and management of myositis. Int Orthop. 2004;28(5):257–60.
55. Garcia J. MRI in inflammatory myopathies. Skeletal Radiol. 2000;29:425–38.
56. Lew DP, Waldvogel FA. Osteomyelitis. Lancet. 2004;364(9431):369–79.
57. Bosse MJ, MacKenzie EJ, Kellam JF, et al. An analysis of outcomes of reconstruction or amputation of leg-threatening injuries. N Engl J Med. 2002;247(24):1924–31.
58. Spellberg B, Lipsky BA. Systemic antibiotic therapy for chronic osteomyelitis in adults. Clin Infect Dis. 2012;54(3):393–407.
59. May AK. Skin and soft tissue infections. Surg Clin North Am. 2009;89(2):403–20.

60. Chan T, Yaghoubian A, Rosing D, et al. Low sensitivity of physical examination findings in necrotizing soft tissue infection is improved with laboratory values: a prospective study. Am J Surg. 2008;196(6):926–30. discussion 930.
61. Wall DB, deVirgilio C, Black S, Klein SR. Objective criteria may assist in distinguishing necrotizing fasciitis from non-necrotizing soft tissue infection. Am J Surg. 2000;179(11):17–21.
62. Wall DB, Klein SR, Black S, deVirgilio C. A simple model to help distinguish necrotizing fasciitis from non-necrotizing soft tissue infections. J Am Coll Surg. 2000;191(3):227–31.
63. Kollef MH, Sherman G, Ward S, Fraser VJ. Inadequate antimicrobial treatment of infections: a risk factor for hospital mortality among critically ill patients. Chest. 1999;115:462–74.
64. Iregui M, Ward S, Sherman G, Fraser VJ, Kollef MH. Clinical importance of delays in the initiation of appropriate antibiotic treatment for ventilator-associated pneumonia. Chest. 2002;122:262–8.
65. Garnacho-Montero J. Impact of adequate empirical antibiotic therapy on the outcome of patients admitted to the intensive care unit with sepsis. Crit Care Med. 2003;31:2742–51.
66. Alvarez-Lerma F. the ICU-Acquired Pneumonia Study Group. Modification of empiric antibiotic treatment in patients with pneumonia acquired in the intensive care unit. Intensive Care Med. 1996;22:387–94.
67. Ruhe JJ, Smith N, Bradsher RW, Menon A. Community-onset methicillin-resistant Staph*ylococcus aureus* skin and soft-tissue infections: impact of antimicrobial therapy on outcome. Clin Infect Dis. 2007;44:777–84.
68. Schramm GE, Johnson JA, Doherty JA, Micek ST, Kollef MH. Methicillin-resistant Staphylococcus aureus sterile-site infection: the importance of appropriate initial antimicrobial treatment. Crit Care Med. 2006;34(8):2069–74.
69. Chuck EA, Frazee BW, Lambert L, McCabe R. The benefit of empiric treatment for methicillin-resistant Staphylococcus aureus. J Emerg Med. 2010;38(5):567–71.
70. Gorwitz RJ, Jernigan DB, Powers JH, Jernigan JA and Participants in the CDC-Convened Experts' Meeting on Management of MRSA in the Community. Strategies for clinical management of MRSA in the community: summary of an experts' meeting convened by the Centers for Disease Control and Prevention. 2006. Available at: http://198.246.98.21/ncidod/dhqp/pdf/ar/CAMRSA_ExpMtgStrategies.pdf.
71. Dellinger RP, Levy MM, Rhodes A, Annane D, Gerlach H, Opal SM, Sevransky JE, Sprung CL, Douglas IS, Jaeschke R, Osborn TM, Nunnally ME, Townsend SR, Reinhart K, Kleinpell RM, Angus DC, Deutschman CS, Machado FR, Rubenfeld GD, Webb SA, Beale RJ, Vincent JL, Moreno R, Surviving Sepsis Campaign Guidelines Committee including the Pediatric Subgroup. Surviving sepsis campaign: international guidelines for management of severe sepsis and septic shock: 2012. Crit Care Med. 2013;41(2):580–637.
72. Dellinger RP, Levy MM, Rhodes A, Annane D, Gerlach H, Opal SM, Sevransky JE, Sprung CL, Douglas IS, Jaeschke R, Osborn TM, Nunnally ME, Townsend SR, Reinhart K, Kleinpell RM, Angus DC, Deutschman CS, Machado FR, Rubenfeld GD, Webb S, Beale RJ, Vincent JL, Moreno R, Surviving Sepsis Campaign Guidelines Committee including The Pediatric Subgroup. Surviving Sepsis Campaign: international guidelines for management of severe sepsis and septic shock, 2012. Intensive Care Med. 2013;39(2):165–228.
73. Kumar A, Roberts D, Wood KE, et al. Duration of hypotension before initiation of effective antimicrobial therapy is the critical determinant of survival in human septic shock. Crit Care Med. 2006;34(6):1589–96.
74. Brook I. Microbiology and management of soft tissue and muscle infections. Int J Surg. 2008;6(4):328–38.
75. Naimi TS, LeDell KH, Como-Sabetti K, et al. Comparison of community- and health care-associated methicillin-resistant Staphylococcus aureus infection. JAMA. 2003;290:2976–84.
76. Crum NF, Lee RU, Thornton SA, et al. Fifteen year study of the changing epidemiology of methicillin-resistant Staphylococcus aureus. Am J Med. 2006;119:943–51.
77. Moran GJ, Krishnadasan A, Gorwitz RJ, et al. Methicillin-resistant S aureus infections among patients in the emergency department. N Engl J Med. 2006;355(7):666–74.
78. Frazee BW, Lynn J, Charlebois ED, et al. High prevalence of methicillin-resistant in emergency department skin and soft tissue infection. Ann Emerg Med. 2005;45:311–20.
79. King MD, Humphrey BJ, Wang YF. Emergence of community acquired methicillin-resistant staphylococcus aureus USA 300 Clone as the prdominant cause of skin and soft-tissue infections. Ann Intern Med. 2006;144:309–17.
80. Ray GT, Suaya JA, Baxter R. Microbiology of skin and soft tissue infections in the age of community-acquired methicillin-resistant Staphylococcus aureus. Diagn Microbiol Infect Dis. 2013;76(1):24–30.
81. Moran GJ, Abrahamian FM, Lovecchio F, Talan DA. Acute bacterial skin infections: developments since the 2005 Infectious Diseases Society of America (IDSA) guidelines. J Emerg Med. 2013;44(6):e397–412.
82. Tattevin P, Schwartz BS, Graber CJ, Volinski J, Bhukhen A, Bhukhen A, Mai TT, Vo NH, Dang DN, Phan TH, Basuino L, Perdreau-Remington F, Chambers HF, Diep BA. Concurrent epidemics of skin and soft tissue infection and bloodstream infection due to community-associated methicillin-resistant Staphylococcus aureus. Clin Infect Dis. 2012;55(6):781–8.
83. Awad SS, Elhabash SI, Lee L, et al. Increasing incidence of methicillin-resistant Staphylococcus aureus skin and soft tissue infections: reconsideration of empiric antimicrobial therapy. Am J Surg. 2007;194:606–10.
84. Hidayat LK, Hsu DI, Quist R, et al. High-dose vancomycin for methicillin-resistant Staphylococcus

aureus infections. Arch Intern Med. 2006;166: 2138–44.
85. Howden BP, Ward PB, Charles PGP, et al. Treatment outcomes for serious infections caused by methicillin-resistant Staphylococcus aureus with reduced vancomycin susceptibility. Clin Infect Dis. 2004;38: 521–8.
86. Lipsky BA, Weigelt JA, Gupta V, et al. Skin, soft tissue, bone and joint infections in hospitalized patients: epidemiology and microbiological, clinical and economic outcomes. Infect Control Hosp Epidemiol. 2007;28(11):1290–8.
87. Stevens DL, Bisno AL, Chambers HF, Infectious Diseases Society of America, et al. Practice guidelines for the diagnosis and management of skin and soft tissue infections. Clin Infect Dis. 2005;41(10):1373–406.
88. Reber A, Moldovan A, Dunkel N, Emonet S, Rohner P, Tahintzi P, Hoffmeyer P, Harbarth S, Uçkay I. Should the methicillin-resistant Staphylococcus aureus carriage status be used as a guide to treatment for skin and soft tissue infections? J Infect. 2012;64(5):513–9.
89. Napolitano LM. Early appropriate parenteral antimicrobial treatment of complicated skin and soft tissue infections caused by methicillin-resistant Staphylococcus aureus. Surg Infect (Larchmt). 2008;9 Suppl 1:S15–27.
90. Liu C, Bayer A, Cosgrove SE, et al. Clinical Practice Guidelines by the Infectious Diseases Society of America for the treatment of methicillin-resistant staphylococcus aureus infections in adults and children. Clin Infect Dis. 2011;52:1–38.
91. Prokocimer P, De Anda C, Fang E, Mehra P, Das A. Tedizolid phosphate vs linezolid for treatment of acute bacterial skin and skin structure infections: the ESTABLISH-1 randomized trial. JAMA. 2013;309(6):559–69.
92. Gurusamy KS, Koti R, Toon CD, Wilson P, Davidson BR. Antibiotic therapy for the treatment of methicillin-resistant Staphylococcus aureus (MRSA) infections in surgical wounds. Cochrane Database Syst Rev. 2013;8:CD009726.
93. Gurusamy KS, Koti R, Toon CD, Wilson P, Davidson BR. Antibiotic therapy for the treatment of methicillin-resistant Staphylococcus aureus (MRSA) in non surgical wounds. Cochrane Database Syst Rev. 2013;11:CD010427.
94. Stevens DL, Ma Y, Salmi DB, McIndoo E, Wallace RJ, Bryant AE. Impact of antibiotics on expression of virulence-associated exotoxin genes in methicillin-sensitive and methicillin-resistant *Staphylococcus aureus*. J Infect Dis. 2007;195:202–11.
95. Wang R, et al. Identification of novel cytolytic peptides as key virulence determinants for community-associated MRSA. Nat Med. 2007;13(12):1510–4.
96. Filbin MR, Ring DC, Wessels MR, et al. Case 2-2009: A 25 year-old man with pain and swelling of the right hand and hypotension. N Engl J Med. 2009;360:281–90.
97. Marshall JC, Maier RV, Jimenez M, Dellinger EP. Source control in the management of severe sepsis and septic shock: an evidence-based review. Crit Care Med. 2004;32(11 Suppl):S513–26.
98. Laterre PF. Progress in medical management of intra-abdominal infections. Curr Opin Infect Dis. 2008;21(4):393–8.
99. Kuo DC, Chasm RM, Witting MD. Emergency department physician ability to predict methicillin-resistant Staphylococcus aureus skin and soft tissue infections. J Emerg Med. 2010;39(1):17–20.
100. Peltola H, Pääkkönen M. Acute osteomyelitis in children. N Engl J Med. 2014;370:352–60.
101. Grayson ML. The treatment triangle for staphylococcal infections. N Engl J Med. 2006;355:724–7.
102. Horan TC, et al. CDC/NHSN surveillance definition of health care–associated infection and criteria for specific types of infections in the acute care setting. Am J Infect Control. 2008;36:309–32.
103. Luchette FA, Bone LB, Born CT, et al. EAST Practice management guidelines for prophylactic antibiotic use in open fractures. www.east.org/tpg/openfrac.pdf.
104. Okike K, Bhattacharyya T. Trends in the management of open fractures. A critical analysis. J Bone Joint Surg Am. 2006;88:2739–48.
105. Holtom PD. Antibiotic prophylaxis: current recommendations. J Am Acad Orthop Surg. 2006; 14:S98–100.
106. Gustilo RB, Anderson JT. Prevention of infection in the treatment of one thousand and twenty-five open fractures of long bones: retrospective and prospective analyses. J Bone Joint Surg Am. 1976;58(4):453–8.
107. Perlroth J, Kuo M, Tan J, Bayer AS, Miller LG. Adjunctive use of rifampin for the treatment of Staphylococcus aureus infections: a systematic review of the literature. Arch Intern Med. 2008;168(8):805–19.
108. Dombrowski JC, Winston LG. Clinical failures of appropriately-treated methicillin-resistant Staphylococcus aureus infections. J Infect. 2008;57(2): 110–5.
109. Livorsi DJ, Daver NG, Atmar RL, Shelburne SA, White Jr AC, Musher DM. Outcomes of treatment for hematogenous Staphylococcus aureus vertebral osteomyelitis in the MRSA ERA. J Infect. 2008;57(2):128–31.

Posttraumatic Acute and Chronic Osteomyelitis

John K. Sontich

Contents

26.1	Introduction	371
26.2	The Causes of Posttraumatic Infections	373
26.3	Acute Posttraumatic Osteomyelitis	373
26.4	Chronic Posttraumatic Osteomyelitis	379
26.4.1	Infected Nonunions	383
References		385

J.K. Sontich, MD
Orthopaedic Surgery,
Case Western Reserve, MetroHealth Medical Center,
2500 MetroHealth Drive, Cleveland,
OH 44109, USA
e-mail: jsontich@metrohealth.org

26.1 Introduction

Posttraumatic osteomyelitis is an unfortunate consequence of injuries resulting from the treatment of open fractures and open treatment of closed fractures. Often the soft tissue is significantly compromised, producing poor blood supply and limiting surgical access to the bone. Cierny et al. [1, 2] defined stages of osteomyelitis with type I (medullary), type II (superficial), type III (localized), and type IV (diffuse) (Fig. 26.1). The host was defined in the classification by their medical fitness. A hosts are healthy patients. B hosts are systemically and/or locally compromised. C hosts are not surgical salvage candidates because of the risks of curative treatment are greater than the potential benefits. Host factors have been shown to influence acute and chronic outcomes. Some examples of comorbidities that increase the risk for treatment failure of infection and osteomyelitis are diabetes, tobacco use, malnutrition, and a poor soft tissue envelope locally. B hosts have a 20 % greater risk of failure in treatment than A host. Conversely, if, preoperatively, a B host can be converted to an A host through medical means, then complications rates are proportionally reduced. Many of these comorbidities can be optimized prior to definitive reconstruction to improve the chance of a successful outcome.

The osteomyelitis patient according to Cierny et al. can be staged by combining the type of

Fig. 26.1 The Cierny classification of adult osteomyelitis

Anatomic Classification of Adult Osteomyelitis

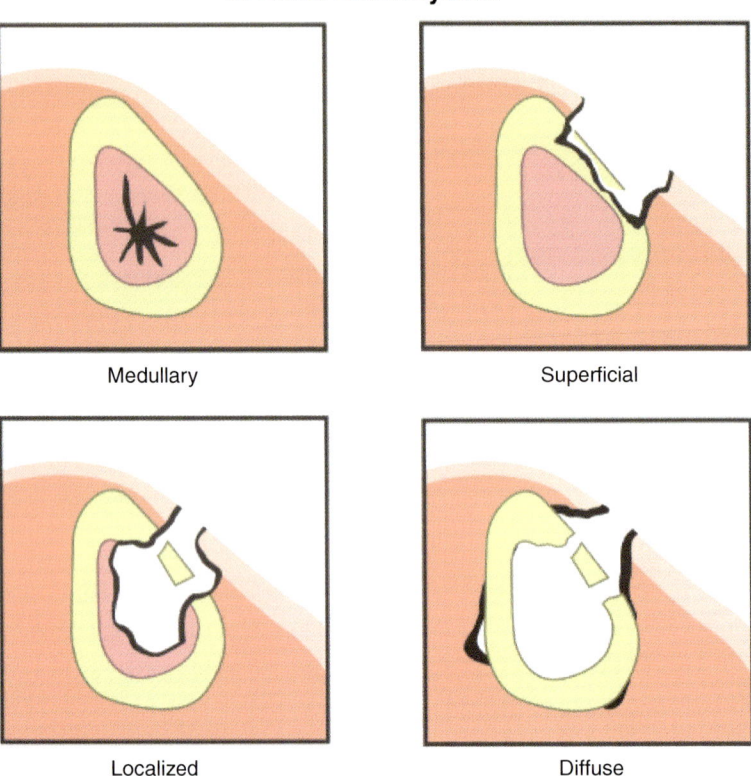

Medullary

Superficial

Localized

Diffuse

osteomyelitis (I–IV) with the host comorbidities. For example, a patient with diabetes and a full-thickness but isolated cortical defect of the tibia would be a stage III B patient. An otherwise healthy patient with an infected nonunion of the tibia with full-thickness permeating circumferential bone involvement is a stage IV A patient.

For the purpose of simplicity, acute osteomyelitis will be defined as an infected fracture with some degree of bone involvement and extends up to 3 months from the traumatic event. Generally speaking, acute fracture infections and acute osteomyelitis are synonymous terms. On the other hand, chronic posttraumatic osteomyelitis develops over time allowing the infection to permeate deeply into the substance of the bone. It often results from the infection of necrotic bone in the acute phase or the lack of removal of metal associated with biofilm. A section of avascular necrotic bone forms the sequestrum, which harbors bacteria and prevents antibiotic penetration.

The Cierny classification is useful to the treating surgeon because it provides an anatomic road map and helps guide the surgical debridement of necrotic bone. Although the original intent of the classification did not define acute from chronic posttraumatic osteomyelitis, the timing of the infection often plays a role in the type of osteomyelitis. For example, an acute intramedullary nail infection is a type I infection because it involves the intramedullary canal but has not penetrated the cortex. Acute open fracture infections, which have incomplete cortical penetration, may be best described as type II infections. Many chronic infection trauma cases have converted to type III or IV as the bacteria penetrate the full depth of the cortex. Most chronic infected nonunions are defined as type IV osteomyelitis because the infection penetrates both sides of the nonunions and complete radical removal of both sides of nonunion is required to eliminate the infection.

26.2 The Causes of Posttraumatic Infections

Fractures become infected for a variety of reasons but can be broken down to five basic causes:

1. Infection occurs when bacteria quantity exceeds the ability of immune system to eradicate it. This may seem obvious, but can be prevented by thorough excisional debridement at the time of open fracture surgery. The understanding of when to delay closure or close primarily is also important. This may be best addressed by the protocols seen at most level one trauma centers. The time between the injury and admission to the definitive level one trauma center was an independent predictor of the likelihood of infection by the LEAP study group [3]. No significant differences were found between patients who had development of an infection and those who did not when the groups were compared with regard to the time from the injury to the first debridement or the time from admission to the first debridement. Preoperative antibiotics are important in the prevention of infection, but the length of treatment of 24 h in open fractures was found just as effective when compared to 2–3 days, 4–5 days, or >5 days [4]. Greater length of treatment of antibiotics did not show any significant differences in the infection risk. The duration of surgery in closed ankle fractures of greater than 90 min was found to be a predictor of infection [5], which suggests more bacterial contamination with extended time of fracture exposure.

2. Infection occurs when the soft tissue damage results in poor blood supply and delayed soft tissue healing. The most direct example of this is revealed in the data produced in combat injuries with infections after damage-controlled orthopedics and delayed IM nailing [6]. Forty percent of the nails became infected at an average of 15 days post nailing. Ninety-one percent of the infections were in soldiers with blast injuries as apposed to non-blast injuries, highlighting the importance of healthy soft tissue and local blood supply in the prevention of fracture site infections.

3. Infections occur when the general health of the patient cannot adequately resists infection. This occurs in patients with diabetes, preexisting vascular disease, smoking history, malnutrition, etc. Bowen et al. [7] evaluated 174 patients with open long bone fractures and divided patients into three categories based on their comorbidities. Class A had no comorbidities, class B one to two, and class C patients three or more. The types of fractures were similar in all groups, but the infection rates were 4 % for class A, 15 % for class B, and 30 % for class C patients based solely on their general health status pre-injury.

4. Infections occur when the bone remains unstable. Stability of the fracture site allows the soft tissue to heal and reestablish a vascular supply to the area and permits the patient's immune system to resists infection locally. Basic science research has demonstrated that lab hamsters with bacterially contaminated femurs developed infection at a much lower rate if they underwent IM nailing of femurs when compared to hamsters with non-stabilized similarly contaminated femur fractures [8]. Temporizing external fixator for pilon and plateau fractures stabilizes the bone and soft tissue without compromising the soft tissue envelope. This allows the soft tissue to heal and has clearly reduced the risks of infection when definite surgical stabilization is performed [9, 10].

5. Infections occur when internal hardware becomes exposed to bacteria. Studies have shown that the minimal infective dose is 10,000 times lower in the presence of metal [11, 12]. A glycocalyx from the bacteria forms a biofilm or slime layer on the implant. Once the biofilm is formed, the implant must be removed or the infection cannot be eradicated [13, 14].

26.3 Acute Posttraumatic Osteomyelitis

The physical exam findings of acute posttraumatic infection (less than 3 month) are erythema at the fracture site, continued drainage over an extended time, additional tissue necrosis, increased warmth, lymphadenopathy, unexpected increased in pain, and elevated body temperature. In many aggressive bacterial infections, like *Staphylococcus aureus*, the clinical symptoms of acute infection are obvious. However, in less aggressive bacteria, the signs and symptoms can be subtle. Radiographic findings may be helpful. The lack of any callus formation along with radiolucency around screws can be suggestive of infection. Hardware, particular non-locking screws, loosens and even back out around the fracture (Fig. 26.2). Medullary osteolysis around an intramedullary nail with pain extending to the insertion site is a typical finding. Screws are sometimes pushed out of the nail (Fig. 26.3). Previous x-rays should be used in the comparison. The erythrocyte sedimentation rate and the C-reactive protein are independently accurate predictors of infection [15]. An additional use of the white blood cell/sulfur colloid scan was not found to be predictive for infection. It is also not cost-effective.

The diagnosis of acute and chronic infection can be made with the evaluation of symptoms, physical exam, radiograph comparison, ESR, and CRP alone 90 % of the time. Complex secondary studies are rarely needed but can be helpful in borderline cases. SPET/CT is useful in differentiating between soft tissue infection and osteomyelitis as well as between cortical, medullary, and subperiosteal infections [16]. Sensitivity ranges from 90 to 100 % and specificity from 80 to 90 % for diagnosing osteomyelitis. The most sensitive and specific test for pure osteomyelitis is the gadolinium MRI scan. Unfortunately, most posttraumatic cases have retained metal making MRI much less helpful.

The true diagnosis of acute or chronic infection requires multiple cultures (at least three) from deeply around or part of the bone in a patient off antibiotics for at least 1 week. Soft tissue or fistula specimens are much less reliable for diagnosis of the true infection [17]. Biopsy of the bone can be diagnostic for osteomyelitis, and excised bone should always be sent to pathology for permanent sections (Fig. 26.4). This can be very valuable in a patient that has been partially treated with

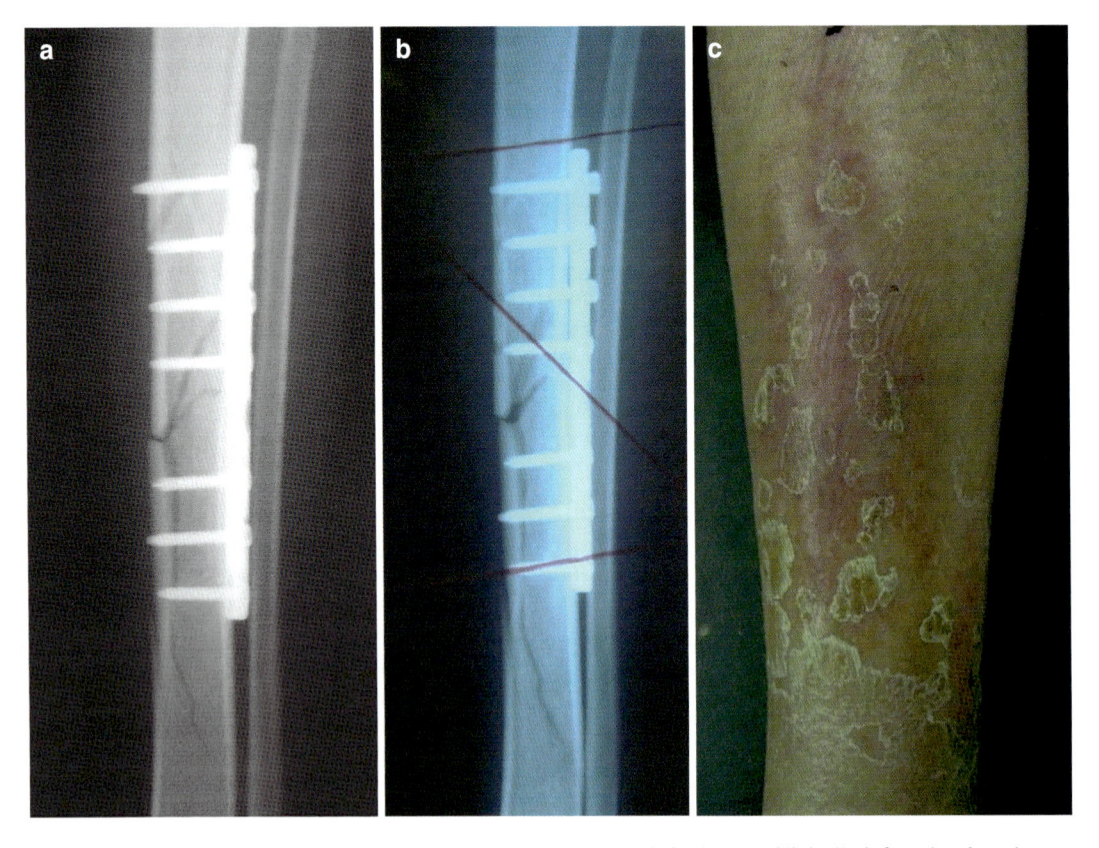

Fig. 26.2 (**a**) Midshaft tibia plating. (**b**) Loosening of screws in early infection. (**c**) Clinically infected at 6 weeks

26 Posttraumatic Acute and Chronic Osteomyelitis

Fig. 26.3 (a) Three months after nailing in open tibia fracture in B host with diabetes, now with distal radiolucency and screw backing out due to gross infection. (b) Nail removed, canal reamed, and antibiotic nail placed. (c) Antibiotic nail removed at 5 months and tibia was healed

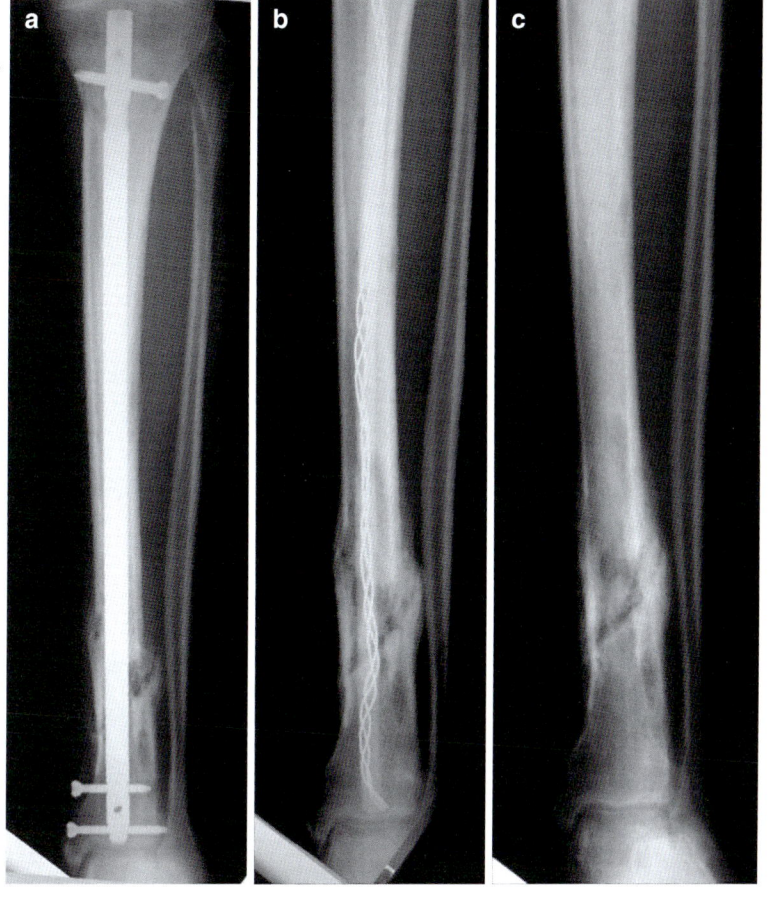

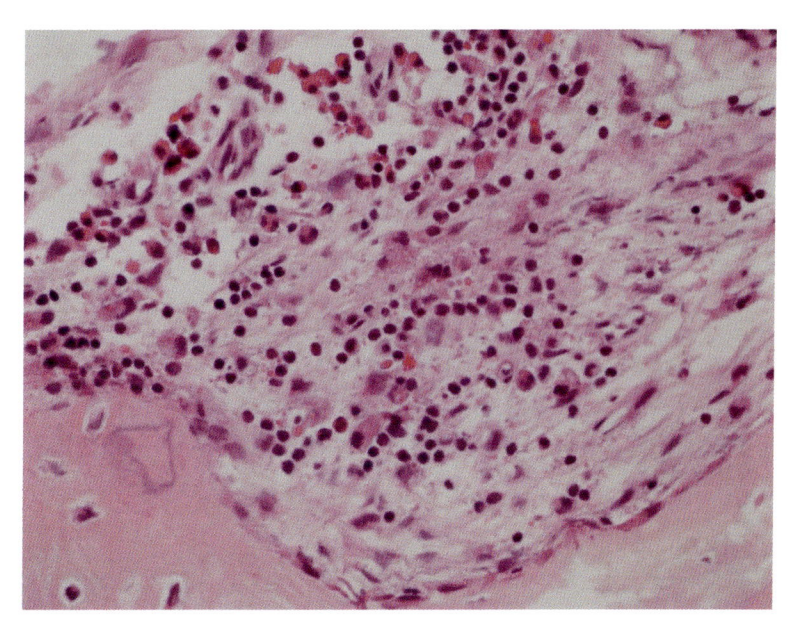

Fig. 26.4 Microscopic specimens demonstrating the invasion of acute and chronic inflammatory cells confirming infiltrating osteomyelitis

antibiotics and the cultures may be negative. Once the diagnosis of acute osteomyelitis or infected fracture has been made, the treatment requires surgical debridement along with appropriate long-term, generally 6 weeks, IV antibiotic treatment for all A and B hosts. Most of the examples

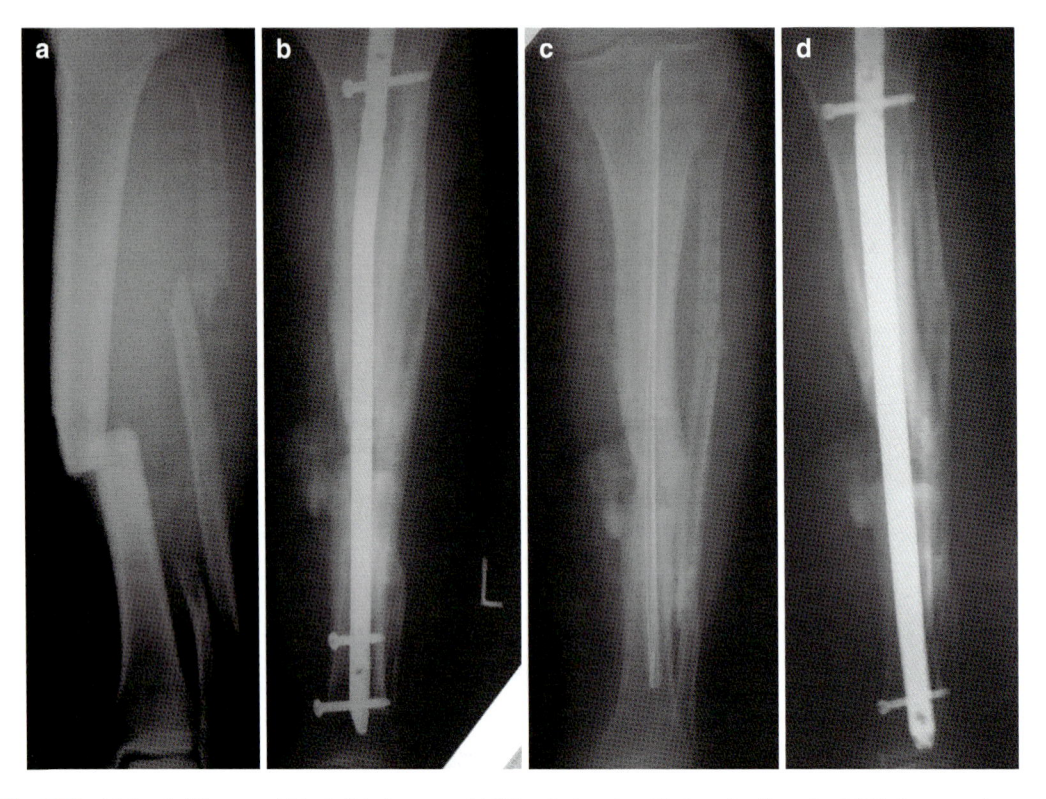

Fig. 26.5 (**a**) Grade III open midshaft tibia fracture. (**b**) IM nail placed and infected at 8 weeks. (**c**) Nail removed, canal reamed, and antibiotic nail placed. (**d**) Renailed at 16 weeks

of acute osteomyelitis can be classified as type I (intramedullary) or type II (partial cortical). The volume of bone involved (osteomyelitis) varied mainly with the length of time that had elapsed before diagnosis and treatment [18]. The real question the surgeon needs to ask is "How aggressively should we treat an infected fracture?"

The answer is that almost all acute posttraumatic infections require surgical debridement along with appropriate, culture-directed antibiotics. Unless you have proven this to be cellulitis ONLY (very rare) or the patient has turned into a C host, antibiotics alone are generally not the best therapy. It appears that surgical debridement and retention of hardware are warranted up to 6 weeks post-op. The success rate of a healed fracture and no infection after this technique is approximately 70 % [19, 20]. High failure rates were seen in patients with open fracture, IM tibia nails, smokers, and pseudomonas infections. In these later cases, the surgeon may consider removing the hardware with debridement even before 6 weeks. After 6 weeks from the definitive internal fixation, the infection has most likely produced a biofilm on the hardware and retention will most likely not cure the problem. The hardware needs to be removed, the infection of the bone debrided, and the bone somehow stabilized in a different manner.

In the case of infected IM tibia nails (type I osteomyelitis), there are several ways of dealing with this problem. Some have recommended exchanged nailing acutely, reaming, and renailing [18]. This works because reaming debrides the intramedullary bone infection and the biofilm is eliminated by the removal of the original nail. Others have recommended removal of the nail, reaming, and external fixation [21]. Still others have recommended removal of the nail, reaming, placement of temporary antibiotic rod, and renailing at the second stage [22, 23] (Fig. 26.5). All these techniques have excellent results if the patient has not progressed to type III or IV osteomyelitis at the original fracture site. Permeative osteomyelitis (III, IV) tends to occur in chronic situations and is the most common reason exchange nail fails with reinfection at times. Renailing DOES NOT adequately debride permeative osteomyelitis at the fracture site in type III or IV osteomyelitis. Radical excisional debridement and bone transport are required for these more extensive infections (see Chronic Infected Nonunions).

At the site of an open tibia fracture, when the soft tissue fails acutely and is not replaced, the bone can develop partial cortical involvement (type II osteomyelitis) requiring direct and aggressive debridement at the fracture site. This must include partial cortical debridement to eliminate the exposed boney surface infection. Some of the bone fragments that were left during the original excisional debridement may not be viable and have developed partial-thickness osteomyelitis. In those cases the deep fragments are removed. If the original trauma or debridement creates a significant bone defect, antibiotic methyl methacrylate beads or calcium sulfate (absorbable) beads can be used to dispense local antibiotics and produce a space filler [24]. The most important additional component of successful treatment is to provide adequate soft tissue coverage [25]. The muscle flap provides not only protection from outside contamination and recontamination, but increases the local blood supply (antibiotics) and a reliable surgical window for future bone grafting. The soft tissue covering the area of bone trauma must heal or the existing bone infection will persist or the patient will have new bone infections develop [26]. A clinical example would be an acute infected proximal tibia fracture that was plated 10 weeks early, requiring removal of hardware, partial cortical debridement (type II osteomyelitis), gastrocnemius rotational muscle flap, and definitive ring fixator for stability (Fig. 26.6).

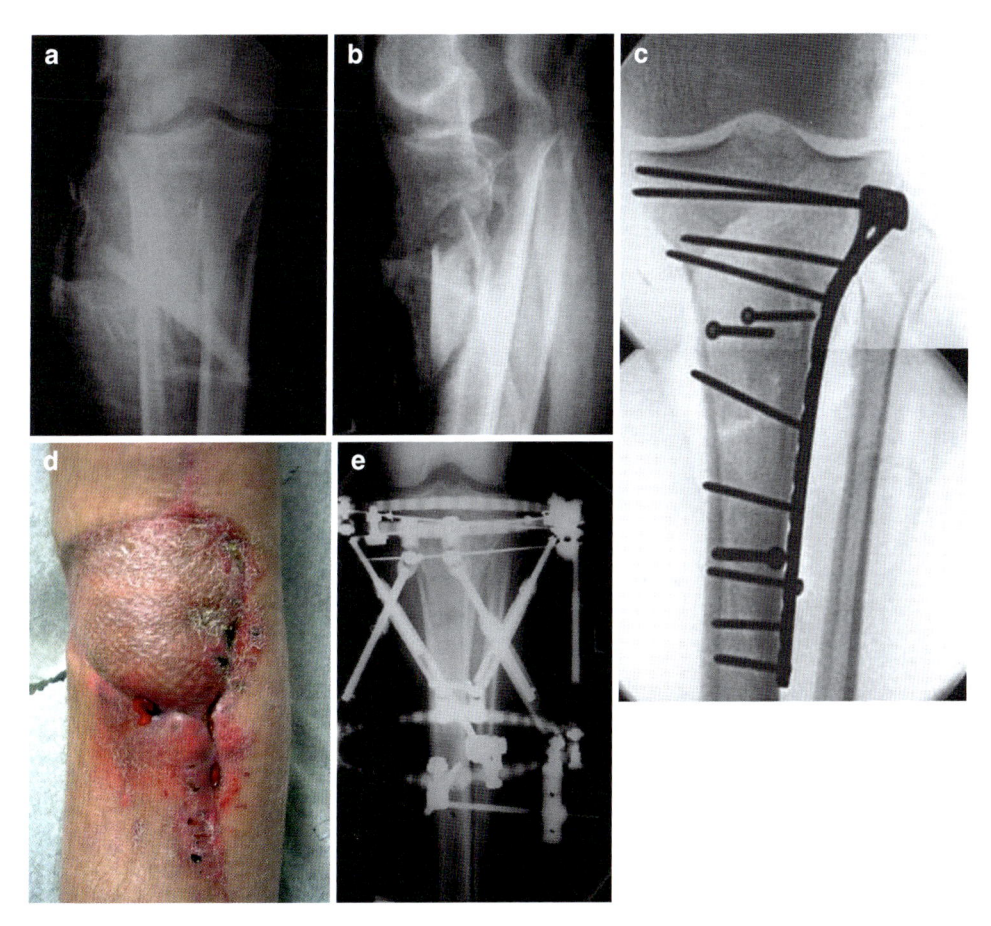

Fig. 26.6 (**a**, **b**) AP and lateral x-rays of tibial plateau fracture, B host. (**c**) Open reduction and internal fixation. (**d**) Infected at 3 months but not healed. (**e**, **f**) All hardware removed, fracture debrided, and TSF applied for type II osteomyelitis. (**g**, **h**) Healed plateau fracture at 6 months, no infection. (**i**) Clinical result and alignment

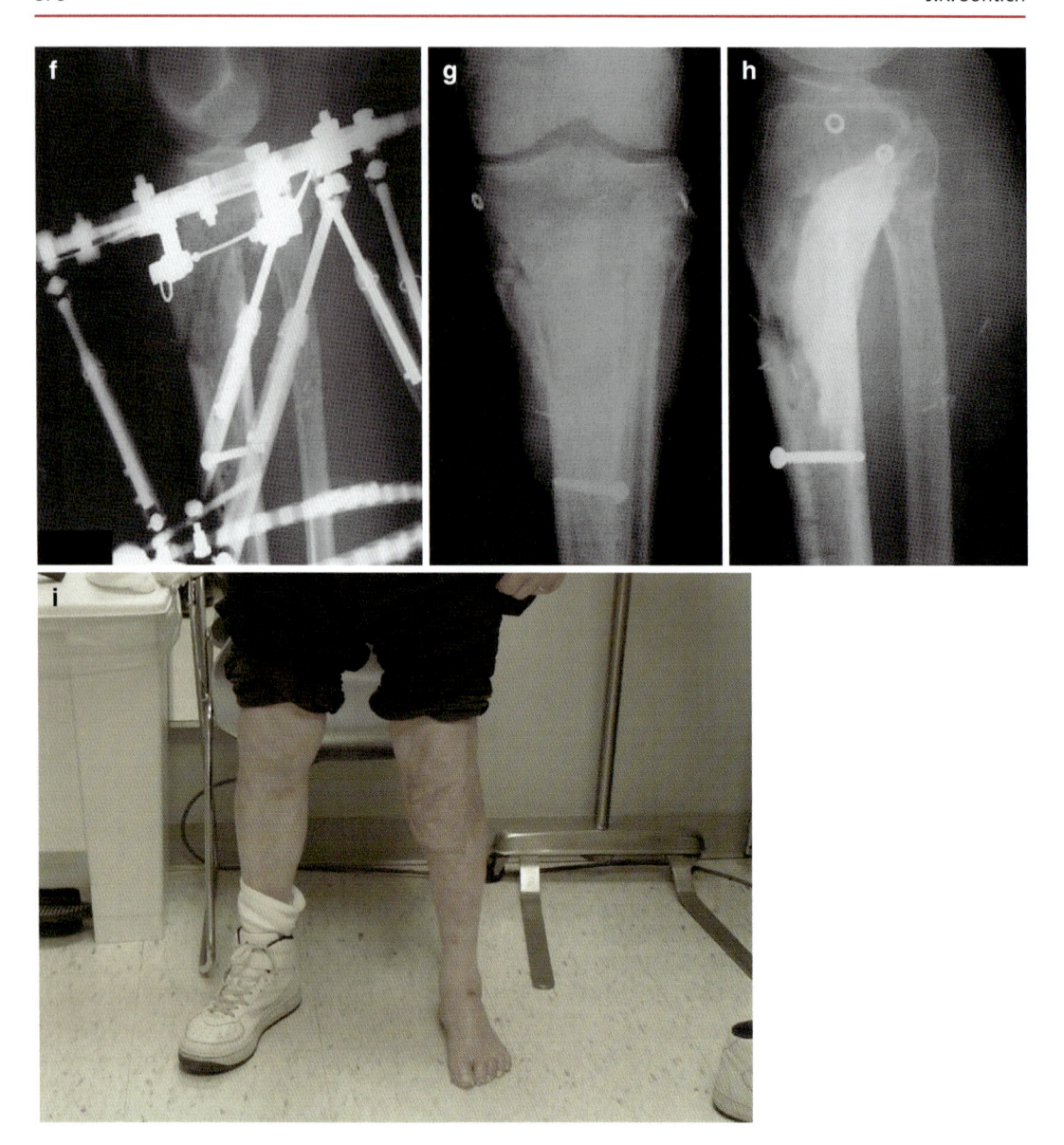

Fig. 26.6 (continued)

The static wire fixator is an important tool in the treatment of acute osteomyelitis, particularly after 6 weeks when the hardware often needs to be removed; yet the fracture still needs stability. The ring fixator (particularly Taylor Spatial Frames) can extend into the metaphysis, provide compression in the oblique plane (compress fractures), can be stable enough for definitive treatment, and does NOT add additional metal directly into the infected area. On the other hand, if the patient had an infected IM nail after 6 weeks, with type I or II osteomyelitis, my preferential treatment strategy is to remove the nail, debride the bone, and use the antibiotic nail temporary for stability, followed by second stage renailing when the infection has been eliminated. Healing and complete elimination of infection should be pursued in the most direct, expedient, and cost-effective strategy possible in acute osteomyelitis. This is a surgical disease and requires a surgical treatment (along with antibiotics) to appropriately address the problem and reduce the risk of the development of chronic osteomyelitis.

26.4 Chronic Posttraumatic Osteomyelitis

Posttraumatic osteomyelitis is an unfortunate consequence of injuries resulting from the treatment of open fractures and open treatment of closed fractures, which develops after at least 3 months. Often the soft tissue envelope is significantly compromised, producing poor blood supply and limiting surgical access to the bone. Cierny et al.'s [1, 2] type III (complete single cortex) and stage IV (permeative) are most often seen. Type IV osteomyelitis is by far more common because it is seen, but not limited to, patients with infected nonunions.

The basic steps to treating patients with chronic osteomyelitis are debridement of infected bone, obliteration of dead space, restoration of the blood supply, obtaining adequate soft tissue coverage, stabilization of the bone, and reconstruction of the defect.

Patients with type III can be treated without segmental reconstruction because the osteomyelitis has only penetrated one cortex and into the medullary canal (Figs. 26.7a, b). The illustrated case demonstrates removal of the affected cortex, collection of deep culture, and debridement into the canal until the bone bleeds. A positive "paprika sign" demonstrates punctate bleeding of the edges of the cortical bone after resection indicating the surgeon has achieved viable bone margins at the resection [2]. This can be done with a high-speed burr and irrigation to prevent thermal necrosis. Antibiotic beads are placed into the defect, the soft tissue is closed primarily (Fig. 26.7c), and monolateral external fixator is added to protect the remaining cortex. Adjunctive IV antibiotics are given for 6 weeks and the simple external fixator allows some weight bearing.

Stage two occurs when the antibiotic course is complete, the site looks noninfective, and the

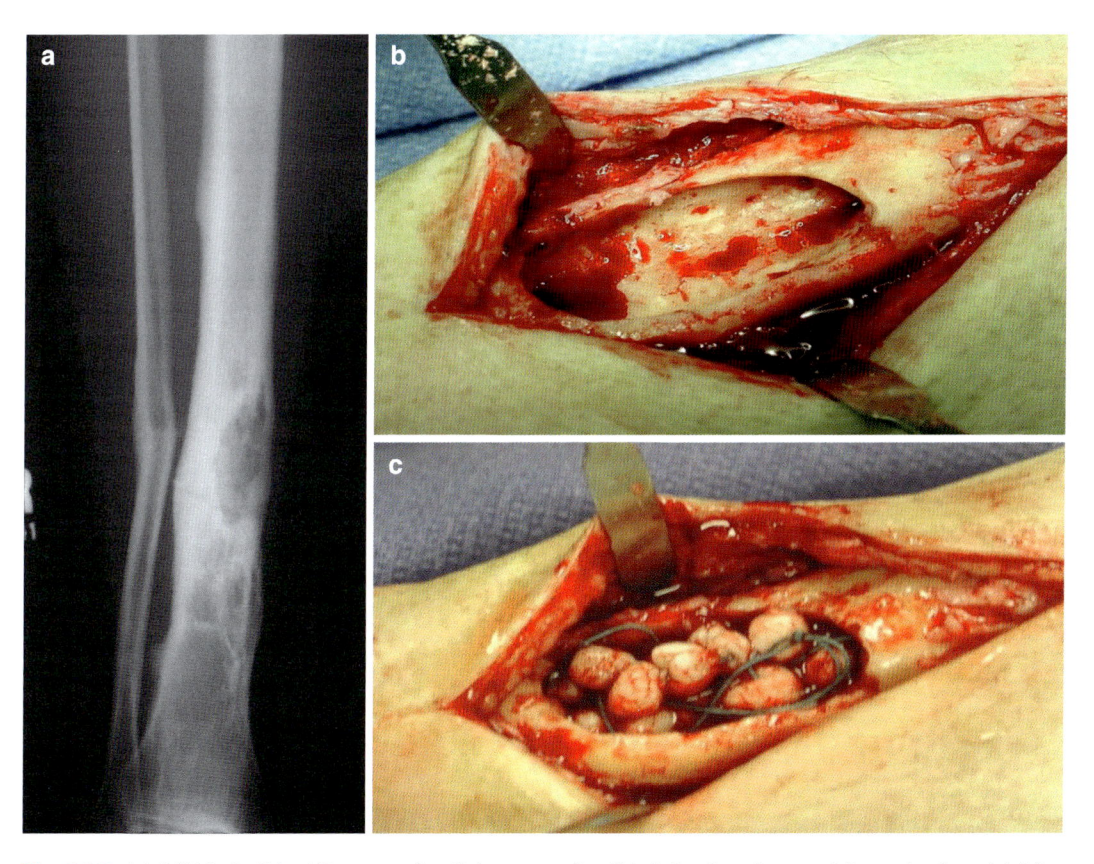

Fig. 26.7 (**a**) Midshaft tibia AP x-ray of radiolucent osteomyelitis. (**b**, **c**) Surgical dissection of type III osteomyelitis and placement of antibiotic beads. (**d**) AP x-rays of antibiotic beads and external fixator in place. (**e**) After fixator removed, defect bone grafted and tibia healed

Fig. 26.7 (continued)

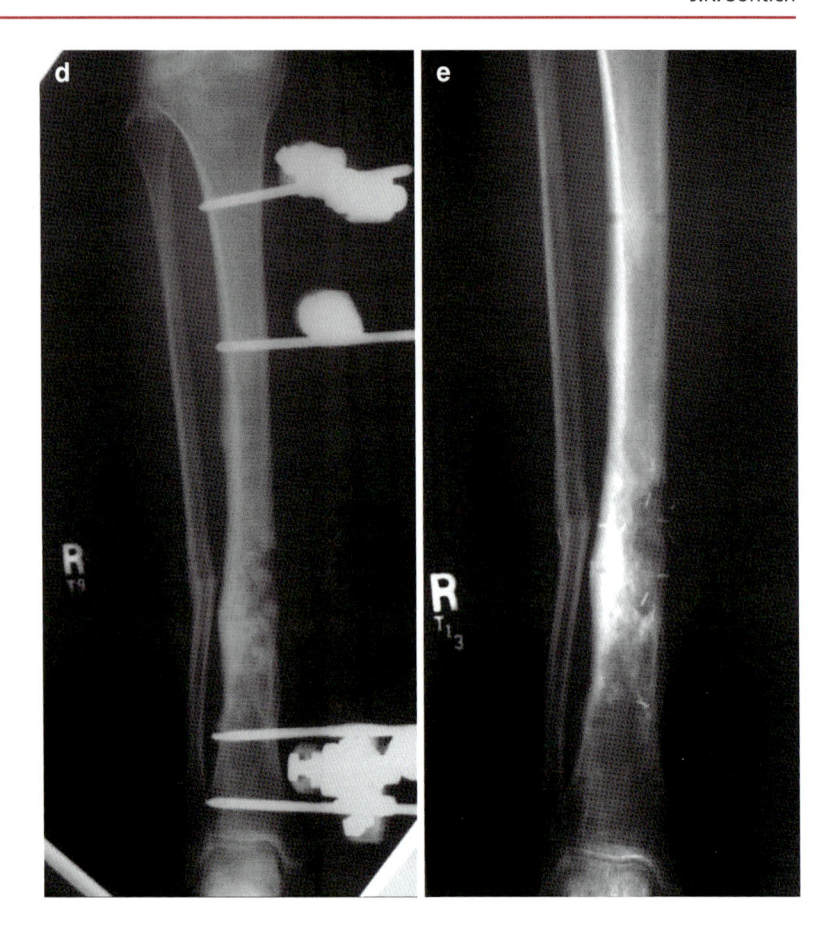

acute phase reactants have normalized. Stage two consists of removal of the beads, re-culturing, and placement of iliac crest bone graft (ICBG) into the defect (Fig. 26.7d). Additional antibiotics are given if the cultures are positive. As a technical note, the ICBG should NOT be contaminated by the osteomyelitis debridement, assuming the osteomyelitis defect may still be infected. This will keep the ICBG site from being infected by osteomyelitis. If the surgeon chooses to harvest the ICBG first, before opening the defect, this is not a problem. However, from time to time, the surgeon may open the defect and find gross pustulant tissue, with the ICBG already harvested. If this should occur, it is recommended that the ICBG be frozen, the defect is redebrided, new antibiotic beads are placed, and the ICBG be saved for another surgical day.

Stage IV disease represents bone with multicortical permeative osteomyelitis which requires segmental resection of bone to eliminate infection. This creates a segmental defect, which is mechanically unstable. The treatment starts with complete eradication of the infected bone with en bloc resection. Preoperative gadolinium MRI scan is often very helpful to access the osteomyelitis and determine the extent of bone resection needed to best eradicate the infection [27, 28]. However, the ultimate decision of how much bone should be resected is determined on the basis of the intraoperative inspection of the exposed bone. The viability of the bone can be seen where the destruction of the periosteum stops and the underlying bone has normal vascularity. The resection can be done with a power saw (never the corticotomy) but should be done with irrigation to avoid thermal necrosis at the margins. It is best to resect the bone perpendicular to the long axis to produce maximum compression at the docking site. The technique of resection to produce vascular bone margins is crucial, but it is also important to establish a clean and biologically viable soft tissue envelope from the onset.

Planning ahead to achieve adequate closure of the soft tissue after bone resection must be part of the preoperative plan. Sometimes by excising the segmental osteomyelitis, there is enough redundancy in the skin to place antibiotic beads and to close the skin. Other strategies include acute shortening of the leg to allow the soft tissue to approximate so that closure can occur.

The ideal coverage for all posttraumatic deformities remains muscle flap coverage. This provides an excellent vascularized soft tissue envelope that permits repeat exposure later in the course of treatment for redebridement and bone grafting. It protects and seals a sterile space and promotes blood supply to the area, which helps eliminate infection and heal bone at the docking site.

It is important to cover the defect early after resection of the osteomyelitis to establish a sterile space. Placement of antibiotic spacers into the defect serves to bathe the local area with antibiotics and prevent invagination of the newly applied soft tissue. At the same time, the orthopedic surgery team should stabilize the bone with an external fixator and try not to impede plastic surgeries access to the area. Placing a simple uniplanar external fixator rather than a ring fixator early allows better access for the micro vascular plastic surgeons reducing the degree of difficulty for an already complicated surgery.

Bone transport, for deficits greater than 2 cm, is the most reliable procedure to reconstruct the limb. Other techniques, which can reconstruct segmental bone defects, include the Masquelet procedure [29–32], vascularized fibular grafts [33, 34], and shortening procedures [35, 36].

The Masquelet technique has shown promise as method to heal defects up to 15 cm and does not require lengthening. Stage I of the technique involves segmental resection of the infected bone, placement of a cement spacer, and stabilizing the bone with a monolateral external fixator in addition to soft tissue coverage. Other stabilizing methods include plates and screwing and intramedullary nailing across the defect [29, 35, 36]. The reoccurrence of infection does appear to be higher with the internal fixation stabilization, which is not totally unexpected. Stage II of the Masquelet technique involves placing large amounts of autologous bone graft inside the self-induced periosteal membrane after removal of the spacer. Others have used Reamer/Irrigator/Aspirator (RIA) as the primary source of grafting and had successful salvage and no harvest site complications [37]. The disadvantages the Masquelet technique are the lack of solid cortical bone formation, the need for massive autologous grafts, and prolonged healing times. It has also been my experience and others that bone transport for the femur is less desirable than in the tibia because of complications related to knee stiffness and soft tissue scarring and prolonged course of treatment [38, 39]. The Masquelet technique is a very attractive reconstruction tool for stage IV osteomyelitis and is probably the first choice in the femur.

Vascularized fibular grafts have been another method of reconstructing segmental defects caused by stage IV osteomyelitis. Unfortunately, the complications of these reconstructions can be extensive [40]. Problems with donor site morbidity, delayed weight bearing, and late mechanical failure of the graft have made this technique less desirable in most trauma reconstruction centers.

Shortening procedures to heal a bone defect of less than 3 cm can be useful in patients with extensive comorbidities and willing to accept a leg length inequality. However, there are problems associated with acute shortening for segmental osteomyelitis. The surgeon should be aware of the risk posed by this technique, which can cause arteriolar kinking and result in the loss of blood supply to the extremity, particularly in chronic deformities. It is particularly important in these cases to check the pulses in the extremity during and after the procedure. If the blood supply is compromised, the leg should be acutely re-lengthened at the defect site until the pulses return. Other problems that arise with acute shortening include the need to resect healthy fibula and reduction in muscle length, causing redundancy and loss of mechanical advantage at the ankle.

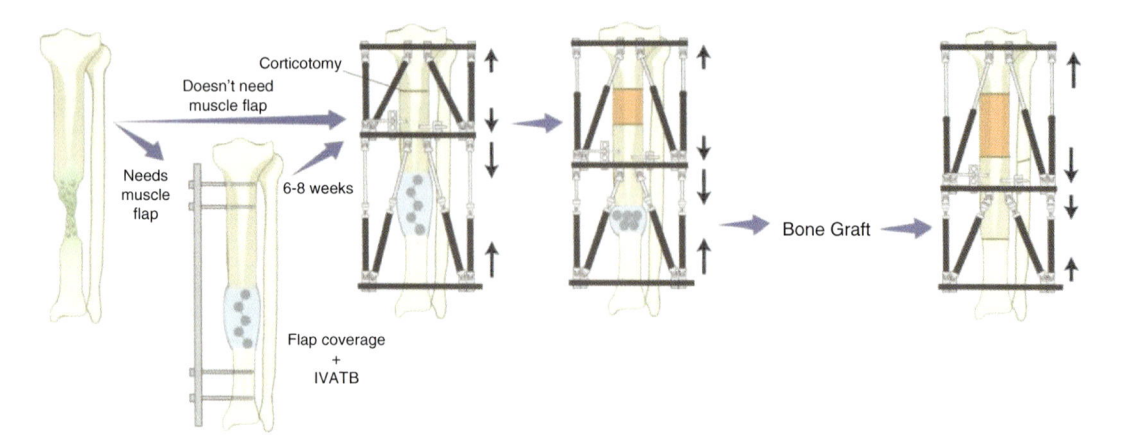

Fig. 26.8 The treatment of stage IV osteomyelitis begins with complete resection of infected bone and placement of antibiotic beads. Muscle flap coverage, if needed, should be accomplished with a uniplanar external fixator. Stacked transport commences when the soft tissue is closed. Prior to docking, beads are removed, ICBG inserted, and fibula osteotomized. New TRC programs are generated at the docking and distraction site

The most reliable technique for reconstruction of segmental defects of the tibia remains bone transport [41–45] (Fig. 26.8). The reconstruction usually commences once the soft tissue envelope is healed and there is no sign of infection. Usually, the definite wire ring transport external fixator can be applied, and the active distraction or transport can commence at 6 weeks after the muscle flap has been placed. Except for the area of the anastomosis, wires and half pins can then be placed through the flap at that time without a significant risk of flap necrosis.

Stacked Taylor Spatial Frame is my preferred setup for transport by placing one frame on top of another. The ring can be of different sizes and strut types, and lengths are individualized depending on the size and location of the defect and the length and width of the leg. When the defect is midshaft or distal, it is best to proceed with proximal to distal bone movement, called the antegrade transport. When setting up an antegrade transport frame in an average-sized leg, medium regular struts are maximally shortened between the proximal two rings, and medium fast struts are maximally lengthened connecting the distal two rings. In this way the frame is preloaded to distract the corticotomy proximally and shortening at the docking site distally. By beginning the transport with the proximal struts maximally shortened and the distal struts maximally lengthened, the necessity for strut changeouts during transport is reduced. The medium regular struts are used proximally because the proximal corticotomy requires no acute adjusts. By contrast, the defect site distally may require acute adjustments on the operating table to line up the future docking site more accurately. This can be done with the fast struts by unlocking the sleeves, lining up the bone under fluoroscopy, and relocking the struts in a better starting position.

A total residual correct (TRC) can then be formulated for the upper and lower stacked frames. If the extremity is severely deformed or the overall length of the leg is short, a fibula osteotomy should be performed after the frame has been applied. A fibula osteotomy allows the upper program to be completely separated from the lower program as long as the fibula remains unhealed. If the fibula heals prematurely or is not osteotomized, then total residual correction (TRC) has to coincide with equal distance and proceeding at the same rate. There are advantages to not osteotomizing the fibula when no other deformity exists except for segmental bone loss. The fibula provides internal mechanical stability to the construct. The fibula can be captured by the proximal ring and distal ring, bridging the defect and providing a bone track for the tibia to transport. Minor adjustment in angulation, translation, and rotation may be accomplished with the fibula intact, but this is restrictive. Major corrections in stacked transport require a fibular osteotomy.

If the leg is short with a segmental defect, then the lengthening program proximally needs to be greater than the shortening program distally. Obviously, a fibula osteotomy is required and the lengthening should be at a faster rate than the shortening to decrease the risk of the fibula healing during the transport. In posttraumatic deformities in adults, the maximum rate should be 0.75 mm/d. In this case the shortening of the segmental defect might progress at 0.5 mm/d, producing lengthening of the fibula at 0.25 mm/d, usually reducing the risk of preconsolidation of the fibula during transport. If one prefers to shorten and lengthen at the same rate during the transport, excising a small segment of the fibula (5 mm) during the osteotomy can also prevent the fibula from healing prematurely. Occasionally this may result in a nonunion of the fibula, but a midshaft fibula nonunion is rarely symptomatic if the tibia heals appropriately.

Unfortunately, nonunions of the docking site of the tibia have been shown to be problematic in transport literature [41–45]. Preventing invagination of the soft tissue can be accomplished by placing an antibiotic string of beads in the defect and allowing the beads to compress at the docking site as the transport proceeds. The beads will compress to the extent they occupy the potential space of the docking site. Just before the beads prevent mechanical advancement of the leading bone segment, the beads are removed, repeat cultures are taken, and autologous cancellous iliac crest bone graft (ICBG) is placed into the defect during a staged surgery. This serves to remove the fibrous scar from the bone ends, and the ICBG provides biological stimulation and promotes healing. A new TRC program is then created for the docking site and the distraction site to fine-tune the transport and allow ideal docking site compression and extremity alignment. Antibiotics are reinstituted if cultures are positive (this maybe a different organism than originally cultured). The transport is continued until the docking site is compressed with the ICBG and the extremity has achieved the desired length. Recurrence of osteomyelitis after stacked transport has been less than 5 %, but appropriate initial debridement is the key eliminated infection.

Retrograde transport is indicated when the segmental osteomyelitis is located in the proximal one-third of the tibia. This requires a distal corticotomy and a proximal docking site, and the fast struts are placed proximally and the regular struts distally. The distal to proximal transport does not create regenerate as easily as antegrade transport, so the rate usually needs to be reduced (between 0.5 and 0.7 mm/d). No more than 70 mm of new bone can be consistently created with retrograde transport.

26.4.1 Infected Nonunions

Infected nonunions represent one of the worst problems that the orthopedic trauma surgeon confronts. These types of nonunions are a combination of stage IV osteomyelitis, bone loss, and/or gap and an unhealed fracture. AO techniques have difficulty addressing large bone deficits and often rely on placing internal fixation in an area with marginal soft tissue coverage and previous infection. The recurrence of infection is greater when metal is required at the nonunion site, probably related to the affinity of bacteria for metal and the creation of a biofilm by 6 weeks. The Taylor Spatial Frame (TSF) offers many advantages in the treatment of infected nonunions and may represent the best available technique to help treat this difficult problem. Even using the TSF, infected cases still have a higher risk of initial failure and persistent nonunions than those cases without infection [46].

Stacked TSF appears to have advantages even over the standard Ilizarov transport frame when comparing the two groups on several variables [46]. A large cohort of 45 patients with TSF stacked transport for infected nonunions was followed for 7 years and compared with a similar cohort of 25 patients treated with standard Ilizarov transport all done by a single surgeon. The TSF transport group displayed better alignment of the extremity (3 vs. 8° of angular deformity) because of the ability to fine-tune the regenerate and the docking sight using the TRC

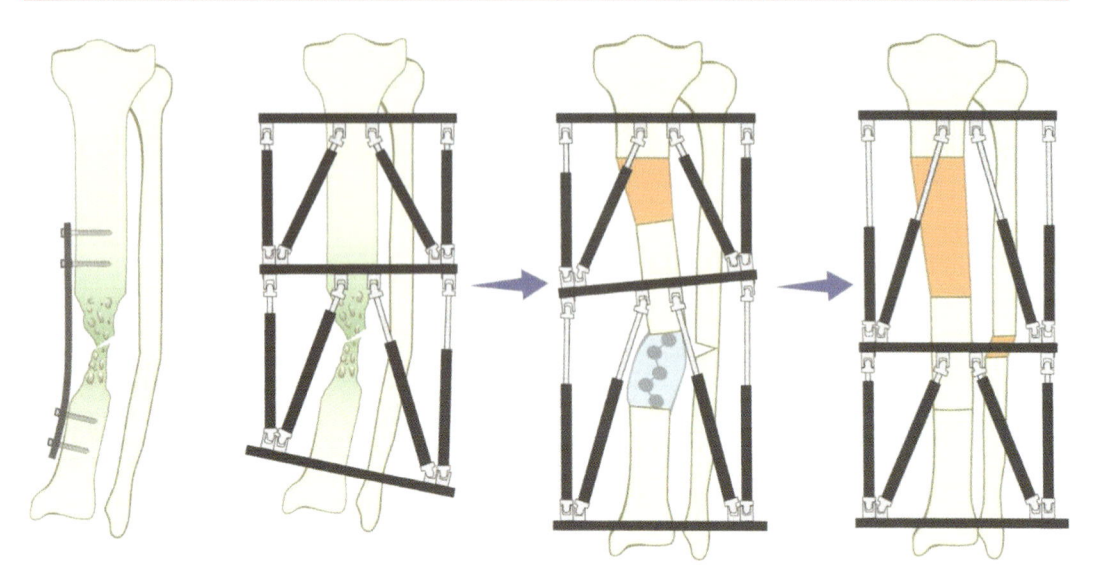

Fig. 26.9 Infected nonunions require excision of infected and necrotic bone, deep cultures, and placement of beads. The stacked TSF can be placed as long as the soft tissue can be closed. Docking site and regenerate site can be realigned, as long as the fibula osteotomy is still mobile

program. There were fewer surgeries required for healing in the TSF group when compared with the standard Ilizarov transport group. Fewer surgeries were required because the frame could be adjusted outside the operating room theater. The TSF group had slightly greater healing times for the regenerate, which was attributed to the single daily turning schedule of the TSF.

However, the transport process for infected nonunions is not without complications, even with the TSF adjustability. There were 29 major complications but no residual infections after healing. Nine patients experienced delayed unions at the docking site, four delayed unions of the regenerate, five soft tissue problems at the docking site, and two amputations.

The musculoskeletal functional outcome scores for the TSF group improved from 1 year [22] to 3 years after removal, approaching a normal range. It is important for the surgeon and the patient to understand that TSF transport for infected nonunions is a limb salvage technique in which the patients spend an average of 10 months in the frame and may not reach potential maximum functional outcome for 3 years.

The standard Ilizarov transport with the use of TSF rings still has a role in the treatment of infected nonunions with bone loss. The standard technique should be used in those infected nonunions that only have a well-aligned defect after resection, no leg length inequality, and preferably an intact well-aligned fibula. In these uncomplicated cases, the Ilizarov transport is simple, is less expensive, and may produce better regenerate. If the patient develops malalignment of the regenerate or the docking site, the threaded rods can be switched out for TSF struts and a TRC program performed to realign the bone.

This author prefers to use the TSF stacked transport for most infected nonunions (Fig. 26.9). Antibiotics are held for 10 days before surgery. The first stage of the surgery, as was the case for stage IV osteomyelitis, is to remove all the infected and avascular bone and hardware at the site. Three sets of deep culture are accompanied by this debridement, and an antibiotic string of beads is loosely placed into the gap. The incision is closed. It is important to have a durable soft tissue envelope that will keep the defect sterile, prevent recurrence of infection, and allow for future exposure to remove the beads and add iliac crest bone to the docking site. This can be done in a single stage if the soft tissue is viable after hardware removal and bone resection. The TSF transport frame can then be applied, the corticotomy placed in the virgin area of bone, and the transport started postoperatively (after a 5–10 day latency period). Infectious disease consultation should be obtained and appropriate antibiotics started.

When the defect has reduced to 1–2 cm during the transport with beads compressed, the patient is taken back to the operating room, the docking site is reopened, antibiotic beads are removed, the defect is recultured, and an ICBG is placed into the site. The docking site is closed. The fibula may need to be re-osteotomized if length of the extremity or major realignment of the docking site is required. Generally, a new TRC program is calculated for the docking site and the regenerate site to fine-tune the reduction. The transport is continued until the docking site is fully compressed and the limb is at the desire length. The consolidation phase is generally longer than the distraction phase. The frame should be dynamized 2 months before removal. The frame is removed in the operating room once the regenerate and docking site are healed.

References

1. Cierny III G, Mader JT, Penninck JJ. A clinical staging for adult osteomyelitis. Cont Orthop. 1985;10:5.
2. Cierny III G. Infected tibial nonunions 1981–95. The evolution of change. Clin Orthop Relat Res. 1999;360:97–105.
3. Pollak AN, Jones AL, Castillo RC. The relationship between timing of surgical debridement and the incidences of infection after open high-energy lower extremity trauma. J Bone Joint Surg Am. 2010;92(1):7–15.
4. Dunkel N, Pittet D, Tovmirzaeva L. Short duration of antibiotic prophylaxis in open fractures does not enhance risk of subsequent infection. J Bone Joint Surg Br. 2013;95(6):831–7.
5. Ovaska MT, Makinen TJ, Huotari K. Risk factors for deep surgical site infection following operative treatment of ankle fractures. J Bone Joint Surg Am. 2013;95(4):348–53.
6. Mody RM, Zapor M, Hartzell JD. Infectious complications of damaged control orthopaedics in war trauma. J Trauma. 2009;67(4):758–61.
7. Bowen TR, Widmaier JC. Host classification predicts infection after open fracture. Clin Orthop Relat Res. 2005;433:205–11.
8. Merrit K, Dowd JP. Role of internal fixation in infection of open fractures: studies with staphylococcus aureus and proteus mirabilis. J Orthop Res. 1987;5(1):23–8.
9. Sirkin M, Sanders R, Dipasquale T, et al. A staged protocol for soft tissue management in the treatment of complex pilon fractures. J Orthop Trauma. 1999;13(2):78–84.
10. Barei DP, Nork SE, Mills WJ, et al. Complications associated with internal fixation of high energy bicondylar tibial plateau fractures utilizing a two incision technique. J Orthop Trauma. 2004;18(10):649–57.
11. Elek S, et al. The Virulence of Staphylococcus aureus for man. Br J Exp Pathol. 1957;38:573–86.
12. Baser J, et al. In Vivo verification of in vitro model of antibiotic treatment of device related infection. Antimicrob Agents Chemother. 1995;39:1134–9.
13. Donlan RM, et al. Biofilms: survival mechanisms of clinically relevant microorganisms. Clin Microbiol Rev. 2002;15:167–93.
14. Costerson JW, et al. Microbial biofilms. Annu Rev Microbiol. 2002;49:711–74.
15. Stucken C, Olsewski DC, Creevy WR, et al. Preoperative diagnosis of infection in patients with nonunions. J Bone Joint Surg Am. 2013;95(15):1409–12.
16. Horger M, Eschmann SM, Pfannenberg C, et al. The value of SPECT/CT in chronic osteomyelitis. Eur J Nucl Med Image. 2003;30(12):1665–73.
17. Zuluags AF, Galvis W, Saldarriaga JG, et al. Etiology diagnosis of chronic osteomyelitis: a prospective study. Arch Intern Med. 2006;166(1):95–100.
18. Court-Brown CM, Keating JF, Mcqueen MM. Infection after intramedullary of the tibia, incidence and protocol for management. J Bone Joint Surg. 1992;74B(5):770–4.
19. Rightmire E, Zurakowski D, Vrahas M, et al. Acute infection after fracture repair: management with hardware in place. Clin Orthop Relat Res. 2008;466(2):466–72.
20. Berkes M, Obremsky WT, Scannell B, et al. Maintenance of hardware after early postoperative infections following fracture internal fixation. J Bone Joint Surg Am. 2010;92(4):823–8.
21. Zych GA, Huston J. Diagnosis and management of infection after tibial intramedullary nailing. Clin Orthop Relat Res. 1995;315:153–62.
22. Sancineto CF, Barla JD. Treatment of long bone osteomyelitis with a mechanically stable intramedullary antibiotic dispenser: nineteen consecutive cases with a minimum of 12 month follow-up. J Trauma. 2008;65(6):1416–20.
23. Selhi HS, Mahindra P, Yamin M, et al. Outcome of patients with an infected nonunion of the long bone treated with a reinforced antibiotic bone cement rod. J OrthopTrauma. 2012;26(3):184–8.
24. Mckee MD, Li-bland EA, Wild LM, et al. A prospective, randomized trial comparing an antibiotic-impregnated bioabsorbable bone substitute …. J Orthop Trauma. 2010;24(8):483–90.
25. Chim H, Sontich J, Kaufman B. Diagnosis and management of infection after tibial intramedullary nailing. Plast Reconstr Surg. 1995;127(6):2364–72.
26. Mader JT, et al. Adult posttraumatic osteomyelitis of the tibia. Clin Orthop Relat Res. 1999;360:14–21.

27. Gross T, Kaim AH, Regazonni P, et al. Current concepts in posttraumatic osteomyelitis: a diagnostic challenge with new imaging options. J Trauma. 2009;52:1210–9.
28. Sayle BA, Fawcett HD, Wilkey DJ, et al. Indium-111 chloride imaging in chronic osteomyelitis. J Nucl Med. 1985;26:225–9.
29. Apard T, Bigorre N, Cronier P. Two stage reconstruction of post-traumatic segmental tibia bone loss with nailing. Orthop Traumatol Surg Res. 2010;96(5):549–53.
30. Giannoudis PV, Faour O, Kanakaris N, et al. Masquelet technique for the treatment of bone defects: tips and tricks and future directions. Injury. 2011;42(6):591–8.
31. Donegan DJ, Scolaro J, Matuszewski PE, et al. Staged bone grafting following placement of an antibiotic spacer block for the management of segmental long bone defects. Orthopedics. 2011;34(11):730–5.
32. Masquelet AC, Begue T. The concept of induced membrane for reconstruction of long bone defects. Orthop Clin North Am. 2010;41(1):27–37.
33. Halim AS, Imran Y. Recalcitrant post-trauma chronic osteomyelitis/infected nonunions of the tibia following grade III fractures: treatment with vascularized osteocutaneous fibular graft. Med J Malaysia. 2006;61(Suppl A):66–70.
34. Yajima H, Kobata Y, Shigematsu K, et al. Vascularized fibular grafting in the treatment of methicillin-resistant staphylococcus aureus osteomyelitis and infected nonunions. J Reconstr Microsurg. 2004;20(1):13–20.
35. Wu CC. Single-stage surgical treatment of infected nonunion of the distal tibia. J Orthop Trauma. 2011;25(3):156–61.
36. Dhar SA, Butt MF, Mir MR, et al. Draining infected nonunion of the distal third of the tibia. The use of invaginating docking over short distances in the older patients. Orthop Traumatol Rehabil. 2009;11(3):264–70.
37. Stafford PR, Norris BL. Reamer-irrigator-aspirator bone graft and bi masquelet technique for segmental bone defect nonunions: a review of 25 cases. Injury. 2010;41 Suppl 2:S72–7.
38. Blum AL, Bongiovanni JC, Morgan SJ, et al. Complications associated with distraction osteogenesis for infected nonunion of the femoral shaft in the presence of a bone defect: a retrospective series. J Bone Joint Surg Br. 2010;92(4):565–70.
39. Saridis A, Panagiotopoulos E, Tyllianakis M, et al. The use of the ilizarov method as a procedure in infected nonunion of the distal femur with bone loss. J Bone Joint Surg Br. 2006;88(2):232–7.
40. Arai K, Toh S, Tsubo K, et al. Complications of vascularized fibula grafts for reconstruction of long bones. Plast Reconstr Surg. 2002;109(7):2301–6.
41. Pearse M, Watson JT, Morandi M. The management of infected intramedullary tibia nails by the Ilizarov method. J Bone Joint Surg Br. 1997;79:S226.
42. Mahaluxmivala J, Nadarajah R, Allen PW, et al. Ilizarov external fixator: acute shortening and lengthening versus bone transport in the management of tibial non-unions. Injury. 2005;36:662–8.
43. Mekhail AO, Abraham E, Gruber B, et al. Bone transport in the management of posttraumatic bone defects in the lower extremity. J Trauma. 2004;56:368–78.
44. Paley D, Maar DC. Ilizarov bone transport treatment for tibial defects. J Orthop Trauma. 2000;14:76–85.
45. Cattaneo R, Catagni M, Johnson EE. The treatment of infected non-unions and segmental defects of the tibia by the methods of Ilizarov. Clin Orthop Relat Res. 1992;280:143–52.
46. Sontich JK, Billow D, Hart D. Taylor spatial frame, stacked transport for tibial infected nonunions with bone loss, functional outcome with long term follow-up. Paper presented at the 19th Annual Scientific Meeting of the LLRS. Louisville, Kentucky; 2009.

Management of Malunions and Nonunions in Patients with Multiple Injuries

27

Nicholas Greco, Peter Siska, and Ivan S. Tarkin

Contents

27.1	Introduction	387
27.2	Damage Control Orthopedic Surgery and Optimal Timing of Definitive Fracture Care	388
27.3	Stratification of the Poly-traumatized Patient Based on Host Physiology and Local Soft Tissue Environment	388
27.3.1	The External Fixator as a Temporizing Device	389
27.4	Acute Total Orthopedic Care in the Poly-traumatized Patient with Associated Fracture	389
27.5	Averting Malunion with an Optimized Initial Treatment Plan	390
27.6	Nonunion Reconstruction After Poly-trauma	390
27.7	Biologic Supplementation for Nonunion After Poly-trauma	392
27.8	Strategic Bone Grafting Approaches	392
27.9	Open Fracture with Bone Loss: "Expected Nonunions"	393
27.10	Innovative Posttraumatic Reconstructive Strategies for Nonunion/Malunion	395
Conclusion		396
References		396

N. Greco, BS, MD • P. Siska, MD • I.S. Tarkin, MD (✉)
Department of Orthopaedic Surgery,
University of Pittsburgh Medical Center,
3471 Fifth Avenue, Suite 1010,
Pittsburgh, PA 15213, USA
e-mail: Greconj2@upmc.edu; siskpa@upmc.edu; tarkinis@upmc.edu

27.1 Introduction

The multiply injured patient with concurrent skeletal fracture presents an intricate problem for the orthopedic traumatologist because of the challenges of fracture treatment coupled with the interaction of multiple tenuous organ systems. In focusing the attention towards the musculoskeletal injury, the orthopedic traumatologist must first recognize the key factors comprising the damage to the local musculoskeletal environment that may potentiate systemic organ failure in a critically ill patient. Emergent management of a fracture associated with vascular injury or debridement of a grossly contaminated open fracture for example, aids in maintaining the well-being of other injured organ systems and prevents systemic compromise of a critically ill patient; however, more difficult decisions need to be rendered initially to prevent musculoskeletal complications such as malunion and nonunion that manifest later in the subacute phase of trauma, which may have a less clear benefit acutely to the systemic health of the patient.

Prior to the advent of damage control orthopedics in the early 1980s, there were many proponents for conservative treatment of fractures in the poly-traumatized patient. This paradigm was based on the theorized complications arising from local manipulations of the fracture site and systemic vulnerability of the multiply-injured patient [1–3].

Bone et al. published the first prospective randomized study in this area in 1989. The authors noted a profound decrease in the incidence of pneumonia, fat-emboli syndrome, ARDS, and morbidity from pulmonary sepsis in the patients randomized to early surgical fixation as compared to those treated greater than 48 h after injury. Furthermore, patients treated with early operative intervention required a shorter ICU and hospital stay which translated to a significantly decreased cost of medical care relative to the group treated with delayed surgery [4]. Poorer outcomes were correlated with a longer duration of skeletal traction prior to the formal reconstructive surgery [5]. These complications are in part believed to be evoked through scar tissue that forms in and around the fractured bone ends which is difficult to manage at the time of the delayed operation. Furthermore, the development of necrotic tissue around the fracture hematoma can cause an expanding area of devascularized bone and soft tissues. These factors complicate operative techniques, fracture healing and ultimately outcomes. These patients are unable to mobilize major joints which can result in profound stiffness and muscular atrophy, which complicates the physical recovery from such devastating injuries and lengthens the course of therapy that is required [6].

27.2 Damage Control Orthopedic Surgery and Optimal Timing of Definitive Fracture Care

With the knowledge that delayed definitive fracture care resulted in less than optimal musculoskeletal outcomes and the emerging research of the 1980s indicating that early operative treatment in the poly-trauma patient population may be more safe than previously anticipated, orthopedic traumatologists continued to refine the definition of early operative care. This significant change in philosophy was questioned as being too extreme by a series of orthopedists including Pape et al. who demonstrated increased mortality and pulmonary morbidity in patients with lung contusions treated with intramedullary nail placement in the first 24 h following injury [7]. The theory behind these findings was that the insertion of an intramedullary nail may dislodge bone marrow components causing embolization to the lungs. In a patient already suffering from a pulmonary injury, embolized marrow contents were believed to theoretically provide an additional stress or "second hit" that could not be handled by the pulmonary system. This concept formed the basis of "damage control orthopedic surgery," which was first popularized by Pape in Germany in the 1990s [8]. The concept of "damage-control" was subsequently translated into the practice of orthopedic surgery by Pape in Germany and first instituted in North America by Scalea and colleagues in the 1990s. The principal tool of damage control orthopedic surgery is the external fixator which is utilized in trauma patients with associated neurologic, hemodynamic, thoracoabdominal, and/or multiple orthopedic injuries that precluded an initial definitive fixation procedure [9]. Achieving rapid alignment not only served to lessen the risk of malunion and nonunion in these patients but also limited ongoing muscle damage, hemorrhage, and inflammatory stimulation caused by the injured extremity.

27.3 Stratification of the Poly-traumatized Patient Based on Host Physiology and Local Soft Tissue Environment

For those poly-traumatized patients presenting in extremis, temporizing orthopedic treatments should be contemplated in the treatment protocol. In these patients, a prolonged orthopedic procedure has the potential to exacerbate the systemic inflammatory state and impair already tenuous organ systems. It is in this subset of patients that damage control orthopedic surgery principles prove to be most applicable [10]. This entails achieving rapid skeletal stabilization of skeletal fractures to stop the cycle of ongoing musculoskeletal injury and to control hemorrhage. External fixation devices are the primary modality of treatment for extremity and pelvic trauma in this subset of patients.

27.3.1 The External Fixator as a Temporizing Device

The external fixator is an invaluable tool in the management of the poly-traumatized patient. Application of an external fixator is a rapid measure that restores gross skeletal alignment and functions as traveling traction. Skeletal stability decreases ongoing bone hemorrhage and stops cycle of injury to the surrounding soft tissue sleeve. An external fixator, opposed to a splint, provides more rigid skeletal stability and allows access for treating/monitoring the soft tissue environment including open wounds and extremity compartments.

The fixator, however, is rarely a definitive treatment and is most commonly used as a temporizing device awaiting either host resuscitation or local soft tissue trauma resolution. Fractures definitively managed with a simple external fixation scheme are prone to nonunion/and or unacceptable deformity. Thus, revision to definitive internal fixation is the preferred management algorithm.

In order to promote reliable healing without deformity of long bone fracture, conversion to intramedullary nail is safe and effective within a 2-week period. However, resuscitation/optimization of the patient for this minimally invasive procedure is typically achievable within the first 24–48 h. Nailing allows for more rigid internal fixation without the risk of significant deformity when performed by the orthopedic traumatologist.

Periarticular fracture, however, may deserve a more conservative approach. Lengthy reconstructive procedure is typically not warranted in the sick poly-traumatized host. Further, local soft tissue conditions are often prohibitive to safe invasive surgery. Thus, the fixator can be used for a more prolonged period of time to allow for host optimization and local soft tissue injury resolution [11].

High energy fractures involving the leg, foot, and ankle are infamously plagued with infection from wound related complication after injudicious acute operation especially in the compromised host. The concept of staged care using interval external fixation has been popularized for fractures of the tibial plateau and pilon. An extension of this concept has been adapted for the high energy midfoot dislocation as well (Fig. 27.1). Kadow et al. [12] from the University of Pittsburgh reported usage of external fixation in the poly-trauma patient with severe midfoot fracture/dislocation. Promising results were reported using the frame as a rigid splint allowing for maintenance of gross skeletal alignment until definitive surgery was indicated (ORIF or selected fusion) based on host physiology and local soft tissue conditions [12].

27.4 Acute Total Orthopedic Care in the Poly-traumatized Patient with Associated Fracture

The breadth of research examining the poly-traumatized patient with associated long bone fractures has provided a foundation with which to further explore and refine treatment algorithms of orthopedic injuries in this patient population. The current approach to the poly-traumatized patient with skeletal fracture underscores the principle of early total care.

Early total care, when safe and feasible based on host and local conditions, affords the multiply-injured patient the best chances for long-term functionality. Long-term outcomes of poly-trauma survivors are often predicated on outcomes of associated musculoskeletal injury [13]. Thus, avoiding malunion and nonunion are of paramount importance for optimizing long-term outcomes after high energy trauma. Current care paradigm suggests that patients should receive

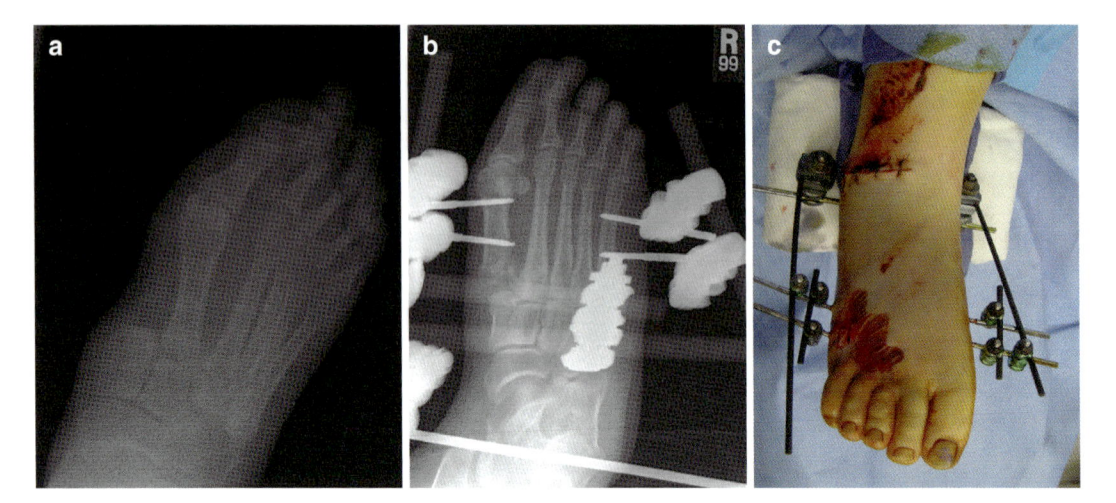

Fig. 27.1 External fixator used as temporizing device awaiting soft tissue resolution in this multiply injured with severe midfoot fracture/dislocation. (**a**) Mid foot fracture/dislocation in multiply injured patient. (**b, c**) External fixator used as a temporizing device awaiting soft tissue resolution

definitive skeletal fixation once their systemic and local physiology can tolerate the required anesthesia, blood loss, and inflammatory response associated with operative fracture reduction and stabilization [14, 15].

27.5 Averting Malunion with an Optimized Initial Treatment Plan

The need for posttraumatic reconstruction is certainly inevitable in the poly-trauma patient population secondary to either nonunion, malunion or both. However, a primary goal of care in the acute setting is to prevent significant deformity leading to malunion. Early total fracture care in the poly-traumatized patient certainly minimizes the risk of developing malunion. Fracture reduction/instrumentation is drastically simplified and more reliable.

Underscoring the principle of averting malunion, symptomatic deformity correction is typically a complex posttraumatic reconstructive process (Fig. 27.2). Angular and/or rotational deformity requires corrective osteotomy. Operation through a contracted soft tissue envelope can lead to muscle imbalance or worse compromise of vital neurovascular anatomy. Further, even in the absence of major complication, outcomes after successful deformity correction can certainly be less than optimal in terms of ultimate patient outcome and satisfaction [16].

27.6 Nonunion Reconstruction After Poly-trauma

When devising a treatment algorithm in the acute phase of injury, preventing skeletal fracture nonunion is of secondary importance relative to avoidance of deformity because of technical factors and physiologic impact of the procedures intended to correct these complications. From a technical standpoint, surgical corrective procedures for nonunion are less demanding for the treating orthopedic traumatologist and thus have a greater propensity to result in successful patient outcomes in contrast to analogous procedures for malunion.

For example, in the lower extremities, diaphyseal fractures of the tibia or femur that have been treated acutely with intra-medullary nailing but which have failed to unite can be treated with exchange nailing using a larger diameter rod [17]. If alignment and rotation of the long bone has already been restored during the index procedure, then this revision operation requires little technical consideration relative to an osteotomy procedure required for correction of malunion. Furthermore, reaming incites a healing response

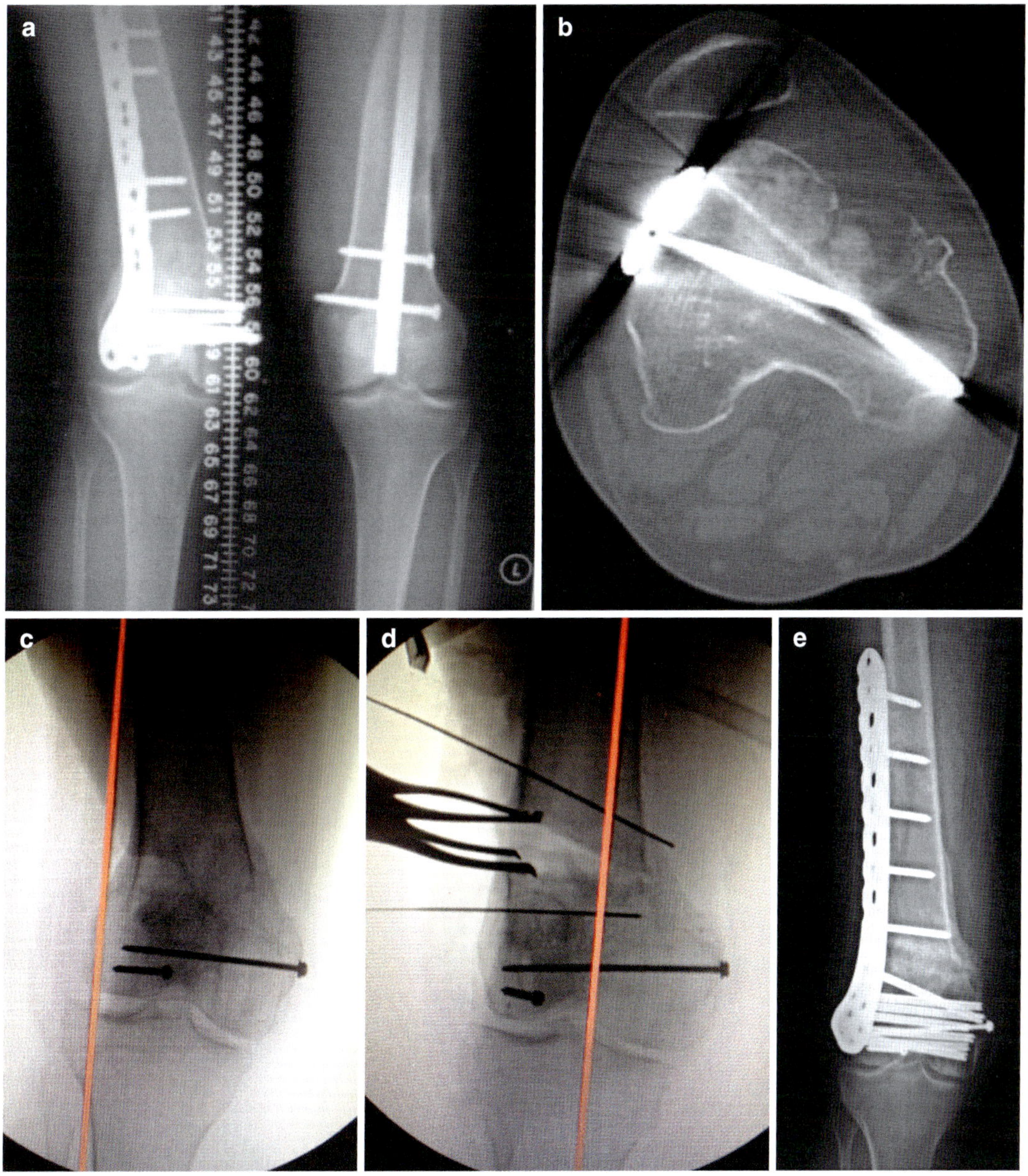

Fig. 27.2 A 20-year-old poly-trauma patient referred with significant malunion causing pain and ambulatory dysfunction. Malunion was associated with valgus and internal rotation deformity as well as shortening. Patella was chronically dislocated laterally. Complex posttraumatic reconstructive ladder was required, highlighting the consequences of marginal acute care. Osteotomy was performed to correct coronal, rotational. and length deformities. Further, soft tissue contracture release was necessary as well. Although the patient healed uneventfully, a 6-month course of healing/rehabilitation was required to achieve acceptable functionality. (**a**) Preoperative AP weightbearing radiograph. (**b**) Axial CT scan of right lower extremity. (**c, d**) Preoperative fluoroscopic image of the right knee with mechanical axis of the right lower extremity depicted in red. Intra-operative fluoroscopic image of the right knee with mechanical axis of the right lower extremity depicted in red. (**e**) Postoperative AP radiograph of right knee

and provides local bone graft at the nonunion site while the minimal surgical exposure needed for surgical execution does not disrupt the local soft tissue vascularity.

For the small percentage of diaphyseal fractures that still do not heal with exchange intramedullary nailing or similarly those periarticular fractures that have not united following the index formal reconstruction with various plating strategies, a revision operation consisting of a more extensive surgical dissection with plate application and utilization of biologic agents will likely be required. The objective under these circumstances is to provide additional mechanical stability as well as biologic factors to improve fracture healing potential.

In the case of nonunion after exchange nailing for a diaphyseal fracture, the plate spans the fracture site with the screws directed around the intra-medullary implant. Conversely, when nonunion is experienced at the site of a periarticular fracture, a supplemental plate must be applied within a new three-dimensional plane. In either situation, the surgical dissection is more extensive, however if the mechanical axis of the fracture is aligned and rotationally controlled then plate application and bone grafting imparts less stress to the local soft tissue environment and remains technically less demanding than correctional osteotomies formal-united fractures [18].

27.7 Biologic Supplementation for Nonunion After Poly-trauma

In addition to adding increased mechanical stability during nonunion reconstruction procedures, the biology within the local soft tissue environment must be enhanced in order to promote fracture healing in aseptic atrophic or oligotrophic cases. Most commonly non-structural cancellous autogenous or allogenic bone graft is incorporated into the fracture site to augment the local biology in these cases. Autogenous cancellous bone graft is osteogenic, osteoconductive, and osteoinductive and is therefore the gold standard grafting material. Bone graft from the iliac crest is preferred due to the large quantity of osseous tissue that can be extracted, and furthermore some researchers believe that iliac crest bone of intramembranous origin may be more osteoconductive than bone of enchondral origin [19]. Shortcomings of harvesting autologous bone graft include the limited amount of bone and complications at the donor site including infection and pain.

Recently, innovative methods have been devised to extract autlogous bone graft from the intra-medullary canal and condyles of long bones. This reamer irrigator aspirator (RIA) technique enables the extraction of significantly large volumes of autologous bone compared to iliac crest bone grafting therefore obviating the need to incorporate bone graft substitutes into nonunion surgery [20]. Additionally, bone graft from the femoral intra-medullary canal is believed to demonstrate similar osteogenic potential to autologous iliac crest bone graft. These devices have characteristics that are unique in comparison to traditional femoral reamers including cutting and suctioning designs that increase the risk for iatrogenic fracture and blood loss making it vital to fully comprehend these capabilities prior to use [21].

Allogenic bone graft functions primarily as an osteoconductive material that is typically used in conjunction with autologous bone graft materials to fill larger skeletal defects. Allogenic bone graft is advantageous because it is obtainable in greater quantity, however, it is less biologically active compared to autologous bone graft. Combining allograft cancellous bone with an osteoinductive agent is a common practice such as recombinanat bone morphogenic protein [22, 23]. However, the relative osteoinductive capability must be weighed against the potential harmful side effects including carcinogenic risk [24].

27.8 Strategic Bone Grafting Approaches

Achieving bone graft incorporation and uneventful nonunion resolution is both art and science. The desirable mechanical environment typically includes rigid nonunion stabilization to allow for bone graft healing via creeping substitution. The soft tissue milieu necessary includes a well vascularized muscular sleeve.

The nonunion site should be debrided to healthy margins removing all interposed scar and necrotic bone ends. Stimulation of the nonunion site is effective by using the feathering method with either osteotome, drill or low speed bur. If fracture geometry is conducive without excessive shortening, bone graft is applied at the nonunion site and mechanical compression is maintained.

Besides direct application of bone graft to the nonunion, autograft and/or allograft/bmp can be used more creatively. For instance, tibia nonunion in the setting of an intact fibula is a desirable indication for creation of a surgical synostosis in an effort to restore long-term skeletal stability via creation of a one-bone leg (Fig. 27.3). Posterolateral grafting is ideally suited for diaphyseal tibial nonunion whereby the cancellous graft is applied along the well vascularized interosseous membrane below the posterior compartment musculature. For distal third fractures, a central bone grafting technique has been popularized [25].

In contrast, creation of a surgical synostosis would not be an effective strategy for radius and/or ulna nonunion. Functionality secondary to restricted forearm rotation would make this strategy less than ideal. Therefore, bone graft containment can be achieved with placement of a rigidly applied tricortical graft.

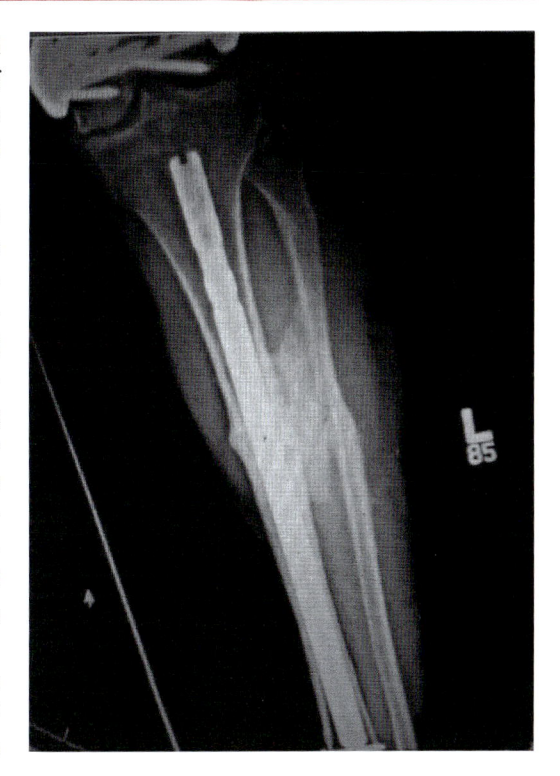

Fig. 27.3 Previous poly-trauma patient with Grade 3b open tibia fracture with subsequent nonunion refractory to exchange nailing. Posterolateral bone grafting strategy performed creating a so-called one bone leg

27.9 Open Fracture with Bone Loss: "Expected Nonunions"

The acute total care principles can be further applied to the poly-traumatized patient sustaining high grade open fracture with bone loss. Aggressive open fracture care is indicated to remove septic nidus yet typically bone defect is a result. The initial aim of the osseous reconstruction is to restore the alignment so as to prevent development of deformity. For diaphyseal fractures of the lower extremities, intra-medullary nailing is the standard care paradigm for rapidly establishing rigid internal splinting. In contrast, periarticular fractures are typically stabilized with fixed angle plate osteosynthesis (Fig. 27.4). Intra-articular fractures are precisely aligned then bridged to the shaft with appropriate length, alignment, and rotation [26]. Rigid stability of the fracture provided by the internal fixation will not only stop the cycle of ongoing injury but also allow for early patient comfort, mobility, and early physiotherapy during the subacute recovery phase from the trauma.

"Expected nonunions" after severe open fractures with bone loss are more easily managed in the absence of concomitant deformity. A proven methodology is usage of the Masquelet technique by initially filling large bone defects with antibiotic beads to encourage wound sterilization and neovascularization [27]. The antibiotic beads or spacer fills the void left by the removal of devitalized bone and prevents scar formation at the site of bone defect. Antibiotics are eluded from the porous cement architecture to aid in preventing infection within the local bone/soft tissue environment. Further, the beads incite an inflammatory process creating a vascular pocket ideally suited for future bone graft incorporation [28].

The staged application of bone graft should be delayed until at least 6–8 weeks after the index procedure in order to allow healing of the host and the local soft tissue environment [29]. Prior to this time point, the host typically has a low physiologic reserve and the traumatized local soft tissue environment displays poor vascularity. In the acute phase following injury, the patient is still trying to heal multiple organ system injuries and will have a high metabolic demand making it difficult to distribute enough nutrients to heal the bone grafting procedure. Furthermore, the traumatized local soft tissue environment is not at an optimal state to support fracture healing as the wound bed has not had time to re-vascularize. For these reasons, a staged bone grafting protocol is believed to permit optimal fracture healing using the concept "Timing is everything!" [30].

After optimization of host physiology including nutrition, the focus can then be shifted towards bone defect reconstruction. Autologous cancellous bone grafting can be extracted from various sites as described above. Harvest from the pelvis or via the RIA technique is most popular. Autograft extenders can be used for large defects such as allograft cancellous chips. Further, osteoinductive agents such as BMP can be added in hosts with severe bone loss and poor fracture healing capability.

In addition to bone grafting the void, a rigid mechanical environment is required to create a biomechanical milieu optimized for fracture consolidation. In the case of a retained nail for diaphyseal fracture, there are two options to create further rigidity. Exchange nailing not only generates bone graft and stimulates a healing response, but also promotes rigidity through passage of a larger diameter rod. Another technique includes nail retention and plating around the intramedullary device. The plate is applied via the exposure used for open bone grafting technique. In the case of the periarticular fracture, supplemental plating of the adjacent or opposite bone column is an effective strategy to promote an optimal mechanical environment for bone graft consolidation.

For example, open supracondylar femur fracture is commonly encountered in the poly-trauma patients typically from high speed motor vehicle trauma as a result of direct impact from dashboard. Dugan et al. reported a staged protocol consisting of aggressive open fracture, locked laterally based plating, as well as antibiotic bead placement at site of bone defect in the acute setting. Bone grafting and parallel plating was then performed after host and local physiology were optimized. Successful outcomes in terms of union and alignment were observed [30].

Clearly, however, the orthopedic traumatologist can employ other care paradigms based on the individual host and injury pattern. The usage of the multi-planar external fixator such as the Ilizarov frame can be invaluable in selected circumstances. Controlled management of deformity in association with nonunion/bone loss is a typical example [31]. These circular frames using tensioned wires can be used to modify alignment while also allowing correction of bone loss through distraction osteogenesis.

Despite the effectiveness in managing the bony reconstruction, this apparatus is infamously fraught with complications especially pin site infection and patient dissatisfaction. These external fixators are burdensome devices that are typically donned for a prolonged period of time. Therefore, in spite of the capability to perform large deformity

Fig. 27.4 Minimally invasive bridge plating of comminuted periarticular fracture restores alignment preventing deformity while preserving fracture biology and encouraging bony union

27　Management of Malunions and Nonunions in Patients with Multiple Injuries

Fig. 27.5 Tibiotalar calcaneal fusion performed as a postreconstructive procedure for talar neck nonunion associated with avascular necrosis and posttraumatic arthrosis (**a**) AP x-ray (**b**) Lateral x-ray

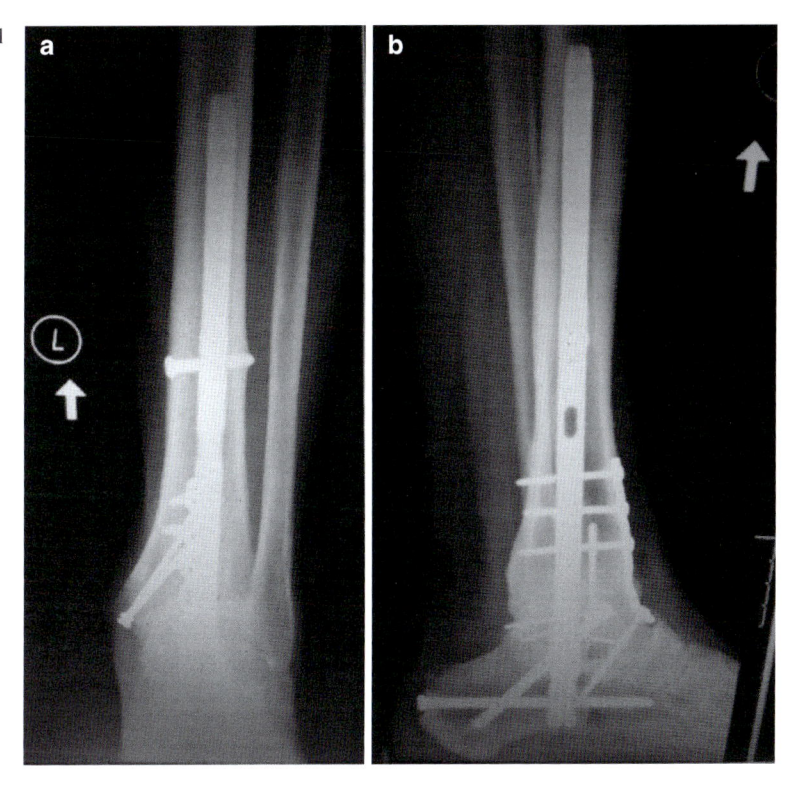

correction with a multi-planar external fixator, if the anatomic and mechanical axes of the bone is restored at the index procedure, using the principles of acute total fracture care, then the primary benefit of the external fixation method is lost.

27.10　Innovative Posttraumatic Reconstructive Strategies for Nonunion/Malunion

Anatomy and joint preserving operation for the management of nonunion/malunion in the previously poly-traumatized patient is typically the preferred care plan especially in the young active patient. Arthroplasty or arthrodesis strategies for periarticular nonunion and/or malunion, however, should certainly be in the treatment armamentarium for selected cases.

Indications for arthroplasty procedure or fusion include nonunion and/or malunion associated with irreversible damage to adjacent joint. Specific conditions include high energy periarticular injury with profound associated chondral damage and/or avascular necrosis. Arthrodesis or joint replacement under these circumstances has the potential to improve functionality/outcomes by concomitantly addressing inevitable posttraumatic arthrosis (Fig. 27.5).

A standard example would include the older patient with femoral neck nonunion/AVN. Total hip arthroplasty has the advantage of addressing the nonunion as well as intra-articular pathology. The combination of nonunion, AVN, and arthrosis after talar neck fracture is a popular indication for arthrodesis. Further, Gruen et al. has recently described optimal union/functional outcomes for addressing pilon fracture nonunion with posterior blade plate arthrodesis [32].

More innovative strategies for periarticular nonunion include the usage of megaprosthesis (Fig. 27.6). Metadiaphyseal nonunion of the proximal and distal femur are examples of scenarios in which megaprosthesis in the form of proximal femoral replacement or distal femoral replacement may be utilized [33, 34]. Despite immediate success, long-term implant reliability, and durability are concerns.

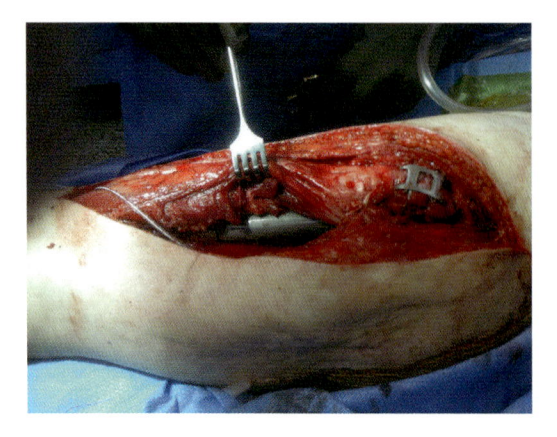

Fig. 27.6 Performance of proximal femoral replacement in geriatric poly-trauma patient to manage peritrochanteric fracture nonunion

Averting deformity is the primary objective of acute total care in an effort to simplify the postreconstructive care plan. On the contrary, nonunion can be tolerated in the subacute phase of recovery assuming that the mechanical alignment and rotation have been restored during the index surgical procedure. Nonunion surgery can subsequently be delayed until the patient has adequately healed other organ injuries and has improved nutrition so as to optimize the potential for fracture healing. These principles can then be applied to diverse patient conditions in order to obtain the most advantageous outcome.

Conclusion

The paradigm of orthopedic fracture care in the poly-traumatized patient has evolved from a period of delayed internal fracture fixation fraught with complications of nonunion and malunion to the theory of acute total care. Over this span, it has been demonstrated that early fracture stabilization tailored to the patient's systemic physiology and local soft tissue conditions does not place the patient at increased risk of systemic compromise but instead halts the cycle of ongoing musculoskeletal injury and fracture associated hemorrhage. Further, early fracture fixation limits the risk of malunion and nonunion while enabling early aggressive rehabilitation thereby maximizing outcomes.

Accordingly, employing damage control orthopedic surgery principles, poly-traumatized patients presenting in extremis or those with severely damaged soft tissue environments should receive rapid stabilization with an external fixator followed by later conversion to definitive fixation. However, all other patients should receive acute total fracture care consisting primarily of internal fixation to restore length, alignment and rotation of the fractured long bone. Interval external fixation still has a prominent role for staged management of high energy periarticular fracture.

References

1. Tscherne H, Schreyer H, Magerl F. Pulmonary and cardiac x-ray findings in traumatic fat embolism. Fortschr Geb Rontgenstr Nuklearmed. 1967;106:703–10.
2. Smith J. The results of early and delayed internal fixation of fractures of the shaft of the femur. J Bone Joint Surg Br. 1964;46:28–31.
3. Riska EB, von Bonsdorff H, Hakkinen S, Jaroma H, Kiviluoto O, Paavilainen T. Prevention of fat embolism by early internal fixation of fractures in patients with multiple injuries. Injury. 1976;8(2):110–6.
4. Bone LB, Johnson KD, Weigelt J, Scheinberg R. Early versus delayed stabilization of femoral fractures: a prospective randomized study. J Bone Joint Surg Am. 1989;71:336–40.
5. Seibel R, LaDuca J, Hassett JM, Babikian G, Mills B, Border DO, Border JR. Blunt multiple trauma (ISS 36), femur traction, and the pulmonary failure-septic state. Ann Surg. 1985;202(3):283–95.
6. Taeger G, Ruchholtz S, Waydhas C, Lewan U, Schmidt B, Nast-Kolb D. Damage control orthopedics in patients with multiple injuries is effective, time saving, and safe. J Trauma. 2005;59(2):409–16.
7. Pape HC, Auf'm'Kolk M, Paffrath T. Pri-mary intramedullary femur fixation in multiple trauma patients with associated lung contusion: a cause of posttraumatic ARDS? J Trauma. 1993;34:540–7.
8. Rotondo MF, Schwab CW, McGonigal MD, Phillips 3rd GR, Fruchterman TM, Kauder DR, Latenser BA, Angood PA. 'Damage control': an approach for improved survival in exsanguinating penetrating abdominal injury. J Trauma. 1993;35(3): 375–82.
9. Scalea TM, Boswell SA, Scott JD, Mitchell KA, Kramer ME, Pollak AN. External fixation as a bridge to intramedullary nailing for patients with multiple injuries and with femur fractures: damage control orthopedics. J Trauma. 2000;48:613–21.

10. Lichte P, Kobbe P, Dombroski D, Pape HC. Damage control orthopedics: current evidence. Curr Opin Crit Care. 2012;18(6):647–50.
11. Tarkin IS, Clare MP, Marcantonio A, Pape HC. An update on the management of high-energy pilon fractures. Injury. 2008;39(2):142–54.
12. Kadow TR, Siska P, Evans A, Taskin IS. Staged treatment of high energy midfoot fracture dislocations. Foot Ankle Int. 2014;35(12):1287–91. doi:10.1177/1071100714552077
13. Zelle BA, Marcantonio A, Tarkin IS. Functional long-term outcomes in poly-trauma patients with orthopaedic injuries. In: Pape HC, Peitzman A, Schwab CW, Giannoudis PV, editors. Damage control management in the polytrauma patient. New York: Springer; 2010. p. 439–51.
14. Schreiber VM, Tarkin IS, Hildebrand F, Darwiche S, Pfeifer R, Chelly J, Giannoudis P, Pape HC. The timing of definitive fixation for major fractures inpolytrauma – a matched-pair comparison between a US and European level I centres: analysis of current fracture management practice in polytrauma. Injury. 2011;42(7):650–4.
15. Pape HC, Tornetta 3rd P, Tarkin I, Tzioupis C, Sabeson V, Olson SA. Timing of fracture fixation in multitrauma patients: the role of early total care and damage control surgery. J Am Acad Orthop Surg. 2009;17(9):541–9.
16. Lasanianos NG, Kanakaris NK, Dimitriou R, Pape HC, Giannoudis PV. Second hit phenomenon: existing evidence of clinical implications. Injury. 2011;42(7):617–29.
17. Brinker MR, O'Connor DP. Exchange nailing of ununited fractures. J Bone Joint Surg Am. 2007;89(1):177–88.
18. Hakeos WM, Richards JE, Obremskey WT. Plate fixation of femoral nonunions over an intramedullary nail with autogenous bone grafting. J Orthop Trauma. 2011;25(2):84–9.
19. Phillips JH, Rahn BA. Fixation effects on membranous and endochondral onlay bone graft resorption. Plast Reconstr Surg. 1988;82:872–7.
20. Kobbe P, Tarkin IS, Pape HC. Use of the 'reamer irrigator aspirator' system for non-infected tibial non-union after failed iliac crest grafting. Injury. 2008;39(7):796–800.
21. Quintero AJ, Tarkin IS, Pape HC. Technical tricks when using the reamer irrigator aspirator technique for autologous bone graft harvesting. J Orthop Trauma. 2010;24(1):42–5.
22. Friedlaender GE, Perry CR, Cole JD, Cook SD, Cierny G, Muschler GF. Osteogenic protein-1 (bone morphogenetic protein-7) in the treatment of tibial nonunions: a prospective, randomized clinical trial comparing rhOP-1 with fresh bone autograft. J Bone Joint Surg Am. 2001;83-A(1 Suppl 2):S151–8.
23. Jones AL, Bucholz RW, Bosse MJ, Mirza SK, Lyon TR, Webb LX, et al. Recombinant human BMP-2 and allograft compared with autogenous bone graft for recon- struction of diaphyseal tibial fractures with cortical defects. J Bone Joint Surg. 2006;88-A(7):1431–41.
24. Carragee EJ, Hurwitz EL, Weinner B. A critical review of recombinant human bone morphogenic protein-2 trials in spinal surgery: emerging safety concerns and lessons learned. Spine J. 2011;11:471–91.
25. Tarkin IS, Siska PA, Zelle BA. Soft tissue and biomechanical challenges encountered with the management of distal tibia nonunions. Orthop Clin North Am. 2010;41(1):119–26.
26. Mitchell SE, Keating JF, Robinson CM. The treatment of open femoral fractures with bone loss. J Bone Joint Surg Br. 2010;92(12):1678–84.
27. Klaue K, Knothe U, Masquelet A. Etret biologique des membranes a corps etranger induites in situ sur la consolidation des greffes d'os spongieux. Rev Chir Orthop Suppl. 1995;70:109–10.
28. Giannoudis PV, Faour O, Goff T, Kanakaris N, Dimitriou R. Masquelet technique for the treatment of bone defects: tips-tricks and future directions. Injury. 2011;42(6):591–8.
29. Donegan DJ, Scolaro J, Matuszewski PE, Mehta S. Staged bone grafting following placement of an antibiotic spacer block for the management of segemental long bone defects. Orthopedics. 2011;34(11):E730–5.
30. Dugan TR, Hubert MG, Siska PA, Pape HC, Tarkin IS. Open supracondylar femur fractures with bone loss in the polytraumatized patient - Timing is everything! Injury. 2013;44(12):1826–31.
31. Ring D, Jupiter JB, Gan BS. Infected nonunion of the tibia. Clin Orthop Relat Res. 1999;369:302–11.
32. Gruen GS, Mears DC. Arthrodesis of the ankle and subtalar joints. Clin Orthop Relat Res. 1991;268:15–20.
33. Chen AF, Choi LE, Colman MW, Goodman MA, Crossett LS, Tarkin IS, McGough RL. Primary versus secondary distal femoral arthroplasty for treatment of total knee arthroplasty periprosthetic femur fractures. J Arthroplasty. 2013;28(9):1580–4.
34. Colman M, Choi L, Chen A, Crossett L, Tarkin I, McGough R. Proximal femoral replacement in the management of acute periprosthetic fractures of the hip: a competing risks survival analysis. J Arthroplasty. 2014;29(2):422–7. Epub 12 Jul 2013.

Psychological Sequelae After Severe Trauma

28

Bianca Voss, Frank Schneider, and Ute Habel

Contents

28.1	Introduction	399
28.2	**Post-traumatic Stress Disorder**	400
28.2.1	Association of PTSD and Traumatic Brain Injuries	401
28.2.2	Risk Factors for the Development of Psychological Sequelae	402
28.2.3	Gender Differences	402
28.2.4	Treatment Options	403
28.3	**Acute Stress Disorder**	403
28.4	**Psychological Sequelae in Different Trauma Settings**	403
28.5	**Psychological Sequelae in Children and Adolescents**	404
Conclusion		404
References		404

B. Voss (✉)
Department of Psychiatry, Psychotherapy, and Psychological Medicine,
RWTH Aachen University, Medical School,
Pauwelsstr. 30, Aachen 52074, NRW, Germany
e-mail: bianca.Voss@jva-hagen.nrw.de

F. Schneider • U. Habel
Department of Psychiatry, Psychotherapy and Psychosomatics, RWTH Aachen University
Pauwelsstr. 30 52074 Aachen

JARA – Translational Brain Medicine, RWTH Aachen University & Research Centre Jülich
e-mail: uhabel@ukaachen.de

28.1 Introduction

Events that bear the risk for traumatization include threat (esp. threat of death) or injuries. Especially, long-term psychological consequences after major traumas (also due to illness-related impairments) can have serious effects on the patient's quality of life resulting in heightened psychological morbidity. As of today, the psychological and psychiatric consequences of major trauma incidences are only superficially explored although psychological consequences of traumatic injuries represent a major public health burden.

A large influence on the current knowledge and literature regarding traumatic injuries to date originate from a specific field of research, the military personnel. Concerning German military personnel a recent study by Wittchen et al. [29] included 1599 German soldiers deployed to Afghanistan during the ISAF mission and demonstrated a significantly heightened risk of traumatic experiences and the development of a post-traumatic stress disorder (PTSD) in deployed soldiers compared to soldiers stationed in Germany. Furthermore, they showed that a high percentage of these persons concerned were not diagnosed nor treated after their return.

In general, research concerning the psychological sequelae of major trauma is limited by methodological constraints, e.g., the frequent lack of a control group in most study settings based in

© Springer-Verlag Berlin Heidelberg 2016
H.-C. Pape et al. (eds.), *The Poly-Traumatized Patient with Fractures:
A Multi-Disciplinary Approach*, DOI 10.1007/978-3-662-47212-5_28

emergency care departments. Therefore, results have to be discussed against the background of general population-wide prevalence rates rather than an injury-specific comparison group.

Among the possible psychological diagnoses following major trauma, post-traumatic stress disorder (PTSD) and acute stress disorder (ASD) are of particular relevance. For PTSD, the three most important symptoms include intrusions, avoidance and arousal.

The fifth edition of the Diagnostic and Statistical Manual of Mental Disorders (DSM 5, [2]) requires a history of exposure to a traumatic event that fulfills symptoms from four clusters: intrusion, aversion, negative alterations in cognition and mood and alterations in arousal and reactivity. Two additional criteria require duration of the symptoms for more than 1 month and the presence of a significant symptom-related distress or functional impairment.

ICD-10 criteria for ASD and PTSD are listed below [30].

F43.0 Acute Stress Reaction
A transient disorder that develops in an individual without any other apparent mental disorder in response to exceptional physical and mental stress and that usually subsides within hours or days. Individual vulnerability and coping capacity play a role in the occurrence and severity of acute stress reactions. The symptoms show a typically mixed and changing picture and include an initial state of "daze" with some constriction of the field of consciousness and narrowing of attention, inability to comprehend stimuli, and disorientation. This state may be followed either by further withdrawal from the surrounding situation (to the extent of a dissociative stupor – F44.2), or by agitation and over-activity (flight reaction or fugue). Autonomic signs of panic anxiety (tachycardia, sweating, flushing) are commonly present. The symptoms usually appear within minutes of the impact of the stressful stimulus or event, and disappear within two to three days (often within hours). Partial or complete amnesia (F44.0) for the episode may be present. If the symptoms persist, a change in diagnosis should be considered.

F43.1 Post-traumatic Stress Disorder
Arises as a delayed or protracted response to a stressful event or situation (of either brief or long duration) of an exceptionally threatening or catastrophic nature, which is likely to cause pervasive distress in almost anyone. Predisposing factors, such as personality traits (e.g. compulsive, asthenic) or previous history of neurotic illness, may lower the threshold for the development of the syndrome or aggravate its course, but they are neither necessary nor sufficient to explain its occurrence. Typical features include episodes of repeated reliving of the trauma in intrusive memories ("flashbacks"), dreams or nightmares, occurring against the persisting background of a sense of "numbness" and emotional blunting, detachment from other people, unresponsiveness to surroundings, anhedonia, and avoidance of activities and situations reminiscent of the trauma. There is usually a state of autonomic hyperarousal with hypervigilance, an enhanced startle reaction, and insomnia. Anxiety and depression are commonly associated with the above symptoms and signs, and suicidal ideation is not infrequent. The onset follows the trauma with a latency period that may range from a few weeks to months. The course is fluctuating but recovery can be expected in the majority of cases. In a small proportion of cases the condition may follow a chronic course over many years, with eventual transition to an enduring personality change (ICD-10, F62.0).

28.2 Post-traumatic Stress Disorder

The lifetime experience of an event that fulfills criteria for a trauma is rather frequent within the general population [12], but the pure exposure to a trauma is not sufficient for the development of a PTSD. Four weeks after a traumatic event, the prevalence of PTSD is estimated as high as 41 % [17]. Lukaschek et al. identified several risk factors which constitute a higher risk for PTSD. Inter alia accidents, nonsexual assault and life-threatening illness are associated with a higher risk for a full and partial PTSD. These results are interpreted as evidence that the impact or consequences following a traumatic event depends on characteristics of the event as well as individual factors of the victim. Regarding clinical and emergency care, victims of motor vehicle accidents or other forms of traumatic injuries (e.g., burn victims, head injuries, orthopedic traumatic injuries) are at heightened risk for development of mental disorders (e.g., PTSD, depression), but studies examining these factors still produce heterogeneous results [26, 20, 24]. Regarding the occurrence of PTSD after a major traumatic experience it is important to keep in mind that PTSD is often associated with other comorbid psychiatric disorders [21]. The most frequent comorbidity is major depressive disorder (MD) followed by generalized anxiety disorder.

Feelings of anxiety and sometimes dissociation in the immediate aftermath of a traumatic experience are common among injury victims. These symptoms are short-term and can be regarded as normal. Although still heterogeneous, literature results show PTSD prevalence rates of 10–30 % after traumatic experiences [24, 10, 20]. Using the Impact of Event Scale (IES, [14]), Haagsma et al. [10] could show that 1 year after a major trauma, 23 % of the participants had probable PTSD, after 2 years 20 % reported symptoms of probable PTSD. Due to the fact that in this study the prevalence of PTSD was exclusively assessed via the IES, the authors use the term "probable PTSD" to indicate the lack of further diagnostic evaluations. Furthermore, female gender and comorbid diseases were strong predictors of probable PTSD after 1 year, whereas head injury and injury of extremities were predictors of PTSD after 2 years. Of the probable PTSD patients after 1 year, 79 % had persistent PTSD after 2 years.

An interesting differentiation was made by Alarcon et al. [1]. In their study, they showed a general prevalence rate of 25 % of PTSD after trauma using the PTSD Checklist-Civilian (PCL-C). Further analyses confirmed an association between the mechanism of injury and the risk of developing symptoms of PTSD with the highest correlation after assault (43 %) and younger age, female gender and motor vehicle collision as significant independent risk factors for subsequent PTSD.

In 2010, an Australian research group published results of a prospective study including over 1000 victims of traumatic injuries [4]. Interested in the extent of psychiatric diagnoses as a consequence of traumatic injuries, they concluded that after 12 month 31 % of their participants developed a psychiatric disorder and 23 % developed a psychiatric disorder that they never had in their life prior to the accident. Psychiatric diagnoses not only included PTSD but a wide range of disorders, most frequently major depression and generalized anxiety disorder next to PTSD.

An Israeli study showed that even noninjured victims of trauma show elevated heart rates at admission to emergency care facilities [25]. According to the authors, these physiological responses in combination with skin conductance (PTSD patients show impaired habituation in skin conductance rates after startling stimuli) and left lateral frontalis electromyogram levels (PTSD patients show increased responses) can distinguish between patients who will develop PTSD later on and patients who will not. The fact that these physiological responses will often recur if a patient re-experiences traumatic memories (flashbacks) during the day can reinforce the memories and thereby impair the treatment of the symptoms.

28.2.1 Association of PTSD and Traumatic Brain Injuries

Against the background of elevated prevalence rates of mental illness after traumatic brain injuries (TBI), Bryant et al. [4] examined the connection between TBI and psychiatric illnesses as a consequence of TBI using a multi-center prospective study design in Australia. Besides the fact that they could repeat an estimated prevalence of 23 % for incidental psychiatric illnesses 12 months after the traumatic event, they documented depression as the most frequent consequence, followed by generalized anxiety disorder and PTSD. These results emphasize the notion that psychological consequences after major traumas can take shape in various diagnoses, not only PTSD. The authors concluded that PTSD may pose a major threat after traumas in connection with violent incidents, while traumatic incidences that involve traumatic brain injuries seem to trigger a broader range of psychiatric consequences.

Recently, Roitman et al. [23] could show that accident victims who suffered head injury in combination with loss of consciousness had a higher prevalence of PTSD 10 days and 8 months after the accident than accident victims without loss of consciousness. In addition, patients with traumatic brain injury and loss of consciousness showed slower recovery rates from early PTSD symptoms. In their discussion, the authors hypothesize that the loss of consciousness may suppress the victims' ability to form a coherent memory of the accident or the traumatic event. Furthermore, they hypothesized that subsequent damage (e.g., headache, confusion) may interfere with processes of fear distinction and therefore act as a reinforcer of traumatic memory processes. Altogether, there seems to be a heightened risk of PTSD if the accident or traumatic event is associated with traumatic brain injury and loss of consciousness, events with a high likelihood regarding polytraumatic injuries. Therefore the medical personnel should be instructed and prepared to consider psychological consequences after polytraumatic incidents as well as physical consequences.

In addition, sport injuries can cause mild TBI and in consequence result in psychiatric problems. Especially acute concussions – injuries common in contact sports, e.g., football or hockey – can lead to prolonged symptoms and to a post-concussive syndrome (PCS). These symptoms can include psychiatric problems like emotional symptoms, behavioral changes or sleep disturbances [18].

28.2.2 Risk Factors for the Development of Psychological Sequelae

The development of psychological disorders after traumatic experiences depends on individual predispositions as well as the characteristics of the traumatic situation. Breslau et al. [3] refuted the rather common assumption of an inverse association between individual predispositions on PTSD and trauma severity with their data. A meta-analysis by Ozer et al. [22] estimated the importance of three antecedent risk factors for PTSD: prior trauma, prior psychological adjustment and family history of psychopathology. These predispositions contributed equally to the development of PTSD after two forms of traumatic experiences: interpersonal violence or accidents. These findings may partially explain the relatively low risk for PTSD after severe polytraumatic experiences as well as the large discrepancy regarding type and assumed severity of traumatic experiences that are supposed to trigger a subsequent PTSD. The authors could not prove that trauma severity or trauma type can be regarded as risk factor for PTSD but as a modifier for the aforementioned risk factors.

28.2.3 Gender Differences

In general, women show a higher prevalence of PTSD than men. The literature furthermore indicates gender differences in risk factors, comorbid disorders (major depression, anxiety) and predispositions for the development of PTSD. This common finding is supposedly mediated by gender differences in the initial responses to trauma [15]. Peritraumatic risk factors seem to be the strongest predictors of PTSD. These risk factors include amongst others peritraumatic dissociation, subjective perception of life threat and general coping appraisals. As Ehlers and Clark [8] pointed out, a main aspect for the development of long-lasting psychological effects after a trauma is the evaluation of the perceived threat during the event and the evaluation of the subjective ability to cope with these extreme experiences. Should a trauma victim appraise the traumatic event as catastrophic, the probability of a sustained feeling of threat and danger after the trauma has passed rises and inhibits a successful treatment of the experiences. This effect increases the possibility of PTSD as a consequence of failed coping strategies.

Symptoms of peritraumatic dissociation include distortions of perceptions during the course of a traumatic event (e.g., feeling that time is running in slow motion or the perception of a disconnection from the own body). This contains in itself the danger that due to these dissociative symptoms the consolidation of the traumatic memory is impaired resulting in impaired adaptation and coping processes [16].

Regarding motor vehicle accidents, Fullerton et al. [9] assessed prevalence rates of PTSD in both genders and differentiated between the four different symptom clusters as demanded in DSM-IV. They could not detect a gender difference regarding the symptom cluster B (re-experiencing) but regarding the symptom cluster C and D (avoidance/numbing and arousal) women were more likely to meet the required symptoms. The most significant result is a gender difference in peritraumatic dissociative symptoms which was associated with a higher risk of PTSD in women as compared to men.

28.2.4 Treatment Options

Treatment options of PTSD include psychotherapeutic interventions (mostly techniques from cognitive-behavioral treatment programs, e.g., exposure-based cognitive therapy) as well as pharmacological interventions (e.g., selective serotonin reuptake inhibitors, SSRI). In addition, some effort has been invested in therapeutic interventions regarding eye movement desensitization and reprocessing therapy (EMDR).

28.3 Acute Stress Disorder

Acute stress disorder (ASD) describes a psychological condition as a response to a traumatic event. The prevalence of ASD following traumatic events is estimated with 13–16 % depending on the nature of the traumatic experience [5]. As opposed to PTSD or major depression, ASD does not constitute a psychiatric disorder but rather a normal response following traumatic exposure and typically transient, characterized by a reduction of the predominant symptoms (e.g., numbness, detachment, dissociative amnesia) hours or days after the traumatic event. ASD represents a particularly "new" diagnostic category, as it was only introduced into the DSM in 1994. The utility of a diagnosis of ASD as predictor for a diagnosis of PTSD is controversial. O'Donnell et al. [21] indicated that a considerable number of accident victims developed PTSD without a prior ASD diagnosis. The authors further stated that affective and anxious symptoms could be a more important predictor for PTSD considering the fact that major depression and generalized anxiety disorders show the highest comorbidity with PTSD. In contrast, other research groups emphasized that very early interventions in patients with ASD can prevent the development of a subsequent PTSD [6]. In this study, the authors compared the efficacy of exposure therapy compared to a trauma-focused cognitive restructuring intervention in participants who were exposed to civilian traumatic events (nonsexual assault or motor vehicle accidents). Results showed a superiority of exposure therapy in the reduction of PTSD occurrence 6 month after the traumatic experience.

28.4 Psychological Sequelae in Different Trauma Settings

Although motor vehicle accidents may be the most common nonpsychological cause of trauma, other sources of injuries can result in additional psychological sequelae as well. In the field of burn injuries, psychosocial consequences seem to predominate after physical recovery. This form of disfiguring and long-lasting alterations of the patient's outward appearance in addition to the traumatic experience and the usually painful treatment process may result in psychopathological secondary damages. Beside high prevalence rates of PTSD and MD after burn injuries, problems in the patients' social and occupational life as well as a reduced quality of life due to potential disfiguration and disability have to be considered [27]. Again, preexisting psychiatric symptoms like depression or anxiety as well as high levels of psychological distress after the injury seem to negatively influence the physical and psychological recovery of the burn victims [28].

Patients with spinal cord injuries resulting in paralysis experience psychological difficulties and high levels of post-traumatic stress symptoms [11]. Due to the life-changing consequences of permanent paralysis the level of acceptance of these changes and constraints have to be considered while planning treatment and rehabilitation programs for the concerned patients.

28.5 Psychological Sequelae in Children and Adolescents

Among children and adolescents, the prevalence of PTSD is likewise common and is as high as 12–46 % 4 months after a road traffic accident and 13–25 % 4–12 months after the accident [19]. Although only a limited number of data exist, these prevalence rates have been reported throughout different studies [13, 7]. The psychological care of these victims becomes even more problematic due to difficulties and heterogeneities in diagnosing symptoms of PTSD in children and adolescents as well as differences in developmental stages. In addition, it is important to keep in mind that particularly children of young age are susceptible to develop psychological symptoms after accidents or other traumatic events even if they were only mildly injured or not injured at all. In consequence of these results, it is important to acknowledge the psychological impact of injuries following accidents in minors and to keep this knowledge in mind while treating physical injuries in a clinical setting.

> **Conclusion**
>
> Traumatic experiences are the leading trigger for trauma-related psychiatric disorders. Being confronted with a probable traumatic event, the majority of patients entering the emergency care system immediately after the event will feel symptoms of acute stress responses. This response is universal and not predictive of following PTSD. While most patients will recover from this temporary feeling of stress, a substantial percentage will not possess the necessary mechanisms or resilience and will develop prolonged psychological sequelae of the traumatic experience, resulting in psychiatric disorders like PTSD or depression. Considering the prevalence of traumatic injuries and the prevalence of subsequent mental diagnoses, this represents a personal as well as a major public health issue. While the treatment of physical injuries is of course of capital importance, the medical personnel – rescue workers as well as emergency department staff and treating physicians – should be aware of these potential impending long-term impairments and be prepared to start the necessary referrals to psychiatric or psychosocial services.
>
> Recovery and rehabilitation after a major traumatic injury are expensive and long-lasting. Furthermore, these processes can be hindered by additional and in the last resort undiagnosed psychiatric illnesses as a consequence of the traumatic event. The prevalence of mental illness after major trauma can be underestimated in the clinical or emergency care context because in general, trauma-related psychiatric illnesses are easier affiliated with psychological trauma (e.g., rape or physical assault).
>
> In addition, it is understandable that the primary concerns are the patient's physical injuries which bear a certain amount of risk that the psychological consequences are "lost" or forgotten during the course of treatment and rehabilitation. On the other hand, it is important for clinicians to keep in mind that poor physical health as a consequence of major trauma can precede subsequent psychological impairments. These impairments in return can prolong the patient's recovery and/or the patient's functional outcome. Therefore, the treating physicians need to be aware of the psychological consequences and of possible treatment facilities and options.

References

1. Alarcon LH, Germain A, Clontz AS, Roach E, Nicholas DH, Zenati MS, Peitzman AB, Sperry JL. Predictors of acute posttraumatic stress disorder

1. symptoms following civilian trauma: highest incidence and severity of symptoms after assault. J Trauma. 2012;72(3):629–37.
2. American Psychiatric Association (APA). Diagnostic and statistical manual of mental disorders. 5th ed. Washington, DC: American Psychiatric Press; 2013.
3. Breslau N, Troost JP, Bohnert K, Luo Z. Influence of predispositions on post-traumatic stress disorder: does it vary by trauma severity? Psychol Med. 2013;43:381–90.
4. Bryant RA, O'Donnell ML, Creamer M, McFarlane AC, Clark CR, Silove D. The psychiatric sequelae of traumatic injury. Am J Psychiatry. 2010;167(3):312–20.
5. Bryant RA, Friedman MJ, Spiegel D, Ursano R, Strain J. A review of acute stress disorder in DSM-5. Depress Anxiety. 2011;28(9):802–17.
6. Bryant RA, Mastrodomenico J, Felmingham KL, Hopwood S, Kenny L, Kandris E, Cahill C, Creamer M. Treatment of acute stress disorder: a randomized controlled trial. Arch Gen Psychiatry. 2008;65(6):659.
7. Daviss WB, Mooney D, Racusin R, Ford JD, Fleischer A, McHugo GJ. Predicting posttraumatic stress after hospitalization for pediatric injury. J Am Acad Child Adolesc Psychiatry. 2000;39:576–83.
8. Ehlers A, Clark DM. A cognitive model of post-traumatic stress disorder. Behav Res Ther. 2000;38:319–45.
9. Fullerton CS, Ursano RJ, Epstein RS, Crowley B, Vance K, Kao TC, Dougall A, Baum A. Gender differences in posttraumatic stress disorder after motor vehicle accidents. Am J Psychiatry. 2001;158(9):1486–91.
10. Haagsma JA, Ringburg AN, van Lieshout EM, van Beeck EF, Patka P, Schipper IB, Polinder S. Prevalence rate, predictors and long-term course of probable posttraumatic stress disorder after major trauma: a prospective cohort study. BMC Psychiatry. 2012;12(1):236.
11. Hatcher MB, Whitaker C, Karl A. What predicts post-traumatic stress following spinal cord injury. Br J Health Psychol. 2009;14(3):541–61.
12. Hauffa R, Rief W, Brähler E, Martin A, Mewes R, Glaesmer H. Lifetime traumatic experiences and post-traumatic stress disorder in the German population: results of a representative population survey. J Nerv Ment Dis. 2011;199(12):934–9.
13. Holbrook TL, Hoyt DB, Coimbra R, Potenza B, Sise M, Anderson JP. Long-term posttraumatic stress disorder persists after major trauma in adolescents: new data on risk factors and functional outcome. J Trauma. 2005;58(4):764–71.
14. Horowitz M, Wilner N, Alvarez W. Impact of Event Scale: a measure of subjective stress. Psychosom Med. 1979;41(3):209–18.
15. Irish LA, Fischer B, Fallon W, Spoonster E, Sledjeski EM, Delahanty DL. Gender differences in PTSD symptoms: an exploration of peritraumatic mechanisms. J Anxiety Disord. 2011;25(2):209–16.
16. Koopman C, Classen C, Spiegel D. Predictors of post-traumatic stress symptoms among survivors of the Oakland/Berkeley, Calif., firestorm. Am J Psychiatry. 1994;151(6):888–94.
17. Lukaschek K, Kruse J, Emeny RT, Lacruz ME, Von Eisenhart Rothe A, Ladwig KH. Lifetime traumatic experiences and their impact on PTSD: a general population study. Soc Psychiatry Psychiatr Epidemiol. 2013;48(4):525–32.
18. McKee AC, Daneshvar DH, Alvarez VE, Stein TD. The neuropathology of sport. Acta Neuropathol. 2014;127(1):29–51.
19. Mehta S, Ameratunga SN. Prevalence of post-traumatic stress disorder among children and adolescents who survive road traffic crashes: a systematic review of the international literature. J Paediatr Child Health. 2012;48(10):876–85.
20. O'Donnell ML, Creamer M, Pattison P. Posttraumatic stress disorder and depression following trauma: understanding comorbidity. Am J Psychiatry. 2004;161(8):1390–6.
21. O'Donnell ML, Creamer M, Pattison P, Atkin C. Psychiatric morbidity following injury. Am J Psychiatry. 2004;161(3):507–14.
22. Ozer EJ, Best SR, Lipsey TL, Weiss DS. Predictors of posttraumatic stress disorder and symptoms in adults: a meta-analysis. Psychol Bull. 2003;129:52–73.
23. Roitman P, Gilad M, Ankri YL, Shalev AY. Head injury and loss of consciousness raise the likelihood of developing and maintaining PTSD symptoms. J Trauma Stress. 2013;26(6):727–34.
24. Schnyder U, Moergeli H, Klaghofer R, Buddeberg C. Incidence and prediction of posttraumatic stress disorder symptoms in severely injured accident victims. Am J Psychiatry. 2001;158(4):594–9.
25. Shalev AY, Orr SP, Pitman RK. Psychophysiologic assessment of traumatic imagery in Israeli civilian patients with posttraumatic stress disorder. Am J Psychiatry. 1993;150(4):620–4.
26. Starr AJ, Smith WR, Frawley WH, Borer DS, Morgan SJ, Reinert CM, Mendoza-Welch M. Symptoms of posttraumatic stress disorder after orthopaedic trauma. J Bone Joint Surg Am. 2004;86(6):1115–21.
27. Van Loey NEE, Maas CJM, Faber AW, Taal LA. Predictors of chronic posttraumatic stress symptoms following burn injury: results of a longitudinal study. J Trauma Stress. 2003;16(4):361–9.
28. Wisely JA, Wilson E, Duncan RT, Tarrier N. Pre-existing psychiatric disorders, psychological reactions to stress and the recovery of burn survivors. Burns. 2010;36(2):183–91.
29. Wittchen HU, Schönfeld S, Kirschbaum C, Thurau C, Trautmann S, Steudte S, Kotsche J, Höfler M, Hauffa R, Zimmermann P. Traumatic experiences and posttraumatic stress disorder in soldiers following deployment abroad: how big is the hidden problem? Dtsch Arztebl Int. 2012;109(35–36):559–68.
30. World Health Organization (2010). International statistical classification of diseases and related health problems - 10th revision, edition 2010. Geneva: World Health Organization.

Outcome After Extremity Injuries

29

Boris A. Zelle

Contents

29.1	Introduction	407
29.2	**Validity of Outcome Data**	408
29.2.1	Level of Evidence	408
29.2.2	Outcome Measures	408
29.2.3	Patient Follow-Up	410
29.3	**Numerical Results**	411
29.3.1	Size of Treatment Effect	411
29.3.2	Precision of the Estimated Treatment Effect	411
29.3.3	Statistical Significance	412
29.4	**Implications for the Clinical Practice**	412
29.5	**Outcomes After Lower Extremity Injuries**	414
29.6	**Outcomes After Upper Extremity Injuries**	416
Conclusion		416
References		417

B.A. Zelle, MD
Division of Orthopaedic Trauma,
Department of Orthopaedic Surgery,
University of Texas Health Science
Center at San Antonio,
Floyd Curl Dr, MC 7774,
San Antonio, TX 78229, USA
e-mail: boris.zelle@gmail.com

29.1 Introduction

Well-designed clinical research remains necessary in order to critically evaluate the quality of orthopaedic trauma care and to advance the field of orthopaedic trauma surgery. Recently, evidence-based medicine has provided valuable insights into clinical research and has emphasized the significance of thoughtful study designs and the importance of a critical appraisal of the orthopaedic literature. In particular, with the growing body of the orthopaedic trauma literature, it is becoming increasingly important for clinicians and researchers to critically evaluate the available literature, to recognize strengths and weaknesses of study designs, and to interpret study results within the clinical context. When assessing orthopaedic trauma outcome studies, important questions to ask include the following:

1. What is the validity of the presented outcome data?
2. What are the numerical results of the outcomes reported?
3. What are the implications for the clinical practice?

In this chapter, these fundamental questions will be discussed in the context of the orthopaedic trauma outcome literature. Moreover, this chapter will summarize the results of the most

© Springer-Verlag Berlin Heidelberg 2016
H.-C. Pape et al. (eds.), *The Poly-Traumatized Patient with Fractures:
A Multi-Disciplinary Approach*, DOI 10.1007/978-3-662-47212-5_29

pertinent outcome studies in the field of orthopaedic trauma and emphasize the lessons learned from these studies.

29.2 Validity of Outcome Data

When assessing the validity of an orthopaedic outcome study, the most pertinent question is whether the study represents an unbiased estimation of treatment outcomes. Bias (or systematic error) is typically linked to the study design and execution of a study. Important variables when assessing the validity of orthopaedic trauma outcome data include the following:

1. Level of evidence
2. Outcome measures used
3. Patient follow-up

29.2.1 Level of Evidence

Evidence-based medicine has recently gained significant prominence in the field of orthopaedic surgery as well as in other areas of medicine. Numerous manuscripts and textbooks in this field have been published and a detailed review of all evidence-based medicine principles is beyond the scope of this chapter. One of the key aspects of evidence-based medicine is the introduction of a hierarchical rating system for the level of evidence whereby the level of evidence is grading the quality of the overall study design. In this context, a higher level of evidence suggests a lower risk of bias. Most rating systems for the level of evidence of therapeutic studies (i.e. the majority of orthopaedic trauma outcome studies) use a five level scale including level 1 (randomized clinical trial), level 2 (prospective cohort study or poor quality randomized clinical trial), level 3 (case control study), level 4 (case series), and level 5 (expert opinion) [1]. Most major orthopaedic journals have adapted this five-level hierarchical rating system and grade the published articles accordingly. When assessing the clinical impact of outcome studies, the hierarchical grading system for the level of evidence plays an important role.

While this rating system provides the reader with important information on potential bias, the level of evidence should also be used cautiously. First, the level of evidence only provides an overall assessment of the study design and further critical assessment of the study methods and study results is necessary. Second, randomized clinical trials are not always possible for each clinical scenario in particular in the orthopaedic trauma population. For instance, the Lower Extremity Assessment Project (LEAP) was designed to evaluate the outcomes of mangled lower extremity injuries to assess lower limb amputation versus salvage [2, 3]. This well-designed study was performed in a non-randomized fashion as randomizing patients with mangled lower extremities into limb salvage versus amputation would not appear feasible [2, 3]. Finally, it must be emphasized that no single study can provide a definitive answer to a study question. Clinical treatment algorithms in orthopaedic trauma should be based on a composite assessment of the entire literature and should consider all levels of evidence from level 1 (randomized clinical trial) to level 5 (expert opinion).

29.2.2 Outcome Measures

The outcome measure is another important variable when assessing the validity of an orthopaedic trauma outcome study. In the orthopaedic trauma literature, numerous outcome scoring systems have been used [4]. In general, outcome measures can be divided into clinician-based and patient-reported outcome measures. Standardized outcome measures may focus on general health, body region-specific function, or disease-specific function. As of today, no general recommendations exist as to which outcome measures should be used in orthopaedic trauma outcome studies. Well-designed outcome studies provide outcome data on the patient's general health in addition to a body region- or disease-specific questionnaire.

When using more than one outcome measure, it is crucial to identify the main outcome measure of the study. The main outcome measure should be according to the main hypothesis that is being tested in the study. Another important consideration is whether the used outcome measure has been validated in prior investigations. An outcome instrument is considered valid if it truly measures what it is supposed to measure. In this context, it is important to emphasize that validation of an outcome measure is not an "all or nothing" concept and validity has several components (e.g. face validity, criterion validity, construct validity, content validity, etc.). A detailed discussion of outcome measure validation procedures is beyond the scope of this chapter. In general, the validity of an outcome measure is typically established by comparison between the tested outcome measure and an established outcome instrument. For instance, the short musculoskeletal function assessment (SMFA) questionnaire has been established for the use of trauma patients, and validation studies used the Medical Outcomes Study 36-Item Short Form (SF-36), a well-established and validated outcome measure, as a reference to test the validity of the SMFA [5].

As of today, the SF-36 can be considered to be one of the most commonly employed outcome instruments in orthopaedic trauma surgery as well as in orthopaedic surgery in general [6]. The SF-36 is a patient-reported outcome measure and measures the overall patient's quality of life. It consists of 36 items and the questionnaire can be completed within 5–10 min. It has been validated and used in numerous studies within the field of orthopaedic surgery as well as in other fields of medicine [7–14]. The SF-36 also allows the comparison of outcomes with normative population data from age- and sex-matched controls. Moreover, the SF-36 has been translated and validated in multiple languages and international normative data have been recorded. Despite these favourable characteristics, the SF-36 also has some limitations that can affect the interpretation of outcome data. First, the items of the SF-36 tend to focus more on lower extremity function than on upper extremity function [15]. This emphasizes the importance of including a body region-specific questionnaire along with a general health questionnaire when performing clinical outcome research in orthopaedic trauma. Moreover, the SF-36 does not incorporate certain basic quality of life domains, such as sexual function or sleep. In some instances, this may lead to the scenario that improvements as well as diminishments in these areas may go undetected. For example, patients undergoing treatment of pelvic fractures may frequently be impaired by sexual dysfunction and recording of SF-36 data may be limited by "ceiling effects" as well as "floor effects".

Besides the assessment of clinical limitations, healthcare utilization and treatment costs represent important outcome measures when evaluating the efficiency of orthopaedic trauma care. In this context, it is important to emphasize that most importantly treatment should be chosen based on the best patient's interest and according to highest level of standard of care. However, in the treatment of patients with musculoskeletal injuries, the orthopaedic trauma surgeon is also mandated to make fiscally sound decisions since cost-efficient treatment is above all in the best interest of the injured patient. The question of cost effectiveness becomes specifically important in areas of complex surgeries and when the effectiveness of treatment remains uncertain. In the orthopaedic trauma literature, a frequently discussed topic is the cost effectiveness of limb salvage versus amputation in patients with mangled lower extremities. A more detailed discussion on the clinical aspects of this topic will be provided in the Sects. 29.5. As of today, the question of limb salvage versus amputation remains controversial and the multiple medical, social, and economical aspects need to be considered when discussing treatment plans with the injured patient. In brief, patients with mangled lower extremities face the situation that attempted limb salvage may offer them the undoubted benefits of keeping their lower limb. In order to achieve this favourable result, these patients may undergo several reconstructive surgical procedures and repeat hospitalizations with the remaining risk of requiring an amputation at a later time point. In contrast, primary amputation may offer the potential benefits of quicker discharge from the hospital, earlier ambulation after prosthesis fitting is completed, and earlier return to work. For these reasons, the medical outcomes of limb salvage versus amputation need to be assessed carefully and this sensitive issue needs to be discussed thoroughly with the patients and their families. Importantly, the financial aspects of treatment have to be included in this discussion as it is clearly in the patient's best interest to be educated about the costs that will incur from medical treatment, hospitalizations, time away from work, as well as lifetime costs from ongoing prostheses needs. Recent investigations have focused on cost utility analyses of amputation versus limb salvage in patients with mangled lower extremities emphasizing the importance of costs as a critical outcome measure [16].

29.2.3 Patient Follow-Up

A critical assessment of study data also requires a careful evaluation of the clinical follow-up that was obtained in the study presented. In an outcome study reporting on patients' recoveries after treatment of extremity injuries, the presented study data may be flawed if subjects who received treatment are not included in the data analysis due to lack of follow-up data. This lack of outcome data may both overestimate as well as underestimate the benefit of treatment effects depending on the outcome of patients not returning for follow-up. Hypothetically, "best case scenarios" and "worst case scenarios" could occur. Thus, patients who did not recover well from their injuries may be upset about their outcome and chose to receive follow-up treatment at a different institution ("I am upset. I am not going back"). If a large number of these patients do not get enrolled in the outcome analysis, the recorded outcomes may be better than the actual real outcomes that have been achieved with the treatment rendered. Vice versa, patients who achieved an excellent recovery potentially may decide to skip their follow-up since they may not feel the necessity to seek any further evaluations ("I feel fine. Why bother?"). If a large number of these patients do not get enrolled in the outcome analysis, the recorded outcomes may be worse than the actual real outcomes of the treatment rendered. For these reasons, any remarkable loss of follow-up carries the risk of skewing the study data and a critical assessment of the study data needs to include the assessment of the loss of follow-up.

Current guidelines of major orthopaedic journals request that any randomized controlled trial with more than 20 % loss of follow-up should be downgraded from an evidence level 1 to an evidence level 2 study [www.jbjs.org]. However, these recommendations are based on traditional postulations and it remains unclear how much loss of follow-up can be considered as acceptable. Recent statistical models using trauma databases have pointed out that even less than 20 % loss of follow-up may frequently yield in a significant change of study results [17]. For these reasons, authors of orthopaedic trauma outcome studies should not only report their loss of follow-up, but should also report which specific attempts were made to minimize loss of follow-up and should record the data available on those patients who did not comply with their final follow-up examinations.

Besides the loss of follow-up, orthopaedic trauma outcome studies need to be assessed for their length of follow-up. Patients with extremity injuries go through different phases in their rehabilitation process. Along the different phases of the recovery process, different outcome variables can be recorded. The immediate postoperative period provides information on early complications, such as surgical site infections, mortality rates, thromboembolic events, and length of hospital stay. Within the first few months after surgery further information, such as fracture healing and return to work, can be recorded. Furthermore, long-term outcome studies provide valuable information on the functional recovery, health-related quality of life, as well as the incidence of posttraumatic arthritis and the need for late reconstructive procedures. Many guidelines have recommended 2-year outcome evaluations for patients with extremity injuries. However, these postulations have recently been challenged.

In patients with mangled lower extremities, comparisons between 1-year follow-up data and 2-year follow-up data have shown that 1-year follow-up data provide sufficient information to test the pertinent study hypotheses while creating significantly less costs than 2-year follow-up evaluations [18]. Current recommendations from the major orthopaedic trauma journal request 6 months of follow-up for pure soft tissue injuries, 1-year patients follow-up for fracture care, and 2-year follow-up data for treatment of arthritic conditions [http://journals.lww.com/jorthotrauma]. These issues emphasize the significant implications of the length of follow-up as an important variable for assessing the validity of outcome data. The length of follow-up provides valuable information as to which specific outcome measures can be addressed in outcome studies on patients with extremity injuries.

29.3 Numerical Results

The numerical results of clinical studies should be scrutinized carefully in order to make appropriate conclusions for the clinical practice. When reviewing the results of orthopaedic trauma outcome studies, pertinent questions include the following:

1. How large was the treatment effect?
2. How precise was the estimate of the treatment effect?
3. What is the statistical significance?

29.3.1 Size of Treatment Effect

With regard to the size of the treatment effect, it is important to distinguish if the main outcome measure was a continuous variable (e.g. SF-36 scores ranging from 0 to 100) or a dichotomous variable (fracture union versus fracture nonunion). For dichotomous variables, several measures of treatment effect size exist. These include odds ratios, relative risk, relative risk reduction, absolute risk reduction, and numbers needed to treat.

In the orthopaedic trauma literature, odds ratios are frequently used to measure treatment effects. The odds ratio is a measure of the association between a risk factor and an outcome. The odds ratio calculates the odds that a particular outcome will occur in association with a particular risk factor as compared to the odds of the outcome occurring in the absence of this particular risk factor. An odds ratio of 1.0 means that the evaluated risk factor does not increase the risk of the recorded outcome. An odds ratio of 1.5 means that the evaluated risk factor increases the odds of the outcome to occur by 50 %. Odds ratios are typically used in case control studies and in logistic regression models.

Another frequently used measure for the size of the treatment effect in the orthopaedic trauma literature is the relative risk reduction. The relative risk reduction plays an important role in the reporting of treatment effects that are observed in prospective controlled trials. The relative risk reduction is expressed as a percentage. A risk reduction of 50 % means that treatment A reduces the risk of a particular outcome by 50 % as compared to treatment B.

29.3.2 Precision of the Estimated Treatment Effect

It is important to realize that the measures of the size of the treatment effect, such as the odds ratio and the relative risk reduction, are point estimates and further information is required in order to measure the precision of these estimates. The confidence interval is the range within which the true treatment effect falls and provides important information on the precision of the estimated size of the treatment effect. By convention, the 95 % confidence interval is used to measure the precision of a point estimate. Thus, a 95 % confidence interval means that if the same study was repeated, there was a 95 % chance that the estimated treatment effect would fall within this interval again. The 95 % confidence interval largely depends on the sample size. With larger sample sizes, the estimated treatment effects become more precise and the 95 % confidence interval becomes smaller. Thus, the clinician can be more confident that the true treatment effect is close to the treatment effect recorded in the outcome study. Vice versa, studies with smaller sample sizes typically result in larger 95 % confidence intervals and with large confidence intervals, the clinician may remain uncertain where the true treatment effect lies.

29.3.3 Statistical Significance

The p value provides another measure for the precision of the results. The p value provides the probability of an α-error. An α-error means that a study observes a difference between two study groups when in fact there is no difference. By convention, a cut-off p value of 0.05 is used in most clinical studies. Thus, a $p<0.05$ means that there is 5 % chance of recording a difference between two study groups when in fact there is no difference between these two groups.

Typically, the p value is given great importance by authors, journals, and clinicians. Thus, a common perception is that reported differences are real whenever there is statistical significance whereas reported differences supposedly are irrelevant when the data is not statistically significant. However, there are several issues with this viewpoint; there has been a recent trend to de-emphasize the importance of the p value [19]. As stated above, the p value only provides information on the size of an α-error and it does not provide any information on the size of the treatment effect. In addition, the 0.05 cut-off is arbitrary and in many scenarios changing only very few events may sometimes change the results from statistically significant to non-significant and vice versa. For this reason, it has been suggested by journal editors to avoid stating "significantly different", but rather providing the exact p value [19]. Moreover, some trauma outcome studies [20–22] have been criticized for artificially creating statistically significant results by deviating from the main hypothesis and performing multiple subgroup analyses with multiple repeat testing procedures that ultimately may yield p values that fall below 0.05 [23]. However, the implications of "statistically significant results" stemming from multiple repeat testing procedures remain questionable since repeat testing naturally increases the likelihood of finding at some point, a p value of less than 0.05 just by chance alone. For these reasons, the interpretation of "statistically significant" versus "statistically non-significant" results should be performed in a cautious fashion. In the interpretation of orthopaedic trauma outcome data, it remains important to review all numerical results including not only the p value, but also the size of the treatment effect as well as the confidence intervals.

29.4 Implications for the Clinical Practice

When interpreting the outcomes of patients with extremity injuries, the foremost question remains how the results can be applied to clinical practice. In order to put an outcome study into a clinical context, it is important to consider several factors including the inclusion criteria, the outcome measures used, the study endpoints, and the distinction between statistical significance and clinical significance.

The inclusion criteria of an outcome study play an important role with regard to the implications for the clinical practice. It is important to be aware which particular patient population was enrolled in the study and recommendations can only be made for this particular type of patient population. For instance, a recently published randomized controlled trial suggested that primary arthrodesis of Lisfranc injuries results in superior outcomes as compared to open reduction and internal fixation [24]. However, these authors only included ligamentous Lisfranc injuries. In addition, patients with associated other orthopaedic injuries were excluded from the study, as well. Moreover, patients with co-morbidities, such as diabetes, peripheral vascular disease, or rheumatoid arthritis were also excluded from this study. Thus, the patient population enrolled in this trial was very specific and probably different from most Lisfranc injuries that typically present to level 1 trauma centres as a result of high energy injuries when associated injuries and co-morbidities are common. Another example would be the investigations performed in the field of mangled lower extremity injuries (LEAP) [2, 3]. These investigations have been performed in civilian trauma patients. In contrast, combat injuries in soldiers represent a completely different scenario with different injury mechanisms (blast injuries versus high speed motorized vehicle collisions) and different rehabilitation resources available to military personnel [25]. Therefore, one must be careful when extrapolating outcome data from civilian patients with mangled lower extremities to combat injuries in soldiers. These issues emphasize that the interpretation of clinical outcomes in patients with extremity injuries requires careful consideration of the patient population in which these outcomes were achieved.

The used outcome measures also play an important role when putting outcome data of patients with extremity injuries into a clinical context. Thus, orthopaedic outcome studies may frequently focus on fracture union as their main outcome measure which certainly appears appropriate since surgical treatment in orthopaedic trauma typically aims at restoring fracture union. However, this may not allow for any definitive conclusions on patient satisfaction, functional outcomes, associated complications, or the need for re-operation. For instance, a well-designed randomized controlled trial investigated the differences of reamed versus non-reamed nailing in patients with tibial shaft fractures [26]. The authors reported that for most parts the outcomes were equivocal. However, there appeared to be a significantly higher rate of required nail dynamization in the non-reamed group. While this outcome suggested superior results in the reamed group, the authors emphasized that the need for nail dynamization was an outcome measure of lower importance and thus, appropriately moderated their conclusions and recommendations.

The clinical implications of an orthopaedic trauma outcome study also strongly depend on the study endpoints. As stated above, patients with extremity injuries go through different stages of recovery. Along the recovery process, different variables become of interest at different time points. Thus, in the early perioperative period, early postoperative complications, length of hospital stay, and mortality rates are typically recorded and important information can be gained during this phase. Within the first few months after surgery, the rates of successful fracture unions and return to work can be recorded efficiently. Long-term outcome studies provide important information on health-related quality of life and late complications from treatment, such as posttraumatic arthritis and the need for late reconstructive procedures. When extracting clinical implications from outcome studies, it is important to recognize which study endpoints were chosen. For instance, a hypothetical clinical study comparing the nonunion rates between operative versus nonoperative treatment of clavicle mid-shaft fractures may use fracture union as an appropriate endpoint of the study. This may provide clinically important information on the union rates and the rate of required subsequent nonunion surgeries following treatment of clavicle fractures. However, when using this information in the clinical setting, it is important to be aware that the study endpoint "fracture union" may not serve as a good source of information for questions about late adverse events, such as the need for hardware removal due to prominent hardware, etc. For these reasons, the study endpoints predict which specific clinical implications can be concluded from outcome studies in patients with extremity injuries.

Another important concept is the distinction between "statistically significant" and "clinically significant". This distinction becomes important when statistically significant results are presented, but the clinical significance of the detected difference remains questionable. For instance, a recently published level 2 study on surgical reconstruction of the anterior cruciate ligament compared the outcomes of two different reconstruction techniques [27]. These authors reported that the postoperative side-to-side difference for anterior tibial translation was significantly higher in one treatment group with a reported p value of 0.001. A closer look at the reported outcomes revealed that one group had a side-to-side difference for anterior tibial translation of 2.2 mm as compared to 1.1 mm in the other group. While statistically significant, these results raise the question if a 1-mm difference for anterior tibial translation represents a clinically significant finding since it can be assumed that most knee surgeons may not be able to clinically detect a 1 mm difference for anterior tibial translation. Moreover, the pertinent question remains if a 1 mm difference of anterior tibial translation results in a remarkable improvement of the patient's perceived health-related quality of life. Thus, in patients with extremity injuries, the results should always be scrutinized carefully in order to assess if the detected difference is not only statistically significant, but also clinically significant.

29.5　Outcomes After Lower Extremity Injuries

Multiple outcome studies in patients with lower extremity injuries have been performed over the last decades. Given the large spectrum of lower extremity fractures and the numerous treatment options, a detailed review of the outcomes of specific fractures or specific treatment options is far beyond the scope of this chapter. The goal of this chapter is to summarize the results of the most pertinent lower extremity outcome studies. Specifically, this review is focused on polytrauma patients with associated lower extremity injuries and patients with mangled lower extremities.

Over the last decades, improved preclinical and clinical emergency care has decreased the mortality and complication rates of polytrauma patients [28, 29]. Given the increased survival rates, the long-term functional outcome and patient satisfaction have gained importance in polytrauma care. It has been shown by several investigations that the incidence of lower extremity injuries has a significant impact on the functional recovery of patients after polytrauma [30–34].

The Hannover Rehab Study has provided important information on the long-term recovery of polytrauma patients [35–44]. Clinical outcome data with a minimum follow-up of years after polytrauma were recorded in this investigation. Detailed data analyses including binary logistic regressions from this study have suggested traumatic amputations and severe spine injuries to be significant independent predictors of poor long-term functional recovery after polytrauma [36]. In addition, it was recorded that patients with lower extremity injuries do significantly worse than patients with upper extremity injuries [36].

With regard to lower extremity injuries, the Hannover Rehab Study revealed that in particular, patients with injuries below the knee joint seem to be significantly limited in their functional recovery as compared to patients with lower extremity injuries above the knee joint [43]. The reasons for this phenomenon remain unclear. However, it can be assumed that the relatively thin soft tissue envelope surrounding the foot and ankle as well as the unfavourable distal circulation may contribute to this problem. Moreover, injuries below the knee joint include a large variety of complex fracture patterns, such as tibial plateau fractures, pilon fractures, talus fractures, and calcaneus fractures, which are also challenging to address from the reconstructive standpoint.

These data clearly emphasize the importance of lower extremity injuries for the long-term functional recovery after polytrauma. In polytrauma patients, injuries to the lower extremities frequently present themselves as mangled lower extremities with significant associated injuries to the surrounding skin, muscles, and neurovascular structures. These injuries frequently require multiple surgical reconstructions and the predominant question remains whether patients will benefit from limb salvage versus amputation. The Lower Extremity Assessment Project (LEAP) study was initiated with the goal to provide answers to this challenging question [2, 3]. The study was performed at eight level 1 trauma centres in North America and represents a milestone in orthopaedic trauma outcome research. A focused summary of the LEAP study will be provided in the following section.

The study focused on patients with severely mangled lower extremity injuries including (1) traumatic amputations; (2) grade 3A open tibia fractures with high degree of nerve, muscle, or bone injury; (3) grade 3B and 3C open tibia fractures; (4) vascular injuries below the distal femur; (5) major soft tissue injuries below the distal femur; (6) grade 3 open pilon fractures; (7) grade 3B open ankle fractures; and (8) severe open hindfoot and midfoot injuries. The rendered treatment of these patients was according to the treating surgeon and the study was performed prospectively, but in a non-randomized fashion. A total of 601 patients were enrolled in this study and the investigators recorded 7-year follow-up data on most patients.

Bosse et al. reported the outcomes of patients undergoing limb salvage versus amputation [45]. At 2 years after injury, there was no significant difference in the outcome scores between the two treatment groups. In both treatment groups, self-efficacy and social support were found to be significant predictors of the functional outcome. Predictors of poor functional outcomes included the incidence of major complications, lower educational level, nonwhite race, low income, lack of health insurance, smoking, and involved litigations. Patients who underwent surgical reconstruction were also significantly more likely to have a secondary re-hospitalization.

Further cohort analyses of the LEAP study data focused on comparing the outcomes of patients with above knee amputation versus knee disarticulation versus below knee amputation [46]. Patients treated with above knee amputation showed no significantly different outcome scores than patients with below knee amputations. However, patients with amputations below the knee joint showed faster walking speeds. Patients with knee disarticulations had significantly worse outcomes than patients with above- or below knee amputations.

A widely used indication for amputation in patients with mangled lower extremities has been the absence of plantar sensation. The LEAP study also investigated the outcomes of patients with absent plantar sensation [47]. This cohort analysis included 29 patients with initially absent plantar sensation who underwent limb salvage. In this cohort, only one patient continued to have absent plantar sensation at 2 years after trauma. The remaining 28 patient showed partial or even full recovery of their plantar sensation. Moreover, patients with initially absent plantar sensation showed no significantly worse functional outcome scores than patients with initially present plantar sensation. Therefore, the LEAP study refuted the widely held belief that absent plantar sensation should be used as a definitive indication for amputation in patients with mangled lower extremities.

The LEAP study also provided important descriptive data on the overall complication rates in patients with mangled lower extremities [48]. In patients undergoing amputation, the revision amputation rate was 5.4 % and the overall complication rate was approximately 25 %. In patients undergoing limb salvage, approximately 4 % required a secondary amputation and the overall complication rate in this cohort was approximately 40 %. This data represents useful information for preoperative patient counselling and patients undergoing limb salvage need to be educated that they are at high risk for complications, re-hospitalization, as well as secondary amputation.

As indicated above, healthcare utilization and treatment costs must be considered important outcome measures in patients with extremity injuries. This question was also addressed in the LEAP study [49]. The cost calculations included hospitalizations, rehabilitation, outpatient visits and therapy, purchase and maintenance of prosthetic devices. At 2 years, limb salvage appeared to be associated with slightly higher costs than amputation ($91,106 versus $81,316). However, the projected life-time costs appeared more than three times higher in the amputation group ($509,275 versus $163,282) which was mostly driven by the required renewal of lower extremity prosthetics. Thus, the LEAP study refuted an additional widely held belief, which is the assumption that amputation is a cheaper solution over limb salvage.

In addition, the LEAP study provided important information on the long-term recovery using 7-year follow-up data [50]. It was found that a total of 58 % of patients with mangled lower extremities had returned to work at 7 years after trauma. Approximately 25 % of patients experienced some degree of limitation with regard to performing their work. There was no significant difference between patients with amputation versus limb salvage. However, factors predicting a successful return to work included younger age, white race, higher education level, non-smoker, high self-efficacy, pre-injury tenure, and absence of litigation.

In summary, the LEAP study has provided a wealth of useful data that may guide the surgeon in counselling the patient. However, it appears that despite the tremendous efforts made by the investigators, the pertinent question remains which patients will benefit from limb salvage versus amputation. Moreover, it appears that the main outcome predictors, such as self-efficacy, age, race, education level, smoking, pre-injury employment, and litigations, cannot be controlled by the surgeon.

29.6 Outcomes After Upper Extremity Injuries

As stated above, injuries to the upper extremity appear to cause fewer limitations in the functional recovery process of polytrauma patients than injuries to the lower extremity [36]. Moreover, the functional recovery of polytrauma patients with upper extremity injuries has gained little attention in the literature and most reports have focused on the outcomes of patients with specific upper extremity injuries. Data from patients with severe upper extremity injuries have suggested that associated brachial plexus injuries significantly limit the functional recovery of patients with severe upper extremity injuries [51, 52]. Further investigations in polytrauma survivors showed that approximately 50 % of patients with shoulder girdle injuries continued to have functional impairments at 5 years after trauma [32]. Displaced and articular fractures were identified to be associated with long-term disability. Moreover, 45 % of patients with shoulder girdle injuries and 62 % of patients with upper extremity fractures complained of chronic pain [32].

Further data on the long-term functional recovery of polytrauma patients with upper extremity injuries has been provided by the Hannover Rehab Study [53]. At approximately 18 years follow-up, polytrauma patients with upper extremity injuries showed significant limitations from their upper extremity injuries with regard to range of motion, muscle weakness, and neurologic impairment. In particular, the combination of associated shaft and articular upper extremity injuries seemed to significantly impact the long-term functional recovery. Thus, decreased range of motion, joint contractures, and muscle weakness were significantly more common in patients with combined articular and shaft injuries as compared to patients with isolated shaft fractures or isolated articular fractures. These data indicate that multiple upper extremity injuries provide significant challenges from the reconstructive and the rehabilitation standpoint.

Conclusion

In summary, the functional long-term outcomes in polytrauma patients require a critical evaluation of the available literature. Lower extremity fractures seem to significantly impact the functional recovery of polytrauma patients. In particular, fractures below the knee joint seem to be associated with significant long-term disability. With regard to mangled lower extremities, it remains unclear which particular subset of patients may benefit from limb salvage versus amputation. Recent research has emphasized that certain patient factors, such as patient self-efficacy, age, race, education level, smoking, pre-injury employment, and litigations, seem to have a significant impact on the patient recovery regardless of the surgical treatment rendered. With regard to upper extremity injuries in polytrauma patients, the currently available literature remains limited. Clinical outcome studies have shown that associated nerve injuries to the upper extremity in particular, such as brachial plexus injuries, seem to limit the functional long-term recovery.

References

1. Wright JG, Swiontkowski MF, Heckman JD. Introducing levels of evidence to the journal. J Bone Joint Surg Am. 2003;85-A(1):1–3.
2. Higgins TF, Klatt JB, Beals TC. Lower Extremity Assessment Project (LEAP): the best available evidence on limb-threatening lower extremity trauma. Orthop Clin North Am. 2010;41(2):233–9.
3. MacKenzie EJ, Bosse MJ. Factors influencing outcome following limb-threatening lower limb trauma: lessons learned from the Lower Extremity Assessment Project (LEAP). J Am Acad Orthop Surg. 2006;14(10 Suppl):S205–10.
4. Suk M, Norvell DC, Hanson B, Dettori JR, Helfet D. Evidence-based orthopaedic surgery: what is evidence without the outcomes? J Am Acad Orthop Surg. 2008;16(3):123–9.
5. Swiontkowski MF, Engelberg R, Martin DP, Agel J. Short musculoskeletal function assessment questionnaire: validity, reliability, and responsiveness. J Bone Joint Surg Am. 1999;81(9):1245–60.
6. Beaton D, Schemitsch E. Measures of health-related quality of life and physical function. Clin Orthop Relat Res. 2003;413:90–105.
7. Brazier JE, Harper R, Jones NM, et al. Validating the SF-36 health survey questionnaire: a new outcome measure for primary care. BMJ. 1992;305:160–4.
8. Bousquet J, Knani J, Dhivert H, et al. Quality of life in asthma: I. Internal consistency and validity of the SF-36 questionnaire. Am J Respir Crit Care Med. 1994;149:371–5.
9. Garratt AM, Ruta DA, Abdalla MI, Buckingham JK, Russell IT. The SF36 health survey questionnaire: an outcome measure suitable for routine use within the NHS? BMJ. 1993;306:1440–4.
10. Jacobson AM, de Groot M, Samson JA. The evaluation of two measures of quality of life in patients with type I and type II diabetes. Diabetes Care. 1994;17:267–74.
11. Marquis P, Fayol C, Joire J, Leplege A. Psychometric properties of a specific quality of life questionnaire in angina pectoris patients. Qual Life Res. 1995;4:540–6.
12. Meyer KB, Espindle DM, DeGiacomo JM, Jenuleson CS, Kurtin PS, Davies AR. Monitoring dialysis patients' health status. Am J Kidney Dis. 1994;24:267–79.
13. Naughton MJ, Anderson RT. Outcomes research in orthopaedics: health-related quality of life and the SF-36. Arthroscopy. 1998;14:127–9.
14. Ware JE, Snow K, Klosinski M, Gandek B. SF-36 health survey: manual and interpretations guide. Boston: The Health Institute, New England Medical Center; 1993.
15. Gartsman GM, Brinker MR, Khan M. Early effectiveness of arthroscopic repair for full-thickness tears of the rotator cuff: an outcome analysis. J Bone Joint Surg Am. 1998;80:33–40.
16. Chung KC, Saddawi-Konefka D, Haase SC, Kaul G. A cost-utility analysis of amputation versus salvage for Gustilo type IIIB and IIIC open tibial fractures. Plast Reconstr Surg. 2009;124(6):1965–73.
17. Zelle BA, Bhandari M, Sanchez AI, Probst C, Pape HC. Loss of follow-up in orthopaedic trauma: is 80% follow-up still acceptable? J Orthop Trauma. 2013;27(3):177–81.
18. Castillo RC, Mackenzie EJ, Bosse MJ, LEAP Study Group. Orthopaedic trauma clinical research: is 2-year follow-up necessary? Results from a longitudinal study of severe lower extremity trauma. J Trauma. 2011;71(6):1726–31.
19. Brand RA. Writing for clinical orthopaedics and related research. Clin Orthop Relat Res. 2008;466(1):239–47.
20. Bracken MB, Shepard MJ, Collins WF, et al. A randomized, controlled trial of methylprednisolone or naloxone in the treatment of acute spinal-cord injury. Results of the Second National Acute Spinal Cord Injury Study. N Engl J Med. 1990;322:1405–11.
21. Bracken MB, Shepard MJ, Holford TR, et al. Administration of methylprednisolone for 24 or 48 h or tirilazad mesylate for 48 h in the treatment of acute spinal cord injury. Results of the Third National Acute Spinal Cord Injury Randomized Controlled Trial. National Acute Spinal Cord Injury Study. JAMA. 1997;277:1597–604.
22. Buckley R, Tough S, McCormack R, Pate G, Leighton R, Petrie D, Galpin R. Operative compared with nonoperative treatment of displaced intra-articular calcaneal fractures: a prospective, randomized, controlled multicenter trial. J Bone Joint Surg Am. 2002;84-A(10):1733–44.
23. Bono CM, Tornetta P. Common errors in the design of orthopaedic studies. Injury. 2006;37:355–60.
24. Ly TV, Coetzee JC. Treatment of primarily ligamentous Lisfranc joint injuries: primary arthrodesis compared with open reduction and internal fixation. A prospective, randomized study. J Bone Joint Surg Am. 2006;88(3):514–20.
25. Rispoli DM, Mackenzie EJ, Extremity War Injuries VII Outcomes Panel. Orthopaedic outcomes: combat and civilian trauma care. J Am Acad Orthop Surg. 2012;20 Suppl 1:S84–7.
26. Study to Prospectively Evaluate Reamed Intramedullary Nails in Patients with Tibial Fractures Investigators, Bhandari M, Guyatt G, Tornetta 3rd P, Schemitsch EH, Swiontkowski M, Sanders D, Walter SD. Randomized trial of reamed and unreamed intramedullary nailing of tibial shaft fractures. J Bone Joint Surg Am. 2008;90(12):2567–78.
27. Branch TP, Siebold R, Freedberg HI, Jacobs CA. Double-bundle ACL reconstruction demonstrated superior clinical stability to single-bundle ACL reconstruction: a matched-pairs analysis of instrumented tests of tibial anterior translation and internal rotation laxity. Knee Surg Sports Traumatol Arthrosc. 2011;19(3):432–40.

28. Pape HC, Hildebrand F, Pertschy S, Zelle B, Garapati R, Grimme K, Krettek C, Reed 2nd RL. Changes in the management of femoral shaft fractures in polytrauma patients: from early total care to damage control orthopedic surgery. J Trauma. 2002;53:452–62.
29. Regel G, Lobenhoffer P, Grotz M, et al. Treatment results of patients with multiple trauma: an analysis of 3406 cases treated between 1972 and 1991 at a German Level I Trauma Center. J Trauma. 1995;38:70–8.
30. Butcher JL, MacKenzie EJ, Cushing B, Jurkovich G, Morris J, Burgess A, McAndrew M, Swiontkowski M. Long-term outcomes after lower extremity trauma. J Trauma. 1996;41(1):4–9.
31. Jurkovich G, Mock C, MacKenzie E, Burgess A, Cushing B, deLateur B, McAndrew M, Morris J, Swiontkowski M. The Sickness Impact Profile as a tool to evaluate functional outcome in trauma patients. J Trauma. 1995;39(4):625–31.
32. Mkandawire NC, Boot DA, Braithwaite IJ, Patterson M. Musculoskeletal recovery 5 years after severe injury: long term problems are common. Injury. 2002;33(2):111–5.
33. O'Toole RV, Castillo RC, Pollak AN, MacKenzie EJ, Bosse MJ, LEAP Study Group. Determinants of patient satisfaction after severe lower-extremity injuries. J Bone Joint Surg Am. 2008;90(6):1206–11.
34. Seekamp A, Regel G, Bauch S, Takacs J, Tscherne H. Long-term results of therapy of polytrauma patients with special reference to serial fractures of the lower extremity. Unfallchirurg. 1994;97(2):57–63.
35. Dienstknecht T, Pfeifer R, Horst K, Sellei RM, Berner A, Zelle BA, Probst C, Pape HC. The long-term clinical outcome after pelvic ring injuries. Bone Joint J. 2013;95-B(4):548–53.
36. Pape HC, Probst C, Lohse R, Zelle BA, Panzica M, Stalp M, Steel JL, Duhme HM, Pfeifer R, Krettek C, Sittaro NA. Predictors of late clinical outcome following orthopedic injuries after multiple trauma. J Trauma. 2010;69(5):1243–51.
37. Pape HC, Zelle B, Lohse R, Stalp M, Hildebrand F, Krettek C, Panzica M, Duhme V, Sittaro NA. Evaluation and outcome of patients after polytrauma–can patients be recruited for long-term follow-up? Injury. 2006;37(12):1197–203.
38. Pfeifer R, Lichte P, Zelle BA, Sittaro NA, Zilkens A, Kaneshige JR, Pape HC. Socio-economic outcome after blunt orthopaedic trauma: implications on injury prevention. Patient Saf Surg. 2011;5(1):9.
39. Pfeifer R, Zelle BA, Kobbe P, Knobe M, Garrison RL, Ohm S, Sittaro NA, Probst C, Pape HC. Impact of isolated acetabular and lower extremity fractures on long-term outcome. J Trauma Acute Care Surg. 2012;72(2):467–72.
40. Probst C, Zelle B, Panzica M, Lohse R, Sitarro NA, Krettek C, Pape HC. Clinical re-examination 10 or more years after polytrauma: is there a gender related difference? J Trauma. 2010;68(3):706–11.
41. Steel J, Youssef M, Pfeifer R, Ramirez JM, Probst C, Sellei R, Zelle BA, Sittaro NA, Khalifa F, Pape HC. Health-related quality of life in patients with multiple injuries and traumatic brain injury 10+ years postinjury. J Trauma. 2010;69(3):523–30.
42. Zeckey C, Hildebrand F, Pape HC, Mommsen P, Panzica M, Zelle BA, Sittaro NA, Lohse R, Krettek C, Probst C. Head injury in polytrauma-Is there an effect on outcome more than 10 years after the injury? Brain Inj. 2011;25(6):551–9.
43. Zelle BA, Brown SR, Panzica M, Lohse R, Sittaro NA, Krettek C, Pape HC. The impact of injuries below the knee joint on the long-term functional outcome following polytrauma. Injury. 2005;36(1):169–77.
44. Zelle BA, Panzica M, Vogt MT, Sittaro NA, Krettek C, Pape HC. Influence of workers' compensation eligibility upon functional recovery 10 to 28 years after polytrauma. Am J Surg. 2005;190(1):30–6.
45. Bosse MJ, MacKenzie EJ, Kellam J, et al. An analysis of outcomes of reconstruction or amputation of leg-threatening injuries. N Engl J Med. 2002;347:1924–31.
46. MacKenzie EJ, Bosse MJ, Castillo RC, et al. Functional outcomes following lower extremity amputation for trauma. J Bone Joint Surg Am. 2004;86-A:1636–45.
47. Bosse MJ, McCarthy ML, Jones AL, Webb LX, Sims SH, Sanders RW, MacKenzie EJ, Lower Extremity Assessment Project (LEAP) Study Group. The insensate foot following severe lower extremity trauma: an indication for amputation? J Bone Joint Surg Am. 2005;87(12):2601–8.
48. Harris AM, Althausen PL, Kellam J, Bosse MJ, Castillo R, Lower Extremity Assessment Project (LEAP) Study Group. Complications following limb-threatening lower extremity trauma. J Orthop Trauma. 2009;23(1):1–6.
49. MacKenzie EJ, Jones AS, Bosse MJ, Castillo RC, Pollak AN, Webb LX, Swiontkowski MF, Kellam JF, Smith DG, Sanders RW, Jones AL, Starr AJ, McAndrew MP, Patterson BM, Burgess AR. Healthcare costs associated with amputation or reconstruction of a limb-threatening injury. J Bone Joint Surg Am. 2007;89(8):1685–92.
50. MacKenzie EJ, Bosse MJ, Kellam JF, Pollak AN, Webb LX, Swiontkowski MF, Smith DG, Sanders RW, Jones AL, Starr AJ, McAndrew MP, Patterson BM, Burgess AR, Travison T, Castillo RC. Early predictors of long-term work disability after major limb trauma. J Trauma. 2006;61(3):688–94.
51. Joshi V, Harding GE, Bottoni DA, Lovell MB, Forbes TL. Determination of functional outcome following upper extremity arterial trauma. Vasc Endovascular Surg. 2007;41(2):111–4.
52. Zelle BA, Pape HC, Gerich TG, Garapati R, Ceylan B, Krettek C. Functional outcome following scapulothoracic dissociation. J Bone Joint Surg Am. 2004;86-A(1):2–8.
53. Pfeifer R, Pape HC. Late outcomes after severe fractures. In: Pape H-C, Sanders R, Borrelli J, editors. The poly-traumatized patient with fractures. 1st ed. Heidelberg/Berlin: Springer; 2011.

Clinical Outcome of Pelvic and Spinal Fractures

30

Roman Pfeifer, Bilal M. Barkatali,
Thomas Dienstknecht, and Hans-Christoph Pape

Contents

30.1	Factors Affecting the Long-Term Outcomes in Polytrauma Patients	419
30.2	Pelvic Fractures	420
30.3	Spinal Fractures	422
Conclusion		423
References		423

Numerous improvements in the delivery of trauma care and rehabilitation have been made over the past decades, such as injury prevention advancements in rescue systems, improvements in hospital diagnostics and surgical techniques, and the development of better treatment strategies. A decrease in the mortality rate (37–18 %) of multiple trauma patients was noted over the past two decades [1–6]. Thus, long-term outcome evaluation and assessment of quality of life and patient satisfaction have gained attention in polytrauma care. Severe musculoskeletal trauma is a life-altering condition leading to prolonged morbidity and numerous repetitive interventions. Trauma is a main contributor to work disability [7–9], impaired long-term psychosocial outcome and persisting disabilities. It has an economic burden to society a lasting impact on the affected individuals and their families [7–12].

R. Pfeifer (✉) • T. Dienstknecht • H.-C. Pape
Department of Trauma, Orthopaedic,
Hand and Plastic Surgery,
University of Aachen Medical Center,
Pauwelsstr. 30, 52074 Aachen, Germany
e-mail: rpfeifer@ukaachen.de;
tdienstknecht@ukaachen.de;
papehc@aol.com

B.M. Barkatali
Department of Trauma,
Royal Bolton Foundation NHS Trust, UK
e-mail: bilalb1@hotmail.com

30.1 Factors Affecting the Long-Term Outcomes in Polytrauma Patients

To look beyond mortality and assess the patients' longitudinal evaluation is a helpful tool for identifying the factors that influence long-term outcome following major injuries and the appropriate beneficial interventions. Several large projects

[7–9, 13–16] have been conducted that focus on patients long-term functional recovery following polytrauma. These studies provide evidence that not only injury-related factors, such as injury severity, injury location, and treatment methods, but also the specific characteristics of the individual, socioeconomic factors, and health habits have a strong impact on outcome [8, 13, 17]. In addition, authors underlined the role of postinjury depression, anxiety, and chronic pain. High incidence rates of posttraumatic stress disorder (24–39 %), anxiety (32–70 %), and depression (35–68 %) have been observed among trauma patients. Additionally, cognitive defects, such as memory impairment, difficulty with concentration and emotional problems have been reported [3, 7, 18–20]. All these factors negatively affect a patient's functional outcome. These studies stress the need for concomitant posttraumatic psychological support. Moreover, self-efficacy was shown to be one of the strongest predictors of the sickness impact profile and return to work [7–9, 13]. It was assumed that persons with low self-efficacy are more likely to be disengaged from the physical rehabilitation and recovery process. To address this issue, it has been suggested that self-efficacy and self-management training should be introduced to polytrauma patients, especially as positive effects have been demonstrated in the treatment of patients with chronic diseases such arthritis, diabetes, and chronic pain [21, 22].

Several groups have demonstrated evidence of gender-related differences after severe injury [16, 23]. The advantages of premenopausal women over men in the acute phase after multiple injuries have been described [16, 23]. However, long-term results demonstrate the opposite. Women showed a higher rate of posttraumatic stress disorder and psychological support, longer duration of rehabilitation, and longer sick leave time [24–27].

Blunt injuries of the trunk are acutely associated with life-threatening complications. Long-term investigations, however, demonstrate that after blunt injuries involving the chest and abdomen, substantial recovery may occur [28]. These injuries were rarely the reason for worse outcome or functional impairments in long-term follow-up studies [28–30].

We previously undertook a long-term outcome study. Its goal was to evaluate functional outcome with a minimum follow-up of 10 years (mean 17.5 years) [14–16]. This study revealed that head and lower extremity injuries accounted for the most frequent causes of long-term disability [16]. At follow-up, 33 % of patients required a medical aid for their disability and 20.1 % reported disability due to their injury. Approximately the same percent of patients (76.5 %) and physicians (69.1 %) reported success from rehabilitation. Moreover, it has to be considered that the results found in isolated fractures are likely to be different from those of severely injured patients who have sustained high-energy trauma and concomitant injuries. In addition, the initial surgical reconstruction of multiple fractures is more complex. A large proportion of these patients require additional operative treatments and reconstructions. The interference of multiple fractures in the rehabilitation process is another factor that negatively affects long-term functional results following severe injury.

30.2 Pelvic Fractures

Approximately 25 % of polytraumatized patients experience pelvic ring injuries [31]. They are known to be associated with a high mortality rates and with multiple concomitant injuries of the lower limb, spine, abdomen, and head [32, 33]. Pelvic fractures are more frequently seen in men than women [34]. However, women appear to have worse outcome results than men [35]. Certainly, the type of fracture fixation and quality of reduction affect the outcome as well. Accordingly, analysis of long-term outcomes may be difficult to interpret because the accompanying injuries may affect the results [36–38]. It has been shown that both the severity of pelvic fracture (stable vs. unstable) and the presence of associated injuries contribute to poor long-term outcome [39]. Incomplete recovery and functional impairments were observed following unstable pelvic ring fractures, while stable pelvic injuries rarely led to major long-term problems [32, 40, 41]. Others could demonstrate an association of sequelae and poor outcomes following open pelvic fractures [42]. Moreover, the clinical outcomes of patients with unstable pelvic ring trauma and associated injuries were less satisfactory than the outcomes of patients with unstable pelvic ring trauma and no associated injuries [38, 40].

Chronic pain syndromes, neurological impairments, and nonunions have been described as determining factors that influence the long-term outcome in patients who have sustained pelvic fracture [41]. An overview of long-term (2 years) pain results was demonstrated by Pohlemann and coauthors [39, 41, 43]. Pain was observed in every fracture classification group; the rate of completely pain-free patients was 55 % after A-type fractures, 41 % after B-type, and 27 % after C-type fractures [39, 41, 43]. Nonanatomical reduction or insufficient fixation can provide poor long-term outcome results resulting in chronic back pain, instability, and mal-unions or nonunions [33, 44, 45].

Moreover, authors described a close correlation between neurological and functional long-term outcome [37]. At follow-up (2.2 years), 21 % of patients with B-Type and 60 % with C-type fractures had at least some neurological impairments [46]. In particular, vertical unstable injuries and trans-foraminal sacral fractures were shown to be associated with severe neurologic disabilities [47, 48]. Among the neurologic sequelae were peripheral nerve lesions, incontinence, and sexual dysfunctions [36, 39–43, 45, 46]. These sequelae are also the main reason for work disability (Table 30.1) [33]. Approximately 50–75 % of previously employed patients with pelvic fractures were able to return to their previous occupation [36, 40, 42].

The evaluation of the Hannover Rehab Study database [14–16] with regard to acetabular and proximal femoral fractures has demonstrated the following long-term outcomes (Table 30.2): A significant percentage of patients (45–50 %) with hip fractures experienced posttraumatic pain and approximately (20–35 %) reported an abnormal gait. High rates of gait abnormality were observed in patients who had sustained acetabular fractures. Moreover, outcome scores, as measured by HASPOC and SF-12, were poor in patients with acetabular fractures and proximal femur injuries. In general, patients with shaft fractures demonstrated significantly more favorable scores than patients with an articular fracture. This might be explained by the degeneration of the affected joint following articular fractures which may lead to functional disabilities and chronic pain [49–53]. The observed rates of arthroplasty were 7.5 % for hip joint.

Table 30.1 Clinical examination of pelvic ring fractures following polytrauma

Study	Fracture type	Follow-up	Patient[b]	Pain	Functional disability	RTW	Neurologic impairments
Pohlemann (1994)	Unstable fractures	2.2 years	58	11–66 %	No data	No data	21–60 %
Miranda (1996)	Pelvic ring fracture	5 years	80	16–35 %	8–21 %	75–81 %	No data
Tornetta (1996)	Unstable fracture	3.7 years	48	37 %	37 %	67 %	35 %
Brenneman (1997)	Open fracture	4 years	27	No data	No data	64 %	18 %
Kabak (2003)	Unstable fracture	3.8 years	36	31 %	No data	72 %	16–31 %[a]
Suzuki (2007)	Unstable fracture	3.9 years	57	No data	No data	84 %	28 %

RTW return to work
[a]Sexual and urinary dysfunction
[b]Skeletally immature patients

Table 30.2 Functional status of the lower extremities following polytrauma with fractures at different localizations

	Acetabulum N=20	Proximal femur N=20
Persistant pain	50.0 %	45.0 %
Abnormal gait	35.0 %	20.0 %
Work disability	27.8 %[a]	10.0 %
Successful rehabilitation	70.0 %	60.0 %
HASPOC-total	78.78	70.07
SF-12 PCU	40.91	40.95

HASPOC Hannover Score for Polytrauma Outcome, *SF-12 PCU* Short-Form 12 items health survey, Physical Component Summary
[a]Significantly worse outcome vs. fractures of the proximal femur ($p<0.05$)

Table 30.3 Long-term results of severely injured patients with: A, isolated anterior pelvic ring injuries; P, posterior pelvic ring injuries; A/P, combined anterior and posterior pelvic ring fractures

	Group A	Group P	Group A/P	p value*
Patients (N)	33	33	43	
Male (n, %)	22 (66.6)	21 (63.6)	35 (81.4)	$p<0.001$
Mean age at follow-up (years)	43	43	46	0.505
ISS (mean)	24 (16–41)	24 (16–43)	22 (16–45)	0.216
SF-12 (mean)				
Physical score	44.4	45.87	38.71	0.004
Mental score	48.68	50.97	48.35	0.476
HASPOC (mean)				
Subjective	47.89	48.7	67.27	0.024
Objective	17.00	25.18	27.78	0.217

* A/P vs A and A/P vs P

When analyzing the outcome of pelvic ring fractures in the same study population, patients with combined anterior and posterior pelvic ring injuries had significantly worse physical SF-12 and HASPOC subjective scores in comparison to patients with isolated anterior or posterior pelvic fractures (Table 30.3) [54]. In addition, patient group with anterior and posterior pelvic ring injuries more frequently reported negative subjective health status. The number of patients reporting failing or bad health status was significantly higher in patients with combined anterior and posterior pelvic injuries than patients with isolated fractures. This might be related to constant pain that has been reported by 32.6 % of patients with combined pelvic injured. Moreover, this patient group reported limping and use of crutches more frequently [54].

Vallier and colleagues analyzed the outcome after pelvic ring injury in women and reported substantial residual dysfunctions [55]. In this study, isolated pelvic ring trauma without concomitant injuries of the lower extremity was associated with a significantly better outcome [55]. On the other hand, patient sustained anteroposterior compression injuries and bladder ruptures showed negative results. Dyspareunia was more frequently reported in women with anteroposterior compression fractures, B-type injuries, bladder ruptures, and symphyseal plate fixation [56]. At follow-up, the caesarean delivery among woman with pelvic injuries was significantly higher performed than in woman without pelvic injuries [57]. However, uncomplicated pregnancy and deliveries are possible even with retained fixation material, [57].

30.3 Spinal Fractures

There are few studies highlighting health-related quality of life in polytrauma patients with spinal fractures. In a retrospective review of 915 polytrauma patients, Tee et al. demonstrated that on arrival, as well as tachycardia, hyperglycemia, and multiple chronic comorbidities; the presence of thoracic spine injuries were early predictive factors for poor outcome [58]. This was substantiated by Reinhold et al. analyzing data from the Spine Study Group of the German Association of Trauma Surgery comparing 733 patients with a 2-year follow-up. They demonstrated that thoracic spine injuries showed less potential for neurological recovery compared to injuries located at other heights [59]. Hofbauer et al. confirmed these findings also for a pediatric cohort [60].

Spinal cord injury in severely injured patients was one of the significant determinants for a decreased health-related quality of life measured with the EuroQol-Score [61]. These findings ware mainly based on decreased autonomous mobility.

In patients with traumatic spinal cord injuries, an initial poor neurological status is among well-known predictors such as traumatic brain injury, high injury severity score and comorbidities that are significantly associated with early mortality [62].

In a 5-year follow-up of 70 patients (38 % polytraumatized) with spinal fractures McLain found a high significance of the neurological injury for unsatisfying recovery. The regained work status also correlated highly with the neurological impairment [63]. The total rate of patients able to return to full-time work was 70 %, but only 12 % in patients with persistent neurological impairment. Holtslag et al. substantiated these findings in a large cohort of severely injured patients, by finding a 21 % return to full-time work rate [64].

It seems natural that outcome of severely injured patients with spinal fractures is mostly determined by the neurological recovery. However, Putz et al. compared two cohorts of multiply injured paraplegic patients with and without neurological recovery after 1 year. Their study was one of the first to look after influence factors of successful recovery other than the initial neurological damage. They could show that functional improvement was independent from neurological improvement [65]. Additionally they suggested that severity of accompanying thoracic trauma is one key factor of successful rehabilitation. None of the patients with a thoracic C-type fracture and an AIS (Impairment scale) type A could show a neurological improvement during the first year. Early decompression and stabilization of spine injuries seems to be beneficial especially in incomplete neurological defects. Length of stay at intensive care unit and days of ventilation seem to be reduced and lower pulmonary complications are seen [66, 67]. However, good prospective clinical trials are still missing on this topic.

Conclusion

Due to improved mortality rates of severely injured patients, long-term follow-up observation studies have gained more attention. Studies emphasized the importance of psychosocial variables on the long-term functional outcome. Early psychological intervention for polytrauma patients has been suggested to address this issue. Large outcome studies have demonstrated that acetabular fractures, anterior and posterior pelvic ring injuries, especially those with concomitant injuries are associated with poor long-term functional results and unfavorable outcome scores. Persisting neurological impairment leads to a decreased health-related quality in spinal injuries. Patients with severe injuries that are associated with poor outcome should be identified earlier in order to improve their rehabilitation results. Social reintegration of patients and return to work were defined as main long-term goals in the treatment of polytrauma patients.

References

1. Bardenheuer M, Obertacke U, Waydhas C, Nast-Kolb D, AG Polytrauma der DGU. Epidemiology of severe multiple trauma- a prospective registration of preclinical and clinical supply. Unfallchirurg. 2000;103:355–63.
2. Ruchholtz S, AG Polytrauma der DGU. The Trauma Registry of the German Society of Trauma Surgery as a basis for interclinical quality mangement. A multicenter study of the German Society of Trauma Surgery. Unfallchirurg. 2000;103:30–7.
3. Nast-Kolb D, Aufmkolk M, Rucholtz S, Obertacke U, Waydhas C. Multiple organ failure still a major cause of morbidity but not mortality in blunt multiple trauma. J Trauma. 2001;51:835–42.
4. Pape HC, Remmers D, Rice J, Ebisch M, Krettek C, Tscherne H. Appraisal of early evaluation of blunt chest trauma: development of a standardized scoring system for initial clinical decision making. J Trauma. 2000;49(496):504.
5. Kuhne CA, Ruchholtz S, Kaiser GM, Nast-Kolb D, AG Polytrauma der DGU. Mortality in severely injured elderly trauma patients—when does age become a risk factor? World J Surg. 2005;29(1476):1482.
6. Regel G, Lobenhoffer P, Grotz M, Pape HC, Lehmann U, Tscherne H. Treatment results of patients with multiple trauma: an analysis of 3406 cases treated between 1972 and 1991 at German Level I Trauma Center. J Trauma. 1995;38(1):70–8.
7. Bosse MJ, MacKenzie EJ, Kellam JF, et al. An analysis of outcomes of reconstruction or amputation

after leg-threatening injuries. N Engl J Med. 2002; 347(24):1924–31.
8. MacKenzie EJ, Bosse MJ, Pollak AN, et al. Long-term persistence of disability following severe lower-limb trauma. Results of seven-year follow-up. J Bone Joint Surg Am. 2005;87(8):1801–9.
9. MacKenzie EJ, Bosse MJ, Kellam JF, et al. Early predictors of long-term work disability after major limb trauma. J Trauma. 2006;61:688–94.
10. Braithwaite IJ, Boot DA, Patterson M, Robinson A. Disability after severe injury: five year follow up of a large cohort. Injury. 1998;29(1):55–9.
11. Mkandawire NC, Boot DA, Braithwaite IJ, Patterson M. Musculoskeletal recovery 5 years after severe injury: long term problems are common. Injury. 2002;33: 111–5.
12. O'toole RV, Castillo RC, Pollak AN, MacKenzie EJ, Bosse MJ, LEAP Study Group. Determinants of patient satisfaction after severe lower-extremity injuries. J Bone Joint Surg Am. 2008;90:1206–11.
13. MacKenzie EJ, Bosse MJ. Factors influencing outcome following limb-threatening lower limb trauma: lessons learned from the Lower Extremity Assessment Project (LEAP). J Am Acad Orthop Surg. 2006;14:S205–10.
14. Zelle BA, Panzica M, Vogt MT, Sittaro NA, Krettek C, Pape HC. Influence of workers' compensation eligibility upon functional recovery 10 to 28 years after polytrauma. Am J Surg. 2005;190(1):30–6.
15. Zelle BA, Brown SR, Panzica M, et al. The impact of injuries below the knee joint on the long-term functional outcome following polytrauma. Injury. 2005;36(1): 169–77.
16. Pape HC, Zelle B, Lohse R, et al. Evaluation and outcome of patients after polytrauma: can patients be recruited for long-term follow-up? Injury. 2006;37(12): 1197–203.
17. MacKenzie EJ, Rivara FP, Jurkovich GJ, et al. The impact of trauma-center care on functional outcomes following major lower-limb trauma. J Bone Joint Surg Am. 2008;90(1):101–9.
18. Michaels AJ, Michaels CE, Smith JS, Moon CH, Peterson C, Long WB. Outcome from injury: general health, work status, and satisfaction 12 months after trauma. J Trauma. 2000;48(5):841–50.
19. Evans SA, Airey MC, Chell SM, Connelly JB, Rigby AS, Tennant A. Disability in young adults following major trauma: 5 year follow up of survivors. BMC Public Health. 2003;3(8):1–8.
20. Piccinelli M, Patterson M, Braithwaite IJ, Boot DA, Wilkinson G. Anxiety and depression disorder 5 years after severe injuries: a prospective follow-p study. J Psychosom Res. 1999;46(5):455–64.
21. Morley S, Eccleston C, Williams A. Systematic review and meta-analysis of randomized controlled trials of cognitive behavior. Pain. 1999;80:1–13.
22. Lorig KR, Sobel DS, Stewart AL, Brown Jr BW, Bandura A, Ritter P, et al. Evidence suggesting that a chronic disease self-management program can improve health status while reducing hospitalisation: a randomized trial. Med Care. 1999;37:5–14.
23. Brenneman FD, Boulanger BR, McLellan BA, Culhane JP. Acute and long-term outcomes of extremely injured blunt trauma victims. J Trauma. 1995;39: 320–4.
24. Probst C, Zelle B, Panzica M, et al. Clinical re-examination 10 or more years after polytrauma: is there a gender related difference? J Trauma. 2010; 68(3):706–11.
25. Holbrook TL, Hoyt DB, Anderson JP. The importance of gender on outcome after major trauma: functional and psychologic outcomes in women versus men. J Trauma. 2001;50:270–3.
26. Holbrook TL, Hoyt DB, Stein MB, Siebert WJ. Gender differences in long-term posttraumatic stress disorder outcomes after major trauma: women are at higher risk of adverse outcomes than men. J Trauma. 2002;53:882–8.
27. Holbrook TL, Hoyt DB. The impact of major trauma: quality-of-life outcomes are worse in women than in men, independent of mechanism and injury severity. J Trauma. 2004;56:284–90.
28. Holbrook TL, Anderson JP, Siebert WJ, Browner D, Hoyt DB. Outcome after major trauma: 12-month and 18-month follow-up results from the Trauma Recovery Project. J Trauma. 1999;46:765–73.
29. Amital A, Shirit D, Fox BD, et al. Long-term pulmonary function after recovery from pulmonary contusion due to blunt chest trauma. Isr Med Assoc J. 2009;11:673–6.
30. Grotz M, Pape HC, Stalp M, van Griensven M, Schreiber TC, Krettek C. Long-term outcome after multiple organ failure following severe trauma. Anaesthesist. 2001;50(4):262–70.
31. Papakostidis C, Kanakaris NK, Kontakis G, Giannoudis PV. Pelvic ring disruptions: treatment modalities and analysis of outcomes. Int Orthop. 2009; 33(2):329–38.
32. Gansslen A, Pohlemann T, Paul C, Lobenhoffer P, Tscherne H. Epidemiology of pelvic ring injuries. Injury. 1996;27 Suppl 1:S-A13–20.
33. Tornetta III P, Matta JM. Outcome of operatively treated unstable posterior pelvic ring disruptions. Clin Orthop Relat Res. 1996;329:186–93.
34. McCarthy ML, MacKenzie EJ, Bosse MJ, Copeland CE, Hash CS, Burgess AR. Functional status following orthopedic trauma in young women. J Trauma. 1995; 39(5):828–36.
35. Borrelli Jr J, Ricci WM, Anglen JO, Gregush R, Engsberg J. Muscle strength recovery and its effects on outcome after open reduction and internal fixation of acetabular fractures. J Orthop Trauma. 2006; 20(6):388–95.
36. Miranda MA, Riemer BL, Butterfield SL, Burke III CJ. Pelvic ring injuries. A long term functional outcome study. Clin Orthop Relat Res. 1996;329:152–9.
37. Suzuki T, Shindo M, Soma K, et al. Long-term functional outcome after unstable pelvic ring fracture. J Trauma. 2007;63(4):884–8.
38. Rommens PM, Hessmann MH. Staged reconstruction of pelvic ring disruption: differences in morbidity, mortality, radiologic results, and functional outcomes

between B1, B2/B3, and C-type lesions. J Orthop Trauma. 2002;16(2):92–8.
39. Pohlemann T, Gansslen A, Schellwald O, Culemann U, Tscherne H. Outcome evaluation after unstable injuries of the pelvic ring. Unfallchirurg. 1996;99(4): 249–59.
40. Kabak S, Halici M, Tuncel M, Avsarogullari L, Baktir A, Basturk M. Functional outcome of open reduction and internal fixation for completely unstable pelvic ring fractures (type C): a report of 40 cases. J Orthop Trauma. 2003;17(8):555–62.
41. Pohlemann T, Tscherne H, Baumgartel F, et al. Pelvic fractures: epidemiology, therapy and long-term outcome. Overview of the multicenter study of the Pelvis Study Group. Unfallchirurg. 1996;99(3):160–7.
42. Brenneman FD, Katyal D, Boulanger BR, Tile M, Redelmeier DA. Long-term outcomes in open pelvic fractures. J Trauma. 1997;42(5):773–7.
43. Pohlemann T, Gansslen A, Schellwald O, Culemann U, Tscherne H. Outcome after pelvic ring injuries. Injury. 1996;27 Suppl 2:B31–8.
44. Kanakaris NK, Angoules AG, Nikolaou VS, Kontakis G, Giannoudis PV. Treatment and outcomes of pelvic malunions and nonunions: a systematic review. Clin Orthop Relat Res. 2009;467(8):2112–24.
45. Eid K, Keel M, Keller A, Ertel W, Trentz O. Influence of sacral fracture on the long-term outcome of pelvic ring injuries. Unfallchirurg. 2005;108(1):35–42.
46. Pohlemann T, Bosch U, Gansslen A, Tscherne H. The Hannover experience in management of pelvic fractures. Clin Orthop Relat Res. 1994;305:69–80.
47. Huittinen VM. Lumbosacral nerve injury in fracture of the pelvis. A postmortem radiographic and pathoanatomical study. Acta Chir Scand Suppl. 1972;429: 3–43.
48. Majeed SA. Neurologic deficits in major pelvic injuries. Clin Orthop Relat Res. 1992;282:222–8.
49. Volpin G, Dowd GS, Stein H, Bentley G. Degenerative arthritis after intra-articular fractures of the knee. Long-term results. J Bone Joint Surg Br. 1990;72(4): 634–8.
50. Harris AM, Patterson BM, Sontich JK, Vallier HA. Results and outcomes after operative treatment of high energy tibial plafond fractures. Foot Ankle Int. 2006;27(4):256–65.
51. Weiss NG, Parvizi J, Trousdale RT, Bryce RD, Lewallen DG. Total knee arthroplasty in patients with a prior fracture of tibial plateau. J Bone Joint Surg Am. 2003;85-A(2):218–21.
52. Letournel E, Judet R. Fractures of the acetabulum. 2nd ed. New York: Springer; 1993.
53. Bhandari M, Matta J, Ferguson T, Matthys G. Predictors of clinical and radiological outcome in patients with fractures of the acetabulum and concomitant posterior dislocation of the hip. J Bone Joint Surg Br. 2006;88(12):1618–24.
54. Dienstknecht T, Pfeifer R, Horst K, et al. The longterm clinical outcome after pelvic ring injuries. Bone Joint J. 2013;95-B(4):548–53.
55. Vallier HA, Cureton BA, Schubeck D, Wang XF. Functional outcomes in women after high-energy pelvic ring injury. J Orthop Trauma. 2012;26(5): 296–301.
56. Vallier HA, Cureton BA, Schubeck D. Pelvic ring injury is associated with sexual dysfunction in women. J Orthop Trauma. 2012;26(5):308–13.
57. Vallier HA, Cureton BA, Schubeck D. Pregnancy outcomes after pelvic ring injury. J Orthop Trauma. 2012;26(5):302–7.
58. Tee JW, Chan CH, Gruen RL, et al. Early predictors of health-related quality of life outcomes in polytrauma patients with spine injuries: a level 1 trauma center study. Global Spine J. 2014;4(1):21–32.
59. Reinhold M, Knop C, Beisse R, et al. Operative treatment of 733 patients with acute thoracolumbar spinal injuries: comprehensive results from the second, prospective, Internet-based multicenter study of the Spine Study Group of the German Association of Trauma Surgery. Eur Spine J. 2010;19(10): 1657–76.
60. Hofbauer M, Jaindl M, Hochtl LL, Ostermann RC, Kdolsky R, Aldrian S. Spine injuries in polytraumatized pediatric patients: characteristics and experience from a Level I trauma center over two decades. J Trauma Acute Care Surg. 2012;73(1):156–61.
61. Holtslag HR, van Beeck EF, Lindeman E, Leenen LP. Determinants of long-term functional consequences after major trauma. J Trauma. 2007;62(4): 919–27.
62. Varma A, Hill EG, Nicholas J, Selassie A. Predictors of early mortality after traumatic spinal cord injury: a population-based study. Spine (Phila Pa 1976). 2010;35(7): 778–83.
63. McLain RF. Functional outcomes after surgery for spinal fractures: return to work and activity. Spine (Phila Pa 1976). 2004;29(4):470–7.
64. Holtslag HR, Post MW, van der Werken C, Lindeman E. Return to work after major trauma. Clin Rehabil. 2007;21(4):373–83.
65. Putz C, Schuld C, Akbar M, et al. Neurological and functional recovery in multiple injured patients with paraplegia: outcome after 1 year. J Trauma. 2011;70(5):1078–85.
66. Bliemel C, Lefering R, Buecking B, et al. Early or delayed stabilization in severely injured patients with spinal fractures? Current surgical objectivity according to the Trauma Registry of DGU: treatment of spine injuries in polytrauma patients. J Trauma Acute Care Surg. 2014;76(2):366–73.
67. Schinkel C, Anastasiadis AP. The timing of spinal stabilization in polytrauma and in patients with spinal cord injury. Curr Opin Crit Care. 2008;14(6):685–9.

Index

A

AAST. *See* American Association for Surgery of Trauma (AAST)
Abbreviated injury scale (AIS), 88, 89, 220, 423
Abdominal injuries
    AAST solid organ injury grading scales, 114, 116
    blunt abdominal trauma
        bowel injury, 121
        genitourinary, 121–122
        liver and spleen, 118–120
        pancreatic injury, 120
    collaboration, 123–124
    contrast extravasation/blush, 115, 116
    CT imaging, 115
    initial evaluation
        primary survey, 111–112
        secondary survey, 112–114
    penetrating injury
        abdominal stab wounds, 117
        GSW, 115–117
        nonoperative management, 118
        SW, 116–117
        triple-contrast CT scan, 118
    postinjury complications, 122–123
Acute bacterial skin and skin structure infections (ABSSSI), 352–353
Acute life-threatening thoracic injuries
    cardiac tamponade, 93
    massive hemothorax, 92
    open pneumothorax, 92
    pulmonary contusion, 92–93
    tension pneumothorax, 91–92
Acute posttraumatic osteomyelitis
    bacterial infections, 374
    C-reactive protein, 374
    diagnosis, 374–376
    erythrocyte sedimentation rate, 374
    fracture site, 373–374
    IM tibia nails, 376
    medullary osteolysis, 374, 375
    non-locking screws, 374
    static wire fixator, 378
    surgical debridement, 376
    type I and II osteomyelitis, 377–378
    type III/IV osteomyelitis, 376–377
Acute respiratory distress syndrome (ARDS)
    features, 102
    mechanical ventilation, 102
    pathogenesis, 102, 103
    PEEP, 103
    risk factors, 102
    treatment, 102, 103
Acute stress disorder (ASD), 400, 403
AIS. *See* Abbreviated injury scale (AIS)
Alarmins, 44
American Association for Surgery of Trauma (AAST), 114, 116, 143–144
Angiography
    orthopedic injuries, 278–279
    vascular injuries, 278–279
Ankylosing spondylitis (AS), 198
Anterior cord syndrome, 193
Anterior pelvic ring fixation technique
    advantage, 139
    bilateral single supra-acetabular pedicle screws, 140
    double plating, 139
    Hohman type retractor, 139
    identify midline raphe, 139
    Jungbluth pelvic reduction clamp, 139
    locked plate fixation, 139
    Pfannenstiel approach, 139
    plate and screw options, 139
    retrograde ramus screw, 140
ARDS. *See* Acute respiratory distress syndrome (ARDS)
Articular fractures
    assessment, 290
    floating joint injuries (*see* Floating joint injuries)
    open articular fractures
        bony fragments, 292, 293
        definitive fixation, 292
        imaging, 292
        initial care, 291–292
        joint spanning external fixator, 292
    polytrauma patients, 290–291
    traumatic knee dislocation

Articular fractures (*cont.*)
  acute setting, 303
  assessment, 301–302
  avulsion fractures, 301
  classification, 300–301
  complications, 304
  definitive management, 303
  evaluation, 302–303
  outcomes, 304
  peroneal nerve damage, 301
  rehabilitation, 303–304
  tibial nerve injury, 301
  vascular injury, 301
    types, 289–290

B
Beck Depression Inventory (BDIA), 2–3
Below-knee deep venous thrombosis (BKDVT), 69
Bladder trauma
  AAST, 146, 147
  blunt traumas, 146
  imaging, 148–150
  signs and symptoms, 146–148
  treatment
    in adults, 148, 151
    antibiotic prophylaxis, 149
    extraperitoneal rupture, 148–149
    intraperitoneal rupture, 148, 152
    vascularized omentum flap, 149–150
Blast injuries
  civilian and military setting, 332
  London attacks, 332, 334
  Madrid train attacks, 332, 334
  mechanisms
    primary effect, 334
    quaternary effect, 337
    secondary effects, 337
    tertiary injury, 337
    tympanic membrane injury, 334, 336–337
  suicide bombing, 332, 334
  surgical management, 334
  treatment, 338–339
  triage and primary resuscitation, 337–338
Blunt abdominal trauma
  aortic injury, 95
  bladder ruptur, 146
  bowel injury, 121
  elderly patients, 209
  genitourinary, 121–122
  liver and spleen, 118–120
  Nexus chest decision, 90
  pancreatic injury, 120
  plexus injury, 79
  renal trauma, 143
  TBI, 59
  tertiary blast injury, 337
  tracheobronchial tree, 94
Blunt cardiac injury (BCI), 100

Bone injuries
  initial management, 344
  rib fractures, 97–98
  scapular fractures, 98
  sternal fractures, 98
Bowel injury, 114, 119, 121, 122
Brown-Séquard syndrome, 193
Bucholz classification, 131

C
Canadian Orthopaedic Trauma Society (COTS), 19–21
Cardiac injuries
  cardiac rupture, 101
  coronary artery injury, 100–101
  myocardial contusion, 100
Cardiac tamponade, 93
Cauda equina syndrome, 193
Centers for Disease Control (CDC), 353, 354
Central cord syndrome, 193
Cerebral perfusion pressure (CCP), 235
Chest trauma
  acute life-threatening thoracic injuries
    cardiac tamponade, 93
    massive hemothorax, 92
    open pneumothorax, 92
    pulmonary contusion, 92–93
    tension pneumothorax, 91–92
  bone injuries
    rib fractures, 97–98
    scapular fractures, 98
    sternal fractures, 98
  cardiac injury
    cardiac rupture, 101
    coronary artery injury, 100–101
    myocardial contusion, 100
  classification, 88–89
  complications
    ARDS, 102–103
    chylothorax, 103
    persistent air leakage, 104
    pleural empyema, 104
  diagnostics
    AP chest radiography, 89
    bronchoscopy, 91
    CT, 89–90
    TEE, 91
    transthoracic ultrasonography, 90–91
  fracture treatment, 101
  incidence, 88
  lung injury
    pulmonary contusion, 99–100
    pulmonary herniation, 100
    pulmonary lacerations, 98–99
  mechanism and pathophysiology, 88
  operative techniques
    anterolateral thoracotomy, 104
    clamshell thoracotomy, 105
    Hilar cross clamping, 105

potentially life-threatening injuries, 88
  esophageal rupture, 96
  hemothorax, 94
  simple pneumothorax, 93–94
  tracheobronchial tree rupture, 94–95
  traumatic aortic disruption, 95, 96
  traumatic diaphragmatic injury, 95–96
Chronic posttraumatic osteomyelitis
  antibiotic beads, 379
  antibiotic course, 379–380
  antibiotic spacers, 381
  bone defect, 381
  bone resection, 381
  bone transport, 381–382
  debridement, 379
  docking site, 383
  ICBG, 380
  infected nonunions, 383–385
  Masquelet technique, 381
  medullary canal, 379
  posttraumatic deformities, 381
  preoperative gadolinium, 380–381
  retrograde transport, 383
  segmental defect, 383
  Stacked Taylor Spatial Frame, 382
  TRC, 382–383
  type III and IV osteomyelitis, 379
  vascularized fibular grafts, 381
Clostridium perfringens, 358–359
Coagulopathy, 26–27
Cognitive dysfunction disorders, 224
Collaborative care (CC), 2
Community-associated MRSA (CA-MRSA), 355, 363
Complementopathy, 27–29
Conus medullaris, 193
COTS. *See* Canadian Orthopaedic Trauma Society (COTS)
Counter anti-inflammatory response syndrome (CARS), 42
C-reactive protein (CRP), 25
Cytokines
  anti-inflammatory, 30–31
  interleukin 1, 44, 46
  pro-inflammatory, 29–30

D
Damage associated molecular patterns (DAMPs), 24, 44, 45
Damage control orthopedics (DCO), 101–102, 169, 183
Debridement, 238, 331, 357
Demineralized bone matrix (DBM), 347
Diagnostic and Statistical Manual of Mental Disorders (DSM), 400, 403
Diffuse idiopathic skeletal hyperostosis (DISH), 198
Digital subtraction angiography (DSA), 279
Dilutional coagulopathy, 43

E
Early total care (ETC), 42, 59–61, 101, 102, 157, 183, 389
EBM. *See* Evidence-based medicine (EBM)
Elderly management
  aging
    cardiovascular, 220–221
    neurologic, 221–222
    pulmonary, 221
    renal, 222
  anticoagulation, 226–227
  clinical impact, 220
  DNR orders, 227
  frailty, 225–226
  perioperative management
    cognitive impairment, 224
    comorbidity, 222–223
    functional status, 225
    medications and supplements, 223
    nutrition, 223–224
  preexisting illness, 220
  single-system injury, 219
Erectile dysfunction, 154
ETC. *See* Early total care (ETC)
Evidence-based medicine (EBM), 15, 16, 21
Evidence cycle, 17–18
Extended spectrum beta-lactamases (ESBLs), 355
Extremity injuries
  clinical practice
    implications, 413
    Lisfranc injuries, 412–413
    open reduction and internal fixation, 412
    reamed *vs.* non-reamed nailing, 413
    significance, 413–414
  evidence-based medicine, 408
  lower extremity injuries
    Hannover Rehab Study, 414
    LEAP study, 414–416
    limb salvage *vs.* amputation, 414–415
    preclinical and clinical emergency care, 414
  numerical results
    precision, 411–412
    statistical significance, 412
    treatment effect, 411
  outcome measures
    assessment, 409–410
    body region/disease-specific questionnaire, 408
    SF-36, 409
    SMFA, 409
    standardization, 408
  patient follow-up, 410–411
  upper extremity injuries, 416
Eye movement desensiti-zation and reprocessing therapy (EMDR), 403

F
Fall-induced injury
  analysis, 207, 213
  average age, 211–212

Fall-induced injury (*cont.*)
    distal radius, 213–214
    epidemiological data, 213–214
    incidence, 210, 212, 213
    Orthopaedic Trauma Unit, 211
    pelvic fracture, 213
    proximal femoral, 213–214
    proximal humeral, 213–214
    upper and lower limbs, 212–213
First hit theory, 33–34
Floating elbow injuries
    adults and children, 299–300
    classification, 300
    complications, 300
    management, 300
    outcomes, 300
Floating joint injuries
    elbow injuries
        adults and children, 299–300
        classification, 300
        complications, 300
        management, 300
        outcomes, 300
    knee injury
        antegrade femoral nailing, 294, 296
        classification, 292, 294
        complications, 292, 293, 296
        isolation, 293–294
        life-threatening injuries, 294
        ligamentous injuries, 295–296
        limb ischaemia, 294
        open fractures, 294–295
        post-operative period, 294
        retrograde femoral nailing, 294–295
        surgical fixation, 293
    shoulder injuries
        classification, 297, 298
        clinical presentation, 298
        complications, 299
        diagnosis, 298
        inherent bony instability, 297
        management, 298–299
Floating knee injury
    antegrade femoral nailing, 294, 296
    classification, 292, 294
    complications, 292, 293, 296
    isolation, 293–294
    life-threatening injuries, 294
    ligamentous injuries, 295–296
    limb ischaemia, 294
    open fractures, 294–295
    post-operative period, 294
    retrograde femoral nailing, 294–295
    surgical fixation, 293
Floating shoulder injuries
    classification, 297, 298
    clinical presentation, 298
    complications, 299
    diagnosis, 298
    inherent bony instability, 297
    management, 298–299

Focused abdominal sonography for trauma (FAST), 90–91, 112, 113, 137–138, 221
Focused assessment with sonography (c), 249
Fracture management
    assessment
        closed fractures, 159
        open fractures, 158–159
    polytrauma
        acute phase, 161
        borderline condition, 162, 163
        in extremis condition, 162, 164
        mangled extremity, 183–184
        multiply injured elderly patient, 208–210
        patient assessment, 161–162, 164–165
        primary phase, 161
        SDS, 162, 163
        secondary phase, 161
        stable condition, 162
        systematic approach, 160
        tertiary phase, 161
        unstable condition, 162
    pregnancy (*see* Pregnant patient)
    SDS, 157
    surgical priorities
        chest injury, 165–166
        head injury, 165
        pelvic ring injuries, 166
        trauma system, 166
    treatment
        serial extremity, 160
        upper *vs.* lower extremity, 159–160
Frailty, 225–226
Frankel grading system, 192
Fresh-frozen plasma (FFP), 226, 227

G

GAS. *See* Group A streptococci (GAS)
Gerota's fascia, 144
Glasgow Outcome Scale (GOS), 56
Glomerular filtration rate (GFR), 222
Gross domestic product (GDP), 307
Group A streptococci (GAS), 359, 363
Gunshot wounds
    early clinical reports, 330
    fracture care, 331–332
        joint reconstruction and fixation, 332, 334
        long bones, 332–335
    gelatin blocks, 330
    high-velocity *vs.* low-velocity, 330
    mechanism, 330
    spinal fractures, 197–198
    treatment, 331
    vascular injuries, 330–331

H

Hannover Fracture Scale (HFS)
    classification, 67–68
    limb-threatening injuries, 318, 323
    lower-extremity injury-severity scores, 315, 318

Index

Head injury
    abnormal neurological status, 58
    clinical assessment and management, 58–59
    consciousness, 58
    damage control surgery, 59
    diagnosis, 58
    hypoxia and hypotension, 56–58
    ICP monitoring, 58
    immunological pathophysiology, 57, 59
    intoxicated patients, 58
    intramedullary reaming and nailing, 59
    neurologic status, 58
    pathophysiology, 56
    pupil size/reactivity, 58
    risks and benefits, 60
    suspected skull fracture, 58
Hemothorax, 92

I

Iliac crest bone graft (ICBG), 380, 382, 383
Incremental cost-effectiveness ratio (ICER), 11–12
Injury severity score (ISS), 201–202
Integra®, 72
Interim management phase, 345
Ipsilateral diaphyseal fractures. *See* Floating elbow injuries
Ischemia/reperfusion (I/R) injury, 34–35

K

Kallikrein-kinin system, 28
Kidney function, 222

L

Laboratory Risk Indicator for Necrotizing Fasciitis (LRINEC) score, 357–358
LEAP study. *See* Lower Extremity Assessment Project (LEAP) study
Letournel classification, 131
Limb salvage
    limb-threatening injuries (*see* Limb-threatening injuries)
    subtotal amputation injury (*see* Subtotal amputation injury)
    traumatic primary amputations (*see* Traumatic primary amputations)
Limb Salvage Index (LSI)
    limb-threatening injuries, 317, 323
    lower-extremity injury-severity scores, 315, 317
    mangled extremity, 171
Limb-threatening injuries, 308
    components, 316, 323
    evaluation, 324
    foot plantar sensation, 315, 323
    HFS, 318, 323
    LEAP study, 311, 323
    LSI, 317, 323
    NISSSA, 319, 323
    open type III tibia fractures, 324
    sensate-salvage group, 323–324
Lisfranc injuries, 412–413
Low-and middle-income countries (LMICs), 4–5
Lower Extremity Assessment Project (LEAP) study
    limb salvage *vs.* amputation, 271
    limb-threatening injuries, 311, 323
    mangled extremity, 181–182
    subtotal amputation injury, 318–320
    traumatic primary amputations, 309–311
Lower extremity injuries
    Hannover Rehab Study, 414
    LEAP study, 414–416
    limb salvage *vs.* amputation, 414–415
    preclinical and clinical emergency care, 414
    vascular and orthopedic injuries
        intimal tears, 282, 285
        knee dislocations, 278, 285–286
        lesions, 286–287
        saphenous vein, 285–286
        temporary stents, 281, 284–285
LSI. *See* Limb Salvage Index (LSI)
Lung contusion, 92, 165, 236, 388
Lung injuries
    pulmonary contusion, 99–100
    pulmonary herniation, 100
    pulmonary lacerations, 98–99

M

Macrophages, 33
Major depressive disorder (MD), 401, 403
Major Trauma Outcome Study (MTOS), 202
Mangled extremity
    complications, 175, 180
    DCO, 169
    emergent management, 169–170
    health-care costs, 182
    injury patterns, 170
    limb salvage *vs.* amputation, 180–182
    management
        ATLS protocol, 171, 173
        BMP-2, 175
        external fixation, 174
        heterotopic ossification, 175
        plantar soft tissue, 172, 173
        sterile dressing, 174
        surgical fixation, 175–180
        thorough irrigation and debridement, 172, 174
        trans-tibial amputations, 172
        wound dressing, 171
    mechanism of injury, 170
    numerous reports, 169
    pediatric patients, 183
    polytrauma, 183–184
    scoring systems
        LSI, 171
        MESS, 171, 172
        predict outcome, 180
        PSI, 171
    upper and lower extremity, 182–183

Mangled extremity severity score (MESS), 171, 315, 316
MAST. *See* Military antishock trousers (MAST)
MatriDerm®, 72
Methicillin-resistant *S. aureus* (MRSA), 353–355, 363–364
Military antishock trousers (MAST), 133–134
Monocytes, 33
MRSA. *See* Methicillin-resistant *S. aureus* (MRSA)
Multi-organ dysfunction syndrome (MODS), 24
Multiple injured patient
   acute-phase reaction, 24–26
   acute total care, 389–390
   blunt trauma, 202
   bone grafting, 392–393
   cell implication
      monocytes/macrophages, 33
      neutrophils, 32
      NK cells, 33
   cytokines
      anti-inflammatory, 30–31
      pro-inflammatory, 29–30
   damage control orthopedics, 388
   DAMPs, 24
   definitive fracture care, 388
   epidemiology
      accident prevention, 204
      elderly population, 210, 211
      incidence, 203
      orthopaedic trauma, 204
      severe injury prevalence, 204
   external fixator, 388–390
   fall-related fractures
      analysis, 207, 213
      average age, 211–212
      distal radius, 213–214
      epidemiological data, 213–214
      incidence, 210, 212, 213
      Orthopaedic Trauma Unit, 211
      pelvic fracture, 213
      proximal femoral, 213–214
      proximal humeral, 213–214
      upper and lower limbs, 212–213
   high-energy injury, 206
   immune response, 26
   injury patterns, 204
   ISS, 201–202
   malunion
      anatomy and joint preserving operation, 395
      indications, 395
      total hip arthroplasty, 395
      treatment plan, 390–391
   mortality, 207–208
   motor vehicle accidents, 205–206
   MTOS, 202
   non-survivors, 202
   nonunion
      anatomy and joint preserving operation, 395
      angle plate osteosynthesis, 393, 394
      biologic supplementation, 392
      concomitant deformity, 393
      indications, 395
      intra-medullary implant, 392
      lower extremities, 390–392
      mechanical environment, 394
      nutrition, 394
      open supracondylar femur fracture, 394
      peritrochanteric fracture, 395, 396
      pin site infection and patient dissatisfaction, 394–395
      staged application, 393–394
      surgical corrective procedures, 390
      total hip arthroplasty, 395
   open fractures, 214–216
   organ dysfunction
      broken barriers, 35
      initial injury (first hit), severity of, 33–34
      I/R injury, 34–35
      two-hit theory, 34
   polytrauma, 208–210
   prevalence, 211, 212
   pro-and anti-inflammatory response, 23
   ROS, 31–32
   serine protease systems
      coagulopathy, 26–27
      complementopathy, 27–29
      kallikrein-kinin system, 28
   SIRS
      diagnostic criteria, 23, 24
      induce trauma, 25
   suicide, 206–207
   treatment, 207

N
National Healthcare Safety Network (NHSN), 353, 354
Natural killer (NK) cells, 33
Necrotizing soft tissue infections (NSTIs)
   CT examination, 358
   diagnosis, 357
   LRINEC score, 357–358
   microbiology, 358–359
   mortality risk calculation, 357
   MR imaging, 358
   vs. non-necrotizing infections, 361–362
   single-institution series, 357
   surgical debridement, 357
Nerve injury, ischemia, soft-tissue injury, skeletal, shock, and age of patient score (NISSSA)
   limb-threatening injuries, 319, 323
   lower-extremity injury-severity scores, 315, 319
Neurological injury
   classification
      ASIA, 192
      incomplete spinal cord injury, 193
      sacral sparing, 192
      spinal shock, 193
   examination, 190–192
Neutrophils, 32

Index   433

NSTIs. *See* Necrotizing soft tissue infections (NSTIs)
Nutation, 330

O

Open fractures
   antibiotic-coated implants, 271
   antibiotic treatment, 265–266
   diagnostics, 262–263
   high-energy modes, 214, 215
   incidence, 214, 215, 261
   limb salvage *vs.* amputation, 271
   patient evaluation, 262–263
   stabilization
     algorithm, 268, 269
     definitive treatment, 267, 268
     head and thoracic injuries, 267
     insofar, 267
     primary plate osteosynthesis, 268
     SPRINT, 268
   Staphylococcus aureus infection, 271
   surgical treatment
     debridement and irrigation, 263–265
     hair removal, 263
   tibial fracture, 261
   wound closure timing
     delayed closure, 270
     emergency free tissue transfer, 268–269
     negative-pressure wound therapy, 270
     prospective study, 269
     retrospective study, 269
     VAC-systems, 270
Open pneumothorax, 92
Organ dysfunction
   broken barriers, 35
   initial injury, severity of, 33–34
   I/R injury, 34–35
   two-hit theory, 34
Orthopaedic trauma care
   biannual meetings, 21
   challenges, 19
   commitment, 20
   COTS, 19–21
   EBM, 15–16
   evidence cycle
     Acquire, 17–18
     Act, 18
     Apply, 18
     Appraise, 18
     Ask, 17
   formalization and funding, 20
   functionally smaller, multicentre and international trials, 19
   hierarchy of evidence, 15–16
   model of success, 19
   operationalization, 21–22
   present state, 16
   research coordinators, 20–21
Orthopedic injuries
   signs and symptoms
     angiography, 278–279
     clinical evaluation, 277–278
     CT angiogram, 279
     Doppler evaluation, 278
     DSA, 279
     MR angiography, 279
   treatment
     extremity bleeding, 280–282, 286
     lower extremity (*see* Lower extremity injuries)
     neck, 282–283
     optimal care, 279–280
     pelvic bleeding, 283, 285
     severe torso bleeding, 280
     upper extremity, 279, 283, 284
Orthopedic Trauma Association Open Fracture Classification (OTA-OFC), 262–263
Osteomyelitis
   diagnosis and treatment, 359–361
   microbiology, 361

P

Pathogen associated molecular patterns (PAMPs)
   inflammatory molecules, 44
   NK cells, 33
Pathophysiology
   classification, 42
   clinical course
     appropriate actions, 46, 48, 49
     immunomarkers, 48, 50, 51
   hyperdynamic flow, 41
   hypodynamic flow, 41
   inflammatory response
     alarmins, 44
     aseptic SIRS and pathogenesis, 44
     cell activation, 43
     danger sensing mechanisms, 44, 45
     firm adhesion, 43
     gender and sex hormones, 46
     interleukin 1 (IL-1), 44
     interleukin 6 (IL-6), 44, 46
     microenvironment theory, 43
     specific gene polymorphism, 46, 47
     TNF, 44
   initial response
     dilutional coagulopathy, 43
     permissive/damage control resuscitation, 43
     resuscitation efforts, 42
     scoop and run perspective, 42
     TIC, 43
     trauma network and effective collaboration, 42
   recovery, 41
Pelvic fractures
   chronic pain syndromes, 421
   clinical examination, 421
   clinical outcomes, 420–421
   evaluation, 421–422
   fracture fixation and quality, 420
   long-term outcomes, 420
   posterior pelvic fractures, 422

Pelvic ring injuries
    anatomy, 127–129
    Bucholz classification, 131
    definitive treatment
        anterior pelvic ring, 139–140
        posterior pelvic ring, 140–141
    emergent treatment/bony stabilization
        anterior external fixation, 134, 135
        C-clamp, 134–135
        MAST, 133–134
        pelvic binders, 133, 134
    hemorrhage control
        angiography, 136
        pelvic packing, 136–137
    Letournel classification, 131
    Pennal classification, 131
    permissive hypotension, 137
    physical examination, 133
    posttraumatic coagulopathy, 137
    Tile classification, 130
    treatment algorithm, 137–139
    Young and Burgess classification, 131–132
Penetrating injuries
    abdominal stab wounds, 117
    GSW, 115–117
    nonoperative management, 118
    SW, 116–117
    triple-contrast CT scan, 118
Penetrating trauma
    blast injuries (see Blast injuries)
    gunshot wounds (see Gunshot wounds)
    spinal cord injury, 195
Pennal classification, 131
Peripheral nerve injury, 79–81
Phenol-soluble modulin (PSM) peptides, 365
Plexus injury, 79
Post-surgery anti-coagulative therapy, 69
Posttraumatic osteomyelitis, 371, 372
    acute posttraumatic infection
        bacterial infections, 374
        C-reactive protein, 374
        diagnosis, 374–376
        erythrocyte sedimentation rate, 374
        fracture site, 373–374
        IM tibia nails, 376
        medullary osteolysis, 374, 375
        non-locking screws, 374
        static wire fixator, 378
        surgical debridement, 376
        type I and II osteomyelitis, 377–378
        type III/IV osteomyelitis, 376–377
    causes, 373
    chronic posttraumatic infection
        antibiotic beads, 379
        antibiotic course, 379–380
        antibiotic spacers, 381
        bone defect, 381
        bone resection, 381
        bone transport, 381–382
        debridement, 379
        docking site, 383
        ICBG, 380
        infected nonunions, 383–385
        Masquelet technique, 381
        medullary canal, 379
        posttraumatic deformities, 381
        preoperative gadolinium, 380–381
        retrograde transport, 383
        segmental defect, 383
        Stacked Taylor Spatial Frame, 382
        TRC, 382–383
        type III and IV osteomyelitis, 379
        vascularized fibular grafts, 381
Post-traumatic stress disorder (PTSD), 400
    anxiety, 401
    diagnosis, 401, 404
    diagnosis of, 2
    gender differences, 402–403
    Israeli study, 401
    prevalence rate, 401, 403–404
    psychological disorders, 402
    risk factors, 400–401
    symptoms, 401
    TBI, 401–402
    treatment options, 403
Potentially life-threatening injuries, 88
    esophageal rupture, 96
    hemothorax, 94
    simple pneumothorax, 93–94
    tracheobronchial tree rupture, 94–95
    traumatic aortic disruption, 95, 96
    traumatic diaphragmatic injury, 95–96
Predictive Salvage Index (PSI), 171, 315, 318
Pregnant patient
    anatomic and physiologic changes, 247
        cardiovascular system, 246
        coagulation factors, 246
        genitourinary system, 248
        obvious and dramatic change, 246
        respiratory system, 248
        stomach cephalad, 248
    clinical outcomes, 256
    epidemiology, 245–246
    initial assessment, 248–249
    radiological assessment
        ALARA principle, 249
        benchmarking, 249–250
        CT, 252–253
        deterministic effects, 250–251
        intrauterine irradiation, 251–252
        intrauterine life, 251
        natural background radiation, 250
        stochastic effects, 251
    surgical intervention
        anaesthesia, 253–254
        intraoperative radiology, 254–255
        orthopaedics, 255–256
Procalcitonin (PCT), 26

Index

Proximal humerus, 283, 284
Psoas-Hitch technique, 148–149
PTSD. *See* Post-traumatic stress disorder (PTSD)
Pulmonary contusion, 92–93
Pyomyositis, 359

Q
Quality of life (QOL), 2–3

R
Reactive oxygen species (ROS), 31–32
Reamer irrigator aspirator (RIA), 346
Renal trauma
    AAST, 143–144
    clinical symptoms, 144
    imaging, 144
    mechanisms, 143
    treatment, 144–145
Research coordinators, 20–21
RIA. *See* Reamer irrigator aspirator (RIA)

S
Safe definitive surgery (SDS), 157, 162, 163
Serine protease systems
    coagulopathy, 26–27
    complementopathy, 27–29
    kallikrein-kinin system, 28
Serum amyloid A (SAA), 25
Short musculoskeletal function assessment (SMFA), 2–3, 409
SIRS. *See* Systemic inflammatory response syndrome (SIRS)
Skin and soft tissue infections (SSTIs)
    classification
        ABSSSI, 352–353
        clinical trials, 351–352
        complicated patients, 352
        uncomplicated patients, 352
    closed long bone fractures, 355–356
    NSTIs
        CT examination, 358
        diagnosis, 357
        LRINEC score, 357–358
        microbiology, 358–359
        mortality risk calculation, 357
        MR imaging, 358
        vs. non-necrotizing infections, 361–362
        single-institution series, 357
        surgical debridement, 357
    open fractures, 356
    optimal management, 351, 352
    osteomyelitis
        diagnosis and treatment, 359–361
        microbiology, 361
    pyomyositis, 359
    SSIs
        CDC/NHSN classification, 353, 354
        microbiology, 354–355
        prevention, 353–354
        scoring system, 353
    traumatic wound infections, 353
    treatment
        antimicrobial therapy, 362, 364–365
        bloodstream infections, 363
        CA-MRSA, 363
        cSSTIs, 363–364
        health care–associated infection, 363
        monomicrobial etiology, 363
        Panton-Valentine leukocidin toxin gene, 363
        pathogen identification, 365, 366
        PSM peptides, 365
        source control, 365, 366
SMFA. *See* Short musculoskeletal function assessment (SMFA)
Socioeconomic factors, 419–420
Soft tissue injury, 159
    acute surgical management, 70
    classifications, 67–68
    early-intent defect coverage
        decision making, 71
        flaps, 72–74
        fracture management and soft tissue coverage, 75–79
        granulation tissue formation, 71–72
        human vascular anatomy, 72, 74
        muscle flaps, 75
        perforator flaps, 74–75
        reconstructive ladder concept, 71
        reconstructive triangle concept, 71
        tissue block, 72
        two-timed approaches, 71
    hypoxia and perfusion-related complications
        compartment syndrome, 68
        decubitus, 69
        tissue hypoxia, 68
        venous tissue congestion and thrombosis, 68–69
    initial inflammatory phase, 65, 66
    life threatening infections, 69–70
    necrotic and microbial load, 65
    peripheral nerve system, 79–81
    phospholipid mediators, 66
    post-acute surgical phase, 70–71
    principles, 67
    surgical debridement, 65
    systemic acute-phase-response, 65
Spinal fractures
    associated injuries and premorbid factors, 187–188
    cervical traction, 195–196
    clinical outcomes, 198–199, 422–423
    AS and DISH, 198
    geriatric patients, 198
    gunshot/open injury, 197–198
    hospital resuscitation, 195
    imaging techniques

Spinal fractures (*cont.*)
    computed tomography, 194
    delayed/missed diagnosis, 193
    MRI, 194–195
    plain film radiography, 193–194
  incidence, 187
  in-hospital management
    ATLS principles, 189, 190
    circulation, 189–190
    history, 189, 190
    neurological injury (*see* Neurological injury)
    physical examination, 190–192
    ventilation, 189
  mortality, 188
  nonsurgical treatment, 196
  pediatric patients, 198
  prehospital management
    clinical history and examination, 188
    immobilization, 188–189
    initial goals, 188
    transport management, 189
  surgical treatment, 196–197
SSIs. *See* Surgical site infections (SSIs)
SSTIs. *See* Skin and soft tissue infections (SSTIs)
Stacked Taylor Spatial Frame, 382
Staphylococcus aureus (S. aureus), 71, 354–355
Subtotal amputation injury, 311–312
  factors, 311–312, 315
  LEAP study, 318–320
  longer-term follow-up, 319
  lower-extremity injury-severity scores
    anterior-posterior and lateral radiographic view, 311, 313
    clinical and functional outcomes, 317–318
    clinical follow-up, 311, 314
    complications, 321–322
    components, 315, 316
    decision-making process, 316–317
    early amputation, 312–313
    external fixation, 311, 313
    HFS, 315, 318
    LSI, 315, 317
    mangled foot and ankle, 320–321
    MESS, 315, 316
    NISSSA, 315, 319
    post-injury radiographs, 311, 313
    PSI, 315, 318
    psychological distress, 322–323
    sensitivity and specificity, 315–316, 320
    severe motorcycle crash, 311, 314–315
  psychosocial and vocational rehabilitation, 319
  self-efficacy, 319
  social support network, 319
  societal costs, 323
Surgical site infections (SSIs)
  CDC/NHSN classification, 353, 354
  microbiology, 354–355
  prevention, 353–354
  scoring system, 353
Systemic inflammatory response syndrome (SIRS)
  diagnostic criteria, 23, 24
  induce trauma, 25
  stress reaction, 41

T
Taylor spatial frame (TSF), 383–385
TBI. *See* Traumatic brain injury (TBI)
Tension pneumothorax, 91–92
Thoracic Trauma Severity Score (TTS), 89
Tile classification, 130
Total residual correction (TRC), 382–383
Transesophageal echocardiography (TEE), 91
Trauma
  chronic/ongoing pain, 3
  disability, 3
  economic aspects
    cost of illness, 10–11
    ICER, 11–12
    multiple/severe injuries, 12–13
    prevention, 11
  LMICs, 4–5
  psychological implications, 2–3
  return to work, 3–4
Trauma-induced coagulopathy (TIC), 43
Traumatic bone defects
  contemporary techniques, 343
  distraction osteogenesis therapy, 343
  initial management, 344–345
  reconstitution
    autologous harvest techniques, 346
    cylindrical titanium cages, 347
    DBM, 347
    Ilizarov distraction osteogenesis, 347, 348
    induced membrane technique, 347
    RIA harvest, 346
    spine fractures, 346
  skeletal fixation, 345
  soft tissue coverage, 345
Traumatic brain injury (TBI), 235, 401–402
Traumatic primary amputations
  indications, 309–310
  life-threatening issues, 308
  lower-extremity amputations, 308–309
  outcomes
    chronic refractory osteomyelitis, 310
    LEAP study, 310–311
    lower-extremity amputation, 310
    posttraumatic fracture nonunion, 310
    SF-36 scores, 310
  severe upper extremity injuries, 309
  Syme amputation, 308, 309
  zone of injury, 308
Triplane fracture, 240
Two-hit theory, 34
Two-plane fracture, 240

U

Upper extremity injuries, 279, 283, 284, 416
Ureteral trauma
    classification, 145–146
    clinical symptom, 146
    imaging, 146, 147
    management, 146, 147
Urethral trauma
    AAST, 151, 152
    bulbomembranous junction, 150
    in children, 150–151
    clinical symptoms, 152
    pendulous and bulbar segments, 151–152
    radiographic examination, 152–153
    treatment
        in females, 154
        in males, 153–154
Urological injuries
    bladder trauma (*see* Bladder trauma)
    incidence, 143
    renal trauma
        AAST, 143–144
        clinical symptoms, 144
        imaging, 144
        mechanisms, 143
        treatment, 144–145
    ureteral trauma
        classification, 145–146
        clinical symptom, 146
        imaging, 146, 147
        management, 146, 147
    urethral trauma
        AAST, 151, 152
        bulbomembranous junction, 150
        in children, 150–151
        clinical symptoms, 152
        pendulous and bulbar segments, 151–152
        radiographic examination, 152–153
        treatment, 153–154

V

Vacuum-assisted closure (VAC) therapy, 269, 270, 291
Vascular compromise, 283, 284
Vascular injuries
    signs and symptoms
        angiography, 278–279
        clinical evaluation, 277–278
        CT angiogram, 279
        Doppler evaluation, 278
        DSA, 279
        MR angiography, 279
    treatment
        extremity bleeding, 280–282, 286
        lower extremity (*see* Lower extremity injuries)
        neck, 282–283
        optimal care, 279–280
        pelvic bleeding, 283, 285
        severe torso bleeding, 280
        upper extremity, 279, 283, 284
Vibrio vulnificus, 358–359

W

Wound closure
    delayed closure, 270
    emergency free tissue transfer, 268–269
    negative-pressure wound therapy, 270
    prospective study, 269
    retrospective study, 269
    VAC-systems, 270

Y

Young and Burgess classification, 131–132
Young children
    abdominal trauma, 236–237
    anatomic and physiologic changes, 232
    bones
        articular and methapyseal fractures, 238–239, 242
        corrective potential, 238, 240
        diaphyseal fractures, 238, 241
        extremities, 238, 239
        growth plate, 239, 242
        localization, 237
        long-term morbidity, 237
        pelvis, 237
        spine, 237–238
        transitional fractures, 240
    definition, 231–232
    epidemiology, 231–232
    first-line treatment, 233–235
    head trauma, 235–236
    patterns of injury, 232–233
    scoring, 233, 234
    thoracic trauma, 236